Probability and Mathematical Statistics (Continued)
  WILKS · Mathematical Statistics
  WILLIAMS · Diffusions, Markov Processes, aume I: Foundations
  ZACKS · The Theory of Statistical Inference

Applied Probability and Statistics
  BAILEY · The Elements of Stochastic Processes with Applications to the Natural Sciences
  BAILEY · Mathematics, Statistics and Systems for Health
  BARNETT and LEWIS · Outliers in Statistical Data
  BARTHOLOMEW · Stochastic Models for Social Processes, *Second Edition*
  BARTHOLOMEW and FORBES · Statistical Techniques for Manpower Planning
  BECK and ARNOLD · Parameter Estimation in Engineering and Science
  BENNETT and FRANKLIN · Statistical Analysis in Chemistry and the Chemical Industry
  BHAT · Elements of Applied Stochastic Processes
  BLOOMFIELD · Fourier Analysis of Time Series: An Introduction
  BOX · R. A. Fisher, The Life of a Scientist
  BOX and DRAPER · Evolutionary Operation: A Statistical Method for Process Improvement
  BOX, HUNTER, and HUNTER · Statistics for Experimenters: An Introduction to Design, Data Analysis, and Model Building
  BROWN and HOLLANDER · Statistics: A Biomedical Introduction
  BROWNLEE · Statistical Theory and Methodology in Science and Engineering, *Second Edition*
  BURY · Statistical Models in Applied Science
  CHAMBERS · Computational Methods for Data Analysis
  CHATTERJEE and PRICE · Regression Analysis by Example
  CHERNOFF and MOSES · Elementary Decision Theory
  CHOW · Analysis and Control of Dynamic Economic Systems
  CLELLAND, deCANI, BROWN, BURSK, and MURRAY · Basic Statistics with Business Applications, *Second Edition*
  COCHRAN · Sampling Techniques, *Third Edition*
  COCHRAN and COX · Experimental Designs, *Second Edition*
  COX · Planning of Experiments
  COX and MILLER · The Theory of Stochastic Processes, *Second Edition*
  DANIEL · Biostatistics: A Foundation for Analysis in the Health Sciences, *Second Edition*
  DANIEL · Application of Statistics to Industrial Experimentation
  DANIEL and WOOD · Fitting Equations to Data
  DAVID · Order Statistics
  DEMING · Sample Design in Business Research
  DODGE and ROMIG · Sampling Inspection Tables, *Second Edition*
  DRAPER and SMITH · Applied Regression Analysis
  DUNN · Basic Statistics: A Primer for the Biomedical Sciences, *Second Edition*
  DUNN and CLARK · Applied Statistics: Analysis of Variance and Regression
  ELANDT-JOHNSON · Probability Models and Statistical Methods in Genetics

continued on back

An Introduction to
Probability Theory
and Mathematical Statistics

An Introduction to Probability Theory and Mathematical Statistics

V. K. ROHATGI
Bowling Green State University

JOHN WILEY AND SONS
New York • Chichester • Brisbane • Toronto

Copyright © 1976 by John Wiley & Sons, Inc.

All rights reserved. Published simultaneously in Canada.

Reproduction or translation of any part of this work beyond that permitted by Sections 107 or 108 of the 1976 United States Copyright Act without the permission of the copyright owner is unlawful. Requests for permission or further information should be addressed to the Permissions Department, John Wiley & Sons, Inc.

Library of Congress Cataloging in Publication Data:
Rohatgi, V   K   1939–
   An introduction to probability theory and mathematical statistics.
   (Wiley series in probability and mathematical statistics)
Includes bibliographical references and indexes.
1. Probabilities.   2. Mathematical statistics.
I. Title.
QA273.R56   519.2   75-14378
ISBN 0-471-73135-8

Printed in the United States of America

10 9 8 7 6

Preface

This book on probability theory and mathematical statistics is designed for a three-quarter course meeting four hours per week or a two-semester course meeting three hours per week. It is designed primarily for advanced seniors and beginning graduate students in mathematics, but it can also be used by students in physics and engineering with strong mathematical backgrounds. Let me emphasize that this is a mathematics text and not a "cookbook." It should not be used as a text for service courses.

The mathematics prerequisites for this book are modest. It is assumed that the reader has had basic courses in set theory and linear algebra and a solid course in advanced calculus. No prior knowledge of probability and/or statistics is assumed.

My aim is to provide a solid and well-balanced introduction to probability theory and mathematical statistics. It is assumed that students who wish to do graduate work in probability theory and mathematical statistics will be taking, concurrently with this course, a measure-theoretic course in analysis if they have not already had one. These students can go on to take advanced-level courses in probability theory or mathematical statistics after completing this course.

This book consists of essentially three parts, although no such formal divisions are designated in the text. The first part consists of Chapters 1 through 6, which form the core of the probability portion of the course. The second part, Chapters 7 through 11, covers the foundations of statistical inference. The third part consists of the remaining three chapters on special topics. For course sequences that separate probability and mathematical statistics, the first part of the book can be used for a course in probability theory, followed by a course in mathematical statistics based on the second part and, possibly, one or more chapters on special topics.

The reader will find here a wealth of material. Although the topics covered are fairly conventional, the discussions and special topics included are not. Many presentations give far more depth than is usually the case in a book at this level. Some special features of the book are the following:

1. A well-referenced chapter on the preliminaries.
2. About 550 problems, over 350 worked-out examples, about 200 remarks, and about 150 references.
3. An advance warning to the reader wherever the details become too involved. He can skip the later portion of the the section in question on first reading without destroying his continuity in any way.
4. Many results on characterizations of distributions (Chapter 5).
5. Proof of the central limit theorem by the method of operators and proof of the strong law of large numbers (Chapter 6).
6. A section on minimal sufficient statistics (Chapter 8).
7. A chapter on special tests (Chapter 10).
8. A careful presentation of the theory of confidence intervals, including Bayesian intervals and shortest-length confidence intervals (Chapter 11).
9. A chapter on the general linear hypothesis, which carries linear models through to their use in basic analysis of variance (Chapter 12).
10. Sections on nonparametric estimation and robustness (Chapter 13).
11. Two sections on sequential estimation (Chapter 14).

The contents of this book were used in a one-year (two-semester) course that I taught three times at the Catholic University of America and once in a three-quarter course at Bowling Green State University. In the fall of 1973 my colleague, Professor Eugene Lukacs, taught the first quarter of this same course on the basis of my notes, which eventually became this book. I have always been able to cover this book (with few omissions) in a one-year course, lecturing three hours a week. An hour-long problem session every week is conducted by a senior graduate student.

In a book of this size there are bound to be some misprints, errors, and ambiguities of presentation. I shall be grateful to any reader who brings these to my attention.

V. K. ROHATGI

Bowling Green, Ohio
February 1975

Acknowledgments

I am indebted to my colleague Eugene Lukacs for encouraging me to write this book. Professor Lukacs followed the first four chapters of this book in his lectures in the fall of 1973 and offered many suggestions for improvements. Robert Tortora read the first draft very carefully and, in addition to detecting many errors, made several helpful suggestions. Thanks are also to Professors R. A. Bradley and J. Sethuraman who read the first draft of this book critically and made numerous suggestions for improvements. I would also like to thank Carol Alf, Ralph Baty, Eugene Lukacs, Thomas O'Connor, M. S. Scott Jr., J. Spielman, and Robert Tortora for their assistance in reading the proofs.

I wish to thank the publishers of the *American Mathematical Monthly*, the *Siam Review*, and the *American Statistician* for permission to include many examples and problems that appeared in these journals. Thanks are also due to the following for permission to include tables: Professors E. S. Pearson and L. R. Verdooren (Table 11), Harvard University Press (Table 1), Hafner Press (Table 3), Iowa State University Press (Table 5), Rand Corporation (Table 6), the American Statistical Association (Tables 7 and 10), the Institute of Mathematical Statistics (Tables 8 and 9), Charles Griffin & Co., Ltd. (Tables 12 and 13), and John Wiley & Sons (Tables 1, 2, 4, 10, and 11). I also wish to express my indebtedness to the National Science Foundation for financially supporting this project.

Kae Lea Main deserves special thanks for undertaking the monumental typing job, which she performed superbly. Finally I acknowledge with thanks my indebtedness to Ms. Beatrice Shube of John Wiley & Sons, Inc. for her cooperation in this endeavor.

V. K. R.

Enumeration of Theorems and References

This book is divided into 14 chapters, numbered 1 through 14, and an unnumbered chapter on the preliminaries. Each chapter is divided into several sections. Lemmas, theorems, equations, definitions, remarks, figures, and so on are numbered consecutively within each section. Thus Theorem $i.j.k$ refers to the kth theorem in Section j of Chapter i, Section $i.j$ refers to the jth section of Chapter i, and so on. Theorem j refers to the jth theorem of the section in which it appears. A similar convention is used for equations except that equation numbers are enclosed in parentheses. Each section is followed by a set of problems for which the same numbering system is used.

References to the chapter on preliminaries begin with the letter P. This chapter consists of three sections: P.1, P.2, and P.3. Theorems and equations are numbered within each section as explained above. Thus Theroem P.2.14 refers to Theorem 14 of Section 2 of the chapter on preliminaries.

References are given at the end of the book and are denoted in the text by numbers enclosed in square brackets, [ ]. If a citation is to a book, the notation ($[i], j$) refers to the jth page of the reference numbered $[i]$.

V. K. R.

Contents

Introduction

P.1	Sets and Classes	1
P.2	Calculus	3
P.3	Linear Algebra	15

1 Probability

1.1	Introduction	18
1.2	Sample Space	19
1.3	Probability Axioms	24
1.4	Combinatorics: Probability on Finite Sample Spaces	35
1.5	Conditional Probability and Bayes Theorem	40
1.6	Independence of Events	45

2 Random Variables and Their Probability Distributions

2.1	Introduction	52
2.2	Random Variables	52
2.3	Probability Distribution of a Random Variable	56
2.4	Discrete and Continuous Random Variables	61
2.5	Functions of a Random Variable	67

3 Moments and Generating Functions

3.1	Introduction	78
3.2	Moments of a Distribution Function	78
3.3	Generating Functions	93
3.4	Some Moment Inequalities	100

4 Random Vectors

4.1	Introduction	105
4.2	Random Vectors	105
4.3	Independent Random Variables	119
4.4	Functions of Random Vectors	127

	4.5	Order Statistics and Their Distributions	149
	4.6	Moments and Moment Generating Functions	154
	4.7	Conditional Expectation	167
	4.8	The Principle of Least Squares	172

5 Some Special Distributions

5.1	Introduction	181
5.2	Some Discrete Distributions	181
5.3	Some Continuous Distributions	202
5.4	The Bivariate and Multivariate Normal Distributions	227
5.5	The Exponential Family of Distributions	236

6 Limit Theorems

6.1	Introduction	240
6.2	Modes of Convergence	240
6.3	The Weak Law of Large Numbers	257
6.4	The Strong Law of Large Numbers	263
6.5	Limiting Moment Generating Functions	276
6.6	The Central Limit Theorem	280

7 Sample Moments and Their Distributions

7.1	Introduction	296
7.2	Random Sampling	297
7.3	Sample Characteristics and Their Distributions	299
7.4	Chi-Square, t-, and F-Distributions: Exact Sampling Distributions	311
7.5	The Distribution of $(\bar{X}, S^2)$ in Sampling from a Normal Population	321
7.6	Sampling from a Bivariate Normal Distribution	325

8 The Theory of Point Estimation

8.1	Introduction	333
8.2	The Problem of Point Estimation	333
8.3	Properties of Estimates	335
8.4	Unbiased Estimation	350
8.5	Unbiased Estimation (Continued): A Lower Bound for the Variance of an Estimate	361
8.6	The Method of Moments	373
8.7	Maximum Likelihood Estimates	375
8.8	Bayes and Minimax Estimation	388
8.9	Minimal Sufficient Statistic	399

CONTENTS

9 Neyman-Pearson Theory of Testing of Hypotheses
- 9.1 Introduction — 404
- 9.2 Some Fundamental Notions of Hypotheses Testing — 404
- 9.3 The Neyman-Pearson Lemma — 412
- 9.4 Families with Monotone Likelihood Ratio — 418
- 9.5 Unbiased and Invariant Tests — 425

10 Some Further Results on Hypotheses Testing
- 10.1 Introduction — 435
- 10.2 The Likelihood Ratio Tests — 435
- 10.3 The Chi-Square Tests — 444
- 10.4 The t-tests — 452
- 10.5 The F-tests — 457
- 10.6 Bayes and Minimax Procedures — 459

11 Confidence Estimation
- 11.1 Introduction — 466
- 11.2 Some Fundamental Notions of Confidence Estimation — 466
- 11.3 Shortest-Length Confidence Intervals — 479
- 11.4 Relation Between Confidence Estimation and Hypotheses Testing — 485
- 11.5 Unbiased Confidence Intervals — 490
- 11.6 Bayes Confidence Intervals — 494

12 The General Linear Hypothesis
- 12.1 Introduction — 497
- 12.2 The General Linear Hypothesis — 497
- 12.3 The Regression Model — 506
- 12.4 One-Way Analysis of Variance — 513
- 12.5 Two-Way Analysis of Variance with One Observation per Cell — 518
- 12.6 Two-Way Analysis of Variance with Interaction — 524

13 Nonparametric Statistical Inference
- 13.1 Introduction — 530
- 13.2 Nonparametric Estimation — 531
- 13.3 Some Single-Sample Problems — 538
- 13.4 Some Two-Sample Problems — 553
- 13.5 Tests of Independence — 564
- 13.6 Some Uses of Order Statistics — 575
- 13.7 Robustness — 580

14 Sequential Statistical Inference

14.1	Introduction	589
14.2	Some Fundamental Ideas of Sequential Sampling	589
14.3	Sequential Unbiased Estimation	596
14.4	Sequential Estimation of the Mean of a Normal Population	602
14.5	The Sequential Probability Ratio Test	612
14.6	Some Properties of the Sequential Probability Ratio Test	620
14.7	The Fundamental Identity of Sequential Analysis and Its Applications	633

References 641

Tables 648

Glossary of Some Frequently Used Symbols and Abbreviations 668

Author Index 671

Subject Index 675

An Introduction to
Probability Theory
and Mathematical Statistics

Introduction

In this chapter we state some basic results from set theory, advanced calculus, and linear algebra that we will have occasion to refer to quite frequently in the rest of the book. The following discussion is not meant to be self-contained and is provided mainly to establish the notation used in subsequent chapters.

P.1 SETS AND CLASSES

In this book, whenever the word *set* is used it is assumed to designate a *subset* of a given set Ω unless otherwise specified. As is usual, the symbols $\cup$ and $\cap$ are used for set-theoretic union and intersection. A^c denotes the complement of set A (in Ω). $A - B$ is the *difference* set $A \cap B^c$ of two sets, A and B. $A \subseteq B$ means that A is a subset of B, and $A = B$ means $A \subseteq B$ and $B \subseteq A$. The *empty* or *null* set is denoted by ϕ. For the union $A \cup B$ of disjoint sets A, B we write $A + B$.

In general, capital letters (A, B, C, etc.) denote sets, and lower case letters (x, y, z, etc.) represent *points* or *elements* of sets. Classes of sets are usually denoted by German or English script letters ($\mathfrak{A}$, $\mathfrak{B}$, $\mathscr{S}$, etc.). We deal only with nonempty classes of sets.

We write $\cap \mathfrak{A} = \cap_{A \in \mathfrak{A}} A$ for the intersection and $\cup \mathfrak{A} = \cup_{A \in \mathfrak{A}} A$ for the union of a class of sets $\mathfrak{A}$. A *disjoint class* $\mathfrak{A}$ of sets is a class such that every two distinct sets of $\mathfrak{A}$ are disjoint. The following identities are known as *DeMorgan's Laws*:

(1) $$\left(\bigcup \mathfrak{A}\right)^c = \bigcap_{A \in \mathfrak{A}} A^c,$$

(2) $$\left(\bigcap \mathfrak{A}\right)^c = \bigcup_{A \in \mathfrak{A}} A^c.$$

Let $\{A_n\}$ be a sequence of sets. The set of all points $\omega \in \Omega$ that belong to A_n for infinitely many values of n is known as the *limit superior* of the sequence and is denoted by

$$\limsup_{n\to\infty} A_n \quad \text{or} \quad \overline{\lim_{n\to\infty}} A_n.$$

The set of all points that belong to A_n for all but a finite number of values of n is known as the *limit inferior* of the sequence $\{A_n\}$ and is denoted by

$$\liminf_{n\to\infty} A_n \quad \text{or} \quad \underline{\lim_{n\to\infty}} A_n.$$

If

$$\underline{\lim_{n\to\infty}} A_n = \overline{\lim_{n\to\infty}} A_n,$$

we say that the limit exists and write $\lim_{n\to\infty} A_n$ for the common set and call it the *limit set*.

We have

(3) $$\underline{\lim_{n\to\infty}} A_n = \bigcup_{n=1}^{\infty} \bigcap_{k=n}^{\infty} A_k \subseteq \bigcap_{n=1}^{\infty} \bigcup_{k=n}^{\infty} A_k = \overline{\lim_{n\to\infty}} A_n.$$

If the sequence $\{A_n\}$ is such that $A_n \subseteq A_{n+1}$, for $n = 1, 2, \cdots$, it is called *nondecreasing*; if $A_n \supseteq A_{n+1}$, $n = 1, 2, \cdots$, it is called *nonincreasing*. If the sequence A_n is nondecreasing, we write $A_n \updownarrow$; if A_n is nonincreasing, we write $A_n \updownarrow$. Clearly, if $A_n \updownarrow$ or $A_n \updownarrow$, the limit exists and we have

(4) $$\lim_n A_n = \bigcup_{n=1}^{\infty} A_n \quad \text{if } A_n \updownarrow$$

and

(5) $$\lim_n A_n = \bigcap_{n=1}^{\infty} A_n \quad \text{if } A_n \updownarrow .$$

A nonempty class of subsets of Ω which is closed under the formation of countable unions and complement and contains ϕ is known as a *σ-field* (or a *σ-algebra*).

It is easily verified that the class of all subsets of Ω is a σ-field and that, if $\mathscr{S}$ is a σ-field, it is closed under the formation of countable intersections and contains the set Ω. Moreover, given a nonempty class of sets $\mathfrak{A}$, there exists a unique σ-field $\mathscr{S}_0$ which contains $\mathfrak{A}$ such that, if $\mathscr{S}$ is any other σ-field containing $\mathfrak{A}$, then $\mathscr{S} \supseteq \mathscr{S}_0$. In other words, $\mathscr{S}_0$ is the *smallest σ-field* containing $\mathfrak{A}$; $\mathscr{S}_0$ is known as the *σ-field generated* by $\mathfrak{A}$.

A σ-field of great interest in the study of probability is the *Borel σ-field* of subsets of the real line, $\mathscr{R}$. It is the σ-field generated by the class of all bounded semiclosed intervals of the form $(a, b]$ and will be denoted by $\mathfrak{B}$.

The sets of $\mathfrak{B}$ are called *Borel sets*. The following results are of key importance.

Theorem 1 (Halmos [45], 62). Every countable set of real numbers is a Borel set.

Theorem 2 (Halmos [45], 63). $\mathfrak{B}$ coincides with the σ-field generated by the class of all open intervals or all intervals on the real line.

It follows that every Borel set of real numbers can be obtained by a countable number of operations of unions, differences, and intersections performed on intervals.

Let A and B be two sets (not necessarily subsets of the same set Ω). We write $A \times B$ for the *Cartesian product* which is the set of all ordered pairs (a, b), where $a \in A$ and $b \in B$. The best-known example of a Cartesian product is the n-dimensional Euclidean space $\mathcal{R}_n$ which is the Cartesian product of n sets, each equal to $\mathcal{R}$. The σ-field of subsets of $\mathcal{R}_n$ generated by rectangles of the form

$$\{(x_1, x_2, \cdots, x_n): a_i < x_i \le b_i, \quad i = 1, 2, \cdots, n\}$$

is known as the *Borel σ-field* on $\mathcal{R}_n$ and will be denoted by $\mathfrak{B}_n$. This σ-field is also generated by product sets of the form $B_1 \times B_2 \times \ldots \times B_n$, where $B_i \in \mathfrak{B}$, $i = 1, 2, \ldots, n$.

Let Ω be a nonempty set, and let $\mathfrak{A}$ be a class of subsets of Ω. A function P that has domain $\mathfrak{A}$ and range contained in $\mathcal{R}$ is known as a *real-valued set function*.

P. 2 CALCULUS

In this section we state some results from calculus. The order in which these results appear below is not necessarily the order in which these topics are covered in a conventional course of lectures on calculus.

The reader who is unfamiliar with the theory of Lebesgue integration may assume that the integral $\int_A f(x)\,dx$, whenever it appears, is defined in the sense of Riemann. To ensure that this is the case, he may assume that f is defined and continuous at all but a finite number of points and that A is either an interval or the union of a finite number of (disjoint) intervals.

1. The Mean-Value Theorem of Integral Calculus

Theorem 1 (Courant and John [16], 143). If f and g are continuous in

$[a, b]$ and g is positive in $[a, b]$, then there exists an x_0 in $[a, b]$ such that

$$\int_a^b f(x) \, g(x) \, dx = f(x_0) \int_a^b g(x) \, dx.$$

In particular, there exists an x_0 in $[a, b]$ such that

$$\int_a^b f(x) \, dx = (b-a) f(x_0).$$

2. L'Hospital's Rule

Theorem 2 (Courant and John [16], 464). Let f and g have continuous first derivatives f', and g', respectively, and let $f(a) = g(a) = 0$. If $\lim_{x \to a} f'(x)/g'(x)$ exists, so does $\lim_{x \to a} f(x)/g(x)$ and

$$\lim_{x \to a} \frac{f(x)}{g(x)} = \lim_{x \to a} \frac{f'(x)}{g'(x)}.$$

If $f'(a) = g'(a) = 0$ and f'' and g'' exist and are continuous, the same result can be applied. This procedure can be continued. (Here f'' denotes the second derivative.)

3. Convexity and Extrema

Convexity. A real-valued function f is said to be *convex* if inequality

(1) $$f\left(\frac{x_1 + x_2}{2}\right) \leq \frac{1}{2}\left[f(x_1) + f(x_2)\right]$$

holds for all values of x_1 and x_2. If $-f$ is convex, we call f *concave*. If the inequality in (1) is strict for $x_1 \neq x_2$, we say that f is *strictly convex*. We shall be interested only in continuous convex functions. If the second derivative exists, the criterion for convexity reduces to $f''(x) \geq 0$.

A function f defined on an interval I has *support* at $x_0 \in I$ if there exists a function $h(x)$, defined by $h(x) = f(x_0) + m(x - x_0)$, m real such that $h(x) \leq f(x)$ for every $x \in I$. The graph of h is called a *line of support* of f at x_0.

Extrema. Let f be defined on $D \subseteq \mathscr{R}$. We say that f has a *relative maximum* at $x_0 \in D$ if there exists an open interval I_{x_0} containing x_0 such that $f(x_0) \geq f(x)$ whenever $x \in D \cap I_{x_0}$. If $f(x_0) \geq f(x)$ for all $x \in D$, we say that f has a *maximum* at x_0. Similar definitions are given for *relative minimum* and *minimum*. The term *relative extremum* is used to cover both terms, relative minimum and relative maximum, and the term *absolute maximum (minimum)* is used for the maximum (minimum) of f in D.

Calculus does not provide any direct method to locate the extrema of a function. It only permits us to locate relative extremum points.

Theorem 3 (Courant and John [16], 240). If f has a relative extremum at point x_0 in the interior of D, and if $f'(x_0)$ exists, then $f'(x_0) = 0$.

Theorem 4 (Courant and John [16], 242). Let f be continuous on an interval D. Let x_0 be an interior point of D where either $f'(x_0)$ does not exist or $f'(x_0) = 0$.

(a) If there exists an interval (a, b) with $x_0 \in (a, b) \subseteq D$ such that

$$f'(x) > 0 \quad \text{for } x \in (a, x_0), \quad \text{and} \quad f'(x) < 0 \quad \text{for } x \in (x_0, b),$$

then f has a relative maximum at x_0. In the case $f'(x) < 0$ for $x \in (a, x_0)$ and $f'(x) > 0$ for $x \in (x_0, b)$, f has a relative minimum at x_0.

(b) If there exists an interval $(a, b) \subseteq D$ with $x_0 \in (a, b)$ such that

$$f'(x) > 0 \quad \text{for } x \in (a, x_0) \quad \text{and} \quad x \in (x_0, b),$$

or

$$f'(x) < 0 \quad \text{for } x \in (a, x_0) \cup (x_0, b),$$

then f does not have a relative extremum at x_0.

Theorem 5 (Courant and John [16], 243). Let f be defined on D. Let $f'(x)$ be defined for $x \in (a, b) \subseteq D$, and $f'(x_0) = 0$ for $x_0 \in (a, b)$.

(a) If $f''(x_0) < 0$, then f has a relative maximum at x_0.
(b) If $f''(x_0) > 0$, then f has a relative minimum at x_0.

4. Quadratic Forms

A real-valued function Q defined on $\mathscr{R}_n$ by

$$Q(\mathbf{x}) = Q(x_1, x_2, \cdots, x_n) = \sum_{i=1}^{n} \sum_{j=1}^{n} a_{ij} x_i x_j, \quad \mathbf{x} = (x_1, x_2, \cdots, x_n) \in \mathscr{R}_n,$$

where a_{ij} are real numbers, is known as a *quadratic form* in n variables. We say that Q is *symmetric* if $a_{ij} = a_{ji}$ for all i and j, *positive semidefinite* if $Q(\mathbf{x}) \geq 0$ for $\mathbf{x} \neq 0$, and *positive definite* if $Q(\mathbf{x}) > 0$ for $\mathbf{x} \neq 0$. Similar definitions are given for *negative semidefinite* and *negative definite* quadratic forms. Let $\varDelta = \det[a_{ij}]$, and let $\varDelta_{n-k}$ denote the determinant with $n - k$ rows and columns, obtained by deleting the last k rows and columns of $\varDelta$. Take $\varDelta_0 = 1$.

Theorem 6 (Apostol [3], 151–152). A necessary and sufficient condition for a symmetric quadratic form Q to be positive definite is that the $n + 1$ numbers $\varDelta_0, \varDelta_1, \cdots, \varDelta_n$ be positive.

In the $n = 2$ case, the form $Q(x, y) = Ax^2 + 2Bxy + Cy^2$ is known as a *binary quadratic form*, and the number $\Delta = AC - B^2$, is the *discriminant* of Q. Q is positive definite if and only if $A > 0$ and $\Delta > 0$, and it is positive semidefinite if and only if we change the sign > 0 to ≥ 0 (not both $>$). See also Section P. 3 below.

5. Orders of Magnitude—The "O" and "o" Notation

Let f and g be two functions, and assume that $g(x) > 0$ for sufficiently large x. We say that $f(x)$ *is at most of the order* of $g(x)$ as $x \to \infty$, and we write $f(x) = O[g(x)]$ as $x \to \infty$ if there exist an x_0 and a constant $c > 0$ such that $|f(x)| < cg(x)$ for all $x \geq x_0$. Thus $f(x) = O[g(x)]$ means that $|f(x)|/g(x)$ is bounded for large x. We write $f(x) = O(1)$ to express the fact that f is bounded for large x. It is easily checked that

(a) if $f_1(x) = O[g_1(x)]$, $f_2(x) = O[g_2(x)]$, then $f_1(x) + f_2(x) = O[g_1(x) + g_2(x)]$;
(b) if $\alpha > 0$ is constant, $f(x) = O[\alpha\, g(x)]$ implies that $f(x) = O[g(x)]$;
(c) if $f_1(x) = O[g_1(x)]$ and $f_2(x) = O[g_2(x)]$, then $f_1(x)f_2(x) = O[g_1(x)g_2(x)]$.

Let f and g be both defined and positive for large x. We say that $f(x)$ is *of smaller order than* $g(x)$ as $x \to \infty$, and we write $f(x) = o[g(x)]$ as $x \to \infty$ if $\lim_{x \to \infty} f(x)/g(x) = 0$. We write $f(x) = o(1)$ to express the fact that $f(x) \to 0$ as $x \to \infty$. It is easily checked that

(a) $f(x) = o[g(x)] \Rightarrow f(x) = O[g(x)]$;
(b) $f_1(x) = O[g_1(x)]$, $f_2(x) = o[g_2(x)] \Rightarrow f_1(x)f_2(x) = o[g_1(x)g_2(x)]$.

The symbols O and o are also used if x tends to a finite value.

6. Taylor's Expansion

Theorem 7 (Courant and John [16], 451–452). Let n be a nonnegative integer, and let $f^{(n+1)}(x)$ denote the $(n+1)$st derivative of f. If $f^{(n+1)}(x)$ is continuous in $|x - a| \leq h$, then

$$(2) \qquad f(x) = f(a) + \sum_{k=1}^{n} \frac{(x-a)^k}{k!} f^{(k)}(a) + R_n,$$

where the remainder R_n which depends on n, a, and h, is given by

$$(3) \qquad R_n = \frac{1}{n!} \int_a^x (x-t)^n f^{(n+1)}(t)\, dt.$$

The *Lagrange form* of the remainder is given by

(4) $R_n = f^{(n+1)}(x_0) \dfrac{(x-a)^{n+1}}{(n+1)!}$ for some unspecified $x_0 \in (a, x)$;

and the *Cauchy form*, by

(5) $R_n = f^{(n+1)}(x_0') \dfrac{(x-x_0')^n}{n!} (x-a)$ for some unspecified $x_0' \in (a, x)$.

7. Power Series

A series of the type

(6) $\qquad c_0 + c_1 x + c_2 x^2 + \cdots + c_n x^n + \cdots$

is known as a *power series*.

Theorem 8 (Courant and John [16], 541). If the power series (6) converges for $x = x_0 \neq 0$, it converges absolutely for x in the interval $(-|x_0|, |x_0|)$, and uniformly in every closed interval $[-\eta, \eta]$, where $0 < \eta < |x_0|$.

Theorem 9 (Courant and John [16], 542–543). Let $f(x) = \sum_{k=0}^{\infty} c_k x^k$ be a power series that converges for $x = x_0 \neq 0$. Then the power series may be integrated and differentiated term by term any number of times in any closed interval $[-\eta, \eta]$, where $0 < \eta < |x_0|$. The resulting power series is uniformly and absolutely convergent in $[-\eta, \eta]$. Moreover, the kth integral or the jth derivative of the power series equals, respectively, the kth integral or jth derivative of f in $[-\eta, \eta]$.

Theorem 10 (Courant and John [16], 544–545). The representation of a function f by a power series, say $f(x) = \sum_{k=0}^{\infty} c_k x^k$ with a nonzero radius of convergence is unique and is given by

(7) $\qquad f(x) = \sum_{k=0}^{\infty} f^{(k)}(0) \dfrac{x^k}{k!},$

where $f^{(0)}(0) = f(0)$.

The expansion (7) is sometimes referred to as a *Maclaurin expansion*. An important expansion is that of the function $f(x) = (1+x)^\alpha$ into a power series. Let us write

(8) $\qquad \dbinom{\alpha}{\nu} = \dfrac{\alpha(\alpha-1)\cdots(\alpha-\nu+1)}{1 \cdot 2 \cdots \cdot \nu},$

where ν is an integer (> 0) and α is an arbitrary real number. Set $\binom{\alpha}{0} = 1$; $\binom{\alpha}{\nu}$ are known as *binomial coefficients*. For $|x| < 1$ we have

(9) $$(1 + x)^\alpha = \sum_{\nu=0}^{\infty} \binom{\alpha}{\nu} x^\nu,$$

which is known as a *binomial series*. If α is a positive integer,

$$\binom{\alpha}{\nu} = \frac{\alpha(\alpha-1)\cdots(\alpha-\nu+1)}{1\cdot 2\cdots\cdot\nu} = \frac{\alpha!}{(\alpha-\nu)!\,\nu!}$$

is the ordinary binomial coefficient. Here $\binom{\alpha}{\nu} = 0$ if $\alpha < \nu$, and (9) reduces to

(10) $$(1 + x)^\alpha = \sum_{\nu=0}^{\alpha} \binom{\alpha}{\nu} x^\nu.$$

For the numerator on the right-hand side of (8) we write $_\alpha P_\nu$. It is easily verified that

(11) $$\binom{\alpha}{\nu} + \binom{\alpha}{\nu-1} = \binom{\alpha+1}{\nu}$$

for any real number α and positive integer ν. Also, for any $\alpha > 0$

(12) $$\binom{-\alpha}{\nu} = (-1)^\nu \binom{\alpha+\nu-1}{\nu}.$$

8. Stirling's Approximation

Theorem 11 (Courant and John [16], 504). As $n \to \infty$,

(13) $$n! \, \{\sqrt{2\pi}\, n^{n+1/2}\, e^{-n}\}^{-1} \to 1 \;;$$

more precisely,

(14) $$(2\pi)^{1/2} n^{n+1/2} e^{-n} < n! < (2\pi)^{1/2} n^{n+1/2} \left(1 + \frac{1}{4n}\right) e^{-n}.$$

9. Infinite Products

Let $P_n = \prod_{k=1}^{n} p_k$. If $P_n \to P \, (\neq 0)$, we say that the *infinite product* $\prod_{k=1}^{\infty} p_k$ *converges* to the limit P. If P_n does not converge to a finite nonzero limit, we say that the product $\prod_{k=1}^{\infty} p_k$ is *divergent*. If $P_n \to 0$, we say that $\prod_{k=1}^{\infty} p_k$ *diverges to* 0. If each of a finite number of factors has the value 0, the product is convergent if it converges when these factors are removed. In such cases the product has the value 0.

Theorem 12 (Courant and John [16], 561–562). A necessary and sufficient condition for $\prod_{k=1}^{\infty} p_k$, $p_k > 0$, to converge is that $\sum_{k=1}^{\infty} \log p_k$ converges.

Theorem 13 (Apostol [3], 381–382).

(a) Let each $p_n > 0$. Then the product $\prod_{n=1}^{\infty} (1 + p_n)$ converges if and only if $\sum_{n=1}^{\infty} p_n$ converges.

(b) Let each $p_n \geq 0$. Then the product $\prod_{n=1}^{\infty} (1 - p_n)$ converges if and only if the series $\sum_{n=1}^{\infty} p_n$ converges.

10. Monotonic Functions

Let f be defined on $D \subseteq \mathscr{R}$. We say that f is *nondecreasing* (*nonincreasing*) if $x < y \Rightarrow f(x) \leq f(y)$ ($f(x) \geq f(y)$) for $x, y \in D$. Also f is said to be (strictly) *increasing* if $x < y \Rightarrow f(x) < f(y)$. Similarly, f is *decreasing* if $x < y \Rightarrow f(x) > f(y)$. We say that f is *monotonic* on D if f is either increasing or decreasing. If f is increasing (decreasing), we write $f \uparrow (\downarrow)$.

Theorem 14 (Apostol [3], 78). If f is nondecreasing on $\mathscr{R}$, then $f(x_0 +) = \lim_{h>0}^{h\to 0} f(x_0 + h)$ and $f(x_0 -) = \lim_{h>0}^{h\to 0} f(x_0 - h)$ both exist for each $x_0 \in \mathscr{R}$ and are finite. The limits at infinity $\lim_{t \downarrow -\infty} f(t) = f(-\infty)$ and $\lim_{t \uparrow \infty} f(t) = f(+\infty)$ exist; the former may be $-\infty$ and the latter may be $+\infty$.

At each point $x_0 \in \mathscr{R}$,

$$f(x_0 -) \leq f(x_0) \leq f(x_0 +)$$

and f is continuous at x_0 if and only if $f(x_0 -)$ and $f(x_0 +)$ are equal. We say that f has a *jump* at x_0 if and only if $f(x_0 -)$ and $f(x_0 +)$ are unequal. It follows that for a nondecreasing function f the only possible kind of discontinuity is a jump discontinuity. If there is a jump at x_0, we say that x_0 is a *jump point* of f and call $f(x_0 +) - f(x_0 -)$ the *size of the jump* or simply the *jump* of f at x_0.

Let f be a continuous monotonic function on $\mathscr{R}$. We shall be interested in the inverse function. Let $f(-\infty)$ and $f(+\infty)$ be the limits at infinity (which may be infinite), and let

$$\alpha = \min \{f(-\infty), f(+\infty)\}, \qquad \beta = \max \{f(-\infty), f(+\infty)\}.$$

Theorem 15 (Courant and John [16], 207). Let f be differentiable at every $x \in \mathscr{R}$ and suppose that either $f'(x) > 0$ for all x or $f'(x) < 0$ for all x. Then

(a) f is increasing if $f' > 0$ and decreasing if $f' < 0$,
(b) f has an inverse function that is continuous and monotone in (α, β),
(c) the inverse function $x = f^{-1}(y)$ is differentiable and satisfies

$$\frac{dx}{dy} \cdot \frac{dy}{dx} = 1;$$

that is,

(15) $$\frac{df^{-1}(y)}{dy} = \frac{1}{dy/dx} = \left(\frac{d}{dx} f(x) \bigg|_{x=f^{-1}(y)}\right)^{-1}.$$

11. The Change of Variable(s)

Theorem 16 (Apostol [3], 216). Let g have a continuous derivative on an interval S with end points c and d, and let f be continuous on $g(S)$. Let $a = g(c)$ and $b = g(d)$, and define F by the equation

$$F(x) = \int_a^x f(t)dt, \qquad x \in g(S)\dagger.$$

Then, for each $x \in S$, $\int_c^x f[g(t)]\, g'(t)\, dt$ exists and has the value $F[g(x)]$. In particular,

(16) $$\int_a^b f(x)\, dx = \int_c^d f[g(t)]\, g'(t)\, dt.$$

If g' never vanishes on S and $g' > 0$ on $S = [c, d]$, say, then $a < b$ and $\mathfrak{X} = g(S) = [a, b]$. Thus (16) becomes

$$\int_{\mathfrak{X}} f(x)\, dx = \int_{g^{-1}(\mathfrak{X})} f[g(t)]\, g'(t)\, dt.$$

If $g' < 0$ on S, then $\mathfrak{X} = g(S) = [b, a]$ and (16) becomes

$$\int_{\mathfrak{X}} f(x)\, dx = -\int_{g^{-1}(\mathfrak{X})} f[g(t)]\, g'(t)\, dt.$$

In either case

(17) $$\int_{\mathfrak{X}} f(x)\, dx = \int_{g^{-1}(\mathfrak{X})} f[g(t)]\, |g'(t)|\, dt.$$

Result (17) can be easily extended to cover the case in which g is not always monotonic but its domain can be divided into a finite number of subintervals in each of which g is monotonic.

It is in form (17) that we generalize the result of change of variables to multiple integrals. For ease in notation we state the result in the two-variable case.

Theorem 17 (Apostol [3], 271). Let

$$u = h_1(x, y), \qquad v = h_2(x, y),$$

together with the inverse functions

$$x = g_1(u, v), \qquad y = g_2(u, v),$$

define a one-to-one mapping of open sets $A^* \subseteq \mathcal{R}_2$ onto $B^* \subseteq \mathcal{R}_2$. Let $A \subseteq A^*$, and $B = \mathbf{h}(A) \subseteq B^*$, where $\mathbf{h} = (h_1, h_2)$. Let the first-order partial derivatives of g_1, g_2 be continuous, and the Jacobian

---

†For definitions of image of set S under function g, and inverse image of set $\mathfrak{X}$ under g, see P. 2.16 below.

CALCULUS

$$J_g(u, v) = \begin{vmatrix} \dfrac{\partial g_1}{\partial u} & \dfrac{\partial g_1}{\partial v} \\ \dfrac{\partial g_2}{\partial u} & \dfrac{\partial g_2}{\partial v} \end{vmatrix}, \quad g = (g_1, g_2)$$

be never zero on B^*. Let f be a real-valued function defined and continuous on A. Then

(18) $$\iint_A f(x, y) \, dx \, dy = \int \int_{g^{-1}(A)} f[g_1(u, v), g_2(u, v)] \, |J_g(u, v)| \, du \, dv.$$

Theorem 17 can also be extended to the case in which A^* can be written as a disjoint union of a finite number of sets, $A_1, A_2, \cdots, A_k$, say, such that the transformation from each A_i to B^* is one to one onto.

12. Interchange of Limit Operations

Let

$$g_1(x) + g_2(x) + \cdots + g_n(x) + \cdots$$

be an infinite series of functions which converges uniformly to $g(x)$. If every $g_n(x)$, $n = 1, 2, \cdots$, is defined on $[a, b]$ and continuous at $x_0 \in [a, b]$, then $g(x)$ is continuous at x_0.

Theorem 18 (Courant and John [16], 537–538).

(a) Let $\{g_n(x)\}$ be a sequence of integrable functions on $[a, b]$ which converges uniformly to a function $g(x)$. Then g is integrable, and

$$\lim_{n \to \infty} \int_a^b g_n(x) \, dx = \int_a^b g(x) \, dx.$$

(b) Let

$$g_1(x) + g_2(x) + \cdots + g_n(x) + \cdots$$

be an infinite series of functions which converges uniformly to $g(x)$. If every $g_n(x)$, $n = 1, 2, \cdots$, is defined and integrable on $[a, b]$, then g is integrable on $[a, b]$; $\sum_{n=1}^{\infty} \int_a^b g_n(x) \, dx$ converges and

$$\int_a^b \sum_{n=1}^{\infty} g_n(x) \, dx = \sum_{n=1}^{\infty} \int_a^b g_n(x) \, dx.$$

Theorem 19 (Courant and John [16], 539). If, on differentiating a convergent infinite series $\sum_{n=1}^{\infty} g_n(x) = g(x)$ of differentiable functions on $[a, b]$

term by term, we obtain a uniformly convergent series of continuous terms $\sum_{n=1}^{\infty} g_n'(x)$, then $\sum_{n=1}^{\infty} g_n(x)$ converges uniformly and

$$\frac{d}{dx} \sum_{n=1}^{\infty} g_n(x) = \sum_{n=1}^{\infty} \frac{d}{dx} g_n(x).$$

13. Interchange of Limit Operations in Infinite Integrals

Integrals over infinite intervals are usually referred to as *infinite integrals*, and integrals with unbounded integrands are known as *improper integrals*. If f is integrable on every finite closed subinterval of $[a, \infty)$ and $\lim_{\alpha \to \infty} \int_a^{\alpha} f(x) \, dx$ exists, we say that the integral $\int_a^{\infty} f(x) \, dx$ converges. Similarly, if f is integrable in every finite closed subinterval of $(-\infty, a]$ and $\lim_{b \to -\infty} \int_b^a f(x) \, dx$ exists, we say that $\int_{-\infty}^a f(x) \, dx$ converges. If both $\int_a^{\infty} f(x) \, dx$ and $\int_{-\infty}^a f(x) \, dx$ exist, then $\int_{-\infty}^a f(x) \, dx + \int_a^{\infty} f(x) \, dx$ is denoted by $\int_{-\infty}^{\infty} f(x) \, dx$. It should be noted that $\lim_{\alpha \to \infty} \int_{-\alpha}^{\alpha} f(x) \, dx$ may exist but $\int_{-\infty}^{\infty} f(x) \, dx$ may not converge. If $\int_{-\infty}^{\infty} f(x) \, dx$ converges, then

$$\int_{-\infty}^{\infty} f(x) \, dx = \lim_{a \to \infty} \int_{-a}^{a} f(x) \, dx.$$

If f is bounded and integrable in $[a, b - \varepsilon]$ for every $\varepsilon > 0$ and the limit $\lim_{\substack{\varepsilon \to 0 \\ \varepsilon > 0}} \int_a^{b-\varepsilon} f(x) \, dx$ exists, we write

$$\int_a^b f(x) \, dx = \lim_{\substack{\varepsilon \to 0 \\ \varepsilon > 0}} \int_a^{b-\varepsilon} f(x) \, dx,$$

and call $\int_a^b f(x) \, dx$ an *improper integral*.

Similarly, we may consider integrals of the type $\int_a^b f(x, t) \, dx$ or $\int_c^d \int_a^b f(x, t) \, dx \, dt$, where any or all of the numbers a, b, c, d may be infinite.

Theorem 20 (Miller [82], 157). Let $f(x, t)$ be continuous for t in an interval I and for all $x \geq a$. If $\int_a^{\infty} f(x, t) \, dx$ converges uniformly for $t \in I$ to the function $\psi(t)$, then ψ is continuous on I.

Theorem 21 (Miller [82], 157). Let $f(x, t)$ and $\partial/\partial t \, f(x, t)$ be continuous for t in an interval I and for all $x \geq a$. Let $\int_a^{\infty} f(x, t) \, dx$ and $\int_a^{\infty} \partial/\partial t \, f(x, t) \, dx$ converge uniformly for $t \in I$. Then

$$\frac{d}{dt} \int_a^{\infty} f(x, t) \, dx = \int_a^{\infty} \frac{d}{dt} f(x, t) \, dx$$

for all $t \in I$.

Theorem 22 (Miller [82], 161). Let $f(x, t)$ be continuous for all $x \geq a$ and all $t \geq c$. Let $\int_a^{\infty} f(x, t) \, dx$ converge uniformly for t in every closed, bounded subinterval of $[c, \infty)$, and $\int_c^{\infty} f(x, t) \, dt$ converge uniformly for x in every closed, bounded subinterval of $[a, \infty)$. Let $\int_a^\infty \int_c^t f(x, y) \, dy \, dx$ converge uniformly for $t \geq c$. Then $\int_a^\infty \int_c^\infty f(x, t) \, dt \, dx$ and $\int_c^\infty \int_a^\infty f(x, t) \, dx \, dt$ exist and are equal.

Theorem 23 (Apostol [3], 451). Let f_n be a sequence of functions defined on $[a, \infty]$ such that $f_n(x) \geq 0$ for $x \geq a$ and $n = 1, 2, \cdots$. If

$$\int_a^b \sum_{n=1}^\infty f_n(x) \, dx = \sum_{n=1}^\infty \int_a^b f_n(x) \, dx$$

holds for every $b \geq a$, then

(19) $$\int_a^\infty \sum_{n=1}^\infty f_n(x) \, dx = \sum_{n=1}^\infty \int_a^\infty f_n(x) \, dx,$$

provided that either side of (19) is convergent.

Theorem 24 (Apostol [3], 373–374). If either of the series $\sum_{m=1}^\infty \sum_{n=1}^\infty a_{mn}$, $\sum_{n=1}^\infty \sum_{m=1}^\infty a_{mn}$ is absolutely convergent, then so is the other and their sums are the same.

14. The Laplace Transform

If the integral $\int_0^\infty e^{-st} f(s) \, ds$ converges for some value of t, we say that the function defined by

(20) $$M(t) = \int_0^\infty e^{-st} f(s) \, ds$$

is the *unilateral Laplace transform* of f. In probability theory we are interested in the integral

(21) $$M(t) = \int_{-\infty}^\infty e^{-st} f(s) \, ds,$$

which, if it converges for some value of t, is known as the *bilateral Laplace transform* of f. Since, however,

$$\int_{-\infty}^\infty e^{-st} f(s) \, ds = \int_0^\infty e^{-st} f(s) \, ds + \int_0^\infty e^{st} f(-s) \, ds,$$

the study of a bilateral transform is reduced to that of the sum of two unilateral transforms in one of which the variable t has been replaced by $-t$.

Theorem 25 (Widder [137], 442). If the integral in (20) converges at $t = t_0$, it converges for $t > t_0$.

Theorem 26 (Widder [137], 446–447). Let $M(t)$ be the Laplace transform of f for $t > t_0$. Then we can differentiate the integral in (20) under the sign of integration any number of times, that is,

$$M^{(k)}(t) = \int_0^\infty (-t)^k e^{-st} f(s)\, ds, \qquad t > t_0, \quad k = 1, 2, \cdots.$$

We may therefore expand $M(t)$ in a Maclaurin series,

$$M(t) = \sum_{k=0}^\infty M^{(k)}(0)\, \frac{t^k}{k!},$$

in the interval of convergence.

Theorem 27 (Widder [137], 460). The Laplace transform M of a continuous function f is unique.

15. Absolute Continuity

A real-valued function f defined on $[a, b]$ is said to be *absolutely continuous* on $[a, b]$ if, for every $\varepsilon > 0$, there exists a $\delta > 0$ such that

$$\sum_{k=1}^n |f(x_k') - f(x_k)| < \varepsilon$$

for every finite collection $\{(x_k, x_k')\}$ of disjoint subintervals with

$$\sum_{k=1}^n |x_k' - x_k| < \delta.$$

It is clear that an absolutely continuous function is continuous. Some other properties of absolutely continuous functions are given in Theorems 28 and 29.

Theorem 28 (Royden [107], 90). If f is absolutely continuous, then $f'(x)$ exists almost everywhere.

Theorem 29 (Royden [107], 91). A function F is an indefinite integral if and only if it is absolutely continous. Thus every absolutely continuous function is the indefinite integral of its derivative.

16. Borel-Measurable Functions

Let f be a function from a set Ω into a set R. Let $A \subseteq \Omega$. By the *image* of A under f we mean the set of elements y of R such that $y = f(\omega)$ for some $\omega \in A$. We denote this image by $f(A)$. If $B \subseteq R$, the *inverse image* of B under f is the set of those $\omega \in \Omega$ for which $f(\omega) \in B$. We write

$$f^{-1}(B) = \{\omega \in \Omega : f(\omega) \in B\}.$$

It can be shown that f^{-1} has the following properties:

(22) $$f^{-1}\left(\bigcup_{\lambda \in \Lambda} B_\lambda\right) = \bigcup_{\lambda \in \Lambda} f^{-1}(B_\lambda),$$

(23) $$f^{-1}\left(\bigcap_{\lambda \in \Lambda} B_\lambda\right) = \bigcap_{\lambda \in \Lambda} f^{-1}(B_\lambda),$$

(24) $$f^{-1}(B^c) = (f^{-1}(B))^c.$$

Let $\Omega = \mathscr{R}_n$, and $\mathfrak{B}_n$ be the Borel σ-field on $\mathscr{R}_n$. A function $f: \mathscr{R}_n \to \mathscr{R}_m$ ($1 \leq m \leq n$) is said to be *Borel measurable* if $f^{-1}(B_m) \in \mathfrak{B}_n$ for every m-dimensional Borel set $B_m \in \mathfrak{B}_m$. The case $m = n = 1$ is of particular importance. A real-valued function f of a real variable x is Borel measurable if the set $\{x: -\infty < f(x) \leq y\}$ is a Borel set for every real number y.

P.3 LINEAR ALGEBRA

In the following, $\mathbf{x} = (x_1, x_2, \cdots, x_n)$ will denote a *row vector* of real numbers, and set of all vectors $\mathbf{x}$ will be denoted by $V_n (= \mathscr{R}_n)$. $\mathbf{x}'$ denotes the *column vector* $(x_1, x_2, \cdots, x_n)$, $\mathbf{0}_n$ is the *zero vector* $(0, 0, \cdots, 0)$ in V_n and $\mathbf{1}_n$ is the *unit vector* $(1, 1, \cdots, 1)$ in V_n. The *scalar* (or *inner*) *product* of $\mathbf{x}$ and $\mathbf{y}$ in V_n is the scalar $\sum_{i=1}^n x_i y_i$ and we write $\mathbf{xy}' = \sum_{i=1}^n x_i y_i$. Clearly $\mathbf{xy}' = \mathbf{yx}'$. Two nonzero vectors $\mathbf{x}$, $\mathbf{y}$ in V_n are *orthogonal* if and only if $\mathbf{xy}' = 0$.

For a set of s vectors $\{\mathbf{a}_1, \mathbf{a}_2, \cdots, \mathbf{a}_s\}$ in V_n, the *vector space* V spanned by $\{\mathbf{a}_1, \mathbf{a}_2, \cdots, \mathbf{a}_s\}$ is the set of all vectors that are linear combinations of the $\mathbf{a}_j$'s. V is also known as a *linear subspace* of V_n. The set of vectors $\{\mathbf{a}_1, \mathbf{a}_2, \cdots, \mathbf{a}_n\}$ is said to be *linearly dependent* if there exists a set of scalars $\{a_1, a_2, \cdots, a_n\}$ not all zero and such that $a_1 \mathbf{a}_1 + a_2 \mathbf{a}_2 + \cdots + a_n \mathbf{a}_n = \mathbf{0}_n$. If no such scalars exist, we say that the set is *linearly independent*.

Let V be a vector space. A set of linearly independent vectors that span V is known as a *basis* for V, and the *dimension* of V is the number of vectors in any basis for V. Every vector space has a basis, and any two bases for a vector space contain the same number of vectors (Scheffé [111], 378–379).

If V_r is an r-dimensional vector space of n-tuples contained in V_n, we write $V_r \subset V_n$. Let $\{\mathbf{a}_1, \mathbf{a}_2, \cdots, \mathbf{a}_n\}$ be a basis for V_n and $\mathbf{x} \in V_n$. The coefficient a_i ($i = 1, 2, \cdots, n$) of $\mathbf{a}_i$ in the unique linear representation $\mathbf{x} = \sum_{i=1}^n a_i \mathbf{a}_i$ of $\mathbf{x}$ in terms of vectors of the basis is called the ith *coordinate of* $\mathbf{x}$ *with respect to the basis* $\{\mathbf{a}_1, \mathbf{a}_2, \cdots, \mathbf{a}_n\}$. A basis $\{\mathbf{a}_1, \mathbf{a}_2, \cdots, \mathbf{a}_r\}$ for $V_r \subset V_n$ is said to be *orthonormal* if the r vectors $\mathbf{a}_i$ are pairwise *orthogonal* and have unit *norm* (or *length*), namely, $\|\mathbf{a}_i\| = (\mathbf{a}_i \mathbf{a}_i')^{1/2} = 1$.

Theorem 1 (Scheffé [111], 381).

(a) If the vectors $\alpha_1, \alpha_2, \cdots, \alpha_r$ are pairwise orthogonal (and nonzero), they are linearly independent.
(b) Any linearly independent set of r vectors in $V_r \subset V_n$ is a basis for V_r.

It follows that any set of r (nonzero) orthogonal vectors in $V_r \subset V_n$ is a basis for V_r.

Theorem 2 (Scheffé [111], 382). Let $\{\alpha_1, \alpha_2, \cdots, \alpha_r\}$ be an arbitrary basis for V_r. Then there exists an orthonormal basis $\{\gamma_1, \gamma_2, \cdots, \gamma_r\}$ for V_r such that each γ_i is a linear combination of the α_i's.

Theorem 3 (Scheffé [111], 382). If $\{\alpha_1, \alpha_2, \cdots, \alpha_r\}$ is an orthonormal basis for $V_r \subset V_n$, it can always be extended to an orthonormal basis $\{\alpha_1, \alpha_2, \cdots, \alpha_r, \alpha_{r+1}, \cdots, \alpha_n\}$ for V_n.

Capital letters will denote matrices. Let $\mathbf{A}^{m \times n}$ be an $m \times n$ (m rows, n columns) *matrix* of real numbers a_{ij}, $i = 1, 2, \cdots, m$; $j = 1, 2, \cdots, n$. We write $\mathbf{A} = (a_{ij})$ and suppress the superscript $m \times n$. The *transpose* of $\mathbf{A}$ will be denoted by $\mathbf{A}'$. The *zero matrix* will be denoted by $\mathbf{0}$, and the $n \times n$ *identity matrix* by $\mathbf{I}_n$. We recall that the *matrix product* of an $m \times n$ matrix $\mathbf{A} = (a_{ij})$ by an $n \times r$ matrix $\mathbf{B} = (b_{jk})$ is an $n \times r$ matrix $\mathbf{C} = (c_{ik})$, where $c_{ik} = \sum_{j=1}^{n} a_{ij} b_{jk}$, and we write $\mathbf{C} = \mathbf{AB}$. We note that the matrix product is not commutative.

A *square matrix* of *order* n (that is, an $n \times n$ matrix) $\mathbf{A}$ is said to be *singular* if its *determinant* $|\mathbf{A}| = 0$; otherwise we call it *nonsingular*. If there exists a matrix $\mathbf{B}$ such that $\mathbf{AB} = \mathbf{BA} = \mathbf{I}_n$, we call $\mathbf{B}$ the *inverse* of $\mathbf{A}$ and write $\mathbf{B} = \mathbf{A}^{-1}$. It is well-known (Scheffé [111], 393) that a square matrix $\mathbf{A}$ has an inverse if and only if $\mathbf{A}$ is nonsingular. Then $\mathbf{A}^{-1}$ is unique. If $\mathbf{A}$ is nonsingular, $(\mathbf{A}')^{-1} = (\mathbf{A}^{-1})'$; and if $\mathbf{A}$ and $\mathbf{B}$ are both nonsingular, so is $\mathbf{AB}$, and $(\mathbf{AB})^{-1} = \mathbf{B}^{-1} \mathbf{A}^{-1}$.

Let $\mathbf{A}^{n \times m} = (\alpha'_1, \alpha'_2, \cdots, \alpha'_m)$. Then the column vectors α'_j may be considered to be vectors in V_n. The *rank* of $\mathbf{A}$ is the maximum number of linearly independent vectors in $\{\alpha'_1, \alpha'_2, \cdots, \alpha'_m\}$, that is, it is the dimension of the vector space spanned by the columns of $\mathbf{A}$. It is well-known (Scheffé [111], 394) that rank $\mathbf{A}^{m \times n}$ = maximum number of linearly independent rows $\leq$ min (m, n) and rank $\mathbf{AB} \leq$ min (rank A, rank B). Moreover, if $\mathbf{A}$ is $m \times n$ and $\mathbf{P}^{m \times m}$ and $\mathbf{Q}^{n \times n}$ are nonsingular, rank $\mathbf{PAQ}$ = rank $\mathbf{A}$. If rank $\mathbf{A}^{m \times n}$ = min (m, n), we say that $\mathbf{A}$ has *full rank*.

An $n \times n$ matrix $\mathbf{P}$ is said to be *orthogonal* if $\mathbf{PP}' = \mathbf{I}_n$; the transformation $\mathbf{x} = \mathbf{yP}'$ is then known as an *orthogonal transformation*.

Theorem 4 (Scheffé [111], 397). Let $\mathbf{A}^{n \times n}$ be a symmetric matrix. Then there exists an orthogonal matrix $\mathbf{P}^{n \times n}$ such that $\mathbf{P}'\mathbf{AP}$ is a diagonal matrix; that is, the off-diagonal elements of $\mathbf{P}'\mathbf{AP}$ are all 0.

Let $\mathbf{A}$ be $n \times n$. Then $|\mathbf{A} - \lambda \mathbf{I}_n|$, a polynomial in λ of degree n, is known as the *characteristic polynomial* of $\mathbf{A}$. The roots of $|\mathbf{A} - \lambda \mathbf{I}_n|$ are termed the *characteristic roots* of $\mathbf{A}$. We say that $\mathbf{A}^{n \times n}$ is *positive definite* if all the characteristic roots of $\mathbf{A}$ are positive, and *positive semidefinite* if all the characteristic roots are nonnegative.

Regardless of what orthogonal $\mathbf{P}$ is used to diagonalize $\mathbf{A}$ in Theorem 4, the elements $\{\lambda_i\}$ of the diagonal matrix $\mathbf{P}'\mathbf{AP}$ are always the same except for order. These $\{\lambda_i\}$ are the roots of the characteristic polynomial.

Theorem 5 (Scheffé [111], 399). For any matrix $\mathbf{A}$, the matrix $\mathbf{AA}'$ is symmetric and positive semidefinite and has the same rank as $\mathbf{A}$.

CHAPTER 1

Probability

1.1 INTRODUCTION

The theory of probability had its origin in gambling and games of chance. It owes much to the curiosity of gamblers who pestered their friends in the mathematical world with all sorts of questions. Unfortunately this association with gambling contributed to a very slow and sporadic growth of probability theory as a mathematical discipline. The mathematicians of the day took little or no interest in the development of any theory but looked only at the combinatorial reasoning involved in each problem.

The first attempt at some mathematical rigor is credited to Laplace. In his monumental work, *Theorie analytique des probabilités* (1812), Laplace gave the classical definition of the probability of an event that can occur only in a finite number of ways as the proportion of the number of favorable outcomes to the total number of all possible outcomes, provided that all the outcomes are *equally likely*. According to this definition, the computation of the probability of events was reduced to combinatorial counting problems. Even in those days, this definition was found inadequate. In addition to being circular and restrictive, it did not answer the question of what probability is, it only gave a practical method of computing the probabilities of some simple events.

An extension of the classical definition of Laplace was used to evaluate the probabilities of sets of events with infinite outcomes. The notion of *equal likelihood* of certain events played a key role in this development. According to this extension, if Ω is some region with a well-defined measure (length, area, volume, etc.), the probability that a point chosen *at random* lies in a subregion A of Ω is the ratio measure (A) / measure (Ω). Many problems of geometric probability were solved using this extension. The trouble is that one can define "at random" in any way one pleases, and different definitions therefore lead to different answers. Joseph Bertrand, for example, in his book *Calcul des probabilités* (Paris, 1889) cited a number of problems in geometric probability where the result depended on the method of solution. In Example 1.3.9 we will discuss the famous Bertrand paradox and show that in reality there is nothing paradoxical about Bertrand's paradoxes; once we define "probability spaces" carefully, the paradox is resolved. Nevertheless difficulties encountered in the field of geometric probability have been largely responsible for the slow growth of probability theory and its tardy acceptance by mathematicians as a mathematical discipline.

The mathematical theory of probability, as we know it today, is of comparatively recent origin. It was A. N. Kolmogorov who axiomatized probability in his fundamental work, *Foundations of the Theory of Probability* (Berlin), in 1933. According to this development, random events are represented by sets and probability is just a *normed measure* defined on these sets. This measure-theoretic development not only provided a logically consistent foundation for probability theory but also, at the same time, joined it to the mainstream of modern mathematics.

In this book we follow Kolmogorov's axiomatic development. In Section 2 we introduce the notion of a sample space. In Section 3 we state Kolmogorov's axioms of probability and study some simple consequences of these axioms. Section 4 is devoted to the computation of probability on finite sample spaces. Section 5 deals with conditional probability and Bayes's rule, while Section 6 examines the independence of events.

1.2 SAMPLE SPACE

In most branches of knowledge experiments are a way of life. In probability and statistics, too, we concern ourselves with special types of experiments. Consider the following examples.

Example 1. A coin is tossed. Assuming that the coin does not land on the side, there are two possible outcomes of the experiment: heads and tails. On any performance of this experiment one does not know what the outcome will be. The coin can be tossed as many times as desired.

Example 2. A roulette wheel is a circular disk divided into 38 equal sectors numbered from 0 to 36 and 00. A ball is rolled on the edge of the wheel, and the wheel is rolled in the opposite direction. One bets on any of the 38 numbers or some combinations of them. One can also bet on a color, red or black. If the ball lands in the sector numbered 32, say, anybody who bet on 32 or combinations including 32 wins, and so on. In this experiment, all possible outcomes are known in advance, namely 00, 0, 1, 2, $\cdots$, 36, but on any performance of the experiment there is uncertainty as to what the outcome will be, provided, of course, that the wheel is not rigged in any manner. Clearly, the wheel can be rolled any number of times.

Example 3. A manufacturer produces footrules. The experiment consists in measuring the length of a footrule produced by the manufacturer as accurately as possible. Because of errors in the production process one does not know what the true length of the footrule selected will be. It is clear, however, that the length will be, say, between 11 and 13 inches, or, if one wants to be safe, between 6 and 18 inches.

Example 4. The length of life of a light bulb produced by a certain manufacturer is recorded. In this case one does not know what the length of life will be for the light bulb selected, but clearly one is aware in advance that it will be some number between 0 and ∞ hours.

The experiments described above have certain common features. For each experiment, we know in advance all possible outcomes, that is, there are no surprises in store after any performance of the experiment. On any performance of the experiment, however, we do not know what the specific outcome will be, that is, there is uncertainty about the outcome on any performance of the experiment. Moreover, the experiment can be repeated under identical conditions. These features describe a *random* (or a *statistical*) *experiment*.

Definition 1. A random (or a statistical) experiment is an experiment in which

(a) all outcomes of the experiment are known in advance,
(b) any performance of the experiment results in an outcome that is not known in advance, and
(c) the experiment can be repeated under identical conditions.

In probability theory we study this uncertainty of a random experiment. It is convenient to associate with each such experiment a set Ω, the set of all possible outcomes of the experiment. To engage in any meaningful discussion about the experiment, we associate with Ω a σ-field $\mathscr{S}$, of subsets of Ω.

Definition 2. The sample space of a statistical experiment is a pair $(\Omega, \mathscr{S})$, where

(a) Ω is the set of all possible outcomes of the experiment, and
(b) $\mathscr{S}$ is a σ-field of subsets of Ω.

We recall that a σ-field is a nonempty class of subsets of Ω that is closed under the formation of countable unions and complements and contains the null set ϕ.

The elements of Ω are called *sample points*. Any set $A \in \mathscr{S}$ is known as an *event*. Clearly A is a collection of sample points. We say that an event A happens if the outcome of the experiment corresponds to a point in A. Each one-point set is known as a *simple* or an *elementary event*. If the set Ω contains only a finite number of points, we say that $(\Omega, \mathscr{S})$ is a *finite sample space*. If Ω contains at most a countable number of points, we call $(\Omega, \mathscr{S})$ a *discrete sample space*. If, however, Ω contains uncountably many points, we say that $(\Omega, \mathscr{S})$ is an *uncountable sample space*. In particular, if $\Omega = \mathscr{R}_k$ or some rectangle in $\mathscr{R}_k$, we call it a *continuous sample space*.

Remark 1. The choice of $\mathscr{S}$ is an important one, and some remarks are in order. If Ω contains at most a countable number of points, we can always take $\mathscr{S}$ to be the class of all subsets of Ω. This is certainly a σ-field. Each one point set is a member of $\mathscr{S}$ and is the fundamental object of interest. Every subset of Ω is an event. If Ω has uncountably many points, the class of all subsets of Ω is still a σ-field, but it is much too large a class of sets to be of interest. It may not be possible to choose the class of all subsets of Ω as $\mathscr{S}$. One of the most important examples of an uncountable sample space is the case in which $\Omega = \mathscr{R}$ or Ω is an interval in $\mathscr{R}$. In this case we would like all one-point subsets of Ω and all intervals (closed, open, or semiclosed) to be events. We use our knowledge of analysis to specify $\mathscr{S}$. We will not go into details here except to recall (Theorem P.1.2) that the class of all semiclosed intervals $(a, b]$ generates a class $\mathfrak{B}_1$ which is a σ-field on $\mathscr{R}$. This class contains all one-point sets and all intervals (finite or infinite). We take $\mathscr{S} = \mathfrak{B}_1$. There are many subsets of $\mathscr{R}$ that are not in $\mathfrak{B}_1$, but we will not demonstrate this fact here. We refer the reader to Halmos [45], Royden [107], or Kolmogorov and Fomin [61] for further details.

We will frequently use the fact that every Borel set $B \in \mathfrak{B}_1$ can be obtained by a countable number of operations of unions, intersections, and differences on intervals. The following relations will be frequently used in subsequent sections:

$$\{x\} = \bigcap_{n=1}^{\infty} (x - \frac{1}{n}, x],$$

$$(x, y) = (x, y] - \{y\},$$

$$[x, y] = (x, y] + \{x\},$$

$$[x, y) = \{x\} + (x, y] - \{y\},$$

$$(x, +\infty) = (-\infty, x]^c,$$

and so on. Similar remarks apply when $\Omega = \mathscr{R}_k$ or a rectangle in $\mathscr{R}_k$ and

$\mathscr{S} = \mathfrak{B}_k$. Since we will be dealing mostly with the one-dimensional case, we will write $\mathfrak{B}$ instead of $\mathfrak{B}_1$.

Example 5. Let us toss a coin. The set Ω is the set of symbols H and T, where H denotes head and T represents tail. Also, $\mathscr{S}$ is the class of all subsets of Ω, namely, $\{\{H\}, \{T\}, \{H, T\}, \phi\}$. If the coin is tossed two times, then

$\Omega = \{(H, H), (H, T), (T, H), (T, T)\}, \quad \mathscr{S} = \{\varnothing, \{(H, H)\},$
$\{(H, T)\}, \{(T, H)\}, \{(T, T)\}, \{(H, H), (H, T)\}, \{(H, H), (T, H)\},$
$\{(H, H), (T, T)\}, \{(H, T), (T, H)\}, \{(T, T), (T, H)\}, \{(T, T),$
$(H, T)\}, \{(H, H), (H, T), (T, H)\}, \{(H, H), (H, T), (T, T)\},$
$\{(H, H), (T, H), (T, T)\}, \{(H, T), (T, H), (T, T)\}, \Omega\}$

where the first element of a pair denotes the outcome of the first toss, and the second element, the outcome of the second toss. The event *at least one head* consists of sample points (H, H), (H, T), (T, H). The event *at most one head* is the collection of sample points (H, T), (T, H), (T, T).

Example 6. A die is rolled n times. The sample space is the pair $(\Omega, \mathscr{S})$, where Ω is the set of all n-tuples $(x_1, x_2, \cdots, x_n)$, $x_i \in \{1, 2, 3, 4, 5, 6\}$, $i = 1, 2, \cdots, n$, and $\mathscr{S}$ is the class of all subsets of Ω. Ω contains 6^n elementary events. The event A that 1 shows at least once is the set

$A = \{(x_1, x_2, \cdots, x_n): \text{ at least one of } x_i\text{'s is } 1\}$
$ = \Omega - \{(x_1, x_2, \cdots, x_n): \text{ none of the } x_i\text{'s is } 1\}$
$ = \Omega - \{(x_1, x_2, \cdots, x_n): x_i \in \{2, 3, 4, 5, 6\}, i = 1, 2, \cdots, n\}.$

Example 7. A coin is tossed until the first head appears. Then

$\Omega = \{H, (T, H), (T, T, H), (T, T, T, H), \cdots\},$

and $\mathscr{S}$ is the class of all subsets of Ω. An equivalent way of writing Ω would be to look at the number of tosses required for the first head. Clearly, this number can take values 1, 2, 3, $\cdots$, so that Ω is the set of all positive integers. Then $\mathscr{S}$ is the class of all subsets of positive integers.

Example 8. Consider a pointer that is free to spin about the center of a circle. If the pointer is spun by an impulse, it will finally come to rest at some point. On the assumption that the mechanism is not rigged in any manner, each point on the circumference is a possible outcome of the experiment. The set Ω consists of all points $0 \leq x < 2\pi r$, where r is the radius of the circle. Every one-point set $\{x\}$ is a simple event, namely, that the pointer will come to rest at x. The events of interest are those in which the pointer stops at a point belonging to a specified arc. Here $\mathscr{S}$ is taken to be the Borel σ-field of subsets of $[0, 2\pi r)$.

SAMPLE SPACE

Example 9. A rod of length l is thrown onto a flat table, which is ruled with parallel lines at distance $2l$. The experiment consists in noting whether or not the rod intersects one of the ruled lines.

Let r denote the distance from the center of the rod to the nearest ruled line, and let θ be the angle that the axis of the rod makes with this line (Fig. 1). Every outcome of this experiment corresponds to a point (r, θ) in the plane. As Ω we take the set of all points (r, θ) in $\{(r, \theta): 0 \leq r \leq l, 0 \leq \theta < \pi\}$. For $\mathscr{S}$ we take the Borel σ-field, $\mathfrak{B}_2$, of subsets of Ω, that is, the smallest σ-field generated by rectangles of the form

$$\{(x, y): a < x \leq b, \quad c < y \leq d, \quad 0 \leq a < b \leq l, \quad 0 \leq c < d < \pi\}.$$

Clearly the rod will intersect a ruled line if and only if the center of the rod lies in the area enclosed by the locus of the center of the rod (while one end touches the nearest line) and the nearest line (shaded area in Fig. 2).

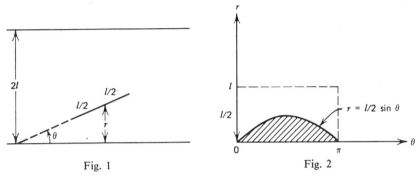

Fig. 1 Fig. 2

Remark 2. From the discussion above it should be clear that in the discrete case there is really no problem. Every one-point set is also an event, and $\mathscr{S}$ is the class of all subsets of Ω. The problem, if there is any, arises only in regard to uncountale sample spaces. The reader has to remember only that in this case not all subsets of Ω are events. The case of interest is the one in which $\Omega = \mathscr{R}_k$. In this case roughly all sets that have a well-defined volume (or area or length) are events. Not every set has the property in question, but sets that lack it are not easy to find and one does not encounter them in practice.

PROBLEMS 1.2

1. A club has five members A, B, C, D, and E. It is required to select a chairman and a secretary. Assuming that one member cannot occupy both positions, write the sample space associated with these selections. What is the event that member A is an office holder?

2. In each of the following experiments, what is the sample space?
(a) In a survey of families with three children, the sexes of the children are recorded in increasing order of age.
(b) The experiment consists of selecting four items from a manufacturer's output and observing whether or not each item is defective.
(c) A given book is opened to any page, and the number of misprints is counted.
(d) Two cards are drawn   (i) with replacement,   (ii) without replacement from an ordinary deck of cards.

3. Let A, B, C be three arbitrary events on a sample space $(\Omega, \mathscr{S})$. What is the event that only A occurs? What is the event that at least two of A, B, C occur? What is the event that both A and C, but not B, occur? What is the event that at most one of A, B, C occurs?

1.3 PROBABILITY AXIOMS

Let $(\Omega, \mathscr{S})$ be the sample space associated with a statistical experiment. In this section we define a probability set function and study some of its properties.

Definition 1. Let $(\Omega, \mathscr{S})$ be a sample space. A set function P defined on $\mathscr{S}$ is called a probability measure (or simply probability) if it satisfies the following conditions:

(i) $P(A) \geq 0$ for all $A \in \mathscr{S}$.
(ii) $P(\Omega) = 1$.
(iii) Let $\{A_j\}$, $A_j \in \mathscr{S}$, $j = 1, 2, \cdots$, be a disjoint sequence of sets, that is, $A_j \cap A_k = \phi$ for $j \neq k$. Then

(1) $$P\left(\sum_{j=1}^{\infty} A_j\right) = \sum_{j=1}^{\infty} P(A_j).$$

We call $P(A)$ the *probability of event A*. If there is no confusion, we write PA instead of $P(A)$. Property (iii) is called *countable additivity*. That $P\phi = 0$ and P is also finitely additive follows from it.

Remark 1. If Ω is discrete and contains at most $n (< \infty)$ points, each single-point set $\{\omega_j\}$, $j = 1, 2, \cdots, n$, is an elementary event, and it is sufficient to assign probability to each $\{\omega_j\}$. Then, if $A \in \mathscr{S}$, where $\mathscr{S}$ is the class of all subsets of Ω, $PA = \sum_{\omega \in A} P\{\omega\}$. One such assignment is the *equally likely* assignment or the assignment of *uniform* probabilities. According to this assignment, $P\{\omega_j\} = 1/n$, $j = 1, 2, \cdots, n$. Thus $PA = m/n$ if A contains m elementary events, $1 \leq m \leq n$.

Remark 2. If Ω is discrete and contains a countable number of points, one cannot make an equally likely assignment of probabilities. It suffices to make the assignment for each elementary event. If $A \in \mathscr{S}$, where $\mathscr{S}$ is the class of all subsets of Ω, define $PA = \sum_{\omega \in A} P\{\omega\}$.

Remark 3. If Ω contains uncountably many points, each one-point set is an elementary event, and again one cannot make an equally likely assignment of probabilities. Indeed, one cannot assign positive probability to each elementary event without violating the axiom $P\Omega = 1$. In this case one assigns probabilities to compound events consisting of intervals. For example, if $\Omega = [0, 1]$, and $\mathscr{S}$ is the Borel σ-field of all subsets of Ω, the assignment $P[I] = $ length of I, where I is a subinterval of Ω, defines a probability.

Definition 2. The triple $(\Omega, \mathscr{S}, P)$ is called a probability space.

Definition 3. Let $A \in \mathscr{S}$; we say that the odds for A are a to b if $PA = a/(a+b)$, and then the odds against A are b to a.

Example 1. Let us toss a coin. The sample space is $(\Omega, \mathscr{S})$, where $\Omega = \{H, T\}$, and $\mathscr{S}$ is the σ-field of all subsets of Ω. Let us define P on $\mathscr{S}$ as follows:

$$P\{H\} = 1/2, \quad P\{T\} = 1/2.$$

Then P clearly defines a probability. Similarly, $P\{H\} = 2/3$, $P\{T\} = 1/3$, and $P\{H\} = 1$, $P\{T\} = 0$ are probabilities defined on $\mathscr{S}$. Indeed,

$$P\{H\} = p \quad \text{and} \quad P\{T\} = 1 - p \quad (0 \leq p \leq 1)$$

defines a probability on $(\Omega, \mathscr{S})$.

Example 2. Let $\Omega = \{1, 2, 3, \cdots\}$ be the set of positive integers, and let $\mathscr{S}$ be the class of all subsets of Ω. Define P on $\mathscr{S}$ as follows:

$$P\{i\} = \frac{1}{2^i}, \quad i = 1, 2, \cdots.$$

Then $\sum_{i=1}^{\infty} P\{i\} = 1$, and P defines a probability.

Example 3. Let $\Omega = (0, \infty)$ and $\mathscr{S} = \mathfrak{B}$, the Borel σ-Field on Ω. Define P as follows: for each interval $I \subseteq \Omega$,

$$PI = \int_I e^{-x} \, dx.$$

Clearly $PI \geq 0$, $P\Omega = 1$, and P is countably additive by properties of integrals.

Theorem 1. P is monotone and subtractive; that is, if $A, B \in \mathscr{S}$ and $A \subseteq B$, then $PA \leq PB$ and $P(B - A) = PB - PA$.

Proof. If $A \subseteq B$, then
$$B = (A \cap B) + (B - A) = A + (B - A),$$
and it follows that $PB = PA + P(B - A)$.

Corollary. For all $A \in \mathscr{S}$, $0 \leq PA \leq 1$.

Remark 4. We wish to emphasize at this point that, if $PA = 0$ for some $A \in \mathscr{S}$, we call A an event with *zero probability* or a *null* event. However, it does not follow that $A = \phi$. Similarly, if $PB = 1$ for some $B \in \mathscr{S}$, we call B a *certain event* but it does not follow that $B = \Omega$.

Theorem 2 (The Addition Rule). If $A, B \in \mathscr{S}$, then

(2) $\qquad P(A \cup B) = PA + PB - P(A \cap B)$.

Proof.
$$A \cup B = (A - B) + (B - A) + (A \cap B),$$
and
$$A = (A \cap B) + (A - B), \quad B = (A \cap B) + (B - A).$$
The result follows by countable additivity of P.

Corollary 1. P is subadditive, that is, if $A, B \in \mathscr{S}$, then

(3) $\qquad P(A \cup B) \leq PA + PB$.

Corollary 1 can be extended to an arbitrary number of events $\cup A_j$,

(4) $\qquad P(\bigcup_j A_j) \leq \sum_j PA_j$.

Corollary 2. If $B = A^c$, then A and B are disjoint and

(5) $\qquad PA = 1 - PA^c$.

The following generalization of (2) is left as an exercise.

Theorem 3 (The Principle of Inclusion-Exclusion). Let $A_1, A_2, \cdots, A_n \in \mathscr{S}$. Then

(6) $$P(\bigcup_{k=1}^{n} A_k) = \sum_{k=1}^{n} PA_k - \sum_{k_1<k_2}^{n} P(A_{k_1} \cap A_{k_2})$$
$$+ \sum_{k_1<k_2<k_3}^{n} P(A_{k_1} \cap A_{k_2} \cap A_{k_3})$$
$$+ \cdots + (-1)^{n+1} P(\bigcap_{k=1}^{n} A_k).$$

Example 4. A die is rolled twice. Let all the elementary events in $\Omega = \{(i, j): i, j = 1, 2, \cdots, 6\}$ be assigned the same probability. Let A be the event that the first throw shows a number ≤ 2, and B, the event that the second throw shows at least 5. Then

$A = \{(i, j): 1 \leq i \leq 2, j = 1, 2, \cdots, 6\},$
$B = \{(i, j): 6 \geq j \geq 5, i = 1, 2, \cdots, 6\},$
$A \cap B = \{(1, 5), (1, 6), (2, 5), (2, 6)\};$

$$P(A \cup B) = PA + PB - P(A \cap B)$$
$$= \tfrac{1}{3} + \tfrac{1}{3} - \tfrac{4}{36} = \tfrac{5}{9}.$$

Example 5. A coin is tossed three times. Let us assign equal probability to each of the 2^3 elementary events in Ω. Let A be the event that at least one head shows up in three throws. Then

$$P(A) = 1 - P(A^c)$$
$$= 1 - P(\text{no heads})$$
$$= 1 - P(\text{TTT}) = \tfrac{7}{8}.$$

We next derive two useful inequalities.

Theorem 4 (Bonferroni's Inequality). Given $n (> 1)$ events $A_1, A_2, \cdots, A_n$,

(7) $$\sum_{i=1}^{n} PA_i - \sum_{i<j} P(A_i \cap A_j) \leq P(\bigcup_{i=1}^{n} A_i) \leq \sum_{i=1}^{n} PA_i.$$

Proof. In view of (4) it suffices to prove the left side of (7). The proof is by induction. The inequality on the left is true for $n = 2$ since

$$PA_1 + PA_2 - P(A_1 \cap A_2) = P(A_1 \cup A_2).$$

For $n = 3$,

$$P(\bigcup_{i=1}^{3} A_i) = \sum_{i=1}^{3} PA_i - \sum\sum_{i<j} P(A_i \cap A_j) + P(A_1 \cap A_2 \cap A_3),$$

and the result holds. Assuming that (7) holds for $3 < m \leq n - 1$, we show that it holds also for $m + 1$:

$$P(\bigcup_{i=1}^{m+1} A_i) = P((\bigcup_{i=1}^{m} A_i) \cup A_{m+1})$$

$$= P(\bigcup_{i=1}^{m} A_i) + PA_{m+1} - P(A_{m+1} \cap (\bigcup_{1}^{m} A_i))$$

$$\geq \sum_{i=1}^{m+1} PA_i - \sum\sum_{i<j}^{m} P(A_i \cap A_j) - P(\bigcup_{i=1}^{m} (A_i \cap A_{m+1}))$$

$$\geq \sum_{i=1}^{m+1} PA_i - \sum\sum_{i<j}^{m} P(A_i \cap A_j) - \sum_{i=1}^{m} P(A_i \cap A_{m+1})$$

$$= \sum_{i=1}^{m+1} PA_i - \sum\sum_{i<j}^{m+1} P(A_i \cap A_j).$$

Theorem 5 (**Boole's Inequality**). For any two events, A and B,

(8) $\qquad P(A \cap B) \geq 1 - PA^c - PB^c.$

Proof. The proof is simple.

Corollary 1. Let $\{A_j\}$, $j = 1, 2, \cdots$, be a countable sequence of events; then

(9) $\qquad P(\bigcap A_j) \geq 1 - \sum P(A_j^c).$

Proof. Take

$$B = \bigcap_{j=2}^{\infty} A_j \quad \text{and} \quad A = A_1$$

in (8).

Corollary 2 (**The Implication Rule**). If $A, B, C \in \mathscr{S}$ and A and B imply C, then

(10) $\qquad PC^c \leq PA^c + PB^c.$

Proof. Since $A \cap B \subseteq C$, $A^c \cup B^c \supseteq C^c$, we have from (3)

$$PC^c \leq P(A^c \cup B^c) \leq PA^c + PB^c.$$

Theorem 6. Let $\{A_n\}$ be a nondecreasing sequence of events in $\mathscr{S}$, that is, $A_n \in \mathscr{S}$, $n = 1, 2, \cdots$, and

$$A_n \supseteq A_{n-1}, \quad n = 2, 3, \cdots.$$

Then

(11) $$\lim_{n\to\infty} PA_n = P(\lim_{n\to\infty} A_n) = P(\bigcup_{n=1}^{\infty} A_n).$$

Proof. Let

$$A = \bigcup_{j=1}^{\infty} A_j.$$

Then

$$A = A_n + \sum_{j=n}^{\infty} (A_{j+1} - A_j).$$

By countable additivity we have

$$PA = PA_n + \sum_{j=n}^{\infty} P(A_{j+1} - A_j),$$

and, letting $n \to \infty$, we see that

$$PA = \lim_{n\to\infty} PA_n + \lim_{n\to\infty} \sum_{j=n}^{\infty} P(A_{j+1} - A_j).$$

The second term on the right tends to zero as $n \to \infty$ since the sum $\sum_{j=1}^{\infty} P(A_{j+1} - A_j) \leq 1$, and the result follows.

Corollary. Let $\{A_n\}$ be a nonincreasing sequence of events in $\mathscr{S}$. Then

(12) $$\lim_{n\to\infty} PA_n = P(\lim_{n\to\infty} A_n) = P(\bigcap_{n=1}^{\infty} A_n).$$

Proof. Consider the nondecreasing sequence of events $\{A_n^c\}$. Then

$$\lim_{n\to\infty} A_n^c = \bigcup_{j=1}^{\infty} A_j^c = A^c.$$

It follows from Theorem 6 that

$$\lim_{n\to\infty} PA_n^c = P(\lim_{n\to\infty} A_n^c) = P(\bigcup_{j=1}^{\infty} A_j^c) = P(A^c).$$

In other words,

$$\lim_{n\to\infty} (1 - PA_n) = 1 - PA,$$

as asserted.

Remark 5. Theorem 6 and its corollary will be used quite frequently in subsequent chapters. Property (11) is called the *continuity of P from below*, and (12) is known as the *continuity of P from above*. Thus Theorem 6 and its corollary assure us that the set function P is continuous from above and below.

We conclude this section with some remarks concerning the use of the word "random" in this book. In probability theory "random" has essentially three meanings. First, in sampling from a finite population a sample is said to be a *random sample* if at each draw all members available for selection have the same probability of being included. We will discuss sampling from a finite population in Section 1.4. Second, we speak of a *random sample from a probability distribution*. This notion is formalized in Section 7.2. The third meaning arises in the context of geometric probability, where statements like "a point is randomly chosen from the interval (a, b)" and "a point is picked randomly from a unit square" are frequently encountered. Once we have studied random variables and their distributions, problems involving geometric probabilities may be formulated in terms of problems involving independent uniformly distributed random variables, and these statements can be given appropriate interpretations.

Roughly speaking, these statements involve a certain assignment of probability. The word "random" expresses our desire to assign equal probability to sets of equal lengths, areas, or volumes. Let $\Omega \subseteq \mathscr{R}_n$ be a given set, and A be a subset of Ω. We are interested in the probability that a "randomly chosen point" in Ω falls in A. Here "randomly chosen" means that the point may be any point of Ω and that the probability of its falling in some subset A of Ω is proportional to the measure of A (independently of the location and shape of A). Assuming that both A and Ω have well defined finite measures (length, area, volume, etc.), we define

$$PA = \frac{\text{measure}(A)}{\text{measure}(\Omega)}.$$

(In the language of measure theory we are assuming that Ω is a measurable subset of $\mathscr{R}_n$ that has a finite, positive Lebesgue measure. If A is any measurable set, $PA = \mu(A)/\mu(\Omega)$, where μ is the n-dimensional Lebesgue measure.) Thus, if a point is chosen at random from the interval (a, b), the probability that it lies in the interval (c, d), $a \leq c < d \leq b$, is $(d-c)/(b-a)$.

We present some examples.

Example 6. A point is picked "at random" from a unit square. Let $\Omega = \{(x, y) : 0 \leq x \leq 1, 0 \leq y \leq 1\}$. It is clear that all rectangles and their unions must be in $\mathscr{S}$. So too should all circles in the unit square, since the area of a circle is also well defined. Indeed every set that has a well-defined area has to be in $\mathscr{S}$. We choose $\mathscr{S} = \mathfrak{B}_2$, the Borel σ-field generated by rectangles in Ω. As for the probability assignment, if $A \in \mathscr{S}$, we assign PA to A, where PA is the area of the set A. If $A = \{(x, y): 0 \leq x \leq 1/2, 1/2 \leq y \leq 1\}$, then $PA = 1/4$. If B is a circle with center $(1/2, 1/2)$ and radius $1/2$, then $PB = \pi(1/2)^2 = \pi/4$. If C is the set of all points which are

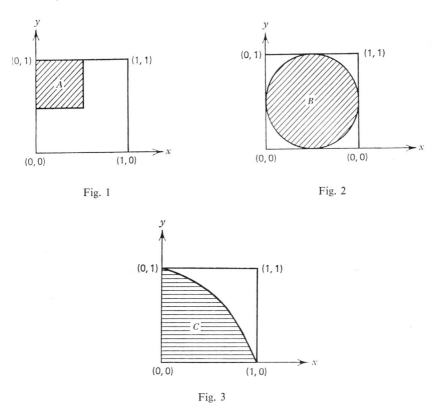

Fig. 1

Fig. 2

Fig. 3

at most a unit distant from the origin, then $PC = \pi/4$. (See Figs. 1 to 3.)

Example 7 (**Buffon's Needle Problem**). We return to Example 1. 2. 9. A needle (rod) of length l is tossed at random on a plane that is ruled with a series of parallel lines at distance $2l$ apart. We wish to find the probability that the needle will intersect one of the lines. Denoting by r the distance from the center of the needle to the closest line and by θ the angle that the needle forms with this line, we see that a necessary and sufficient condition for the needle to intersect the line is that $r \leq (l/2) \sin \theta$. The needle will intersect the nearest line if and only if its center falls in the shaded region in Fig. 1.2.2. We assign probability to an event A as follows:

$$PA = \frac{\text{area of set } A}{l\pi}.$$

Thus the required probability is

$$\frac{1}{l\pi} \int_0^\pi \frac{l}{2} \sin \theta \, d\theta = \frac{1}{\pi}.$$

Here we have interpreted "at random" to mean that the position of the needle is characterized by a point (r, θ) which lies in the rectangle $0 \leq r \leq l$, $0 \leq \theta \leq \pi$. We have assumed that the probability that the point (r, θ) lies in any arbitrary subset of this rectangle is proportional to the area of this set. Roughly, this means that "all positions of the midpoint of the needle are assigned the same weight and all directions of the needle are assigned the same weight."

Example 8. An interval of length 1, say (0, 1), is divided into three intervals by choosing two points at random. What is the probability that the three line segments form a triangle?

It is clear that a necessary and sufficient condition for the three segments to form a triangle is that the length of any one of the segments be less than the sum of the other two. Let x, y be the abscissas of the two points chosen at random. Then we must have either

$$0 < x < \frac{1}{2} < y < 1 \quad \text{and} \quad y - x < \frac{1}{2}$$

or

$$0 < y < \frac{1}{2} < x < 1 \quad \text{and} \quad x - y < \frac{1}{2}.$$

This is precisely the shaded area in Fig. 4. It follows that the required probability is 1/4.

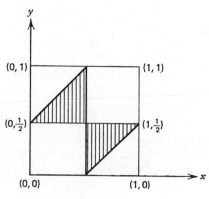

Fig. 4

If it is specified in advance that the point x is chosen at random from $(0, 1/2)$, and the point y at random from $(1/2, 1)$, we must have

$$0 < x < \frac{1}{2}, \qquad \frac{1}{2} < y < 1,$$

and

$$y - x < x + 1 - y \qquad \text{or} \qquad 2(y - x) < 1.$$

In this case the area bounded by these lines is the shaded area in Fig. 5, and it follows that the required probability is $1/2$.

Note the difference in sample spaces in the two computations made above.

Example 9 (Bertrand's Paradox). A chord is drawn at random in the unit circle. What is the probability that the chord is longer than the side of the equilateral triangle inscribed in the circle?

We present here two of several solutions to this problem, depending on

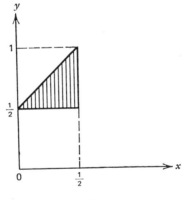

Fig. 5

how we interpret the phrase "at random." The paradox is resolved once we define the probability spaces carefully.

Solution 1. Since the length of a chord is uniquely determined by the position of its midpoint, choose a point C at random in the circle and draw a line through C and O, the center of the circle (Fig. 6). Draw the chord through C perpendicular to the line OC. If l_1 is the length of the chord with C as midpoint, $l_1 > \sqrt{3}$ if and only if C lies inside the circle with center O and radius $1/2$. Thus $PA = \pi(1/2)^2/\pi = 1/4$.

In this case Ω is the circle with center O and radius one, and the event A is the concentric circle with centre O and radius $\frac{1}{2}$. $\mathscr{S}$ is the usual Borel σ-field of subsets of Ω.

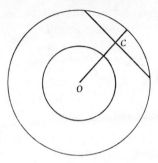

Fig. 6

Solution 2. Because of symmetry, we may fix one end point of the chord at some point P and then choose the other end point P_1 at random. Let the probability that P_1 lies on an arbitrary arc of the circle be proportional to the length of this arc. Now the inscribed equilateral triangle having P as one of its vertices divides the circumference into three equal parts. A chord drawn through P will be longer than the side of the triangle if and only if the other end point P_1 (Fig. 7) of the chord lies on that one third of the circumference that is opposite to P. It follows that the required probability is $1/3$. In this case $\Omega = [0, 2\pi]$, $\mathscr{S} = \mathfrak{B}_1 \cap \Omega$ and $A = [2\pi/3, 4\pi/3]$.

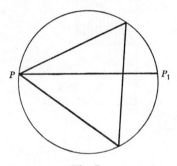

Fig. 7

PROBLEMS 1.3

1. Let Ω be the set of all nonnegative integers, and $\mathscr{S}$ the class of all subsets of Ω. In each of the following cases does P define a probability measure on $(\Omega, \mathscr{S})$?

(a) For $A \in \mathscr{S}$, let

$$PA = \sum_{x \in A} \frac{e^{-\lambda}\lambda^x}{x!}, \qquad \lambda > 0.$$

(b) For $A \in \mathscr{S}$, let

$$PA = \sum_{x \in A} p(1-p)^x, \qquad 0 < p < 1.$$

(c) For $A \in \mathscr{S}$, let $PA = 1$ if A has a finite number of elements, and $PA = 0$ otherwise.

2. Let $\Omega = \mathscr{R}$ and $\mathscr{S} = \mathscr{B}$. In each of the following cases does P define a probability measure on $(\Omega, \mathscr{S})$?

(a) For each interval I, let
$$PI = \int_I \frac{1}{\pi} \frac{1}{1+x^2} \, dx.$$

(b) For each interval I, let $PI = 1$ if I is an interval of finite length, and $PI = 0$ if I is an infinite interval.

(c) For each interval I, let $PI = 0$ if $I \subseteq (-\infty, 1)$, and $PI = \int_I (1/2) \, dx$ if $I \subseteq [1, \infty]$. (If $I = I_1 + I_2$, where $I_1 \subseteq (-\infty, 1)$ and $I_2 \subseteq [1, \infty)$, then $PI = PI_2$.)

3. Let A and B be two events such that $B \supseteq A$. What is $P(A \cup B)$? What is $P(A \cap B)$? What is $P(A - B)$?

4. In Problems 1(a) and 1(b), let $A = \{$all integers $> 2\}$, $B = \{$all integers $< 3\}$, and $C = \{$all integers x, $3 < x < 6\}$. Find PA, PB, PC, $P(A \cap B)$, $P(A \cup B)$, $P(B \cup C)$, $P(A \cap C)$, and $P(B \cap C)$.

5. In Problem 2(a) let A be the event $A = \{x : x \geq 0\}$. Find PA.

6. A box contains 1000 light bulbs. The probability that there is at least 1 defective bulb in the box is .1, and the probability that there are at least 2 defective bulbs is .05. Find the probability in each of the following cases:

(a) The box contains no defective bulbs.
(b) The box contains exactly 1 defective bulb.
(c) The box contains at most 1 defective bulb.

7. Two points are chosen at random on a line of unit length. Find the probability that each of the three line segments so formed will have a length $> 1/4$.

8. Find the probability that the sum of two randomly chosen positive numbers (both ≤ 1) will not exceed 1, and that their product will be $\leq 1/4$.

9. Prove Theorem 3.

10. Let $\{A_n\}$ be a sequence of events such that $A_n \to A$ as $n \to \infty$. Show that $PA_n \to PA$ as $n \to \infty$.

1.4 COMBINATORICS: PROBABILITY ON FINITE SAMPLE SPACES

In this section we restrict our attention to sample spaces that have at most a finite number of points. Let $\Omega = \{\omega_1, \omega_2, \cdots, \omega_n\}$ and $\mathscr{S}$ be the σ-field of all subsets of Ω. For any $A \in \mathscr{S}$,

$$PA = \sum_{\omega_j \in A} P\{\omega_j\}.$$

Definition 1. An assignment of probability is said to be equally likely (or uniform) if each elementary event in Ω is assigned the same probability. Thus, if Ω contains n points ω_j, $P\{\omega_j\} = 1/n$, $j = 1, 2, \cdots, n$.

With this assignment

(1) $$PA = \frac{\text{number of elementary events in } A}{\text{total number of elementary events in } \Omega}.$$

Example 1. A coin is tossed twice. The sample space consists of four points. Under the uniform assignment, each of four elementary events is assigned probability $1/4$.

Example 2. Three dice are rolled. The sample space consists of 6^3 points. Each one-point set is assigned probability $1/6^3$.

In games of chance we usually deal with finite sample spaces where uniform probability is assigned to all simple events. The same is the case in sampling schemes. In such instances the computation of the probability of an event A reduces to a combinatorial counting problem. We therefore consider some rules of counting.

Rule 1
Given a collection of n_1 elements $a_{11}, a_{12}, \cdots, a_{1n_1}$, n_2 elements $a_{21}, a_{22}, \cdots, a_{2n_2}$, and so on, up to n_k elements $a_{k1}, a_{k2}, \cdots, a_{kn_k}$, it is possible to form $n_1 \cdot n_2 \cdot \cdots \cdot n_k$ ordered k-tuplets $(a_{1j_1}, a_{2j_2}, \cdots, a_{kj_k})$ containing one element of each kind, $1 \leq j_i \leq n_i$, $i = 1, 2, \cdots, k$.

Proof. The proof is easy and is left as an exercise.

Example 3. Here r balls are to be placed in n cells. This amounts to choosing one cell for each ball. The sample space consists of n^r r-tuples $(i_1, i_2, \ldots, i_r)$, where i_j is the cell number of the jth ball, $j = 1, 2, \cdots, r$, $(1 \leq i_j \leq n)$.

Consider r tossings with a coin. There are 2^r possible outcomes. The probability that no heads will show up in r throws is $(1/2)^r$. Similarly, the probability that no 6 will turn up in r throws of a die is $(5/6)^r$.

Rule 2 is concerned with *ordered samples*. Consider a set of n elements $a_1, a_2, \cdots, a_n$. Any ordered arrangement $(a_{i_1}, a_{i_2}, \cdots, a_{i_r})$ of r symbols is called an *ordered sample* of size r. If elements are selected one by one, there are two possibilities:

(a) *Sampling with replacement:* In this case repetitions are permitted, and we can draw samples of arbitrary size. Clearly there are n^r samples of size r.

(b) *Sampling without replacement:* In this case an element once chosen is not replaced, so that there can be no repetitions. Clearly the sample size cannot exceed n, the size of the population. There are $n(n-1)\cdots(n-r+1) = {}_nP_r$, say, possible samples of size r. Clearly ${}_nP_r = 0$ for integers $r > n$. If $r = n$, then ${}_nP_r = n!$.

Rule 2
If ordered samples of size r are drawn from a population of n elements, there are n^r different samples with replacement and ${}_nP_r$ samples without replacement.

Corollary. The number of permutations of n objects is $n!$.

Remark. We will frequently use the term "random sample" in this book to describe the equal assignment of probability to all possible samples in sampling from a finite population. Thus, when we speak of a random sample of size r from a population of n elements, it means that each of n^r samples, in sampling with replacement, has the same probability $1/n^r$ or that each of ${}_nP_r$ samples, in sampling without replacement, is assigned probability $1/{}_nP_r$.

Example 4. Consider a set of n elements. A sample of size r is drawn at random with replacement. Then the probability that no element appears more than once is clearly ${}_nP_r/n^r$.

Thus, if n balls are to be randomly placed in n cells, the probability that each cell will be occupied is $n!/n^n$.

Example 5. Consider a class of r students. The birthdays of these r students form a sample of size r from the 365 days in the year. Then the probability that all r birthdays are different is ${}_{365}P_r/(365)^r$. One can show that this probability is $< 1/2$ if $r = 23$.

Next suppose that each of the r students is asked for his birth date in order, with the instruction that as soon as a student hears his birth date he is to raise his hand. Let us compute the probability that a hand is first raised when the kth ($k = 1, 2, \cdots, r$) student is asked his birth date. Let p_k be the probability that the procedure terminates at the kth student. Then

$$p_1 = 1 - \left(\frac{364}{365}\right)^{r-1}$$

and

$$p_k = \frac{{}_{365}P_{k-1}}{(365)^{k-1}}\left(1 - \frac{k-1}{365}\right)^{r-k+1}\left[1 - \left(\frac{365-k}{365-k+1}\right)^{r-k}\right], \quad k = 2, 3, \cdots.$$

Example 6. Let Ω be the set of all permutations of n objects. Let A_i be the set of all permutations that leave the ith object unchanged. Then the set $\cup_{i=1}^{n} A_i$ is the set of permutations with at least one fixed point. Clearly

$$PA_i = \frac{(n-1)!}{n!}, \qquad i = 1, 2, \cdots, n,$$

$$P(A_i \cap A_j) = \frac{(n-2)!}{n!}, \qquad i < j;\ i, j = 1, 2, \cdots, n, \text{ etc.}$$

By Theorem 1.3.3 we have

$$P(\bigcup_{i=1}^{n} A_i) = \left(1 - \frac{1}{2!} + \frac{1}{3!} - \cdots \pm \frac{1}{n!}\right).$$

As an application consider an absent-minded secretary who places n letters in n envelopes at random. Then the probability that she will misplace every letter is

$$1 - \left(1 - \frac{1}{2!} + \frac{1}{3!} - \cdots \pm \frac{1}{n!}\right).$$

Rule 3
There are $\binom{n}{r}$ different subpopulations of size $r \leq n$ from a population of n elements, where

(2) $$\binom{n}{r} = \frac{n!}{r!(n-r)!}.$$

Proof. The proof of this result is left as an exercise.

Example 7. Consider the random distribution of r balls in n cells. Let A_k be the event that a specified cell has exactly k balls, $k = 0, 1, 2, \cdots, r$; k balls can be chosen in $\binom{r}{k}$ ways. We place k balls in the specified cell and distribute the remaining $r - k$ balls in the $n - 1$ cells in $(n-1)^{r-k}$ ways. Thus

$$PA_k = \binom{r}{k}\frac{(n-1)^{r-k}}{n^r} = \binom{r}{k}\left(\frac{1}{n}\right)^k\left(1 - \frac{1}{n}\right)^{r-k}.$$

Example 8. There are $\binom{52}{13} = 635{,}013{,}559{,}600$ different hands at bridge, and $\binom{52}{5} = 2{,}598{,}960$ hands at poker.

The probability that all 13 cards in a bridge hand have different face values is $4^{13}/\binom{52}{13}$.

The probability that a hand at poker contains five different face values is $\binom{13}{5} 4^5 / \binom{52}{5}$.

FINITE SAMPLE SPACES

Rule 4

Consider a population of n elements. The number of ways in which the population can be partitioned into k subpopulations of sizes $r_1, r_2, \cdots, r_k$, respectively, $r_1 + r_2 + \cdots + r_k = n$, $0 \leq r_i \leq n$, is given by

$$(3) \qquad \binom{n}{r_1, r_2, \cdots, r_k} = \frac{n!}{r_1! r_2! \cdots r_k!}.$$

The numbers defined in (3) are known as *multinomial coefficients*.

Proof. For the proof of Rule 4 one uses Rule 3 repeatedly. Note that

$$(4) \qquad \binom{n}{r_1, r_2, \cdots, r_k} = \binom{n}{r_1}\binom{n-r_1}{r_2}\cdots\binom{n-r_1\cdots-r_{k-2}}{r_{k-1}}.$$

Example 9. In a game of bridge the probability that a hand of 13 cards contains 2 spades, 7 hearts, 3 diamonds, and 1 club is

$$\frac{\binom{13}{2}\binom{13}{7}\binom{13}{3}\binom{13}{1}}{\binom{52}{13}}.$$

Example 10. An urn contains 5 red, 3 green, 2 blue, and 4 white balls. A sample of size 8 is selected at random without replacement. The probability that the sample contains 2 red, 2 green, 1 blue, and 3 white balls is

$$\frac{\binom{5}{2}\binom{3}{2}\binom{2}{1}\binom{4}{3}}{\binom{14}{8}}.$$

PROBLEMS 1.4

1. How many different words can be formed by permuting letters of the word "Mississippi"? How many of these start with the letters "Mi"?

2. An urn contains R red and W white marbles. Marbles are drawn from the urn one after another without replacement. Let A_k be the event that a red marble is drawn for the first time on the kth draw. Show that

$$PA_k = \left(\frac{R}{R+W-k+1}\right) \prod_{j=1}^{k-1}\left(1 - \frac{R}{R+W-j+1}\right).$$

Let p be the proportion of red marbles in the urn before the first draw. Show that $PA_k \to p(1-p)^{k-1}$ as $R + W \to \infty$. Is this to be expected?

40   PROBABILITY

3. In a population of N elements, R are red and $W = N - R$ are white. A group of n elements is selected at random. Find the probability that the group so chosen will contain exactly r red elements.

4. Each permutation of the digits 1, 2, 3, 4, 5, 6 determines a six-digit number. If the numbers corresponding to all possible permutations are listed in increasing order of magnitude, find the 319th number on this list.

5. The numbers $1, 2, \cdots, n$ are arranged in random order. Find the probability that the digits $1, 2, \cdots, k$ $(k < n)$ appear as neighbors in that order.

6. A pin table has seven holes through which a ball can drop. Five balls are played. Assuming that at each play a ball is equally likely to go down any one of the seven holes, find the probability that more than one ball goes down at least one of the holes.

7. If $2n$ boys are divided into two equal subgroups, find the probability that the two tallest boys will be (a) in different subgroups, and (b) in the same subgroup.

8. In a movie theater that can accommodate $n + k$ people, n people are seated. What is the probability that $r \leq n$ given seats are occupied?

9. Waiting in line for a Sunday morning Disney show, are $2n$ children. Tickets are priced at a quarter each. Find the probability that nobody will have to wait for change if, before a ticket is sold to the first customer, the cashier has $2k$ $(k < n)$ quarters. Assume that it is equally likely that each ticket is paid for with a quarter or a half-dollar coin.

10. Each box of a certain brand of breakfast cereal contains a small charm, with k distinct charms forming a set. Assuming that the chance of drawing any particular charm is equal to that of drawing any other charm, show that the probability of finding at least one complete set of charms in a random purchase of $N \geq k$ boxes equals

$$1 - \binom{k}{1}\left(\frac{k-1}{k}\right)^N + \binom{k}{2}\left(\frac{k-2}{k}\right)^N - \binom{k}{3}\left(\frac{k-3}{k}\right)^N + \cdots + (-1)^{k-1}\binom{k}{k-1}\left(\frac{1}{k}\right)^N$$

[Hint: Use (1.3.6).]

11. Prove Rules 1 through 4.

1.5 CONDITIONAL PROBABILITY AND BAYES THEOREM

So far, we have computed probabilities of events on the assumption that no information was available about the experiment other than the sample space. Sometimes, however, it is known that an event H has happened. How do we use this information in making a statement concerning the outcome of another event A?

Consider the following examples.

Example 1. Let urn 1 contain one white and two black balls, and urn 2, one black and two white balls. A fair coin is tossed. If a head turns up, a ball is drawn at random from urn 1; otherwise, from urn 2. Let E be the event that the ball drawn is black. The sample space is $\Omega = \{Hb_{11}, Hb_{12}, Hw_{11}, Tb_{21}, Tw_{21}, Tw_{22}\}$, where H denotes head, T denotes tail, b_{ij} denotes jth black ball in ith urn, $i = 1, 2$, and so on. Then

$$PE = P\{Hb_{11}, Hb_{12}, Tb_{21}\} = \tfrac{3}{6} = \tfrac{1}{2}.$$

If, however, it is known that the coin showed a head, the ball could not have been drawn from urn 2. Thus the probability of E, conditional on information H, is $\tfrac{2}{3}$. Note that this probability equals the ratio $P\{\text{Head and ball drawn black}\}/P\{\text{Head}\}$.

Example 2. Let us toss two fair coins. Then the sample space of the experiment is $\Omega = \{HH, HT, TH, TT\}$. Let event $A = \{\text{both coins show same face}\}$ and $B = \{\text{at least one coin shows H}\}$. Then $PA = 2/4$. If B is known to have happened, this information assures that TT cannot happen, and $P\{A \text{ conditional on the information that } B \text{ has happened}\} = \tfrac{1}{3} = \tfrac{1}{4}/\tfrac{3}{4} = P(AB)/PB$.

Definition 1. Let $(\Omega, \mathscr{S}, P)$ be a probability space, and let $H \in \mathscr{S}$ with $PH > 0$. For an arbitrary $A \in \mathscr{S}$ we shall write

(1) $$P\{A \mid H\} = \frac{P(A \cap H)}{PH}$$

and call the quantity so defined the conditional probability of A, given H. Conditional probability remains undefined when $PH = 0$.

Theorem 1. Let $(\Omega, \mathscr{S}, P)$ be a probability space, and let $H \in \mathscr{S}$ with $PH > 0$. Then $(\Omega, \mathscr{S}, P_H)$, where $P_H(A) = P\{A|H\}$ for all $A \in \mathscr{S}$, is a probability space.

Proof. Clearly $P_H(A) = P\{A|H\} \geq 0$ for all $A \in \mathscr{S}$. Also, $P_H(\Omega) = P(\Omega \cap H)/PH = 1$. If $A_1, A_2, \cdots$ is a disjoint sequence of sets in $\mathscr{S}$, then

$$P_H(\sum_{i=1}^{\infty} A_i) = P\{\sum_{i=1}^{\infty} A_i | H\} = \frac{P\{(\sum_{1}^{\infty} A_i) \cap H\}}{PH}$$

$$= \frac{\sum_{i=1}^{\infty} P(A_i \cap H)}{PH}$$

$$= \sum_{i=1}^{\infty} P_H(A_i).$$

Remark 1. What we have done is to consider a new sample space consisting of the basic set H and the σ-field $\mathscr{S}_H = \mathscr{S} \cap H$, of subsets $A \cap H$, $A \in \mathscr{S}$, of H. On this space we have defined a set function P_H by multiplying the probability of each event by $(PH)^{-1}$. Indeed, $(H, \mathscr{S}_H, P_H)$ is a probability space.

Let A and B be two events with $PA > 0$, $PB > 0$. Then it follows from (1) that

(2)
$$P(A \cap B) = PA \cdot P\{B \mid A\},$$
$$P(A \cap B) = PB \cdot P\{A \mid B\}.$$

Equations (2) may be generalized to any number of events. Let $A_1, A_2, \ldots, A_n \in \mathscr{S}$, $n \geq 2$, and assume that $P(\bigcap_{j=1}^{n-1} A_j) > 0$. Since

$$A_1 \supset (A_1 \cap A_2) \supset (A_1 \cap A_2 \cap A_3) \supset \cdots \supset (\bigcap_{j=1}^{n-2} A_j) \supset (\bigcap_{j=1}^{n-1} A_j),$$

we see that

$$PA_1 > 0, \quad P(A_1 \cap A_2) > 0, \cdots, \quad P(\bigcap_{j=1}^{n-2} A_j) > 0.$$

It follows that $P\{A_j \mid \bigcap_{j=1}^{k-1} A_j\}$ are well defined for $k = 2, 3, \cdots, n$.

Theorem 2 (**The Multiplication Rule**). Let $(\Omega, \mathscr{S}, P)$ be a probability space and $A_1, A_2, \cdots, A_n \in \mathscr{S}$, with $P(\bigcap_{j=1}^{n-1} A_j) > 0$. Then

(3) $$P\{\bigcap_{j=1}^{n} A_j\} = P(A_1)\, P\{A_2 | A_1\}\, P\{A_3 | A_1 \cap A_2\} \cdots P\{A_n \mid \bigcap_{j=1}^{n-1} A_j\}.$$

Proof. The proof is simple.

Let us suppose that $\{H_j\}$ is a countable collection of events in $\mathscr{S}$ such that $H_j \cap H_k = \phi$, $j \neq k$, and $\sum_{j=1}^{\infty} H_j = \Omega$. Suppose that $PH_j > 0$ for all j. Then

(4) $$PB = \sum_{j=1}^{\infty} P(H_j)\, P\{B \mid H_j\} \qquad \text{for all } B \in \mathscr{S}.$$

For the proof we note that

$$B = \sum_{j=1}^{\infty} (B \cap H_j),$$

and the result follows. Equation (4) is called the *total probability rule*.

CONDITIONAL PROBABILITY   43

Example 3. Consider a hand of five cards in a game of poker. If the cards are dealt at random, there are $\binom{52}{5}$ possible hands of five cards each. Let
$A = \{$at least 3 cards of spades$\}$, $B = \{$all 5 cards of spades$\}$.
Then
$$P(A \cap B) = P\{\text{All 5 cards of spades}\}$$
$$= \frac{\binom{13}{5}}{\binom{52}{5}},$$
and
$$P\{B|A\} = \frac{P(A \cap B)}{PA}$$
$$= \frac{\binom{13}{5} / \binom{52}{5}}{\left[\binom{13}{3}\binom{39}{2} + \binom{13}{4}\binom{39}{1} + \binom{13}{5}\right] / \binom{52}{5}}.$$

Example 4. Urn 1 contains one white and two black marbles, urn 2 contains one black and two white marbles, and urn 3 contains three black and three white marbles. A die is rolled. If 1, 2, or 3 shows up, urn 1 is selected; if 4 shows up, urn 2 is selected; and if 5 or 6 shows up, urn 3 is selected. A marble is then drawn at random from the selected urn. Let A be the event that the marble drawn is white. If U, V, W, respectively, denote the events that the urn selected is 1, 2, 3, then

$$A = (A \cap U) + (A \cap V) + (A \cap W),$$
$$P(A \cap U) = P(U) \cdot P\{A|U\} = \tfrac{3}{6} \cdot \tfrac{1}{3},$$
$$P(A \cap V) = P(V) \cdot P\{A|V\} = \tfrac{1}{6} \cdot \tfrac{2}{3},$$
$$P(A \cap W) = P(W) \cdot P\{A|W\} = \tfrac{2}{6} \cdot \tfrac{3}{6}.$$

It follows that
$$PA = \tfrac{1}{6} + \tfrac{1}{9} + \tfrac{1}{6} = \tfrac{4}{9}.$$

A simple consequence of the total probability rule is the Bayes theorem, which we now prove.

Theorem 3 (Bayes Rule). Let $\{H_n\}$ be a disjoint sequence of events such that $PH_n > 0$, $n = 1, 2, \cdots$, and $\sum_{n=1}^{\infty} H_n = \Omega$. Let $B \in \mathscr{S}$ with $PB > 0$. Then

(5) $$P\{H_j|B\} = \frac{P(H_j) P\{B|H_j\}}{\sum_{i=1}^{\infty} P(H_i) P\{B|H_i\}}, \qquad j = 1, 2, \cdots.$$

Proof. From (2)
$$P\{B \cap H_j\} = P(B) P\{H_j|B\} = PH_j P\{B|H_j\},$$
and it follows that
$$P\{H_j|B\} = \frac{PH_j P\{B|H_j\}}{PB}.$$
The result now follows on using (4).

Remark 2. Suppose that H_1, H_2, $\cdots$ are all the "causes" that lead to the outcome of a random experiment. Let H_j be the set of outcomes corresponding to the jth cause. Assume that the probabilities PH_j, $j = 1, 2, \cdots$, called the *prior* probabilities, can be assigned. Now suppose that the experiment results in an event B of positive probability. This information leads to a reassesment of the prior probabilities. The conditional probabilities $P\{H_j|B\}$ are called the *posterior* probabilities. Formula (5) can be interpreted as a rule giving the probability that observed event B was due to cause or hypothesis H_j.

Example 5. In Example 4 let us compute the conditional probability $P\{V|A\}$. We have

$$P\{V|A\} = \frac{PV\, P\{A|V\}}{PU\, P\{A|U\} + PV\, P\{A|V\} + PW\, P\{A|W\}}$$

$$= \frac{\frac{1}{6} \cdot \frac{2}{3}}{\frac{3}{6} \cdot \frac{1}{3} + \frac{1}{6} \cdot \frac{2}{3} + \frac{2}{6} \cdot \frac{3}{6}} = \frac{\frac{1}{9}}{\frac{4}{9}} = \frac{1}{4}.$$

PROBLEMS 1.5

1. Let A and B be two events such that $PA = p_1 > 0$, $PB = p_2 > 0$, and $p_1 + p_2 > 1$. Show that $P\{B|A\} \geq 1 - [(1 - p_2)/p_1]$.

2. Two digits are chosen at random without replacement from the set of integers $\{1, 2, 3, 4, 5, 6, 7, 8\}$.

(a) Find the probability that both digits are greater than 5.
(b) Show that the probability that the sum of the digits will be equal to 5 is the same as the probability that their sum will exceed 13.

3. The probability of a family chosen at random having exactly k children is αp^k, $0 < p < 1$. Suppose that the probability that any child has blue eyes is b, $0 < b < 1$, independently of others. What is the probability that a family chosen at random has exactly r ($r \geq 0$) children with blue eyes?

4. In Problem 3 let us write

p_k = probability of a randomly chosen family having exactly k children = αp^k, $k = 1, 2, \cdots$,

$$p_0 = 1 - \frac{\alpha p}{(1-p)}.$$

Suppose that all sex distributions of k chidren are equally likely. Find the probability that a family has exactly r boys, $r \geq 1$. Find the conditional probability that a family has at least two boys, given that it has at least one boy.

5. Each of $(N+1)$ identical urns marked $0, 1, 2, \cdots, N$ contains N balls. The kth urn contains k black and $N - k$ white balls, $k = 0, 1, 2, \cdots, N$. An urn is chosen at random, and n random drawings are made from it, the ball drawn being always replaced. If all the n draws result in black balls, find the probability that the $(n + 1)$th draw will also produce a black ball. How does this probability behave as $N \to \infty$?

6. Each of n urns contains four white and six black balls, while another urn contains five white and five black balls. An urn is chosen at random from the $(n + 1)$ urns, and two balls are drawn from it, both being black. The probability that five white and three black balls remain in the chosen urn is $1/7$. Find n.

7. In answering a question on a multiple choice test, a candidate either knows the answer with probability p ($0 \leq p < 1$) or does not know the answer with probability $1 - p$. If he knows the answer, he puts down the correct answer with probability .99, whereas if he guesses, the probability of his putting down the correct result is $1/k$ (k choices to the answer). Find the conditional probability that the candidate knew the answer to a question, given that he has made the correct answer. Show that this probability tends to 1 as $k \to \infty$.

8. An urn contains five white and four black balls. Four balls are transferred to a second urn. A ball is then drawn from this urn, and it happens to be black. Find the probability of drawing a white ball from among the remaining three.

9. Prove Theorem 2.

1.6 INDEPENDENCE OF EVENTS

Let $(\Omega, \mathscr{S}, P)$ be a probability space, and let $A, B \in \mathscr{S}$, with $PB > 0$. By the multiplication rule we have

$$P(A \cap B) = P(B)\, P\{A|B\}.$$

In many experiments the information provided by B does not affect the probability of event A, that is, $P\{A|B\} = P\{A\}$.

Example 1. Let two fair coins be tossed, and let
$A = \{\text{head on the second throw}\}$, $B = \{\text{head on the first throw}\}$. Then

$$P(A) = P\{HH, TH\} = \tfrac{1}{2}, \qquad P(B) = \{HH, HT\} = \tfrac{1}{2},$$

and

$$P\{A|B\} = \frac{P(A \cap B)}{P(B)} = \frac{\tfrac{1}{4}}{\tfrac{1}{2}} = \tfrac{1}{2} = P(A).$$

Thus

$$P(A \cap B) = P(A) P(B).$$

In the following we will write $A \cap B = AB$.

Definition 1. Two events, A and B, are said to be independent if and only if

(1) $$P(AB) = P(A) P(B).$$

Note that we have not placed any restriction on $P(A)$ or $P(B)$. Thus conditional probability is not defined when $P(A)$ or $P(B) = 0$, but independence is. Clearly, if $P(A) = 0$, then A is independent of every $E \in \mathscr{S}$ with $PE > 0$. Also, any event $A \in \mathscr{S}$ with $PA > 0$ is independent of ϕ and Ω.

Theorem 1. If A and B are independent events, then

and
$$P\{A|B\} = P(A) \quad \text{if } P(B) > 0,$$
$$P\{B|A\} = P(B) \quad \text{if } P(A) > 0.$$

Theorem 2. If A and B are independent, so are A and B^c, A^c and B, and A^c and B^c.

Proof.

$$\begin{aligned} P(A^c B) &= P(B - (A \cap B)) \\ &= P(B) - P(A \cap B) \quad \text{since } B \supseteq (A \cap B) \\ &= P(B)\{1 - P(A)\} \\ &= P(A^c) P(B). \end{aligned}$$

Similarly, one proves that A^c and B^c, and A and B^c, are independent.

We wish to emphasize that independence of events is not to be confused with disjoint or mutually exclusive events. If two events, each with nonzero probability, are mutually exclusive, they are obviously dependent since the occurrence of one will automatically preclude the occurrence of the other. Similarly, if A and B are independent and $PA > 0$, $PB > 0$, then A and B cannot be mutually exclusive.

Example 2. A card is chosen at random from a deck of 52 cards. Let A be the event that the card is an ace, and B, the event that it is a club. Then

$$P(A) = \tfrac{4}{52} = \tfrac{1}{13}, \qquad P(B) = \tfrac{13}{52} = \tfrac{1}{4},$$

$$P(AB) = P\{\text{Ace of clubs}\} = \tfrac{1}{52},$$

so that A and B are independent.

Example 3. Consider families with two children, and assume that all four possible distributions of sex: BB, BG, GB, GG, where B stands for boy and G for girl, are equally likely. Let E be the event that a randomly chosen family has at most one girl, and F, the event that the family has children of both sexes. Then

$$P(E) = \tfrac{3}{4}, \qquad P(F) = \tfrac{1}{2}, \quad \text{and} \quad P(EF) = \tfrac{1}{2},$$

so that E and F are not independent.

Now consider families with three children. Assuming that each of the eight possible sex distributions is equally likely, we have

$$P(E) = \tfrac{4}{8}, \qquad P(F) = \tfrac{6}{8}, \qquad P(EF) = \tfrac{3}{8},$$

so that E and F are independent.

An obvious extension of the concept of independence between two events A and B to a given collection $\mathfrak{A}$ of events is to require that any two distinct events in $\mathfrak{A}$ be independent.

Definition 2. Let $\mathfrak{A}$ be a family of events from $\mathscr{S}$. We say that the events $\mathfrak{A}$ are *pairwise independent* if and only if, for every pair of distinct events $A, B \in \mathfrak{A}$,

$$P(AB) = PA\,PB.$$

A much stronger and more useful concept is *mutual* or *complete independence*.

Definition 3. A family of events $\mathfrak{A}$ is said to be a *mutually* or *completely independent* family if and only if, for every finite subcollection $\{A_{i_1}, A_{i_2}, \cdots, A_{i_k}\}$ of $\mathfrak{A}$, the following relation holds:

(2) $$P(A_{i_1} \cap A_{i_2} \cap \ldots \cap A_{i_k}) = \prod_{j=1}^{k} PA_{i_j}.$$

In what follows we will omit the adjective "mutual" or "complete" and speak of independent events. It is clear from Definition 3 that in order to check the independence of n events $A_1, A_2, \cdots, A_n \in \mathscr{S}$ we must check the following $2^n - n - 1$ relations.

$$P(A_i A_j) = PA_i\, PA_j, \qquad i \neq j;\, i, j = 1, 2, \cdots, n,$$
$$P(A_i A_j A_k) = PA_i\, PA_j\, PA_k, \qquad i \neq j \neq k;\, i, j, k = 1, 2, \cdots, n,$$
$$\vdots$$
$$P(A_1 A_2 \cdots A_n) = PA_1\, PA_2 \cdots PA_n.$$

The first of these requirements is pairwise independence. Independence therefore implies pairwise independence, but not conversely.

Example 4 (Wong [144]). Take four identical marbles. On the first, write symbols $A_1 A_2 A_3$. On each of the other three, write A_1, A_2, A_3, respectively. Put the four marbles in an urn and draw one at random. Let E_i denote the event that the symbol A_i appears on the drawn marble. Then

$$P(E_1) = P(E_2) = P(E_3) = \tfrac{1}{2},$$
$$P(E_1 E_2) = P(E_2 E_3) = P(E_1 E_3) = \tfrac{1}{4},$$

and

$$P(E_1 E_2 E_3) = \tfrac{1}{4}.$$

It follows that, although events E_1, E_2, E_3, are not independent, they are pairwise independent.

Example 5 (Kac [54], 22–23). In this example $P(E_1 E_2 E_3) = P(E_1)\, P(E_2)\, P(E_3)$, but E_1, E_2, E_3 are not pairwise independent and hence not independent. Let $\Omega = \{1, 2, 3, 4\}$, and let p_i be the probability assigned to $\{i\}$, $i = 1, 2, 3, 4$. Let $p_1 = \frac{\sqrt{2}}{2} - \frac{1}{4}$, $p_2 = \frac{1}{4}$, $p_3 = \frac{3}{4} - \frac{\sqrt{2}}{2}$, $p_4 = \frac{1}{4}$. Let $E_1 = \{1, 3\}$, $E_2 = \{2, 3\}$, $E_3 = \{3, 4\}$. Then

$$P(E_1 E_2 E_3) = P\{3\} = \frac{3}{4} - \frac{\sqrt{2}}{2} = \frac{1}{2}\left(1 - \frac{\sqrt{2}}{2}\right)\left(1 - \frac{\sqrt{2}}{2}\right)$$
$$= (p_1 + p_3)(p_2 + p_3)(p_3 + p_4)$$
$$= P(E_1)\, P(E_2)\, P(E_3).$$

But $P(E_1 E_2) = \frac{3}{4} - \frac{\sqrt{2}}{2} \neq PE_1\, PE_2$, and it follows that E_1, E_2, E_3 are not independent.

Example 6. A die is rolled repeatedly until a 6 turns up. We will show that event A that "a 6 will eventually show up" is certain to occur. Let A_k be the event that a 6 will show up for the first time on the kth throw. Then

$$A = \sum_{k=1}^{\infty} A_k \quad \text{and} \quad PA_k = \frac{1}{6}\left(\frac{5}{6}\right)^{k-1}, \quad k = 1, 2, \cdots.$$

Thus

$$PA = \frac{1}{6} \sum_{k=1}^{\infty} \left(\frac{5}{6}\right)^{k-1} = \frac{1}{6} \frac{1}{1 - \frac{5}{6}} = 1.$$

Alternatively, we can use the corollary to Theorem 1.3.6. Let B_n be the event that a 6 does not show up on the first n trials. Clearly $B_{n+1} \subseteq B_n$, and we have $A^c = \bigcap_{n=1}^{\infty} B_n$. Thus

$$1 - PA = PA^c = P(\bigcap_{n=1}^{\infty} B_n) = \lim_{n \to \infty} P(B_n) = \lim_{n \to \infty} \left(\frac{5}{6}\right)^n = 0.$$

Example 7. A slip of paper is given to person A, who marks it with either a plus or a minus sign; the probability of his writing a plus sign is 1/3. A passes the slip to B, who may either leave it alone or change the sign before passing it to C. Next, C passes the slip to D after perhaps changing the sign; finally, D passes it to a referee after perhaps changing the sign. The referee sees a plus sign on the slip. It is known that B, C, and D each change the sign with probability 2/3. We shall compute the probability that A originally wrote a plus.

Let N be the event that A wrote a plus sign, and M, the event that he wrote a minus sign. Let E be the event that the referee saw a plus sign on the slip. We have

$$P\{N|E\} = \frac{P(N)\, P\{E|N\}}{P(M)\, P\{E|M\} + P(N)\, P\{E|N\}}.$$

Now

$$P\{E|N\} = P\{\text{The plus sign was either not changed or changed exactly twice}\}$$
$$= \left(\tfrac{1}{3}\right)^3 + 3\left(\tfrac{2}{3}\right)^2 \left(\tfrac{1}{3}\right),$$

and

$$P\{E|M\} = P\{\text{The minus sign was changed either once or three times}\}$$
$$= 3\left(\tfrac{2}{3}\right)\left(\tfrac{1}{3}\right)^2 + \left(\tfrac{2}{3}\right)^3.$$

It follows that

$$P\{N|E\} = \frac{\left(\tfrac{1}{3}\right)[\left(\tfrac{1}{3}\right)^3 + 3\left(\tfrac{2}{3}\right)^2 \left(\tfrac{1}{3}\right)]}{\left(\tfrac{1}{3}\right)[\left(\tfrac{1}{3}\right)^3 + 3\left(\tfrac{2}{3}\right)^2 \left(\tfrac{1}{3}\right)] + \left(\tfrac{2}{3}\right)[3\left(\tfrac{2}{3}\right)\left(\tfrac{1}{3}\right)^2 + \left(\tfrac{2}{3}\right)^3]}$$
$$= \frac{\frac{13}{81}}{\frac{41}{81}} = \frac{13}{41}.$$

PROBLEMS 1.6

1. A biased coin is tossed until a head appears for the first time. Let p be the probability of a head, $0 < p < 1$. What is the probability that the number of tosses required is odd? Even?

2. Let A and B be two independent events defined on some probability space, and let $PA = 1/3$, $PB = 3/4$. Find (a) $P(A \cup B)$, (b) $P\{A|A \cup B\}$, (c) $P\{B|A \cup B\}$.

3. Let A_1, A_2, A_3 be three independent events. Show that A_1^c, A_2^c, and A_3^c are independent.

4. A biased coin with probability p, $0 < p < 1$, of success (heads) is tossed until for the first time the same result occurs three times in succession (that is, three heads or three tails in succession). Find the probability that the game will end at the seventh throw.

5. A box contains 20 black and 30 green balls. One ball at a time is drawn at random, its color is noted, and the ball is then replaced in the box for the next draw.

(a) Find the probability that the first green ball is drawn on the fourth draw.

(b) Find the probability that the third and fourth green balls are drawn on the sixth and ninth draws, respectively.

(c) Let N be the trial at which the fifth green ball is drawn. Find the probability that the fifth green ball is drawn on the nth draw. (Note that N take values $5, 6, 7, \cdots$.)

6. An urn contains four red and four black balls. A sample of two balls is drawn at random. If both balls drawn are of the same color, these balls are set aside and a new sample is drawn. If the two balls drawn are of different colors, they are returned to the urn and another sample is drawn. Assume that the draws are independent and that the same sampling plan is pursued at each stage until all balls are drawn.

(a) Find the probability that at least n samples are drawn before two balls of the same color appear.

(b) Find the probability that after the first two samples are drawn four balls are left, two black and two red.

7. Let A, B, and C be three boxes with three, four, and five cells respectively. There are three yellow balls numbered 1 to 3, four green balls numbered 1 to 4, and five red balls numbered 1 to 5. The yellow balls are placed at random in box A, the green in B, and the red in C, with no cell receiving more than one ball. Find the probability that only one of the boxes will show no matches.

8. A pond contains red and golden fish. There are 3000 red and 7000 golden fish, of which 200 and 500, respectively, are tagged. Find the probability that a random sample of 100 red and 200 golden fish will show 15 and 20 tagged fish, respectively.

9. Let $(\Omega, \mathscr{S}, P)$ be a probability space. Let $A, B, C \in \mathscr{S}$ with PB and $PC > 0$. If B and C are independent show that

$$P\{A|B\} = P\{A|B \cap C\}PC + P\{A|B \cap C^c\}PC^c.$$

Conversely, if this relation holds, $P\{A|BC\} \neq P\{A|B\}$, and $PA > 0$, then B and C are independent. (Strait [128])

10. Show that the converse of Theorem 2 also holds. Thus A and B are independent if, and only if A and B^c are independent, and so on.

CHAPTER 2

Random Variables and Their Probability Distributions

2.1 INTRODUCTION

In Chapter 1 we dealt essentially with random experiments which can be described by finite sample spaces. We studied the assignment and computation of probabilities of events. In practice, one observes a function defined on the space of outcomes. Thus, if a coin is tossed n times, one is not interested in knowing which of the 2^n n-tuples in the sample space has occurred. Rather, one would like to know the number of heads in n tosses. In games of chance one is interested in the net gain or loss of a certain player. Actually in Chapter 1 we were concerned with such functions without defining the term *random variable*. Here we study the notion of a random variable and examine some of its properties.

In Section 2 we define a random variable, while in Section 3 we study the notion of probability distribution of a random variable. Section 4 deals with some special types of random variables, and Section 5 considers functions of a random variable and their induced distributions.

The fundamental difference between a random variable and a real-valued function of a real variable is the associated notion of a probability distribution. Nevertheless our knowledge of advanced calculus or real analysis is the basic tool in the study of random variables and their probability distributions.

2.2 RANDOM VARIABLES

In Chapter 1 we studied properties of a set function P defined on a sample space $(\Omega, \mathscr{S})$. Since P is a set function, it is not very easy to handle. Moreover, in practice one frequently observes some function of elementary events. When a coin is tossed repeatedly, which replication resulted in heads is not of much interest. Rather one is interested in the number of heads, and consequently the number of tails, that appear in, say, n tossings of the coin. It is therefore desirable to introduce a point function on the sample space. We can then use our knowledge of advanced calculus.

Definition 1. Let $(\Omega, \mathscr{S})$ be a sample space. A finite, single-valued function which maps Ω into $\mathscr{R}$ is called a random variable (rv) if the inverse images under X of all Borel sets in $\mathscr{R}$ are events, that is, if

(1) $\qquad X^{-1}(B) = \{\omega : X(\omega) \in B\} \in \mathscr{S} \qquad$ for all $B \in \mathfrak{B}$.

Let $x \in \mathscr{R}$, and consider the semiclosed interval $(-\infty, x]$. Since $(-\infty, x] \in \mathfrak{B}$, it follows that if X is an rv, then $X^{-1}(-\infty, x] = \{X(\omega) \leq x\}$ is an event in $\mathscr{S}$. Also, if B is a Borel set in $\mathscr{R}$, then B can be obtained by a countable number of operations of unions, intersections, and differences of semiclosed intervals. The following result is obtained, using the properties of inverse images under X (see P. 2. 16).

Theorem 1. X is an rv if and only if for each $x \in \mathscr{R}$

(2) $\qquad \{\omega : X(\omega) \leq x\} = \{X \leq x\} \in \mathscr{S}.$

Remark 1. Note that the notion of probability does not enter into the definition of an rv.

Remark 2. If X is an rv, the sets $\{X = x\}$, $\{a < X \leq b\}$, $\{X < x\}$, $\{a \leq X < b\}$, $\{a < X < b\}$, $\{a \leq X \leq b\}$ are all events. Indeed, we could have defined an rv in the following equivalent manner: X is an rv if and only if

(3) $\qquad \{\omega : X(\omega) < x\} \in \mathscr{S} \qquad$ for all $x \in \mathscr{R}$.

We have

(4) $\qquad \{X < x\} = \bigcup_{n=1}^{\infty} \{X \leq x - \tfrac{1}{n}\}$

and

(5) $\qquad \{X \leq x\} = \bigcap_{n=1}^{\infty} \{X < x + \tfrac{1}{n}\}.$

Theorem 2. Let X be an rv defined on $(\Omega, \mathscr{S})$, and a, b be constants. Then $aX + b$ is also an rv on $(\Omega, \mathscr{S})$.

Proof.
$$\{\omega: aX(\omega) + b \leq x\} = \{aX \leq x - b\}.$$

If $a > 0$, then
$$\{aX \leq x - b\} = \left\{X \leq \frac{x-b}{a}\right\} \in \mathscr{S}.$$

If $a < 0$, then
$$\{aX \leq x - b\} = \left\{X \geq \frac{x-b}{a}\right\} = \left\{X < \frac{x-b}{a}\right\}^c \in \mathscr{S}.$$

If $a = 0$, then
$$\{aX \leq x - b\} = \begin{cases} \Omega & \text{if } x - b \geq 0, \\ \phi & \text{if } x - b < 0. \end{cases}$$

The proof is complete.

Example 1. For any set $A \subseteq \Omega$, define
$$I_A(\omega) = \begin{cases} 0, & \omega \notin A, \\ 1, & \omega \in A. \end{cases}$$

$I_A(\omega)$ is called the *indicator function* of set A. I_A is an rv if and only if $A \in \mathscr{S}$.

Example 2. Let $\Omega = \{H, T\}$, and $\mathscr{S}$ be the class of all subsets of Ω. Define X by $X(H) = 1$, $X(T) = 0$. Then
$$X^{-1}(-\infty, x] = \begin{cases} \phi & \text{if } x < 0, \\ \{T\} & \text{if } 0 \leq x < 1, \\ \{H, T\} & \text{if } 1 \leq x, \end{cases}$$

and we see that X is an rv.

Example 3. Let $\Omega = \{HH, TT, HT, TH\}$, and $\mathscr{S}$ be the class of all subsets of Ω. Define X by
$$X(\omega) = \text{number of H's in } \omega.$$

Then $X(HH) = 2$, $X(HT) = X(TH) = 1$, and $X(TT) = 0$.

$$X^{-1}(-\infty, x] = \begin{cases} \phi, & x < 0, \\ \{TT\}, & 0 \le x < 1, \\ \{TT, HT, TH\}, & 1 \le x < 2, \\ \Omega, & 2 \le x. \end{cases}$$

Thus X is an rv.

Remark 3. Let $(\Omega, \mathscr{S})$ be a discrete sample space; that is, let Ω be a countable set of points, and $\mathscr{S}$ be the class of all subsets of Ω. Then every numerical valued function defined on $(\Omega, \mathscr{S})$ is an rv.

Example 4. Let $\Omega = [0, 1]$ and $\mathscr{S} = \mathscr{B} \cap [0, 1]$, be the σ-field of Borel sets on $[0, 1]$. Define X on Ω by

$$X(\omega) = \omega, \quad \omega \in [0, 1].$$

Clearly X is an rv. Any Borel subset of Ω is an event.

Remark 4. Let X be an rv. Then X^2 is an rv and $1/X$ is also an rv, provided that $\{X = 0\} = \phi$. For $\{X^2 \le x\} = \phi$ if $x < 0$ and if $x \ge 0$, then $\{X^2 \le x\} = \{-\sqrt{x} \le X \le \sqrt{x}\} \in \mathscr{S}$. Similarly,

$$\left\{\frac{1}{X} \le x\right\} = \left\{\frac{1}{X} \le x, X < 0\right\} + \left\{\frac{1}{X} \le x, X > 0\right\} + \left\{\frac{1}{X} \le x, X = 0\right\}$$

$$= \{xX \le 1\} \cap \{X < 0\} + \{xX \ge 1\} \cap \{X > 0\}$$

$$= \begin{cases} \{X < 0\} & \text{if } x = 0, \\ \left\{X \le \frac{1}{x}\right\} \cap \{X < 0\} + \left\{X \ge \frac{1}{x}\right\} \cap \{X > 0\} & \text{if } x > 0, \\ \left\{X \ge \frac{1}{x}\right\} \cap \{X < 0\} + \left\{X \le \frac{1}{x}\right\} \cap \{X > 0\} & \text{if } x < 0. \end{cases}$$

For a general result see Theorem 2.5.1.

PROBLEMS 2.2

1. Let X be the number of heads in three tosses of a coin. What is Ω? What are the values that X assigns to points of Ω? What are the events $\{X \le 2.75\}$, $\{.5 \le X \le 1.72\}$?

2. A die is tossed two times. Let X be the sum of face values on the two tosses, and Y be the absolute value of the difference in face values. What is Ω? What values do X and Y assign to points of Ω? Check to see whether X and Y are random variables.

3. Let X be an rv. Is $|X|$ also an rv? If X is an rv that takes only nonnegative values, is $\sqrt{X}$ also an rv?

4. A die is rolled five times. Let X be the sum of face values. Write the events $\{X = 4\}$, $\{X = 6\}$, $\{X = 30\}$, $\{X \geq 29\}$.

5. Let $\Omega = [0,1]$, and $\mathscr{S}$ be the Borel σ-field of subsets of Ω. Define X on Ω as follows: $X(\omega) = \omega$ if $0 \leq \omega \leq 1/2$, and $X(\omega) = \omega - 1/2$ if $1/2 < \omega \leq 1$. Is X an rv? If so, what is the event $\{\omega: X(\omega) \in (1/4, 1/2)\}$?

2.3 PROBABILITY DISTRIBUTION OF A RANDOM VARIABLE

In Section 2.2 we introduced the concept of an rv and noted that the concept of probability on the sample space was not involved in this definition. Let $(\Omega, \mathscr{S}, P)$ be a probability space, and let X be an rv defined on our probability space.

Theorem 1. The rv X defined on the probability space $(\Omega, \mathscr{S}, P)$ induces a probability space $(\mathscr{R}, \mathfrak{B}, Q)$ by means of the correspondence

(1) $\qquad Q(B) = P\{X^{-1}(B)\} = P\{\omega: X(\omega) \in B\} \qquad$ for all $B \in \mathfrak{B}$.

We write $Q = PX^{-1}$ and call Q or PX^{-1} the (probability) *distribution* of X.

Proof. Clearly $Q(B) \geq 0$ for all $B \in \mathfrak{B}$, and also $Q(\mathscr{R}) = P\{X \in \mathscr{R}\} = P(\Omega) = 1$.

Let $B_i \in \mathfrak{B}$, $i = 1, 2, \cdots$ with $B_i \cap B_j = \phi$, $i \neq j$. Since the inverse image of a disjoint union of Borel sets is the disjoint union of their inverse images, we have

$$Q(\sum_{i=1}^{\infty} B_i) = P\{X^{-1}(\sum_{i=1}^{\infty} B_i)\}$$
$$= P\{\sum_{i=1}^{\infty} X^{-1}(B_i)\}$$
$$= \sum_{i=1}^{\infty} PX^{-1}(B_i) = \sum_{i=1}^{\infty} Q(B_i).$$

It follows that $(\mathscr{R}, \mathfrak{B}, Q)$ is a probability space, and the proof is complete.

Since Q is a set function and set functions are not easy to handle, let us introduce a point function on $\mathscr{R}$.

Definition 1. A real-valued function F defined on $(-\infty, \infty)$ that is nondecreasing, right continous and satisfies

DISTRIBUTION OF A RANDOM VARIABLE 57

$$F(-\infty) = 0 \quad \text{and} \quad F(+\infty) = 1$$

is called a distribution function (df).

Remark 1. From our knowledge of calculus (see P.2. 10) we see that if F is a nondecreasing function on $\mathscr{R}$, then $F(x-) = \lim_{t\uparrow x} F(t)$, $F(x+) = \lim_{t\downarrow x} F(t)$ exist and are finite. Also, $F(+\infty)$ and $F(-\infty)$ exist as $\lim_{t\uparrow+\infty} F(t)$ and $\lim_{t\downarrow-\infty} F(t)$, respectively. In general,

$$F(x-) \le F(x) \le F(x+),$$

and x is a jump point of F if and only if $F(x+)$ and $F(x-)$ exist but are unequal. Thus a nondecreasing function F has only jump discontinuities. If we define

$$F^*(x) = F(x+) \quad \text{for all } x,$$

we see that F^* is nondecreasing and right continuous on $\mathscr{R}$. Thus in Definition 1 the nondecreasing part is very important. Some authors demand left continuity in the definition of a df instead of right continuity.

Theorem 2. *The set of discontinuity points of a df F is at most countable.*

Proof. Let $(a, b]$ be a finite interval with at least n discontinuity points:

$$a < x_1 < x_2 < \cdots < x_n \le b.$$

Then

$$F(a) \le F(x_1-) < F(x_1) \le \cdots \le F(x_n-) < F(x_n) \le F(b).$$

Let $p_k = F(x_k) - F(x_k-)$, $k = 1, 2, \cdots, n$. Clearly

$$\sum_{k=1}^{n} p_k \le F(b) - F(a),$$

and it follows that the number of points x in $(a, b]$ with jump $p(x) > \varepsilon > 0$ is at most $\varepsilon^{-1}\{F(b) - F(a)\}$. Thus, for every integer N, the number of discontinuity points with jump greater than $1/N$ is finite. It follows that there are no more than a countable number of discontinuity points in every finite interval $(a, b]$. Since $\mathscr{R}$ is a countable union of such intervals, the proof is complete.

Definition 2. Let X be an rv defined on $(\Omega, \mathscr{S}, P)$. Define a point function $F(.)$ on $\mathscr{R}$ by

(2) $$F(x) = P\{\omega : X(\omega) \leq x\} \quad \text{for all} \quad x \in \mathscr{R}.$$

The function F is called the distribution function of rv X.

If there is no confusion, we will write
$$F(x) = P\{X \leq x\}.$$
The following result justifies our calling F as defined by (2) a df.

Theorem 3. The function F defined in (2) is indeed a df.

Proof. Let $x_1 < x_2$. Then $(-\infty, x_1] \subset (-\infty, x_2]$, and we have
$$F(x_1) = P\{X \leq x_1\} \leq P\{X \leq x_2\} = F(x_2).$$
Since F is nondecreasing, it is sufficient to show that for any sequence of numbers $x_n \downarrow x$, $x_1 > x_2 > \cdots > x_n > \cdots > x$, $F(x_n) \to F(x)$. Let $A_k = \{\omega : X(\omega) \in (x, x_k]\}$. Then $A_k \in \mathscr{S}$ and $A_k \downarrow$. Also,
$$\lim_{k \to \infty} A_k = \bigcap_{k=1}^{\infty} A_k = \phi,$$
since none of the intervals $(x, x_k]$ contains x. It follows that $\lim_{k \to \infty} P(A_k) = 0$. But
$$P(A_k) = P\{X \leq x_k\} - P\{X \leq x\}$$
$$= F(x_k) - F(x),$$
so that
$$\lim_{k \to \infty} F(x_k) = F(x),$$
and F is right continuous.

Finally, let x_n be a sequence of numbers decreasing to $-\infty$. Then
$$\{X \leq x_n\} \supseteq \{X \leq x_{n+1}\} \quad \text{for each } n$$
and
$$\lim_{n \to \infty} \{X \leq x_n\} = \bigcap_{n=1}^{\infty} \{X \leq x_n\} = \phi.$$
Therefore
$$F(-\infty) = \lim_{n \to \infty} P\{X \leq x_n\} = P\{\lim_{n \to \infty} \{X \leq x_n\}\} = 0.$$
Similarly,

$$F(+\infty) = \lim_{x_n \to \infty} P\{X \le x_n\} = 1,$$

and the proof is complete.

The next result, which we state without proof, establishes a correspondence between the induced probability Q on $(\mathcal{R}, \mathcal{B})$ and a point function F defined on $\mathcal{R}$.

Theorem 4. Given a probability Q on $(\mathcal{R}, \mathcal{B})$, there exists a distribution function F satisfying

(3) $\qquad Q(-\infty, x] = F(x) \qquad \text{for all} \quad x \in \mathcal{R},$

and, conversely, given a df F, there exists a unique probability Q defined on $(\mathcal{R}, \mathcal{B})$ that satisfies (3).

For proof see Chung [15], pages 23–24.

Theorem 5. Every df is the df of an rv on some probability space.

Proof. Let F be a df. From Theorem 4 it follows that there exists a unique probability Q defined on $\mathcal{R}$ that satisfies

$$Q(-\infty, x] = F(x) \qquad \text{for all} \quad x \in \mathcal{R}.$$

Let $(\mathcal{R}, \mathcal{B}, Q)$ be the probability space on which we define

$$X(\omega) = \omega, \qquad \omega \in \mathcal{R}.$$

Then

$$Q\{\omega : X(\omega) \le x\} = Q(-\infty, x] = F(x),$$

and F is the df of rv X.

Remark 2. If X is an rv on $(\Omega, \mathcal{S}, P)$, we have seen (Theorem 3) that $F(x) = P\{X \le x\}$ is a df associated with X. Theorem 5 assures us that to every df F we can associate some rv. Thus, given an rv, there exists a df, and conversely. In this book when we speak of an rv we will assume that it is defined on some probability space.

Example 1. Let X be defined on $(\Omega, \mathcal{S}, P)$ by

$$X(\omega) = c \qquad \text{for all} \quad \omega \in \Omega.$$

Then

$$P\{X = c\} = 1,$$
$$F(x) = Q(-\infty, x] = P\{X^{-1}(-\infty, x]\} = 0 \quad \text{if} \quad x < c$$

and

$$F(x) = 1 \quad \text{if} \quad x \geq c.$$

Example 2. Let $\Omega = \{H, T\}$, and X be defined by

$$X(H) = 1, \qquad X(T) = 0.$$

If P assigns equal mass to $\{H\}$ and $\{T\}$, then

$$P\{X = 0\} = \tfrac{1}{2} = P\{X = 1\},$$

and

$$F(x) = Q(-\infty, x] = \begin{cases} 0, & x < 0, \\ \tfrac{1}{2}, & 0 \leq x < 1, \\ 1, & 1 \leq x. \end{cases}$$

Example 3. Let $\Omega = \{(i, j): i, j \in \{1, 2, 3, 4, 5, 6\}\}$, and $\mathscr{S}$ be the set of all subsets of Ω. Let $P\{(i, j)\} = 1/6^2$ for all 6^2 pairs (i, j) in Ω. Define

$$X(i, j) = i + j, \qquad 1 \leq i, \quad j \leq 6.$$

Then

$$F(x) = Q(-\infty, x] = P\{X \leq x\} = \begin{cases} 0, & x < 2, \\ \tfrac{1}{36}, & 2 \leq x < 3, \\ \tfrac{3}{36}, & 3 \leq x < 4, \\ \tfrac{6}{36}, & 4 \leq x < 5, \\ \vdots & \vdots \\ \tfrac{35}{36}, & 11 \leq x < 12, \\ 1, & 12 \leq x. \end{cases}$$

Example 4. We return to Example 2.2.4. For every subinterval I of $[0, 1]$ let $P(I)$ be the length of the interval. Then $(\Omega, \mathscr{S}, P)$ is a probability space, and the df of rv $X(\omega) = \omega$, $\omega \in \Omega$, is given by $F(x) = 0$ if $x < 0$, $F(x) = P\{\omega: X(\omega) \leq x\} = P([0, x]) = x$ if $x \in [0, 1]$, and $F(x) = 1$ if $x \geq 1$.

PROBLEMS 2.3

1. Write the df of rv X defined in Problem 2.2.1, assuming that the coin is fair.

2. What is the df of rv Y defined in Problem 2.2.2, assuming that the die is not loaded?

3. Do the following functions define df's?
   (a) $F(x) = 0$ if $x < 0$, $= x$ if $0 \le x < 1/2$, and $= 1$ if $x \ge \frac{1}{2}$.
   (b) $F(x) = (1/\pi) \tan^{-1} x$, $-\infty < x < \infty$.
   (c) $F(x) = 0$ if $x \le 1$, and $= 1 - (1/x)$ if $1 < x$.
   (d) $F(x) = 1 - e^{-x}$ if $x \ge 0$, and $= 0$ if $x < 0$.
4. Let X be an rv with df F.
   (a) If F is the df defined in Problem 3(a), find $P\{X > \frac{1}{4}\}$, $P\{\frac{1}{3} < X \le \frac{3}{8}\}$.
   (b) If F is the df defined in Problem 3(d), find $P\{-\infty < X < 2\}$.

2.4 DISCRETE AND CONTINUOUS RANDOM VARIABLES

Let X be an rv defined on some fixed, but otherwise arbitrary, probability space $(\Omega, \mathcal{S}, P)$, and let F be the df of X. In this book we shall restrict ourselves to two types of rv's, namely, the case in which the rv assumes at most a countable number of values, and that in which the df F is absolutely continuous (see P.2.15).

Definition 1. An rv X defined on $(\Omega, \mathcal{S}, P)$ is said to be of the discrete type, or simply discrete, if there exists a countable set $E \subseteq \mathcal{R}$ such that $P\{X \in E\} = 1$. The points of E which have positive mass are called jump points or points of increase of the df of X, and their probabilities are called jumps of the df.

Note that $E \in \mathcal{B}$ since every one-point set is in $\mathcal{B}$. Indeed, if $x \in \mathcal{R}$, then

(1) $$\{x\} = \bigcap_{n=1}^{\infty} \left\{ \left(x - \frac{1}{n} < x \le x + \frac{1}{n} \right) \right\}.$$

Thus $\{X \in E\}$ is an event. Let X take on the value x_i with probability $p_i (i = 1, 2, \cdots)$. We have
$$P\{\omega : X(\omega) = x_i\} = p_i, \quad i = 1, 2, \cdots, \quad p_i \ge 0 \quad \text{for all } i.$$
Then $\sum_{i=1}^{\infty} p_i = 1$.

Definition 2. The collection of numbers $\{p_i\}$ satisfying $P\{X = x_i\} = p_i \ge 0$, for all i and $\sum_{i=1}^{\infty} p_i = 1$, is called the probability mass function (pmf) of rv X.

The df F of X is given by

(2) $$F(x) = P\{X \le x\} = \sum_{x_i \le x} p_i.$$

If I_A denotes the indicator function of the set A, we may write

(3) $$X(\omega) = \sum_{i=1}^{\infty} x_i I_{[X=x_i]}(\omega).$$

Let us define a function $\varepsilon(x)$ as follows:

$$\varepsilon(x) = \begin{cases} 1, & x \geq 0, \\ 0, & x < 0. \end{cases}$$

Then we have

(4) $$F(x) = \sum_{i=1}^{\infty} p_i \varepsilon(x - x_i).$$

Example 1. The simplest example is that of an rv X *degenerate* at c, $P\{X = c\} = 1$:

$$F(x) = \varepsilon(x - c) = \begin{cases} 0, & x < c, \\ 1, & x \geq c. \end{cases}$$

Example 2. A box contains good and defective items. If an item drawn is good, we assign the number 1 to the drawing; otherwise, the number 0. Let p be the probability of drawing at random a good item. Then

$$P\left\{X = \begin{matrix} 0 \\ 1 \end{matrix}\right\} = \begin{cases} 1 - p, \\ p, \end{cases}$$

and

$$F(x) = P\{X \leq x\} = \begin{cases} 0, & x < 0, \\ 1 - p, & 0 \leq x < 1, \\ 1, & 1 \leq x. \end{cases}$$

Example 3. Let X be an rv with pmf

$$P\{X = k\} = \frac{6}{\pi^2} \cdot \frac{1}{k^2}, \quad k = 1, 2, \cdots.$$

Then

$$F(x) = \frac{6}{\pi^2} \sum_{k=1}^{\infty} \frac{1}{k^2} \varepsilon(x - k).$$

The following result is obvious.

Theorem 1. Let $\{p_k\}$ be a collection of nonnegative real numbers such that $\sum_{k=1}^{\infty} p_k = 1$. Then $\{p_k\}$ is the pmf of some rv X.

We next consider rv's associated with df's that have no jump points. The df of such rv's is continuous. We shall restrict our attention to a special subclass of such rv's.

Definition 3. Let X be an rv defined on $(\Omega, \mathscr{S}, P)$ with df F. Then X is said to be of the continuous type (or, simply, continuous) if F is absolutely continuous, that is, if there exists a nonnegative function $f(x)$ such that for every real number x we have

$$(5) \qquad F(x) = \int_{-\infty}^{x} f(t) \, dt.$$

The function f is called the probability density function (pdf) of the rv X.

Note that $f \geq 0$ and satisfies $\lim_{x \to +\infty} F(x) = F(+\infty) = \int_{-\infty}^{\infty} f(t) \, dt = 1$. Let a and b be any two real numbers with $a < b$. Then

$$P\{a < X \leq b\} = F(b) - F(a)$$
$$= \int_{a}^{b} f(t) \, dt.$$

Let B be a Borel set of the real line. Since B can be obtained by a countable number of operations of unions, intersections, and differences on intervals, the following result holds.

Theorem 2. Let X be an rv of the continuous type with pdf f. Then for every Borel set $B \in \mathfrak{B}$

$$(6) \qquad P(B) = \int_{B} f(t) \, dt.$$

If F is absolutely continuous and f is continuous at x, we have

$$(7) \qquad F'(x) = \frac{dF(x)}{dx} = f(x).$$

Theorem 3. Every nonnegative real function f that is integrable over $\mathscr{R}$ and satisfies

$$\int_{-\infty}^{\infty} f(x) \, dx = 1$$

is the pdf of some continuous type rv X.

Proof. In view of Theorem 2.3.5 it suffices to show that there corresponds a df F to f. Define

$$F(x) = \int_{-\infty}^{x} f(t) \, dt, \qquad x \in \mathscr{R}.$$

Then $F(-\infty) = 0$, $F(+\infty) = 1$, and, if $x_2 > x_1$,

$$F(x_2) = \left(\int_{-\infty}^{x_1} + \int_{x_1}^{x_2}\right)f(t)\,dt \geq \int_{-\infty}^{x_1} f(t)\,dt = F(x_1).$$

Finally, F is (absolutely) continuous and hence continuous from the right.

Remark 1. In the discrete case, $P\{X = a\}$ is the probability that X takes the value a. In the continuous case, $f(a)$ is not the probability that X takes the value a. Indeed, if X is of the continuous type, it assumes every value with probability 0.

Theorem 4 Let X be any rv. Then
(8) $$P\{X = a\} = \lim_{\substack{t \to a \\ t < a}} P\{t < X \leq a\}.$$

Proof. Let $t_1 < t_2 < \cdots < a$, $t_n \to a$, and write

$$A_n = \{t_n < X \leq a\}.$$

Then A_n is a nonincreasing sequence of events which converges to $\bigcap_{n=1}^{\infty} A_n = \{X = a\}$. It follows that $\lim_{n \to \infty} PA_n = P\{X = a\}$.

Remark 2. Since $P\{t < X \leq a\} = F(a) - F(t)$, it follows that

$$\lim_{\substack{t \to a \\ t < a}} P\{t < X \leq a\} = P\{X = a\} = F(a) - \lim_{\substack{t \to a \\ t < a}} F(t)$$
$$= F(a) - F(a-).$$

Thus F has a jump disontinuity at a if and only if $P\{X = a\} > 0$, that is, F is continuous at a if and only if $P\{X = a\} = 0$. If X is an rv of the continuous type, $P\{X = a\} = 0$ for all $a \in \mathscr{R}$. Moreover,

$$P\{X \in \mathscr{R} - \{a\}\} = 1.$$

This justifies Remark 4 in Section 1.3.

Example 4. Let X be an rv with df F given by (Fig. 1)

$$F(x) = \begin{cases} 0, & x \leq 0, \\ x, & 0 < x \leq 1, \\ 1, & 1 < x. \end{cases}$$

Differentiating F with respect to x at continuity points of f, we get

$$f(x) = F'(x) = \begin{cases} 0, & x < 0 \text{ or } x > 1, \\ 1, & 0 < x < 1. \end{cases}$$

The function f is not continuous at $x = 0$, or at $x = 1$ (Fig. 2). We may define $f(0)$ and $f(1)$ in any manner. Choosing $f(0) = f(1) = 0$, we have

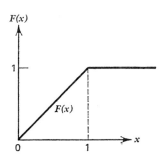

Fig. 1   Fig. 2

$$f(x) = \begin{cases} 1, & 0 < x < 1, \\ 0, & \text{otherwise.} \end{cases}$$

Then
$$P\{.4 < X \le .6\} = F\{.6\} - F\{.4\} = .2.$$

Example 5. Let X have the *triangular* pdf (Fig. 3)

$$f(x) = \begin{cases} x, & 0 < x \le 1, \\ 2 - x, & 1 \le x \le 2, \\ 0, & \text{otherwise.} \end{cases}$$

It is easy to check that f is a pdf. For the df F of X we have (Fig. 4)

$$F(x) = 0 \qquad \text{if } x \le 0,$$

$$F(x) = \int_0^x t \, dt = \frac{x^2}{2} \qquad \text{if } 0 < x \le 1,$$

$$F(x) = \int_0^1 t \, dt + \int_1^x (2 - t) \, dt = 2x - \frac{x^2}{2} - 1 \qquad \text{if } 1 < x \le 2,$$

and
$$F(x) = 1 \qquad \text{if } x \ge 2.$$

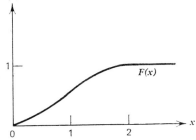

Fig. 3 Graph of f.   Fig. 4 Graph of F.

Then
$$P\{.3 < X \le 1.5\} = P\{X \le 1.5\} - P\{X \le .3\}$$
$$= .83.$$

Example 6. Let $k > 0$ be a constant, and
$$f(x) = \begin{cases} kx(1-x), & 0 < x < 1, \\ 0, & \text{otherwise.} \end{cases}$$

Then $\int_0^1 f(x)\,dx = k/6$. It follows that $f(x)$ defines a pdf if $k = 6$. We have
$$P\{X > .3\} = 1 - 6 \int_0^{.3} x(1-x)\,dx = .784.$$

We conclude this discussion by emphasizing that the two types of rv's considered above form only a small part of the class of all rv's. These two classes, however, contain practically all the random variables that arise in practice. We note without proof (see Chung [15], 9) that every df F can be decomposed into two parts according to

(9) $$F(x) = aF_d(x) + (1-a)F_c(x).$$

Here F_d and F_c are both df's; F_d is the df of a discrete rv, while F_c is a continuous (not necessarily absolutely continuous) df. In fact, F_c can be further decomposed, but we will not go into that. (See Chung [15], 11.)

Example 7. Let X be an rv with df
$$F(x) = \begin{cases} 0, & x < 0, \\ \dfrac{1}{2}, & x = 0, \\ \dfrac{1}{2} + \dfrac{x}{2}, & 0 < x < 1, \\ 1, & 1 \le x. \end{cases}$$

Note that the df F has a jump at $x = 0$ and F is continuous (in fact, absolutely continuous) in the interval $(0, 1)$. F is the df of an rv X that is neither discrete nor continuous. We can write
$$F(x) = \frac{1}{2} F_d(x) + \frac{1}{2} F_c(x),$$
where
$$F_d(x) = \begin{cases} 0, & x = 0, \\ 1, & x \ge 0; \end{cases}$$

and

$$F_d(x) = \begin{cases} 0, & x \le 0, \\ x, & 0 < x < 1, \\ 1, & 1 \le x. \end{cases}$$

Here $F_d(x)$ is the df of the rv degenerate at $x = 0$, and $F_c(x)$ is the df with pdf

$$f_c(x) = \begin{cases} 1, & 0 < x < 1, \\ 0, & \text{otherwise}. \end{cases}$$

PROBLEMS 2.4

1. Let

$$p_k = p(1-p)^k, \quad k = 0, 1, 2, \cdots, \quad 0 < p < 1.$$

Does $\{p_k\}$ define the pmf of some rv? What is the df of this rv? If X is an rv with pmf $\{p_k\}$, what is $P\{n \le X \le N\}$, where n, $N(N > n)$ are positive integers?

2. In Problem 2.3.3, find the pdf associated with the df's of parts (a),(c), and (d).

3. Does the function $f_\theta(x) = \theta^2 x \, e^{-\theta x}$ if $x > 0$, and $= 0$ if $x \le 0$, where $\theta > 0$, define a pdf? Find the df associated with $f_\theta(x)$; if X is an rv with pdf $f_\theta(x)$, find $P\{X \ge 1\}$.

4. Does the function $f_\theta(x) = (x+1)/[\theta(\theta+1)]e^{-x/\theta}$ if $x > 0$, and $= 0$ otherwise, where $\theta > 0$ define a pdf? Find the corresponding df.

5. For what values of K do the following functions define the pmf of some rv?
(a) $f(x) = K(\lambda^x/x!)$, $x = 0, 1, 2, \cdots$, $\lambda > 0$.
(b) $f(x) = K/N$, $x = 1, 2, \cdots, N$.

6. Show that the function

$$f(x) = \frac{1}{2}e^{-|x|}, \quad -\infty < x < \infty,$$

is a pdf.

7. For the pdf $f(x) = x$ if $0 \le x < 1$, and $= 2 - x$ if $1 \le x < 2$, find $P\{1/6 < X \le 7/4\}$.

8. Prove Theorems 1 and 2.

2.5 FUNCTIONS OF A RANDOM VARIABLE

Let X be an rv with a known distribution, and let g be a function defined on the real line. We seek the distribution of $Y = g(X)$, provided that Y is also an rv. We first prove the following result.

Theorem 1. Let X be an rv defined on $(\Omega, \mathscr{S}, P)$. Also, let g be a Borel-measurable function on $\mathscr{R}$. Then $g(X)$ is also an rv.

Proof.
$$\{g(X) \leq y\} = \{X \in g^{-1}(-\infty, y]\},$$
and since g is Borel-measurable, $g^{-1}(-\infty, y]$ is a Borel set. It follows that $\{g(X) \leq y\} \in \mathscr{S}$, and the proof is complete.

Theorem 2. Given an rv X with a known df, the distribution of the rv $Y = g(X)$, where g is a Borel-measurable function, is determined.

Proof. We have

(1) $$P\{Y \leq y\} = P\{X \in g^{-1}(-\infty, y]\}.$$

In what follows, we will always assume that the functions under consideration are Borel-measurable.

Example 1. Let X be an rv with df F. Then $|X|$, $aX + b$ (where $a \neq 0$ and, b are constants), X^k (where $k \geq 0$ is an integer), and $|X|^\alpha$ ($\alpha > 0$) are all rv's. Define

$$X^+ = \begin{cases} X, & X \geq 0, \\ 0, & X < 0, \end{cases}$$

and

$$X^- = \begin{cases} X, & X \leq 0, \\ 0, & X > 0. \end{cases}$$

Then X^+, X^- are also rv's. We have

$$P\{|X| \leq y\} = P\{-y \leq X \leq y\} = P\{X \leq y\} - P\{X < -y\}$$
$$= F(y) - F(-y) + P\{X = -y\}, \quad y > 0;$$

$$P\{aX + b \leq y\} = P\{aX \leq y - b\}$$
$$= \begin{cases} P\left\{X \leq \dfrac{y-b}{a}\right\} & \text{if } a > 0, \\ P\left\{X \geq \dfrac{y-b}{a}\right\} & \text{if } a < 0; \end{cases}$$

$$P\{X^+ \leq y\} = \begin{cases} 0 & \text{if } y < 0, \\ P\{X \leq 0\} & \text{if } y = 0, \\ P\{X < 0\} + P\{0 \leq X \leq y\} & \text{if } y > 0. \end{cases}$$

Similarly,

$$P\{X^- \le y\} = \begin{cases} 1 & \text{if } y \ge 0, \\ P\{X \le y\} & \text{if } y < 0. \end{cases}$$

Let X be an rv of the discrete type, and A be the countable set such that $P\{X \in A\} = 1$ and $P\{X = x\} > 0$ for $x \in A$. Let $Y = g(X)$ be a one-to-one mapping from A onto some set B. Then the inverse map, g^{-1}, is a single-valued function of y. To find $P\{Y = y\}$, we note that

$$P\{Y = y\} = P\{g(X) = y\} = P\{X = g^{-1}(y)\}, \quad y \in B,$$
$$\text{and} \quad P\{Y = y\} = 0, \quad y \in B^c.$$

Example 2. Let X be a *Poisson* rv with pmf

$$P\{X = k\} = \begin{cases} e^{-\lambda} \dfrac{\lambda^k}{k!}, & k = 0, 1, 2, \cdots; \lambda > 0, \\ 0, & \text{otherwise.} \end{cases}$$

Let $Y = X^2 + 3$. Then $y = x^2 + 3$ maps $A = \{0, 1, 2, \cdots\}$ onto $B = \{3, 4, 7, 12, 19, 28, \cdots\}$. The inverse map is $x = \sqrt{(y-3)}$, and since there are no negative values in A we take the positive square root of $y - 3$. We have

$$P\{Y = y\} = P\{X = \sqrt{y - 3}\} = \frac{e^{-\lambda} \lambda^{\sqrt{y-3}}}{\sqrt{(y-3)!}}, \quad y \in B,$$

and $P\{Y = y\} = 0$ elsewhere.

Actually the restriction of a single-valued inverse on g is not necessary. If g has a finite (or even a countable) number of inverses for each y, from countable additivity of P we have

$$P\{Y = y\} = P\{g(X) = y\} = P\{\bigcup_a [X = a, g(a) = y]\}$$
$$= \sum_a P\{X = a, g(a) = y\}.$$

Example 3. Let X be an rv with pmf

$$P\{X = -2\} = \tfrac{1}{5}, \quad P\{X = -1\} = \tfrac{1}{6}, \quad P\{X = 0\} = \tfrac{1}{5},$$
$$P\{X = 1\} = \tfrac{1}{15}, \quad \text{and} \quad P\{X = 2\} = \tfrac{11}{30}.$$

Let $Y = X^2$. Then

$$A = \{-2, -1, 0, 1, 2\}, \quad \text{and} \quad B = \{0, 1, 4\}.$$

We have

$$P\{Y = y\} = \begin{cases} \frac{1}{5} & y = 0, \\ \frac{1}{6} + \frac{1}{15} = \frac{7}{30}, & y = 1, \\ \frac{1}{5} + \frac{11}{30} = \frac{17}{30}, & y = 4. \end{cases}$$

The case in which X is an rv of the continuous type is not as simple. First we note that, if X is a continuous type rv and g is some Borel-measurable function, $Y = g(X)$ may not be an rv of the continuous type.

Example 4. Let X be an rv with *uniform* distribution on $[-1, 1]$, that is, the pdf of X is $f(x) = 1/2$, $-1 \leq x \leq 1$, and $= 0$ elsewhere. Let $Y = X^+$. Then, from Example 1,

$$P\{Y \leq y\} = \begin{cases} 0, & y < 0, \\ \frac{1}{2}, & y = 0, \\ \frac{1}{2} + \frac{1}{2}y, & 1 \geq y > 0, \\ 1, & y > 1. \end{cases}$$

We see that the df of Y has a jump at $y = 0$ and that Y is neither discrete nor continuous. Note that all we require is that $P\{X < 0\} > 0$.

Example 4 shows that we need some conditions on g to ensure that $g(X)$ is also an rv of the continuous type whenever X is continuous. This will be the case when g is a continuous monotonic function (see P. 2.10). A sufficient condition is given in the following theorem.

Theorem 3. Let X be an rv of the continuous type with pdf f. Let $y = g(x)$ be differentiable for all x and either $g'(x) > 0$ for all x or $g'(x) < 0$ for all x. Then $Y = g(X)$ is also an rv of of the continuous type with pdf given by

(2) $$h(y) = \begin{cases} f[g^{-1}(y)] \left| \frac{d}{dy} g^{-1}(y) \right|, & \alpha < y < \beta, \\ 0, & \text{otherwise,} \end{cases}$$

where $\alpha = \min \{g(-\infty), g(+\infty)\}$ and $\beta = \max \{g(-\infty), g(+\infty)\}$.

Proof. If g is differentiable for all x and $g'(x) > 0$ for all x, then g is continuous and strictly increasing, the limits α, β exist (may be infinite), and the inverse function $x = g^{-1}(y)$ exists, is strictly increasing, and is differentiable (see P.2.10). The df of Y for $\alpha < y < \beta$ is given by

$$P\{Y \leq y\} = P\{X \leq g^{-1}(y)\}.$$

The pdf of g is obtained on differentiation. We have

$$h(y) = \frac{d}{dy} P\{Y \le y\}$$
$$= f[g^{-1}(y)] \frac{d}{dy} g^{-1}(y).$$

Similarly, if $g' < 0$, then g is strictly decreasing and we have

$$P\{Y \le y\} = P\{X \ge g^{-1}(y)\}$$
$$= 1 - P\{X \le g^{-1}(y)\} \quad (X \text{ is a continuous type rv})$$

so that

$$h(y) = -f[g^{-1}(y)] \cdot \frac{d}{dy} g^{-1}(y).$$

Since g and g^{-1} are both strictly decreasing, $d/dy \, g^{-1}(y)$ is negative and (2) follows.

Note that (see Theorem P.2.15)

$$\frac{d}{dy} g^{-1}(y) = \frac{1}{d \, g(x)/dx} \bigg|_{x=g^{-1}(y)},$$

so that (2) may be rewritten as

(3) $$h(y) = \frac{f(x)}{|d \, g(x)/dx|} \bigg|_{x=g^{-1}(y)}, \quad \alpha < y < \beta.$$

Remark 1. The key to computation of the induced distribution of $Y = g(X)$ from the distribution of X is (1). If the conditions of Theorem 3 are satisfied, we are able to identify the set $\{X \in g^{-1}(-\infty, y]\}$ as $\{X \le g^{-1}(y)\}$ or $\{X \ge g^{-1}(y)\}$, according to whether g is increasing or decreasing. In practice Theorem 3 is quite useful, but whenever the conditions are violated one should return to (1) to compute the induced distribution. This is the case, for example, in Examples 7 and 8 and Theorem 4 below.

Remark 2. If the pdf f of X vanishes outside an interval $[a, b]$ of finite length, we need only to assume that g is differentiable in (a, b), and either $g'(x) > 0$ or $g'(x) < 0$ throughout the interval. Then we take

$$\alpha = \min \{g(a), g(b)\} \quad \text{and} \quad \beta = \max \{g(a), g(b)\}$$

in Theorem 3.

Example 5. Let X have the density $f(x) = 1$, $0 < x < 1$, and $= 0$ otherwise. Let $Y = e^X$. Then $X = \log Y$, and we have

$$h(y) = \left|\frac{1}{y}\right| \cdot 1, \qquad 0 < \log y < 1,$$

that is,

$$h(y) = \begin{cases} \dfrac{1}{y}, & 1 < y < e, \\ 0, & \text{otherwise.} \end{cases}$$

If $Y = -2 \log X$, then $x = e^{-y/2}$ and

$$h(y) = \left|-\tfrac{1}{2} e^{-y/2}\right| \cdot 1, \qquad 0 < e^{-y/2} < 1,$$
$$= \begin{cases} \tfrac{1}{2} e^{-y/2}, & 0 < y < \infty, \\ 0, & \text{otherwise.} \end{cases}$$

Example 6. Let X be a nonnegative rv of the continuous type with pdf f, and let $\alpha > 0$. Let $Y = X^\alpha$. Then

$$P\{X^\alpha \leq y\} = \begin{cases} P\{X \leq y^{1/\alpha}\} & \text{if } y \geq 0, \\ 0 & \text{if } y < 0. \end{cases}$$

The pdf of Y is given by

$$h(y) = f(y^{1/\alpha}) \left|\frac{d}{dy} y^{1/\alpha}\right|$$
$$= \begin{cases} \dfrac{1}{\alpha} y^{1/\alpha - 1} f(y^{1/\alpha}), & y > 0, \\ 0, & y \leq 0. \end{cases}$$

Example 7. Let X be an rv with pdf

$$f(x) = \frac{1}{\sqrt{2\pi}} e^{-x^2/2}, \qquad -\infty < x < \infty.$$

Let $Y = X^2$. In this case, $g'(x) = 2x$ which is > 0 for $x > 0$, and < 0 for $x < 0$, so that the conditions of Theorem 3 are not satisfied. But for $y > 0$

$$P\{Y \leq y\} = P\{-\sqrt{y} \leq X \leq \sqrt{y}\}$$
$$= F(\sqrt{y}) - F(-\sqrt{y}),$$

where F is the df of X. Thus the pdf of Y is given by

$$h(y) = \begin{cases} \dfrac{1}{2\sqrt{y}} \{f(\sqrt{y}) + f(-\sqrt{y})\}, & y > 0, \\ 0, & y \leq 0. \end{cases}$$

Thus

$$h(y) = \begin{cases} \dfrac{1}{\sqrt{2\pi y}} e^{-y/2}, & 0 < y, \\ 0, & y \le 0. \end{cases}$$

Example 8. Let X be an rv with pdf

$$f(x) = \begin{cases} \dfrac{2x}{\pi^2}, & 0 < x < \pi, \\ 0, & \text{otherwise.} \end{cases}$$

Let $Y = \sin X$. In this case $g'(x) = \cos x > 0$ for x in $(0, \pi/2)$ and < 0 for x in $(\pi/2, \pi)$, so that the conditions of Theorem 3 are not satisfied. To compute the pdf of Y we return to (1) and see that the df of Y is given by

$$P\{Y \le y\} = P\{\sin X \le y\}, \qquad 0 < y < 1,$$
$$= P\{[0 \le X \le x_1] \cup [x_2 \le X \le \pi]\},$$

where $x_1 = \sin^{-1} y$ and $x_2 = \pi - \sin^{-1} y$. Thus

$$P\{Y \le y\} = \int_0^{x_1} f(x)\, dx + \int_{x_2}^{\pi} f(x)\, dx$$
$$= \left(\frac{x_1}{\pi}\right)^2 + 1 - \left(\frac{x_2}{\pi}\right)^2,$$

and the pdf of Y is given by

$$h(y) = \frac{d}{dy}\left(\frac{\sin^{-1} y}{\pi}\right)^2 + \frac{d}{dy}\left[1 - \left(\frac{\pi - \sin^{-1} y}{\pi}\right)^2\right]$$
$$= \begin{cases} \dfrac{2}{\pi\sqrt{1 - y^2}}, & 0 < y < 1, \\ 0, & \text{otherwise.} \end{cases}$$

In Examples 7 and 8 the function $y = g(x)$ can be written as the sum of two monotone functions. We applied Theorem 3 to each of these monotonic summands. These two examples are special cases of the following result, the proof of which the reader is asked to construct.

Theorem 4. Let X be an rv of the continuous type with pdf f. Let $y = g(x)$ be differentiable for all x, and assume that $g'(x)$ is continuous and nonzero at all but a finite number of values of x. Then, for every real number y,

(a) there exist a positive integer $n = n(y)$ and real numbers (inverses) $x_1(y), x_2(y), \cdots, x_n(y)$ such that

$$g[x_k(y)] = y, \qquad g'[x_k(y)] \neq 0, \quad k = 1, 2, \cdots, n(y),$$

or

(b) there does not exist any x such that $g(x) = y$, $g'(x) \neq 0$, in which case we write $n(y) = 0$.

Then Y is a continuous rv with pdf given by

$$h(y) = \begin{cases} \sum_{k=1}^{n} f[x_k(y)] |g'[x_k(y)]|^{-1} & \text{if } n > 0, \\ 0 & \text{if } n = 0. \end{cases}$$

Example 9. Let X be an rv with pdf f, and let $Y = |X|$. Here $n(y) = 2$, $x_1(y) = y$, $x_2(y) = -y$ for $y > 0$, and

$$h(y) = \begin{cases} f(y) + f(-y), & y > 0, \\ 0, & y \leq 0. \end{cases}$$

Thus, if $f(x) = 1/2$, $-1 \leq x \leq 1$, and $= 0$ otherwise, then

$$h(y) = \begin{cases} 1, & 0 \leq y \leq 1, \\ 0, & \text{otherwise}. \end{cases}$$

If $f(x) = (1/\sqrt{2\pi}) e^{-(x^2/2)}$, $-\infty < x < \infty$, then

$$h(y) = \begin{cases} \dfrac{2}{\sqrt{2\pi}} e^{-(y^2/2)}, & y > 0, \\ 0, & \text{otherwise}. \end{cases}$$

Example 10. Let X be an rv of the continuous type with pdf f, and let $Y = X^{2m}$, where m is a positive integer. In this case $g(x) = x^{2m}$, $g'(x) = 2mx^{2m-1} > 0$ for $x > 0$ and $g'(x) < 0$ for $x < 0$. Writing $n = 2m$, we see that, for any $y > 0$, $n(y) = 2$, $x_1(y) = -y^{1/n}$, $x_2(y) = y^{1/n}$. It follows that

$$h(y) = f[x_1(y)] \cdot \frac{1}{ny^{1-1/n}} + f[x_2(y)] \frac{1}{ny^{1-1/n}}$$

$$= \begin{cases} \dfrac{1}{ny^{1-1/n}} \{f(y^{1/n}) + f(-y^{1/n})\} & \text{if } y > 0, \\ 0 & \text{if } y \leq 0. \end{cases}$$

In particular, if f is the pdf given in Example 7, then

$$h(y) = \begin{cases} \dfrac{2}{\sqrt{2\pi} \, ny^{1-1/n}} \exp\left\{-\dfrac{y^{2/n}}{2}\right\} & \text{if } y > 0, \\ 0 & \text{if } y \leq 0. \end{cases}$$

Remark 3. The basic formula (1) and the countable additivity of probability allow us to compute the distribution of $Y = g(X)$ in some instances even if g has a countable number of inverses. Let $A \subseteq \mathcal{R}$ and g map A into $B \subseteq \mathcal{R}$. Suppose that A can be represented as a countable union of disjoint

FUNCTIONS OF A RANDOM VARIABLE 75

sets A_k, $k = 1, 2, \cdots$. Then the df of Y is given by

$$\begin{aligned} P\{Y \le y\} &= P\{X \in g^{-1}(-\infty, y]\} \\ &= P\{X \in \sum_{k=1}^{\infty} [\{g^{-1}(-\infty, y]\} \cap A_k]\} \\ &= \sum_{k=1}^{\infty} P\{X \in A_k \cap \{g^{-1}(-\infty, y]\}\}. \end{aligned}$$

If the conditions of Theorem 3 are satisfied by the restriction of g to each A_k, we may obtain the pdf of Y on differentiating the df of Y. We remind the reader that term-by-term differentiation is permissible if the differentiated series is uniformly convergent (see Theorem P.2.19).

Example 11. Let X be an rv with pdf

$$f(x) = \begin{cases} \theta e^{-\theta x}, & x > 0, \\ 0, & x \le 0, \end{cases} \quad \theta > 0.$$

Let $Y = \sin X$, and let $\sin^{-1} y$ be the principal value. Then, for $0 < y < 1$,

$P\{\sin X \le y\}$
$= P\{0 < X \le \sin^{-1} y \text{ or } (2n-1)\pi - \sin^{-1} y \le X \le 2n\pi + \sin^{-1} y$
$\qquad\qquad\qquad\qquad\qquad\qquad\qquad\qquad\qquad \text{for all integers } n \ge 1\}$
$= P\{0 < X \le \sin^{-1} y\} + \sum_{n=1}^{\infty} P\{(2n-1)\pi - \sin^{-1} y \le X \le 2n\pi + \sin^{-1} y\}$
$= 1 - e^{-\theta \sin^{-1} y} + \sum_{n=1}^{\infty} [e^{-\theta[(2n-1)\pi - \sin^{-1} y]} - e^{-\theta(2n\pi + \sin^{-1} y)}]$
$= 1 - e^{-\theta \sin^{-1} y} + (e^{\theta \pi + \theta \sin^{-1} y} - e^{-\theta \sin^{-1} y}) \sum_{n=1}^{\infty} e^{-(2\theta\pi)n}$
$= 1 - e^{-\theta \sin^{-1} y} + (e^{\theta\pi + \theta \sin^{-1} y} - e^{-\theta \sin^{-1} y}) \left(\dfrac{e^{-2\theta\pi}}{1 - e^{-2\theta\pi}} \right)$
$= 1 + \dfrac{e^{-\theta\pi + \theta \sin^{-1} y} - e^{-\theta \sin^{-1} y}}{1 - e^{-2\pi\theta}}.$

A similar computation can be made for $y < 0$. If follows that the pdf of Y is given by

$$h(y) = \begin{cases} \theta e^{-\theta\pi} (1 - e^{-2\theta\pi})^{-1} (1 - y^2)^{-1/2} [e^{\theta \sin^{-1} y} + e^{-\theta\pi - \theta \sin^{-1} y}] & \text{if } -1 < y < 0, \\ \theta (1 - e^{-2\theta\pi})^{-1} (1 - y^2)^{-1/2} [e^{-\theta \sin^{-1} y} + e^{-\theta\pi + \theta \sin^{-1} y}] & \text{if } 0 < y < 1, \\ 0 & \text{otherwise.} \end{cases}$$

PROBLEMS 2.5

1. Let X be a random variable with probability mass function

$$P\{X = r\} = \binom{n}{r} p^r (1-p)^{n-r}, \quad r = 0, 1, 2, \cdots, n, \quad 0 \le p \le 1.$$

Find the pmf's of the rv's (a) $Y = aX + b$, (b) $Y = X^2$, (c) $Y = \sqrt{X}$.

2. Let X be an rv with pdf

$$f(x) = \begin{cases} 0 & \text{if } x \le 0, \\ \dfrac{1}{2} & \text{if } 0 < x \le 1, \\ \dfrac{1}{2x^2} & \text{if } 1 < x < \infty. \end{cases}$$

Find the pdf of the rv $1/X$.

3. Let X be a positive rv of the continuous type with pdf $f(\cdot)$. Find the pdf of the rv $U = X/(1 + X)$. If, in particular, X has the pdf

$$f(x) = \begin{cases} 1, & 0 \le x \le 1, \\ 0, & \text{otherwise,} \end{cases}$$

what is the pdf of U?

4. Let X be an rv with pdf f defined in Example 11. Let $Y = \cos X$, and $Z = \tan X$. Find the df's of Y and Z.

5. Let X be an rv with pdf

$$f_\theta(x) = \begin{cases} \theta e^{-\theta x} & \text{if } x \ge 0, \\ 0 & \text{otherwise.} \end{cases}$$

Let $Y = [X - 1/\theta]^2$. Find the pdf of Y.

6. A point is chosen at random on the circumference of a circle of radius r with center at the origin, that is, the polar angle θ of the point chosen has the pdf

$$f(\theta) = \frac{1}{2\pi}, \qquad \theta \in (-\pi, \pi).$$

Find the pdf of the abscissa of the point selected.

7. For the rv X of Example 7 find the pdf of the following rv's: (a) $Y_1 = e^X$, (b) $Y_2 = 2X^2 + 1$, (c) $Y_3 = g(X)$, where $g(x) = 1$ if $x > 0$, $= 1/2$ if $x = 0$, and $= -1$ if $x < 0$.

8. Suppose that a projectile is fired at an angle θ above the earth with a velocity V. Assuming that θ is an rv with pdf

$$f(\theta) = \begin{cases} \dfrac{12}{\pi} & \text{if } \dfrac{\pi}{6} < \theta < \dfrac{\pi}{4}, \\ 0 & \text{otherwise,} \end{cases}$$

find the pdf of the range R of the projectile, where $R = V^2 \sin 2\theta/g$, g being the gravitational constant.

9. Let X be an rv with pdf $f(x) = 1/(2\pi)$ if $0 < x < 2\pi$, and $= 0$ otherwise. Let $Y = \sin X$. Find the df and pdf of Y.

10. Let X be an rv with pdf $f(x) = 1/3$ if $-1 < x < 2$, and $= 0$ otherwise. Let $Y = |X|$. Find the pdf of Y.

11. Let X be an rv with pdf $f(x) = (1/2\theta)$ if $-\theta \leq x \leq \theta$, and $= 0$ otherwise. Let $Y = 1/X^2$. Find the pdf of Y.

12. Let X be an rv of the continuous type, and let $Y = g(X)$ be defined as follows:
(a) $g(x) = 1$ if $x > 0$, and $= -1$ if $x \leq 0$.
(b) $g(x) = b$ if $x \geq b$, $= x$ if $|x| < b$, and $= -b$ if $x \leq -b$.
(c) $g(x) = x$ if $|x| \geq b$, and $= 0$ if $|x| < b$.

Find the distribution of Y in each case.

13. Prove Theorem 4.

CHAPTER 3

Moments and Generating Functions

3.1 INTRODUCTION

The study of the probability distributions of a random variable is essentially the study of some numerical characteristics associated with them. These so-called *parameters of the distribution* play a key role in mathematical statistics. In Section 2 we introduce some of these parameters, namely, moments and order parameters, and investigate their properties. In Section 3 the idea of generating functions is introduced. In particular, we study probability generating functions and moment generating functions. Section 4 deals with some moment inequalities.

3.2 MOMENTS OF A DISTRIBUTION FUNCTION

In this section we investigate some numerical characteristics, called *parameters*, associated with the distribution of an rv X. These parameters are (a) *moments* and their functions and (b) *order parameters*. We will concentrate mainly on moments and their properties.

Let X be a random variable of the discrete type with probability mass function $p_k = P\{X = x_k\}$, $k = 1, 2, \cdots$. If

(1) $$\sum_{k=1}^{\infty} |x_k| p_k < \infty,$$

we say that the *expected value* (or the *mean* or the *mathematical expectation*) of X exists and write

(2) $$\mu = EX = \sum_{k=1}^{\infty} x_k p_k.$$

Note that the series $\sum_{k=1}^{\infty} x_k p_k$ may converge but the series $\sum_{k=1}^{\infty} |x_k| p_k$ may not. In that case we say that EX does not exist.

Example 1. Let X have the pmf given by

$$p_j = P\left\{X = (-1)^{j+1} \frac{3^j}{j}\right\} = \frac{2}{3^j}, \qquad j = 1, 2, \cdots.$$

Then

$$\sum_{j=1}^{\infty} |x_j| p_j = \sum_{j=1}^{\infty} \frac{2}{j} = \infty,$$

and EX does not exist, although the series

$$\sum_{j=1}^{\infty} x_j p_j = \sum_{j=1}^{\infty} (-1)^{j+1} \frac{2}{j}$$

is convergent.

If X is of the continuous type and has pdf f, we say that EX exists and equals $\int x f(x)\, dx$, provided that

$$\int |x| f(x)\, dx < \infty.$$

Similar definition is given for the mean of any Borel-measurable function $h(X)$ of X.

We emphasize that the condition $\int |x| f(x)\, dx < \infty$ must be checked before it can be concluded that EX exists and equals $\int x f(x)\, dx$. Moreover, it is worthwhile to recall at this point that the integral $\int_{-\infty}^{\infty} \varphi(x)\, dx$ exists, provided that the limit $\lim_{\substack{a\to-\infty \\ b\to\infty}} \int_{-b}^{a} \varphi(x)\, dx$ exists. It is quite possible for the limit $\lim_{a\to\infty} \int_{-a}^{a} \varphi(x)\, dx$ to exist without the existence of $\int_{-\infty}^{\infty} \varphi(x)\, dx$. As an example consider the Cauchy pdf:

$$f(x) = \frac{1}{\pi} \cdot \frac{1}{1 + x^2}, \qquad -\infty < x < \infty.$$

Clearly

$$\lim_{a\to\infty} \int_{-a}^{a} \frac{x}{\pi} \frac{1}{1 + x^2}\, dx = 0.$$

However, EX does not exist since the integral $(1/\pi) \int_{-\infty}^{\infty} |x| / (1 + x^2)\, dx$ diverges.

Remark 1. Let $X(\omega) = I_A(\omega)$ for some $A \in \mathscr{S}$. Then $EX = P(A)$.

Remark 2. If we write $h(X) = |X|$, we see that EX exists if and only if $E|X|$

does.

Remark 3. We say that an rv X is *symmetric* about a point α if
$$P\{X \geq \alpha + x\} = P\{X \leq \alpha - x\} \qquad \text{for all } x.$$
In terms of df F of X, this means that, if
$$F(\alpha - x) = 1 - F(\alpha + x) + P\{X = \alpha + x\}$$
holds for all $x \in \mathcal{R}$, we say that the df F (or the rv X) is symmetric with α as the *center* of *symmetry*. If $\alpha = 0$, then for every x
$$F(-x) = 1 - F(x) + P\{X = x\}.$$
In particular, if X is an rv of the continuous type, X is symmetric with center α if and only if the pdf f of X satisfies
$$f(\alpha - x) = f(\alpha + x) \qquad \text{for all } x.$$
If $\alpha = 0$, we will say simply that X is symmetric (or that F is symmetric).

As an immediate consequence of this definition we see that, if X is symmetric with α as the center of symmetry and $E|X| < \infty$, then $EX = \alpha$. Examples of symmetric df's are easy to construct, and we will encounter many such distributions in this book.

Remark 4. If a and b are constants and X is an rv with $E|X| < \infty$, then $E|aX + b| < \infty$ and $E\{aX + b\} = aEX + b$. In particular, $E\{X - \mu\} = 0$, a fact that should not come as a surprise.

Remark 5. If X is bounded, that is, $P\{|X| < M\} = 1$, $0 < M < \infty$, then EX exists.

Remark 6. If $P\{X \geq 0\} = 1$, and EX exists, then $EX \geq 0$.

Theorem 1. Let X be an rv, and g be a Borel-measurable function on $\mathcal{R}$. Let $Y = g(X)$. Then

(3) $$EY = \sum_{j=1}^{\infty} g(x_j) P\{X = x_j\}$$

in the sense that, if either side of (3) exists, so does the other, and then the two are equal.

Remark 7. Let X be a discrete rv. Then Theorem 1 says that
$$\sum_{j=1}^{\infty} g(x_j) P\{X = x_j\} = \sum_{k=1}^{\infty} y_k P\{Y = y_k\}$$

in the sense that, if either of the two series converges absolutely, so does the other, and the two sums are equal. If X is of the continuous type with pdf f, let $h(y)$ be the pdf of $Y = g(X)$. Then, according to Theorem 1,

$$\int g(x) f(x)\, dx = \int y\, h(y)\, dy,$$

provided that $E|g(X)| < \infty$.

Proof of Theorem 1. Let X be discrete, and suppose that $P\{X \in A\} = 1$. If $y = g(x)$ is a one-to-one mapping of A onto some set B, then

$$P\{Y = y\} = P\{X = g^{-1}(y)\}, \qquad y \in B.$$

We have

$$\sum_{x \in A} g(x)\, P\{X = x\} = \sum_{y \in B} y\, P\{Y = y\}.$$

If X is of the continuous type with pdf f, and g satisfies the conditions of Theorem 2.5.3, then

$$\int g(x) f(x)\, dx = \int_\alpha^\beta y\, f[g^{-1}(y)]\, \left| \frac{d}{dy} g^{-1}(y) \right| dy$$

by changing the variable to $y = g(x)$. Thus

$$\int g(x) f(x)\, dx = \int_\alpha^\beta y\, h(y)\, dy.$$

In the general case, including the case where g may not be one-to-one, we refer the reader to Loève [71], page 166.

The functions $h(x) = x^n$, where n is a positive integer, and $h(x) = |x|^\alpha$, where α is a positive real number, are of special importance. If EX^n exists for some positive integer n, we call EX^n the *n*th moment of (the distribution function of) X *about the origin*. If $E|X|^\alpha < \infty$ for some positive real number α, we call $E|X|^\alpha$ the αth absolute moment of X. We shall use the following notation:

(4) $$m_n = EX^n, \qquad \beta_\alpha = E|X|^\alpha,$$

whenever the expectations exist.

Example 2. Let X have the *uniform* distribution on the first N natural numbers, that is, let

$$P\{X = k\} = \frac{1}{N}, \qquad k = 1, 2, \cdots, N.$$

Clearly moments of all order exist:

$$EX = \sum_{k=1}^{N} k \cdot \frac{1}{N} = \frac{N+1}{2},$$

$$EX^2 = \sum_{k=1}^{N} k^2 \cdot \frac{1}{N} = \frac{(N+1)(2N+1)}{6}.$$

Example 3. Let X be an rv with pdf

$$f(x) = \begin{cases} \dfrac{2}{x^3}, & x \geq 1, \\ 0, & x < 1. \end{cases}$$

Then

$$EX = \int_1^\infty \frac{2}{x^2}\,dx = 2.$$

But

$$EX^2 = \int_1^\infty \frac{2}{x}\,dx$$

does not exist. Indeed, it is easily possible to construct examples of random variables for which all moments of a specified order exist but no higher-order moments do.

Example 4. Two players, A and B, play a coin-tossing game. A gives B one dollar if a head turns up; otherwise, B pays A one dollar. If the probability that the coin shows a head is p, find the gain of A.

Let X denote the expected gain of A. Then

$$P\{X = 1\} = P\{\text{Tails}\} = 1 - p, \qquad P\{X = -1\} = p$$

and

$$EX = 1 - p - p = 1 - 2p \begin{cases} > 0 & \text{if and only if } p < \tfrac{1}{2}, \\ = 0 & \text{if and only if } p = \tfrac{1}{2}. \end{cases}$$

Thus $EX = 0$ if and only if the coin is *fair*.

Theorem 2. *If the moment of order t exists for an rv X, moments of order $0 < s < t$ exist.*

Proof. Let X be of the continuous type with pdf f. We have

$$E|X|^s = \int_{|x|^s \leq 1} |x|^s f(x)\,dx + \int_{|x|^s > 1} |x|^s f(x)\,dx$$

$$\leq P\{|X|^s \leq 1\} + E|X|^t < \infty.$$

A similar proof can be given when X is a discrete rv.

Theorem 3. Let X be an rv on a probability space $(\Omega, \mathscr{S}, P)$. Let $E|X|^k < \infty$ for some $k > 0$. Then

$$n^k P\{|X| > n\} \to 0 \quad \text{as} \quad n \to \infty.$$

Proof. We provide the proof for the case in which X is of the continuous type with density f. We have

$$\infty > \int |x|^k f(x)\, dx = \lim_{n \to \infty} \int_{|x| \leq n} |x|^k f(x)\, dx.$$

It follows that

$$\lim_{n \to \infty} \int_{|x| > n} |x|^k f(x)\, dx \to 0 \quad \text{as} \quad n \to \infty.$$

But

$$\int_{|x| > n} |x|^k f(x)\, dx \geq n^k P\{|X| > n\},$$

completing the proof.

Remark 8. Probabilities of the type $P\{|X| > n\}$ or either of its components, $P\{X > n\}$ or $P\{X < -n\}$, are called *tail probabilities*. The result of Theorem 3, therefore, gives the rate at which $P\{|X| > n\}$ converges to 0 as $n \to \infty$.

Remark 9. The converse of Theorem 3 does not hold in general, that is,

$$n^k P\{|X| > n\} \to 0 \quad \text{as } n \to \infty \text{ for some } k$$

does not necessarily imply that $E|X|^k < \infty$, for consider the rv

$$P\{X = n\} = \frac{c}{n^2 \log n}, \quad n = 2, 3, \cdots,$$

where c is a constant determined from

$$\sum_{n=2}^{\infty} \frac{c}{n^2 \log n} = 1.$$

We have

$$P\{X > n\} \approx c \int_n^\infty \frac{1}{x^2 \log x} dx \approx cn^{-1} (\log n)^{-1}$$

and $nP\{X > n\} \to 0$ as $n \to \infty$. (Here and subsequently $\approx$ means that the ratio of two sides $\to 1$ as $n \to \infty$.) But

$$EX = \sum \frac{c}{n \log n} = \infty .$$

In fact, we need

$$n^{k+\delta} P\{|X| > n\} \to 0 \quad \text{as} \quad n \to \infty$$

for some $\delta > 0$ to ensure that $E|X|^k < \infty$. A condition such as this is called a *moment condition*.

For the proof we need the following lemma.

Lemma 1. Let X be a nonnegative rv with distribution function F. Then

(5) $$EX = \int_0^\infty [1 - F(x)] dx,$$

in the sense that, if either side exists, so does the other and the two are equal.

Proof. If X is of the continuous type with density f and $EX < \infty$, then

$$EX = \int_0^\infty x f(x) dx = \lim_{n \to \infty} \int_0^n x f(x) dx.$$

On integration by parts we obtain

$$\int_0^n x f(x) dx = n F(n) - \int_0^n F(x) dx$$
$$= -n[1 - F(n)] + \int_0^n [1 - F(x)] dx.$$

But

$$n[1 - F(n)] = n \int_n^\infty f(x) dx$$
$$< \int_n^\infty x f(x) dx,$$

and, since $E|X| < \infty$, it follows that

$$n[1 - F(n)] \to 0 \quad \text{as} \quad n \to \infty.$$

We have

$$EX = \lim_{n\to\infty} \int_0^n x f(x)\, dx = \lim_{n\to\infty} \int_0^n [1 - F(x)]\, dx$$
$$= \int_0^\infty [1 - F(x)]\, dx.$$

If $\int_0^\infty [1 - F(x)]\, dx < \infty$, then

$$\int_0^n x f(x)\, dx \le \int_0^n [1 - F(x)]\, dx \le \int_0^\infty [1 - F(x)]\, dx,$$

and it follows that $E|X| < \infty$.

If, on the other hand, X is a discrete rv, let us write $P\{X = x_j\} = p_j$. Then

$$EX = \sum_{j=1}^\infty x_j p_j.$$

Let $I = \int_0^\infty [1 - F(x)]\, dx$. Then

$$I = \sum_{k=1}^\infty \int_{(k-1)/n}^{k/n} P\{X > x\}\, dx,$$

and since $P\{X > x\}$ is nonincreasing, we have

$$\frac{1}{n}\sum_{k=1}^\infty P\left\{X > \frac{k}{n}\right\} \le I \le \frac{1}{n}\sum_{k=1}^\infty P\left\{X > \frac{k-1}{n}\right\}$$

for every positive integer n. Let

$$U_n = \frac{1}{n}\sum_{k=1}^\infty P\left\{X > \frac{k-1}{n}\right\}, \quad \text{and} \quad L_n = \frac{1}{n}\sum_{k=1}^\infty P\left\{X > \frac{k}{n}\right\}.$$

We have

$$L_n = \frac{1}{n}\sum_{k=1}^\infty \sum_{j=k}^\infty P\left\{\frac{j}{n} < X \le \frac{j+1}{n}\right\}$$
$$= \frac{1}{n}\sum_{k=2}^\infty (k-1) P\left\{\frac{k-1}{n} < X \le \frac{k}{n}\right\}$$

on rearranging the series. Thus

$$L_n = \sum_{k=2}^\infty \frac{k}{n} P\left\{\frac{k-1}{n} < X \le \frac{k}{n}\right\} - \sum_{k=2}^\infty \frac{1}{n} P\left\{\frac{k-1}{n} < X \le \frac{k}{n}\right\}$$
$$= \sum_{k=2}^\infty \frac{k}{n} \sum_{(k-1)/n < x_j \le k/n} p_j - \frac{1}{n} P\left\{X > \frac{1}{n}\right\}$$
$$\ge \sum_{k=1}^\infty \sum_{(k-1)/n < x_j \le k/n} x_j p_j - \frac{1}{n}$$
$$= EX - \frac{1}{n}.$$

Similarly we can show that for every positive integer n

$$U_n \leq EX + \frac{1}{n}.$$

Thus we have

$$EX - \frac{1}{n} \leq \int_0^\infty [1 - F(x)]\, dx \leq EX + \frac{1}{n}$$

for every positive integer n. Taking the limit as $n \to \infty$, we see that

$$EX = \int_0^\infty [1 - F(x)]\, dx.$$

Corollary. For any rv X, $E|X| < \infty$ if and only if the integrals $\int_{-\infty}^0 P\{X \leq x\}\, dx$ and $\int_0^\infty P\{X > x\}\, dx$ both converge, and in that case

$$EX = \int_0^\infty P\{X > x\}\, dx - \int_{-\infty}^0 P\{X \leq x\}\, dx.$$

Actually we can get a little more out of Lemma 1 than the above corollary. In fact,

$$E|X|^\alpha = \int_0^\infty P\{|X|^\alpha > x\}\, dx = \alpha \int_0^\infty x^{\alpha-1} P\{|X| > x\}\, dx,$$

and we see that *an rv X possesses* an *absolute moment of order $\alpha > 0$ if and and only if $|x|^{\alpha-1} P\{|X| > x\}$ is integrable over $(0, \infty)$*.

A simple application of the integral test (see Apostol [3], 361) leads to the following *moments lemma*.

Lemma 2.

(6) $$E|X|^\alpha < \infty \Leftrightarrow \sum_{n=1}^\infty P\{|X| > n^{1/\alpha}\} < \infty.$$

In Section 6.4 we will construct another proof of (6). Note that an immediate consequence of Lemma 2 is Theorem 3. We are now ready to prove the following result.

Theorem 4. Let X be an rv with a distribution satisfying $n^\alpha P\{|X| > n\} \to 0$ as $n \to \infty$ for some $\alpha > 0$. Then $E|X|^\beta < \infty$ for $0 < \beta < \alpha$.

Proof. Given $\varepsilon > 0$, we can choose an $N = N(\varepsilon)$ such that

$$P\{|X| > n\} < \frac{\varepsilon}{n^\alpha} \qquad \text{for all } n \geq N.$$

It follows that for $0 < \beta < \alpha$

$$E|X|^\beta = \beta \int_0^N x^{\beta-1} P\{|X| > x\} \, dx + \beta \int_N^\infty x^{\beta-1} P\{|X| > x\} \, dx$$
$$\leq N^\beta + \beta \varepsilon \int_N^\infty x^{\beta-\alpha-1} \, dx$$
$$< \infty.$$

Remark 10. Using Theorems 3 and 4, we demonstrate the existence of random variables for which moments of any order do not exist, that is, for which $E|X|^\alpha = \infty$ for every $\alpha > 0$. For such an rv $n^\alpha \, P\{|X| > n\} \not\to 0$ as $n \to \infty$ for any $\alpha > 0$. Consider, for example, the rv X with pdf

$$f(x) = \begin{cases} \dfrac{1}{2|x| (\log |x|)^2} & \text{for } |x| > e \\ 0 & \text{otherwise.} \end{cases}$$

The df of X is given by

$$F(x) = \begin{cases} \dfrac{1}{2 \log |x|} & \text{if } x \leq -e \\ \dfrac{1}{2} & \text{if } -e < x < e, \\ 1 - \dfrac{1}{2 \log x} & \text{if } x \geq e. \end{cases}$$

Then for $x > e$

$$P\{|X| > x\} = 1 - F(x) + F(-x)$$
$$= \frac{1}{2 \log x},$$

and $x^\alpha \, P\{|X| > x\} \to \infty$ as $x \to \infty$ for any $\alpha > 0$. If follows that $E|X|^\alpha = \infty$ for every $\alpha > 0$. In this example we see that $P\{|X| > cx\}/P\{|X| > x\} \to 1$ as $x \to \infty$ for every $c > 0$. A positive function $L(\cdot)$ defined on $(0, \infty)$ is said to be a function of *slow variation* if and only if $L(cx)/L(x) \to 1$ as $x \to \infty$ for every $c > 0$. For such a function $x^\alpha L(x) \to \infty$ for every $\alpha > 0$ (see Feller [29], 275–279). It follows that, if $P\{|X| > x\}$ is slowly varying, $E|X|^\alpha = \infty$ for every $\alpha > 0$. Functions of slow variation play an important role in the theory of probability. We again refer the reader to Feller [29].

Random variables for which $P\{|X| > x\}$ is slowly varying are clearly excluded from the domain of the following result.

Theorem 5. Let X be an rv satisfying

(7) $$\frac{P\{|X| > \alpha k\}}{P\{|X| > k\}} \to 0 \quad \text{as } k \to \infty \text{ for all } \alpha > 1;$$

then X possesses moments of all orders. (Note that, if $\alpha = 1$, the limit in (7) is 1, whereas if $\alpha < 1$, the limit will not go to 0 since $P\{|X| > \alpha k\} \geq P\{|X| > k\}$.)

Proof. Let $\varepsilon > 0$ (we will choose ε later), choose K_0 so large that

(8) $$\frac{P\{|X| > \alpha k\}}{P\{|X| > k\}} < \varepsilon \quad \text{for all} \quad k \geq K_0,$$

and choose K_1 so large that

(9) $$P\{|X| > k\} < \varepsilon \quad \text{for all} \quad k \geq K_1.$$

Let $N = \max(K_0, K_1)$. We have, for a fixed positive integer r,

(10) $$\frac{P\{|X| > \alpha^r k\}}{P\{|X| > k\}} = \prod_{p=1}^{r} \frac{P\{|X| > \alpha^p k\}}{P\{|X| > \alpha^{p-1} k\}} \leq \varepsilon^r$$

for $k \geq N$. Thus for $k \geq N$ we have, in view of (9),

(11) $$P\{|X| > \alpha^r k\} \leq \varepsilon^{r+1}.$$

Next note that, for any fixed positive integer n,

(12) $$\begin{aligned} E|X|^n &= n \int_0^\infty x^{n-1} P\{|X| > x\} \, dx \\ &= n \int_0^N x^{n-1} P\{|X| > x\} \, dx + n \int_N^\infty x^{n-1} P\{|X| > x\} \, dx. \end{aligned}$$

Since the first integral in (12) is finite, we need only show that the second integral is also finite. We have

$$\begin{aligned} \int_N^\infty x^{n-1} P\{|X| > x\} \, dx &= \sum_{r=1}^{\infty} \int_{\alpha^{r-1} N}^{\alpha^r N} x^{n-1} P\{|X| > x\} \, dx \\ &\leq \sum_{r=1}^{\infty} (\alpha^r N)^{n-1} \varepsilon^r \cdot 2\alpha^r N \\ &= 2N^n \sum_{r=1}^{\infty} (\varepsilon \alpha^n)^r \\ &= 2N^n \frac{\varepsilon \alpha^n}{1 - \varepsilon \alpha^n} < \infty, \end{aligned}$$

provided that we choose ε such that $\varepsilon \alpha^n < 1$. It follows that $E|X|^n < \infty$ for $n = 1, 2, \cdots$. Actually we have shown that (7) implies $E|X|^\delta < \infty$ for all $\delta > 0$.

Theorem 6. If $h_1, h_2, \cdots, h_n$ are Borel-measurable functions of an rv X and $Eh_i(X)$ exists for $i = 1, 2, \cdots, n$, then $E\{\sum_{i=1}^n h_i(X)\}$ exists and equals $\sum_{i=1}^n Eh_i(X)$.

Proof. The proof is simple.

Definition 1. Let k be a positive integer, and c be a constant. If $E(X - c)^k$ exists, we call it the moment of order k about the point c. If we take $c = EX = \mu$, which exists since $E|X|^k < \infty$, we call $E(X - \mu)^k$ the central moment of order k or the moment of order k about the mean. We shall write

$$\mu_k = E\{X - \mu\}^k.$$

If we know $m_1, m_2, \cdots, m_k$, we can compute $\mu_1, \mu_2, \cdots, \mu_k$, and conversely. We have

(13) $\quad \mu_k = E\{X - \mu\}^k = m_k - \binom{k}{1}\mu m_{k-1} + \binom{k}{2}\mu^2 m_{k-2} - \cdots + (-1)^k \mu^k$

and

(14) $\quad m_k = E\{X - \mu + \mu\}^k = \mu_k + \binom{k}{1}\mu \mu_{k-1} + \binom{k}{2}\mu^2 \mu_{k-2} + \cdots + \mu^k.$

The case $k = 2$ is of special importance.

Definition 2. If EX^2 exists, we call $E\{X - \mu\}^2$ the variance of X, and we write $\sigma^2 = \text{var}(X) = E(X - \mu)^2$. The quantity σ is called the standard deviation (SD) of X.

From Theorem 6 we see that

(15) $\qquad\qquad\qquad \sigma^2 = \mu_2 = EX^2 - (EX)^2.$

Variance has some important properties.

Theorem 7. Var $(X) = 0$ if and only if X is degenerate.

Theorem 8. Var $(X) < E(X - c)^2$ for any $c \neq EX$.

Proof. We have

$$\text{var}(X) = E\{X - \mu\}^2 = E\{X - c\}^2 + (c - \mu)^2.$$

Note that

$$\text{var}(aX + b) = a^2 \text{var}(X).$$

Let $E|X|^2 < \infty$. Then we define

$$\text{(16)} \qquad Z = \frac{X - EX}{\sqrt{\text{var}(X)}} = \frac{X - \mu}{\sigma}$$

and see that $EZ = 0$ and var $(Z) = 1$. We call Z a *standardized* rv.

Example 5. Let X be an rv with *binomial* pmf

$$P\{X = k\} = \binom{n}{k} p^k (1-p)^{n-k}, \qquad k = 0, 1, 2, \cdots, n; \quad 0 < p < 1.$$

Then

$$EX = \sum_{k=0}^{n} k \binom{n}{k} p^k (1-p)^{n-k}$$
$$= np \sum \binom{n-1}{k-1} p^{k-1}(1-p)^{n-k}$$
$$= np;$$

$$EX^2 = E\{X(X-1)\} + X\}$$
$$= \sum k(k-1) \binom{n}{k} p^k (1-p)^{n-k} + np$$
$$= n(n-1) p^2 + np;$$

$$\text{var}(X) = n(n-1) p^2 + np - n^2 p^2$$
$$= np(1-p);$$

$$EX^3 = E\{X(X-1)(X-2) + 3X(X-1) + X\}$$
$$= n(n-1)(n-2) p^3 + 3n(n-1) p^2 + np;$$

$$\mu_3 = m_3 - 3\mu m_2 + 2\mu^3$$
$$= n(n-1)(n-2) p^3 + 3n(n-1) p^2 + np - 3np[n(n-1) p^2 + np] + 2n^3 p^3$$
$$= np(1-p)(1-2p).$$

We have seen that for some distributions even the mean does not exist. We next consider some parameters, called *order parameters*, which always exist.

Definition 3. A number x satisfying

$$\text{(17)} \qquad P\{X \leq x\} \geq p, \qquad P\{X \geq x\} \geq 1 - p, \quad 0 < p < 1,$$

is called a quantile of order p [or $(100p)$th percentile] for the rv X (or for the df F of X). We write $\zeta_p(X)$ for a quantile of order p for the rv X.

If x is a quantile of order p for an rv X with df F, then

$$\text{(18)} \qquad p \leq F(x) \leq p + P\{X = x\}.$$

If $P\{X = x\} = 0$, as is the case—in particular, if X is of the continuous type— a quantile of order p is a solution of the equation

(19) $$F(x) = p.$$

If F is strictly increasing, (19) has a unique solution. Otherwise there may be many (even uncountably many) solutions of (19), each one of which is then called a quantile of order p.

Definition 4. Let X be an rv with df F. A number x satisfying

(20) $$\frac{1}{2} \le F(x) \le \frac{1}{2} + P\{X = x\}$$

or, equivalently,

(21) $$P\{X \le x\} \ge \frac{1}{2} \quad \text{and} \quad P\{X \ge x\} \ge \frac{1}{2}$$

is called a median of X (or F).

Again we note that there may be many values that satisfy (20) or (21). Thus a median is not necessarily unique.

If F is a symmetric df, the center of symmetry is clearly the median of the df F. The median is an important centering constant especially in cases where the mean of the distribution does not exist.

Example 6. Let X be an rv with *Cauchy* pdf

$$f(x) = \frac{1}{\pi} \frac{1}{1 + x^2}, \quad -\infty < x < \infty.$$

Then $E|X|$ is not finite. The median of the rv X is clearly $x = 0$.

Example 7. Let X be an rv with pmf

$$P\{X = -2\} = P\{X = 0\} = \tfrac{1}{4}, \quad P\{X = 1\} = \tfrac{1}{3}, \quad P\{X = 2\} = \tfrac{1}{6}.$$

Then

$$P\{X \le 0\} = \tfrac{1}{2} \quad \text{and} \quad P\{X \ge 0\} = \tfrac{3}{4} > \tfrac{1}{2}.$$

In fact, if x is any number such that $0 < x < 1$, then

$$P\{X \le x\} = P\{X = -2\} + P\{X = 0\} = \tfrac{1}{2}$$

and

$$P\{X \ge x\} = P\{X = 1\} + P\{X = 2\} = \tfrac{1}{2},$$

and it follows that every x, $0 \leq x < 1$, is a median of the rv X.
If $p = .2$, the quantile of order p is $x = -2$, since

$$P\{X \leq -2\} = \frac{1}{4} > p \quad \text{and} \quad P\{X \geq -2\} = 1 > 1 - p.$$

PROBLEMS 3.2

1. Find the expected number of throws of a fair die until a 6 is obtained.

2. From a box containing N identical tickets numbered 1 through N, n tickets are drawn with replacement. Let X be the largest number drawn. Find EX.

3. Let X be an rv with pdf

$$f(x) = \frac{c}{(1+x^2)^m}, \quad -\infty < x < \infty, \quad m \geq 1,$$

where $c = \Gamma(m)/[\Gamma(1/2)\Gamma(m-1/2)]$. Show that EX^{2r} exists if and only if $2r < 2m-1$. What is EX^{2r} if $2r < 2m - 1$?

4. Let X be an rv with pdf

$$f(x) = \begin{cases} \dfrac{ka^k}{(x+a)^{k+1}} & \text{if } x \geq 0, \\ 0 & \text{otherwise } (a > 0). \end{cases}$$

Show that $E|X|^\alpha < \infty$ for $\alpha < k$. Find the quantile of order p for the rv X.

5. Let X be an rv such that $E|X| < \infty$. Show that $E|X-c|$ is minimized if we choose c equal to the median of the distribution of X.

6. *Pareto's distribution* with parameters α and β (both α and β positive) is defined by the pdf

$$f(x) = \begin{cases} \dfrac{\beta \alpha^\beta}{x^{\beta+1}} & \text{if } x \geq \alpha, \\ 0 & \text{if } x < \alpha. \end{cases}$$

Show that the moment of order n exists if and only if $n < \beta$. Let $\beta > 2$. Find the mean and the variance of the distribution.

7. For an rv X with pdf

$$f(x) = \begin{cases} \frac{1}{2}x & \text{if } 0 \leq x < 1, \\ \frac{1}{2} & \text{if } 1 < x \leq 2, \\ \frac{1}{2}(3-x) & \text{if } 2 < x \leq 3, \end{cases}$$

show that moments of all order exist. Find the mean and the variance of X.

8. For the pmf of Example 5 show that

$$EX^4 = np + 7n(n-1)p^2 + 6n(n-1)(n-2)p^3 + n(n-1)(n-2)(n-3)p^4$$

and

$$\mu_4 = 3(npq)^2 + npq(1 - 6pq),$$

where $0 \leq p \leq 1$, $q = 1 - p$.

9. For the Poisson rv X with pmf

$$P\{X = x\} = e^{-\lambda}\frac{\lambda^x}{x!}, \quad x = 0, 1, 2, \cdots,$$

show that $EX = \lambda$, $EX^2 = \lambda + \lambda^2$, $EX^3 = \lambda + 3\lambda^2 + \lambda^3$, $EX^4 = \lambda + 7\lambda^2 + 6\lambda^3 + \lambda^4$, and $\mu_2 = \mu_3 = \lambda$, $\mu_4 = \lambda + 3\lambda^2$.

10. For any rv X with $E|X|^4 < \infty$ define

$$\alpha_3 = \frac{\mu_3}{(\mu_2)^{3/2}}, \quad \alpha_4 = \frac{\mu_4}{\mu_2^2}.$$

Here α_3 is known as the *coefficient of skewness* and is sometimes used as a measure of asymmetry, and α_4 is known as *kurtosis* and is used to measure the peakedness ("flatness of the top") of a distribution.

Compute α_3 and α_4 for the pmf's of Problems 8 and 9.

11. For a positive rv X define the negative moment of order n by EX^{-n}, where $n > 0$ is an integer. Find $E\{1/(X + 1)\}$ for the pmf's of Example 5 and Problem 9.

12. Prove Theorem 6.
13. Prove Theorem 7.

3.3 GENERATING FUNCTIONS

In this section we consider some functions that generate probabilities or moments of an rv. The simplest type of generating function in probability theory is the one associated with integer-valued rv's. Let X be an rv, and let

$$p_k = P\{X = k\}, \quad k = 0, 1, 2, \cdots$$

with $\sum_{k=0}^{\infty} p_k = 1$.

Definition 1. The function defined by

(1) $$P(s) = \sum_{k=0}^{\infty} p_k s^k,$$

which surely converges for $|s| \leq 1$, is called the probability generating function (pgf) of X.

Example 1. Consider the Poisson rv

$$P\{X = k\} = e^{-\lambda}\frac{\lambda^k}{k!}, \quad k = 0, 1, 2, \cdots.$$

We have

$$P(s) = \sum_{k=0}^{\infty} (s\lambda)^k \frac{e^{-\lambda}}{k!} = e^{-\lambda} e^{s\lambda} = e^{-\lambda(1-s)}, \qquad |s| \le 1.$$

Example 2. Let X be an rv with *geometric* distribution, that is, let

$$P\{X=k\} = pq^k, \qquad k = 0, 1, 2, \cdots; \quad 0 < p < 1, q = 1 - p.$$

Then

$$P(s) = \sum_{k=0}^{\infty} s^k pq^k = p\frac{1}{1 - sq}, \qquad |s| \le 1.$$

Remark 1. Since $P(1) = 1$, series (1) is uniformly and absolutely convergent in $|s| \le 1$ and the pgf P is a continuous function of s. It determines the pgf uniquely, since $P(s)$ can be represented in a unique manner as a power series.

Remark 2. The moments of the rv X, if they exist, can be determined by the derivative at the point $s = 1$ of the function $P(s)$. Thus

$$P'(s) = \sum_{k=1}^{\infty} kp_k s^{k-1}, \qquad \text{so that } P'(1) = EX \qquad \text{if } EX < \infty.$$

$$P''(s) = \sum_{k=2}^{\infty} k(k-1)p_k s^{k-2}, \qquad \text{so that } P''(1) = E\{X(X-1)\} \qquad \text{if } EX^2 < \infty,$$

and so on.

Example 3. In Example 1 we found that $P(s) = e^{-\lambda(1-s)}$, $|s| \le 1$, for a Poisson rv. Thus

$$P'(s) = \lambda e^{-\lambda(1-s)},$$
$$P''(s) = \lambda^2 e^{-\lambda(1-s)}.$$

Also, $EX = \lambda$, $E\{X^2 - X\} = \lambda^2$, so that var $(X) = EX^2 - (EX)^2 = \lambda^2 + \lambda - \lambda^2 = \lambda$.

In Example 2 we computed $P(s) = p/(1 - sq)$, so that

$$P'(s) = \frac{pq}{(1 - sq)^2} \qquad \text{and} \qquad P''(s) = \frac{2pq^2}{(1 - sq)^3}.$$

Thus

$$EX = \frac{q}{p}, \qquad EX^2 = \frac{q}{p} + \frac{2pq^2}{p^3}, \qquad \text{var } (X) = \frac{q^2}{p^2} + \frac{q}{p} = \frac{q}{p^2}.$$

Next we consider the important concept of a moment generating function.

GENERATING FUNCTIONS

Definition 2. Let X be an rv defined on $(\Omega, \mathscr{S}, P)$. The function

(2) $$M(s) = E\, e^{sX}$$

is known as the moment generating function (mgf) of the rv X if the expectation on the right side of (2) exists in some neighborhood of the origin.

Example 4. Let X have the pmf

$$f(k) = \begin{cases} \dfrac{6}{\pi^2} \cdot \dfrac{1}{k^2}, & k = 1, 2, \cdots, \\ 0, & \text{otherwise.} \end{cases}$$

Then $(1/\pi^2) \sum_{k=1}^{\infty} e^{sk}/k^2$, is infinite for every $s > 0$. We see that the mgf of X does not exist. In fact, $EX = \infty$.

Example 5. Let X have the pdf

$$f(x) = \begin{cases} \tfrac{1}{2} e^{-x/2}, & x > 0, \\ 0, & \text{otherwise.} \end{cases}$$

Then

$$M(s) = \frac{1}{2} \int_0^{\infty} e^{(s-1/2)x}\, dx$$
$$= \frac{1}{1 - 2s}, \quad s < \frac{1}{2}.$$

Example 6. Let X have pmf

$$P\{X = k\} = \begin{cases} e^{-\lambda} \dfrac{\lambda^k}{k!}, & k = 0, 1, 2, \cdots, \\ 0, & \text{otherwise.} \end{cases}$$

Then

$$M(s) = E e^{sX} = e^{-\lambda} \sum_{k=0}^{\infty} e^{sk} \frac{\lambda^k}{k!}$$
$$= e^{-\lambda(1-e^s)} \quad \text{for all } s.$$

The following result will be quite useful in what follows.

Theorem 1. The mgf uniquely determines a df and, conversely, if the mgf exists, it is unique.

For the proof we refer the reader to Widder [137], page 460, or Curtiss [20]. See also P.2.14. Theorem 2 explains why we call $M(s)$ an mgf.

Theorem 2. If the mgf $M(s)$ of an rv X exists for s in $(-s_0, s_0)$ say, $s_0 > 0$, the derivatives of all order exist at $s = 0$ and can be evaluated under the integral sign, that is,

(3) $$M^{(k)}(s)\big|_{s=0} = EX^k \text{ for positive integral } k.$$

For the proof of Theorem 2 we refer to Widder [137], pages 446–447. See also P.2.14 and Problem 9.

Remark 3. Alternatively, if the mgf $M(s)$ exists for s in $(-s_0, s_0)$ say, $s_0 > 0$, one can express $M(s)$ (uniquely) in a Maclaurin series expansion:

(4) $$M(s) = M(0) + \frac{M'(0)}{1!} s + \frac{M''(0)}{2!} s^2 + \cdots,$$

so that EX^k is the coefficient of $s^k/k!$ in expansion (4).

Example 7. Let X be an rv with pdf $f(x) = (1/2)e^{-x/2}$, $x > 0$. From Example 5, $M(s) = 1/(1 - 2s)$ for $s < 1/2$. Thus

$$M'(s) = \frac{2}{(1-2s)^2} \quad \text{and} \quad M''(s) = \frac{4 \cdot 2}{(1-2s)^3}, \quad s < \frac{1}{2}.$$

It follows that

$$EX = 2, \quad EX^2 = 8, \quad \text{and} \quad \text{var}(X) = 4.$$

Example 8. Let X be an rv with pdf $f(x) = 1$, $0 \le x \le 1$, and $= 0$ otherwise. Then

$$M(s) = \int_0^1 e^{sx} \, dx = \frac{e^s - 1}{s}, \quad \text{all } s,$$

$$M'(s) = \frac{e^s \cdot s - (e^s - 1) \cdot 1}{s^2},$$

$$EX = M'(0) = \lim_{s \to 0} \frac{se^s - e^s + 1}{s^2} = \frac{1}{2}.$$

Remark 4. Since there exist rv's for which the mgf may not exist, its utility is somewhat limited. It is much more convenient to work with the *characteristic function* of an rv X, which is defined as $E(e^{itX})$, where $i = \sqrt{(-1)}$, the imaginary unit, and t is any real number. $E(e^{itX})$ exists for every distribution. Moreover, it uniquely determines the distribution of rv X. Since we do not assume a knowledge of complex variables in this book, we will deal with only mgf's whenever they exist.

We next consider the problem of characterizing a distribution from its moments. Let X be an rv with mgf $M(s)$. Since $n! e^{|sx|} \geq |sx|^n$ for $n > 0$, n integral, we see that $E|X|^n < \infty$ for any n. Given the mgf $M(s)$, we can determine EX^n for any n (positive integer) with the help of Theorem 2. Suppose now that moments of all orders exist for an rv X. It does not follow that the mgf exists.

Example 9. Let X be an rv with pdf

$$f(x) = c e^{-|x|^\alpha}, \quad 0 < \alpha < 1, \quad -\infty < x < \infty,$$

where c is a constant determined from

$$c \int_{-\infty}^{\infty} e^{-|x|^\alpha} dx = 1.$$

Let $s > 0$. Then

$$\int_{0}^{\infty} e^{sx} e^{-x^\alpha} dx = \int_{0}^{\infty} e^{x(s - x^{\alpha-1})} dx$$

and since $\alpha - 1 < 0$, $\int_0^\infty e^{sx} e^{-x^\alpha} dx$ is not finite for any $s > 0$. Hence the mgf does not exist. But

$$E|X|^n = c \int_{-\infty}^{\infty} |x|^n e^{-|x|^\alpha} dx = 2c \int_{0}^{\infty} x^n e^{-x^\alpha} dx < \infty \quad \text{for each } n,$$

as is easily checked by substituting $y = x^\alpha$.

Theorem 3. Let $\{m_k\}$ be the moment sequence of an rv X. If the series

(5) $$\sum_{k=1}^{\infty} \frac{m_k}{k!} s^k$$

converges absolutely for some $s > 0$, then $\{m_k\}$ uniquely determines the df F of X.

The proof of this result is much too complicated to be included here, and we refer the reader to original papers by Hamburger [46]. It should be noted that condition (5) is not necessary (see Dharmadhikari [25]).

In particular if for some constant c

$$|m_k| \leq c^k, \quad k = 1, 2, \cdots,$$

then

$$\sum_{k=1}^{\infty} \frac{|m_k|}{k!} s^k \leq \sum_{1}^{\infty} \frac{(cs)^k}{k!} < e^{cs} \quad \text{for } s > 0,$$

and the df of X is uniquely determined.

If not all the moments of an rv X exist, there is no chance of determining the df of X. The df of X is surely not determined uniquely by the moments that do exist.

Example 10. Let X be an rv with pmf

$$P\left\{X = \frac{3^k}{k^2}\right\} = \frac{2}{3^k}, \quad k = 1, 2, \cdots.$$

Then

$$EX = \sum_{k=1}^{\infty} \frac{2}{k^2} < \infty, \quad \text{and} \quad EX^2 = \sum_{k=1}^{\infty} \frac{2 \cdot 3^k}{k^4} = \infty.$$

Let Y be an rv with pmf

$$P\{Y = 0\} = \frac{1}{3}, \quad \text{and} \quad P\left\{Y = \frac{3^{k+1}}{k^2}\right\} = \frac{2}{3^{k+1}}, \quad k = 1, 2, \cdots.$$

Then

$$EY = \sum_{k=1}^{\infty} \frac{2}{k^2} < \infty, \quad \text{and} \quad EX = EY.$$

But

$$EY^2 = 2 \sum_{k=1}^{\infty} \frac{3^{k+1}}{k^4} = \infty,$$

and X and Y do not have the same distribution.

Finally we mention some sufficient conditions for a moment sequence to determine a unique df.

(i) The range of the rv is finite.
(ii) (Carleman) $\sum_{k=1}^{\infty} (m_{2k})^{-1/2k} = \infty$ when the range of the rv is $(-\infty, \infty)$. If the range is $(0, \infty)$, a sufficient condition is $\sum_{k=1}^{\infty} (m_k)^{-1/2k} = \infty$.
(iii) $\overline{\lim}_{n \to \infty} \{(m_{2n})^{1/2n}/2n\}$ is finite.

PROBLEMS 3.3

1. Find the pgf of the rv's with the following pmf's:

(a) $P\{X = k\} = \binom{n}{k} p^k (1-p)^{n-k}$, $k = 0, 1, 2, \cdots, n; 0 \le p \le 1$.
(b) $P\{X = k\} = [e^{-\lambda}/(1 - e^{-\lambda})](\lambda^k/k!)$, $k = 1, 2, \cdots; \lambda > 0$.
(c) $P\{X = k\} = pq^k(1 - q^{N+1})^{-1}$, $k = 0, 1, 2, \cdots, N; 0 < p < 1, q = 1 - p$.

2. Let X be an integer-valued rv with pgf $P(s)$. Let a and b be nonnegative integers, and write $Y = aX + b$. Find the pgf of Y.

3. Let X be an integer-valued rv with pgf $P(s)$, and suppose that the mgf $M(s)$ exists for $s \in (-s_0, s_0)$, $s_0 > 0$. How are $M(s)$ and $P(s)$ related? Using $M^{(k)}(s)|_{s=0} = EX^k$ for positive integral k, find EX^k in terms of the derivatives of $P(s)$ for values of $k = 1, 2, 3, 4$.

4. For the Cauchy pdf
$$f(x) = \frac{1}{\pi} \frac{1}{1+x^2}, \quad -\infty < x < \infty,$$
does the mgf exist?

5. Let X be an rv with pmf
$$P\{X = j\} = p_j, \quad j = 0, 1, 2, \cdots.$$
Set $P\{X > j\} = q_j$, $j = 0, 1, 2, \cdots$. Clearly $q_j = p_{j+1} + p_{j+2} + \cdots$, $j \geq 0$. Write $Q(s) = \sum_{j=0}^{\infty} q_j s^j$. Then the series for $Q(s)$ converges in $|s| < 1$. Show that
$$Q(s) = \frac{1 - P(s)}{1 - s} \quad \text{for } |s| < 1,$$
where $P(s)$ is the pgf of X. Find the mean and the variance of X (when they exist) in terms of Q and its derivatives.

6. For the pmf
$$P\{X = j\} = \frac{a_j \theta^j}{f(\theta)}, \quad j = 0, 1, 2, \cdots, \theta > 0,$$
where $a_j \geq 0$ and $f(\theta) = \sum_{j=0}^{\infty} a_j \theta^j$, find the pgf and the mgf in terms of f.

7. For the Laplace pdf
$$f(x) = \frac{1}{2\lambda} e^{-|x-\mu|/\lambda}, \quad -\infty < x < \infty; \lambda > 0, -\infty < \mu < \infty,$$
show that the mgf exists and equals
$$M(t) = (1 - \lambda^2 t^2)^{-1} e^{\mu t}, \quad t < \frac{1}{\lambda}.$$

8. For any integer-valued rv X, show that
$$\sum_{n=0}^{\infty} s^n P\{X \leq n\} = (1-s)^{-1} P(s),$$
where P is the pgf of X.

9. Let X be an rv with mgf $M(t)$, which exists for $t \in (-t_0, t_0)$, $t_0 > 0$. Show that
$$E|X|^n < n! \, s^{-n}[M(s) + M(-s)]$$
for any fixed s, $0 < s < t_0$, and for each integer $n \geq 1$. Expanding e^{tx} in a power series, show that, for $t \in (-s, s)$, $0 < s < t_0$,
$$M(t) = \sum_{n=0}^{\infty} t^n \frac{EX^n}{n!}.$$
(Since a power series can be differentiated term by term within the interval of convergence, it follows that for $|t| < s$,

$$M^{(k)}(t)|_{t=0} = EX^k$$

for each integer $k \geq 1$.)

(Roy, LePage, and Moore [106])

3.4 SOME MOMENT INEQUALITIES

In this section we derive some inequalities for moments of an rv. The main result of this section is Theorem 1 (and its corollary), which gives a bound for tail probability in terms of some moment of the random variable.

Theorem 1. Let $h(X)$ be a nonnegative Borel-measurable function of an rv X. If $Eh(X)$ exists, then, for every $\varepsilon > 0$,

(1) $$P\{h(X) \geq \varepsilon\} \leq \frac{Eh(X)}{\varepsilon}.$$

Proof. We prove the result when X is discrete. Let $P\{X = x_k\} = p_k$, $k = 1, 2, \cdots$. Then

$$Eh(X) = \sum_k h(x_k) p_k$$
$$= \left(\sum_A + \sum_{A^c}\right) h(x_k) p_k,$$

where

$$A = \{k : h(x_k) \geq \varepsilon\}.$$

Then

$$Eh(X) \geq \sum_A h(x_k) p_k \geq \varepsilon \sum_A p_k$$
$$= \varepsilon P\{h(X) \geq \varepsilon\}.$$

Corollary. Let $h(X) = |X|^r$ and $\varepsilon = K^r$, where $r > 0$ and $K > 0$. Then

(2) $$P\{|X| \geq K\} \leq \frac{E|X|^r}{K^r},$$

which is *Markov's inequality*. In particular, if we take $h(X) = (X - \mu)^2$, $\varepsilon = K^2 \sigma^2$, we get *Chebychev's inequality*:

(3) $$P\{|X - \mu| > K\sigma\} \leq \frac{1}{K^2},$$

where $EX = \mu$, $\text{var}(X) = \sigma^2$.

For rv's with finite second-order moments one cannot do better than the inequality in (3).

Example 1.
$$P\{X = 0\} = 1 - \frac{1}{K^2}, \quad K > 1, \text{ constant,}$$
$$P\{X = \mp 1\} = \frac{1}{2K^2}$$
$$EX = 0, \quad EX^2 = \frac{1}{K^2}, \quad \sigma = \frac{1}{K},$$
$$P\{|X| \geq K\sigma\} = P\{|X| \geq 1\} = \frac{1}{K^2},$$

so that equality is achieved.

Example 2. Let X be distributed with pdf $f(x) = 1$ if $0 < x < 1$, and $= 0$ otherwise. Then
$$EX = \tfrac{1}{2}, \quad EX^2 = \tfrac{1}{3}, \text{ var}(X) = \tfrac{1}{3} - \tfrac{1}{4} = \tfrac{1}{12},$$
$$P\{|X - \tfrac{1}{2}| \leq 2\sqrt{\tfrac{1}{12}}\} = P\{\tfrac{1}{2} - \tfrac{1}{\sqrt{3}} < X < \tfrac{1}{2} + \tfrac{1}{\sqrt{3}}\} = 1.$$

From Chebychev's inequality
$$P\{|X - \tfrac{1}{2}| \leq 2\sqrt{\tfrac{1}{12}}\} \geq 1 - \tfrac{1}{4} = .75.$$

It is possible to improve upon Chebychev's inequality, at least in some cases, if we assume the existence of higher-order moments. We need the following lemma.

Lemma 1. Let X be an rv with $EX = 0$ and var$(X) = \sigma^2$. Then

(4) $$P\{X > x\} \leq \frac{\sigma^2}{\sigma^2 + x^2} \quad \text{if } x > 0,$$

(5) $$P\{X > x\} \geq \frac{x^2}{\sigma^2 + x^2} \quad \text{if } x < 0.$$

Proof. Let $h(t) = (t + c)^2$, $c > 0$. Then $h(t) \geq 0$ for all t and
$$h(t) \geq (x + c)^2 \quad \text{for } t > x > 0.$$

It follows that

(6) $$P\{X > x\} \leq P\{h(X) \geq (x + c)^2\}$$
$$\leq \frac{E(X + c)^2}{(x + c)^2} \quad \text{for all } c > 0, x > 0.$$

Since $EX = 0$, $EX^2 = \sigma^2$, and the right side of (6) is minimum when $c = \sigma^2/x$. We have

$$P\{X > x\} \leq \frac{\sigma^2}{\sigma^2 + x^2}, \qquad x > 0.$$

Similar proof holds for (5).

Remark 1. Inequalities (4) and (5) cannot be improved (Problem 3).

Theorem 2. Let $E|X|^4 < \infty$, and let $EX = 0$, $EX^2 = \sigma^2$. Then

(7) $$P\{|X| \geq K\alpha\} \geq \frac{\mu_4 - \sigma^4}{\mu_4 + \sigma^4 K^4 - 2K^2\sigma^4} \qquad \text{for } K > 1,$$

where $\mu_4 = EX^4$.

Proof. For the proof let us substitute $(X^2 - \sigma^2)/(K^2\sigma^2 - \sigma^2)$ for X and take $x = 1$ in (4). Then

$$P\{X^2 - \sigma^2 \geq K^2\sigma^2 - \sigma^2\} \leq \frac{\text{var}\{(X^2 - \sigma^2)/(K^2\sigma^2 - \sigma^2)\}}{1 + \text{var}\{(X^2 - \sigma^2)/(K^2\sigma^2 - \sigma^2)\}}$$

$$= \frac{\mu_4 - \sigma^4}{\sigma^4(K^2 - 1)^2 + \mu_4 - \sigma^4}$$

$$= \frac{\mu_4 - \sigma^4}{\mu_4 + \sigma^4 K^4 - 2K^2\sigma^4}, \qquad K > 1,$$

as asserted.

Remark 2. Bound (7) is better than bound (3) if $K^2 \geq \mu_4/\sigma^4$ and worse if $1 \leq K^2 < \mu_4/\sigma^4$ (Problem 5).

Example 3. Let X have the uniform density

$$f(x) = \begin{cases} 1 & \text{if } 0 < x < 1, \\ 0 & \text{otherwise.} \end{cases}$$

Then

$$EX = \tfrac{1}{2}, \qquad \text{var}(X) = \tfrac{1}{12}, \qquad \mu_4 = E\{X - \tfrac{1}{2}\}^4 = \tfrac{1}{80},$$

and

$$P\{|X - \tfrac{1}{2}| \geq 2\sqrt{\tfrac{1}{12}}\} \leq \frac{\tfrac{1}{80} - \tfrac{1}{144}}{\tfrac{1}{80} + \tfrac{1}{144} \cdot 16 - 8\tfrac{1}{144}} = \tfrac{4}{49},$$

that is

$$P\{|X - \tfrac{1}{2}| \leq 2\sqrt{\tfrac{1}{12}}\} \geq \tfrac{45}{49} \approx .92,$$

which is much better than the bound given by Chebychev's inequality (Example 2).

Theorem 3 (Lyapunov Inequality). Let $\beta_n = E|X|^n < \infty$. Then for arbitrary k, $2 \leq k \leq n$, we have

(8) $$\beta_{k-1}^{1/(k-1)} \leq \beta_k^{1/k}.$$

Proof. Consider the quadratic form:

$$Q(u, v) = \int_{-\infty}^{\infty} (u|x|^{(k-1)/2} + v|x|^{(k+1)/2})^2 f(x)\, dx,$$

where we have assumed that X is continuous with pdf f. We have

$$Q(u, v) = u^2 \beta_{k-1} + 2uv\beta_k + \beta_{k+1} v^2.$$

Clearly $Q \geq 0$ for all u, v real. It follows (see P. 2.4) that

$$\begin{vmatrix} \beta_{k-1} & \beta_k \\ \beta_k & \beta_{k+1} \end{vmatrix} \geq 0,$$

implying that

$$\beta_k^{2k} \leq \beta_{k-1}^k \beta_{k+1}^k.$$

Thus

$$\beta_1^2 \leq \beta_0^1 \beta_2^1, \quad \beta_2^4 \leq \beta_1^2 \beta_3^2, \quad \cdots, \quad \beta_{n-1}^{2(n-1)} \leq \beta_{n-2}^{n-1} \beta_n^{n-1},$$

where $\beta_0 = 1$. Multiplying successive $k - 1$ of these, we have

$$\beta_{k-1}^k \leq \beta_k^{k-1} \quad \text{or} \quad \beta_{k-1}^{1/(k-1)} \leq \beta_k^{1/k}.$$

It follows that

$$\beta_1 \leq \beta_2^{1/2} \leq \beta_3^{1/3} \leq \cdots \leq \beta_n^{1/n}.$$

The equality holds if and only if

$$\beta_k^{1/k} = \beta_{k+1}^{1/(k+1)} \quad \text{for } k = 1, 2, \cdots,$$

that is, $\{\beta_k^{1/k}\}$ is a constant sequence of numbers, which happens if and only if $|X|$ is degenerate.

PROBLEMS 3.4

1. For the rv with pdf

$$f(x; \lambda) = \frac{e^{-x} x^\lambda}{\lambda!}, \quad x > 0,$$

where $\lambda \geq 0$ is an integer, show that

$$P\{0 < X < 2(\lambda + 1)\} > \frac{\lambda}{\lambda+1}.$$

2. Let X be any rv, and suppose that the mgf of X, $M(t) = Ee^{tX}$, exists for every $t > 0$. Then for any $t > 0$

$$P\{tX > s^2 + \log M(t)\} < e^{-s^2}.$$

3. Construct an example to show that inequalities (4) and (5) cannot be improved.

4. Let $g(.)$ be a function satisfying $g(x) > 0$ for $x > 0$, $g(x)$ increasing for $x > 0$, and $g(|X|) < \infty$. Show that

$$P\{|X| > \varepsilon\} \leq \frac{Eg(|X|)}{g(\varepsilon)} \quad \text{for every } \varepsilon > 0.$$

5. Let X be an rv with $EX = 0$, var $(X) = \sigma^2$, and $EX^4 = \mu_4$. Let K be any positive real number. Show that

$$P\{|X| \geq K\sigma\} \leq \begin{cases} 1 & \text{if } K^2 < 1, \\ \dfrac{1}{K^2} & \text{if } 1 \leq K^2 < \dfrac{\mu_4}{\sigma^4}, \\ \dfrac{\mu_4 - \sigma^4}{\mu_4 + \sigma^4 K^4 - 2K^2\sigma^4} & \text{if } K^2 \geq \dfrac{\mu_4}{\sigma^4}. \end{cases}$$

In other words, show that bound (7) is better than bound (3) if $K^2 \geq \mu_4/\sigma^4$ and worse if $1 \leq K^2 < \mu_4/\sigma^4$. Construct an example to show that the last inequalities cannot be improved.

6. (a) Let X be an rv with df F, and let g be a strictly convex function on the range of F. Let φ be a Borel-measurable function. Suppose that EX, $E\varphi(X)$, and $Eg(X)$ all exist, and write $EX = \mu$. Let $t(x) = g(\mu) + K(x - \mu)$ be a line of support for g at $x = \mu$. Also, let $h(x) = g(x) - t(x)$. Then for every $\varepsilon > 0$ we have

$$E|\varphi(X)| \leq \sup_{|x-\mu|<\varepsilon} |\varphi(x)| + [\sup_{|x-\mu|\geq\varepsilon} \frac{|\varphi(x)|}{h(x)}] Eh(X).$$

(This inequality is due to M. Riesz; see, for example, Lukacs [76]. See P.2.3 for definitions of strictly convex functions and line of support.)

(b) Derive Markov's inequality from Riesz's inequality. (c) Show that for $a > 0$ and $\varepsilon > 0$

$$P\{|X - \mu| > \varepsilon\} \leq Me^{-a\mu}[Ee^{aX} - e^{a\mu}],$$

where $M = \{e^{-a\varepsilon} - 1 + a\varepsilon\}^{-1}$.
(*Hint*: Take $g(x) = e^{ax}$, $\varphi(x) = 0$ if $|x - \mu| < \varepsilon$, and $= 1$ if $|x - \mu| \geq \varepsilon$; $h(x) = e^{a\mu}[e^{a(x-\mu)} - 1 - a(x - \mu)]$.)

7. For any rv X, show that

$$P\{X \geq 0\} \leq \inf \{\varphi(t): t \geq 0\} \leq 1,$$

where $\varphi(t) = Ee^{tX}$, $0 < \varphi(t) \leq \infty$.

CHAPTER 4

Random Vectors

4.1 INTRODUCTION

In many experiments an observation is expressible, not as a single numerical quantity, but as a family of several separate numerical quantities. Thus, for example, if a pair of distinguishable dice is tossed, the outcome is a pair (x, y), where x denotes the face value on the first die, and y, the face value on the second die. Similarly, to record the height and weight of every person in a certain community we need a pair (x, y), where the components represent, respectively, the height and the weight of a particular individual. To be able to describe such experiments mathematically we must study the *multidimensional random variables* or *random vectors*.

In Section 2 we introduce the basic notions involved and study joint, marginal, and conditional distributions. In Section 3 we examine independent random variables and investigate some consequences of independence. Sections 4 and 5 deal with functions of random vectors and their induced distributions. Sections 6 and 7 consider moments and their generating functions, and in Section 8 we study the functional relationship between two dependent random variables.

4.2 RANDOM VECTORS

In this section we study multidimensional rv's. Let $(\Omega, \mathscr{S}, P)$ be a fixed but otherwise arbitrary probability space.

Definition 1. The collection $\mathbf{X} = (X_1, X_2, \cdots, X_n)$ defined on $(\Omega, \mathscr{S}, P)$ into $\mathscr{R}_n$ by

$$\mathbf{X}(\omega) = (X_1(\omega), X_2(\omega), \cdots, X_n(\omega)), \qquad \omega \in \Omega,$$

is called an *n*-dimensional rv (or a random vector) if the inverse image of every *n*-dimensional interval

$$I = \{(x_1, x_2, \cdots, x_n): -\infty < x_i \leq a_i, \ a_i \in \mathcal{R}, \ i = 1, 2, \cdots, n\}$$

is also in $\mathcal{S}$, that is, if

$$\mathbf{X}^{-1}(I) = \{\omega: X_1(\omega) \leq a_1, \cdots, X_n(\omega) \leq a_n\} \in \mathcal{S} \quad \text{for } a_i \in \mathcal{R}.$$

Theorem 1. Let $X_1, X_2, \cdots, X_n$ be n rv's. Then $\mathbf{X} = (X_1, X_2, \cdots, X_n)$ is an *n*-dimensional rv.

Proof. Let $I = \{(x_1, x_2, \cdots, x_n): -\infty < x_i \leq a_i, \ i = 1, 2, \cdots, n\}$. Then

$$\{(X_1, X_2, \cdots, X_n) \in I\} = \{\omega: X_1(\omega) \leq a_1, X_2(\omega) \leq a_2, \cdots, X_n(\omega) \leq a_n\}$$
$$= \bigcap_{k=1}^{n} \{\omega: X_k(\omega) \leq a_k\} \in \mathcal{S},$$

as asserted.

From now on we will restrict our attention to two-dimensional random variables. The discussion for the *n*-dimensional ($n > 2$) case is similar except when indicated. The development follows closely the one-dimensional case.

Definition 2. The function $F(\cdot, \cdot)$, defined by

(1) $$F(x, y) = P\{X \leq x, Y \leq y\}, \quad \text{all } (x, y) \in \mathcal{R}_2,$$

is known as the df of the rv (X, Y).

Following the discussion in Section 2.3, it is easily shown that

(i) $F(x, y)$ is nondecreasing and continuous from the right with respect to each coordinate, and

(ii) $$\lim_{\substack{x \to +\infty \\ y \to +\infty}} F(x, y) = F(+\infty, +\infty) = 1,$$
$$\lim_{y \to -\infty} F(x, y) = F(x, -\infty) = 0 \quad \text{for all } x,$$
$$\lim_{x \to -\infty} F(x, y) = F(-\infty, y) = 0 \quad \text{for all } y.$$

But (i) and (ii) are not sufficient conditions to make any function $F(\cdot, \cdot)$ a df.

Example 1. Let F be a function of two variables defined by

$$F(x, y) = \begin{cases} 0, & x < 0 \text{ or } x + y < 1 \text{ or } y < 0, \\ 1, & \text{otherwise.} \end{cases}$$

Then F satisfies both (i) and (ii) above. However, F is not a df since

$$P\{\tfrac{1}{3} < X \le 1, \tfrac{1}{3} < Y \le 1\} = F(1,1) + F(\tfrac{1}{3}, \tfrac{1}{3}) - F(1, \tfrac{1}{3}) - F(\tfrac{1}{3}, 1)$$
$$= 1 + 0 - 1 - 1 = -1 \not\ge 0.$$

Let $x_1 < x_2$ and $y_1 < y_2$. We have

$$P\{x_1 < X \le x_2, y_1 < Y \le y_2\}$$
$$= P\{X \le x_2, Y \le y_2\} + P\{X \le x_1, Y \le y_1,\}$$
$$\qquad - P\{X \le x_1, Y \le y_2\} - P\{X \le x_2, Y \le y_1\}$$
$$= F(x_2, y_2) + F(x_1, y_1) - F(x_1, y_2) - F(x_2, y_1)$$
$$\ge 0$$

for all pairs (x_1, y_1), (x_2, y_2) with $x_1 < x_2$, $y_1 < y_2$.

Theorem 2. A function F of two variables is a df of some two-dimensional rv if and only if it satisfies the following conditions

(i) F is nondecreasing and right continuous with respect to both arguments,

(ii) $F(-\infty, y) = F(x, -\infty) = 0$ and $F(+\infty, +\infty) = 1$, and

(iii) for every (x_1, y_1), (x_2, y_2) with $x_1 < x_2$ and $y_1 < y_2$ the inequality

(2) $\qquad F(x_2, y_2) - F(x_2, y_1) + F(x_1, y_1) - F(x_1, y_2) \ge 0$

holds.

The "if" part of the theorem has already been established. The "only if" part will not be proved here. (See Tucker [132], 26.)

Theorem 2 can be generalized to the n-dimensional case in the following manner.

Theorem 3. A function $F(x_1, x_2, \cdots, x_n)$ is the df of some n-dimensional rv if and only if F is nondecreasing and continuous from the right with respect to all the arguments $x_1, x_2, \cdots, x_n$ and satisfies the following conditions:

(i) $F(-\infty, x_2, \cdots, x_n) = F(x_1, -\infty, x_3, \cdots, x_n) = \cdots$
$$= F(x_1, \cdots, x_{n-1}, -\infty) = 0,$$
$$F(+\infty, +\infty, \cdots, +\infty) = 1,$$

and

(ii) for every $(x_1, x_2, \cdots, x_n) \in \mathscr{R}_n$ and all $\varepsilon_i > 0$ $(i = 1, 2, \cdots, n)$ the inequality

$$F(x_1 + \varepsilon_1, x_2 + \varepsilon_2, \cdots, x_n + \varepsilon_n)$$
$$- \sum_{i=1}^{n} F(x_1 + \varepsilon_1, \cdots, x_{i-1} + \varepsilon_{i-1}, x_i, x_{i+1} + \varepsilon_{i+1}, \cdots, x_n + \varepsilon_n)$$
(3)
$$+ \sum_{\substack{i,j=1 \\ i<j}}^{n} F(x_1 + \varepsilon_1, \cdots, x_{i-1} + \varepsilon_{i-1}, x_i, x_{i+1} + \varepsilon_{i+1}, \cdots, x_{j-1} + \varepsilon_{j-1},$$
$$x_j, x_{j+1} + \varepsilon_{j+1}, \cdots, x_n + \varepsilon_n)$$
$$+ \cdots$$
$$+ (-1)^n F(x_1, x_2, \cdots, x_n) \geq 0$$

holds.

We restrict ourselves here to two-dimensional rv's of the discrete or the continuous type, which we now define.

Definition 3. A two-dimensional rv (X, Y) is said to be of the discrete type if it takes on pairs of values belonging to a countable set of pairs A with probability 1. We call every pair (x_i, y_j) that is assumed with positive probability p_{ij} a jump point of the df of (X, Y), and call p_{ij} the jump at (x_i, y_j).

Clearly $\sum_{ij} p_{ij} = 1$. As for the df of (X, Y), we have

$$F(x, y) = \sum_{B} p_{ij}$$

where $B = \{(i, j): x_i \leq x, y_j \leq y\}$.

Definition 4. Let (X, Y) be an rv of the discrete type that takes on pairs of values (x_i, y_j), $i = 1, 2, \cdots$, and $j = 1, 2, \cdots$. We call

$$p_{ij} = P\{X = x_i, Y = y_j\}, \quad i = 1, 2, \cdots, \quad j = 1, 2, \cdots,$$

the joint probability mass function (pmf) of (X, Y).

Example 2. A die is rolled, and a coin is tossed independently. Let X be the face value on the die, and let $Y = 0$ if a tail turns up and $Y = 1$ if a head turns up. Then

$$A = \{(1, 0), (2, 0), \cdots, (6, 0), (1, 1), (2, 1), \cdots, (6, 1)\},$$
$$p_{ij} = \tfrac{1}{12} \quad \text{for } i = 1, 2, \cdots, 6; \quad j = 0, 1.$$

The df of (X, Y) is given by

$$F(x, y) = \begin{cases} 0, & x < 1, -\infty < y < \infty; -\infty < x < \infty, y < 0, \\ \frac{1}{12}, & 1 \le x < 2, 0 \le y < 1, \\ \frac{1}{6}, & 2 \le x < 3, 0 \le y < 1; 1 \le x < 2, 1 \le y, \\ \frac{1}{4}, & 3 \le x < 4, 0 \le y < 1, \\ \frac{1}{3}, & 4 \le x < 5, 0 \le y < 1; 2 \le x < 3, 1 \le y, \\ \frac{5}{12}, & 5 \le x < 6, 0 \le y < 1, \\ \frac{1}{2}, & 6 \le x, 0 \le y < 1; 3 \le x < 4, 1 \le y, \\ \frac{2}{3}, & 4 \le x < 5, 1 \le y, \\ \frac{5}{6}, & 5 \le x < 6, 1 \le y, \\ 1, & 6 \le x, 1 \le y. \end{cases}$$

Theorem 4. A collection of nonnegative numbers $\{p_{ij}: i = 1, 2, \cdots; j = 1, 2, \cdots\}$ satisfying $\sum_{i,j=1}^{\infty} p_{ij} = 1$ is the pmf of some rv.

Proof. The proof of Theorem 4 is easy to construct with the help of Theorem 2.

Definition 5. A two-dimensional rv (X, Y) is said to be of the continuous type if there exists a nonnegative function $f(\cdot, \cdot)$ such that for every pair $(x, y) \in \mathscr{R}_2$ we have

(4) $$F(x, y) = \int_{-\infty}^{x} \left[\int_{-\infty}^{y} f(u, v) \, dv \right] du,$$

where F is the df of (X, Y). The function f is called the (joint) pdf of (X, Y).

Clearly

$$F(+\infty, +\infty) = \lim_{\substack{x \to +\infty \\ y \to +\infty}} \int_{-\infty}^{x} \int_{-\infty}^{y} f(u, v) \, dv \, du$$

$$= \int_{-\infty}^{\infty} \int_{-\infty}^{\infty} f(u, v) \, dv \, du = 1.$$

If f is continuous at (x, y), then

(5) $$\frac{\partial^2 F(x, y)}{\partial x \, \partial y} = f(x, y).$$

Example 3. Let (X, Y) be an rv with joint pdf given by

$$f(x, y) = \begin{cases} e^{-(x+y)}, & 0 < x < \infty, \ 0 < y < \infty, \\ 0, & \text{otherwise.} \end{cases}$$

Then

$$F(x, y) = \begin{cases} (1 - e^{-x})(1 - e^{-y}), & 0 < x < \infty, \ 0 < y < \infty, \\ 0, & \text{otherwise.} \end{cases}$$

Theorem 5. If f is a nonnegative function satisfying $\int_{-\infty}^{\infty} \int_{-\infty}^{\infty} f(x, y) \, dx \, dy = 1$, then f is the joint density function of some rv.

Proof. For the proof define

$$F(x, y) = \int_{-\infty}^{x} \left[\int_{-\infty}^{y} f(u, v) \, dv \right] du$$

and use Theorem 2.

Let (X, Y) be a two-dimensional rv with pmf

$$p_{ij} = P\{X = x_i, Y = y_j\}.$$

Then

(6) $$\sum_{i=1}^{\infty} p_{ij} = \sum_{i=1}^{\infty} P\{X = x_i, Y = y_j\} = P\{Y = y_j\}$$

and

(7) $$\sum_{j=1}^{\infty} p_{ij} = \sum_{j=1}^{\infty} P\{X = x_i, Y = y_j\} = P\{X = x_i\}.$$

Let us write

(8) $$p_{i\cdot} = \sum_{j=1}^{\infty} p_{ij} \quad \text{and} \quad p_{\cdot j} = \sum_{i=1}^{\infty} p_{ij}.$$

Then $p_{i\cdot} \geq 0$ and $\sum_{i=1}^{\infty} p_{i\cdot} = 1$, $p_{\cdot j} \geq 0$ and $\sum_{j=1}^{\infty} p_{\cdot j} = 1$, and $\{p_{i\cdot}\}$, $\{p_{\cdot j}\}$ represent pmf's.

Definition 6. The collection of numbers $\{p_{i\cdot}\}$ is called the marginal pmf of X, and the collection $\{p_{\cdot j}\}$, the marginal pmf of Y.

Example 4. A fair coin is tossed three times. Let X = number of heads in three tossings, and Y = difference, in absolute value, between number of heads and number of tails. The joint pmf of (X, Y) is given in the following table:

Y \ X	0	1	2	3	$P\{Y = y\}$
1	0	$\frac{3}{8}$	$\frac{3}{8}$	0	$\frac{6}{8}$
3	$\frac{1}{8}$	0	0	$\frac{1}{8}$	$\frac{2}{8}$
$P\{X = x\}$	$\frac{1}{8}$	$\frac{3}{8}$	$\frac{3}{8}$	$\frac{1}{8}$	1

The marginal pmf of Y is shown in the column representing row totals, and the marginal pmf of X, in the row representing column totals.

If (X, Y) is an rv of the continuous type with pdf f, then

(9) $$f_1(x) = \int_{-\infty}^{\infty} f(x, y) \, dy$$

and

(10) $$f_2(y) = \int_{-\infty}^{\infty} f(x, y) \, dx$$

satisfy $f_1(x) \geq 0$, $f_2(y) \geq 0$, and $\int_{-\infty}^{\infty} f_1(x) \, dx = 1$, $\int_{-\infty}^{\infty} f_2(y) \, dy = 1$. It follows that $f_1(x)$ and $f_2(y)$ are pdf's.

Definition 7. The functions $f_1(x)$ and $f_2(y)$, defined in (9) and (10), are called the marginal pdf of X and the marginal pdf of Y, respectively.

Example 5. Let (X, Y) be jointly distributed with pdf $f(x, y) = 2$, $0 < x < y < 1$, and $= 0$ otherwise. Then

$$f_1(x) = \int_x^1 2 \, dy = \begin{cases} 2 - 2x, & 0 < x < 1, \\ 0, & \text{otherwise,} \end{cases}$$

and

$$f_2(y) = \int_0^y 2 \, dx = \begin{cases} 2y, & 0 < y < 1, \\ 0, & \text{otherwise,} \end{cases}$$

are the two marginal density functions.

Definition 8. Let (X, Y) be an rv with df F. Then the marginal df of X is defined by

(11) $$F_1(x) = F(x, \infty) = \lim_{y \to \infty} F(x, y)$$

$$= \begin{cases} \sum_{x_i \leq x} p_i. & \text{if } (X, Y) \text{ is discrete,} \\ \int_{-\infty}^{x} f_1(t)\, dt & \text{if } (X, Y) \text{ is continuous.} \end{cases}$$

A similar definition is given for the marginal df of Y.

In general, given a df $F(x_1, x_2, \cdots, x_n)$ of an n-dimensional rv $(X_1, X_2, \cdots, X_n)$, one can obtain any k-dimensional $(1 \leq k \leq n - 1)$ marginal df from it. Thus the marginal df of $(X_{i_1}, X_{i_2}, \cdots, X_{i_k})$, where $1 \leq i_1 < i_2 < \cdots < i_k \leq n$, is given by

$$\lim_{x_i \to \infty} F(x_1, x_2, \cdots, x_n) = F(+\infty, \cdots, +\infty, x_{i_1}, +\infty, \cdots, +\infty, x_{i_k}, +\infty, \cdots, +\infty).$$
$i \neq i_1, i_2, \cdots, i_k$

We now consider the concept of *conditional distributions*. Let (X, Y) be an rv of the discrete type with pmf $p_{ij} = P\{X = x_i, Y = y_j\}$. The marginal pmf's are $p_i. = \sum_{j=1}^{\infty} p_{ij}$ and $p_{.j} = \sum_{i=1}^{\infty} p_{ij}$. Recall that, if $A, B \in \mathscr{S}$ and $PB > 0$, the conditional probability of A, given B, is given by

$$P\{A|B\} = \frac{P(AB)}{P(B)}.$$

Take $A = \{X = x_i\} = \{(x_i, y): -\infty < y < \infty\}$, and $B = \{Y = y_j\} = \{(x, y_j); -\infty < x < \infty\}$, and assume that $PB = P\{Y = y_j\} = p_{.j} > 0$. Then $A \cap B = \{X = x_i, Y = y_j\}$, and

$$P\{A|B\} = P\{X = x_i | Y = y_j\} = \frac{p_{ij}}{p_{.j}}.$$

For fixed j, the function $P\{X = x_i | Y = y_j\} \geq 0$ and $\sum_{i=1}^{\infty} P\{X = x_i | Y = y_j\} = 1$. Thus $P\{X = x_i | Y = y_j\}$, for fixed j, defines a pmf.

Definition 9. Let (X, Y) be an rv of the discrete type. If $P\{Y = y_j\} > 0$, the function

(12) $$P\{X = x_i | Y = y_j\} = \frac{P\{X = x_i, Y = y_j\}}{P\{Y = y_j\}},$$

for fixed j, is known as the conditional pmf of X, given $Y = y_j$. A similar definition is given for $P\{Y = y_j | X = x_i\}$, the conditional pmf of Y, given $X = x_i$, provided that $P\{X = x_i\} > 0$.

Example 6. For the joint pmf of Example 4, we have for $Y = 1$

$$P\{X = i | Y = 1\} = \begin{cases} 0, & i = 0, 3, \\ \frac{1}{2}, & i = 1, 2. \end{cases}$$

Similarly

$$P\{X = i | Y = 3\} = \begin{cases} \frac{1}{2}, & \text{if } i = 0, 3, \\ 0, & \text{if } i = 1, 2, \end{cases}$$

$$P\{Y = j | X = 0\} = \begin{cases} 0, & \text{if } j = 1, \\ 1, & \text{if } j = 3, \end{cases}$$

and so on.

Next suppose that (X, Y) is an rv of the continuous type with joint pdf f. Since $P\{X = x\} = 0$, $P\{Y = y\} = 0$ for any x, y, the probability $P\{X \leq x | Y = y\}$, or $P\{Y \leq y | X = x\}$, is not defined. Let $\varepsilon > 0$, and suppose that $P\{y - \varepsilon < Y \leq y + \varepsilon\} > 0$. For every x and every interval $(y - \varepsilon, y + \varepsilon]$, consider the conditional probability of the event $\{X \leq x\}$, given that $Y \in (y - \varepsilon, y + \varepsilon]$. We have

$$P\{X \leq x | y - \varepsilon < Y \leq y + \varepsilon\} = \frac{P\{X \leq x, y - \varepsilon < Y \leq y + \varepsilon\}}{P\{Y \in (y - \varepsilon, y + \varepsilon]\}}.$$

For any fixed interval $(y - \varepsilon, y + \varepsilon]$, the above expression defines the conditional df of X, given that $Y \in (y - \varepsilon, y + \varepsilon]$, provided that $P\{Y \in (y - \varepsilon, y + \varepsilon]\} > 0$. We shall be interested in the case where the limit

$$\lim_{\varepsilon \to 0+} P\{X \leq x | Y \in (y - \varepsilon, y + \varepsilon]\}$$

exists.

Definition 10. The conditional df of an rv X, given $Y = y$, is defined as the limit

(13) $$\lim_{\varepsilon \to 0+} P\{X \leq x | Y \in (y - \varepsilon, y + \varepsilon]\},$$

provided that the limit exists. If the limit exists, we denote it by $F_{X|Y}(x|y)$, and define the conditional density function of X, given $Y = y$, $f_{X|Y}(x|y)$, as a nonnegative function satisfying

(14) $$F_{X|Y}(x|y) = \int_{-\infty}^{x} f_{X|Y}(t|y)\, dt \quad \text{for all} \quad x \in \mathcal{R}.$$

For fixed y, we see that $f_{X|Y}(x|y) \geq 0$ and $\int_{-\infty}^{\infty} f_{X|Y}(x|y)\, dx = 1$. Thus $f_{X|Y}(x|y)$ is a pdf for fixed y.

Suppose that (X, Y) is an rv of the continuous type with pdf f. At every point (x, y) where f is continuous and the marginal pdf $f_2(y) > 0$ and is continuous, we have

$$F_{X|Y}(x|y) = \lim_{\varepsilon \to 0+} \frac{P\{X \le x, Y \in (y-\varepsilon, y+\varepsilon]\}}{P\{Y \in (y-\varepsilon, y+\varepsilon]\}}$$

$$= \lim_{\varepsilon \to 0+} \frac{\int_{-\infty}^{x} \left\{\int_{y-\varepsilon}^{y+\varepsilon} f(u,v)\, dv\right\} du}{\int_{y-\varepsilon}^{y+\varepsilon} f_2(v)\, dv}.$$

Dividing numerator and denominator by 2ε and passing to the limit as $\varepsilon \to 0+$, we have

$$F_{X|Y}(x|y) = \frac{\int_{-\infty}^{x} f(u, y)\, du}{f_2(y)}$$

$$= \int_{-\infty}^{x} \left\{\frac{f(u, y)}{f_2(y)}\right\} du.$$

It follows that there exists a conditional pdf of X, given $Y = y$, that is expressed by

$$f_{X|Y}(x|y) = \frac{f(x, y)}{f_2(y)}, \qquad f_2(y) > 0.$$

We have thus proved the following theorem.

Theorem 6. Let f be the pdf of an rv (X, Y) of the continuous type, and let f_2 be the marginal pdf of Y. At every point (x, y) at which f is continuous and $f_2(y) > 0$ and is continuous, the conditional pdf of X, given $Y = y$, exists and is expressed by

(15) $$f_{X|Y}(x|y) = \frac{f(x, y)}{f_2(y)}.$$

Note that

$$\int_{-\infty}^{x} f(u, y)\, du = f_2(y)\, F_{X|Y}(x|y),$$

so that

(16) $$F_1(x) = \int_{-\infty}^{\infty} \left\{\int_{-\infty}^{x} f(u, y)\, du\right\} dy = \int_{-\infty}^{\infty} f_2(y)\, F_{X|Y}(x|y)\, dy,$$

where F_1 is the marginal df of X.

It is clear that similar definitions may be made for the conditional df and conditional pdf of the rv Y, given $X = x$, and an analogue of Theorem 6 holds.

In the general case, let $(X_1, X_2, \cdots, X_n)$ be an n-dimensional rv of the continuous type with pdf $f_{X_1, X_2, \cdots, X_n}(x_1, x_2, \cdots, x_n)$. Also, let $\{i_1 < i_2 < \ldots < i_k, j_1 < j_2 \ldots < j_l\}$ be a subset of $\{1, 2, \cdots, n\}$. Then

(17) $$F(x_{i_1}, x_{i_2}, \cdots, x_{i_k} | x_{j_1}, x_{j_2}, \cdots, x_{jl})$$

$$= \frac{\int_{-\infty}^{x_{i_1}} \cdots \int_{-\infty}^{x_{i_k}} f_{X_{i_1}, \cdots, X_{i_k}, X_{j_1}, \cdots, X_{jl}}(x_{i_1}, \cdots, x_{i_k}, x_{j_1}, \cdots, x_{jl}) \prod_{p=1}^{k} dx_{i_p}}{\int_{-\infty}^{\infty} \cdots \int_{-\infty}^{\infty} f_{X_{i_1}, \cdots, X_{i_k}, X_{j_1}, \cdots, X_{jl}}(x_{i_1}, \cdots, x_{i_k}, x_{j_1}, \cdots, x_{jl}) \prod_{p=1}^{k} dx_{i_p}},$$

provided that the denominator exceeds 0. Here $f_{X_{i_1}, \cdots, X_{i_k}, X_{j_1}, \cdots, X_{jl}}$ is the joint marginal pdf of $(X_{i_1}, X_{i_2}, \cdots, X_{i_k}, X_{j_1}, X_{j_2}, \cdots, X_{jl})$. The conditional densities are obtained in a similar manner.

The case in which $(X_1, X_2, \cdots, X_n)$ is of the discrete type is similarly treated.

Example 7. For the joint pdf of Example 5 we have

$$f_{Y|X}(y|x) = \frac{f(x, y)}{f_1(x)} = \frac{1}{1 - x}, \quad x < y < 1,$$

so that the conditional pdf $f_{Y|X}$ is uniform on $(x, 1)$. Also,

$$f_{X|Y}(x|y) = \frac{1}{y}, \quad 0 < x < y,$$

which is uniform on $(0, y)$. Thus

$$P\{Y \geq \tfrac{1}{2} | x = \tfrac{1}{2}\} = \int_{1/2}^{1} \frac{1}{1 - \tfrac{1}{2}} \, dy = 1,$$

$$P\{X \geq \tfrac{1}{3} | y = \tfrac{2}{3}\} = \int_{1/3}^{2/3} \frac{1}{\tfrac{2}{3}} \, dx = \tfrac{1}{2}.$$

Example 8. Let $\{X_n\}$ be a sequence of discrete rv's satisfying

$$P\{X_n = x_{i_n} | X_1 = x_{i_1}, \cdots, X_{n-1} = x_{i_{n-1}}\} = P\{X_n = x_{i_n} | X_{n-1} = x_{i_{n-1}}\}.$$

Such a sequence is called a *Markovian dependent sequence*. The probabilities $P\{X_n = x_{i_n} | X_{n-1} = x_{i_{n-1}}\}$ are called *transition probabilities*.

As an example consider a sequence of trials with a fair coin. Let X_n = number of heads in the first n trials. If $X_{n-1} = k$, then X_n must be either k or $k + 1$. Thus

$$P\{X_n = x_{i_n} | X_1 = x_{i_1}, \cdots, X_{n-2} = x_{i_{n-2}}, X_{n-1} = k\}$$
$$= \begin{cases} \tfrac{1}{2}, & \text{if } x_{i_n} = k \text{ or } k + 1, \\ 0, & \text{otherwise.} \end{cases}$$

It follows that $\{X_n\}$ is a Markov sequence. In this case the conditional probability is independent of n. Such a sequence is said to have *stationary transition* probabilities.

Similar considerations apply if we take $\{X_n\}$ to be rv's of the continuous type. We leave the reader to modify the definition given above. For further details we refer to Feller [28], Chapter 15, and Fisz [32], Chapters 7 and 8.

We conclude this section with a discussion of truncated distributions. Truncation is a very useful device in the study of probability limit theorems.

Definition 11. Let X be an rv on $(\Omega, \mathscr{S}, P)$, and $T \in \mathscr{B}$ such that $0 < P\{X \in T\} < 1$. Then the conditional distribution $P\{X \leq x | X \in T\}$, defined for any real x, is called the truncated distribution of X.

If X is a discrete rv with pmf $p_i = P\{X = x_i\}$, $i = 1, 2, \cdots$, the truncated distribution of X is given by

$$(18) \quad P\{X = x_i | X \in T\} = \frac{P\{X = x_i, X \in T\}}{P\{X \in T\}} = \begin{cases} \dfrac{p_i}{\sum_{x_j \in T} p_j} & \text{if } x_i \in T, \\ 0 & \text{otherwise.} \end{cases}$$

If X is of the continuous type with pdf f, then

$$(19) \quad P\{X \leq x | X \in T\} = \frac{P\{X \leq x, X \in T\}}{P\{X \in T\}} = \frac{\int_{(-\infty, x] \cap T} f(y)\, dy}{\int_T f(y)\, dy}.$$

The pdf of the truncated distribution is given by

$$(20) \quad h(x) = \begin{cases} \dfrac{f(x)}{\int_T f(y)\, dy}, & x \in T, \\ 0, & x \notin T. \end{cases}$$

Example 9. Let X be an rv with *standard normal* pdf

$$f(x) = \frac{1}{\sqrt{2\pi}} e^{-x^2/2}.$$

Let $T = (-\infty, 0]$. Then $P\{X \in T\} = 1/2$, since X is symmetric and continuous. For the truncated pdf, we have

$$h(x) = \begin{cases} 2f(x), & -\infty < x \leq 0, \\ 0, & x > 0. \end{cases}$$

Truncation is specially important in cases where the df F in question does not have a finite mean. If X is an rv, we truncate X at some $c > 0$, where c is finite, by replacing X by $X^c = X$ if $|X| \leq c$, and $= 0$ if $|X| > c$. Then X^c is X truncated at c, and all moments of X^c exist and are finite. In fact, we can always select c sufficiently large so that

$$P\{X \neq X^c\} = P\{|X| > c\}$$

is arbitrarily small. The distribution of X^c is given by

$$P\{X^c \le x\} = P\{X \le x \mid |X| \le c\}$$

(21)
$$= \frac{\int_{(-\infty,x]\cap[-c,c]} f(y)\,dy}{P\{|X| \le c\}},$$

if X is a continuous rv with pdf f, and by

(22)
$$P\{X^c = x_j\} = \begin{cases} \dfrac{p_j}{\sum_{x_i \in [-c,c]} p_i} & \text{if } x_j \in [-c, c], \\ 0 & \text{otherwise,} \end{cases}$$

if X is an rv of the discrete type with pmf p_j, $j = 1, 2, \cdots$.

We have for $\alpha > 0$

(23)
$$E|X^c|^\alpha \le c^\alpha.$$

Example. 10 Let X be an rv with Cauchy pdf

$$f(x) = \frac{1}{\pi}\frac{1}{1+x^2}, \quad -\infty < x < \infty.$$

Then EX does not exist. Let $c > 0$ be finite, and let us truncate X at c by writing

$$X^c = \begin{cases} X & \text{if } |X| \le c, \\ 0 & \text{if } |X| > c. \end{cases}$$

Then

$$P\{|X| \le c\} = \frac{1}{\pi}\int_{-c}^{c} \frac{1}{1+x^2}\,dx$$
$$= \frac{2}{\pi}\tan^{-1} c.$$

The pdf of X^c is given by

$$h(x) = \begin{cases} \dfrac{1}{2}\dfrac{1}{(1+x^2)}\dfrac{1}{\tan^{-1} c} & \text{if } x \in [-c, c], \\ 0 & \text{otherwise,} \end{cases}$$

with

$$EX^c = \frac{1}{2\tan^{-1} c}\int_{-c}^{c} \frac{x}{1+x^2}\,dx = 0,$$

$$E(X^c)^2 = \frac{1}{2\tan^{-1} c}\int_{-c}^{c} \frac{x^2}{1+x^2}\,dx = \frac{c}{\tan^{-1} c} - 1,$$

and so on.

PROBLEMS 4.2

1. Let $F(x, y) = 1$ if $x + 2y \geq 1$, and $= 0$ if $x + 2y < 1$. Does F define a df in the plane?

2. Let T be a closed triangle in the plane with vertices $(0, 0)$, $(0, \sqrt{2})$, and $(\sqrt{2}, \sqrt{2})$. Let $F(x, y)$ denote the elementary area of the intersection of T with $\{(x_1, x_2): x_1 \leq x, x_2 \leq y\}$. Show that F defines a df in the plane, and find its marginal df's.

3. Let (X, Y) have the joint pdf f defined by $f(x, y) = 1/2$ inside the square with corners at the points $(1, 0)$, $(0, 1)$, $(-1, 0)$, and $(0, -1)$ in the (x, y) – plane, and $= 0$ otherwise. Find the marginal pdf's of X and Y and the two conditional pdf's.

4. Let $f(x, y, z) = e^{-x-y-z}$, $x > 0$, $y > 0$, $z > 0$, and $= 0$ otherwise, be the joint pdf of (X, Y, Z). Compute $P\{X < Y < Z\}$ and $P\{X = Y < Z\}$.

5. Let (X, Y) have the joint pdf $f(x, y) = \frac{4}{3}[xy + (x^2/2)]$ if $0 < x < 1$, $0 < y < 2$, and $= 0$ otherwise. Find $P\{Y < 1 | X < 1/2\}$.

6. For df's $F, F_1, F_2, \cdots, F_n$ show that
$$1 - \sum_{i=1}^{n} \{1 - F_i(x_i)\} \leq F(x_1, x_2, \cdots, x_n) \leq \min_{1 \leq i \leq n} F_i(x_i)$$
for all real numbers $x_1, x_2, \cdots, x_n$ if and only if F_i's are the marginal df's of F.

7. For the *bivariate negative binomial* distribution
$$P\{X = x, Y = y\} = \frac{(x + y + k - 1)!}{x! \, y! \, (k - 1)!} p_1^x p_2^y (1 - p_1 - p_2)^k,$$
where $x, y = 0, 1, 2, \cdots$, $k \geq 1$ is an integer, $0 < p_1 < 1$, $0 < p_2 < 1$, and $p_1 + p_2 < 1$, find the marginal pmf's of X and Y and the conditional distributions.

In Problems 8–10 the bivariate distributions considered are not unique generalizations of the corresponding univariate distributions.

8. For the *bivariate Cauchy* rv (X, Y) with pdf
$$f(x, y) = \frac{c}{2\pi}(c^2 + x^2 + y^2)^{-3/2}, \quad -\infty < x < \infty, \; -\infty < y < \infty, \; c > 0,$$
find the marginal pdf's of X and Y. Find the conditional pdf of Y given $X = x$.

9. For the *bivariate beta* rv (X, Y) with pdf
$$f(x, y) = \frac{\Gamma(p_1 + p_2 + p_3)}{\Gamma(p_1)\Gamma(p_2)\Gamma(p_3)} x^{p_1-1} y^{p_2-1} (1 - x - y)^{p_3-1}, \quad x \geq 0, y \geq 0, x + y \leq 1,$$
where p_1, p_2, p_3 are positive real number, find the marginal pdf's of X and Y and the conditional pdf's. Find also the conditional pdf of $Y/(1 - X)$, given $X = x$.

10. For the *bivariate gamma* rv (X, Y) with pdf
$$f(x, y) = \frac{\beta^{\alpha+\gamma}}{\Gamma(\alpha)\Gamma(\gamma)} x^{\alpha-1}(y - x)^{\gamma-1} e^{-\beta y}, \quad 0 < x < y; \alpha, \beta, \gamma > 0,$$

find the marginal pdf's of X and Y and the conditional pdf's. Also, find the conditional pdf of $Y - X$, given $X = x$, and the conditional distribution of X/Y, given $Y = y$.

11. For the *bivariate hypergeometric* rv (X, Y) with pmf

$$P\{X = x, Y = y\} = \binom{N}{n}^{-1} \binom{Np_1}{x} \binom{Np_2}{y} \binom{N - Np_1 - Np_2}{n - x - y}, \quad x, y = 0, 1, 2, \cdots, n,$$

where $x \leq Np_1$, $y \leq Np_2$, $n - x - y \leq N(1 - p_1 - p_2)$, N, n integers with $n \leq N$, and $0 < p_1 < 1$, $0 < p_2 < 1$ so that $p_1 + p_2 \leq 1$, find the marginal pmf's of X and Y and the conditional pmf's.

12. Let X be an rv with pdf $f(x) = 1$ if $0 \leq x \leq 1$, and $= 0$ otherwise. Let $T = \{x : 1/3 < x \leq 1/2\}$. Find the pdf of the truncated distribution of X, its mean, and its variance.

13. Let X be an rv with pmf

$$P\{X = x\} = e^{-\lambda} \frac{\lambda^x}{x!}, \quad x = 0, 1, 2, \cdots, \lambda > 0.$$

Suppose that the value $x = 0$ cannot be observed. Find the pmf of the truncated rv, its mean, and its variance.

4.3 INDEPENDENT RANDOM VARIABLES

We recall that the joint distribution of a random vector uniquely determines the marginal distributions of the component random variables, but, in general, knowledge of marginal distributions is not enough to determine the joint distribution. Indeed, it is quite possible to have an infinite collection of joint densities f_α with given marginal densities.

Example 1 (Gumbel [42]). Let f_1, f_2, f_3 be three pdf's with corresponding df's F_1, F_2, F_3, and let α be a constant, $|\alpha| \leq 1$. Define

$$f_\alpha(x_1, x_2, x_3) = f_1(x_1) f_2(x_2) f_3(x_3) \cdot \{1 + \alpha[2F_1(x_1) - 1][2F_2(x_2) - 1][2F_3(x_3) - 1]\}.$$

We show that f_α is a pdf for each α in $[-1, 1]$ and that the collection of densities $\{f_\alpha; -1 \leq \alpha \leq 1\}$ has the same marginal densities f_1, f_2, f_3. First note that

$$|[2F_1(x_1) - 1][2F_2(x_2) - 1][2F_3(x_3) - 1]| \leq 1,$$

so that

$$1 + \alpha[2F_1(x_1) - 1][2F_2(x_2) - 1][2F_3(x_3) - 1] \geq 0.$$

Also,

$$\iiint f_\alpha(x_1, x_2, x_3)\, dx_1\, dx_2\, dx_3$$
$$= 1 + \alpha\left(\int [2F_1(x_1) - 1] f_1(x_1)\, dx_1\right)\left(\int [2F_2(x_2) - 1] f_2(x_2)\, dx_2\right)$$
$$\cdot \left(\int [2F_3(x_3) - 1] f_3(x_3)\, dx_3\right)$$
$$= 1 + \alpha\{[F_1^2(x_1)]\Big|_{-\infty}^{\infty} - 1][F_2^2(x_2)]\Big|_{-\infty}^{\infty} - 1][F_3^2(x_3)]\Big|_{-\infty}^{\infty} - 1]\}$$
$$= 1.$$

It follows that f_α is a density function. That f_1, f_2, f_3 are the marginal densities of f_α follows similarly.

In this section we deal with a very special class of distributions in which the marginal distributions uniquely determine the joint distribution of a random vector. First we consider the bivariate case.

Let $F(x, y)$ and $F_1(x)$, $F_2(y)$, respectively, be the joint df of (X, Y) and the marginal df's of X and Y.

Definition 1. We say that X and Y are independent if and only if

(1) $\qquad F(x, y) = F_1(x) F_2(y) \quad \text{for all} \quad (x, y) \in \mathcal{R}_2.$

Lemma 1. If X and Y are independent and $a < c$, $b < d$ are real numbers, then

(2) $\qquad P\{a < X \leq c, b < Y \leq d\} = P\{a < X \leq c\} P\{b < Y \leq d\}.$

Proof. The proof is left as an exercise.

Theorem 1. (a) A necessary and sufficient condition for rv's X, Y of the discrete type to be independent is that

(3) $\qquad P\{X = x_i, Y = y_j\} = P\{X = x_i\} P\{Y = y_j\}$

for all pairs (x_i, y_j). (b) Two rv's X and Y, of the continuous type are independent if and only if

(4) $\qquad f(x, y) = f_1(x) f_2(y) \quad \text{for all} \quad (x, y) \in \mathcal{R}_2,$

and where f, f_1, f_2, respectively, are the joint and marginal densities of X and Y, and f is everywhere continuous.

Proof. (a) Let X, Y be independent. Then from Lemma 1, letting $a \to c$ and $b \to d$, we get

$$P\{X = c, Y = d\} = P\{X = c\} P\{Y = d\}.$$

Conversely,
$$F(x, y) = \sum_B P\{X = x_i, Y = y_j\},$$
where
$$B = \{(i, j): x_i \leq x, y_j \leq y\}.$$
Then
$$F(x, y) = \sum_B P\{X = x_i\} P\{Y = y_j\}$$
$$= \sum_{x_i \leq x} [\sum_{y_j \leq y} P\{Y = y_j\}] P\{X = x_i\} = F(x) F(y).$$

The proof of part (b) is left as an exercise.

Corollary. Let X and Y be independent rv's. Then $F_{Y|X}(y|x) = F_Y(y)$ for all y, and $F_{X|Y}(x|y) = F_X(x)$ for all x.

We now return to Lemma 1. Recalling once again that every Borel set on the real line can be obtained by a countable number of operations of unions, intersections, and differences on semiclosed intervals, we can establish the following theorem.

Theorem 2. X and Y are independent if and only if
(5) $$P\{X \in A_1, Y \in A_2\} = P\{X \in A_1\} P\{Y \in A_2\}$$
for all Borel sets A_1 on the x-axis and A_2 on the y-axis.

Theorem 3. Let X and Y be independent rv's and f and g be Borel-measurable functions. Then $f(X)$ and $g(Y)$ are also independent.

Proof. We have
$$P\{f(X) \leq x, g(Y) \leq y\} = P\{X \in f^{-1}(-\infty, x], Y \in g^{-1}(-\infty, y]\}$$
$$= P\{X \in f^{-1}(-\infty, x]\} P\{Y \in g^{-1}(-\infty, y]\}$$
$$= P\{f(X) \leq x\} P\{g(Y) \leq y\}.$$

Note that a degenerate rv is independent of any rv.

Example 2. Let X and Y be jointly distributed with pdf
$$f(x, y) = \begin{cases} \dfrac{1 + xy}{4}, & |x| < 1, |y| < 1, \\ 0, & \text{otherwise.} \end{cases}$$

Then X and Y are not independent since $f_1(x) = 1/2$, $|x| < 1$, and

$f_2(y) = 1/2$, $|y| < 1$, are the marginal densities of X and Y, respectively. However, the rv's X^2 and Y^2 are independent. Indeed,

$$\begin{aligned}
P\{X^2 \leq u, Y^2 \leq v\} &= \int_{-v^{1/2}}^{v^{1/2}} \int_{-u^{1/2}}^{u^{1/2}} f(x, y)\, dx\, dy \\
&= \frac{1}{4} \int_{-v^{1/2}}^{v^{1/2}} \left[\int_{-u^{1/2}}^{u^{1/2}} (1 + xy) dx \right] dy \\
&= u^{1/2} v^{1/2} \\
&= P\{X^2 \leq u\}\, P\{Y^2 \leq v\}.
\end{aligned}$$

Example 3. We return to Buffon's needle problem, discussed in Examples 1.2.9 and 1.3.7. Suppose that the rv R, which represents the distance from the center of the needle to the nearest line, is uniformly distributed on $(0, l]$. Suppose further that Θ, the angle that the needle forms with this line, is uniformly distributed on $[0, \pi)$. If R and Θ are assumed to be independent, the joint pdf is given by

$$f_{R,\Theta}(r, \theta) = f_R(r) f_\Theta(\theta) = \begin{cases} \dfrac{1}{l} \cdot \dfrac{1}{\pi} & \text{if } 0 < r \leq l,\ 0 \leq \theta < \pi, \\ 0 & \text{otherwise.} \end{cases}$$

The needle will intersect the nearest line if and only if

$$\frac{l}{2} \sin \Theta \geq R.$$

Therefore the required probability is given by

$$\begin{aligned}
P\left\{\sin \Theta \geq \frac{2R}{l}\right\} &= \int_0^\pi \int_0^{l/2 \sin \theta} f_{R,\Theta}(r, \theta)\, dr\, d\theta \\
&= \frac{1}{l\pi} \int_0^\pi \frac{l}{2} \sin \theta\, d\theta = \frac{1}{\pi}.
\end{aligned}$$

Definition 2. A collection of rv's $X_1, X_2, \cdots, X_n$, is said to be mutually or completely independent if and only if

(6) $\quad F(x_1, x_2, \cdots, x_n) = \prod_{i=1}^{n} F_i(x_i), \quad$ for all $(x_1, x_2, \cdots, x_n) \in \mathscr{R}_n$,

where F is the joint df of $(X_1, X_2, \cdots, X_n)$, and F_i ($i = 1, 2, \cdots, n$) is the marginal df of X_i. $X_1, \cdots, X_n$ are said to be pairwise independent if and only if any two of them are independent.

It is clear that an analogue of Theorem 1 holds, but we leave the reader to construct it.

Example 4. In Example 1 we cannot write

$$f_\alpha(x_1, x_2, x_3) = f_1(x_1) f_2(x_2) f_3(x_3)$$

except when $\alpha = 0$. It follows that X_1, X_2, and X_3 are not independent except when $\alpha = 0$.

The following result is easy to prove.

Theorem 4. If $X_1, X_2, \cdots, X_n$ are independent, every subcollection $X_{i_1}, X_{i_2}, \cdots, X_{i_k}$ of $X_1, X_2, \cdots, X_n$ is also independent.

Remark 1. It is quite possible for rv's $X_1, X_2, \cdots, X_n$ to be *pairwise independent* without being mutually independent. Let (X, Y, Z) have the joint pmf defined by

$$P\{X = x, Y = y, Z = z\} = \begin{cases} \frac{3}{16} & \text{if } (x, y, z) \in \{(0, 0, 0), (0, 1, 1), \\ & (1, 0, 1), (1, 1, 0)\}, \\ \frac{1}{16} & \text{if } (x, y, z) \in \{(0, 0, 1), (0, 1, 0), \\ & (1, 0, 0), (1, 1, 1)\}. \end{cases}$$

Clearly X, Y, Z, not are independent (Why?). We have

$$P\{X = x, Y = y\} = \tfrac{1}{4}, \quad (x, y) \in \{(0, 0), (0, 1), (1, 0), (1, 1)\},$$
$$P\{Y = y, Z = z\} = \tfrac{1}{4}, \quad (y, z) \in \{(0, 0), (0, 1), (1, 0), (1, 1)\},$$
$$P\{X = x, Z = z\} = \tfrac{1}{4}, \quad (x, z) \in \{(0, 0), (0, 1), (1, 0), (1, 1)\},$$
$$P\{X = x\} = \tfrac{1}{2}, \quad x = 0, x = 1,$$
$$P\{Y = y\} = \tfrac{1}{2}, \quad y = 0, y = 1,$$
$$P\{Z = z\} = \tfrac{1}{2}, \quad z = 0, z = 1.$$

It follows that X and Y, Y and Z, and X and Z are pairwise independent.

Definition 3. A sequence $\{X_n\}$ of rv's is said to be independent if for every $n = 2, 3, 4, \cdots$ the rv's $X_1, X_2, \cdots, X_n$ are independent.

Similarly one can speak of an independent family of rv's.

Definition 4. We say that rv's X and Y are identically distributed if X and Y have the same df, that is,

$$F_X(x) = F_Y(x) \quad \text{for all} \quad , x \in \mathcal{R}$$

where F_X and F_Y are the df's of X and Y, respectively.

Definition 5. We say that $\{X_n\}$ is a sequence of independent, identically distributed (iid) rv's with common law $\mathcal{L}(X)$ if $\{X_n\}$ is an independent sequence of rv's and the distribution of X_n ($n = 1, 2, \cdots$) is the same as that of X.

According to Definition 4, X and Y are identically distributed if and only if they have the same distribution. It does not follow that $X = Y$ with probability 1 (see Problem 8). If $P\{X = Y\} = 1$, we say that X and Y are *equivalent* rv's. All Definition 4 says is that X and Y are identically distributed if and only if

$$P\{X \in A\} = P\{Y \in A\} \qquad \text{for all} \quad A \in \mathcal{B}.$$

Nothing is said about the equality of events $\{X \in A\}$ and $\{Y \in A\}$.

Definition 6. Two random vectors $(X_1, X_2, \cdots, X_m)$ and $(Y_1, Y_2, \cdots, Y_n)$ are said to be independent if

(7)
$$\begin{aligned}F(x_1, x_2, \cdots, x_m, y_1, y_2, \cdots, y_n) \\= F_1(x_1, x_2, \cdots, x_m) F_2(y_1, y_2, \cdots, y_n)\end{aligned}$$

for all $(x_1, x_2, \cdots, x_m, y_1, y_2, \cdots, y_n) \in \mathcal{R}_{m+n}$, where F, F_1, F_2 are the joint distribution functions of $(X_1, X_2, \cdots, X_m, Y_1, Y_2, \cdots, Y_n)$, $(X_1, X_2, \cdots, X_m)$, and $(Y_1, Y_2, \cdots, Y_n)$, respectively.

Of course, the independence of $(X_1, X_2, \cdots, X_m)$ and $(Y_1, Y_2, \cdots, Y_n)$ does not imply the independence of components $X_1, X_2, \cdots, X_m$ of $\mathbf{X}$ or components $Y_1, Y_2, \cdots, Y_n$ of $\mathbf{Y}$.

Example 5. Let (X, Y) be an rv with joint density function

$$f(x, y) = \frac{1}{4}[1 + xy(x^2 - y^2)], \qquad |x| \leq 1, |y| \leq 1,$$

and (U, V) be an rv with joint density

$$g(u, v) = \frac{1}{2\pi\sqrt{1 - \rho^2}} \exp\left\{-\frac{1}{2(1 - \rho^2)}(u^2 - 2\rho uv + v^2)\right\},$$
$$-\infty < u < \infty, \; -\infty < v < \infty,$$

where $|\rho| < 1$ is a given constant. Let the joint density of (X, Y, U, V) be given by

$$h(x, y, u, v) = \frac{1}{8\pi\sqrt{1 - \rho^2}}[1 + xy(x^2 - y^2)]$$
$$\cdot \exp\left\{-\frac{1}{2(1 - \rho^2)}(u^2 - 2\rho uv + v^2)\right\},$$
$$|x| \leq 1, |y| \leq 1, -\infty < u, v < \infty.$$

Then (X, Y) and (U, V) are independent, but X and Y are not independent and neither are U and V.

Theorem 5. Let $\mathbf{X} = (X_1, X_2, \cdots, X_m)$ and $\mathbf{Y} = (Y_1, Y_2, \cdots, Y_n)$ be independent random vectors. Then the component X_j of $\mathbf{X}$ ($j = 1, 2, \cdots, m$) and the component Y_k of $\mathbf{Y}$ ($k = 1, 2, \cdots, n$) are independent rv's. If h and g are Borel-measurable functions, $h(X_1, X_2, \cdots, X_m)$ and $g(Y_1, Y_2, \cdots, Y_n)$ are independent.

Proof. The proof is simple and is left as an exercise.

Remark 2. It is possible that an rv X may be independent of Y and also of Z, but X may not be independent of the random vector (Y, Z). See the example in Remark 1.

We conclude this section by mentioning that the notion of independence of random vectors can be generalized to any number of random vectors. We leave the details to the reader.

PROBLEMS 4.3

1. Let A be a set of k numbers, and Ω be the set of all ordered samples of size n from A with replacement. Also, let $\mathscr{S}$ be the set of all subsets of Ω, and P be a probability defined on $\mathscr{S}$. Let $X_1, X_2, \cdots, X_n$ be rv's defined on $(\Omega, \mathscr{S}, P)$ by setting

$$X_i(a_1, a_2, \cdots, a_n) = a_i, \quad (i = 1, 2, \cdots, n).$$

Show that $X_1, X_2, \cdots, X_n$ are independent if and only if each sample point is equally likely.

2. Let X_1, X_2 be iid rv's with common pmf

$$P\{X = \pm 1\} = \frac{1}{2}.$$

Write $X_3 = X_1 X_2$. Show that X_1, X_2, X_3 are pairwise independent but not independent.

3. Let (X_1, X_2, X_3) be a random vector with joint pmf

$$f(x_1, x_2, x_3) = \frac{1}{4} \quad \text{if } (x_1, x_2, x_3) \in A,$$
$$= 0 \quad \text{otherwise,}$$

where

$$A = \{(1, 0, 0), (0, 1, 0), (0, 0, 1), (1, 1, 1)\}.$$

Are X_1, X_2, X_3 independent? Are X_1, X_2, X_3 pairwise independent? Are $X_1 + X_2$ and X_3 independent?

4. Let X and Y be independent rv's such that XY is a degenerate rv. Show that X and Y are also degenerate.

5. Let $(\Omega, \mathscr{S}, P)$ be a probability space and $A, B \in \mathscr{S}$. Define X and Y so that

$$X(\omega) = I_A(\omega), \qquad Y(\omega) = I_B(\omega) \text{ for all } \omega \in \Omega.$$

Show that X and Y are independent if and only if A and B are independent.

6. Let X and Y be independent, and write $Z = X + Y$. Show that, if Z possesses moments of all orders, so do X and Y.

7. A set of rv's $X_1, X_2, \cdots, X_n$ is said to be *interchangeable* if the joint df of $(X_1, X_2, \cdots, X_n)$ remains unchanged under any permutation of the X_i's. Let $X_1, X_2, \cdots, X_n$ be a set of positive, interchangeable rv's. Then

$$E\left\{\frac{X_1 + X_2 + \cdots + X_k}{X_1 + X_2 + \cdots + X_n}\right\} = \frac{k}{n}, \qquad 1 \le k \le n.$$

(Kullback and Yasaimaibodi [64])

8. Let X and Y be identically distributed. Construct an example to show that X and Y need not be equal, that is, $P\{X = Y\}$ need not equal 1.

9. Let X and Y be independent rv's. Show that, if X and $X - Y$ are independent, X must be degenerate.

10. Let X be a nonnegative rv with $EX < \infty$, and let T be independent of X with uniform distribution on $[0, a]$. Write $Y = a[(X + T)/a]$, where $[x]$ denotes the greatest integer in x. Show that $EY = EX$. (Rowland [104])

11. Let X and Y be independent rv's with pdf's f_λ and f_μ, respectively, where

$$f_\lambda(x) = \begin{cases} \lambda e^{-\lambda x} & \text{if } x > 0 \\ 0 & \text{otherwise} \end{cases}, \quad \lambda > 0,$$

and

$$f_\mu(y) = \begin{cases} \mu e^{-\mu y} & \text{if } y > 0 \\ 0 & \text{otherwise} \end{cases}, \quad \mu > 0.$$

Let $U = \min\{X, Y\}$ and $V = 1$ if $U = X$, and $= 0$ if $U = Y$. Find the joint distribution of (U, V).

12. Let $\{X_n\}$ be a sequence of independent rv's with pmf

$$P\{X_n = 1\} = p, \qquad P\{X_n = -1\} = 1 - p, \qquad 0 < p < 1.$$

Write $S_n = \sum_{k=1}^n X_k$. Show that $\{S_n\}$ is a Markov dependent sequence.

13. Prove Lemma 1.

14. Let $X_1, X_2, \cdots, X_n$ be rv's with joint pdf f, and let f_j be the marginal pdf of $X_j (j = 1, 2, \cdots, n)$. Show that $X_1, X_2, \cdots, X_n$ are independent if and only if

$$f(x_1, x_2, \cdots, x_n) = \prod_{j=1}^n f_j(x_j) \qquad \text{for all } (x_1, x_2, \cdots, x_n) \in \mathscr{R}_n.$$

15. Prove Theorem 4.

16. Prove Theorem 5.

4.4 FUNCTIONS OF RANDOM VECTORS

Here we first consider some simple functions of a random vector. Let X and Y be rv's defined on a probability space $(\Omega, \mathscr{S}, P)$. Define

$(X + Y)(\omega) = X(\omega) + Y(\omega)$ for all $\omega \in \Omega$;
$(XY)(\omega) = X(\omega)Y(\omega)$ for all $\omega \in \Omega$,
$\left(\dfrac{X}{Y}\right)(\omega) = \dfrac{X(\omega)}{Y(\omega)}$ for all $\omega \in \Omega$, provided that $\{Y = 0\} = \phi$.

Theorem 1. Let X and Y be rv's defined on $(\Omega, \mathscr{S}, P)$. Then so are $X + Y$ and XY. If in addition $\{Y = 0\} = \phi$, then X/Y is also an rv.

Proof. We have

(1) $$\{X + Y < z\} = \bigcup_r \{X < r\}\{Y < z - r\},$$

where the union is over the set of all rational numbers r. Since the set of all rational numbers is countable, we see that $A = \bigcup_r \{X < r\} \cap \{Y < z - r\} \in \mathscr{S}$. We need only to show that the sets on the two sides of (1) are the same. We first note that set $A \subseteq B = \{X + Y < z\}$. Let $\omega \in B$. Then $X(\omega) + Y(\omega) < z$. Let r_0 be a rational number satisfying $X(\omega) < r_0 < z - Y(\omega)$. Then $X(\omega) < r_0$ and $Y(\omega) < z - r_0$, so that $\omega \in \{X < r_0\} \cap \{Y < z - r_0\} \in A$. It follows that $B \subseteq A$ and that $X + Y$ is an rv.

Since $X + Y$, $X - Y$, are rv's, $(X+Y)^2$ and $(X-Y)^2$ are also rv's (Theorem 2.5.1), and it follows that

$$XY = \frac{(X+Y)^2 - (X-Y)^2}{4}$$

is also an rv.

Finally, since $\{Y = 0\} = \phi$, we have

$$\left\{\frac{X}{Y} \leq x\right\} = \left\{\frac{X}{Y} \leq x, Y < 0\right\} + \left\{\frac{X}{Y} \leq x, Y > 0\right\}$$
$$= \{X \geq xY, Y < 0\} + \{X \leq xY, Y > 0\}$$
$$= \{X - xY \geq 0\}\{Y < 0\} + \{X - xY \leq 0\}\{Y > 0\} \in \mathscr{S}.$$

Let $X_1, X_2, \cdots, X_n$ be rv's on $(\Omega, \mathscr{S}, P)$. Define

(2) $$M_n = \max\{X_1, X_2, \cdots, X_n\}$$

by

$$M_n(\omega) = \max\{X_1(\omega), X_2(\omega), \cdots, X_n(\omega)\} \quad \text{for all } \omega \in \Omega$$

and

(3) $$N_n = \min\{X_1, X_2, \cdots, X_n\}$$
by
$$N_n(\omega) = \min\{X_1(\omega), X_2(\omega), \cdots, X_n(\omega)\} \quad \text{for all} \quad \omega \in \Omega.$$
Note that
$$N_n = -\max\{-X_1, -X_2, \cdots, -X_n\}.$$

Theorem 2. M_n and N_n are rv's.

Proof. It suffices to show that M_n is an rv. We have
$$\{M_n \leq z\} = \{X_1 \leq z, X_2 \leq z, \cdots, X_n \leq z\} = \bigcap_{i=1}^{n}\{X_i \leq z\} \in \mathcal{S},$$
and the proof is complete.

Theorem 3. Let $X_2, X_2, \cdots, X_n$ be independent rv's. Then for all $z \in \mathcal{R}$

(4) $$P\{M_n \leq z\} = \prod_{i=1}^{n} P\{X_i \leq z\}$$

and

(5) $$P\{N_n \leq z\} = 1 - \prod_{i=1}^{n}[1 - P\{X_i \leq z\}].$$

Proof. We have
$$P\{M_n \leq z\} = P\{X_1 \leq z, X_2 \leq z, \cdots, X_n \leq z\} = \prod_{i=1}^{n} P\{X_i \leq z\},$$
and
$$\begin{aligned}P\{N_n \leq z\} &= 1 - P\{N_n > z\} \\ &= 1 - P\{X_1 > z, X_2 > z, \cdots, X_n > z\} \\ &= 1 - \prod_{i=1}^{n}[1 - P\{X_i \leq z\}].\end{aligned}$$

Corollary. If $X_1, X_2, \cdots, X_n$ are independent rv's with common distribution function F, then

(6) $$P\{M_n \leq z\} = F^n(z), \quad z \in \mathcal{R},$$

and

(7) $$P\{N_n \leq z\} = 1 - [1 - F(z)]^n, \quad z \in \mathcal{R}.$$

If, in addition, F is absolutely continuous and f is the density of X_i, the densities of M_n and N_n are given, respectively, by

(8) $$g(z) = nF^{n-1}(z)f(z)$$

and

(9) $$h(z) = n[1 - F(z)]^{n-1}f(z)$$

at all continuity points of f.

Theorem 4. Let $X_1, X_2, \cdots, X_n$ be independent rv's. Then the joint df of M_n and N_n is given by

(10) $$P\{M_n \leq x, N_n \leq y\} = \begin{cases} \prod_{i=1}^{n} P\{X_i \leq x\} - \prod_{i=1}^{n} [P\{X_i \leq x\} - P\{X_i \leq y\}] & \text{if } x > y, \\ \prod_{i=1}^{n} P\{X_i \leq x\} & \text{if } x \leq y. \end{cases}$$

Proof.

$$P\{M_n \leq x, N_n \leq y\} = P\{M_n \leq x\} - P\{M_n \leq x, N_n > y\}$$
$$= \prod_{i=1}^{n} P\{X_i \leq x\} - P\{X_1 \leq x, X_2 \leq x, \cdots, X_n \leq x;$$
$$X_1 > y, \cdots, X_n > y\}.$$

If $x \leq y$,

$$P\{M_n \leq x, N_n \leq y\} = \prod_{i=1}^{n} P\{X_i \leq x\}.$$

If $x > y$,

$$P\{M_n \leq x, N_n \leq y\} = \prod_{i=1}^{n} P\{X_i \leq x\} - \prod_{i=1}^{n} P\{y < X_i \leq x\}$$
$$= \prod_{i=1}^{n} P\{X_i \leq x\} - \prod_{i=1}^{n} [P\{X_i \leq x\} - P\{X_i \leq y\}].$$

Corollary. If $X_1, X_2, \cdots, X_n$ are independently distributed with common df F, then

(11) $$P\{M_n \leq x, N_n \leq y\} = \begin{cases} F^n(x) - [F(x) - F(y)]^n, & x > y, \\ F^n(x), & x \leq y. \end{cases}$$

If, in addition, F is absolutely continuous with density f, the joint density of M_n and N_n is given by

(12) $$u(x, y) = \begin{cases} 0, & x \leq y, \\ n(n-1)[F(x) - F(y)]^{n-2} f(x)f(y), & x > y. \end{cases}$$

Note that M_n and N_n are not independent.

Example 1. Let $X_1, \cdots, X_n$ be rv's with common density function

$$f(x) = \begin{cases} \dfrac{1}{b-a} & \text{for } x \in [a, b], \\ 0 & \text{otherwise.} \end{cases}$$

Then the density of M_n is

$$g(z) = \begin{cases} n\left(\dfrac{z-a}{b-a}\right)^{n-1} \dfrac{1}{b-a} & \text{for } z \in [a, b] \\ 0 & \text{otherwise.} \end{cases}$$

The density of N_n is

$$h(z) = \begin{cases} n\left(\dfrac{b-z}{b-a}\right)^{n-1} \dfrac{1}{b-a} & \text{for } z \in [a, b], \\ 0 & \text{otherwise.} \end{cases}$$

The joint density of M_n and N_n is

$$u(x, y) = \begin{cases} 0 & \text{if } x \le y, \\ \dfrac{n(n-1)}{(b-a)^n} (x-y)^{n-2} & \text{if } a \le y < x \le b. \end{cases}$$

Example 2. Let X_1, X_2 be iid rv's with common probability function

$$P\{X_i = x\} = e^{-\lambda} \dfrac{\lambda^x}{x!}, \quad x = 0, 1, 2, \cdots; \quad i = 1, 2,$$

where $\lambda > 0$ is a constant. Let $M = \max(X_1, X_2)$ and $N = \min(X_1, X_2)$. We have

$$P\{X_1 = x, M = z\} = \begin{cases} 0, & x > z, \\ P\{X_1 = x, X_2 \le z\}, & x = z, \\ P\{X_1 = x, X_2 = z\}, & x < z. \end{cases}$$

By independence, therefore, the marginal pmf of M is given by

$$P\{M = m\} = \sum_{x=0}^{m} P\{X_1 = x, M = m\}$$

$$= e^{-2\lambda} \dfrac{\lambda^m}{m!} \sum_{x=0}^{m-1} \dfrac{\lambda^x}{x!} + e^{-2\lambda} \dfrac{\lambda^m}{m!} \sum_{j=0}^{m} \dfrac{\lambda^j}{j!} \quad (m = 0, 1, 2, \cdots)$$

$$= 2e^{-2\lambda} \dfrac{\lambda^m}{m!} \sum_{j=0}^{m-1} \dfrac{\lambda^j}{j!} + \dfrac{\lambda^{2m} e^{-2\lambda}}{(m!)^2}.$$

Clearly

$$\sum_{m=0}^{\infty} P\{M = m\} = \left\{\sum_{m} \dfrac{\lambda^m}{m!} e^{-\lambda}\right\}^2 = 1.$$

Similarly

$$P\{X_1 = x, N = n\} = \begin{cases} 0 & \text{if } x < n, \\ P\{X_1 = n, X_2 \ge n\} & \text{if } x = n, \\ P\{X_1 = x, X_2 = n\} & \text{if } x > n, \end{cases}$$

so that

$$P\{N = n\} = \sum_{x=n}^{\infty} P\{X_1 = x, N = n\}$$
$$= P\{X_1 = n, X_2 \geq n\} + \sum_{x=n+1}^{\infty} P\{X_1 = x, X_2 = n\}$$
$$= e^{-2\lambda} \frac{\lambda^n}{n!} \sum_{j=n}^{\infty} \frac{\lambda^j}{j!} + e^{-2\lambda} \frac{\lambda^n}{n!} \sum_{x=n+1}^{\infty} \frac{\lambda^x}{x!}$$
$$= 2e^{-\lambda} \frac{\lambda^n}{n!} \sum_{j=n+1}^{\infty} \frac{\lambda^j}{j!} + e^{-2\lambda} \frac{\lambda^{2n}}{(n!)^2}$$

$$n = 0, 1, 2, \cdots.$$

Theorems 1 and 2 are special cases of the following general result.

Theorem 5. Let $f: \mathcal{R}_n \to \mathcal{R}_m$ be a Borel-measurable function, that is, if $B \in \mathfrak{B}_m$, then $f^{-1}(B) \in \mathfrak{B}_n$. If $\mathbf{X} = (X_1, X_2, \cdots, X_n)$ is an n-dimensional rv ($n \geq 1$), then $f(\mathbf{X})$ is an m-dimensional rv.

Proof. For $B \in \mathfrak{B}_m$

$$\{f(X_1, X_2, \cdots, X_n) \in B\} = \{(X_1, X_2, \cdots, X_n) \in f^{-1}(B)\},$$

and, since $f^{-1}(B) \in \mathfrak{B}_n$, it follows that $\{(X_1, X_2 \cdots, X_n) \in f^{-1}(B)\} \in \mathcal{S}$, which concludes the proof.

In Theorems 3 and 4 we computed the df of some simple functions of $X_1, X_2, \cdots, X_n$ by a combinatorial argument. It is possible to compute the distribution of rv's such as $X + Y$, $X - Y$, XY, and X/Y (where $[Y = 0] = \phi$) directly. It will be convenient, however, to state a general result that will be of subsequent use.

The case in which $\mathbf{X} = (X_1, X_2, \cdots, X_n)$ is a random vector of the discrete type causes no special problem. Let $A \subseteq \mathcal{R}_n$ be the countable set of points such that $P\{\mathbf{X} \in A\} = 1$, and assume that every point in A has positive mass. Let $u_1 = g_1(x_1, x_2, \cdots, x_n)$, $u_2 = g_2(x_1, x_2, \cdots, x_n)$, $\cdots$, $u_n = g_n(x_1, x_2, \cdots, x_n)$ define a one-to-one mapping of A onto B. Then, writing $\mathbf{u} = (u_1, u_2, \cdots, u_n)$, we have

$$P\{\mathbf{U} = \mathbf{u}\} = P\{g_1(\mathbf{X}) = u_1, g_2(\mathbf{X}) = u_2, \cdots, g_n(\mathbf{X}) = u_n\}$$
$$= P\{X_1 = h_1(\mathbf{u}), X_2 = h_2(\mathbf{u}), \cdots, X_n = h_n(\mathbf{u})\} \quad \text{for all } \mathbf{u} \in B,$$

and

$$P\{\mathbf{U} = \mathbf{u}\} = 0 \quad \text{if } \mathbf{u} \notin B,$$

where $x_1 = h_1(\mathbf{u})$, $x_2 = h_2(\mathbf{u})$, $\cdots$, $x_n = h_n(\mathbf{u})$ is the inverse transformation. The marginal pmf of any U_j or the joint marginal pmf of any subcollection of $U_1, U_2, \cdots, U_n$ is now easily computed by summing over the remaining $u_i's$.

Example 3. Let X have the binomial pmf

$$P\{X = k\} = \begin{cases} \binom{n}{k} p^k (1-p)^{n-k}, & k = 0, 1, 2, \cdots, n; 0 < p < 1, \\ 0, & \text{otherwise.} \end{cases}$$

Let Y be independent of X and have the same pmf as X. We wish to find the pmf of $X+Y$. We have

$$P\{X + Y = z\} = \sum_{k=0}^{n} P\{X = k, Y = z - k\}$$
$$= \sum_{k=0}^{n} \binom{n}{k} p^k (1-p)^{n-k} \binom{n}{z-k} p^{z-k} (1-p)^{n-z+k}$$
$$= \sum_{k=0}^{n} \binom{n}{k} \binom{n}{z-k} p^z (1-p)^{2n-z}$$
$$= \binom{2n}{z} p^z (1-p)^{2n-z}, \qquad z = 0, 1, 2, \cdots, 2n.$$

Suppose that $W = X - Y$, then W takes on values $w = -n, -(n-1), \cdots, -1, 0, 1, \cdots, n$. We have

$$P\{W = w\} = \sum_{k=0}^{n} P\{X = k + w\} P\{Y = k\}$$
$$= \sum_{0}^{n} \binom{n}{k+w} \binom{n}{k} p^{2k+w} (1-p)^{2n-2k-w}$$
$$= \left(\frac{p}{1-p}\right)^w \sum_{k=0}^{n} \binom{n}{k+w} \binom{n}{k} p^{2k} (1-p)^{2n-2k},$$

where $w = -n, -(n-1), \cdots, -1, 0, 1, \cdots, n$. Thus

$$P\{W = 0\} = \sum_{k=0}^{n} \binom{n}{k}^2 p^{2k} (1-p)^{2n-2k},$$
$$P\{W = -n\} = \left(\frac{1-p}{p}\right)^n p^{2n} = p^n (1-p)^n,$$

and so on.

Finally, let $U = X/(Y + 1)$ and $V = Y + 1$. Then $x = uv$ and $y = v - 1$. The joint pmf of U and V is given by

$$P\{U = u, V = v\} = \binom{n}{uv} p^{uv} (1-p)^{n-uv} \binom{n}{v-1} p^{v-1} (1-p)^{n+1-v}$$
$$= \binom{n}{uv} \binom{n}{v-1} p^{vu+v-1} (1-p)^{2n+1-v-uv}.$$

Here V takes values $1, 2, \cdots, n + 1$, and UV takes values $0, 1, 2, \cdots, n$. Thus

$$P\{U = n, V = 1\} = p^n (1-p)^n,$$

and

$$P\{U = \frac{k}{j+1}, V = j+1\} = \binom{n}{k}\binom{n}{j}p^{k+j}(1-p)^{2n-k-j},$$
$$k = 0, 1, 2, \cdots, n; j = 0, 1, 2, \cdots, n.$$

Example 4. Consider the bivariate negative binomial distribution with pmf

$$P\{X = x, Y = y\} = \frac{(x+y+k-1)!}{x!\,y!\,(k-1)!}p_1^x p_2^y (1-p_1-p_2)^k,$$

where $x, y = 0, 1, 2, \cdots; k \geq 1$ is an integer; $p_1, p_2 \in (0, 1)$; and $p_1 + p_2 < 1$. Let us find the pmf of $U = X + Y$. We introduce an rv $V = Y$ (see Remark 1 below) so that $u = x + y$, $v = y$ represents a one-to-one mapping of $A = \{(x, y): x, y = 0, 1, 2, \cdots\}$ onto the set $B = \{(u, v): v = 0, 1, 2, \cdots, u; u = 0, 1, 2, \cdots\}$ with inverse map $x = u - v$, $y = v$. It follows that the joint pmf of (U, V) is given by

$$P\{U = u, V = v\} = \begin{cases} \frac{(u+k-1)!}{(u-v)!\,v!\,(k-1)!}p_1^{u-v} p_2^v (1-p_1-p_2)^k \\ \qquad\qquad\qquad\qquad \text{for } (u, v) \in B, \\ 0 \qquad\qquad\qquad\qquad \text{otherwise.} \end{cases}$$

The marginal pmf of U is given by

$$P\{U = u\} = \frac{(u+k-1)!(1-p_1-p_2)^k}{(k-1)!\,u!}\sum_{v=0}^{u}\binom{u}{v}p_1^{u-v}p_2^v$$
$$= \frac{(u+k-1)!(1-p_1-p_2)^k}{(k-1)!\,u!}(p_1+p_2)^u$$
$$= \binom{u+k-1}{u}(p_1+p_2)^u(1-p_1-p_2)^k \qquad (u = 0, 1, 2, \cdots).$$

Next suppose that the mapping of $\mathbf{x}$ to $\mathbf{u}$ is not a one-to-one mapping of A onto B, but that for each $\mathbf{u} \in B$ there are a finite (or even countable) number of inverses $\mathbf{x}_1, \mathbf{x}_2, \cdots, \mathbf{x}_{k(\mathbf{u})}$. In that case

$$P\{\mathbf{U} = \mathbf{u}\} = P\{g_1(\mathbf{X}) = u_1, g_2(\mathbf{X}) = u_2, \cdots, g_n(\mathbf{X}) = u_n\}$$
$$= \sum_{j=1}^{k(\mathbf{u})} P\{g_1(\mathbf{x}_j) = u_1, g_2(\mathbf{x}_j) = u_2, \cdots, g_n(\mathbf{x}_j) = u_n; \mathbf{X} = \mathbf{x}_j\}.$$

Example 5. Let (X, Y) be an rv of the discrete type with joint pmf as follows:

Y \ X	−1	0	1	
−2	$\frac{1}{6}$	$\frac{1}{12}$	$\frac{1}{6}$	$\frac{5}{12}$
1	$\frac{1}{6}$	$\frac{1}{12}$	$\frac{1}{6}$	$\frac{5}{12}$
2	$\frac{1}{12}$	0	$\frac{1}{12}$	$\frac{1}{6}$
	$\frac{5}{12}$	$\frac{1}{6}$	$\frac{5}{12}$	1

Let $U = |X|$ and $V = Y^2$. Then
$$A = \{(x, y): x \in \{-1, 0, 1\}, y \in \{-2, 1, 2\}\}$$
and
$$B = \{(u, v): u = 0 \text{ or } 1 \quad \text{and} \quad v = 1 \text{ or } 4\}.$$

The pmf of (U, V) is computed in the following table:

(u, v)	Inverses (x, y)	$P\{U = u, V = v\}$
$(0, 1)$	$(0, 1)$	$\frac{1}{12}$
$(0, 4)$	$(0, 2), (0, -2)$	$\frac{1}{12}$
$(1, 1)$	$(-1, 1), (1, 1)$	$\frac{1}{3}$
$(1, 4)$	$(-1, -2), (-1, 2), (1, -2), (1, 2)$	$\frac{1}{2}$

The marginal pmf of U is given by
$$P\{U = u\} = \begin{cases} \frac{1}{6}, & u = 0, \\ \frac{5}{6}, & u = 1, \end{cases}$$
and that of V by
$$P\{V = v\} = \begin{cases} \frac{5}{12}, & v = 1, \\ \frac{7}{12}, & v = 4. \end{cases}$$

The case in which $\mathbf{X} = (X_1, X_2, \cdots, X_n)$ is of the continuous type is not as simple. We would like to obtain an analogue of Theorem 2.5.3. To this end, let $\mathbf{X}$ be an rv with joint pdf $f(x_1, x_2, \cdots, x_n)$, and let $\mathbf{U} = \mathbf{g}(X_1, X_2, \cdots, X_n) = (U_1, U_2, \cdots, U_n)$, where
$$u_i = g_i(x_1, x_2, \cdots, x_n), \quad i = 1, 2, \cdots, n,$$
be a mapping of $\mathcal{R}_n$ into itself. Then
$$P\{\mathbf{U} \in B\} = P\{\mathbf{X} \in \mathbf{g}^{-1}(B)\} = \int \cdots \int_{\mathbf{g}^{-1}(B)} f(x_1, x_2, \cdots, x_n) \prod_{i=1}^{n} dx_i,$$
where $\mathbf{g}^{-1}(B) = \{\mathbf{x} = (x_1, x_2, \cdots, x_n) \in \mathcal{R}_n: \mathbf{g}(\mathbf{x}) \in B\}$. Let us choose B to be the n-dimensional interval
$$B = B_{\mathbf{u}} = \{(u_1', u_2', \cdots, u_n'): -\infty < u_i' \leq u_i, i = 1, 2, \cdots, n\}.$$
Then the joint df of $\mathbf{U}$ is given by
$$P\{\mathbf{U} \in B_{\mathbf{u}}\} = G(\mathbf{u}) = P\{g_1(\mathbf{X}) \leq u_1, g_2(\mathbf{X}) \leq u_2, \cdots, g_n(\mathbf{X}) \leq u_n\}$$
$$= \int \cdots \int_{\mathbf{g}^{-1}(B_{\mathbf{u}})} f(x_1, x_2, \cdots, x_n) \prod_{i=1}^{n} dx_i,$$
and (if G is absolutely continuous) the pdf of $\mathbf{U}$ is given by

$$w(\mathbf{u}) = \frac{\partial^n G(\mathbf{u})}{\partial u_1 \, \partial u_2 \cdots \partial u_n}$$

at every continuity point $\mathbf{u}$ of w. Under certain conditions it is possible to write w in terms of f by making a change of variable in the multiple integral (see Theorem P.2.17 for a statement of this result).

Theorem 6. Let $(X_1, X_2, \cdots, X_n)$ be an n-dimensional rv of the continuous type with pdf $f(x_1, x_2, \cdots, x_n)$.

(a) Let
$$\begin{aligned} u_1 &= g_1(x_1, x_2, \cdots, x_n), \\ u_2 &= g_2(x_1, x_2, \cdots, x_n), \\ &\vdots \\ u_n &= g_n(x_1, x_2, \cdots, x_n), \end{aligned}$$

be a one-to-one mapping of $\mathcal{R}_n$ into itself, that is, there exists the inverse transformation

$$x_1 = h_1(u_1, u_2, \cdots, u_n), \quad x_2 = h_2(u_1, u_2, \cdots, u_n), \cdots,$$
$$x_n = h_n(u_1, u_2, \cdots, u_n)$$

defined over the range of the transformation.

(b) Assume that both the mapping and its inverse are continuous.
(c) Assume that the partial derivatives

$$\frac{\partial x_i}{\partial u_j}, \quad 1 \le i \le n, \, 1 \le j \le n,$$

exist and are continuous.

(d) Assume that the Jacobian J of the inverse transformation

$$J = \frac{\partial(x_1, \cdots, x_n)}{\partial(u_1, \cdots, u_n)} = \begin{vmatrix} \frac{\partial x_1}{\partial u_1} & \frac{\partial x_1}{\partial u_2} & \cdots & \frac{\partial x_1}{\partial u_n} \\ \frac{\partial x_2}{\partial u_1} & \frac{\partial x_2}{\partial u_2} & \cdots & \frac{\partial x_2}{\partial u_n} \\ \vdots & \vdots & & \vdots \\ \frac{\partial x_n}{\partial u_1} & \frac{\partial x_n}{\partial u_2} & \cdots & \frac{\partial x_n}{\partial u_n} \end{vmatrix}$$

is different from zero for $(u_1, u_2, \cdots, u_n)$ in the range of the transformation.

Then the random vector $(U_1, U_2, \cdots, U_n)$ has a joint absolutely continuous df with pdf given by

(13) $\quad w(u_1, u_2, \cdots, u_n) = |J| f(h_1(u_1, \cdots, u_n), \cdots, h_n(u_1, \cdots, u_n)).$

Proof. For $(u_1, u_2, \cdots, u_n) \in \mathcal{R}_n$, let

$$B = \{(u'_1, u'_2, \cdots, u'_n) \in \mathcal{R}_n : -\infty < u'_i \le u_i, \, i = 1, 2, \cdots, n\}.$$

Then
$$\mathbf{g}^{-1}(B) = \{\mathbf{x} \in \mathcal{R}_n : \mathbf{g}(\mathbf{x}) \in B\} = \{(x_1, x_2, \cdots, x_n) : g_i(\mathbf{x}) \leq u_i, i = 1, 2, \cdots, n\}$$
and
$$\begin{aligned}
G_{\mathbf{U}}(\mathbf{u}) &= P\{\mathbf{U} \in B\} = P\{\mathbf{X} \in \mathbf{g}^{-1}(B)\} \\
&= \int_{\mathbf{g}^{-1}(B)} \cdots \int f(x_1, x_2, \cdots, x_n) \, dx_1 \, dx_2 \cdots dx_n \\
&= \int_{-\infty}^{u_1} \cdots \int_{-\infty}^{u_n} f(h_1(\mathbf{u}), \cdots, h_n(\mathbf{u})) \left| \frac{\partial(x_1, x_2, \cdots, x_n)}{\partial(u_1, u_2, \cdots, u_n)} \right| du_1 \cdots du_n.
\end{aligned}$$

Result (13) now follows on differentiation of df $G_{\mathbf{U}}$.

Remark 1. In actual applications we will not know the mapping from $x_1, x_2, \cdots, x_n$ to $u_1, u_2, \cdots, u_n$ completely, but one or more of the functions g_i will be known. If only k, $1 \leq k < n$, of the g_i's are known, we introduce arbitrarily $n - k$ functions such that the conditions of the theorem are satisfied. To find the joint marginal density of these k variables we simply integrate the w function over all the $n - k$ variables that were arbitrarily introduced.

Remark 2. An analogue of Theorem 2.5.4 holds, which we state without proof.

Let $\mathbf{X} = (X_1, X_2, \cdots, X_n)$ be an rv of the continuous type with joint pdf f, and let $u_i = g_i(x_1, x_2, \cdots, x_n)$, $i = 1, 2, \cdots, n$, be a mapping of $\mathcal{R}_n$ into itself. Suppose that for each $\mathbf{u}$ the transformation $\mathbf{g}$ has a finite number $k = k(\mathbf{u})$ of inverses. Suppose further that $\mathcal{R}_n$ can be partitioned into k disjoint sets $A_1, A_2, \cdots, A_k$, such that the transformation $\mathbf{g}$ from $A_i (i = 1, 2, \cdots, n)$ into $\mathcal{R}_n$ is one-to-one with inverse transformation

$$x_1 = h_{1i}(u_1, u_2, \cdots, u_n), \cdots, \quad x_n = h_{ni}(u_1, u_2, \cdots, u_n), \quad i = 1, 2, \cdots, k.$$

Suppose that the first partial derivatives are continuous and that each Jacobian

$$J_i = \begin{vmatrix} \dfrac{\partial h_{1i}}{\partial u_1} & \dfrac{\partial h_{1i}}{\partial u_2} & \cdots & \dfrac{\partial h_{1i}}{\partial u_n} \\ \dfrac{\partial h_{2i}}{\partial u_1} & \dfrac{d h_{2i}}{\partial u_2} & \cdots & \dfrac{\partial h_{2i}}{\partial u_n} \\ \vdots & \vdots & & \vdots \\ \dfrac{d h_{ni}}{d u_1} & \dfrac{\partial h_{ni}}{\partial u_2} & \cdots & \dfrac{\partial h_{ni}}{\partial u_n} \end{vmatrix}$$

is different from zero in the range of the transformation. Then the joint pdf of $\mathbf{U}$ is given by

$$w(u_1, u_2, \cdots, u_n) = \sum_{i=1}^{k} |J_i| f(h_{1i}(u_1, u_2, \cdots, u_n), \cdots, h_{ni}(u_1, u_2, \cdots, u_n)).$$

Example 6. Let X_1, X_2, X_3 be iid rv's with common *exponential* density function

$$f(x) = \begin{cases} e^{-x} & \text{if } x > 0, \\ 0 & \text{otherwise.} \end{cases}$$

Also, let

$$Y_1 = X_1 + X_2 + X_3, \quad Y_2 = \frac{X_1 + X_2}{X_1 + X_2 + X_3}, \quad Y_3 = \frac{X_1}{X_1 + X_2}.$$

Then

$$x_1 = y_1 y_2 y_3, \quad x_2 = y_1 y_2 - x_1 = y_1 y_2 (1 - y_3), \quad \text{and}$$
$$x_3 = y_1 - y_1 y_2 = y_1 (1 - y_2).$$

The Jacobian of transformation is given by

$$J = \begin{vmatrix} y_2 y_3 & y_1 y_3 & y_1 y_2 \\ y_2(1 - y_3) & y_1(1 - y_3) & -y_1 y_2 \\ 1 - y_2 & -y_1 & 0 \end{vmatrix} = -y_1^2 y_2.$$

Note that $0 < y_1 < \infty$, $0 < y_2 < 1$, and $0 < y_3 < 1$. Thus the joint pdf of Y_1, Y_2, Y_3 is given by

$$w(y_1, y_2, y_3) = y_1^2 y_2 \, e^{-y_1}$$
$$= (2y_2)(\tfrac{1}{2} y_1^2 \, e^{-y_1}), \quad 0 < y_1 < \infty, \, 0 < y_2, y_3 < 1.$$

It follows that Y_1, Y_2, and Y_3 are independent.

Example 7. Let X_1, X_2 be independent rv's with common density given by

$$f(x) = \begin{cases} 1 & \text{if } 0 < x < 1, \\ 0 & \text{otherwise.} \end{cases}$$

Let $Y_1 = X_1 + X_2$, $Y_2 = X_1 - X_2$. Then the Jacobian of the transformation is given by

$$J = \begin{vmatrix} \tfrac{1}{2} & \tfrac{1}{2} \\ \tfrac{1}{2} & -\tfrac{1}{2} \end{vmatrix} = -\tfrac{1}{2},$$

and the joint density of Y_1, Y_2 (Fig. 1) is given by

$$f_{Y_1, Y_2}(y_1, y_2) = \tfrac{1}{2} f\!\left(\frac{y_1 + y_2}{2}\right) f\!\left(\frac{y_1 - y_2}{2}\right)$$

$$\text{if } 0 < \frac{y_1 + y_2}{2} < 1, \, 0 < \frac{y_1 - y_2}{2} < 1,$$

$$= \tfrac{1}{2} \quad \text{if } (y_1, y_2) \in \{0 < y_1 + y_2 < 2, \, 0 < y_1 - y_2 < 2\}.$$

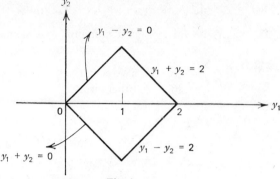

Fig. 1

The marginal pdf's of Y_1 and Y_2 are given by

$$f_{Y_1}(y_1) = \begin{cases} \int_{-y_1}^{y_1} \frac{1}{2}\, dy_2 = y_1, & 0 < y_1 \le 1, \\ \int_{y_1-2}^{2-y_1} \frac{1}{2}\, dy_2 = 2 - y_1, & 1 < y_1 < 2, \\ 0, & \text{otherwise}; \end{cases}$$

$$f_{Y_2}(y_2) = \begin{cases} \int_{-y_2}^{y_2+2} \frac{1}{2}\, dy_1 = y_2 + 1, & -1 < y_2 \le 0, \\ \int_{y_2}^{2-y_2} \frac{1}{2}\, dy_1 = 1 - y_2, & 0 < y_2 < 1, \\ 0, & \text{otherwise}. \end{cases}$$

Example 8. Let X_1, X_2, X_3 be iid rv's with common pdf

$$f(x) = \frac{1}{\sqrt{2\pi}} e^{-x^2/2}, \qquad -\infty < x < \infty.$$

Let $Y_1 = (X_1 - X_2)/\sqrt{2}$, $Y_2 = (X_1 + X_2 - 2X_3)/\sqrt{6}$, and $Y_3 = (X_1 + X_2 + X_3)/\sqrt{3}$. Then

$$x_1 = \frac{y_1}{\sqrt{2}} + \frac{y_2}{\sqrt{6}} + \frac{y_3}{\sqrt{3}},$$

$$x_2 = -\frac{y_1}{\sqrt{2}} + \frac{y_2}{\sqrt{6}} + \frac{y_3}{\sqrt{3}},$$

$$x_3 = -\frac{\sqrt{2}\, y_2}{\sqrt{3}} + \frac{y_3}{\sqrt{3}}.$$

The Jacobian of transformation is given by

$$J = \begin{vmatrix} \frac{1}{\sqrt{2}} & \frac{1}{\sqrt{6}} & \frac{1}{\sqrt{3}} \\ \frac{-1}{\sqrt{2}} & \frac{1}{\sqrt{6}} & \frac{1}{\sqrt{3}} \\ 0 & \frac{-\sqrt{2}}{\sqrt{3}} & \frac{1}{\sqrt{3}} \end{vmatrix} = 1.$$

The joint pdf of X_1, X_2, X_3 is given by

$$g(x_1, x_2, x_3) = \frac{1}{(\sqrt{2\pi})^3} \exp\left\{-\frac{x_1^2 + x_2^2 + x_3^2}{2}\right\}, \qquad x_1, x_2, x_3 \in \mathcal{R}.$$

It is easily checked that

$$x_1^2 + x_2^2 + x_3^2 = y_1^2 + y_2^2 + y_3^2,$$

so that the joint pdf of Y_1, Y_2, Y_3 is given by

$$w(y_1, y_2, y_3) = \frac{1}{(\sqrt{2\pi})^3} \exp\left\{-\frac{y_1^2 + y_2^2 + y_3^2}{2}\right\}.$$

It follows that Y_1, Y_2, Y_3 are also iid rv's with common pdf f.

In Example 8 the transformation used is orthogonal and is known as *Helmert's transformation*. In fact, we will show in Section 5.4 that under orthogonal transformations iid rv's with pdf f defined above are transformed into iid rv's with the same pdf.

Actually it is easily verified that

$$y_1^2 + y_2^2 = \sum_{j=1}^{3}\left(x_j - \frac{x_1 + x_2 + x_3}{3}\right)^2.$$

We have therefore proved that $(X_1 + X_2 + X_3)$ is independent of $\sum_{j=1}^{3}\{X_j - [(X_1 + X_2 + X_3)/3]\}^2$. This is a very important result in mathematical statistics, and we will return to it in Section 7·5.

Example 9. Let (X, Y) be a *bivariate normal* rv with joint pdf

$$f(x, y) = \frac{1}{2\pi\sigma_1\sigma_2(1 - \rho^2)^{1/2}}$$

$$\cdot \exp\left\{-\frac{1}{2(1 - \rho^2)}\left[\frac{(x - \mu_1)^2}{\sigma_1^2} - \frac{2\rho(x - \mu_1)(y - \mu_2)}{\sigma_1\sigma_2} + \frac{(y - \mu_2)^2}{\sigma_2^2}\right]\right\},$$

$$-\infty < x < \infty, \quad -\infty < y < \infty; \; \mu_1 \in \mathcal{R}, \mu_2 \in \mathcal{R};$$
$$\text{and } \sigma_1 > 0, \sigma_2 > 0, |\rho| < 1.$$

Let

$$U_1 = \sqrt{X^2 + Y^2}, \qquad U_2 = \frac{X}{Y}.$$

If $u_1 > 0$, then

$$\sqrt{x^2 + y^2} = u_1 \quad \text{and} \quad \frac{x}{y} = u_2$$

have two solutions:

$$x_1 = \frac{u_1 u_2}{\sqrt{1 + u_2^2}}, \quad y_1 = \frac{u_1}{\sqrt{1 + u_2^2}}, \quad \text{and} \quad x_2 = -x_1, \quad y_2 = -y_1$$

for any $u_2 \in \mathcal{R}$. The Jacobians are given by

$$J_1 = J_2 = \begin{vmatrix} \dfrac{u_2}{\sqrt{1+u_2^2}} & \dfrac{u_1}{(1+u_2^2)^{3/2}} \\ \dfrac{1}{\sqrt{1+u_2^2}} & -\dfrac{u_1 u_2}{(1+u_2^2)^{3/2}} \end{vmatrix} = -\dfrac{u_1}{1+u_2^2}.$$

It follows from the result in Remark 2 that the joint pdf of (U_1, U_2) is given by

$$w(u_1, u_2) = \begin{cases} \dfrac{u_1}{1+u_2^2} \left[f\left(\dfrac{u_1 u_2}{\sqrt{1+u_2^2}}, \dfrac{u_1}{\sqrt{1+u_2^2}}\right) \\ \qquad + f\left(\dfrac{-u_1 u_2}{\sqrt{1+u_2^2}}, \dfrac{-u_1}{\sqrt{1+u_2^2}}\right) \right] & \text{if } u_1 > 0,\ u_2 \in \mathcal{R}, \\ 0 & \text{otherwise.} \end{cases}$$

In the special case where $\mu_1 = \mu_2 = 0$, $\rho = 0$, and $\sigma_1 = \sigma_2 = \sigma$, we have

$$f(x, y) = \dfrac{1}{2\pi\sigma^2} e^{-[(x^2+y^2)/2\sigma^2]}$$

so that X and Y are independent. Moreover,

$$f(x, y) = f(-x, -y),$$

and it follows that

$$w(u_1, u_2) = \begin{cases} \dfrac{1}{2\pi\sigma^2} \dfrac{2u_1}{1+u_2^2} e^{-u_1^2/2\sigma^2}, & u_1 > 0,\ -\infty < u_2 < \infty, \\ 0, & \text{otherwise.} \end{cases}$$

Since

$$w(u_1, u_2) = \dfrac{1}{\pi(1+u_2^2)} \dfrac{u_1}{\sigma^2} e^{-u_1^2/2\sigma^2},$$

it follows that U_1 and U_2 are independent with marginal pdf's given by

$$w_1(u_1) = \begin{cases} \dfrac{u_1}{\sigma^2} e^{-u_1^2/2\sigma^2}, & u_1 > 0, \\ 0, & u_1 \leq 0, \end{cases}$$

and

$$w_2(u_2) = \dfrac{1}{\pi(1+u_2^2)}, \qquad -\infty < u_2 < \infty,$$

respectively.

An important application of the result in Remark 2 will appear in Theorem 4.5.2.

Theorem 7. Let (X, Y) be an rv of the continuous type with pdf f. Let
$$Z = X + Y, \ U = X - Y, \ V = XY;$$
and, if $P\{Y = 0\} = 0$, let $W = X/Y$. Then the pdf's of Z, U, V, and W are, respectively, given by

(14) $$f_Z(z) = \int_{-\infty}^{\infty} f(x, z - x) \, dx,$$

(15) $$f_U(u) = \int_{-\infty}^{\infty} f(u + y, y) \, dy,$$

(16) $$f_V(v) = \int_{-\infty}^{\infty} f\left(x, \frac{v}{x}\right) \frac{1}{|x|} \, dx,$$

(17) $$f_W(w) = \int_{-\infty}^{\infty} f(xw, x) \, |x| \, dx.$$

Proof. The proof is left as an exercise.

Corollary. If X and Y are independent with pdf's f_1 and f_2, respectively, then

(18) $$f_Z(z) = \int_{-\infty}^{\infty} f_1(x) f_2(z - x) \, dx,$$

(19) $$f_U(u) = \int_{-\infty}^{\infty} f_1(u + y) f_2(y) \, dy,$$

(20) $$f_V(v) = \int_{-\infty}^{\infty} f_1(x) f_2\left(\frac{v}{x}\right) \frac{1}{|x|} \, dx,$$

(21) $$f_W(w) = \int_{-\infty}^{\infty} f_1(xw) f_2(x) \, |x| \, dx.$$

Remark 3. Let F and G be two absolutely continuous df's; then
$$H(x) = \int_{-\infty}^{\infty} F(x - y) G'(y) \, dy = \int_{-\infty}^{\infty} G(x - y) F'(y) \, dy$$
is also an absolutely continuous df with pdf
$$H'(x) = \int_{-\infty}^{\infty} F'(x - y) G'(y) \, dy = \int_{-\infty}^{\infty} G'(x - y) F'(y) \, dy.$$
If
$$F(x) = \sum_k p_k \, \varepsilon(x - x_k) \quad \text{and} \quad G(x) = \sum_j q_j \, \varepsilon(x - y_j)$$
are two df's, then
$$H(x) = \sum_k \sum_j p_k q_j \, \varepsilon(x - x_k - y_j)$$

is also a df of an rv of the discrete type. The df H is called the *convolution* of F and G, and we write $H = F * G$. Clearly the operation is commutative and associative; that is, if F_1, F_2, F_3 are df's, $F_1 * F_2 = F_2 * F_1$ and $(F_1 * F_2) * F_3 = F_1 * (F_2 * F_3)$. In this terminology, if X and Y are independent rv's with df's F and G, respectively, $X + Y$ has the convolution df $H = F * G$. Extension to an arbitrary number of independent rv's is obvious.

Example 10. Let $X_1, X_2, \cdots, X_n$ be iid rv's with common density function f. Also, let
$$M_n = \max(X_1, X_2, \cdots, X_n), \qquad N_n = \min(X_1, X_2, \cdots, X_n),$$
and write
$$R_n = M_n - N_n.$$
Then R_n is also an rv called the *range* of $X_1, X_2, \cdots, X_n$. By the corollary to Theorem 4 the density of R_n is given by
$$\begin{aligned} f_{R_n}(u) &= \int_{-\infty}^{\infty} f_{M_n, N_n}(u + y, y) \, dy \\ &= \int_{-\infty}^{\infty} f_{M_n, N_n}(x, x - u) \, dx \\ &= n(n-1) \int_{-\infty}^{\infty} [F(x) - F(x-u)]^{n-2} f(x) f(x-u) \, dx \end{aligned} \qquad (u > 0),$$
where F is the df of X_i. If, in particular,
$$f(x) = \begin{cases} 1 & \text{if } 0 \le x \le 1, \\ 0 & \text{otherwise,} \end{cases}$$
we see that
$$f_{R_n}(u) = \begin{cases} 0 & \text{for } u < 0 \text{ or } u > 1, \\ n(n-1) u^{n-2}(1-u) & \text{for } 0 \le u \le 1. \end{cases}$$

Theorem 8. Let X and Y be independent and identically distributed. Then $X - Y$ is a symmetric rv.

Proof. We prove the theorem for the case in which X and Y are of the discrete type and leave the reader to supply the proof for the continuous case. We have
$$\begin{aligned} P\{X - Y \le x\} &= \sum_{j=1}^{\infty} P\{-Y \le x - x_j\} P\{X = x_j\} \\ &= \sum_j P\{-X \le x - x_j\} P\{X = x_j\} \\ &= 1 - \sum_j P\{-X > x - x_j\} P\{X = x_j\}. \end{aligned}$$

Therefore

$$P\{X - Y \leq -x\} = 1 - \sum_j P\{-X > -x - x_j\} P\{X = x_j\}$$
$$= 1 - \sum_j P\{X < x + x_j\} P\{Y = x_j\}$$
$$= 1 - P\{X - Y < x\}$$
$$= 1 - P\{X - Y \leq x\} + P\{X - Y = x\},$$

and it follows that $X - Y$ is a symmetric rv.

Remark 4. An rv X has a symmetric df if and only if X and $-X$ have the same distribution. If X and Y are identically distributed but not independent, let us write $Z = X - Y$. Since X and Y are identically distributed and $-Z = Y - X$, there is a temptation to conclude that Z and $-Z$ have the same distribution, so that Z is a symmetric rv. That this is not true is seen in the following example.

Example 11. Let (X, Y) have the joint pmf given in the following table:

Y \ X	1	2	3	
1	$\frac{1}{12}$	$\frac{2}{12}$	$\frac{1}{12}$	$\frac{4}{12}$
2	$\frac{1}{12}$	0	$\frac{1}{12}$	$\frac{2}{12}$
3	$\frac{2}{12}$	0	$\frac{4}{12}$	$\frac{6}{12}$
	$\frac{4}{12}$	$\frac{2}{12}$	$\frac{6}{12}$	1

Here X and Y are identically distributed but not independent. The pmf of $Z = X - Y$ is given by

Z	$P\{Z = z\}$
-2	$\frac{1}{6}$
-1	$\frac{1}{12}$
0	$\frac{5}{12}$
1	$\frac{3}{12}$
2	$\frac{1}{12}$

Z and $-Z$ do not have the same distribution, and therefore Z is not symmetric.

It is easy to find examples of jointly distributed rv's X, Y such that X and Y are identically distributed and $X - Y$ is symmetric.

Example 12. Let (X, Y) be jointly distributed with density

$$f(x, y) = \begin{cases} (1 + \alpha) - 2\alpha(x + y - 2xy), & 0 < x < 1, 0 < y < 1, \\ 0, & \text{otherwise,} \end{cases}$$

where $|\alpha| \leq 1$. The marginal pdf's of X and Y are the same, namely,

$$g(x) = \begin{cases} 1 & \text{if } 0 < x < 1, \\ 0 & \text{otherwise.} \end{cases}$$

If we write $U = X - Y$, the density of U is easily shown to be

$$f_U(u) = \begin{cases} (1 + \frac{1}{3}\alpha) + (1 + \alpha)u - 2\alpha \frac{u^3}{3}, & -1 < u \leq 0, \\ (1 + \frac{1}{3}\alpha) - (1 + \alpha)u + 2\alpha \frac{u^3}{3}, & 0 < u < 1, \\ 0, & \text{otherwise.} \end{cases}$$

Clearly U is symmetric.

The following result gives a condition under which rv's X and Y are dependent and $X - Y$ has a symmetric distribution.

Theorem 9. Let (X, Y) be a jointly distributed rv such that the joint distribution is symmetric in the arguments. Then $X - Y$ has a symmetric distribution.

Proof. The proof is left as an exercise.

Corollary. If X and Y are iid, $X - Y$ has a symmetric distribution.

It is possible to improve Theorem 9 in special cases, but we will not do so in detail here. The following example suffices.

Example 13. Let (X, Y) have the joint density

$$f(x, y) = \begin{cases} \frac{1}{4}(1 - x^3 y), & -1 < x < 1, -1 < y < 1, \\ 0, & \text{otherwise.} \end{cases}$$

Then X and Y both have the same pdf:

$$g(x) = \begin{cases} \frac{1}{2} & \text{if } -1 < x < 1, \\ 0 & \text{otherwise.} \end{cases}$$

The pdf of $U = X - Y$ is given by

$$f_U(u) = \begin{cases} \frac{1}{4}\left[\frac{9}{5} - \frac{3}{4}u + \frac{(u-1)^5}{5} - \frac{u(u-1)^4}{4}\right], & 0 < u < 2, \\ \frac{1}{4}\left[\frac{9}{5} + \frac{3}{4}u - \frac{(1+u)^5}{5} + \frac{u(1+u)^4}{4}\right], & -2 < u \leq 0, \\ 0, & \text{otherwise.} \end{cases}$$

Clearly U is symmetric.

FUNCTIONS OF RANDOM VECTORS

Definition. Let X be an rv, and let X' be an rv that is independent of X and has the same distribution as X. We call the rv

$$X^s = X - X'$$

the symmetrized X.

The technique of symmetrization is an important tool in the study of limit theorems.

Remark 5. Clearly X^s is symmetric about the origin so that

$$P\{X^s \geq 0\} \geq \tfrac{1}{2} \quad \text{and} \quad P\{X^s \leq 0\} \geq \tfrac{1}{2}.$$

Remark 6. If EX exists, so does EX^s, since

$$E|X^s| \leq E|X| + E|X'| = 2E|X| < \infty,$$

and we have $EX^s = 0$.

Remark 7. Not every symmetric distribution is the result of a symmetrization procedure. Thus $P\{X = \pm 1\} = \tfrac{1}{2}$ defines a symmetric rv that cannot be obtained by symmetrization.

Theorem 10.

(a) $P\{|X^s| > \varepsilon\} \leq 2P\{|X| > \varepsilon/2\}$.
(b) If $a \geq 0$ such that $P\{X \geq a\} \leq 1 - p$ and $P\{X \leq -a\} \leq 1 - p$, then $P\{|X^s| \geq \varepsilon\} \geq pP\{|X| > a + \varepsilon\}$.

Proof. The reader is asked to construct a proof of Theorem 10.

We conclude this section by noting that the mgf can frequently be used to find distributions of functions of random vectors. The following result is of great importance.

Theorem 11. Let $X_1, X_2, \cdots, X_n$ be independent rv's, and suppose that the mgf of X_i exists for each $i = 1, 2, \cdots, n$. Then the mgf of $S_n = X_1 + X_2 + \cdots + X_n$ exists and satisfies

(22) $$M_{S_n}(t) = \prod_{i=1}^{n} M_{X_i}(t).$$

Proof. We have

$$M_{S_n}(t) = Ee^{tS_n} = E\prod_{i=1}^{n} e^{tX_i} = \prod_{i=1}^{n} Ee^{tX_i},$$

since $e^{tX_1}, \cdots, e^{tX_n}$ are independent.

In particular, if $X_1, X_2, \cdots, X_n$ are iid with common law $\mathscr{L}(X)$, then
$$M_{X_i}(t) = M_X(t) \quad \text{for } i = 1, 2, \cdots, n,$$

and we have
$$M_{S_n}(t) = [M_X(t)]^n.$$

Remark 8. The converse of Theorem 11 does not hold. We leave the reader to construct an example to illustrate this fact.

Example 14. Let $X_1, X_2, \cdots, X_m$ be iid rv's with common pmf
$$P\{X = k\} = \binom{n}{k} p^k (1-p)^{n-k}, \qquad k = 0, 1, 2, \cdots, n; \; 0 < p < 1.$$
Then the mgf of X_i is given by
$$M(t) = (1 - p + pe^t)^n.$$
It follows that the mgf of $S_m = X_1 + X_2 + \cdots + X_m$ is
$$\begin{aligned} M_{S_m}(t) &= \prod_1^m (1 - p + pe^t)^n \\ &= (1 - p + pe^t)^{nm}, \end{aligned}$$
and we see that S_m has the pmf
$$P\{S_m = s\} = \binom{mn}{s} p^s (1-p)^{mn-s}, \qquad s = 0, 1, 2, \cdots, mn.$$

Example 15. Let X have the pdf
$$f(x) = \begin{cases} 1 & \text{if } 0 < x < 1, \\ 0 & \text{otherwise.} \end{cases}$$
The mgf of X is given by
$$M(t) = \frac{e^t - 1}{t}.$$
Let $Y = aX + b$, where a and b are constants. Then
$$\begin{aligned} M_Y(t) &= Ee^{atX+bt} \\ &= e^{bt} M(at) = e^{bt} \frac{e^{at} - 1}{at}. \end{aligned}$$
This is the mgf of the uniform distribution on $(b, a+b)$ with pdf
$$g(y) = \begin{cases} \dfrac{1}{a}, & b < y < a + b, \\ 0, & \text{otherwise,} \end{cases}$$
as can easily be checked.

FUNCTIONS OF RANDOM VECTORS 147

From these examples it is clear that to use this technique effectively one must be able to recognize the mgf of the function under consideration. In Chapter 5 we will study a number of commonly occurring probability distributions and derive their mgf's (whenever they exist). We will have occasion to use Theorem 11 quite frequently.

For integer-valued rv's one can sometimes use pgf's to compute the distribution of certain functions of a random vector. See, for example, Problem 9.

PROBLEMS 4.4

1. Let F be a df and ε be any positive real number. Show that
$$\Psi_1(x) = \frac{1}{\varepsilon} \int_x^{x+\varepsilon} F(x)\, dx$$
and
$$\Psi_2(x) = \frac{1}{2\varepsilon} \int_{x-\varepsilon}^{x+\varepsilon} F(x)\, dx$$
are also distribution functions.

2. Let X, Y be iid rv's with common pdf
$$f(x) = \begin{cases} e^{-x} & \text{if } x > 0, \\ 0 & \text{if } x \leq 0. \end{cases}$$
(a) Find the pdf of rv's $X + Y$, $X - Y$, XY, X/Y, min $\{X, Y\}$, max $\{X, Y\}$, min $\{X, Y\}$/max $\{X, Y\}$, and $X/(X+Y)$.
(b) Let $U = X + Y$ and $V = X - Y$. Find the conditional pdf of V, given $U = u$, for some fixed $u > 0$.
(c) Show that U and $Z = X/(X+Y)$ are independent.

3. Let X and Y be independent rv's defined on the space $(\Omega, \mathscr{S}, P)$. Let X be uniformly distributed on $(-a, a)$, $a > 0$, and Y be an rv of the continuous type with density f, where f is continuous and positive on $\mathscr{R}$. Let F be the df of Y. If $u_0 \in (-a, a)$ is a fixed number, show that
$$f_{Y|X+Y}(y|u_0) = \begin{cases} \dfrac{f(y)}{F(u_0 + a) - F(u_0 - a)} & \text{if } u_0 - a < y < u_0 + a, \\ 0 & \text{otherwise,} \end{cases}$$
where $f_{Y|X+Y}(y|u_0)$ is the conditional density function of Y, given $X + Y = u_0$.

4. Let X and Y be iid rv's with common pdf
$$f(x) = \begin{cases} 1 & \text{if } 0 \leq x \leq 1, \\ 0 & \text{otherwise.} \end{cases}$$
Find the pdf's of rv's $X + Y$, XY, X/Y, min $\{X, Y\}$, max $\{X, Y\}$, min $\{X, Y\}$/max $\{X, Y\}$.

5. Let X_1, X_2, X_3 be iid rv's with common density function

$$f(x) = \begin{cases} 1 & \text{if } 0 \leq x \leq 1; \\ 0 & \text{otherwise.} \end{cases}$$

Show that the pdf of $U = X_1 + X_2 + X_3$ is given by

$$g(u) = \begin{cases} \dfrac{u^2}{2}, & 0 \leq u < 1, \\ 3u - u^2 - \dfrac{3}{2}, & 1 \leq u < 2, \\ \dfrac{(u-3)^2}{2}, & 2 \leq u \leq 3, \\ 0, & \text{elsewhere.} \end{cases}$$

6. Let X and Y be independent rv's with common geometric pmf

$$P\{X = k\} = \pi(1-\pi)^k, \quad k = 0, 1, 2, \cdots; 0 < \pi < 1.$$

Also, let $M = \max\{X, Y\}$. Find the joint distribution of M and X, the marginal distribution of M, and the conditional distribution of X, given M.

7. Let X be a nonnegative rv of the continuous type. The integral part, Y, of X is distributed with pmf $P\{Y = k\} = \lambda^k e^{-\lambda}/k!$, $k = 0, 1, 2, \cdots, \lambda > 0$; and the fractional part, Z, of X has pdf $f_Z(z) = 1$ if $0 \leq z \leq 1$, and $= 0$ otherwise. Find the pdf of X, and the mean and variance of X assuming that Y and Z are independent.

8. Let X and Y be independent rv's. If at least one of X and Y is of the continuous type, show that $X + Y$ is also continuous. What if X and Y are not independent?

9. Let X and Y be independent integral rv's. Show that

$$P(t) = P_X(t)P_Y(t),$$

where P, P_X, and P_Y, respectively, are the pgf's of $X + Y$, X, and Y.

10. Let X and Y be independent nonnegative rv's of the continuous type with pdf's f and g, respectively. Let $f(x) = e^{-x}$ if $x > 0$, and $= 0$ if $x \leq 0$, and let g be arbitrary. Show that the mgf $M(t)$ of Y, which is assumed to exist, has the property that the df of X/Y is $1 - M(-t)$.

11. Let X, Y, Z have the joint pdf

$$f(x, y, z) = \begin{cases} 6(1 + x + y + z)^{-4} & \text{if } 0 < x, 0 < y, 0 < z, \\ 0 & \text{otherwise.} \end{cases}$$

Find the pdf of $U = X + Y + Z$.

12. Prove Theorem 9. In view of Example 10 can you suggest an improvement in Theorem 9?

13. Prove Theorem 10.

14. Let X and Y be iid rv's with common pdf

$$f(x) = \begin{cases} (x\sqrt{2\pi})^{-1} e^{-(1/2)(\log x)^2}, & x > 0, \\ 0, & x \leq 0. \end{cases}$$

Find the pdf of $Z = XY$.

15. Let X and Y be iid rv's with common pdf f defined in Example 8. Find the joint pdf of U and V in the following cases:
(a) $U = \sqrt{X^2 + Y^2}$, $V = \tan^{-1}(X/Y)$, $-(\pi/2) < V \leq (\pi/2)$.
(b) $U = (X + Y)/2$, $V = (X - Y)^2/2$.

16. Prove Theorem 7.

17. Construct an example to show that even when the mgf of $X + Y$ can be written as a product of the mgf of X and the mgf of Y, X and Y need not be independent.

4.5 ORDER STATISTICS AND THEIR DISTRIBUTIONS

Let $(X_1, X_2, \cdots, X_n)$ be an n-dimensional random vector, and $(x_1, x_2, \cdots, x_n)$ be an n-tuple assumed by $(X_1, X_2, \cdots, X_n)$. Then let us arrange $x_1, x_2, \cdots, x_n$ in increasing order of magnitude so that

$$x_{(1)} \leq x_{(2)} \leq \cdots \leq x_{(n)},$$

where $x_{(1)} = \min(x_1, x_2, \cdots, x_n)$, $x_{(2)}$ is the second smallest value in $x_1, x_2, \cdots, x_n$, and so on, $x_{(n)} = \max(x_1, x_2, \cdots, x_n)$. If any two x_i, x_j, are equal, their order does not matter.

Definition. The function $X_{(k)}$ of $(X_1, X_2, \cdots, X_n)$ that takes on the value $x_{(k)}$ in each possible sequence $(x_1, x_2, \cdots, x_n)$ of values assumed by $(X_1, X_2, \cdots, X_n)$ is known as the kth-order statistic or the statistic of order k. $\{X_{(1)}, X_{(2)}, \cdots, X_{(n)}\}$ is called the set of order statistics for $(X_1, X_2, \cdots, X_n)$.

Example 1. Let X_1, X_2, X_3 be three rv's of the discrete type. Also, let X_1, X_3 take on values 0, 1, and X_2 take on values 1, 2, 3. Then the random vector (X_1, X_2, X_3) assumes these triplets of values: (0, 1, 0), (0, 2, 0), (0, 3, 0), (0, 1, 1), (0, 2, 1), (0, 3, 1), (1, 1, 0), (1, 2, 0), (1, 3, 0), (1, 1, 1), (1, 2, 1), (1, 3, 1); $X_{(1)}$ takes on values 0, 1; $X_{(2)}$ takes on values 0, 1; and $X_{(3)}$ takes on values 1, 2, 3.

In Section 4.4 we showed that $X_{(1)}$ and $X_{(n)}$ are themselves rv's.

Theorem 1. Let $(X_1, X_2, \cdots, X_n)$ be an n-dimensional rv. Let $X_{(k)}$, $1 \leq k \leq n$, be the order statistic of order k. Then $X_{(k)}$ is also an rv.

Proof. The proof of Theorem 1 is left as an exercise.

In the following we assume that $X_1, X_2, \cdots, X_n$ are iid rv's of the continuous type with pdf f. Let $\{X_{(1)}, X_{(2)}, \cdots, X_{(n)}\}$ be the set of order statistics for

$X_1, X_2, \cdots, X_n$. Since the X_i are all continuous type rv's, it follows with probability 1 that

$$X_{(1)} < X_{(2)} < \cdots < X_{(n)}.$$

We will compute the joint pdf of $(X_{(1)}, X_{(2)}, \cdots, X_{(n)})$.

Theorem 2. The joint pdf of $(X_{(1)}, X_{(2)}, \cdots, X_{(n)})$ is given by

(1) $$g(x_{(1)}, x_{(2)}, \cdots, x_{(n)}) = \begin{cases} n! \prod_{i=1}^{n} f(x_{(i)}), & x_{(1)} < x_{(2)} < \cdots < x_{(n)}, \\ 0, & \text{otherwise.} \end{cases}$$

Proof. The transformation from $(X_1, X_2, \cdots, X_n)$ to $(X_{(1)}, X_{(2)}, \cdots, X_{(n)})$ is not one-to-one. In fact, there is a total of $n!$ possible arrangements of $x_1, x_2, \cdots, x_n$ in increasing order of magnitude. Thus there are $n!$ inverses to the transformation. For example, one of the $n!$ permutations might be

$$x_4 < x_1 < x_{n-1} < x_3 < \cdots < x_n < x_2.$$

Then the corresponding inverse is

$$x_4 = x_{(1)}, \quad x_1 = x_{(2)}, \quad x_{n-1} = x_{(3)}, \quad x_3 = x_{(4)}, \cdots, x_n = x_{(n-1)},$$
$$x_2 = x_{(n)}.$$

The Jacobian of this transformation is the determinant of an $n \times n$ identity matrix with rows rearranged, since each $x_{(i)}$ equals one and only one of $x_1, x_2, \cdots, x_n$. Therefore $J = \pm 1$, and

$$g(x_{(2)}, x_{(n)}, x_{(4)}, x_{(1)}, \cdots, x_{(3)}, x_{(n-1)}) |J| = \prod_{i=1}^{n} f(x_{(i)}),$$
$$x_{(1)} < x_{(2)} < \cdots < x_{(n)}.$$

The same expression holds for each of the $n!$ arrangements.

It follows (see Remark 4. 4. 2) that

$$g(x_{(1)}, x_{(2)}, \cdots, x_{(n)}) = \sum_{\substack{\text{all } n! \\ \text{inverses}}} \prod_{i=1}^{n} f(x_{(i)})$$
$$= \begin{cases} n! f(x_{(1)}) f(x_{(2)}) \cdots f(x_{(n)}) & \text{if } x_{(1)} < x_{(2)} \cdots < x_{(n)}, \\ 0 & \text{otherwise.} \end{cases}$$

Example 2. Let $X_1, X_2, \cdots, X_n$ be iid rv's with common pdf

$$f(x) = \begin{cases} 1 & \text{if } 0 < x < 1, \\ 0 & \text{otherwise.} \end{cases}$$

Then the joint pdf of $X_{(1)}, X_{(2)}, \cdots, X_{(n)}$ is

$$g(y_1, y_2, \cdots, y_n) = \begin{cases} n!, & 0 < y_1 < y_2 < \cdots < y_n < 1, \\ 0, & \text{otherwise.} \end{cases}$$

Example 3. Let X_1, X_2, X_3, X_4 be iid rv's with pdf f. The joint pdf of $X_{(1)}$, $X_{(2)}, X_{(3)}, X_{(4)}$ is

$$g(y_1, y_2, y_3, y_4) = \begin{cases} 4!f(y_1)f(y_2)f(y_3)f(y_4), & y_1 < y_2 < y_3 < y_4, \\ 0, & \text{otherwise.} \end{cases}$$

Let us compute the marginal pdf of $X_{(2)}$. We have

$$g_2(y_2) = 4! \iiint f(y_1)f(y_2)f(y_3)f(y_4)\, dy_1\, dy_3\, dy_4$$

$$= 4!f(y_2) \int_{-\infty}^{y_2} \int_{y_2}^{\infty} \left[\int_{y_3}^{\infty} f(y_4)\, dy_4\right] f(y_3)f(y_1)\, dy_3\, dy_1$$

$$= 4!f(y_2) \int_{-\infty}^{y_2} \left\{\int_{y_2}^{\infty} [1 - F(y_3)]\, f(y_3)\, dy_3\right\} f(y_1)\, dy_1$$

$$= 4!f(y_2) \int_{-\infty}^{y_2} \frac{[1 - F(y_2)]^2}{2} f(y_1)\, dy_1$$

$$= 4!f(y_2) \frac{[1 - F(y_2)]^2}{2!} F(y_2), \quad y_1 < y_2 < y_3 < y_4.$$

The procedure for computing the marginal pdf of $X_{(r)}$, the rth-order statistic of $X_1, X_2, \cdots, X_n$, is similar. The following theorem summarizes the result.

Theorem 3. The marginal pdf of $X_{(r)}$ is given by

(2) $$g_r(y_r) = \frac{n!}{(r-1)!(n-r)!} [F(y_r)]^{r-1} [1 - F(y_r)]^{n-r} f(y_r),$$

where F is the common df of $X_1, X_2, \cdots, X_n$.

Proof.

$$g_r(y_r) = n!f(y_r) \int_{-\infty}^{y_r} \int_{-\infty}^{y_{r-1}} \cdots \int_{-\infty}^{y_2} \int_{y_r}^{\infty} \int_{y_{r+1}}^{\infty} \cdots \int_{y_{n-1}}^{\infty} \prod_{i \neq r}^{n} f(y_i)\, dy_n \cdots dy_{r+1}$$
$$\cdot dy_1 \cdots dy_{r-1}$$

$$= n!f(y_r) \frac{[1 - F(y_r)]^{n-r}}{(n-r)!} \int_{-\infty}^{y_r} \cdots \int_{-\infty}^{y_2} \prod_{i=1}^{r-1} [f(y_i)\, dy_i]$$

$$= n!f(y_r) \frac{[1 - F(y_r)]^{n-r}}{(n-r)!} \cdot \frac{[F(y_r)]^{r-1}}{(r-1)!},$$

as asserted.

We now compute the joint pdf of $X_{(j)}$ and $X_{(k)}$, $1 \leq j < k \leq n$.

Theorem 4. The joint pdf of $X_{(j)}$ and $X_{(k)}$ is given by

$$\text{(3)} \quad g_{jk}(y_j, y_k) = \begin{cases} \dfrac{n!}{(j-1)!(k-j-1)!(n-k)!} F^{j-1}(y_j)[F(y_k) - \\ F(y_j)]^{k-j-1}[1 - F(y_k)]^{n-k} f(y_j)f(y_k) & \text{if } y_j < y_k, \\ 0 & \text{otherwise.} \end{cases}$$

Proof.

$$g_{jk}(y_j, y_k) = \int_{-\infty}^{y_j} \cdots \int_{-\infty}^{y_2} \int_{y_j}^{y_k} \cdots \int_{y_{k-2}}^{y_k} \int_{y_k}^{\infty} \cdots \int_{y_{n-1}}^{\infty} n! f(y_1) \cdots f(y_n)$$
$$\cdot dy_n \cdots dy_{k+1}\, dy_{k-1} \cdots dy_{j+1}\, dy_1 \cdots dy_{j-1}$$
$$= n! \int_{-\infty}^{y_j} \cdots \int_{-\infty}^{y_2} \int_{y_j}^{y_k} \cdots \int_{y_{k-2}}^{y_k} \frac{[1 - F(y_k)]^{n-k}}{(n-k)!} f(y_1) f(y_2) \cdots f(y_k)$$
$$\cdot dy_{k-1} \cdots dy_{j+1}\, dy_1 \cdots dy_{j-1}$$
$$= n! \frac{[1 - F(y_k)]^{n-k}}{(n-k)!} f(y_k) \int_{-\infty}^{y_j} \cdots \int_{-\infty}^{y_2} \frac{[F(y_k) - F(y_j)]^{k-j-1}}{(k-j-1)!}$$
$$\cdot f(y_1) f(y_2) \cdots f(y_j)\, dy_1 \cdots dy_{j-1}$$
$$= \frac{n!}{(n-k)!(k-j-1)!} [1 - F(y_k)]^{n-k} [F(y_k) - F(y_j)]^{k-j-1}$$
$$\cdot f(y_k) f(y_j) \frac{[F(y_j)]^{j-1}}{(j-1)!}, \qquad y_j < y_k,$$

as asserted.

In a similar manner we can show that the joint pdf of $X_{(j_1)}, \cdots, X_{(j_k)}$, $1 \le j_1 < j_2 < \cdots < j_k \le n$, $1 \le k \le n$, is given by

$$g_{j_1, j_2, \ldots, j_k}(y_1, y_2, \ldots, y_k) = \frac{n!}{(j_1 - 1)!(j_2 - j_1 - 1)! \cdots (n - j_k)!}$$
$$\cdot F^{j_1 - 1}(y_1) f(y_1) [F(y_2) - F(y_1)]^{j_2 - j_1 - 1} f(y_2) \cdots [1 - F(y_k)]^{n - j_k} f(y_k)$$

for $y_1 < y_2 < \cdots < y_k$, and $= 0$ otherwise.

Example 4. Let $X_1, X_2, \cdots, X_n$ be iid rv's with common pdf

$$f(x) = \begin{cases} 1 & \text{if } 0 < x < 1, \\ 0 & \text{otherwise.} \end{cases}$$

Then

$$g_r(y_r) = \begin{cases} \dfrac{n!}{(n-1)!(n-r)!} y_r^{r-1} (1 - y_r)^{n-r}, & 0 < y_r < 1 \\ 0, & \text{otherwise,} \end{cases} \quad (1 \le r \le n).$$

The joint distribution of $X_{(j)}$ and $X_{(k)}$ is given by

$$g_{jk}(y_j, y_k) = \begin{cases} \dfrac{n!}{(j-1)!(k-j-1)!(n-k)!} y_j^{j-1}(y_k - y_j)^{k-j-1}(1-y_k)^{n-k}, \\ \qquad\qquad\qquad\qquad\qquad\qquad\qquad\qquad\qquad\qquad y_j < y_k, \\ 0, \qquad\qquad\qquad\qquad\qquad\qquad\qquad\qquad\qquad\quad \text{otherwise,} \end{cases}$$

where $1 \leq j < k \leq n$.

Example 5. Let $X_{(1)}$, $X_{(2)}$, $X_{(3)}$ be the order statistics of iid rv's X_1, X_2, X_3 with common pdf

$$f(x) = \begin{cases} \beta e^{-x\beta}, & x > 0, \\ 0, & \text{otherwise,} \end{cases} \quad (\beta > 0).$$

Let $Y_1 = X_{(3)} - X_{(2)}$ and $Y_2 = X_{(2)}$. We show that Y_1 and Y_2 are independent. The joint pdf of $X_{(2)}$ and $X_{(3)}$ is given by

$$g_{23}(x, y) = \begin{cases} \dfrac{3!}{1!0!0!} (1 - e^{-\beta x}) \beta e^{-\beta x} \beta e^{-\beta y}, & x < y, \\ 0, & \text{otherwise.} \end{cases}$$

The pdf of (Y_1, Y_2) is
$$f(y_1, y_2) = 3! \, \beta^2 (1 - e^{-\beta y_2}) e^{-\beta y_2} e^{-(y_1 + y_2)\beta}$$
$$= \begin{cases} \{3!\beta e^{-2\beta y_2}(1 - e^{-\beta y_2})\} \{\beta e^{-\beta y_1}\}, & 0 < y_1 < \infty, \, 0 < y_2 < \infty, \\ 0, & \text{otherwise.} \end{cases}$$

It follows that Y_1 and Y_2 are independent.

PROBLEMS 4.5

1. Let $X_{(1)}$, $X_{(2)}$, $\cdots$, $X_{(n)}$ be the set of order statistics of independent rv's X_1, X_2, $\cdots$, X_n with common pdf

$$f(x) = \begin{cases} \beta e^{-x\beta} & \text{if } x \geq 0, \\ 0 & \text{otherwise.} \end{cases}$$

(a) Show that $X_{(r)}$ and $X_{(s)} - X_{(r)}$ are independent for any $s > r$.
(b) Find the pdf of $X_{(r+1)} - X_{(r)}$.
(c) Let $Z_1 = nX_{(1)}$, $Z_2 = (n-1)(X_{(2)} - X_{(1)})$, $Z_3 = (n-2)(X_{(3)} - X_{(2)})$, $\cdots$, $Z_n = (X_{(n)} - X_{(n-1)})$. Show that $(Z_1, Z_2, \cdots, Z_n)$ and $(X_1, X_2, \cdots, X_n)$ are identically distributed.

2. Let X_1, X_2, $\cdots$, X_n be iid with pdf

$$f(x) = \begin{cases} \dfrac{1}{\sigma} e^{-[(x-\theta)/\sigma]} & \text{if } x > \theta, \\ 0 & \text{if } x < \theta. \end{cases}$$

Show that $X_{(1)}$, $X_{(2)} - X_{(1)}$, $X_{(3)} - X_{(2)}$, $\cdots$, $X_{(n)} - X_{(n-1)}$ are independent.

3. Let $X_1, X_2, \cdots, X_n$ be iid with a df
$$F(y) = \begin{cases} y^\alpha & \text{if } 0 < y < 1, \\ 0 & \text{otherwise,} \end{cases} \quad \alpha > 0.$$
Show that $X_{(i)}/X_{(n)}$, $i = 1, 2, \cdots, n-1$, and $X_{(n)}$ are independent.

4. Let X_1 and X_2 be independent rv's with common pmf
$$P\{X = x\} = p(1-p)^{x-1}, \; x = 1, 2, \cdots; \; 0 < p < 1.$$
Show that $X_{(1)}$ and $X_{(2)} - X_{(1)}$ are independent.

5. Let $X_1, X_2 \cdots, X_n$ be iid nonnegative rv's of the continuous type. If $E|X| < \infty$, show that $E|X_{(r)}| < \infty$. Write $M_n = X_{(n)} = \max(X_1, X_2, \cdots, X_n)$. Show that
$$EM_n = EM_{n-1} + \int_0^\infty F^{n-1}(x)[1 - F(x)]\, dx, \quad n = 2, 3, \cdots.$$
Find EM_n in each of the following cases:

(a) X_i have the common df
$$F(x) = 1 - e^{-\beta x}, \quad x \geq 0.$$

(b) X_i have the common df
$$F(x) = x, \; 0 < x < 1.$$

6. Let $X_{(1)}, X_{(2)}, \cdots, X_{(n)}$ be the order statistics of n independent rv's $X_1, X_2, \cdots, X_n$ with common pdf $f(x) = 1$ if $0 < x < 1$, and $= 0$ otherwise. Show that $Y_1 = X_{(1)}/X_{(2)}$, $Y_2 = X_{(2)}/X_{(3)}, \cdots, Y_{n-1} = X_{(n-1)}/X_{(n)}$, and $Y_n = X_{(n)}$ are independent. Find the pdf's of $Y_1, Y_2, \cdots, Y_n$.

4.6 MOMENTS AND MOMENT GENERATING FUNCTIONS

Let (X, Y) be a two-dimensional rv, and g be a Borel-measurable function on $\mathcal{R}_2$.

Definition 1. Let (X, Y) be a two-dimensional rv of the discrete type with pmf $p_{ij} = P\{X = x_i, Y = y_j\}$, and let $g: \mathcal{R}_2 \to \mathcal{R}$ be a Borel-measurable function. If $\sum_{i,j} p_{ij} |g(x_i, y_j)| < \infty$, the series
$$Eg(X, Y) = \sum_{i,j} p_{ij}\, g(x_i, y_j)$$
is called the expected value of $g(X, Y)$.

Definition 2. Let (X, Y) be an rv of the continuous type with joint density $f(x, y)$, and let $g: \mathcal{R}_2 \to \mathcal{R}$ be a Borel-measurable function. If
$$\int_{-\infty}^\infty \int_{-\infty}^\infty |g(x, y)| f(x, y)\, dx\, dy < \infty,$$
the integral

MOMENT GENERATING FUNCTIONS

$$Eg(X, Y) = \int_{-\infty}^{\infty} \int_{-\infty}^{\infty} g(x, y) f(x, y) \, dx \, dy$$

is called the expected value of $g(X, Y)$.

We shall be mainly interested in functions of the type $g(x, y) = x^j y^k$ where j and k are nonnegative integers.

Definition 3. If $E(X^j Y^k)$ exists for nonnegative integers j and k, we call it a moment of order $(j + k)$ of (X, Y) and write

(1) $$m_{jk} = E(X^j Y^k).$$

If moments of order 1 of (X, Y) exist, then clearly

$$m_{10} = EX, \quad m_{01} = EY.$$

Either or both of these moments may not exist. Similarly, there are three possible moments of order 2:

$$m_{20} = EX^2, \quad m_{11} = E(XY), \quad m_{02} = EY^2.$$

Definition 4. If $E\{(X - EX)^j (Y - EY)^k\}$ exists for nonnegative integers j and k, we call it a central moment of order $(j + k)$ and write

(2) $$\mu_{jk} = E\{(X - EX)^j (Y - EY)^k\}.$$

Clearly

$$\mu_{10} = \mu_{01} = 0,$$
$$\mu_{20} = \text{var}(X), \quad \mu_{02} = \text{var}(Y),$$

and

$$\mu_{11} = E\{(X - m_{10})(Y - m_{01})\}.$$

Definition 5. If $E\{(X - EX)(Y - EY)\}$ exists, we call it the covariance between X and Y and write

(3) $$\text{cov}(X, Y) = E\{(X - EX)(Y - EY)\}.$$

We have

$$\text{cov}(X, Y) = E\{(X - m_{10})(Y - m_{01})\}$$
$$= E(XY) - EX \, EY$$
(4) $$= m_{11} - m_{10} m_{01}.$$

Example 1. Let (X, Y) be jointly distributed with density function

$$f(x, y) = \begin{cases} x + y, & 0 < x < 1, 0 < y < 1, \\ 0, & \text{otherwise.} \end{cases}$$

Then

$$EX^l Y^m = \int_0^1 \int_0^1 x^l y^m (x + y)\, dx\, dy$$

$$= \int_0^1 \int_0^1 x^{l+1} y^m\, dx\, dy + \int_0^1 \int_0^1 x^l y^{m+1}\, dx\, dy$$

$$= \frac{1}{(l + 2)(m + 1)} + \frac{1}{(l + 1)(m + 2)},$$

where l and m are positive integers. Thus

$$EX = EY = \tfrac{7}{12},$$
$$EX^2 = EY^2 = \tfrac{5}{12},$$
$$\operatorname{var}(X) = \operatorname{var}(Y) = \tfrac{5}{12} - \tfrac{49}{144} = \tfrac{11}{144},$$
$$\operatorname{cov}(X, Y) = \tfrac{1}{3} - \tfrac{49}{144} = -\tfrac{1}{144}.$$

It is clear that Definitions 1 to 4 can be generalized to the n-dimensional case.

Theorem 1. Let $X_1, X_2, \cdots, X_n$ be rv's such that $E|X_i| < \infty$, $i = 1, 2, \cdots, n$. Let $a_1, a_2, \cdots, a_n$ be real numbers, and write

$$S = a_1 X_1 + a_2 X_2 + \cdots + a_n X_n.$$

Then ES exists, and we have

(5) $$ES = \sum_{j=1}^n a_j\, EX_j.$$

Proof. If $(X_1, X_2, \cdots, X_n)$ is of the discrete type, then

$$ES = \sum_{i_1, i_2, \cdots, i_n} (a_1 x_{i_1} + a_2 x_{i_2} + \cdots + a_n x_{i_n}) P\{X_1 = x_{i_1}, X_2 = x_{i_2}, \cdots, X_n = x_{i_n}\}$$
$$= a_1 \sum_{i_1} x_{i_1} \sum_{i_2, \cdots, i_n} P\{X_1 = x_{i_1}, \cdots, X_n = x_{i_n}\}$$
$$+ \cdots + a_n \sum_{i_n} x_{i_n} \sum_{i_1, \cdots, i_{n-1}} P\{X_1 = x_{i_1}, \cdots, X_n = x_{i_n}\}$$
$$= a_1 \sum_{i_1} x_{i_1} P\{X_1 = x_{i_1}\} + \cdots + a_n \sum_{i_n} x_{i_n} P\{X_n = x_{i_n}\}$$
$$= a_1 EX_1 + \cdots + a_n EX_n.$$

The existence of ES follows easily by replacing each a_j by $|a_j|$ and each x_{ij} by $|x_{ij}|$ and remembering that $E|X_j| < \infty$, $j = 1, 2, \cdots n$. The case of continuous type $(X_1, X_2, \cdots, X_n)$ is similarly treated.

Corollary. Take $a_1 = a_2 = \cdots = a_n = 1/n$. Then

$$E\left(\frac{X_1 + X_2 + \cdots + X_n}{n}\right) = \frac{1}{n} \sum_{i=1}^{n} EX_i,$$

and if $EX_1 = EX_2 = \cdots = EX_n = \mu$, then

$$E\left(\frac{X_1 + X_2 + \cdots X_n}{n}\right) = \mu.$$

Theorem 2. Let $X_1, X_2, \cdots, X_n$ be independent rv's such that $E|X_i| < \infty$, $i = 1, 2, \cdots, n$. Then $E(\prod_{i=1}^{n} X_i)$ exists and

(6) $$E(\prod_{i=1}^{n} X_i) = \prod_{i=1}^{n} EX_i.$$

Proof. The proof is left as an exercise.

Corollary. If X and Y are independent, cov $(X, Y) = 0$.

We recall that

$$\text{cov}(X, Y) = E(XY) - EX\, EY,$$

so that, if X and Y are independent, $EXY = EX\, EY$ and the result follows.

We emphasize the fact that, if X and Y have zero covariance, it does not follow that they are independent.

Example 2. Let X be a symmetric rv with finite third moment, and let $Y = X^2$. Then $E(XY) = EX^3 = 0$, and it follows that X and Y have zero covariance. But X and Y are not independent. In fact, they are strongly dependent.

Let X and Y be independent, and $g_1(\cdot)$ and $g_2(\cdot)$ be Borel-measurable functions. Then we know (Theorem 4.3.3) that $g_1(X)$ and $g_2(Y)$ are independent. If $E\{g_1(X)\}$, $E\{g_2(Y)\}$, and $E\{g_1(X) g_2(Y)\}$ exist, it follows from Theorem 2 that

(7) $$E\{g_1(X) g_2(Y)\} = E\{g_1(X)\}\, E\{g_2(Y)\}.$$

Conversely, if for any Borel sets A_1 and A_2 we take $g_1(X) = 1$ if $X \in A_1$, and $= 0$ otherwise, and $g_2(Y) = 1$ if $Y \in A_2$, and $= 0$ otherwise, then

$$E\{g_1(X) g_2(Y)\} = P\{X \in A_1, Y \in A_2\}$$

and $E\{g_1(X)\} = P\{X \in A_1\}$, $E\{g_2(Y)\} = P\{Y \in A_2\}$. Relation (7) implies that for any Borel sets A_1 and A_2 of real numbers

$$P\{X \in A_1, Y \in A_2\} = P\{X \in A_1\}\, P\{Y \in A_2\}.$$

It follows that X and Y are independent if (7) holds. We have thus proved the following theorem.

Theorem 3. Two rv's X and Y are independent if and only if for every pair of Borel-measurable functions g_1 and g_2 the relation

(8) $$E\{g_1(X)\, g_2(Y)\} = E\{g_1(X)\}\, E\{g_2(Y)\}$$

holds, provided that the expectations on both sides of (8) exist.

Theorem 4. Let X and Y be two rv's with finite variances. Then $\text{cov}(X, Y)$ exists. Moreover,

(9) $$E^2(XY) \leq EX^2\, EY^2$$

with equality if and only if there exists real numbers and β not both zero such that $P\{\alpha X + \beta Y = 0\}$. [Relation (9) is known as the *Cauchy-Schwarz inequality*]

Proof. Since

$$|ab| \leq \frac{a^2 + b^2}{2}$$

for every pair of real numbers a and b, it follows that $E|XY|$ exists if $EX^2 < \infty$ and $EY^2 < \infty$. To prove (9) we have, for any real number α,

$$E(\alpha X + Y)^2 = \alpha^2 EX^2 + 2\alpha E(XY) + EY^2 \geq 0.$$

If $EX^2 = 0$, (9) is trivially true. Assume that $EX^2 > 0$, and choose $\alpha = -[E(XY)/EX^2]$. We have

$$\frac{E^2(XY)}{EX^2} - 2\frac{E^2(XY)}{EX^2} + EY^2 \geq 0,$$

which yields (9). Strict equality holds if and only if there exists real numbers α and β not both zero such that $E(\alpha X + Y\beta) = 0$, which happens if and only if $P\{\alpha X + \beta Y = 0\} = 1$.

Theorem 5. Let $X_1, X_2, \cdots, X_n$ be rv's with $E|X_i|^2 < \infty$ for $i = 1, 2, \cdots, n$. Let $a_1, a_2, \cdots, a_n$ be real numbers and write

$$S = \sum_{i=1}^{n} a_i X_i.$$

Then the variance of S exists and is given by

(10) $$\text{var}(S) = \sum_{i=1}^{n} a_i^2 \text{var}(X_i) + \sum_{\substack{i=1 \\ i \neq j}}^{n} \sum_{j=1}^{n} a_i a_j \text{cov}(X_i, X_j).$$

If, in particular, $X_1, X_2, \cdots, X_n$ are such that $\text{cov}(X_i, X_j) = 0$ for $i, j = 1, 2, \cdots, n, i \neq j$, then

(11) $$\text{var}(S) = \sum_{i=1}^{n} a_i^2 \text{var}(X_i).$$

Proof. We have

$$\text{var}(S) = E\{\sum_{i=1}^{n} a_i X_i - \sum_{i=1}^{n} a_i EX_i\}^2 \quad \text{by Theorem 1}$$

$$= E\{\sum_{i=1}^{n} a_i^2 (X_i - EX_i)^2 + \sum\sum_{i \neq j} a_i a_j (X_i - EX_i)(X_j - EX_j)\}$$

$$= \sum_{i=1}^{n} a_i^2 E(X_i - EX_i)^2 + \sum\sum_{i \neq j} a_i a_j E\{(X_i - EX_i)(X_j - EX_j)\}.$$

If the X_i's satisfy

$$\text{cov}(X_i, X_j) = 0 \quad \text{for } i, j = 1, 2, \cdots, n; i \neq j,$$

the second term on the right side of (10) vanishes, and we have (11).

Corollary. Let $X_1, X_2, \cdots, X_n$ be iid rv's with $\text{var}(X_i) = \sigma^2$, $i = 1, 2, \cdots, n$. Then

$$\text{var}(\sum_{i=1}^{n} a_i X_i) = \sigma^2 \sum_{i=1}^{n} a_i^2.$$

In particular,

$$\text{var}\left(\frac{X_1 + X_2 + \cdots + X_n}{n}\right) = \frac{\sigma^2}{n}.$$

Note that we only need $\text{cov}(X_j, X_k) = 0$, $j, k = 1, 2, \cdots, n, j \neq k$, rather than independence of the X_i's.

Theorem 6. Let $X_1, X_2, \cdots, X_n$ be iid rv's with common variance σ^2. Also, let $a_1, a_2, \cdots, a_n$ be real numbers such that $\sum_{1}^{n} a_i = 1$, and let $S = \sum_{i=1}^{n} a_i X_i$. Then the variance of S is least if we choose $a_i = 1/n$, $i = 1, 2, \cdots, n$.

Proof. We have

$$\text{var}(S) = \sigma^2 \sum_{i=1}^{n} a_i^2,$$

which is least if and only if we choose the a_i's so that $\sum_{i=1}^{n} a_i^2$ is smallest, subject to the condition $\sum_{i=1}^{n} a_i = 1$. We have

$$\sum_{i=1}^{n} a_i^2 = \sum_{i=1}^{n} \left(a_i - \frac{1}{n} + \frac{1}{n}\right)^2$$

$$= \sum_{i=1}^{n} \left(a_i - \frac{1}{n}\right)^2 + \frac{2}{n}\sum_{i=1}^{n}\left(a_i - \frac{1}{n}\right) + \frac{1}{n}$$

$$= \sum_{i=1}^{n} \left(a_i - \frac{1}{n}\right)^2 + \frac{1}{n}$$

which is minimized for the choice $a_i = 1/n$, $i = 1, 2, \cdots, n$.

Note that the result holds if we replace independence by the condition that $\text{cov}(X_i, X_j) = 0$, $i, j = 1, 2, \cdots, n$ $(i \neq j)$.

Example 3. Suppose that r balls are drawn one at a time without replacement from a bag containing n white and m black balls. Let S_r be the number of black balls drawn.

Let us define rv's X_k as follows:

$$X_k = 1 \quad \text{if the } k\text{th ball drawn is black}$$
$$= 0 \quad \text{if the } k\text{th ball drawn is white} \quad k = 1, 2, \cdots, r.$$

Then

$$S_r = X_1 + X_2 + \cdots + X_r.$$

Also

(12) $$P\{X_k = 1\} = \frac{m}{m+n}, \quad P\{X_k = 0\} = \frac{n}{m+n}.$$

Thus $EX_k = m/(m+n)$, and

$$\text{var}(X_k) = \frac{m}{m+n} - \frac{m^2}{(m+n)^2} = \frac{mn}{(m+n)^2}.$$

To compute $\text{cov}(X_j, X_k)$, $j \neq k$, note that the rv $X_j X_k = 1$ if the jth and the kth balls drawn are black, and $= 0$ otherwise. Thus

(13) $$E(X_j X_k) = P\{X_j = 1, X_k = 1\} = \frac{m}{m+n} \frac{m-1}{m+n-1}$$

and

$$\text{cov}(X_j, X_k) = -\frac{mn}{(m+n)^2(m+n-1)}.$$

Thus

$$ES_r = \sum_{k=1}^{r} EX_k = \frac{mr}{m+n},$$

and
$$\operatorname{var}(S_r) = r\frac{mn}{(m+n)^2} - r(r-1)\frac{mn}{(m+n)^2(m+n-1)}$$
$$= \frac{mnr}{(m+n)^2(m+n+1)}(m+n-r).$$

The reader is asked to satisfy himself that (12) and (13) hold.

Example 4. Let $X_1, X_2, \cdots, X_n$ be independent, and $a_1, a_2, \cdots, a_n$ be real numbers such that $\sum a_i = 1$. Assume that $E|X_i^2| < \infty$, $i = 1, 2, \cdots, n$, and let var $(X_i) = \sigma_i^2$, $i = 1, 2, \cdots, n$. Write $S = \sum_{i=1}^n a_i X_i$. Then var $(S) = \sum_{i=1}^n a_i^2 \sigma_i^2 = \sigma$, say. To find weights a_i such that σ is minimum, we write

$$\sigma = a_1^2 \sigma_1^2 + a_2^2 \sigma_2^2 + \cdots + (1 - a_1 - a_2 - \cdots - a_{n-1})^2 \sigma_n^2,$$

and differentiate partially with respect to $a_1, a_2, \cdots, a_{n-1}$, respectively. We get

$$\frac{\partial \sigma}{\partial a_1} = 2a_1 \sigma_1^2 - 2(1 - a_1 - a_2 - \cdots - a_{n-1})\sigma_n^2 = 0,$$
$$\vdots$$
$$\frac{\partial \sigma}{\partial a_{n-1}} = 2a_{n-1} \sigma_{n-1}^2 - 2(1 - a_1 - a_2 - \cdots - a_{n-1})\sigma_n^2 = 0.$$

It follows that

$$a_j \sigma_j^2 = a_n \sigma_n^2, \quad j = 1, 2, \cdots, n - 1,$$

that is, the weights a_j, $j = 1, 2, \cdots, n$, should be chosen proportional to $1/\sigma_j^2$. The minimum value of σ is then

$$\sigma_{\min} = \sum_{i=1}^n \frac{k^2}{\sigma_i^4} \sigma_i^2 = k^2 \sum_{i=1}^n \frac{1}{\sigma_i^2},$$

where k is given by $\sum_{j=1}^n (k/\sigma_j^2) = 1$. Thus

$$\sigma_{\min} = \frac{1}{\sum_{i=1}^n (1/\sigma_j^2)} = \frac{H}{n},$$

where H is the harmonic mean of the variances, σ_i^2.

We now define the moment generating function of a random vector. For notational convenience we restrict ourselves to the bivariate case.

Definition 6. Let (X, Y) be a two-dimensional rv. If $E(e^{t_1 X + t_2 Y})$ exists for $|t_1| \leq h_1$, $|t_2| \leq h_2$, where h_1 and h_2 are positive real numbers, we write

(14) $$M(t_1, t_2) = E(e^{t_1 X + t_2 Y})$$

and call it the moment generating function of the joint distribution of (X, Y) or, simply, the mgf of (X, Y).

Theorem 7. The mgf $M(t_1, t_2)$ uniquely determines the joint distribution of (X, Y), and conversely, if the mgf exists it is unique.

Corollary. The mgf $M(t_1, t_2)$ completely determines the marginal distributions of X and Y. Indeed,

(15) $$M(t_1, 0) = E(e^{t_1 X}) = M_X(t_1),$$
(16) $$M(0, t_2) = E(e^{t_2 Y}) = M_Y(t_2).$$

Proof. For the proofs we refer the reader to Widder [137], page 460 (see also P. 2.14).

Theorem 8. If $M(t_1, t_2)$ exists, the moments of all orders of (X, Y) exist and may be obtained from

(17) $$\left. \frac{\partial^{m+n} M(t_1, t_2)}{\partial t_1^m \partial t_2^n} \right|_{t_1 = t_2 = 0} = E(X^m Y^n).$$

Thus

$$\frac{\partial M(0, 0)}{\partial t_1} = EX, \qquad \frac{\partial M(0, 0)}{\partial t_2} = EY,$$

$$\frac{\partial^2 M(0, 0)}{\partial t_1^2} = EX^2, \qquad \frac{\partial^2 M(0, 0)}{\partial t_2^2} = EY^2,$$

$$\frac{\partial^2 M(0, 0)}{\partial t_1 \partial t_2} = E(XY),$$

and so on.

Proof. See Widder [137], pages 446–447, for the proof (see also P. 2.14).

Theorem 9. X and Y are independent rv's if and only if

(18) $$M(t_1, t_2) = M(t_1, 0) M(0, t_2) \qquad \text{for all } t_1, t_2 \in \mathscr{R}.$$

Proof. Let X and Y be independent. Then

$$M(t_1, t_2) = E\{e^{t_1 X + t_2 Y}\} = E(e^{t_1 X}) E(e^{t_2 Y}) = M(t_1, 0) M(0, t_2).$$

Conversely, if

$$M(t_1, t_2) = M(t_1, 0) M(0, t_2),$$

then, in the continuous case,

$$\iint e^{t_1 x + t_2 y} f(x, y) \, dx \, dy = \left[\int e^{t_1 x} f_1(x) \, dx\right]\left[\int e^{t_2 y} f_2(y) \, dy\right],$$

that is,

$$\iint e^{t_1 x + t_2 y} f(x, y) \, dx \, dy = \iint e^{t_1 x + t_2 y} f_1(x) f_2(y) \, dx \, dy.$$

By the uniqueness of the mgf (Theorem 7) we must have

$$f(x, y) = f_1(x) f_2(y) \quad \text{for all } (x, y) \in \mathscr{R}_2.$$

It follows that X and Y are independent. A similar proof is given in the case where (X, Y) is of the discrete type.

Example 5. Let (X, Y) be jointly distributed with density function

$$f(x, y) = \begin{cases} e^{-x-y}, & 0 < x < \infty, 0 < y < \infty, \\ 0, & \text{otherwise.} \end{cases}$$

Then

$$M(t_1, t_2) = \int_0^\infty \int_0^\infty e^{t_1 x + t_2 y} e^{-x-y} \, dx \, dy$$

$$= \frac{1}{1 - t_1} \frac{1}{1 - t_2}, \quad t_1 < 1, t_2 < 1,$$

$$EX = \frac{\partial M(0, 0)}{\partial t_1} = 1, \quad EY = \frac{\partial M(0, 0)}{\partial t_2} = 1,$$

and

$$\frac{\partial^2 M(t_1, t_2)}{\partial t_1^2} = \frac{2}{(1 - t_1)^3}, \quad \text{so that } EX^2 = 2.$$

Similarly $EY^2 = 2$, so that $\text{var}(X) = \text{var}(Y) = 1$, and

$$EXY = \frac{\partial^2 M(t_1, t_2)}{\partial t_1 \, \partial t_2}\bigg|_{t_1 = 0, t_2 = 0} = \frac{1}{(1 - t_1)^2} \frac{1}{(1 - t_2)^2}\bigg|_{t_1 = 0, t_2 = 0} = 1,$$

so that $\text{cov}(X, Y) = 1 - 1 = 0$. Indeed, X and Y are independent since

$$f(x, y) = e^{-x} e^{-y} = f_1(x) f_2(y) \quad \text{for all } x, y \in \mathscr{R},$$

where f_1, f_2 are the two marginal densities.

Example 6. For the bivariate density of Example 1 we have

$$M(t_1, t_2) = \int_0^1 \int_0^1 (x+y)\, e^{t_1 x}\, e^{t_2 y}\, dx\, dy$$
$$= \frac{1}{t_1^2 t_2}[(t_1-1)e^{t_1}+1](e^{t_2}-1)$$
$$+ \frac{1}{t_1 t_2^2}[(t_2-1)e^{t_2}+1](e^{t_1}-1).$$

Using (17), we can again check that
$$EX = EY = \tfrac{7}{12}, \quad \operatorname{var}(X) = \operatorname{var}(Y) = \tfrac{11}{144},$$
$$\text{and } \operatorname{cov}(X,Y) = -\tfrac{1}{144}.$$

We conclude this section with some important moment inequalities. We begin with the simple inequality

(19) $$|a+b|^r \le c_r(|a|^r + |b|^r),$$

where $c_r = 1$ for $0 \le r \le 1$, and $= 2^{r-1}$ for $r > 1$. For $r = 0$ and $r = 1$, (19) is trivially true.

First note that it is sufficient to prove (19) when $0 < a \le b$. Let $0 < a \le b$, and write $x = a/b$. Then
$$\frac{(a+b)^r}{a^r + b^r} = \frac{(1+x)^r}{1+x^r}.$$

Writing $f(x) = (1+x)^r/(1+x^r)$, we see that
$$f'(x) = \frac{r(1+x)^{r-1}}{(1+x^r)^2}(1 - x^{r-1}),$$

where $0 < x \le 1$. It follows that $f'(x) > 0$ if $r > 1$, $= 0$ if $r = 1$, and < 0 if $r < 1$. Thus
$$\max_{0 \le x \le 1} f(x) = f(0) = 1 \quad \text{if } r \le 1,$$

while
$$\max_{0 \le x \le 1} f(x) = f(1) = 2^{r-1} \quad \text{if } r \ge 1.$$

Note that $|a+b|^r \le 2^r(|a|^r + |b|^r)$ is trivially true since
$$|a+b| \le \max(2|a|, 2|b|).$$

An immediate application of (19) is the following result.

Theorem 10. Let X and Y be rv's and $r > 0$ be a fixed number. If $E|X|^r$, $E|Y|^r$ are both finite, so also is $E|X+Y|^r$.

Proof. Let $a = X$ and $b = Y$ in (19). Taking the expectation on both sides, we see that

$$E|X + Y|^r \le c_r(E|X|^r + E|Y|^r),$$
where $c_r = 1$ if $0 < r \le 1$, and $= 2^{r-1}$ if $r > 1$.

Next we establish Hölder's inequality,

(20) $$|xy| \le \frac{|x|^p}{p} + \frac{|y|^q}{q},$$

where p and q are positive real numbers such that $p > 1$ and $1/p + 1/q = 1$. Note that for $x > 0$ the function $w = \log x$ is concave. It follows that for $x_1, x_2 > 0$

$$\log[tx_1 + (1-t)x_2] \ge t \log x_1 + (1-t) \log x_2.$$

Taking antilogarithms, we get

$$x_1^t x_2^{1-t} \le tx_1 + (1-t)x_2.$$

Now we choose $x_1 = |x|^p$, $x_2 = |y|^q$, $t = 1/p$, $1 - t = 1/q$, where $p > 1$ and $1/p + 1/q = 1$, to get (20).

Theorem 11. Let $p > 1$, $q > 1$ so that $1/p + 1/q = 1$. Then
(21) $$E|XY| \le (E|X|^p)^{1/p} (E|Y|^q)^{1/q}.$$

Proof. By Hölder's inequality, letting $x = X\{E|X|^p\}^{-1/p}$, $y = Y\{E|Y|^q\}^{-1/q}$, we get

$$|XY| \le p^{-1}|X|^p \{E|X|^p\}^{1/p-1} \{E|Y|^q\}^{1/q}$$
$$+ q^{-1}|Y|^q \{E|Y|^q\}^{1/q-1} \{E|X|^p\}^{1/p}.$$

Taking the expectation on both sides leads to (21).

Corollary. Taking $p = q = \frac{1}{2}$, we obtain the Cauchy-Schwarz inquality,

$$E|XY| \le E^{1/2}|X|^2 E^{1/2}|Y|^2.$$

The final result of this section is an inequality due to Minkowski.

Theorem 12. For $p \ge 1$
(22) $$\{E|X + Y|^p\}^{1/p} \le \{E|X|^p\}^{1/p} + \{E|Y|^p\}^{1/p}.$$

Proof. We have, for $p > 1$,

$$|X + Y|^p \le |X| |X + Y|^{p-1} + |Y| |X + Y|^{p-1}.$$

Taking expectations and using Hölder's inequality with Y replaced by $|X + Y|^{p-1}$ ($p > 1$), we have

$$E|X + Y|^p \le \{E|X|^p\}^{1/p} \{E|X + Y|^{(p-1)q}\}^{1/q}$$
$$+ \{E|Y|^p\}^{1/p} \{E|X + Y|^{(p-1)q}\}^{1/q}$$
$$= [\{E|X|^p\}^{1/p} + \{E|Y|^p\}^{1/p}] \cdot \{E|X + Y|^{(p-1)q}\}^{1/q}.$$

Excluding the trivial case in which $E|X + Y|^p = 0$, and noting that $(p - 1)q = p$, we have, after dividing both sides of the last inequality by $\{E|X + Y|^p\}^{1/q}$,

$$\{E|X + Y|^p\}^{1/p} \le \{E|X|^p\}^{1/p} + \{E|Y|^p\}^{1/p}, \qquad p > 1.$$

The case $p = 1$ being trivial, this establishes (22).

PROBLEMS 4.6

1. Suppose that the rv (X, Y) is uniformly distributed over the region $R = \{(x, y): 0 < x < y < 1\}$. Find the mgf of (X, Y). Hence or otherwise find the covariance between X and Y.

2. Let (X, Y) have the joint pdf given by

$$f(x, y) = \begin{cases} x^2 + \dfrac{xy}{3} & \text{if } 0 < x < 1, \ 0 < y < 2, \\ 0 & \text{otherwise.} \end{cases}$$

Find the mgf of (X, Y), and moments of order 2.

3. Let (X, Y) be distributed with joint density

$$f(x, y) = \begin{cases} \frac{1}{4}[1 + xy(x^2 - y^2)] & \text{if } |x| \le 1, \ |y| \le 1, \\ 0 & \text{otherwise.} \end{cases}$$

Find the mgf of (X, Y). Are X, Y independent? If not, find the covariance between X and Y.

4. For a positive rv X with finite first moment show that (1) $E\sqrt{X} \le \sqrt{EX}$, and (2) $E\{1/X\} \ge 1/EX$.

5. If X is a nondegenerate rv with finite expectation and such that $X \ge a > 0$, then
$$E\{\sqrt{X^2 - a^2}\} < \sqrt{(EX)^2 - a^2}.$$
(Kruskal [63])

6. Show that for $x > 0$
$$\left(\int_x^\infty te^{-t^2/2} \, dt\right)^2 \le \int_x^\infty e^{-t^2/2} \, dt \int_x^\infty t^2 e^{-t^2/2} \, dt,$$

and hence that
$$\int_x^\infty e^{-t^2/2} \, dt \ge \tfrac{1}{2}[(4 + x^2)^{1/2} - x] e^{-x^2/2}.$$

7. Given a pdf f that is nondecreasing in the interval $a \le x \le b$, show that for any $s > 0$

$$\int_a^b x^{2s} f(x)\, dx \geq \frac{b^{2s+1} - a^{2s+1}}{(2s+1)(b-a)} \int_a^b f(x)\, dx,$$

with the inequality reversed if f is nonincreasing.

8. Derive the Lyapunov inequality (Theorem 3.4.3)

$$\{E|X|^r\}^{1/r} \leq \{E|X|^s\}^{1/s}, \quad 1 < r < s < \infty,$$

from Hölder's inequality (21).

9. Let X be an rv with $E|X|^r < \infty$ for $r > 0$. Show that the function $\log E|X|^r$ is a convex function of r.

10. Show with the help of an example that Theorem 12 is not true for $p < 1$.

11. Show that the converse of Theorem 10 also holds for independent rv's that is, if $E|X + Y|^r < \infty$ for some $r > 0$ and X and Y are independent, then $E|X|^r < \infty$, $E|Y|^r < \infty$.

(*Hint*: Without loss of generality assume that the median of X is 0. Show that, for any $t > 0$, $P\{|X + Y| > t\} > \frac{1}{2} P\{|X| > t\}$. Now use the remarks preceding Lemma 3.2.2 to conclude that $E|X|^r < \infty$.)

12. Let $(\Omega, \mathscr{S}, P)$ be a probability space, and $A_1, A_2, \cdots, A_n$ be events in $\mathscr{S}$ such that $P(\bigcup_{k=1}^n A_k) > 0$. Show that

$$2 \sum\sum_{1 \leq j < k \leq n} P(A_j A_k) \geq \frac{(\sum_{k=1}^n PA_k)^2 - \sum_{k=1}^n PA_k}{P(\bigcup_{k=1}^n A_k)}.$$

(Chung and Erdös [14])

(*Hint*: Let X_k be the indicator function of A_k, $k = 1, 2, \cdots, n$. Use the Cauchy-Schwarz inequality.)

13. Prove Theorem 2.

4.7 CONDITIONAL EXPECTATION

In Section 2 we defined the conditional distribution of an rv X, given Y. We showed that, if (X, Y) is of the discrete type, the conditional pmf of X, given $Y = y_j$, where $P\{Y = y_j\} > 0$, is a pmf when considered as a function of the x_i's (for fixed y_j). Similarly, if (X, Y) is an rv of the continuous type with pdf $f(x, y)$ and marginal densities f_1 and f_2, respectively, then, at every point (x, y) at which f is continuous and at which $f_2(y) > 0$ and is continuous, a conditional density function of X, given Y, exists and may be defined by

$$f_{X|Y}(x|y) = \frac{f(x, y)}{f_2(y)}.$$

We also showed that $f_{X|Y}(x|y)$, for fixed y, when considered as a function of x is a pdf in its own right. Therefore we can (and do) consider the moments of this conditional distribution.

Definition 1. Let X and Y be rv's defined on a probability space $(\Omega, \mathcal{S}, P)$, and let h be a Borel-measurable function. Assume that $Eh(X)$ exists. Then the conditional expectation of $h(X)$, given Y, written as $E\{h(X)|Y\}$, is an rv that takes the value $E\{h(X)|y\}$, defined by

$$(1) \quad E\{h(X)|y\} = \begin{cases} \sum_x h(x) P\{X = x | Y = y\} & \text{if } (X, Y) \text{ is of the discrete type and } P\{Y = y\} > 0, \\ \int_{-\infty}^{\infty} h(x) f_{X|Y}(x|y) \, dx & \text{if } (X, Y) \text{ is of the continuous type and } f_2(y) > 0, \end{cases}$$

when the rv Y assumes the value y.

Needless to say, a similar definition may be given for the conditional expectation $E\{h(Y)|X\}$, provided that $Eh(Y)$ exists.

It is immediate that the rv $E\{h(X)|Y\}$ satisfies the usual properties of an expectation. For example, one can show easily that

(2) $E\{c|Y\} = c$, where c is a constant;
(3) $E\{aX + b|Y\} = a E\{X|Y\} + b$, a, b constants;
(4) if g_1, g_2 are Borel functions and $Eg_i(X)$ exists for $i = 1, 2$, then

$$E\{[a_1 g_1(X) + a_2 g_2(X)]|Y\} = a_1 E\{g_1(X)|Y\} + a_2 E\{g_2(X)|Y\};$$

(5) if $X \geq 0$, then $E\{X|Y\} \geq 0$;
(6) if $X_1 \geq X_2$, then $E\{X_1|Y\} \geq E\{X_2|Y\}$.

The moments of a conditional distribution are defined in the usual manner. Thus, if EX^r exists for some integer $r \geq 0$, then $E\{X^r|Y\}$ defines the rth moment of the conditional distribution. We can define the central moments of the conditional distribution and, in particular, the variance. There is no difficulty in generalizing these concepts for n-dimensional distributions when $n > 2$. We leave the reader to furnish the details.

Example 1. An urn contains three red and two green balls. A random sample of two balls is drawn (a) with replacement, and (b) without replacement. Let $X = 0$ if the first ball drawn is green, $= 1$ if the first ball drawn is red, and let $Y = 0$ if the second ball drawn is green, $= 1$ if the second ball drawn is red.

The joint pmf of (X, Y) is given in the following tables:

CONDITIONAL EXPECTATION   169

(a) With Replacement				(b) Without Replacement			
Y \ X	0	1		Y \ X	0	1	
0	$\frac{4}{25}$	$\frac{6}{25}$	$\frac{2}{5}$	0	$\frac{2}{20}$	$\frac{6}{20}$	$\frac{2}{5}$
1	$\frac{6}{25}$	$\frac{9}{25}$	$\frac{3}{5}$	1	$\frac{6}{20}$	$\frac{6}{20}$	$\frac{3}{5}$
	$\frac{2}{5}$	$\frac{3}{5}$	1		$\frac{2}{5}$	$\frac{3}{5}$	1

The conditional pmf's and the conditional expectations are as follows:

(a) $P\{X = x|0\} = \begin{cases} \frac{2}{5}, & x = 0, \\ \frac{3}{5}, & x = 1, \end{cases}$ $P\{Y = y|0\} = \begin{cases} \frac{2}{5}, & y = 0, \\ \frac{3}{5}, & y = 1, \end{cases}$

$P\{X = x|1\} = \begin{cases} \frac{2}{5}, & x = 0, \\ \frac{3}{5}, & x = 1, \end{cases}$ $P\{Y = y|1\} = \begin{cases} \frac{2}{5}, & y = 0, \\ \frac{3}{5}, & y = 1, \end{cases}$

$E\{X|Y\} = \begin{cases} \frac{3}{5}, & y = 0, \\ \frac{3}{5}, & y = 1, \end{cases}$ $E\{Y|X\} = \begin{cases} \frac{3}{5}, & x = 0, \\ \frac{3}{5}, & x = 1; \end{cases}$

(b) $P\{X = x|0\} = \begin{cases} \frac{1}{4}, & x = 0, \\ \frac{3}{4}, & x = 1, \end{cases}$ $P\{Y = y|0\} = \begin{cases} \frac{1}{4}, & y = 0, \\ \frac{3}{4}, & y = 1, \end{cases}$

$P\{X = x|1\} = \begin{cases} \frac{1}{2}, & x = 0, \\ \frac{1}{2}, & x = 1, \end{cases}$ $P\{Y = y|1\} = \begin{cases} \frac{1}{2}, & y = 0, \\ \frac{1}{2}, & y = 1, \end{cases}$

$E\{X|Y\} = \begin{cases} \frac{3}{4}, & y = 0, \\ \frac{1}{2}, & y = 1, \end{cases}$ $E\{Y|X\} = \begin{cases} \frac{3}{4}, & x = 0, \\ \frac{1}{2}, & x = 1. \end{cases}$

Example 2. For the rv (X, Y) considered in Examples 4.2.5 and 4.2.7

$$E\{Y|x\} = \int_x^1 y f_{Y|X}(y|x) \, dy = \frac{1}{2} \frac{1-x^2}{1-x} = \frac{1+x}{2} \qquad 0 < x < 1,$$

and

$$E\{X|y\} = \int_0^y x f_{X|Y}(x|y) \, dx = \frac{y}{2}, \qquad 0 < y < 1.$$

Also,

$$E\{X^2|y\} = \int_0^y x^2 \frac{1}{y} \, dx = \frac{y^2}{3}, \qquad 0 < y < 1,$$

and

$$\begin{aligned} \text{var}\{X|y\} &= E\{X^2|y\} - [E\{X|y\}]^2 \\ &= \frac{y^2}{3} - \frac{y^2}{4} = \frac{y^2}{12}, \qquad 0 < y < 1. \end{aligned}$$

Theorem 1. Let $Eh(X)$ exist. Then

(7) $$Eh(X) = E\{E\{h(X)|Y\}\}.$$

Proof. Let (X, Y) be of the discrete type. Then

$$\begin{aligned}
E\{E\{h(X)|Y\}\} &= \sum_y \{\sum_x h(x) P\{X = x | Y = y\}\} P\{Y = y\} \\
&= \sum_y \{\sum_x h(x) P\{X = x, Y = y\}\} \\
&= \sum_x h(x) \sum_y P\{X = x, Y = y\} \\
&= Eh(X).
\end{aligned}$$

If (X, Y) is of the continuous type, then

$$\begin{aligned}
E\{E\{h(X)|Y\}\} &= \int_{-\infty}^{\infty} \left\{\int_{-\infty}^{\infty} h(x) f_{X|Y}(x|y) \, dx\right\} f_Y(y) \, dy \\
&= \int_{-\infty}^{\infty} \left\{\int_{-\infty}^{\infty} h(x) f(x, y) \, dx\right\} dy \\
&= \int_{-\infty}^{\infty} h(x) \left\{\int_{-\infty}^{\infty} f(x, y) \, dy\right\} dx \\
&= Eh(X).
\end{aligned}$$

In Theorem 1 it is important that $E|h(X)| < \infty$ for (7) to hold.

Example 3 (Enis [26]). Let Y be a positive rv with pdf

$$g(y) = \begin{cases} \dfrac{(\frac{1}{2})^{1/2} y^{-1/2} e^{-y/2}}{\Gamma(\frac{1}{2})}, & y > 0, \\ 0, & y \leq 0, \end{cases}$$

and let the conditional pdf of X, given $y(> 0)$, be expressed by

$$h(x|y) = (2\pi)^{-1/2} y^{1/2} e^{-yx^2/2}, \quad -\infty < x < \infty.$$

Then $E\{X|Y\}$ exists, and since $h(x|y)$ is symmetric it follows that $E\{X|y\} = 0$. The marginal pdf of X is given by

$$\begin{aligned}
f(x) &= \int_0^{\infty} g(y) h(x|y) \, dy \\
&= \frac{1}{2\sqrt{\pi} \Gamma(\frac{1}{2})} \int_0^{\infty} e^{-y(1+x^2)/2} \, dy \\
&= \frac{1}{\pi(1 + x^2)}, \quad -\infty < x < \infty.
\end{aligned}$$

This is the Cauchy density, for which we have seen that $E|X| = \infty$. Thus $E\{E\{X|Y\}\}$ exists, but not EX, and (7) does not hold.

Theorem 2. *If $EX^2 < \infty$, then*

(8) $\qquad \text{var}(X) = \text{var}(E\{X|Y\}) + E(\text{var}\{X|Y\}).$

Proof. The right-hand side of (8) equals, by definition,

$$\{E(E^2\{X|Y\}) - [E(E\{X|Y\})]^2\} + E(E\{X^2|Y\} - E^2\{X|Y\})$$
$$= \{E(E^2\{X|Y\}) - (EX)^2\} + EX^2 - E(E^2\{X|Y\})$$
$$= \text{var}(X).$$

Corollary. If $EX^2 < \infty$, then

(9) $$\text{var}(X) \geq \text{var}(E\{X|Y\})$$

with equality if and only if X is a function of Y.

Equation (9) follows immediately from (8). The equality in (9) holds if and only if

$$E(\text{var}\{X|Y\}) = E(X - E\{X|Y\})^2 = 0,$$

which holds if and only if

(10) $$X = E\{X|Y\}.$$

We recall that, if X and Y are independent,

$$F_{X|Y}(x|y) = F_X(x) \quad \text{for all } x,$$

and

$$F_{Y|X}(y|x) = F_Y(y) \quad \text{for all } y.$$

It follows, therefore, that if $E\{h(X)\}$ exists and X and Y are independent, then

(11) $$E\{h(X)|Y\} = E\{h(X)\}.$$

Similarly

(12) $$E\{h(Y)|X\} = E\{h(Y)\},$$

provided that $E\{h(Y)\}$ exists, and X and Y are independent. Note that (8) becomes trivial if X and Y are independent.

PROBLEMS 4.7

1. Let X be an rv with pdf given by

$$f(x) = \frac{1}{\sigma\sqrt{2\pi}} \exp\left\{-\frac{1}{2}\frac{(x-\mu)^2}{\sigma^2}\right\}, \quad -\infty < x < \infty, \; -\infty < \mu < \infty, \; \sigma > 0.$$

Find $E\{X|a < X < b\}$, where a and b are constants.

2. (a) Let (X, Y) be jointly distributed with density

$$f(x, y) = \begin{cases} y(1 + x)^{-4} e^{-y(1+x)^{-1}}, & x, y \geq 0, \\ 0, & \text{otherwise.} \end{cases}$$

Find $E\{Y|X\}$.

(b) Do the same for the joint density

$$f(x, y) = \tfrac{4}{5}(x + 3y)e^{-x-2y}, \quad x, y \geq 0,$$
$$= 0, \quad \text{otherwise.}$$

3. Let (X, Y) be jointly distributed with bivariate normal density

$$f(x, y) = \frac{1}{2\pi\sigma_1\sigma_2\sqrt{1-\rho^2}}$$
$$\cdot \exp\left\{-\frac{1}{2(1-\rho^2)}\left[\left(\frac{x-\mu_1}{\sigma_1}\right)^2 - 2\rho\left(\frac{x-\mu_1}{\sigma_1}\right)\left(\frac{y-\mu_2}{\sigma_2}\right) + \left(\frac{y-\mu_2}{\sigma_2}\right)^2\right]\right\}.$$

Find $E\{X|y\}$ and $E\{Y|x\}$. (Here $\mu_1, \mu_2 \in \mathscr{R}$, $\sigma_1, \sigma_2 > 0$, and $|\rho| < 1$.)

4. Find $E\{Y - E\{Y|X\}\}^2$.

5. Let X and Y be rv's, and $\varphi(X)$ be an rv. Assume that EY and $E\varphi(X)$ exist. Show that (a) $E\{\varphi(X)|X\} = \varphi(X)$, and (b) $E\{\varphi(X)Y|X\} = \varphi(X)E\{Y|X\}$.

[Results (a) and (b) hold in general. We ask the reader to supply the proof in the discrete case only.]

4.8 THE PRINCIPLE OF LEAST SQUARES

Let X, Y be dependent rv's, and suppose that we wish to find the functional relationship between X and Y. Suppose also that this relationship is $y = h(x)$. We assume that both EY^2 and $E[h(X)]^2$ exist. Our object is to find the function $h(x)$. The *principle of least squares* consists of choosing $h(x)$ so that the quantity

$$E\{Y - h(X)\}^2$$

is minimum. If (X, Y) is of the continuous type, this means that we want

$$E\{Y - h(X)\}^2 = \iint [y - h(x)]^2 f(x, y)\, dx\, dy$$
$$= \int_{-\infty}^{\infty} f_X(x)\left\{\int_{-\infty}^{\infty} [y - h(x)]^2 f_{Y|X}(y|x)\, dy\right\} dx$$

to be minimum. Clearly this will be achieved if we minimize

$$\left\{\int_{-\infty}^{\infty} [y - h(x)]^2 f_{Y|X}(y|x)\, dy\right\}.$$

From Theorem 3.2.8 it follows that we must have

(1) $$h(x) = E\{Y|x\}.$$

A similar argument holds when (X, Y) is of the discrete type.

Definition 1. The relation $y = E\{Y|x\}$ is called the *regression of Y on X*, and $x = E\{X|y\}$ is called the *regression of X on Y*.

In practice, one is frequently interested in approximating the relationship between X and Y with the help of a straight line,

$$y = ax + b,$$

say. In that case, the principle of least squares requires us to choose a and b so as to minimize

$$L = E\{Y - aX - b\}^2.$$

We have

$$L = EY^2 + a^2 EX^2 + b^2 - 2a\, EXY + 2ab\, EX - 2b\, EY,$$

where, of course, we have assumed that $EY^2 < \infty$, $EX^2 < \infty$. We solve for a and b from

$$\frac{\partial L}{\partial a} = 2a\, EX^2 - 2EXY + 2b\, EX = 0,$$

$$\frac{\partial L}{\partial b} = 2b + 2a\, EX - 2EY = 0;$$

that is, we solve a and b from the so-called *normal equations*

$$a\, EX + b = EY,$$
$$a\, EX^2 + b\, EX = E(XY).$$

We get

(2) $$a = \frac{E(XY) - EX\, EY}{EX^2 - (EX)^2} = \frac{\text{cov}(X, Y)}{\text{var}(X)},$$

(3) $$b = EY - EX\frac{\text{cov}(X, Y)}{\text{var}(X)}.$$

Thus we have

(4) $$y - EY = \frac{\text{cov}(X, Y)}{\text{var}(X)}\{x - EX\}.$$

We call (4) the *line of regression of Y on X*. Similarly the line

(5) $$x - EX = \frac{\text{cov}(X, Y)}{\text{var}(Y)}\{y - EY\}$$

is the *line of regression of X on Y*. We wish to emphasize that one cannot simply solve (4) for $(x - EX)$ to obtain (5), because the roles of y and x are reversed. Relation (4) was obtained with x as the causal variable, whereas

(5) is obtained by taking y as the causal variable and minimizing $E\{X - \alpha Y - \beta\}^2$.

The procedure for fitting a polynomial of any order is similar. Thus, if we wish to fit

$$y = a + bx + cx^2 + dx^3,$$

the normal equations are

$$a + b\,EX + c\,EX^2 + d\,EX^3 = EY,$$
$$a\,EX + b\,EX^2 + c\,EX^3 + d\,EX^4 = EXY,$$
$$a\,EX^2 + b\,EX^3 + c\,EX^4 + d\,EX^5 = EX^2Y,$$
$$a\,EX^3 + b\,EX^4 + c\,EX^5 + d\,EX^6 = EX^3Y,$$

and so on.

Definition 2. Let EX^2, EY^2 exist. Then the quantity $\operatorname{cov}(X, Y)/\operatorname{var}(X)$ is called the coefficient of regression of Y on X. Similarly, $\operatorname{cov}(X, Y)/\operatorname{var}(Y)$ is called the coefficient of regression of X on Y. We write

$$r_{X|Y} = \frac{\operatorname{cov}(X, Y)}{\operatorname{var}(Y)}, \quad \text{and} \quad r_{Y|X} = \frac{\operatorname{cov}(X, Y)}{\operatorname{var}(X)}.$$

Definition 3. If EX^2, EY^2 exist, we define the correlation coefficient between X and Y as

(6) $$\rho = \frac{\operatorname{cov}(X, Y)}{SD(X)\,SD(Y)} = \frac{E(XY) - EX\,EY}{\sqrt{EX^2 - (EX)^2}\,\sqrt{EY^2 - (EY)^2}}.$$

Note that

$$\rho^2 = r_{X|Y}\,r_{Y|X},$$

and the sign of ρ is the same as that of $\operatorname{cov}(X, Y)$.

Definition 4. We say that two rv's X and Y are uncorrelated if and only if $\rho = 0$.

Clearly, $\rho = 0$ if and only if $\operatorname{cov}(X, Y) = 0$. If X and Y are independent, then $\operatorname{cov}(X, Y) = 0$ and X and Y are uncorrelated. We emphasize that, if X and Y are uncorrelated [and hence $\operatorname{cov}(X, Y) = 0$], X and Y are not necessarily independent.

Example 1. Let U and V be two rv's with the same mean and the same variance. Write $X = U + V$, $Y = U - V$. Then

$$\text{cov}(X, Y) = E(U^2 - V^2) - E(U - V) E(U + V) = 0,$$

so that X and Y are uncorrelated but not necessarily independent.

Some basic properties of the correlation coefficient are given in the following theorems.

Theorem 1.

(a) The correlation coefficient between two rv's X and Y satisfies
$$|\rho| \leq 1.$$

(b) The equality $\rho = \pm 1$ holds if and only if there exist constants a and b such that
$$P\{Y = aX + b\} = 1.$$

Proof. By the Cauchy-Schwarz inequality
$$\text{cov}(X, Y) \leq \text{var}(X) \text{var}(Y)$$
with equality if and only if there exist constants a and b such that
$$P\{Y = aX + b\} = 1.$$

Corollary. The two lines of regression (4) and (5) coincide if and only if $\rho = +1$ or $\rho = -1$. Thus the case $\rho = \pm 1$ is the opposite of independence.

Theorem 2. Let $EX^2 < \infty$, $EY^2 < \infty$, and let $U = aX + b$, $V = cY + d$. Then
$$\rho_{X,Y} = \pm \rho_{U,V},$$
where $\rho_{X,Y}$ and $\rho_{U,V}$, respectively, are the correlation coefficients between X and Y and U and V.

Proof. The proof is simple and is left as an exercise.

Example 2. Let X, Y be identically distributed with common pmf
$$P\{X = k\} = \frac{1}{N}, \quad k = 1, 2, \cdots, N \ (N > 1).$$
Then
$$EX = EY = \frac{N+1}{2}, \quad EX^2 = EY^2 = \frac{(N+1)(2N+1)}{6},$$
so that

$$\operatorname{var}(X) = \operatorname{var}(Y) = \frac{N^2 - 1}{12}.$$

Also,
$$E(XY) = \tfrac{1}{2}\{EX^2 + EY^2 - E(X - Y)^2\}$$
$$= \frac{(N + 1)(2N + 1)}{6} - \frac{E(X - Y)^2}{2}.$$

Thus
$$\operatorname{cov}(X, Y) = \frac{(N + 1)(2N + 1)}{6} - \frac{E(X - Y)^2}{2} - \frac{(N + 1)^2}{4}$$
$$= \frac{(N + 1)(N - 1)}{12} - \tfrac{1}{2} E(X - Y)^2,$$

and
$$\rho_{X,Y} = \frac{(N^2 - 1)/12 - E(X - Y)^2/2}{(N^2 - 1)/12}$$
$$= 1 - \frac{6E(X - Y)^2}{N^2 - 1}.$$

If $P\{X = Y\} = 1$, then $\rho = 1$, and conversely. If $P\{Y = N + 1 - X\} = 1$, then
$$E(X - Y)^2 = E(2X - N - 1)^2$$
$$= 4\frac{(N + 1)(2N + 1)}{6} - 4\frac{(N + 1)^2}{2} + (N + 1)^2,$$

and it follows that $\rho_{XY} = -1$. Conversely, if $\rho_{X,Y} = -1$, from Theorem 1 it follows that $Y = -aX + b$ with probability 1 for some $a > 0$ and some real number b. To find a and b, we note that $EY = -aEX + b$ so that $b = [(N + 1)/2](1 + a)$. Also $EY^2 = E(b - aX)^2$, which yields
$$(1 - a^2)EX^2 + 2abEX - b^2 = 0.$$

Substituting for b in terms of a and the values of EX^2 and EX, we see that $a^2 = 1$, so that $a = 1$. Hence $b = N + 1$, and it follows that $Y = N + 1 - X$ with probability 1.

Example 3. Let (X, Y) have the joint pdf given by
$$f(x, y) = \begin{cases} 1 & \text{if } |y| < x, \ 0 < x < 1, \\ 0 & \text{otherwise.} \end{cases}$$

Then
$$f_1(x) = \begin{cases} 2x, & 0 < x < 1, \\ 0, & \text{otherwise,} \end{cases}$$

and

$$f_2(y) = \begin{cases} 1 - y, & 0 < y < 1, \\ 1 + y, & -1 < y \le 0, \\ 0, & \text{otherwise,} \end{cases}$$

are the marginal pdf's. The conditional densities are given by

$$f_{X|Y}(x|y) = \begin{cases} \dfrac{1}{1-y}, & 0 < y < 1 \\ \dfrac{1}{1+y}, & -1 < y \le 0 \\ 0, & \text{otherwise;} \end{cases} \quad (0 < x < 1)$$

$$f_{Y|X}(y|x) = \begin{cases} \dfrac{1}{2x}, & |y| < x, \ 0 < x < 1, \\ 0, & \text{otherwise;} \end{cases}$$

$$E\{Y|x\} = \int_{-x}^{x} \frac{1}{2x} y \, dy = 0;$$

and

$$E\{X|y\} = \int x f_{X|Y}(x|y) \, dx = \begin{cases} \dfrac{1}{2(1-y)}, & 0 < y < 1, \\ \dfrac{1}{2(1+y)}, & -1 < y \le 0. \end{cases}$$

Also,

$$\text{cov}(X, Y) = \iint xy \, dx \, dy = \int_0^1 x \int_{-x}^{x} y \, dy \, dx = 0,$$

so that

$$\rho_{XY} = 0.$$

Example 4. Let (X, Y) be jointly distributed with density

$$f(x, y) = \begin{cases} e^{-y}, & 0 < x < y < \infty, \\ 0, & \text{otherwise.} \end{cases}$$

Then the marginal densities are

$$f_1(x) = \begin{cases} e^{-x}, & 0 < x < \infty, \\ 0, & \text{otherwise;} \end{cases}$$

$$f_2(y) = \begin{cases} ye^{-y}, & 0 < y < \infty, \\ 0, & \text{otherwise;} \end{cases}$$

and the conditional densities are

$$f_{X|Y}(x|y) = \frac{1}{y}, \quad 0 < x < y,$$

$$f_{Y|X}(y|x) = e^{x-y}, \quad 0 < x < y < \infty;$$

$$E\{X|y\} = \int_0^y x \frac{1}{y} dx = \frac{y}{2},$$

$$E\{Y|x\} = \int_x^\infty y\, e^{x-y}\, dy = \int_0^\infty (u+x) e^{-u}\, du = x+1.$$

It follows that the coefficient of regression of X on Y is $\frac{1}{2}$, and that of Y on X is 1. Therefore

$$\rho^2 = \tfrac{1}{2},$$

hence

$$\rho = +\sqrt{\tfrac{1}{2}}.$$

PROBLEMS 4.8

1. Let (X, Y) have the joint pdf

$$f(x, y) = \begin{cases} x + y, & 0 < x < 1, 0 < y < 1, \\ 0, & \text{otherwise.} \end{cases}$$

Find the correlation coefficient between X and Y.

2. Show that the acute angle, θ, between the two lines of regression is given by

$$\tan \theta = \frac{1 - \rho^2}{\rho} \cdot \frac{\sigma_1 \sigma_2}{\sigma_1^2 + \sigma_2^2},$$

where σ_1, σ_2 are the standard deviations of X and Y, respectively, and ρ is the correlation coefficient.

3. Let $(\Omega, \mathscr{S}, P)$ be a probability space, and $A, B, \in \mathscr{S}$ with $0 < PA < 1$, $0 < PB < 1$. Define $\rho(A, B)$ by $\rho(A, B) = $ correlation coefficient between rv's I_A, and I_B, where I_A, I_B, are the indicator functions of A and B, respectively. Express $\rho(A,B)$ in terms of PA, PB and $P(AB)$, and conclude that $\rho(A, B) = 0$ if and only if A and B are independent. What happens if $A = B$ or if $A = B^c$?

(a) Show that

$$\rho(A, B) > 0 \Leftrightarrow P\{A|B\} > P(A) \Leftrightarrow P\{B|A\} > P(B),$$

and

$$\rho(A, B) < 0 \Leftrightarrow P\{A|B\} < PA \Leftrightarrow P\{B|A\} < PB.$$

(b) Show that

$$\rho(A, B) = \frac{P(AB)\, P(A^c B^c) - P(AB^c)\, P(A^c B)}{(PA\, PA^c - PB\, PB^c)^{1/2}}.$$

4. Let $X_1, X_2, \cdots, X_n$ be iid rv's, and define

$$\bar{X} = \frac{\sum_{i=1}^{n} X_i}{n}, \qquad S^2 = \frac{\sum_{i=1}^{n}(X_i - \bar{X})^2}{(n-1)}.$$

Suppose that the common distribution is symmetric. Assuming the existence of moments of appropriate order, show that cov $(\bar{X}, S^2) = 0$.

5. Let X, Y be iid rv's with common standard normal density

$$f(x) = \frac{1}{\sqrt{2\pi}} e^{-x^2/2}, \qquad -\infty < x < \infty.$$

Let $U = X + Y$ and $V = X^2 + Y^2$. Find the mgf of the random variable (U, V). Also, find the correlation coefficient between U and V. Are U and V independent?

6. Let X and Y be two discrete rv's:

$$P\{X = x_1\} = p_1, \qquad P\{X = x_2\} = 1 - p_1;$$

and

$$P\{Y = y_1\} = p_2, \qquad P\{Y = y_2\} = 1 - p_2.$$

Show that X and Y are independent if and only if the correlation coefficient between X and Y is zero.

7. Two fair coins, each with faces numbered 1 and 2, are thrown. Let X denote the sum of the two numbers obtained, and Y, the maximum of the two numbers obtained. Find the correlation coefficient between X and Y.

8. Let X and Y be dependent rv's with common means 0, variances 1, and correlation coefficient ρ. Show that

$$E\{\max(X^2, Y^2)\} \leq 1 + \sqrt{1 - \rho^2}.$$

9. Let X_1, X_2 be independent normal rv's with density functions

$$f_i(x) = \frac{1}{\sigma_i \sqrt{2\pi}} \exp\left\{-\tfrac{1}{2}\left(\frac{x - \mu_i}{\sigma_i}\right)^2\right\}, \qquad -\infty < x < \infty;\ i = 1, 2.$$

Also let

$$Z = X_1 \cos\theta + X_2 \sin\theta \quad \text{and} \quad W = X_2 \cos\theta - X_1 \sin\theta.$$

Find the correlation coefficient between Z and W, and show that

$$0 \leq \rho^2 \leq \left(\frac{\sigma_1^2 - \sigma_2^2}{\sigma_1^2 + \sigma_2^2}\right)^2,$$

where ρ denotes the correlation coefficient between Z and W.

10. Find $E\{Y - aX - b\}^2$, where a and b are given by (2) and (3), respectively.

11. Let $(X_1, X_2, \cdots, X_n)$ be an rv such that the correlation coefficient between each pair X_i, X_j, $i \neq j$, is ρ. Show that $-(n-1)^{-1} \leq \rho \leq 1$.

12. Let $X_1, X_2, \cdots, X_{m+n}$ be iid rv's with finite second moment. Let $S_k = \sum_{j=1}^{k} X_j$, $k = 1, 2, \cdots, m + n$. Find the correlation coefficient between S_n and $S_{m+n} - S_m$, where $n > m$.

13. Let f be the pdf of a positive rv, and write

$$g(x, y) = \begin{cases} \dfrac{f(x+y)}{x+y} & \text{if } x > 0,\ y > 0, \\ 0 & \text{otherwise.} \end{cases}$$

Show that g is a density function in the plane. If the mth moment of f exists for some positive integer m, find EX^m. Compute the means and variances of X and Y and the correlation coefficient between X and Y in terms of moments of f.

(Adapted from Feller [29], 100)

14. A die is thrown $n + 2$ times. After each throw a $+$ sign is recorded for 4, 5, or 6, and a $-$ sign for 1, 2, or 3, the signs forming an ordered sequence. Each sign, except the first and the last, is attached a characteristic rv that assumes the value 1 if both the neighboring signs differ from the one between them and 0 otherwise. Let $X_1, X_2, \cdots, X_n$ be these characteristic rv's, where X_i corresponds to the $(i + 1)$st sign ($i = 1, 2, \cdots, n$) in the sequence. Show that

$$E\left\{\sum_{1}^{n} X_i\right\} = \frac{n}{4}, \quad \text{and} \quad \operatorname{var}\left\{\sum_{1}^{n} X_i\right\} = \frac{5n - 2}{16}.$$

15. Let (X, Y) be jointly distributed with pdf f defined by $f(x, y) = \frac{1}{2}$ inside the square with corners at the points $(0, 1)$, $(1, 0)$, $(-1, 0)$, $(0, -1)$ in the (x, y) – plane, and $f(x, y) = 0$ otherwise. Are X, Y independent? Are they uncorrelated?

16. Predict the length L of a word selected at random from the sentence

IT IS TOO GOOD TO BE TRUE.

Let X be the number of O's in the word selected. Find the joint pmf of X and L. Compute the best least square predictor of L, given X, and compute the expected loss if you lose the square of your error.

17. Let $(\Omega, \mathscr{S}, P)$ be a probability space, and A and $B \in \mathscr{S}$ be such that

$$P(A) = \tfrac{1}{4}, \qquad P\{B|A\} = \tfrac{1}{2}, \qquad P\{A|B\} = \tfrac{1}{4}.$$

Define $X(\omega) = I_A(\omega)$, $Y(\omega) = I_B(\omega)$ for all $\omega \in \Omega$. Find EX, EY, $\operatorname{var}(X)$, $\operatorname{var}(Y)$, and the correlation coefficient between X and Y. Are X and Y independent?

CHAPTER 5

Some Special Distributions

5.1 INTRODUCTION

In preceding chapters we studied probability distributions in general. In this chapter we study some commonly occurring probability distributions and investigate their basic properties. The results of this chapter will be of considerable use in theoretical as well as practical applications. We begin with some discrete distributions in Section 2 and follow with some continuous models in Section 3. Section 4 deals with bivariate and multivariate normal distributions, while in Section 5 we discuss the exponential family of distributions.

5.2 SOME DISCRETE DISTRIBUTIONS

In this section we study some well-known univariate and multivariate discrete distributions and describe their important properties.

a. The Degenerate Distribution

The simplest distribution is that of an rv X *degenerate* at point k, that is, $P\{X = k\} = 1$, and $= 0$ elsewhere. If we define

$$\varepsilon(x) = \begin{cases} 0 & \text{if } x < 0, \\ 1 & \text{if } x \geq 0, \end{cases} \tag{1}$$

the df of the rv X is $\varepsilon(x - k)$. Clearly, $EX^l = k^l$, $l = 1, 2, \cdots$, and $M(t) = e^{tk}$. In particular, var $(X) = 0$. This property characterizes a degenerate rv. As we shall see, the degenerate rv plays an important role in the study of limit theorems.

b. The Two-Point Distribution

We say that an rv X has a *two-point distribution* if it takes two values, x_1 and x_2, with probabilities

$$P\{X = x_1\} = p, \quad P\{X = x_2\} = 1 - p, \quad 0 < p < 1.$$

We may write

(2) $$X = x_1 I_{[X=x_1]} + x_2 I_{[X=x_2]}.$$

The df of X is given by

(3) $$F(x) = p\varepsilon(x - x_1) + (1 - p)\varepsilon(x - x_2).$$

Also,

(4) $$EX^k = px_1^k + (1 - p)x_2^k, \quad k = 1, 2, \cdots,$$

(5) $$M(t) = pe^{tx_1} + (1 - p)e^{tx_2} \quad \text{for all } t.$$

In particular,

(6) $$EX = px_1 + (1 - p)x_2.$$

and

(7) $$\text{var}(X) = p(1 - p)(x_1 - x_2)^2.$$

If $x_1 = 1$, $x_2 = 0$, we get the important *Bernoulli* rv:

(8) $$P\{X = 1\} = p, \quad P\{X = 0\} = 1 - p, \quad 0 < p < 1.$$

For a Bernoulli rv X with parameter p, we write $X \sim b(1, p)$ and have

(9) $$EX = p, \quad \text{var}(X) = p(1 - p), \quad M(t) = 1 + p(e^t - 1), \quad \text{all } t.$$

Bernoulli rv's occur in practice, for example, in coin-tossing experiments. Suppose that $P\{H\} = p$, $0 < p < 1$, and $P\{T\} = 1 - p$. Define rv X so that $X(H) = 1$ and $X(T) = 0$. Then $P\{X = 1\} = p$ and $P\{X = 0\} = 1 - p$. Each repetition of the experiment will be called a *trial*. More generally, any nontrivial experiment can be dichotomized to yield a Bernoulli model. Let $(\Omega, \mathscr{S}, P)$ be the sample space of an experiment, and let $A \in \mathscr{S}$ with $P(A) = p > 0$. Then $P(A^c) = 1 - p$. Each performance of the experiment is a Bernoulli trial. It will be convenient to call the occurrence of event A a *success*, and the occurrence of A^c, a *failure*.

Example 1 (Sabharwal [108]). In a sequence of n Bernoulli trials with constant probability p of success (S), and $1 - p$ of failure (F), let Y_n denote the number of times the combination SF occurs. To find EY_n and var (Y_n), let X_i represent the event that occurs on the ith trial, and define rv's

$$f(X_i, X_{i+1}) = \begin{cases} 1 & \text{if } X_i = S, X_{i+1} = F, \\ 0 & \text{otherwise.} \end{cases} \quad (i = 1, 2, \ldots, n-1)$$

Then
$$Y_n = \sum_{i=1}^{n-1} f(X_i, X_{i+1})$$

and
$$EY_n = (n-1)p(1-p).$$

Also,
$$EY_n^2 = E\{\sum_{i=1}^{n-1} f^2(X_i, X_{i+1})\} + E\{\sum\sum_{i \neq j} f(X_i, X_{i+1}) f(X_j, X_{j+1})\}$$
$$= (n-1)p(1-p) + (n-2)(n-3)p^2(1-p)^2,$$

so that
$$\text{var}(Y_n) = p(1-p)\{n - 1 + p(1-p)(5 - 3n)\}.$$

If $p = 1/2$, then
$$EY_n = \frac{n-1}{4}, \quad \text{and} \quad \text{var}(Y_n) = \frac{n+1}{16}.$$

c. The Uniform Distribution on n Points

X is said to have a *uniform distribution* on n points $\{x_1, x_2, \ldots, x_n\}$ if its pmf is of the form

(10) $$P\{X = x_i\} = \frac{1}{n}, \quad i = 1, 2, \ldots, n.$$

Thus we may write
$$X = \sum_{i=1}^{n} x_i I_{[X=x_i]}, \quad \text{and} \quad F(x) = \frac{1}{n} \sum_{i=1}^{n} \varepsilon(x - x_i),$$

(11) $$EX = \frac{1}{n} \sum_{i=1}^{n} x_i,$$

(12) $$EX^l = \frac{1}{n} \sum_{i=1}^{n} x_i^l, \quad l = 1, 2, \ldots,$$

and

(13) $$\text{var}(X) = \frac{1}{n} \sum_{i=1}^{n} x_i^2 - (\frac{1}{n} \sum x_i)^2 = \frac{1}{n} \sum_{i=1}^{n} (x_i - \bar{x})^2,$$

if we write $\bar{x} = \sum_{i=1}^{n} x_i/n$. Also,

(14) $$M(t) = \frac{1}{n} \sum_{i=1}^{n} e^{tx_i} \quad \text{for all } t.$$

If, in particular, $x_i = i$, $i = 1, 2, \ldots, n$,

(15) $$EX = \frac{n+1}{2}, \quad EX^2 = \frac{(n+1)(2n+1)}{6},$$

(16) $$\operatorname{var}(X) = \frac{n^2 - 1}{12}.$$

Example 2. A box contains tickets numbered 1 to N. Let X be the largest number drawn in n random drawings with replacement.

Then $P\{X \leq k\} = (k/N)^n$, so that
$$\begin{aligned} P\{X = k\} &= P\{X \leq k\} - P\{X \leq k - 1\} \\ &= \left(\frac{k}{N}\right)^n - \left(\frac{k-1}{N}\right)^n. \end{aligned}$$

Also,
$$\begin{aligned} EX &= N^{-n} \sum_{1}^{N} [k^{n+1} - (k-1)^{n+1} - (k-1)^n] \\ &= N^{-n} [N^{n+1} - \sum_{1}^{N} (k-1)^n]. \end{aligned}$$

d. The Binomial Distribution

We say that X has a *binomial distribution* with parameter p if its pmf is given by

(17) $\quad p_k = P\{X = k\} = \binom{n}{k} p^k (1-p)^{n-k}, \quad k = 0, 1, 2, \ldots, n; 0 \leq p \leq 1.$

Since $\sum_{k=0}^{n} p_k = [p + (1-p)]^n = 1$, the p_k's indeed define a pmf. If X has pmf (17), we will write $X \sim b(n, p)$. This is consistent with the notation for a Bernoulli rv. We have

$$F(x) = \sum_{k=0}^{n} \binom{n}{k} p^k (1-p)^{n-k} \varepsilon(x - k).$$

In Example 3.2.5 we showed that

(18) $\quad EX = np,$

(19) $\quad EX^2 = n(n-1)p^2 + np,$

and

(20) $\quad \operatorname{var}(X) = np(1-p) = npq,$

where $q = 1 - p$. Also,

$$M(t) = \sum_{k=0}^{n} e^{tk}\binom{n}{k} p^k(1-p)^{n-k}$$

(21)
$$= (q + pe^t)^n \quad \text{for all } t.$$

The pgf of $X \sim b(n, p)$ is given by $P(s) = \{1 - p(1 - s)\}^n$, $|s| \leq 1$.

Binomial distribution can also be considered as the distribution of the sum of n independent, identically distributed $b(1, p)$ random variables. If we toss a coin, with constant probability p of heads and $1 - p$ of tails, n times, the distribution of the number of heads is given by (17). Alternatively, if we write

$$X_k = \begin{cases} 1 & \text{if } k\text{th toss results in a head,} \\ 0 & \text{otherwise,} \end{cases}$$

the number of heads in n trials is the sum $S_n = X_1 + X_2 + \cdots + X_n$. Also

$$P\{X_k = 1\} = p, \quad P\{X_k = 0\} = 1 - p, \quad k = 1, 2, \cdots, n.$$

Thus

$$ES_n = \sum_{1}^{n} EX_i = np,$$
$$\text{var}(S_n) = \sum_{1}^{n} \text{var}(X_i) = np(1 - p),$$

and

$$M(t) = \prod_{i=1}^{n} Ee^{tX_i}$$
$$= (q + pe^t)^n.$$

Theorem 1. Let X_i ($i = 1, 2, \cdots, k$) be independent rv's with $X_i \sim b(n_i, p)$. Then $S_k = \sum_{i=1}^{k} X_i$ has a $b(n_1 + n_2 + \cdots + n_k, p)$ distribution.

Proof.
$$M_{S_k}(t) = \prod_{i=1}^{k} M_{X_i}(t) = \prod_{i=1}^{k} (q + pe^t)^{n_i},$$

and the result follows from uniqueness of the mgf.

Corollary. If X_i ($i = 1, 2, \cdots, k$) are iid rv's with common pmf $b(n, p)$, then S_k has a $b(nk, p)$ distribution.

Actually the additive property described in Theorem 1 characterizes the binomial distribution in the following sense. Let X and Y be two independent, nonnegative, finite integer-valued rv's and let $Z = X + Y$. Then Z is a binomial rv with parameter p if and only if X and Y are binomial rv's with the same parameter p. The "only if" part is due to Shanbhag and Basawa [113] and will not be proved here.

Example 3. A fair die is rolled n times. The probability of obtaining exactly one 6 is $n(\frac{1}{6})(\frac{5}{6})^{n-1}$, the probability of obtaining no 6 is $(\frac{5}{6})^n$, and the probability of obtaining at least one 6 is $1 - (\frac{5}{6})^n$.

The number of trials needed for the probability of at least one 6 to be $\geq 1/2$ is given by the smallest integer n such that

$$1 - (\tfrac{5}{6})^n \geq \tfrac{1}{2}$$

or

$$n \log (\tfrac{5}{6}) \leq -\log 2.$$

Therefore

$$n \geq \frac{\log 2}{\log 1.2} \approx 3.8.$$

Example 4. Here r balls are distributed in n cells so that each of n^r possible arrangements has probability n^{-r}. We are interested in the probability p_k that a specified cell has exactly k balls ($k = 0, 1, 2, \cdots, r$). Then the distribution of each ball may be considered as a trial. A success results if the ball goes to the specified cell (with probability $1/n$); otherwise the trial results in a failure (with probability $1 - 1/n$). Let X denote the number of successes *in r trials.* Then

$$p_k = P\{X = k\} = \binom{r}{k}\left(\frac{1}{n}\right)^k \left(1 - \frac{1}{n}\right)^{r-k}, \qquad k = 0, 1, 2, \cdots, n.$$

e. The Negative Binomial Distribution (Pascal or Waiting Time Distribution)

Let $(\Omega, \mathscr{L}, P)$ be a probability space of a given statistical experiment, and let $A \in \mathscr{S}$ with $P(A) = p$. On any performance of the experiment, if A happens we call it a success, otherwise a failure. Consider a succession of trials of this experiment, and let us compute the probability of observing exactly r successes, where $r \geq 1$ is a fixed integer. If X denotes the number of failures that precede rth success, $X + r$ is the total number of replications needed to produce r successes. This will happen if and only if the last trial results in a success and among the previous $(r + X - 1)$ trials there are exactly X failures. It follows by independence that

$$(22) \qquad P\{X = x\} = \binom{x + r - 1}{x} p^r (1 - p)^x, \qquad x = 0, 1, 2, \cdots.$$

Rewriting (22) in the form (see P. 2.7)

(23) $$P\{X = x\} = \binom{-r}{x} p^r (-q)^x, \quad x = 0, 1, 2, \cdots; q = 1 - p,$$

we see that

(24) $$\sum_{x=0}^{\infty} \binom{-r}{x}(-q)^x = (1-q)^{-r} = p^{-r}.$$

It follows that

$$\sum_{x=0}^{\infty} P\{X = x\} = 1.$$

Definition 1. For fixed positive integer $r \geq 1$ and $0 < p < 1$, an rv with pmf given by (22) is said to have a negative binomial distribution. We will use the notation $X \sim NB(r; p)$.

We may write

$$X = \sum_{x=0}^{\infty} x I_{[X=x]}, \quad \text{and} \quad F(x) = \sum_{k=0}^{\infty} \binom{k+r-1}{k} p^r (1-p)^k \varepsilon(x-k).$$

For the mgf of X we have

$$M(t) = \sum_{x=0}^{\infty} \binom{x+r-1}{x} p^r (1-p)^x e^{tx}$$

$$= p^r \sum_{x=0}^{\infty} (qe^t)^x \binom{x+r-1}{x} \quad (q = 1 - p)$$

(25) $$= p^r (1 - qe^t)^{-r} \quad \text{for} \quad qe^t < 1.$$

The pgf is given by $P(s) = p^r(1 - sq)^{-r}$, $|s| \leq 1$. Also,

$$EX = \sum_{x=0}^{\infty} x \binom{x+r-1}{x} p^r q^x$$

$$= rp^r \sum_{x=0}^{\infty} \binom{x+r}{x} q^{x+1}$$

(26) $$= rp^r q(1-q)^{-r-1} = \frac{rq}{p}.$$

Similarly, we can show that

(27) $$\text{var}(X) = \frac{rq}{p^2}.$$

If, however, we are interested in the distribution of the number of trials required to get r successes, we have, writing $Y = X + r$,

(28) $$P\{Y = y\} = \binom{y-1}{r-1} p^r (1-p)^{y-r}, \quad y = r, r+1, \cdots,$$

(29) $$\begin{cases} EY = EX + r = \dfrac{r}{p}, \\ \text{var}(Y) = \text{var}(X) = \dfrac{rq}{p^2}, \end{cases}$$

and

(30) $$M_Y(t) = (p\,e^t)^r (1 - qe^t)^{-r} \quad \text{for } qe^t < 1.$$

Let X be a $b(n, p)$ rv, and let Y be the rv defined in (28). If there are r or more successes in the first n trials, at most n trials were required to obtain the first r of these successes. We have

(31) $$P\{X \geq r\} = P\{Y \leq n\}$$

and also

(32) $$P\{X < r\} = P\{Y > n\}.$$

In the special case when $r = 1$, the distribution of X is given by

(33) $$P\{X = x\} = pq^x, \quad x = 0, 1, 2, \cdots.$$

An rv X with pmf (33) is said to have a *geometric distribution*. Clearly, for the geometric distribution, we have

(34) $$\begin{cases} M(t) = p(1 - qe^t)^{-1}, \\ EX = \dfrac{q}{p}, \\ \text{var}(X) = \dfrac{q}{p^2}. \end{cases}$$

Example 5 (Banach's matchbox problem). A mathematician carries one matchbox each in his right and left pockets. When he wants a match, he selects the left pocket with probability p and the right pocket with probability $1 - p$. Suppose that initially each box contains N matches. Consider the moment when the mathematician discovers that a box is empty. At that time the other box may contain $0, 1, 2, \ldots, N$ matches. Let us identify success with the choice of the left pocket. The left-pocket box will be empty at the moment when the right-pocket box contains exactly r matches if and only if exactly $N - r$ failures precede the $(N + 1)$st success. A similar argument applies to the right pocket, and we have

P_r = Probability that the mathematician discovers a box empty while the other contains r matches

$$= \binom{2N - r}{N - r} p^{N+1} q^{N-r} + \binom{2N - r}{N - r} q^{N+1} p^{N-r}.$$

Example 6. A die is rolled repeatedly. Let us compute the probability of event A that a 2 will show up before a 5. Let A_j be the event that a 2 shows up on the jth trial ($j = 1, 2, \cdots$) for the first time, and a 5 does not show up on the previous $j - 1$ trials. Then $PA = \sum_{j=1}^{\infty} PA_j$, where $PA_j = \frac{1}{6}(\frac{4}{6})^{j-1}$. It follows that

$$P(A) = \sum_{j=1}^{\infty} \left(\frac{1}{6}\right)\left(\frac{4}{6}\right)^{j-1} = \frac{1}{2}.$$

Similarly the probability that a 2 will show up before a 5 or a 6 is 1/3, and so on.

Theorem 2. Let $X_1, X_2, \cdots, X_k$ be independent $NB(r_i; p)$ rv's, $i = 1, 2, \cdots, k$, respectively. Then $S_k = \sum_{i=1}^{k} X_i$ is distributed as $NB(r_1 + r_2 + \cdots + r_k; p)$.

Proof. The proof is left as an exercise.

Corollary. If $X_1, X_2, \cdots, X_k$ are iid geometric rv's, then S_k is an $NB(k; p)$ rv.

Theorem 3. Let X and Y be independent rv's with pmf's $NB(r_1; p)$ and $NB(r_2; p)$, respectively. Then the conditional pmf of X, given $X + Y = t$, is expressed by

$$P\{X = x \mid X + Y = t\} = \frac{\binom{x + r_1 - 1}{x}\binom{t + r_2 - x - 1}{t - x}}{\binom{t + r_1 + r_2 - 1}{t}}.$$

If, in particular, $r_1 = r_2 = 1$, the conditional distribution is uniform on $t + 1$ points.

Proof. By Theorem 2, $X + Y$ is an $NB(r_1 + r_2; p)$ rv. Thus

$$P\{X = x \mid X + Y = t\} = \frac{P\{X = x, Y = t - x\}}{P\{X + Y = t\}}$$

$$= \frac{\binom{x + r_1 - 1}{x} p^{r_1}(1-p)^x \binom{t - x + r_2 - 1}{t - x} p^{r_2}(1-p)^{t-x}}{\binom{t + r_1 + r_2 - 1}{t} p^{r_1 + r_2}(1-p)^t}$$

$$= \frac{\binom{x + r_1 - 1}{x}\binom{t + r_2 - x - 1}{t - x}}{\binom{t + r_1 + r_2 - 1}{t}}, \quad t = 0, 1, 2, \cdots.$$

If $r_1 = r_2 = 1$, that is, if X and Y are independent geometric rv's, then

(35) $\qquad P\{X = x \mid X + Y = t\} = \dfrac{1}{t+1}, \qquad x = 0, 1, 2, \cdots, t; t = 0, 1, 2, \cdots.$

Theorem 4 (Chatterji [12]). Let X and Y be iid rv's, and let
$$P\{X = k\} = p_k > 0, \quad k = 0, 1, 2, \cdots.$$
If
(36) $\quad P\{X = t \mid X + Y = t\} = P\{X = t - 1 \mid X + Y = t\} = \dfrac{1}{t+1}, \quad t \geq 0,$

then X and Y are geometric rv's.

Proof. We have

(37) $\quad P\{X = t \mid X + Y = t\} = \dfrac{p_t p_0}{\sum_{k=0}^{t} p_k p_{t-k}} = \dfrac{1}{t+1}$

and

(38) $\quad P\{X = t - 1 \mid X + Y = t\} = \dfrac{p_{t-1} p_1}{\sum_{k=0}^{t} p_k p_{t-k}} = \dfrac{1}{t+1}.$

It follows that
$$\frac{p_t}{p_{t-1}} = \frac{p_1}{p_0}$$
and by iteration $p_t = (p_1/p_0)^t p_0$. Since $\sum_{t=0}^{\infty} p_t = 1$, we must have $(p_1/p_0) < 1$. Moreover,
$$1 = p_0 \frac{1}{1 - (p_1/p_0)},$$
so that $(p_1/p_0) = 1 - p_0$, and the proof is complete.

Remark 1. It is possible to drop the requirement of identical distribution in Theorem 4. (See Chatterji [12].)

Theorem 5. If X has a geometric distribution, then, for any two positive integers m and n,

(39) $\quad P\{X > m + n \mid X > m\} = P\{X \geq n\}.$

Proof. The proof is left as an exercise.

Remark 2. Theorem 5 says that the geometric distribution has *no memory*, that is, the information of no successes in m trials is forgotten in subsequent calculations.

The converse of Theorem 5 is also true.

Theorem 6. Let X be a nonnegative integer-valued rv satisfying
$$P\{X > m + n \mid X > m\} = P\{X \geq n\}$$
for any two positive integers m and n. Then X must have a geometric distribution.

Proof. Let the pmf of X be written as
$$P\{X = k\} = p_k, \quad k = 0, 1, 2, \cdots.$$
Then
$$P\{X \geq n\} = \sum_{k=n}^{\infty} p_k,$$
and
$$P\{X > m\} = \sum_{m+1}^{\infty} p_k = q_m, \quad \text{say,}$$
$$P\{X > m + n \mid X > m\} = \frac{P\{X > m + n\}}{P\{X > m\}} = \frac{q_{m+n}}{q_m}.$$
Thus
$$q_{m+n} = q_m q_{n-1}, \quad q_{m+1} = q_m q_0,$$
where $q_0 = P\{X > 0\} = p_1 + p_2 + \cdots = 1 - p_0$. It follows that $q_k = (1 - p_0)^{k+1}$, and hence $p_k = q_{k-1} - q_k = (1 - p_0)^k p_0$, as asserted.

Theorem 7. Let $X_1, X_2, \ldots, X_n$ be independent geometric rv's with parameters $p_1, p_2, \cdots, p_n$, respectively. Then $N_n = \min(X_1, X_2, \cdots, X_n)$ is also a geometric rv with parameter
$$p = 1 - \prod_{i=1}^{n} (1 - p_i).$$

Proof. The proof is left as an exercise.

Corollary. Iid rv's $X_1, X_2, \cdots, X_n$ are $NB(1; p)$ if and only if $N_n = \min(X_1, X_2, \cdots, X_n)$ is a geometric rv with parameter $1 - (1 - p)^n$.

Proof. The necessity follows from Theorem 7. For the sufficiency part of the proof let
$$P\{N_n \leq k\} = 1 - P\{N_n > k\} = 1 - (1 - p)^{n(k+1)}.$$
But
$$P\{N_n \leq k\} = 1 - P\{X_1 > k, X_2 > k, \ldots, X_n > k\}$$
$$= 1 - [1 - F(k)]^n,$$

where F is the common df of $X_1, X_2, \ldots, X_n$. It follows that
$$[1 - F(k)] = (1 - p)^{k+1},$$
so that $P\{X_1 > k\} = (1 - p)^{k+1}$, which completes the proof.

f. The Hypergeometric Distribution

A box contains N marbles. Of these, M are taken out at random, marked, and returned to the box. The contents of the box are then thoroughly mixed. Next, n marbles are drawn at random from the box, and the marked marbles are counted. If X denotes the number of marked marbles, then

(40) $$P\{X = x\} = \binom{N}{n}^{-1}\binom{M}{x}\binom{N - M}{n - x}.$$

Since x cannot exceed M or n, we must have

(41) $$x \leq \min(M, n).$$

Also $x \geq 0$ and $N - M \geq n - x$, so that

(42) $$x \geq \max(0, M + n - N).$$

Note that
$$\sum_{k=0}^{n}\binom{a}{k}\binom{b}{n - k} = \binom{a + b}{n}$$
for arbitrary numbers a, b and positive integer n. It follows that
$$\sum_{x} P\{X = x\} = \binom{N}{n}^{-1} \sum_{x}\binom{M}{x}\binom{N - M}{n - x} = 1.$$

It is easy to write the df of the rv X.

Definition 2. An rv X with pmf given by (40) is called a hypergeometric rv.

It is easy to check that

(43) $$EX = \frac{n}{N} M,$$

(44) $$EX^2 = \frac{M(M - 1)}{N(N - 1)} n(n - 1) + \frac{nM}{N},$$

and

(45) $$\text{var}(X) = \frac{nM}{N^2(N - 1)} (N - M)(N - n).$$

Example 7. A lot consisting of 50 bulbs is inspected by taking at random

10 bulbs and testing them. If the number of defective bulbs is at most 1, the lot is accepted; otherwise it is rejected. If there are, in fact, 10 defective bulbs in the lot, the probability of accepting the lot is

$$\frac{\binom{10}{1}\binom{40}{9}}{\binom{50}{10}} + \frac{\binom{40}{10}}{\binom{50}{10}}.$$

Example 8. Suppose that an urn contains b white and c black balls, $b + c = N$. A ball is drawn at random, and before drawing the next ball, $s + 1$ balls of the same color are added to the urn. The procedure is repeated n times. Let X be the number of white balls drawn in n draws, $X = 0$, 1, 2, $\cdots$, n. We shall find the pmf of X.

First note that the probability of drawing k white balls in successive draws is

$$\left(\frac{b}{N}\right)\left(\frac{b+s}{N+s}\right)\left(\frac{b+2s}{N+2s}\right)\cdots\left[\frac{b+(k-1)s}{N+(k-1)s}\right],$$

and the probability of drawing k white balls in the first k draws and then $n - k$ black balls in the next $n - k$ draws is

(46) $$p_k = \left(\frac{b}{N}\right)\left(\frac{b+s}{N+s}\right)\cdots\left[\frac{b+(k-1)s}{N+(k-1)s}\right]\left(\frac{c}{N+ks}\right)\left[\frac{c+s}{N+(k+1)s}\right]$$
$$\cdots\left[\frac{c+(n-k-1)s}{N+(n-1)s}\right].$$

Here p_k also gives the probability of drawing k white and $n - k$ black balls in any given order. It follows that

(47) $$P\{X = k\} = \binom{n}{k}p_k.$$

An rv X with pmf given by (47) is said to have a *Polya distribution*. Let us write

(48) $$Np = b, \quad N(1 - p) = c, \quad \text{and } N\alpha = s.$$

Then, with $q = 1 - p$, we have

$$P\{X = k\} = \binom{n}{k}\frac{p(p + \alpha)\cdots[p + (k-1)\alpha]q(q + \alpha)\cdots[q + (n-k-1)\alpha]}{1(1 + \alpha)\cdots[1 + (n-1)\alpha]}.$$

Let us take $s = -1$. This means that the ball drawn at each draw is not replaced in the urn before drawing the next ball. In this case $\alpha = -1/N$, and we have

$$P\{X=k\} = \binom{n}{k}\frac{Np(Np-1)\ldots[Np-(k-1)]\;c(c-1)\ldots[c-(n-k-1)]}{N(N-1)\ldots[N-(n-1)]}$$

(49)
$$= \frac{\binom{Np}{k}\binom{Nq}{n-k}}{\binom{N}{n}},$$

which is a hypergeometric distribution. Here

(50) $$\max(0, n - Nq) \le k \le \min(n, Np).$$

Theorem 3. Let X and Y be independent rv's with pmf's $b(m, p)$ and $b(n, p)$, respectively. Then the conditional distribution of X, given $X + Y$, is hypergeometric.

Proof. The proof is left as an exercise.

For a characterization of the hypergeometric distribution we refer the reader to Skibinsky [118].

g. The Poisson Distribution

Definition 3. An rv X is said to be a Poisson rv with parameter $\lambda > 0$ if its pmf is given by

(51) $$P\{X = k\} = \frac{e^{-\lambda}\lambda^k}{k!}, \quad k = 0, 1, 2, \cdots.$$

We first check to see that (51) indeed defines a pmf. We have

$$\sum_{k=0}^{\infty} P\{X = k\} = \frac{e^{-\lambda}\sum_{k=0}^{\infty}\lambda^k}{k!} = e^{-\lambda}e^{\lambda} = 1.$$

If X has the pmf given by (51), we will write $X \sim P(\lambda)$. Clearly

$$X = \sum_{k=0}^{\infty} kI_{[X=k]}$$

and

$$F(x) = \sum_{k=0}^{\infty} e^{-\lambda}\frac{\lambda^k}{k!}\varepsilon(x - k).$$

The mean and the variance are given by (see Problem 3.2.9)

(52) $$EX = \lambda, \quad EX^2 = \lambda + \lambda^2,$$

and

(53) $$\text{var}(X) = \lambda.$$

The mgf of X is given by (see Example 3.3.6)

(54) $$Ee^{tX} = \exp\{\lambda(e^t - 1)\},$$

and the pgf is given by $P(s) = e^{-\lambda(1-s)}$, $|s| \leq 1$.

Theorem 9. Let $X_1, X_2, \cdots, X_n$ be independent Poisson rv's with $X_k \sim P(\lambda_k)$, $k = 1, 2, \cdots, n$. Then $S_n = X_1 + X_2 + \cdots + X_n$ is a $P(\lambda_1 + \lambda_2 + \cdots + \lambda_n)$ rv.

Proof. The proof is left as an exercise.

The converse of Theorem 9 is also true. Indeed, Raikov [93] showed that if $X_1, X_2, \cdots, X_n$ are independent and $S_n = \sum_{i=1}^n X_i$ has a Poisson distribution, each of the rv's $X_1, X_2, \cdots, X_n$ has a Poisson distribution.

Example 9. The number of female insects in a given region follows a Poisson distribution with mean λ. The number of eggs laid by each insect is a $P(\mu)$ rv. We are interested in the probability distribution of the number of eggs in the region.

Let F be the number of female insects in the given region. Then

$$P\{F = f\} = \frac{e^{-\lambda} \lambda^f}{f!}, \quad f = 0, 1, 2, \cdots.$$

Let Y be the number of eggs laid by each insect. Then

$$P\{Y = y, F = f\} = P\{F = f\} \cdot P\{Y = y | F = f\}$$
$$= \frac{e^{-\lambda} \lambda^f}{f!} \frac{(f\mu)^y e^{-\mu f}}{y!}.$$

Thus

$$P\{Y = y\} = \frac{e^{-\lambda} \mu^y}{y!} \sum_{f=0}^{\infty} \frac{(\lambda e^{-\mu})^f f^y}{f!}.$$

The mgf of Y is given by

$$M(t) = \sum_{f=0}^{\infty} \frac{\lambda^f e^{-\lambda}}{f!} \sum_{y=0}^{\infty} \frac{e^{yt}(f\mu)^y}{y!} e^{-\mu f}$$
$$= \sum_{f=0}^{\infty} \frac{\lambda^f e^{-\lambda}}{f!} \exp\{f\mu(e^t - 1)\}$$
$$= e^{-\lambda} \sum_{f=0}^{\infty} \frac{\{\lambda e^{\mu(e^t-1)}\}^f}{f!}$$
$$= e^{-\lambda} \exp\{\lambda e^{\mu(e^t-1)}\}.$$

Theorem 10. Let X and Y be independent rv's with pmf's $P(\lambda_1)$ and $P(\lambda_2)$, respectively. Then the conditional distribution of X, given $X + Y$, is binomial.

Proof. For positive integers m and n, $m < n$, we have

$$\begin{aligned}
P\{X = m \mid X + Y = n\} &= \frac{P\{X = m, Y = n - m\}}{P\{X + Y = n\}} \\
&= \frac{e^{-\lambda_1}(\lambda_1^m / m!) \, e^{-\lambda_2}(\lambda_2^{n-m} / (n-m)!)}{e^{-(\lambda_1 + \lambda_2)}(\lambda_1 + \lambda_2)^n / n!} \\
&= \binom{n}{m} \frac{\lambda_1^m \lambda_2^{n-m}}{(\lambda_1 + \lambda_2)^n} \\
&= \binom{n}{m} \left(\frac{\lambda_1}{\lambda_1 + \lambda_2}\right)^m \left(1 - \frac{\lambda_1}{\lambda_1 + \lambda_2}\right)^{n-m}, \\
m &= 0, 1, 2, \cdots, n,
\end{aligned}$$

and the proof is complete.

Remark 3. The converse of this result is also true in the following sense. If X and Y are independent nonnegative integer-valued rv's such that $P\{X = k\} > 0$, $P\{Y = k\} > 0$, for $k = 0, 1, 2, \cdots$, and the conditional distribution of X, given $X + Y$, is binomial, both X and Y are Poisson. This result is due to Chatterji [12]. For the proof see Problem 13.

Theorem 11. If $X \sim P(\lambda)$ and the conditional distribution of Y, given $X = x$, is $b(x, p)$, then Y is a $P(\lambda p)$ rv.

Proof. The proof is left as an exercise.

Example 10 (Lamperti and Kruskal [67]). Let N be a nonnegative integer-valued rv. Independently of each other, N balls are placed either in urn A with probability $p \, (0 < p < 1)$ or in urn B with probability $1 - p$, resulting in N_A balls in urn A and $N_B = N - N_A$ balls in urn B. We will show that the rv's N_A and N_B are independent if and only if N has a Poisson distribution. We have

$$P\{N_A = a \text{ and } N_B = b \mid N = a + b\} = \binom{a + b}{a} p^a (1 - p)^b,$$

where a, b, are integers ≥ 0. Thus

$$P\{N_A = a, N_B = b\} = \binom{a + b}{a} p^a q^b \, P\{N = n\}, \quad q = 1 - p, \, n = a + b.$$

If N has a Poisson (λ) distribution, then

$$P\{N_A = a, N_B = b\} = \frac{(a+b)!}{a!\,b!} p^a q^b \frac{e^{-\lambda} \lambda^{a+b}}{(a+b)!}$$
$$= \left(\frac{p^a \lambda^a e^{-\lambda/2}}{a!}\right)\left(\frac{q^b \lambda^b}{b!} e^{-\lambda/2}\right)$$

so that N_A and N_B are independent.

Conversely, if N_A and N_B are independent, then

$$P\{N = n\}\, n! = f(a)\, g(b)$$

for some functions f and g. Clearly, $f(0) \neq 0$, $g(0) \neq 0$ because $P\{N_A = 0, N_B = 0\} > 0$. Thus there is a function h such that $h(a+b) = f(a)\,g(b)$ for all nonnegative integers a, b. It follows that

$$h(1) = f(1)\,g(0) = f(0)\,g(1),$$
$$h(2) = f(2)\,g(0) = f(1)\,g(1) = f(0)\,g(2),$$

and so on. By induction,

$$f(a) = f(1)\left[\frac{g(1)}{g(0)}\right]^{a-1}, \qquad g(b) = g(1)\left[\frac{f(1)}{f(0)}\right]^{b-1}.$$

We may write, for some $\alpha_1, \alpha_2, \lambda$,

$$f(a) = \alpha_1 e^{a\lambda}, \qquad g(b) = \alpha_2 e^{b\lambda},$$
$$P\{N = n\} = \alpha_1 \alpha_2 \frac{e^{\lambda(a+b)}}{(a+b)!}$$

so that N is a Poisson rv.

h. The Multinomial Distribution

The binomial distribution is generalized in the following natural fashion. Suppose that an experiment is repeated n times. Each replication of the experiment terminates in one of k mutually exclusive and exhaustive events $A_1, A_2, \cdots, A_k$. Let p_j be the probability that the experiment terminates in A_j, $j = 1, 2, \cdots, k$, and suppose that $p_j (j = 1, 2, \cdots, k)$ remains constant for all n replications. We assume that the n replications are independent.

Let $x_1, x_2, \cdots, x_{k-1}$ be nonnegative integers such that $x_1 + x_2 + \cdots + x_{k-1} \leq n$. Then the probability that exactly x_i trials terminate in A_i, $i = 1, 2, \cdots, k-1$, and hence that $x_k = n - (x_1 + x_2 + \cdots + x_{k-1})$ trials terminate in A_k, is clearly

$$\frac{n!}{x_1!\,x_2!\cdots x_k!} p_1^{x_1} p_2^{x_2} \cdots p_k^{x_k}.$$

If $(X_1, X_2, \cdots, X_k)$ is a random vector such that $X_j = x_j$ means that event A_j has occurred x_j times, $x_j = 0, 1, 2, \cdots, n$, the joint pmf of $(X_1, X_2, \cdots, X_k)$ is given by

(55) $P\{X_1 = x_1, X_2 = x_2, \cdots, X_k = x_k\}$
$$= \begin{cases} \dfrac{n!}{x_1!\, x_2! \cdots x_k!} p_1^{x_1} p_2^{x_2} \cdots p_k^{x_k} & \text{if } n = \sum_1^k x_i, \\ 0 & \text{otherwise.} \end{cases}$$

Definition 4. An rv $(X_1, X_2, \cdots, X_{k-1})$ with joint pmf given by

(56) $P\{X_1 = x_1, X_2 = x_2, \cdots, X_{k-1} = x_{k-1}\}$
$$= \begin{cases} \dfrac{n!}{x_1!\, x_2! \cdots (n - x_1 - \cdots - x_{k-1})!} p_1^{x_1} p_2^{x_2} \cdots p_k^{n - x_1 - \cdots - x_{k-1}} \\ \qquad \text{if } x_1 + x_2 + \cdots + x_{k-1} \le n, \\ 0 \qquad \text{otherwise,} \end{cases}$$

is said to have a multinomial distribution.

For the mgf of $(X_1, X_2, \cdots, X_{k-1})$ we have

$$M(t_1, t_2, \cdots, t_{k-1}) = E e^{t_1 X_1 + t_2 X_2 + \cdots + t_{k-1} X_{k-1}}$$
$$= \sum_{\substack{x_1, x_2, \cdots, x_{k-1}=0 \\ x_1 + x_2 + \cdots + x_{k-1} \le n}}^{n} e^{t_1 x_1 + \cdots + t_{k-1} x_{k-1}} \frac{n!\, p_1^{x_1} p_2^{x_2} \cdots p_k^{x_k}}{x_1!\, x_2! \cdots x_k!}$$
$$= \sum_{\substack{x_1, x_2, \cdots, x_{k-1}=0 \\ x_1 + x_2 + \cdots + x_{k-1} \le n}}^{n} \frac{n!}{x_1!\, x_2! \cdots x_k!} (p_1 e^{t_1})^{x_1} (p_2 e^{t_2})^{x_2} \cdots$$
$$\cdot (p_{k-1} e^{t_{k-1}})^{x_{k-1}} p_k^{x_k}$$
(57) $\quad = (p_1 e^{t_1} + p_2 e^{t_2} + \cdots + p_{k-1} e^{t_{k-1}} + p_k)^n$
$$\text{for all } t_1, t_2, \cdots, t_{k-1} \in \mathscr{R}.$$

Clearly
$$M(t_1, 0, 0, \cdots, 0) = (p_1 e^{t_1} + p_2 + \cdots + p_k)^n = (1 - p_1 + p_1 e^{t_1})^n,$$
which is binomial. Indeed, the marginal pmf of each X_i, $i = 1, 2, \cdots, k-1$, is binomial. Similarly, the joint mgf of X_i, X_j, $i, j = 1, 2, \cdots, k-1$ ($i \ne j$), is

$$M(0, 0, \cdots, 0, t_i, 0, \cdots, 0, t_j, 0, \cdots, 0) = [p_i e^{t_i} + p_j e^{t_j} + (1 - p_i - p_j)]^n,$$

which is the mgf of a *trinomial distribution* with pmf

(58) $\quad f(x_i, x_j) = \dfrac{n!}{x_i!\, x_j!(n - x_i - x_j)!} p_i^{x_i} p_j^{x_j} p_k^{n - x_i - x_j}, \quad p_k = 1 - p_i - p_j.$

Note that the rv's $X_1, X_2, \cdots, X_{k-1}$ are dependent.
From the mgf of $(X_1, X_2, \cdots, X_{k-1})$ or directly from the marginal pmf's we can compute the moments. Thus

(59) $EX_j = np_j$ and var $(X_j) = np_j(1 - p_j)$, $j = 1, 2, \cdots, k - 1$,

and for $j = 1, 2, \cdots, k - 1$, and $i \neq j$,

(60) $\operatorname{cov}(X_i, X_j) = E\{(X_i - np_i)(X_j - np_j)\} = -np_i p_j$.

It follows that the correlation coefficient between X_i and X_j is given by

(61) $\rho_{ij} = -\left[\dfrac{p_i p_j}{(1 - p_i)(1 - p_j)}\right]^{1/2}$, $i, j = 1, 2, \cdots, k - 1 (i \neq j)$.

Example 11. Consider the trinomial distribution with pmf

$$P\{X = x, Y = y\} = \dfrac{n!}{x! y! (n - x - y)!} p_1^x p_2^y p_3^{n-x-y},$$

where x, y are nonnegative integers such that $x + y \leq n$, and $p_1, p_2, p_3 > 0$ with $p_1 + p_2 + p_3 = 1$. The marginal pmf of X is given by

$$P\{X = x\} = \binom{n}{x} p_1^x (1 - p_1)^{n-x}, \quad x = 0, 1, 2, \cdots, n.$$

It follows that

(62) $P\{Y = y \mid X = x\}$

$$= \begin{cases} \dfrac{(n - x)!}{y!(n - x - y)!}\left(\dfrac{p_2}{1 - p_1}\right)^y\left(\dfrac{p_3}{1 - p_1}\right)^{n-x-y} & \text{if } y = 0, 1, 2, \cdots, n - x, \\ 0 & \text{otherwise,} \end{cases}$$

which is $b(n - x, p_2/(1 - p_1))$. Thus

(63) $$E\{Y \mid x\} = (n - x)\dfrac{p_2}{1 - p_1}.$$

Similarly,

(64) $$E\{X \mid y\} = (n - y)\dfrac{p_1}{1 - p_2}.$$

Finally we note that, if $\mathbf{X} = (X_1, X_2, \cdots, X_k)$ and $\mathbf{Y} = (Y_1, Y_2, \cdots, Y_k)$ are two independent multinomial rv's with common parameter $(p_1, p_2, \cdots, p_k)$, then $\mathbf{Z} = \mathbf{X} + \mathbf{Y}$ is also a multinomial rv with probabilities $(p_1, p_2, \cdots, p_k)$. This follows easily if one employs the mgf technique, using (57). Actually this property characterizes the multinomial distribution. If $\mathbf{X}$ and $\mathbf{Y}$ are k-dimensional, nonnegative, independent random vectors, and if $\mathbf{Z} = \mathbf{X} + \mathbf{Y}$ is a multinomial random vector with parameter $(p_1, p_2, \cdots, p_k)$, then $\mathbf{X}$ and $\mathbf{Y}$ also have multinomial distribution with the same parameter. This result is due to Shanbhag and Basawa [113] and will not be proved here.

PROBLEMS 5.2

1. (a) Let us write

$$b(k; n, p) = \binom{n}{k} p^k (1-p)^{n-k}, \quad k = 0, 1, 2, \cdots, n.$$

Show that, as k goes from 0 to n, $b(k; n, p)$ first increases monotonically and then decreases monotonically. The greatest value is assumed when $k = m$, where m is an integer such that

$$(n+1)p - 1 < m \leq (n+1)p$$

except that $b(m-1; n, p) = b(m; n, p)$ when $m = (n+1)p$.

(b) If $k \geq np$, then

$$P\{X \geq k\} \leq b(k; n, p) \frac{(k+1)(1-p)}{k+1-(n+1)p};$$

and if $k \leq np$, then

$$P\{X \leq k\} \leq b(k; n, p) \frac{(n-k+1)p}{(n+1)p - k}.$$

2. Generalize the result in Theorem 10 to n independent Poisson rv's, that is, if $X_1, X_2, \cdots, X_n$ are independent rv's with $X_i \sim P(\lambda_i)$, $i = 1, 2, \cdots, n$, the conditional distribution of $X_1, X_2, \cdots, X_n$, given $\sum_{i=1}^{n} X_i = t$, is multinomial with parameters $t, \lambda_1/\Sigma_1^n \lambda_i, \cdots, \lambda_n/\Sigma_1^n \lambda_i$.

3. Let X_1, X_2 be independent rv's with $X_i \sim b(n_i, \frac{1}{2})$, $i = 1, 2$. What is the pmf of $X_1 - X_2 + n_2$?

4. A box contains N identical balls numbered 1 through N. Of these balls, n are drawn at a time. Let $X_1, X_2, \cdots, X_n$ denote the numbers on the n balls drawn. Let $S_n = \sum_{i=1}^{n} X_i$. Find var (S_n).

5. From a box containing N identical balls marked 1 through N, M balls are drawn one after another without replacement. Let X_i denote the number on the ith ball drawn, $i = 1, 2, \cdots, M$, $1 \leq M \leq N$. Let $Y = \max(X_1, X_2, \cdots, X_M)$. Find the df and the pmf of Y. Also find the conditional distribution of $X_1, X_2, \cdots, X_M$, given $Y = y$. Find EY and var (Y).

6. Let $f(x; r, p)$, $x = 0, 1, 2, \cdots$, denote the pmf of an $NB(r; p)$ rv. Show that the terms $f(x; r, p)$ first increase monotonically and then decrease monotonically. When is the greatest value assumed?

7. Show that the terms

$$P_\lambda\{X = k\} = e^{-\lambda} \frac{\lambda^k}{k!}, \quad k = 0, 1, 2, \cdots,$$

of the Poisson pmf reach their maxima when k is the largest integer $\leq \lambda$.

8. Show that

$$\binom{n}{k} p^k (1-p)^{n-k} \to e^{-\lambda} \frac{\lambda^k}{k!} \quad \text{(uniformly in } k\text{)}$$

as $n \to \infty$ and $p \to 0$, so that $np = \lambda$ remains constant.

(*Hint*: Use Stirling's approximation, namely, $n! \approx \sqrt{2\pi}\, n^{n+1/2} e^{-n}$ as $n \to \infty$.)

9. A biased coin is tossed indefinitely. Let p ($0 < p < 1$) be the probability of success (heads). Let Y_1 denote the length of the first run, and Y_2, the length of the second run. Find the pmf's of Y_1 and Y_2, and show that $EY_1 = q/p + p/q$, $EY_2 = 2$. If Y_n denotes the length of the nth run, $n \geq 1$, what is the pmf of Y_n? Find EY_n.

10. Show that
$$\binom{N}{n}^{-1} \binom{Np}{k} \binom{N(1-p)}{n-k} \to \binom{n}{k} p^k (1-p)^{n-k}$$
as $N \to \infty$.

11. Show that
$$\binom{r+k-1}{k} p^r (1-p)^k \to e^{-\lambda} \frac{\lambda^k}{k!}$$
as $p \to 1$ and $r \to \infty$ in such a way that $r(1-p) = \lambda$ remains fixed.

12. Let X and Y be independent geometric rv's. Show that min (X, Y) and $X - Y$ are independent.

13. Let X and Y be independent rv's with pmfs $P\{X = k\} = p_k$, $P\{Y = k\} = q_k$, $k = 0, 1, 2, \cdots$, where $p_k, q_k > 0$ and $\sum_{k=0}^{\infty} p_k = \sum_{k=0}^{\infty} q_k = 1$. Let
$$P\{X = k \mid X + Y = t\} = \binom{t}{k} \alpha_t^k (1 - \alpha_t)^{t-k}, \qquad 0 \leq k \leq t.$$
Then $\alpha_t = \alpha$ for all t, and
$$p_k = \frac{e^{-\theta\beta}(\theta\beta)^k}{k!}, \qquad q_k = \frac{e^{-\theta}\theta^k}{k!},$$
where $\beta = \alpha/(1 - \alpha)$, and $\theta > 0$ is arbitrary.

(Chatterji [12])

14. Generalize the result of Example 10 to the case of k urns, $k \geq 3$.

15. Let $(X_1, X_2, \cdots, X_{k-1})$ have a multinomial distribution with parameters $n, p_1, p_2, \cdots, p_{k-1}$. Write
$$Y = \sum_{i=1}^{k} \frac{(X_i - np_i)^2}{np_i},$$
where $p_k = 1 - p_1 - \cdots - p_{k-1}$, and $X_k = n - X_1 - \cdots - X_{k-1}$. Find EY and var (Y).

16. Let X_1, X_2 be iid rv's with common df F, having positive mass at $0, 1, 2, \cdots$. Also, let $U = \max(X_1, X_2)$ and $V = X_1 - X_2$. Then
$$P\{U = j, V = 0\} = P\{U = j\} P\{V = 0\}$$
for all j if and only if F is a geometric distribution.

(Srivastava [122])

17. Let X and Y be mutually independent rv's, taking nonegative integer values. Then
$$P\{X \leq n\} - P\{X + Y \leq n\} = \alpha P\{X + Y = n\}$$
holds for $n = 0, 1, 2, \cdots$ and some $\alpha > 0$ if and only if
$$P\{Y = n\} = \frac{1}{1 + \alpha}\left(\frac{\alpha}{1 + \alpha}\right)^n, \quad n = 0, 1, 2, \cdots.$$
(Puri [92])

(*Hint:* Use Problem 3.3.8.)

18. Let $X_1, X_2, \cdots$ be a sequence of independent $b(1, p)$ rv's with $0 < p < 1$. Also, let $Z_N = \sum_{i=1}^{N} X_i$, where N is a $P(\lambda)$ rv. Show that Z_N and $N - Z_N$ are independent.

19. Prove Theorems 5 and 7.

20. Prove Theorem 8.

21. Prove Theorem 11.

5.3 SOME CONTINUOUS DISTRIBUTIONS

In this section we study some most frequently used absolutely continuous distributions and describe their important properties.

a. The Uniform Distribution (Rectangular Distribution)

Definition 1. An rv X is said to have a uniform distribution on the interval $[a, b]$, $-\infty < a < b < \infty$, if its pdf is given by

(1) $$f(x) = \begin{cases} \dfrac{1}{b - a}, & a \leq x \leq b, \\ 0, & \text{otherwise.} \end{cases}$$

We will write $X \sim U[a, b]$ if X has a uniform distribution on $[a, b]$.

The end point a or b or both end points may be excluded. Clearly,
$$\int_{-\infty}^{\infty} f(x) = 1,$$
so that (1) indeed defines a pdf. The df of X is given by

(2) $$F(x) = \begin{cases} 0, & x < a, \\ \dfrac{x - a}{b - a}, & a \leq x < b, \\ 1, & b \leq x; \end{cases}$$

(3) $$EX = \frac{a+b}{2}, \quad EX^k = \frac{b^{k+1} - a^{k+1}}{(k+1)(b-a)}, \quad k > 0 \text{ is an integer;}$$

(4) $$\text{var}(X) = \frac{(b-a)^2}{12};$$

(5) $$M(t) = \frac{1}{t(b-a)}(e^{tb} - e^{ta}), \quad t \neq 0.$$

Example 1. Let X have the density given by

$$f(x) = \begin{cases} \lambda e^{-\lambda x}, & \infty > x > 0, \; \lambda > 0, \\ 0, & \text{otherwise.} \end{cases}$$

Then

$$F(x) = \begin{cases} 0 & x \leq 0, \\ 1 - e^{-\lambda x}, & x > 0. \end{cases}$$

Let $Y = F(X) = 1 - e^{-\lambda X}$. The pdf of Y is given by

$$f_Y(y) = \frac{1}{\lambda} \cdot \frac{1}{1-y} \lambda e^{-\lambda(-\frac{1}{\lambda})\log(1-y)} \quad 0 \leq y < 1,$$
$$= 1, \quad 0 \leq y < 1.$$

Let us define $f_Y(y) = 1$ at $y = 1$. Then we see that Y has the density function

$$f_Y(y) = \begin{cases} 1, & 0 \leq y \leq 1, \\ 0, & \text{otherwise,} \end{cases}$$

which is the $U[0, 1]$ distribution. That this is not a mere coincidence is shown in the following theorem.

Theorem 1 (**Probability Integral Transformation**). Let X be an rv with a continuous df F. Then $F(X)$ has the uniform distribution on $[0,1]$.

Proof. The proof is left as an exercise.

The reader is asked to consider what happens in the case where F is the df of a discrete rv. In the converse direction the following result holds.

Theorem 2. Let F be any df, and let X be a $U[0,1]$ rv. Then there exists a function h such that $h(X)$ has df F, that is,

(6) $$P\{h(X) \leq x\} = F(x) \quad \text{for all } x \in (-\infty, \infty).$$

Proof. If F is the df of a discrete rv Y, let

$$P\{Y = y_k\} = p_k, \quad k = 1, 2, \cdots.$$

Define h as follows:

$$h(x) = \begin{cases} y_1 & \text{if } 0 \le x < p_1, \\ y_2 & \text{if } p_1 \le x < p_1 + p_2, \\ \vdots & \vdots \end{cases}$$

Then

$$P\{h(X) = y_1\} = P\{0 \le X < p_1\} = p_1,$$
$$P\{h(X) = y_2\} = P\{p_1 \le X < p_1 + p_2\} = p_2,$$

and, in general,

$$P\{h(X) = y_k\} = p_k, \quad k = 1, 2, \cdots.$$

Thus $h(X)$ is a discrete rv with df F.

If F is continuous and strictly increasing F^{-1} is well defined, and we take $h(X) = F^{-1}(X)$. We have

$$\begin{aligned} P\{h(X) \le x\} &= P\{F^{-1}(X) \le x\} \\ &= P\{X \le F(x)\} \\ &= F(x), \end{aligned}$$

as asserted.

In general, define

(7) $$F^{-1}(y) = \inf\{x : F(x) \ge y\},$$

and let $h(X) = F^{-1}(X)$. Then we have

(8) $$\{F^{-1}(y) \le x\} = \{y \le F(x)\}.$$

$F^{-1}(y) \le x$ implies, that, for every $\varepsilon > 0$, $y \le F(x + \varepsilon)$. Since $\varepsilon > 0$ is arbitrary and F is continuous on the right, we let $\varepsilon \to 0$ and conclude that $y \le F(x)$. Since $y \le F(x)$ implies $F^{-1}(y) \le x$ by definition (7), it follows that (8) holds generally. Thus

$$P\{F^{-1}(X) \le x\} = P\{X \le F(x)\} = F(x).$$

Theorem 2 is quite useful in generating samples with the help of the uniform distribution.

Example. 2 Let F be the df defined by

$$F(x) = \begin{cases} 0, & x \le 0 \\ 1 - e^{-x}, & x > 0. \end{cases}$$

Then the inverse to $y = 1 - e^{-x}$, $x > 0$, is $x = -\log(1 - y)$, $0 < y < 1$. Thus

$$h(y) = -\log(1 - y),$$

and $-\log(1 - X)$ has the required distribution, where X is a $U[0, 1]$ rv.

SOME CONTINUOUS DISTRIBUTIONS

Theorem 3. Let X be an rv defined on $[0, 1]$. If $P\{x < X \leq y\}$ depends only on the length $y - x$ for all $0 \leq x \leq y \leq 1$, then X is $U[0, 1]$. ($P\{X=0\}=0$.)

Proof. Let $P\{x < X \leq y\} = f(y - x)$; then $f(x + y) = P\{0 < X \leq x + y\}$ $= P\{0 < X \leq x\} + P\{x < X \leq x + y\} = f(x) + f(y)$. Note that f is continuous from the right. We have

$$f(x) = f(x) + f(0),$$

so that

$$f(0) = 0.$$

Also

$$0 = f(0) = f(x - x) = f(x) + f(-x),$$

so that

$$f(-x) = -f(x).$$

We will show that $f(x) = cx$ for some constant c. It suffices to prove the result for positive x. Let m be an integer; then

$$f(x + x + \cdots + x) = f(x) + \cdots + f(x) = mf(x).$$

Letting $x = n/m$, we get

$$f\left(m \cdot \left(\frac{n}{m}\right)\right) = mf\left(\frac{n}{m}\right),$$

so that

$$f\left(\frac{n}{m}\right) = \frac{1}{m} f(n) = \frac{n}{m} f(1),$$

for positive integers n and m. Letting $f(1) = c$, we have proved that

$$f(x) = cx$$

for rational numbers x.

To complete the proof we consider the case where x is a positive irrational number. Then we can find a decreasing sequence of positive rationals $x_1, x_2, \cdots$ such that $x_n \to x$. Since f is right continuous,

$$f(x) = \lim_{x_n \downarrow x} f(x_n) = \lim_{x_n \downarrow x} cx_n = cx.$$

Now, for $0 \leq x \leq 1$,

$$\begin{aligned} F(x) &= P\{X \leq 0\} + P\{0 < X \leq x\} \\ &= F(0) + P\{0 < X \leq x\} \\ &= f(x) \\ &= cx, \quad 0 \leq x \leq 1. \end{aligned}$$

Since $F(1) = 1$, we must have $c = 1$, so that
$$F(x) = x, \quad 0 \leq x \leq 1.$$
This completes the proof.

b. The Gamma Distribution

For $\alpha > 0$, $\Gamma(\alpha)$ is defined by

(9) $$\Gamma(\alpha) = \int_0^\infty x^{\alpha-1} e^{-x} \, dx.$$

In particular, if $\alpha = 1$, $\Gamma(1) = 1$. If $\alpha > 1$, integration by parts yields

(10) $$\Gamma(\alpha) = (\alpha - 1) \int_0^\infty x^{\alpha-2} e^{-x} \, dx = (\alpha - 1) \Gamma(\alpha - 1).$$

If $\alpha = n$ is a positive integer, then

(11) $$\Gamma(n) = (n-1)!.$$

Let us write $x = y/\beta$, $\beta > 0$, in the integral in (9). Then

(12) $$\Gamma(\alpha) = \int_0^\infty \frac{y^{\alpha-1}}{\beta^\alpha} e^{-y/\beta} \, dy,$$

so that

(13) $$\int_0^\infty \frac{1}{\Gamma(\alpha) \beta^\alpha} y^{\alpha-1} e^{-y/\beta} \, dy = 1.$$

Since the integrand in (13) is positive for $y > 0$, it follows that the function

(14) $$f(y) = \begin{cases} \dfrac{1}{\Gamma(\alpha)\beta^\alpha} y^{\alpha-1} e^{-y/\beta} & 0 < y < \infty, \\ 0, & y \leq 0. \end{cases}$$

defines a pdf for $\alpha > 0$, $\beta > 0$.

Definition 2. An rv X with pdf defined by (14) is said to have a gamma distribution with parameters α and β. We will write $X \sim G(\alpha, \beta)$.

The df of a $G(\alpha, \beta)$ rv is given by

(15) $$F(x) = \begin{cases} 0, & x \leq 0, \\ \dfrac{1}{\Gamma(\alpha)\beta^\alpha} \int_0^x y^{\alpha-1} e^{-y/\beta} \, dy, & 0 < x. \end{cases}$$

The mgf of X is easily computed. We have

$$M(t) = \frac{1}{\Gamma(\alpha)\beta^\alpha} \int_0^\infty e^{x(t-1/\beta)} x^{\alpha-1} dx$$

$$= \left(\frac{1}{1-\beta t}\right)^\alpha \int_0^\infty \frac{y^{\alpha-1} e^{-y}}{\Gamma(\alpha)} dy, \quad t < \frac{1}{\beta}$$

(16)
$$= (1-\beta t)^{-\alpha}, \quad t < \frac{1}{\beta}.$$

It follows that

(17) $$EX = M'(t)|_{t=0} = \alpha\beta,$$

(18) $$EX^2 = M''(t)|_{t=0} = \alpha(\alpha+1)\beta^2,$$

so that

(19) $$\operatorname{var}(X) = \alpha\beta^2.$$

Indeed, we can compute the moment of order n directly from the density. We have

$$EX^n = \frac{1}{\Gamma(\alpha)\beta^\alpha} \int_0^\infty e^{-x/\beta} x^{\alpha+n-1} dx$$

$$= \beta^n \frac{\Gamma(\alpha+n)}{\Gamma(\alpha)}.$$

(20) $$= \beta^n(\alpha+n-1)(\alpha+n-2)\cdots\alpha.$$

The special case in which $\alpha = 1$ leads to the *exponential distribution with parameter* β. The pdf of an exponentially distributed rv is therefore

(21) $$f(x) = \begin{cases} \beta^{-1} e^{-x/\beta}, & x > 0, \\ 0, & \text{otherwise.} \end{cases}$$

Note that we can talk of the exponential distribution on $(-\infty, 0)$. The pdf of such an rv is

(22) $$f(x) = \begin{cases} \beta^{-1} e^{x/\beta}, & x < 0, \\ 0, & x \geq 0. \end{cases}$$

Clearly, if $X \sim G(1, \beta)$, we have

(23) $$EX^n = n!\,\beta^n$$

(24) $$EX = \beta \quad \text{and} \quad \operatorname{var}(X) = \beta^2,$$

(25) $$M(t) = (1-\beta t)^{-1} \quad \text{for } t < \beta^{-1}.$$

Another special case of importance is that in which $\alpha = n/2$, $n > 0$ (an integer) and $\beta = 2$.

Definition 3. An rv X is said to have a chi-square distribution (χ^2-distribution) with n degrees of freedom if its pdf is given by

(26) $$f(x) = \begin{cases} \dfrac{1}{\Gamma(n/2)\, 2^{n/2}} e^{-x/2} x^{n/2-1}, & 0 < x < \infty, \\ 0, & x \leq 0. \end{cases}$$

We will write $X \sim \chi^2(n)$ for a χ^2 rv with n degrees of freedom (d.f.). [Note the difference in the abbreviations of distribution function (df) and degrees of freedom (d.f.).]

If $X \sim \chi^2(n)$, then

(27) $$EX = n, \quad \mathrm{var}(X) = 2n,$$

(28) $$EX^k = \frac{2^k \Gamma[(n/2) + k]}{\Gamma(n/2)},$$

and

(29) $$M(t) = (1 - 2t)^{-n/2} \quad \text{for } t < \tfrac{1}{2}.$$

Theorem 4. Let $X_1, X_2, \cdots, X_n$ be independent rv's such that $X_j \sim G(\alpha_j, \beta)$, $j = 1, 2, \cdots, n$. Then $S_n = \sum_{k=1}^{n} X_k$ is a $G(\sum_{j=1}^{n} \alpha_j, \beta)$ rv.

Corollary 1. Let $X_1, X_2, \cdots, X_n$ be iid rv's, each with an exponential distribution with parameter β. Then S_n is a $G(n, \beta)$ rv.

Corollary 2. If $X_1, X_2, \cdots, X_n$ are independent rv's such that $X_j \sim \chi^2(r_j)$, $j = 1, 2, \cdots, n$, then S_n is a $\chi^2(\sum_{i=1}^{n} r_j)$ rv.

The proof of Theorem 4 is simple, and Corollaries 1 and 2 follow immediately.

Theorem 5. Let $X \sim U(0, 1)$. Then $Y = -2 \log X$ is $\chi^2(2)$.

Corollary. Let $X_1, X_2, \cdots, X_n$ be iid rv's with common distribution $U(0, 1)$. Then $-2 \sum_{i=1}^{n} \log X_i = 2 \log(1 / \prod_{i=1}^{n} X_i)$ is $\chi^2(2n)$.

For proof see Example 2.5.5. The corollary follows from Corollary 2 to Theorem 4.

Theorem 6. Let $X \sim G(\alpha_1, \beta)$ and $Y \sim G(\alpha_2, \beta)$ be independent rv's. Then $X + Y$ and X/Y are independent.

Corollary. Let $X \sim G(\alpha_1, \beta)$ and $Y \sim G(\alpha_2, \beta)$ be independent rv's. Then $X + Y$ and $X/(X + Y)$ are independent.

Proof. The proof is straightforward and is left as an exercise.

The converse of Theorem 6 is also true. The result is due to Lukacs [73], and we state it without proof.

Theorem 7. Let X and Y be two nondegenerate rv's that take only positive values. Suppose that $U = X + Y$ and $V = X/Y$ are independent. Then X and Y have gamma distribution with the same parameter β.

Theorem 8. Let $X \sim G(1, \beta)$. Then the rv X has "no memory," that is,

(30) $$P\{X > r + s \mid X > s\} = P\{X > r\},$$

for any two positive real numbers r and s.

Proof. The proof is left as an exercise.

The converse of Theorem 8 is also true in the following sense.

Theorem 9. Let F be a df such that $F(x) = 0$ if $x < 0$, $F(x) < 1$ if $x > 0$, and

(31) $$\frac{1 - F(x + y)}{1 - F(y)} = 1 - F(x) \qquad \text{for all } x, y > 0.$$

Then there exists a constant $\beta > 0$ such that

(32) $$1 - F(x) = e^{-x/\beta}, \qquad x > 0.$$

Proof. Equation (31) is equivalent to

$$g(x + y) = g(x) + g(y)$$

if we write $g(x) = \log\{1 - F(x)\}$. From the proof of Theorem 3 it is clear that the only right continuous solution is $g(x) = cx$. Hence $F(x) = 1 - e^{cx}$, $x \geq 0$. Since $F(x) \to 1$ as $x \to \infty$, it follows that $c < 0$ and the proof is complete.

Theorem 10. Let $X_1, X_2, \cdots, X_n$ be iid rv's. Then $X_i \sim G(1, n\beta)$, $i = 1, 2, \cdots, n$, if and only if $N_n = \min\{X_1, X_2, \cdots, X_n\}$ is $G(1, \beta)$.

Proof. The proof is left as an exercise.

Note that, if $X_1, X_2, \cdots, X_n$ are independent with $X_i \sim G(1, \beta_i)$, $i = 1, 2, \cdots, n$, then N_n is a $G(1, 1/\sum_1^n \beta_i^{-1})$ rv.

Theorem 11 (Desu [23]). Let $X_1, X_2, \cdots, X_n$ be iid nondegenerate rv's with common df F, and let $N_n = \min(X_1, X_2, \cdots, X_n)$. Then X_1 and nN_n are identically distributed if and only if

(33) $$F(x) = 1 - e^{-\lambda x} \quad \text{for } x \geq 0,$$

where λ is a positive constant.

Proof. If
$$F(x) = \begin{cases} 1 - e^{-\lambda x} & \text{if } x \geq 0, \\ 0 & \text{if } x < 0, \end{cases}$$

then

(34) $$F_{nN_n}(y) = P\left\{N_n \leq \frac{y}{n}\right\} = 1 - \left\{1 - F\left(\frac{y}{n}\right)\right\}^n$$
$$= 1 - e^{-\lambda y} = F(y).$$

Conversely, let

(35) $$F_{nN_n}(y) = F(y) \quad \text{for real } y.$$

Let $G(y) = 1 - F(y)$. Then
$$1 - F(y) = \left[1 - F\left(\frac{y}{n}\right)\right]^n,$$

and it follows that

(36) $$G(ny) = G^n(y) \quad \text{for real } y \text{ and all integers } n \geq 1.$$

Thus $G(0) = G^n(0)$, and we must have $G(0) = 0$ or $G(0) = 1$. We show that $G(0) = 1$.

Let $y_0 < 0$ so that $G(y_0) < 1$. Then $G^n(y_0) \to 0$ and $G(ny_0) \to 1$ as $n \to \infty$, which is a contradiction. Thus, for all $y < 0$, $G(y) \geq 1$, and hence $G(y) = 1$ for $y < 0$. Since F is nondegenerate, we cannot have $G(y) = 0$ for all $y \geq 0$, so that $G(y_1) > 0$ for some $y_1 \geq 0$. If $y_1 = 0$, then $G(0) > 0$, and hence $G(0) = 1$. On the other hand, if $G(y_1) > 0$ for $y_1 > 0$, then $G(y) > 0$ for $0 \leq y \leq y_1$, since G is nonincreasing. Since G is right continuous, it follows that $G(0) = 1$. Thus

(37) $$G(y) \begin{cases} = 1 & \text{for } y \leq 0, \\ > 0 & \text{for } 0 < y \leq y_1. \end{cases}$$

Suppose that there is a constant $y_2 > y_1$ for which $G(y_2) = 0$. Then, from (36),

(38) $$G(y_2) = G^n\left(\frac{y_2}{n}\right) = 0 \quad \text{for } n \geq 1.$$

It follows that $G(y) = 0$ for $y > y_2/n$; therefore, for sufficiently large n,

$G(y) = 0$ when $y_2/n \leq y \leq y_1$, which contradicts (37). Hence we have shown that

(39) $\qquad G(y) > 0 \qquad$ for $y > 0$.

From (36), we have
$$G^n(1) = G(n) = G\left(m \cdot \frac{n}{m}\right) = G^m\left(\frac{n}{m}\right),$$

so that

(40) $\qquad G\left(\frac{n}{m}\right) = G^{n/m}(1) \qquad$ for $n, m \geq 1$ (n, m integers).

Since $G(y)$ and $\{G(1)\}^y$ are both nonincreasing and coincide for positive rational numbers y, and since $\{G(1)\}^y$ is continuous, it follows that

(41) $\qquad G(y) = \{G(1)\}^y = e^{y \log G(1)} \qquad$ for all real $y > 0$.

Since F is a nondegenerate df,

(42) $\qquad \lim_{y \to \infty} G(y) = 0,$

so that $G(1) < 1$. Hence

(43) $\qquad G(y) = \begin{cases} e^{-\lambda y}, & y > 0, \\ 1, & y \leq 0, \end{cases}$

where $\lambda = -\log G(1)$.

Corollary. Let Y be a nondegenerate rv with df H, where $H(0) = 0$. Also, let $Y_1, Y_2, \cdots, Y_n$ be iid rv's with common df H. Then Y and $M_n = \max\{Y_1^n, \cdots, Y_n^n\}$ have the same df H (for each integer $n \geq 2$) if and only if
$$H(y) = y^\lambda \qquad 0 < y < 1 \qquad \text{for some } \lambda > 0.$$

For proof take $X = -\log Y$ in Theorem 11.

The following result describes the relationship between exponential and Poisson rv's.

Theorem 12. Let $X_1, X_2, \cdots$ be a sequence of iid rv's having common exponential density with parameter $\beta > 0$. Let $S_n = \sum_{k=1}^n X_k$ be the nth partial sum, $n = 1, 2, \cdots$, and suppose that $t > 0$. If $Y = $ number of $S_n \in [0, t]$, then Y is a $P(t/\beta)$ rv.

Proof. We have
$$P\{Y = 0\} = P\{S_1 > t\} = \frac{1}{\beta} \int_t^\infty e^{-x/\beta} \, dx = e^{-t/\beta},$$

so that the assertion holds for $Y = 0$. Let n be a positive integer. Since the X_i's are nonnegative, S_n is nondecreasing, and

(44) $$P\{Y = n\} = P\{S_n \le t, S_{n+1} > t\}.$$

Now

(45) $$P\{S_n \le t\} = P\{S_n \le t, S_{n+1} > t\} + P\{S_{n+1} \le t\}.$$

It follows that

(46) $$P\{Y = n\} = P\{S_n \le t\} - P\{S_{n+1} \le t\},$$

and, since $S_n \sim G(n, \beta)$, we have

$$P\{Y = n\} = \int_0^t \frac{1}{\Gamma(n)\beta^n} x^{n-1} e^{-x/\beta} \, dx - \int_0^t \frac{1}{\Gamma(n+1)\beta^{n+1}} x^n e^{-x/\beta} \, dx$$
$$= \frac{t^n e^{-t/\beta}}{\beta^n n!},$$

as asserted.

Theorem 13. If X and Y are independent exponential rv's with parameter β, then $Z = X/(X + Y)$ has a $U(0, 1)$ distribution.

Proof. The proof is essentially contained in Problem 4.4.2.

Note that, in view of Theorem 7, Theorem 13 characterizes the exponential distribution in the following sense. Let X and Y be independent rv's that are nondegenerate and take only positive values. Suppose that $X + Y$ and X/Y are independent. If $X/(X + Y)$ is $U(0, 1)$, X and Y both have the exponential distribution with parameter β. This follows since, by Theorem 7, X and Y must have the gamma distribution with parameter β. Thus $X/(X + Y)$ must have (see Theorem 15) the pdf

$$f(x) = \frac{\Gamma(\alpha_1 + \alpha_2)}{\Gamma(\alpha_1)\Gamma(\alpha_2)} x^{\alpha_1 - 1}(1 - x)^{\alpha_2 - 1}, \quad 0 < x < 1,$$

and this is the uniform density on $(0, 1)$ if and only if $\alpha_1 = \alpha_2 = 1$. Thus X and Y both have the $G(1, \beta)$ distribution.

Theorem 14. Let X be a $P(\lambda)$ rv. Then

(47) $$P\{X \le K\} = \frac{1}{K!} \int_\lambda^\infty e^{-x} x^K \, dx$$

expresses the df of X in terms of an incomplete gamma function.

Proof.
$$\frac{d}{d\lambda} P\{X \le K\} = \sum_{j=0}^{K} \frac{1}{j!} \{je^{-\lambda}\lambda^{j-1} - \lambda^j e^{-\lambda}\}$$
$$= \frac{-\lambda^K e^{-\lambda}}{K!},$$

and it follows that

$$P\{X \le K\} = \frac{1}{K!} \int_{\lambda}^{\infty} e^{-x} x^K \, dx,$$

as asserted.

An alternative way of writing (47) is the following:

$$P\{X \le K\} = P\{Y \ge 2\lambda\},$$

where $X \sim P(\lambda)$, and $Y \sim \chi^2(2K+2)$.

c. The Beta Distribution

For $\alpha > 0$, $\beta > 0$, $B(\alpha, \beta)$ is defined by

(48) $$B(\alpha, \beta) = \int_{0+}^{1-} x^{\alpha-1}(1-x)^{\beta-1} \, dx.$$

It follows that

(49) $$f(x) = \begin{cases} \dfrac{x^{\alpha-1}(1-x)^{\beta-1}}{B(\alpha, \beta)}, & 0 < x < 1, \\ 0, & \text{otherwise}, \end{cases}$$

defines a pdf.

Definition 4. An rv X with pdf given by (49) is said to have a beta distribution with parameters α and β, $\alpha > 0$, $\beta > 0$. We will write $X \sim B(\alpha, \beta)$ for a beta variable with density (49).

The df of a $B(\alpha, \beta)$ rv is given by

(50) $$F(x) = \begin{cases} 0, & x \le 0, \\ [B(\alpha, \beta)]^{-1} \int_{0+}^{x} y^{\alpha-1}(1-y)^{\beta-1} \, dy, & 0 < x < 1, \\ 1, & x \ge 1. \end{cases}$$

If n is a positive number, then

$$EX^n = \frac{1}{B(\alpha, \beta)} \int_0^1 x^{n+\alpha-1}(1-x)^{\beta-1} \, dx$$

(51) $$= \frac{B(n+\alpha, \beta)}{B(\alpha, \beta)} = \frac{\Gamma(n+\alpha)\,\Gamma(\alpha+\beta)}{\Gamma(\alpha)\,\Gamma(n+\alpha+\beta)},$$

using the fact that $B(\alpha, \beta) = \dfrac{\Gamma(\alpha)\,\Gamma(\beta)}{\Gamma(\alpha + \beta)}$. In particular,

$$(52) \qquad EX = \frac{\alpha}{\alpha + \beta}$$

and

$$(53) \qquad \mathrm{var}\,(X) = \frac{\alpha\beta}{(\alpha + \beta)^2(\alpha + \beta + 1)}.$$

For the mgf of $X \sim B(\alpha, \beta)$, we have

$$(54) \qquad M(t) = \frac{1}{B(\alpha, \beta)} \int_0^1 e^{tx} x^{\alpha-1}(1 - x)^{\beta-1}\,dx.$$

Since moments of all order exist, and $E|X|^j < 1$ for all j, we have

$$M(t) = \sum_{j=0}^{\infty} \frac{t^j}{j!}\,EX^j$$

$$(55) \qquad = \sum_{j=0}^{\infty} \frac{t^j}{\Gamma(j+1)} \frac{\Gamma(\alpha+j)\,\Gamma(\alpha+\beta)}{\Gamma(\alpha+\beta+j)\Gamma(\alpha)}.$$

Remark 1. Note that in the special case where $\alpha = \beta = 1$ we get the uniform distribution on $(0, 1)$.

Remark. 2 If X is a beta rv with parameters α and β, then $1 - X$ is a beta variate with parameters β and α. In particular, X is $B(\alpha, \alpha)$ if and only if $1 - X$ is $B(\alpha, \alpha)$. A special case is the uniform distribution on $(0, 1)$. If X and $1 - X$ have the same distribution, it does not follow that X has to be $B(\alpha, \alpha)$. All this entails is that the pdf satisfies

$$f(x) = f(1 - x), \qquad 0 < x < 1.$$

Take

$$f(x) = \frac{1}{B(\alpha, \beta) + B(\beta, \alpha)}\,[x^{\alpha-1}(1-x)^{\beta-1} + (1-x)^{\alpha-1} x^{\beta-1}], \quad 0 < x < 1.$$

Example 3. Let X be distributed with pdf

$$f(x) = \begin{cases} \frac{1}{12} x^2(1 - x), & 0 < x < 1, \\ 0, & \text{otherwise.} \end{cases}$$

Then $X \sim B(3, 2)$ and

$$EX^n = \frac{\Gamma(n+3)\Gamma(5)}{\Gamma(3)\Gamma(n+5)} = \frac{4!}{2!} \cdot \frac{(n+2)!}{(n+4)!} = \frac{12}{(n+4)(n+3)},$$

$$EX = \tfrac{12}{20}, \qquad \mathrm{var}\,(X) = \frac{6}{5^2 \cdot 6} = \frac{1}{25},$$

$$M(t) = \sum_{j=0}^{\infty} \frac{t^j}{j!} \cdot \frac{(j+2)!}{(j+4)!} \frac{4!}{2!}$$
$$= \sum_{j=0}^{\infty} \frac{12}{(j+4)(j+3)} \cdot \frac{t^j}{j!},$$

and

$$P\{.2 < X < .5\} = \frac{1}{12} \int_{.2}^{.5} (x^2 - x^3)\, dx$$
$$= .023.$$

Theorem 15. Let X and Y be independent $G(\alpha_1, \beta)$ and $G(\alpha_2, \beta)$, respectively, rv's. Then $X/(X + Y)$ is a $B(\alpha_1, \alpha_2)$ rv.

Proof. The proof is left as an exercise.

Theorem 16. Let $X \sim B(\alpha_1, \beta_1)$ and $Y \sim B(\alpha_2, \beta_2)$, and let X and Y be independent. Suppose that $\alpha_1 = \alpha_2 + \beta_2$. Then XY is a $B(\alpha_2, \beta_1 + \beta_2)$ rv. If $\alpha_2 = \alpha_1 + \beta_1$, then XY is $B(\alpha_1, \beta_1 + \beta_2)$.

Proof. For the pdf of $XY = Z$, say, we have

$$f(z) = \int_z^1 \frac{1}{x} \cdot \frac{x^{\alpha_2+\beta_2-1}(1-x)^{\beta_1-1}}{B(\alpha_1, \beta_1)} \cdot \frac{(z/x)^{\alpha_2-1}[1-(z/x)]^{\beta_2-1}}{B(\alpha_2, \beta_2)}\, dx$$
$$= \frac{z^{\alpha_2-1}}{B(\alpha_1, \beta_1)} \frac{1}{B(\alpha_2, \beta_2)} \int_z^1 (1-x)^{\beta_1-1}(x-z)^{\beta_2-1}\, dx$$
$$= \frac{z^{\alpha_2-1}(1-z)^{\beta_1+\beta_2-1}}{B(\alpha_1, \beta_1) B(\alpha_2, \beta_2)} \int_0^1 (1-u)^{\beta_1-1} u^{\beta_2-1}\, du,$$

where we substituted $u = (x - z)/(1 - z)$ in the integral. Thus

$$f(z) = \frac{z^{\alpha_2-1}(1-z)^{\beta_1+\beta_2-1}}{B(\alpha_1, \beta_1)} \frac{B(\beta_2, \beta_1)}{B(\alpha_2, \beta_2)}$$
$$= \frac{z^{\alpha_2-1}(1-z)^{\beta_1+\beta_2-1}}{B(\alpha_2, \beta_1+\beta_2)},$$

as asserted. The second assertion follows similarly.

We remark that a partial converse to Theorem 16 holds; that is, if XY is $B(\alpha, \beta)$, and X and Y are independent $B(\alpha_1, \beta_1)$ and $B(\alpha_2, \beta_2)$, respectively, we have $\beta = \beta_1 + \beta_2$, and $\alpha = \alpha_1$ or $\alpha = \alpha_2$. This is easy to see, since, for all $t \geq 0$, $EX^t\, EY^t = E(XY)^t$, so that

$$\left(\frac{\alpha_1 + \beta_1 + t}{\alpha_1 + t}\right)\left(\frac{\alpha_2 + \beta_2 + t}{\alpha_2 + t}\right) = \frac{\alpha + \beta + t}{\alpha + t} \quad \text{holds for all } t \geq 0.$$

Cross-multiplying and equating coefficients of t^k, $k = 0, 1, 2$, we get $\beta = \beta_1 + \beta_2$ and either $\alpha = \alpha_1$ or $\alpha = \alpha_2$. (See Ramachandaran [94].)

Theorem 17. Let $X_1, X_2, \cdots, X_k$ be independent beta rv's with $X_i \sim B(\alpha_i, \beta_i)$, $i = 1, 2, \cdots, k$. Let $\alpha_i = \alpha_{i+1} + \beta_{i+1}$, $i = 1, 2, \cdots, k - 1$. Then $\prod_{i=1}^{k} X_i$ is also a beta rv with parameters $(\alpha_k, \beta_1 + \beta_2 + \cdots + \beta_k)$.

Let $X_1, X_2, \cdots, X_n$ be iid rv's with the uniform distribution on $[0, 1]$. Let $X_{(k)}$ be the kth-order statistic.

Theorem 18. The rv $X_{(k)}$ has a beta distribution with parameters $\alpha = k$ and $\beta = n - k + 1$.

Proof. Let X be the number of X_i's that lie in $[0, t]$. Then X is $b(n, t)$. We have

$$P\{X_{(k)} \leq t\} = P\{X \geq k\}$$
$$= \sum_{j=k}^{n} \binom{n}{j} t^j (1 - t)^{n-j}.$$

Also

$$\frac{d}{dt} P\{X \geq k\} = \sum_{j=k}^{n} \binom{n}{j} \{jt^{j-1}(1 - t)^{n-j} - (n - j)t^j (1 - t)^{n-j-1}\}$$
$$= \sum_{j=k}^{n} \{n\binom{n-1}{j-1} t^{j-1}(1 - t)^{n-j} - n\binom{n-1}{j} t^j (1 - t)^{n-j-1}\}$$
$$= n\binom{n-1}{k-1} t^{k-1}(1 - t)^{n-k}.$$

On integration, we get

$$P\{X_{(k)} \leq t\} = n\binom{n-1}{k-1} \int_0^t x^{k-1}(1 - x)^{n-k} dx,$$

as asserted.

Remark 3. Note that we have shown that, if X is $b(n, p)$, then

(56) $$1 - P\{X < k\} = n\binom{n-1}{k-1} \int_0^p x^{k-1}(1 - x)^{n-k} dx,$$

which expresses the df of X in terms of an incomplete beta function.

Theorem 19. Let $X_1, X_2, \cdots, X_n$ be independent rv's, and write $M_n = \max(X_1, X_2, \cdots, X_n)$. Then $X_1, X_2, \cdots, X_n$ are iid $B(\alpha, 1)$ rv's if and only if $M_n \sim B(\alpha n, 1)$.

Proof. The proof is simple.

For a more general result see the corollary to Theorem 11.

d. The Cauchy Distribution

Definition 5. An rv X is said to have a Cauchy distribution with parameters μ and θ if its pdf is given by

$$(57) \qquad f(x) = \frac{\mu}{\pi} \frac{1}{\mu^2 + (x - \theta)^2}, \qquad -\infty < x < \infty, \quad \mu > 0.$$

We will write $X \sim \mathscr{C}(\mu, \theta)$ for a Cauchy rv with density (57).

We first check that (57) in fact defines a pdf. Substituting $y = (x - \theta)/\mu$, we get

$$\int_{-\infty}^{\infty} f(x)\,dx = \frac{1}{\pi} \int_{-\infty}^{\infty} \frac{dy}{1 + y^2} = \frac{2}{\pi} (\tan^{-1} y)\Big|_0^{\infty} = 1.$$

The df of a $\mathscr{C}(1, 0)$ rv is given by

$$(58) \qquad F(x) = \frac{1}{2} + \frac{1}{\pi} \tan^{-1} x, \qquad -\infty < x < \infty.$$

Theorem 20. Let X be a Cauchy rv with parameters μ and θ. The moments of order < 1 exist, but the moments of order ≥ 1 do not exist for the rv X.

Proof. It suffices to consider the pdf

$$f(x) = \frac{1}{\pi} \frac{1}{1 + x^2}, \qquad -\infty < x < \infty.$$

$$E|X|^\alpha = \frac{2}{\pi} \int_0^\infty x^\alpha \frac{1}{1 + x^2} dx,$$

and, letting $z = 1/(1 + x^2)$ in the integral, we get

$$E|X|^\alpha = \frac{1}{\pi} \int_0^1 z^{(1-\alpha)/2 - 1} (1 - z)^{[(\alpha+1)/2]-1} dz,$$

which converges for $\alpha < 1$ and diverges for $\alpha \geq 1$ (Widder [137], 374). This completes the proof of the theorem.

It follows from Theorem 20 that the mgf of a Cauchy rv does not exist. This creates some manipulative problems.

Theorem 21. Let $X \sim \mathscr{C}(\mu_1, \theta_1)$, and $Y \sim \mathscr{C}(\mu_2, \theta_2)$ be independent rv's. Then $X + Y$ is a $\mathscr{C}(\mu_1 + \mu_2, \theta_1 + \theta_2)$ rv.

Proof. For notational convenience we will prove the result in the special case where $\mu_1 = \mu_2 = 1$ and $\theta_1 = \theta_2 = 0$, that is, where X and Y have the common pdf

$$f(x) = \frac{1}{\pi} \cdot \frac{1}{1+x^2}, \quad -\infty < x < \infty.$$

The proof in the general case follows along the same lines. If $Z = X + Y$, the pdf of Z is given by

$$f_Z(z) = \frac{1}{\pi^2} \int_{-\infty}^{\infty} \frac{1}{1+x^2} \cdot \frac{1}{1+(z-x)^2} \, dx.$$

Now

$$\frac{1}{(1+x^2)[1+(z-x)^2]} = \frac{1}{z^2(z^2+4)} \left[\frac{2zx}{1+x^2} + \frac{z^2}{1+x^2} + \frac{2z^2 - 2zx}{1+(z-x)^2} + \frac{z^2}{1+(z-x)^2} \right],$$

so that

$$f_Z(z) = \frac{1}{\pi^2} \frac{1}{z^2(z^2+4)} \left[z \log \frac{1+x^2}{1+(z-x)^2} + z^2 \tan^{-1} x + z^2 \tan^{-1}(x-z) \right]_{-\infty}^{\infty}$$

$$= \frac{1}{\pi} \frac{2}{z^2 + 2^2}, \quad -\infty < z < \infty.$$

It follows that, if X and Y are iid $\mathscr{C}(1, 0)$ rv's, then $X + Y$ is a $\mathscr{C}(2, 0)$ rv.

Corollary. Let $X_1, X_2, \cdots, X_n$ be independent Cauchy rv's, $X_k \sim \mathscr{C}(\mu_k, \theta_k)$, $k = 1, 2, \cdots, n$. Then $S_n = \sum_1^n X_k$ is a $\mathscr{C}(\sum_1^n \mu_k, \sum_1^n \theta_k)$ rv.

In particular, if $X_1, X_2, \cdots, X_n$ are iid $\mathscr{C}(1, 0)$ rv's, $n^{-1} S_n$ is also a $\mathscr{C}(1, 0)$ rv. This is a remarkable result, the importance of which will become clear in Chapter 6. Actually this property uniquely characterizes the Cauchy distribution. If F is a nondegenerate df with the property that $n^{-1} S_n$ also has df F, then F must be a Cauchy distribution. (See Thompson [130], 112.)

The proof of the following result is simple.

Theorem 22. Let X be $\mathscr{C}(\mu, 0)$. Then λ/X, where λ is a constant, is a $\mathscr{C}(|\lambda|/\mu, 0)$ rv.

Corollary. X is $\mathscr{C}(1, 0)$ if and only $1/X$ is $\mathscr{C}(1, 0)$.

We emphasize that, if X and $1/X$ have the same pdf on $(-\infty, \infty)$, it does not follow[†] that X is $\mathscr{C}(1, 0)$, for let X be an rv with pdf

$$f(x) = \tfrac{1}{4} \quad \text{if } |x| \leq 1,$$

[†]Menon [80] has shown that we need the condition that both X and $1/X$ be *stable* to conclude that X is Cauchy.

$$= \frac{1}{4x^2} \quad \text{if } |x| > 1.$$

Then X and $1/X$ have the same pdf, as can be easily checked.

Theorem 23. Let X be a $U(-\pi/2, \pi/2)$ rv. Then $Y = \tan X$ is a Cauchy rv.

Many important properties of the Cauchy distribution can be derived from this result. (See Pitman and Williams [89].)

e. The Normal Distribution (the Gaussian Law)

One of the most important distributions in the study of probability and mathematical statistics is the *normal distribution*, which we will examine presently.

Definition 6. An rv X is said to have a standard normal distribution if its pdf is given by

(59) $$\varphi(x) = \frac{1}{\sqrt{2\pi}} e^{-(x^2/2)}, \quad -\infty < x < \infty.$$

We first check that f defines a pdf. Let

$$I = \int_{-\infty}^{\infty} e^{-x^2/2} \, dx.$$

Then

$$0 < e^{-x^2/2} < e^{-|x|+1}, \quad -\infty < x < \infty,$$

$$\int_{-\infty}^{\infty} e^{-|x|+1} \, dx = 2e,$$

and it follows that I exists. We have

$$I = \int_0^{\infty} y^{-1/2} e^{-y/2} \, dy$$
$$= \Gamma(\tfrac{1}{2}) \, 2^{1/2}$$
$$= \sqrt{2\pi}.$$

Thus $\int_{-\infty}^{\infty} \varphi(x) \, dx = 1$, as required.

A nondegenerate distribution function F is said to be stable if, for two iid rv's X_1, X_2 with common df F, and given constants $a_1, a_2 > 0$, we can find $\alpha > 0$ and $\beta(a_1, a_2)$ such that the rv

$$X_3 = \alpha^{-1}(a_1 X_1 + a_2 X_2 - \beta)$$

again has the same distribution F. Examples are the Cauchy (see the corollary to Theorem 21) and normal (discussed in Section 5.3e) distributions. See also Section 6.6.

Let us write $Y = \sigma X + \mu$, where $\sigma > 0$. Then the pdf of Y is given by

$$\psi(y) = \frac{1}{\sigma} \varphi\left(\frac{y-\mu}{\sigma}\right)$$

(60)
$$= \frac{1}{\sigma\sqrt{2\pi}} e^{-[(y-\mu)^2/2\sigma^2]}, \quad -\infty < y < \infty; \sigma > 0, -\infty < \mu < \infty.$$

Definition 7. An rv X is said to have a normal distribution with parameters $\mu (-\infty < \mu < \infty)$ and $\sigma (> 0)$ if its pdf is given by (60).

If X is a normally distributed rv with parameters μ and σ, we will write $X \sim \mathcal{N}(\mu, \sigma^2)$. In this notation φ defined by (59) is the pdf of an $\mathcal{N}(0, 1)$ rv. The df of an $\mathcal{N}(0, 1)$ rv will be denoted by $\Phi(x)$, where

(61)
$$\Phi(x) = \frac{1}{\sqrt{2\pi}} \int_{-\infty}^{x} e^{-u^2/2} \, du.$$

Clearly, if $X \sim \mathcal{N}(\mu, \sigma^2)$, then $Z = (X-\mu)/\sigma \sim \mathcal{N}(0, 1)$. Z is called a *standard normal rv*. For the mgf of an $\mathcal{N}(\mu, \sigma^2)$ rv, we have

$$M(t) = \frac{1}{\sqrt{2\pi}\,\sigma} \int_{-\infty}^{\infty} \exp\left\{\frac{-x^2}{2\sigma^2} + x\frac{(t\sigma^2 + \mu)}{\sigma^2} - \frac{\mu^2}{2\sigma^2}\right\} dx$$

$$= \frac{1}{\sqrt{2\pi}\,\sigma} \int_{-\infty}^{\infty} \exp\left\{\frac{-(x-\mu-\sigma^2 t)^2}{2\sigma^2} + \mu t + \frac{\sigma^2 t^2}{2}\right\} dx$$

(62)
$$= \exp\left(\mu t + \frac{\sigma^2 t^2}{2}\right),$$

for all real values of t. Moments of all order exist and may be computed from the mgf. Thus

(63)
$$EX = M'(t)\big|_{t=0} = (\mu + \sigma^2 t) M(t)\big|_{t=0} = \mu,$$

and

$$EX^2 = M''(t)\big|_{t=0} = [M(t)\sigma^2 + (\mu + \sigma^2 t)^2 M(t)]_{t=0}$$

(64)
$$= \sigma^2 + \mu^2.$$

Thus

(65)
$$\text{var}(X) = \sigma^2.$$

Clearly, the central moments of odd order are all zero. The central moments of even order are as follows:

$$E\{X - \mu\}^{2n} = \frac{1}{\sigma\sqrt{2\pi}} \int_{-\infty}^{\infty} x^{2n} e^{-x^2/2\sigma^2} \, dx \quad (n \text{ is a positive integer})$$

$$= \frac{\sigma^{2n}}{\sqrt{2\pi}} 2^{n+1/2} \Gamma(n + \tfrac{1}{2})$$

(66)
$$= [(2n - 1)(2n - 3) \cdots 3 \cdot 1]\sigma^{2n}.$$

As for the absolute moment of order α, for a standard normal rv Z we have

$$E|Z|^\alpha = \frac{1}{\sqrt{2\pi}} 2 \int_0^\infty z^\alpha e^{-z^2/2} \, dz$$

$$= \frac{1}{\sqrt{2\pi}} \int_0^\infty y^{[(\alpha+1)/2]-1} e^{-y/2} \, dy$$

(67)
$$= \frac{\Gamma[(\alpha + 1)/2] \, 2^{\alpha/2}}{\sqrt{\pi}}.$$

As remarked earlier, the normal distribution is one of the most important distributions in probability and statistics, and for this reason the standard normal distribution is available in tabular form. Table 2 on page 651 gives the probability $P\{Z > z\}$ for various values of $z(> 0)$ in the tail of an $\mathcal{N}(0, 1)$ rv. In this book we will write z_α for the value of Z that satisfies $\alpha = P\{Z > z_\alpha\}$, $0 \le \alpha \le 1$.

Example 4. By Chebychev's inequality, if $E|X|^2 < \infty$, $EX = \mu$, and var $(X) = \sigma^2$, then

$$P\{|X - \mu| > K\sigma\} \le \frac{1}{K^2}.$$

For $K = 2$, we get $P\{|X - \mu| > K\sigma\} \le .25$, and for $K = 3$, we have $P\{|X - \mu| > K\sigma\} \le \tfrac{1}{9}$. If X is, in particular, $\mathcal{N}(\mu, \sigma^2)$, then

$$P\{|X - \mu| > K\sigma\} = P\{|Z| > K\},$$

where Z is $\mathcal{N}(0, 1)$. From Table 2 on page 651

$$P\{|Z| > 1\} = .318, \quad P\{|Z| > 2\} = .046, \quad \text{and} \quad P\{|Z| > 3\} = .002.$$

Thus practically all the distribution is concentrated within three standard deviations of the mean.

Example 5. Let $X \sim \mathcal{N}(3, 4)$. Then

$$P\{2 < X \le 5\} = P\left\{\frac{2-3}{2} < \frac{X-3}{2} \le \frac{5-3}{2}\right\} = P\{-.5 < Z \le 1\}$$
$$= P\{Z \le 1\} - P\{Z \le -.5\}$$
$$= .841 - P\{Z \ge .5\}$$
$$= 0.841 - .309 = .532.$$

Theorem 24 (Feller [28], 175). Let Z be a standard normal rv. Then

(68) $$P\{Z > x\} \approx \frac{1}{\sqrt{2\pi}\, x} e^{-x^2/2} \quad \text{as} \quad x \to \infty.$$

More precisely, for every $x > 1/\sqrt{2}$

(69) $$\frac{1}{\sqrt{2\pi}} e^{-x^2/2}\left(\frac{1}{x} - \frac{1}{x^3}\right) < P\{Z > x\} < \frac{1}{x\sqrt{2\pi}} e^{-x^2/2}.$$

Proof. We have

(70) $$\frac{1}{\sqrt{2\pi}} \int_x^\infty e^{-(1/2)y^2}\left(1 - \frac{3}{y^4}\right) dy = \frac{1}{\sqrt{2\pi}} e^{-x^2/2}\left(\frac{1}{x} - \frac{1}{x^3}\right)$$

and

(71) $$\frac{1}{\sqrt{2\pi}} \int_x^\infty e^{-y^2/2}\left(1 + \frac{1}{y^2}\right) dy = \frac{1}{\sqrt{2\pi}} e^{-x^2/2} \frac{1}{x},$$

as can be checked on differentiation. Equation (69) follows immediately.

Remark 4. The lower bound in (69) can be improved quite easily. Indeed, from Problem 4.6.6, we have for $x > 0$

(72) $$P\{Z > x\} \geq \frac{1}{2\sqrt{2\pi}} e^{-x^2/2} (\sqrt{4 + x^2} - x).$$

Theorem 25. Let $X_1, X_2, \cdots, X_n$ be independent rv's with $X_k \sim \mathcal{N}(\mu_k, \sigma_k^2)$, $k = 1, 2, \cdots, n$. Then $S_n = \sum_{k=1}^n X_k$ is an $\mathcal{N}(\sum_{k=1}^n \mu_k, \sum_1^n \sigma_k^2)$ rv.

Corollary 1. If $X_1, X_2, \cdots, X_n$ are iid $\mathcal{N}(\mu, \sigma^2)$ rv's, then S_n is an $\mathcal{N}(n\mu, n\sigma^2)$ rv and $n^{-1} S_n$ is an $\mathcal{N}(\mu, \sigma^2/n)$ rv.

Corollary 2. If $X_1, X_2, \cdots, X_n$ are iid $\mathcal{N}(0, 1)$ rv's, then $n^{-1/2} S_n$ is also an $\mathcal{N}(0, 1)$ rv.

We remark that if $X_1, X_2, \cdots, X_n$ are iid rv's with $EX = 0$, $EX^2 = 1$ such that $n^{-1/2} S_n$ also has the same distribution for each $n = 1, 2, \cdots$, that distribution can only be $\mathcal{N}(0, 1)$. This characterization of the normal distribution will become clear when we study the central limit theorem in Chapter 6.

Theorem 26. Let X and Y be independent rv's. Then $X + Y$ is normally distributed if and only if X and Y are both normal.

If X and Y are independent normal rv's, $X + Y$ is normal by Theorem 25. The converse is due to Cramér [17] and will not be proved here.

Theorem 27. Let X and Y be independent rv's with common $\mathcal{N}(0, 1)$ distribution. Then $X + Y$ and $X - Y$ are independent.

Proof. The proof is left as an exercise.

The converse is due to Bernstein [6] and is stated here without proof.

Theorem 28. If X and Y are independent rv's with the same distribution, which has finite variance, and if $Z_1 = X + Y$ and $Z_2 = X - Y$ are independent, all rv's X, Y, Z_1 and Z_2 are normally distributed.

The following result generalizes Theorem 27.

Theorem 29. If $X_1, X_2, \cdots, X_n$ are independent normal rv's and $\sum_{i=1}^{n} a_i b_i \operatorname{var}(X_i) = 0$, then $L_1 = \sum_{i=1}^{n} a_i X_i$ and $L_2 = \sum_{i=1}^{n} b_i X_i$ are independent. Here $a_1, a_2, \cdots, a_n$ and $b_1, b_2, \cdots, b_n$ are fixed (non-zero) real numbers.

Proof. Let $\operatorname{var}(X_i) = \sigma_i^2$, and assume without loss of generality that $EX_i = 0$, $i = 1, 2, \cdots, n$. For any real numbers α, β, and t

$$Ee^{(\alpha L_1 + \beta L_2)t} = E \exp \left\{ t \sum_{1}^{n} (\alpha a_i + \beta b_i) X_i \right\}$$

$$= \prod_{i=1}^{n} \exp \left\{ \frac{t^2}{2} (\alpha a_i + \beta b_i)^2 \sigma_i^2 \right\}$$

$$= \exp \left\{ \frac{\alpha^2 t^2}{2} \sum_{1}^{n} a_i^2 \sigma_i^2 + \frac{\beta^2 t^2}{2} \sum_{1}^{n} b_i^2 \sigma_i^2 \right\} \text{ (since } \sum_{1}^{n} a_i b_i \sigma_i^2 = 0\text{)}$$

$$= \prod_{i=1}^{n} \exp \left\{ \frac{t^2 \alpha^2}{2} a_i^2 \sigma_i^2 \right\} \cdot \prod_{i=1}^{n} \exp \left\{ \frac{t^2 \beta^2}{2} b_i^2 \sigma_i^2 \right\}$$

$$= \prod_{1}^{n} Ee^{t\alpha a_i X_i} \cdot \prod_{1}^{n} Ee^{t\beta b_i X_i}$$

$$= E \exp (t\alpha \sum a_i X_i) \cdot E \exp (t\beta \sum_{1}^{n} b_i X_i)$$

$$= Ee^{\alpha t L_1} Ee^{\beta t L_2}.$$

Thus we have shown that

$$M(\alpha t, \beta t) = M(\alpha t, 0) M(0, \beta t) \quad \text{for all} \quad \alpha, \beta, t.$$

It follows that L_1 and L_2 are independent.

Corollary. If X_1, X_2 are independent $\mathcal{N}(\mu_1, \sigma^2)$ and $\mathcal{N}(\mu_2, \sigma^2)$ rv's, then $X_1 - X_2$ and $X_1 + X_2$ are independent. (This gives Theorem 27.)

Darmois [22] and Skitovitch [119] provided the converse of Theorem 29, which we state without proof.

Theorem 30. If $X_1, X_2, \cdots, X_n$ are independent rv's, $a_1, a_2, \cdots, a_n, b_1, b_2, \cdots, b_n$ are real numbers none of which equals zero, and if the linear forms

$$L_1 = \sum_{i=1}^{n} a_i X_i, \qquad L_2 = \sum_{i=1}^{n} b_i X_i$$

are independent, then all the rv's are normally distributed.

Corollary. If X and Y are independent rv's such that $X + Y$ and $X - Y$ are independent, $X, Y, X + Y$ and $X - Y$ are all normal.

Yet another result of this type is the following theorem.

Theorem 31. Let $X_1, X_2, \cdots, X_n$ be iid rv's with finite variance. Then the common distribution is normal if and only if

$$S_n = \sum_{k=1}^{n} X_k \quad \text{and} \quad Y_n = \sum_{i=1}^{n} (X_i - n^{-1} S_n)^2$$

are independent.

In Chapter 7 we will prove the necessity part of this result, which is basic to the theory of t-tests in statistics (Chapter 10). See also Example 4.4.8. The sufficiency part was proved by Lukacs[72], and we will not prove it here.

Theorem 32. $X \sim \mathcal{N}(0, 1) \Rightarrow X^2 \sim G(\frac{1}{2}, 2)$.

Proof. See Example 2.5.7 for the proof.

Corollary 1. If $X \sim \mathcal{N}(\mu, \sigma^2)$, the rv $Z^2 = (X - \mu)^2/\sigma^2$ is $\chi^2(1)$.

Corollary 2. If $X_1, X_2, \cdots, X_n$ are independent rv's and $X_k \sim \mathcal{N}(\mu_k, \sigma_k^2)$, $k = 1, 2, \cdots, n$, then $\sum_{k=1}^{n} (X_k - \mu_k)^2/\sigma_k^2$ is $\chi^2(n)$.

Theorem 33. Let X and Y be iid $\mathcal{N}(0, \sigma^2)$ rv's. Then X/Y is $\mathscr{C}(1, 0)$.

Proof. For the proof see Example 4.4.9.

We remark that the converse of this result does not hold; that is, if $Z = X/Y$ is the quotient of two iid rv's and Z has a $\mathscr{C}(1, 0)$ distribution, it does not follow that X and Y are normal, for take X and Y to be iid with pdf

$$f(x) = \frac{\sqrt{2}}{\pi} \frac{1}{1+x^4}, \qquad -\infty < x < \infty.$$

We leave the reader to verify that $Z = X/Y$ is $\mathscr{C}(1, 0)$.

PROBLEMS 5.3

1. Prove Theorem 1.

2. Let X be an rv with pmf $p_k = P\{X = k\}$ given below. If F is the corresponding df, find the distribution of $F(X)$, in the following cases.

   (a) $p_k = \binom{n}{k} p^k (1-p)^{n-k}$, $k = 0, 1, 2, \cdots, n$; $0 < p < 1$.

   (b) $p_k = e^{-\lambda}(\lambda^k/k!)$, $k = 0, 1, 2, \cdots$; $\lambda > 0$.

3. Let $Y_1 \sim U[0, 1]$, $Y_2 \sim U[0, Y_1]$, $\cdots$, $Y_n \sim U[0, Y_{n-1}]$. Show that
$$Y_1 \sim X_1, \quad Y_2 \sim X_1 X_2, \quad \cdots, \quad Y_n \sim X_1 X_2 \cdots X_n,$$
where $X_1, X_2, \cdots, X_n$ are iid $U[0, 1]$ rv's. If U is the number of $Y_1, Y_2, \cdots, Y_n$ in $[t, 1]$, where $0 < t < 1$, show that U has a Poisson distribution with parameter $-\log t$.

4. Let $X_1, X_2, \cdots, X_n$ be iid $U[0, 1]$ rv's. Prove by induction or otherwise that $S_n = \sum_{k=1}^n X_k$ has the pdf
$$f_n(x) = [(n-1)!]^{-1} \sum_{k=0}^n (-1)^k \binom{n}{k} [\varepsilon(x-k)]^n \cdot (x-k)^{n-1}.$$

5. N numbers are chosen independently at random, one from each of the N intervals $[0, L_i]$, $i = 1, 2, \cdots, N$. If the distribution of each random number is uniform with respect to the length of the interval from which it is chosen, find the expected value of the smallest of the N numbers chosen. (Klamkin [59])

6. Prove (a) Theorem 6 and its corollary, and (b) Theorem 10.

7. Let X be a nonnegative rv of the continuous type, and let $Y \sim U(0, X)$. Also, let $Z = X - Y$. Then the rv's Y and Z are independent if and only if X is $G(2, 1/\lambda)$ for some $\lambda > 0$. (Lamperti [66])

8. Let X and Y be independent rv's with common pdf $f(x) = \beta^{-\alpha} \alpha x^{\alpha-1}$ if $0 < x < \beta$, and $= 0$ otherwise; $\alpha \geq 1$. Let $U = \min(X, Y)$ and $V = \max(X, Y)$. Find the joint pdf of U and V and the pdf of $U + V$. Show that U/V and V are independent.

9. Prove Theorem 15.

10. Prove Theorem 8.

11. Prove Theorems 22 and 23.

12. Let $X_1, X_2, \cdots, X_n$ be independent rv's with $X_i \sim \mathscr{C}(\mu_i, \lambda_i)$, $i = 1, 2, \cdots, n$. Show that the rv $X = 1/\sum_{i=1}^n X_i^{-1}$ is also a Cauchy rv with parameters $\mu/(\lambda^2 + \mu^2)$ and $\lambda/(\lambda^2 + \mu^2)$ where
$$\lambda = \sum_{i=1}^n \frac{\lambda_i}{\lambda_i^2 + \mu_i^2} \quad \text{and} \quad \mu = \sum_{i=1}^n \frac{\mu_i}{\lambda_i^2 + \mu_i^2}.$$

13. Let $X_1, X_2, \cdots, X_n$ be iid $\mathscr{C}(1, 0)$ rv's and $a_i \neq 0$, b_i, $i = 1, 2, \cdots, n$, be any real numbers. Find the distribution of $\sum_{i=1}^n 1/(a_i X_i + b_i)$.

14. Suppose that the load of an airplane wing is a random variable X with $\mathscr{N}$ (1000, 14400) distribution. The maximum load that the wing can withstand is an

rv Y, which is $\mathcal{N}(1260, 2500)$. If X and Y are independent, find the probability that the load encountered by the wing is less than its critical load.

15. Let $X \sim \mathcal{N}(0, 1)$. Find the pdf of $Z = 1/X^2$. If X and Y are iid $\mathcal{N}(0, 1)$, deduce that $U = XY/\sqrt{X^2 + Y^2}$ is $\mathcal{N}(0, 1/\sqrt{2})$.

16. In Problem 15 let X and Y be independent normal rv's with zero means. Show that $U = XY/\sqrt{(X^2 + Y^2)}$ is normal. If, in addition, var (X) = var (Y) show that $V = (X^2 - Y^2)/\sqrt{(X^2 + Y^2)}$ is also normal. Moreover, U and V are independent.
(Shepp [114])

17. Let X_1, X_2, X_3, X_4 be independent $\mathcal{N}(0, 1)$. Show that $Y = X_1 X_2 + X_3 X_4$ has the pdf $f(y) = \frac{1}{2} e^{-|y|}$, $-\infty < y < \infty$.

18. Let $X \sim \mathcal{N}(15, 16)$. Find (a) $P\{X \le 12\}$, (b) $P\{10 \le X \le 17\}$, (c) $P\{10 \le X \le 19 | X \le 17\}$, (d) $P\{|X - 15| \ge .5\}$.

19. Let $X \sim \mathcal{N}(-1, 9)$. Find x such that $P\{X > x\} = .38$. Also, find x such that $P\{|X + 1| < x\} = .4$.

20. Let X be an rv such that $\log (X - a)$ is $\mathcal{N}(\mu, \sigma^2)$. Show that X has pdf

$$f(x) = \begin{cases} \dfrac{1}{\sigma(x - a)\sqrt{2\pi}} \exp\left\{-\dfrac{[\log(x - a) - \mu]^2}{2\sigma^2}\right\} & \text{if } x > a, \\ 0 & \text{if } x \le a. \end{cases}$$

If m_1, m_2 are the first two moments of this distribution and $\alpha_3 = \mu_3/\mu_2^{3/2}$ is the coefficient of skewness, show that a, μ, σ are given by

$$a = m_1 - \frac{\sqrt{m_2 - m_1^2}}{\eta}, \qquad \sigma^2 = \log(1 + \eta^2),$$

and

$$\mu = \log(m_1 - a) - \tfrac{1}{2}\sigma^2,$$

where η is the real root of the equation $\eta^3 + 3\eta - \alpha_3 = 0$.

21. Prove Theorem 27.

22. Let X and Y be iid $\mathcal{N}(0, 1)$ rv's. Find the pdf of $X/|Y|$. Also, find the pdf of $|X|/|Y|$.

23. Construct a continuous analogue of Problem 5.2.17.

24. Let $X_1, X_2, \cdots, X_n$ be iid $\mathcal{N}(\mu, \sigma^2)$ rv's. Find the distribution of

$$Y_n = \frac{\sum\limits_{k=1}^{n} k X_k - \mu \sum\limits_{k=1}^{n} k}{(\sum\limits_{k=1}^{n} k^2)^{1/2}}.$$

25. Let $F_1, F_2, \cdots, F_n$ be n df's. Show that min $\{F_1(x_1), F_2(x_2), \cdots, F_n(x_n)\}$ is an n-dimensional df with marginal df's $F_1, F_2, \cdots F_n$. (Kemp [55])

26. Let $X \sim NB(1; p)$ and $Y \sim G(1, 1/\lambda)$. Show that X and Y are related by the equation

$$P\{X \le x\} = P\{Y \le [x]\} \quad \text{for } x > 0, \quad \lambda = \log\left(\frac{1}{1-p}\right),$$

where $[x]$ is the largest integer $\le x$. Equivalently, show that

$$P\{Y \in (n-1, n]\} = P_\theta\{X = n\},$$

where $\theta = 1 - e^{-\lambda}$. (Prochaska[91])

5.4 THE BIVARIATE AND MULTIVARIATE NORMAL DISTRIBUTIONS

In this section we introduce the bivariate and multivariate normal distributions and investigate some of their important properties.

Definition. A two-dimensional rv (X, Y) is said to have a bivariate normal distribution if the joint pdf is of the form

(1) $$f(x, y) = \frac{1}{2\pi\sigma_1\sigma_2\sqrt{1-\rho^2}} e^{-Q(x,y)/2},$$
$$-\infty < x < \infty, \quad -\infty < y < \infty,$$

where $\sigma_1 > 0$, $\sigma_2 > 0$, $|\rho| < 1$, and Q is the quadratic form

(2) $$Q(x, y) = \frac{1}{1-\rho^2}\left[\left(\frac{x-\mu_1}{\sigma_1}\right)^2 - 2\rho\left(\frac{x-\mu_1}{\sigma_1}\right)\left(\frac{y-\mu_2}{\sigma_2}\right) + \left(\frac{y-\mu_2}{\sigma_2}\right)^2\right].$$

We first show that (1) indeed defines a joint pdf. In fact, we prove the following result.

Theorem 1. The function defined by (1) and (2) with $\sigma_1 > 0$, $\sigma_2 > 0$, $|\rho| < 1$ is a joint pdf. The marginal pdf's of X and Y are respectively, $\mathcal{N}(\mu_1, \sigma_1^2)$ and $\mathcal{N}(\mu_2, \sigma_2^2)$, and ρ is the correlation coefficient between X and Y.

Proof. Let $f_1(x) = \int_{-\infty}^{\infty} f(x, y)\, dy$. Note that

$$(1-\rho^2) Q(x, y) = \left(\frac{y-\mu_2}{\sigma_2} - \rho\frac{x-\mu_1}{\sigma_1}\right)^2 + (1-\rho^2)\left(\frac{x-\mu_1}{\sigma_1}\right)^2$$
$$= \left\{\frac{y - [\mu_2 + \rho(\sigma_2/\sigma_1)(x-\mu_1)]}{\sigma_2}\right\}^2 + (1-\rho^2)\left(\frac{x-\mu_1}{\sigma_1}\right)^2.$$

It follows that

(3) $$f_1(x) = \frac{1}{\sigma_1\sqrt{2\pi}} \exp\left\{\frac{-(x-\mu_1)^2}{2\sigma_1^2}\right\} \int_{-\infty}^{\infty} \frac{\exp\{-(y-\beta_x)^2/[2\sigma_2^2(1-\rho^2)]\}}{\sigma_2\sqrt{1-\rho^2}\sqrt{2\pi}}\, dy,$$

where we have written

(4) $$\beta_x = \mu_2 + \rho\left(\frac{\sigma_2}{\sigma_1}\right)(x-\mu_1).$$

The integrand is the pdf of an $\mathcal{N}(\beta_x, \sigma_2^2(1-\rho^2))$ rv, so that

$$f_1(x) = \frac{1}{\sigma_1\sqrt{2\pi}} \exp\left\{-\frac{1}{2}\left(\frac{x-\mu_1}{\sigma_1}\right)^2\right\}, \qquad -\infty < x < \infty.$$

Thus

$$\int_{-\infty}^{\infty} \left\{\int_{-\infty}^{\infty} f(x, y)\, dy\right\} dx = \int_{-\infty}^{\infty} f_1(x)\, dx = 1,$$

and $f(x, y)$ is a joint pdf of two rv's of the continuous type. It also follows that f_1 is the marginal pdf of X, so that X is $\mathcal{N}(\mu_1, \sigma_1^2)$. In a similar manner we can show that Y is $\mathcal{N}(\mu_2, \sigma_2^2)$.

Furthermore, we have

(5) $$\frac{f(x, y)}{f_1(x)} = \frac{1}{\sigma_2\sqrt{1-\rho^2}\sqrt{2\pi}} \exp\left\{\frac{-(y-\beta_x)^2}{2\sigma_2^2(1-\rho^2)}\right\},$$

where β_x is given by (4). It is clear, then, that the conditional pdf $f_{Y|X}(y|x)$ given by (5) is also normal, with parameters β_x and $\sigma_2^2(1-\rho^2)$. We have

(6) $$E\{Y|x\} = \beta_x = \mu_2 + \rho\frac{\sigma_2}{\sigma_1}(x - \mu_1).$$

Since $\rho(\sigma_2/\sigma_1)$ is the coefficient of regression of Y on X and σ_1, σ_2 are the standard deviations, it follows immediately that ρ is the correlation coefficient between X and Y.

In a similar manner, we can show that

(7) $$E\{X|y\} = \beta_y = \mu_1 + \rho\frac{\sigma_1}{\sigma_2}(y - \mu_2)$$

and that the conditional pdf of X, given Y, is $\mathcal{N}(\beta_y, \sigma_1^2(1-\rho^2))$. The proof is now complete.

Remark 1. If $\rho^2 = 1$, then (1) becomes meaningless. But in that case we know (Theorem 4.8.1) that there exist constants a and b such that $P\{Y = aX + b\} = 1$. We thus have a univariate distribution, which is called the *bivariate degenerate* (or *singular*) *normal* distribution. The bivariate degenerate normal distribution does not have a pdf but corresponds to an rv (X, Y) whose marginal distributions are normal or degenerate and are such that (X, Y) falls on a fixed line with probability 1. It is for this reason that degenerate distributions are considered as normal distributions with variance 0.

Next we compute the mgf $M(t_1, t_2)$ of a bivariate normal rv (X, Y). We have, if $f(x, y)$ is the pdf given in (1) and f_1 is the marginal pdf of X,

$$M(t_1, t_2) = \int_{-\infty}^{\infty} \int_{-\infty}^{\infty} e^{t_1 x + t_2 y} f(x, y) \, dx \, dy,$$

$$= \int_{-\infty}^{\infty} \left\{ \int_{-\infty}^{\infty} f_{Y|X}(y|x) \, e^{t_2 y} dy \right\} e^{t_1 x} f_1(x) \, dx$$

$$= \int_{-\infty}^{\infty} e^{t_1 x} f_1(x) \left\{ \exp \left[\tfrac{1}{2} \sigma_2^2 t_2^2 (1 - \rho^2) + t_2(\mu_2 + \rho \tfrac{\sigma_2}{\sigma_1}(x - \mu_1)) \right] \right\} dx$$

$$= \exp \left[\tfrac{1}{2} \sigma_2^2 t_2^2 (1 - \rho^2) + t_2 \mu_2 - \rho t_2 \tfrac{\sigma_2}{\sigma_1} \mu_1 \right] \int_{-\infty}^{\infty} e^{t_1 x} e^{(\rho \sigma_2/\sigma_1) x t_2} f_1(x) \, dx.$$

Now

$$\int_{-\infty}^{\infty} e^{(t_1 + \rho t_2 \sigma_2/\sigma_1) x} f_1(x) \, dx = \exp \left[\mu_1 \left(t_1 + \rho \tfrac{\sigma_2}{\sigma_1} t_2 \right) + \tfrac{1}{2} \sigma_1^2 \left(t_1 + \rho t_2 \tfrac{\sigma_2}{\sigma_1} \right)^2 \right].$$

Therefore

(8) $$M(t_1, t_2) = \exp \left(\mu_1 t_1 + \mu_2 t_2 + \frac{\sigma_1^2 t_1^2 + \sigma_2^2 t_2^2 + 2\rho \sigma_1 \sigma_2 t_1 t_2}{2} \right).$$

The following result is an immediate consequence of (8).

Theorem 2. If (X, Y) has a bivariate normal distribution, X and Y are independent if and only if $\rho = 0$.

Remark 2. It is quite possible for an rv (X, Y) to have a bivariate density such that the marginal densities of X and Y are normal and the correlation coefficient is 0, yet X and Y are not independent. Indeed, if the marginal densities of X and Y are normal, it does not follow that the joint density of (X, Y) is a bivariate normal. Let

(9) $$f(x, y) = \tfrac{1}{2} \left\{ \frac{1}{2\pi(1 - \rho^2)^{1/2}} \exp \left[\frac{-1}{2(1 - \rho^2)} (x^2 - 2\rho xy + y^2) \right] \right.$$
$$\left. + \frac{1}{2\pi(1 - \rho^2)^{1/2}} \exp \left[\frac{-1}{2(1 - \rho^2)} (x^2 + 2\rho xy + y^2) \right] \right\}.$$

Here $f(x, y)$ is a joint pdf such that both marginal densities are normal, $f(x, y)$ is not bivariate normal, and X and Y have zero correlation. But X and Y are not independent. We have

$$f_1(x) = \frac{1}{\sqrt{2\pi}} e^{-x^2/2}, \quad -\infty < x < \infty,$$

$$f_2(y) = \frac{1}{\sqrt{2\pi}} e^{-y^2/2}, \quad -\infty < y < \infty,$$

$$EXY = 0.$$

Example 1 (Rosenberg [103]). Let f and g be pdf's with corresponding df's F and G. Also, let

(10) $$h(x, y) = f(x)g(y)[1 + \alpha(2F(x) - 1)(2G(y) - 1)],$$

where $|\alpha| \leq 1$ is a constant. It was shown in Example 4.3.1 that h is a bivariate density function with given marginal densities f and g.

In particular, take f and g to be the pdf of $\mathcal{N}(0, 1)$, that is,

(11) $$f(x) = g(x) = \frac{1}{\sqrt{2\pi}} e^{-x^2/2}, \quad -\infty < x < \infty,$$

and let (X, Y) have the joint pdf $h(x, y)$. We will show that $X + Y$ is not normal except in the trivial case $\alpha = 0$, when X and Y are independent.

Let $Z = X + Y$. Then

$$EZ = 0, \quad \text{var}(Z) = \text{var}(X) + \text{var}(Y) + 2 \text{cov}(X, Y).$$

It is easy to show (Problem 2) that cov $(X, Y) = \alpha/\pi$, so that var $(Z) = 2[1 + (\alpha/\pi)]$. If Z is normal, its mgf must be

(12) $$M_Z(t) = e^{t^2[1+(\alpha/\pi)]}.$$

Next we compute the mgf of Z directly from the joint pdf (10). We have

$$M_1(t) = E\{e^{tX+tY}\}$$
$$= e^{t^2} + \alpha \int_{-\infty}^{\infty} \int_{-\infty}^{\infty} e^{tx+ty}[2F(x) - 1][2F(y) - 1] f(x) f(y) \, dx \, dy$$
$$= e^{t^2} + \alpha \left[\int_{-\infty}^{\infty} e^{tx}[2F(x) - 1] f(x) \, dx\right]^2.$$

Now

$$\int_{-\infty}^{\infty} e^{tx}[2F(x) - 1] f(x) \, dx = -2 \int_{-\infty}^{\infty} e^{tx}[1 - F(x)] f(x) \, dx + e^{t^2/2}$$
$$= e^{t^2/2} - 2 \int_{-\infty}^{\infty} \int_{x}^{\infty} \frac{1}{2\pi} \exp\{-\tfrac{1}{2}(x^2 + u^2 - 2tx)\} \, du \, dx$$
$$= e^{t^2/2} - \int_{-\infty}^{\infty} \int_{0}^{\infty} \frac{\exp\{-\tfrac{1}{2}[x^2 + (v+x)^2 - 2tx]\}}{\pi} \, dv \, dx$$
$$= e^{t^2/2} - \int_{0}^{\infty} \frac{\exp\{-v^2/2 + (v-t)^2/4\}}{\sqrt{\pi}} \int_{-\infty}^{\infty} \frac{\exp\{-[x+(v-t)/2]^2\}}{\sqrt{\pi}} \, dx \, dv$$
$$= e^{t^2/2} - 2e^{t^2/2} \int_{0}^{\infty} \frac{\exp\{-\tfrac{1}{2}[(v+t)^2/2]\}}{2\sqrt{\pi}} \, dv$$

(13) $$= e^{t^2/2} - 2e^{t^2/2} P\left\{Z_1 > \frac{t}{\sqrt{2}}\right\},$$

where Z_1 is an $\mathcal{N}(0, 1)$ rv.

It follows that

THE MULTIVARIATE NORMAL DISTRIBUTION   231

$$M_1(t) = e^{t^2} + \alpha\left[e^{t^2/2} - 2e^{t^2/2} P\left\{Z_1 > \frac{t}{\sqrt{2}}\right\}\right]^2$$

(14)
$$= e^{t^2}\left[1 + \alpha\left(1 - 2P\left\{Z_1 > \frac{t}{\sqrt{2}}\right\}\right)^2\right].$$

If Z were normally distributed, we must have $M_Z(t) = M_1(t)$ for all t and all $|\alpha| \leq 1$, that is,

(15)
$$e^{t^2} e^{(\alpha/\pi)t^2} = e^{t^2}\left[1 + \alpha\left(1 - 2P\left\{Z_1 > \frac{t}{\sqrt{2}}\right\}\right)^2\right].$$

For $\alpha = 0$, the equality clearly holds. The expression within the brackets on the right side of (15) is bounded by $1 + \alpha$, whereas the expression $e^{(\alpha/\pi)t^2}$ is unbounded, so the equality cannot hold for all t and α.

Next we investigate the multivariate normal distribution of dimension n, $n \geq 2$. Let $\mathbf{M}$ be an $n \times n$ real, symmetric, and positive definite matrix. Let $\mathbf{x}$ denote the $1 \times n$ row vector of real numbers $(x_1, x_2, \cdots, x_n)$, and let $\boldsymbol{\mu}$ denote the row vector $(\mu_1, \mu_2, \cdots, \mu_n)$, where μ_i ($i = 1, 2, \cdots, n$) are real constants.

Theorem 3. The nonnegative function

(16) $f(\mathbf{x}) = c \exp\left\{-\frac{(\mathbf{x} - \boldsymbol{\mu})\mathbf{M}(\mathbf{x} - \boldsymbol{\mu})'}{2}\right\}$ $-\infty < x_i < \infty,$

$i = 1, 2, \cdots, n,$

defines the joint pdf of some random vector $\mathbf{X} = (X_1, X_2, \cdots, X_n)$, provided that the constant c is chosen appropriately. The mgf of $\mathbf{X}$ exists and is given by

(17) $M(t_1, t_2, \cdots, t_n) = \exp\left\{\mathbf{t}\boldsymbol{\mu}' + \frac{\mathbf{t}\mathbf{M}^{-1}\mathbf{t}'}{2}\right\},$

where $\mathbf{t} = (t_1, t_2, \cdots, t_n)$, and $t_1, t_2, \cdots, t_n$ are arbitrary real numbers.

Proof. Let

(18) $I = c \int_{-\infty}^{\infty} \cdots \int_{-\infty}^{\infty} \exp\left\{\mathbf{t}\mathbf{x}' - \frac{(\mathbf{x} - \boldsymbol{\mu})\mathbf{M}(\mathbf{x} - \boldsymbol{\mu})'}{2}\right\} \prod_{i=1}^{n} dx_i.$

Changing the variables of integration to $y_1, y_2, \cdots, y_n$ by writing $x_i - \mu_i = y_i$, $i = 1, 2, \cdots, n$, and $\mathbf{y} = (y_1, y_2, \cdots, y_n)$, we have $\mathbf{x} - \boldsymbol{\mu} = \mathbf{y}$ and

(19) $I = c \exp(\mathbf{t}\boldsymbol{\mu}') \int_{-\infty}^{\infty} \cdots \int_{-\infty}^{\infty} \exp\left(\mathbf{t}\mathbf{y}' - \frac{\mathbf{y}\mathbf{M}\mathbf{y}'}{2}\right) \prod_{i=1}^{n} dy_i.$

Since **M** is positive definite, it follows (see Section P. 3) that all the n characteristic roots of **M**, say $m_1, m_2, \cdots, m_n$, are positive. Moreover, since **M** is symmetric there exists (Theorem P. 3.4) an $n \times n$ orthogonal matrix **L** such that **L'ML** is a diagonal matrix with diagonal elements $m_1, m_2, \cdots, m_n$. Let us change the variables to $z_1, z_2, \cdots, z_n$ by writing $\mathbf{y'} = \mathbf{Lz'}$, where $\mathbf{z} = (z_1, z_2, \cdots, z_n)$, and note that the Jacobian of this orthogonal transformation is $|\mathbf{L}|$. Since $\mathbf{L'L} = \mathbf{I}_n$, where $\mathbf{I}_n$ is an $n \times n$ unit matrix, $|\mathbf{L}| = 1$ and we have

(20) $$I = c \exp(\mathbf{t}\boldsymbol{\mu}') \int_{-\infty}^{\infty} \cdots \int_{-\infty}^{\infty} \exp\left(\mathbf{tLz'} - \frac{\mathbf{zL'MLz'}}{2}\right) \prod_{i=1}^{n} dz_i.$$

If we write $\mathbf{tL} = \mathbf{u} = (u_1, u_2, \cdots, u_n)$, then $\mathbf{tLz'} = \sum_{i=1}^{n} u_i z_i$. Also $\mathbf{L'ML} = \text{diag}(m_1, m_2, \cdots, m_n)$ so that $\mathbf{zL'MLz'} = \sum_{i=1}^{n} m_i z_i^2$. The integral in (20) can therefore be written as

$$\prod_{i=1}^{n}\left[\int_{-\infty}^{\infty} \exp\left(u_i z_i - \frac{m_i z_i^2}{2}\right) dz_i\right] = \prod_{i=1}^{n}\left[\sqrt{\frac{2\pi}{m_i}} \exp\left(\frac{u_i^2}{2m_i}\right)\right].$$

It follows that

(21) $$I = c \exp(\mathbf{tu'}) \frac{(2\pi)^{n/2}}{(m_1 m_2 \cdots m_n)^{1/2}} \exp\left(\sum_{i=1}^{n} \frac{u_i^2}{2m_i}\right).$$

Setting $t_1 = t_2 = \cdots = t_n = 0$, we see from (18) and (21) that

$$\int_{-\infty}^{\infty} \cdots \int_{-\infty}^{\infty} f(x_1, x_2, \cdots, x_n) \, dx_1 \, dx_2 \cdots dx_n = \frac{c(2\pi)^{n/2}}{(m_1 m_2 \cdots m_n)^{1/2}}.$$

By choosing

(22) $$c = \frac{(m_1 m_2 \cdots m_n)^{1/2}}{(2\pi)^{n/2}}$$

we see that f is a joint pdf of some random vector **X**, as asserted.

Finally, since

$$(\mathbf{L'ML})^{-1} = \text{diag}(m_1^{-1}, m_2^{-1}, \cdots, m_n^{-1}),$$

we have

$$\sum_{i=1}^{n} \frac{u_i^2}{m_i} = \mathbf{u}(\mathbf{L'M^{-1}L})\mathbf{u'} = \mathbf{tM^{-1}t'}.$$

Also

$$|\mathbf{M}^{-1}| = |\mathbf{L'M^{-1}L}| = (m_1 m_2 \cdots m_n)^{-1}.$$

It follows from (21) and (22) that the mgf of **X** is given by (17), and we may write

$$(23) \qquad c = \frac{1}{\{(2\pi)^n |\mathbf{M}^{-1}|\}^{1/2}}.$$

This completes the proof of Theorem 3.

Let us write $\mathbf{M}^{-1} = ((\sigma_{ij}))_{i,j=1,2,\cdots,n}$. Then

$$M(0, 0, \cdots, 0, t_i, 0, \cdots, 0) = \exp\left(t_i \mu_i + \sigma_{ii} \frac{t_i^2}{2}\right)$$

is the mgf of X_i, $i = 1, 2, \cdots, n$. Thus each X_i is $\mathcal{N}(\mu_i, \sigma_{ii})$, $i = 1, 2, \cdots, n$. For $i \neq j$, we have for the mgf of X_i and X_j

$$M(0, 0, \cdots, 0, t_i, 0, \cdots, 0, t_j, 0, \cdots, 0) =$$
$$\exp\left(t_i \mu_i + t_j \mu_j + \frac{\sigma_{ii} t_i^2 + 2\sigma_{ij} t_i t_j + t_j^2 \sigma_{jj}}{2}\right).$$

This is the mgf of a bivariate normal distribution with means μ_i, μ_j, variances σ_{ii}, σ_{jj}, and covariance σ_{ij}. Thus we see that

$$(24) \qquad \boldsymbol{\mu} = (\mu_1, \mu_2, \cdots, \mu_n)$$

is the mean vector of $\mathbf{X} = (X_1, \cdots, X_n)$,

$$(25) \qquad \sigma_{ii} = \sigma_i^2 = \text{var}(X_i), \qquad i = 1, 2, \cdots, n,$$

and

$$(26) \qquad \sigma_{ij} = \rho \sigma_i \sigma_j, \qquad i \neq j; i, j = 1, 2, \cdots, n.$$

The matrix $\mathbf{M}^{-1}$ is called the *dispersion (variance-covariance)* matrix of the multivariate normal distribution.

If $\sigma_{ij} = 0$ for $i \neq j$, the matrix $\mathbf{M}^{-1}$ is a diagonal matrix, and it follows that the rv's $X_1, X_2, \cdots, X_n$ are independent. Thus we have the following analogue of Theorem 2.

Theorem 4. The components $X_1, X_2, \cdots, X_n$ of a normally distributed random vector $\mathbf{X}$ are independent if and only if the covariances $\sigma_{ij} = 0$ for all $i \neq j$ $(i, j = 1, 2, \cdots, n)$.

The following result is stated without proof. The proof is similar to the two-variate case (Theorem 4.8.1) except that now we consider the quadratic form in n variables: $E\{\sum_{i=1}^{n} t_i(X_i - \mu_i)\}^2 \geq 0$. (See P. 2.4).

Theorem 5. The probability that the rv's $X_1, X_2, \cdots, X_n$ with finite variances satisfy at least one linear relationship is 1 if and only if $|\mathbf{M}| = 0$.

Accordingly, if $|\mathbf{M}| = 0$ all the probability mass is concentrated on a hyperplane of dimension $< n$.

Theorem 6. Let $\mathbf{X} = (X_1, X_2, \cdots, X_n)$ be an n-dimensional rv with a normal distribution. Let $Y_1, Y_2, \cdots, Y_k, k \leq n$, be linear functions of $X_j (j = 1, 2, \cdots, n)$. Then $(Y_1, Y_2, \cdots, Y_k)$ also has a multivariate normal distribution.

Proof. Without loss of generality let us assume that $EX_i = 0, i = 1, 2, \cdots, n$. Let

$$(27) \qquad Y_p = \sum_{j=1}^{n} A_{pj} X_j, \quad p = 1, 2, \cdots, k; k \leq n.$$

Then $EY_p = 0, p = 1, 2, \cdots, k$, and

$$(28) \qquad \operatorname{cov}(Y_p, Y_q) = \sum_{i,j=1}^{n} A_{pi} A_{qj} \sigma_{ij},$$

where $E(X_i X_j) = \sigma_{ij}, i, j = 1, 2, \cdots, n$.

The mgf of $(Y_1, Y_2, \cdots, Y_k)$ is given by

$$M^*(t_1, t_2, \cdots, t_k) = E\{\exp(t_1 \sum_{j=1}^{n} A_{1j} X_j + \cdots + t_k \sum_{j=1}^{n} A_{kj} X_j)\}.$$

Writing $u_j = \sum_{p=1}^{k} t_p A_{pj}, j = 1, 2, \cdots, n$, we have

$$M^*(t_1, t_2, \cdots, t_k) = E\{\exp(\sum_{i=1}^{n} u_i X_i)\}$$

$$= \exp\left(\frac{1}{2} \sum_{i,j=1}^{n} \sigma_{ij} u_i u_j\right) \quad \text{by (17)}$$

$$= \exp\left(\frac{1}{2} \sum_{i,j=1}^{n} \sigma_{ij} \sum_{l,m=1}^{k} t_l t_m A_{li} A_{mj}\right)$$

$$= \exp\left(\frac{1}{2} \sum_{l,m=1}^{k} t_l t_m \sum_{i,j=1}^{n} A_{li} A_{mj} \sigma_{ij}\right)$$

$$(29) \qquad = \exp\left\{\frac{1}{2} \sum_{l,m=1}^{k} t_l t_m \operatorname{cov}(Y_l, Y_m)\right\}.$$

When (17) and (29) are compared, the result follows.

Corollary 1. Every marginal distribution of an n-dimensional normal distribution is univariate normal. Moreover, any linear function of $X_1, X_2, \cdots, X_n$ is univariate normal.

Corollary 2. If $X_1, X_2, \cdots, X_n$ are iid $\mathcal{N}(\mu, \sigma^2)$, and $\mathbf{A}$ is an $n \times n$ orthogonal transformation matrix, the components $Y_1, Y_2, \cdots, Y_n$ of $\mathbf{Y} = \mathbf{XA}'$, where $\mathbf{X} = (X_1, \cdots, X_n)$, are independent rv's, each normally distributed with the same variance σ^2.

We have from (27) and (28)

$$\text{cov}(Y_p, Y_q) = \sum_{i=1}^{n} A_{pi} A_{qi} \sigma_{ii} + \sum_{i \neq j} A_{pi} A_{qj} \sigma_{ij}$$

$$= \begin{cases} 0 & \text{if } p \neq q, \\ \sigma^2 & \text{if } p = q, \end{cases}$$

since $\sum_{i=1}^{n} A_{pi} A_{qi} = 0$ and $\sum_{j=1}^{n} A_{pj}^2 = 1$. It follows that

$$M^*(t_1, t_2, \cdots, t_n) = \exp\left(\tfrac{1}{2} \sum_{l=1}^{n} t_l^2 \sigma^2 \right),$$

and Corollary 2 follows.

Theorem 7. Let $\mathbf{X} = (X_1, X_2, \cdots, X_n)$. Then $\mathbf{X}$ has an n-dimensional normal distribution if and only if every linear function of $\mathbf{X}$

$$\mathbf{Xt'} = t_1 X_1 + t_2 X_2 + \cdots + t_n X_n$$

has a univariate normal distribution.

Proof. Suppose that $\mathbf{Xt'}$ is normal for any $\mathbf{t}$. Then the mgf of $\mathbf{Xt'}$ is given by

(30) $$M(s) = \exp(bs + \tfrac{1}{2} \sigma^2 s^2).$$

Here $b = E\{\mathbf{Xt'}\} = \sum_1^n t_i \mu_i = \mathbf{t}\boldsymbol{\mu}'$, where $\boldsymbol{\mu} = (\mu_1, \cdots, \mu_n)$, and $\sigma^2 = \text{var}(\mathbf{Xt'}) = \text{var}(\sum t_i X_i) = \mathbf{t}\mathbf{M}^{-1}\mathbf{t'}$, where $\mathbf{M}^{-1}$ is the dispersion matrix of $\mathbf{X}$. Thus

(31) $$M(s) = \exp(\mathbf{t'}\boldsymbol{\mu} s + \tfrac{1}{2} \mathbf{t}\mathbf{M}^{-1}\mathbf{t'} s^2).$$

Let $s = 1$; then

(32) $$M(1) = \exp(\mathbf{t}\boldsymbol{\mu}' + \tfrac{1}{2} \mathbf{t}\mathbf{M}^{-1}\mathbf{t'}),$$

and since the mgf is unique, it follows that $\mathbf{X}$ has a multivariate normal distribution. The converse follows from Corollary 1 to Theorem 6.

Many characterization results for the multivariate normal distribution are now available. We refer the reader to Lukacs and Laha [74], page 79.

PROBLEMS 5.4

1. For a bivariate normal rv (X, Y) does the the conditional probability density function of (X, Y), given $X + Y = t$, exist? If so, find it. If not, why not?
2. In Example 1 show that $\text{cov}(X, Y) = \alpha/\pi$.

3. Let (X, Y) be a bivariate normal rv with parameters μ_1, μ_2, σ_1^2, σ_2^2, and ρ. What is the distribution of $X + Y$? Compare your result with that of Example 1.

4. Let (X, Y) be a bivariate normal rv with parameters μ_1, μ_2, σ_1^2, σ_2^2, and ρ, and let $U = aX + b$, $a \neq 0$, and $V = cY + d$, $c \neq 0$. Find the joint distribution of (U, V).

5. Let (X, Y) be a bivariate normal rv with parameters $\mu_1 = 5$, $\mu_2 = 8, \sigma_1^2 = 16$, $\sigma_2^2 = 9$, and $\rho = .6$. Find $P\{5 < Y < 11 | X = 2\}$.

6. Let X and Y be jointly normal with means 0. Also, let

$$W = X \cos \theta + Y \sin \theta, \quad Z = X \cos \theta - Y \sin \theta.$$

Find θ such that W and Z are independent.

7. Let (X, Y) be a normal rv with parameters μ_1, μ_2, σ_1^2, σ_2^2, and ρ. Find a necessary and sufficient condition for $X + Y$ and $X - Y$ to be independent.

8. Let (X, Y) be a bivariate normal rv with parameters μ_1, μ_2, σ_1^2, σ_2^2, and ρ. Find the mean square error of the best predictor (in the sense of least squares) of Y given X.

9. Show that every variance-covariance matrix is positive semidefinite and conversely. If the variance-covariance matrix is not positive definite, then with probability 1 the random vector **X** lies in some hyperplane $\mathbf{cX}' = a$ with $\mathbf{c} \neq \mathbf{0}$.

10. Let (X, Y) be a bivariate normal rv with $EX = EY = 0$, var $(X) =$ var$(Y) = 1$, and cov $(X, Y) = \rho$. Show that the rv $Z = Y/X$ has a Cauchy distribution.

5.5 THE EXPONENTIAL FAMILY OF DISTRIBUTIONS

Most of the distributions that we have so far encountered belong to a general family of distributions that we now study. Let Θ be an interval on the real line, and let $\{f_\theta, \theta \in \Theta\}$ be a family of pdf's (pmf's). We will assume that the set $\{f_\theta(x) > 0\}$ is independent of θ. Here and in what follows we write $\mathbf{x} = (x_1, x_2, \cdots, x_n)$ unless otherwise specified.

Definition 1. If there exist real-valued functions $Q(\theta)$ and $D(\theta)$ on Θ and Borel-measurable functions $T(X_1, X_2, \cdots, X_n)$ and $S(X_1, X_2, \cdots, X_n)$ on $\mathcal{R}_n$ such that

(1) $\qquad f_\theta(x_1, x_2, \cdots, x_n) = \exp\{Q(\theta) T(\mathbf{x}) + D(\theta) + S(\mathbf{x})\},$

we say that the family $\{f_\theta, \theta \in \Theta\}$ is a one-parameter exponential family.

Let $\mathbf{X}_1, \mathbf{X}_2, \cdots, \mathbf{X}_m$ be iid with pmf (pdf) f_θ. Then the joint distribution of $\mathbf{X} = (\mathbf{X}_1, \mathbf{X}_2, \cdots, \mathbf{X}_m)$ is given by

$$g_\theta(\mathbf{x}) = \prod_{i=1}^{m} f_\theta(\mathbf{x}_i) = \prod_{i=1}^{m} \exp\{Q(\theta) T(\mathbf{x}_i) + D(\theta) + S(\mathbf{x}_i)\}$$
$$= \exp\{Q(\theta) \sum_{i=1}^{m} T(\mathbf{x}_i) + mD(\theta) + \sum_{i=1}^{m} S(\mathbf{x}_i)\},$$

where $\mathbf{x} = (\mathbf{x}_1, \mathbf{x}_2, \cdots, \mathbf{x}_m)$, $\mathbf{x}_j = (x_{j1}, x_{j2}, \cdots, x_{jn})$, $j = 1, 2, \cdots, m$, and it follows that $\{g_\theta: \theta \in \Theta\}$ is again a one-parameter exponential family.

Example 1. Let $X \sim \mathcal{N}(\mu_0, \sigma^2)$, where μ_0 is known and σ^2 unknown. Then
$$f_{\sigma^2}(x) = \frac{1}{\sigma\sqrt{2\pi}} \exp\left\{-\frac{(x-\mu_0)^2}{2\sigma^2}\right\}$$
$$= \exp\left\{-\log(\sigma\sqrt{2\pi}) - \frac{(x-\mu_0)^2}{2\sigma^2}\right\}$$

is a one-parameter exponential family with
$$Q(\sigma^2) = -\frac{1}{2\sigma^2}, \quad T(x) = (x - \mu_0)^2, \quad S(x) = 0, \quad \text{and}$$
$$D(\sigma^2) = -\log(\sigma\sqrt{2\pi}).$$

If $X \sim \mathcal{N}(\mu, \sigma_0^2)$, where σ_0 is known but μ is unknown, then
$$f_\mu(x) = \frac{1}{\sigma_0\sqrt{2\pi}} \exp\left\{-\frac{(x-\mu)^2}{2\sigma_0^2}\right\}$$
$$= \frac{1}{\sigma_0\sqrt{2\pi}} \exp\left(-\frac{x^2}{2\sigma_0^2} + \frac{\mu x}{\sigma_0^2} - \frac{\mu^2}{2\sigma_0^2}\right)$$

is a one-parameter exponential family with
$$Q(\mu) = \frac{\mu}{\sigma_0^2}, \quad D(\mu) = -\frac{\mu^2}{2\sigma_0^2}, \quad T(x) = x,$$

and
$$S(x) = -\left[\frac{x^2}{2\sigma_0^2} + \tfrac{1}{2}\log(2\pi\sigma_0^2)\right].$$

Example 2. Let $X \sim P(\lambda)$, $\lambda > 0$ unknown. Then
$$P_\lambda\{X = x\} = e^{-\lambda}\frac{\lambda^x}{x!} = \exp\{-\lambda + x\log\lambda - \log(x!)\},$$

and we see that the family of Poisson pmf's with parameter λ is a one-parameter exponential family.

Some other important examples of one-parameter exponential families are binomial, $G(\alpha, \beta)$ (provided that one of α, β is fixed), $B(\alpha, \beta)$ (provided that one of α, β is fixed), negative binomial, and geometric. The Cauchy family of densities and the uniform distribution on $[0, \theta]$ do not belong to this class.

Theorem. Let $\{f_\theta : \theta \in \Theta\}$ be a one-parameter family of exponential pdf's (pmf's) given in (1). Then the family of distributions of $T(X)$ is also a one-parameter exponential family of pdf's (pmf's), given by

$$g_\theta(t) = \exp\{tQ(\theta) + D(\theta) + S^*(t)\}$$

for suitable $S^*(t)$.

Proof. The proof of Theorem 1 is a simple application of the transformation of variables technique studied in Section 4.4 and is left as an exercise, at least for the cases considered in Section 4.4. For the general case we refer to Lehmann [70], page 52.

Let us now consider the k-parameter exponential family, $k \geq 2$. Let $\Theta \subseteq \mathcal{R}_k$ be a k-dimensional interval, and assume that $\{f_\theta > 0\}$ does not depend on θ

Definition 2. If there exist real-valued functions $Q_1, Q_2, \cdots, Q_k, D$ defined on Θ, and Borel-measurable functions $T_1, T_2, \cdots, T_k, S$ on $\mathcal{R}_n$ such that

(2) $$f_\theta(\mathbf{x}) = \exp\left\{\sum_{i=1}^k Q_i(\boldsymbol{\theta}) T_i(\mathbf{x}) + D(\boldsymbol{\theta}) + S(\mathbf{x})\right\},$$

we say that the family $\{f_\theta, \boldsymbol{\theta} \in \Theta\}$ is a k-parameter exponential family.

Once again, if $\mathbf{X} = (X_1, X_2, \cdots, X_m)$ and X_j are iid with common distribution (2), the joint distributions of $\mathbf{X}$ form a k-parameter exponential family. An analogue of Theorem 1 also holds for the k-parameter exponential family.

Example 3. The most important example of a k-parameter exponential family is $\mathcal{N}(\mu, \sigma^2)$ when both μ and σ^2 are unknown. We have

$$\boldsymbol{\theta} = (\mu, \sigma^2), \quad \Theta = \{(\mu, \sigma^2): -\infty < \mu < \infty, \sigma^2 > 0\}$$

and

$$f_\theta(x) = \frac{1}{\sigma\sqrt{2\pi}} \exp\left(-\frac{x^2 - 2\mu x + \mu^2}{2\sigma^2}\right)$$

$$= \exp\left\{-\frac{x^2}{2\sigma^2} + \frac{\mu}{\sigma^2}x - \frac{1}{2}\left[\frac{\mu^2}{\sigma^2} + \log(2\pi\sigma^2)\right]\right\}.$$

It follows that f_θ is a two-parameter exponential family with

$$Q_1(\theta) = -\frac{1}{2\sigma^2}, \qquad Q_2(\theta) = \frac{\mu}{\sigma^2}, \qquad T_1(x) = x^2, \qquad T_2(x) = x,$$

$$D(\theta) = -\frac{1}{2}\left[\frac{\mu^2}{\sigma^2} + \log(2\pi\sigma^2)\right], \qquad \text{and} \qquad S(x) = 0.$$

Other examples are the $G(\alpha, \beta)$ and $B(\alpha, \beta)$ distributions when both α, β are unknown, and the multinomial distribution. $U[\alpha, \beta]$ does not belong to this family, nor does $\mathscr{C}(\alpha, \beta)$.

Some general properties of exponential families will be studied in Chapter 8, and the importance of these families will then become evident.

PROBLEMS 5.5

1. Show that the following families of distributions are one-parameter exponential families:
   (a) $X \sim b(n, p)$.
   (b) $X \sim G(\alpha, \beta)$, (i) if α is known, (ii) if β is known.
   (c) $X \sim B(\alpha, \beta)$, (i) if α is known, (ii) if β is known.
   (d) $X \sim NB(r; p)$, where r is known, p unknown.

2. Let $X \sim \mathscr{C}(1, \theta)$. Show that the family of distributions of X is not a one-parameter exponential family.

3. Let $X \sim U[0, \theta]$, $\theta \in [0, \infty)$. Show that the family of distributions of X is not an exponential family.

4. Is the family of pdf's

$$f_\theta(x) = \tfrac{1}{2}e^{-|x-\theta|}, \qquad -\infty < x < \infty,\ \theta \in (-\infty, \infty),$$

an exponential family?

5. Show that the following families of distributions are two-parameter exponential families.
   (a) $X \sim G(\alpha, \beta)$, both α and β unknown.
   (b) $X \sim B(\alpha, \beta)$, both α and β unknown.

6. Show that the families of distributions $U[\alpha, \beta]$ and $\mathscr{C}(\alpha, \beta)$ do not belong to the exponential families.

7. Show that the multinomial distribution forms an exponential family.

CHAPTER 6

Limit Theorems

6.1 INTRODUCTION

In this chapter we investigate convergence properties of sequences of random variables. The three limit results proved here, namely, the two laws of large numbers and the central limit theorem, are of considerable importance in the study of probability and statistics. Just as in analysis, we distinguish among several types of convergence. The various modes of convergence are introduced in Section 2. Sections 3 and 4 deal with the laws of large numbers, and the central limit theorem is proved in Section 6.

The reader may find some parts of this chapter difficult, at least on the first reading. These have been identified with a dagger (†) and include the concept of almost sure convergence (Section 2), the strong law of large numbers (Section 4), and the proof of the central limit theorem (Theorem 6.6.1). An alternative proof of the central limit theorem is provided in Problem 6.6.4, which the reader may find simpler although the conditions are more restrictive. Since the central limit result is basic and will be used repeatedly in the rest of the book, it is important for the reader to familiarize himself with this result and its application and to understand its significance. He can pick up the proof of Theorem 6.6.1 later if he so desires. Similarly, on the first reading it will suffice to know the strong law of large numbers and to understand its significance.

6.2 MODES OF CONVERGENCE

In this section we consider several modes of convergence and investigate their interrelationships. We begin with the weakest mode.

Definition 1. Let $\{F_n\}$ be a sequence of distribution functions. If there exists a df F such that, as $n \to \infty$,

(1) $F_n(x) \to F(x)$

at every point x at which F is continuous, we say that F_n converges in law (or, weakly,) to F, and we write $F_n \xrightarrow{w} F$.

If $\{X_n\}$ is a sequence of rv's and $\{F_n\}$ is the corresponding sequence of df's, we say that X_n converges in distribution (or law) to X if there exists an rv X with df F such that $F_n \xrightarrow{w} F$. We write $X_n \xrightarrow{L} X$.

It must be remembered that it is quite possible for a given sequence df's to converge to a function that is not a df.

Example 1. Consider the sequence of df's

$$F_n(x) = \begin{cases} 0, & x < n, \\ 1, & x \geq n. \end{cases}$$

Here $F_n(x)$ is the df of the rv X_n degenerate at $x=n$. We see that $F_n(x)$ converges to a function F that is identically equal to 0, and hence is not a df.

Example 2. Let $X_1, X_2, \cdots, X_n$ be iid rv's with common density function

$$f(x) = \begin{cases} \dfrac{1}{\theta} & 0 < x < \theta, \\ 0 & \text{otherwise}, \end{cases} \quad (0 < \theta < \infty).$$

Let $M_n = \max(X_1, X_2, \cdots, X_n)$. Then the density function of M_n is

$$f_{M_n}(x) = \begin{cases} \dfrac{nx^{n-1}}{\theta^n} & 0 < x < \theta, \\ 0 & \text{otherwise}, \end{cases}$$

and the df of M_n is

$$F_n(x) = \begin{cases} 0 & x < 0, \\ (x/\theta)^n & 0 \leq x < \theta, \\ 1, & x \geq \theta. \end{cases}$$

We see that, as $n \to \infty$,

$$F_n(x) \to F(x) = \begin{cases} 0, & x < \theta, \\ 1, & x \geq \theta, \end{cases}$$

which is a df. Thus $F_n \xrightarrow{w} F$.

The following example shows that convergence in distribution does not imply convergence of moments.

Example 3. Let F_n be a sequence of df's defined by

$$F_n(x) = \begin{cases} 0, & x < 0, \\ 1 - \dfrac{1}{n}, & 0 \leq x < n, \\ 1, & n \leq x. \end{cases}$$

Clearly $F_n \xrightarrow{w} F$, where F is the df given by

$$F(x) = \begin{cases} 0, & x < 0, \\ 1, & x \geq 0. \end{cases}$$

Note that F_n is the df of the rv X_n with pmf

$$P\{X_n = 0\} = 1 - \frac{1}{n}, \qquad P\{X_n = n\} = \frac{1}{n},$$

and F is the df of the rv X degenerate at 0. We have

$$EX_n^k = n^k \left(\frac{1}{n}\right) = n^{k-1},$$

where k is a positive integer. Also $EX^k = 0$. So that

$$EX_n^k \not\to EX^k \qquad \text{for any } k.$$

We next give an example to show that weak convergence of distribution functions does not imply the convergence of corresponding pmf's or pdf's.

Example 4. Let $\{X_n\}$ be a sequence of rv's with pmf

$$f_n(x) = P\{X_n = x\} = \begin{cases} 1 & \text{if } x = 2 + 1/n, \\ 0 & \text{otherwise.} \end{cases}$$

Note that none of the f_n's assigns any probability to the point $x = 2$. It follows that

$$f_n(x) \to f(x) \qquad \text{as } n \to \infty,$$

where $f(x) = 0$ for all x. However, the sequence of df's $\{F_n\}$ of rv's X_n converges to the function

$$F(x) = \begin{cases} 0, & x < 2, \\ 1, & x \geq 2, \end{cases}$$

at all continuity points of F. Since F is the df of the rv degenerate at $x = 2$, $F_n \xrightarrow{w} F$.

The following result is easy to prove.

Theorem 1. Let X_n be a sequence of integer-valued rv's. Also, let $f_n(k) = P\{X_n = k\}$, $k = 0, 1, 2, \cdots$, be the pmf of X_n, $n = 1, 2, \cdots$, and $f(k) = P\{X = k\}$ be the pmf of X. Then

$$f_n(x) \to f(x) \quad \text{for all } x \Leftrightarrow X_n \xrightarrow{L} X.$$

In the continuous case we state the following result of Scheffé [110] without proof.

Theorem 2. Let X_n, $n = 1, 2, \cdots$, and X be continuous rv's such that

$$f_n(x) \to f(x) \quad \text{for (almost) all } x \text{ as } n \to \infty.$$

Here f_n and f are the pdf's of X_n and X, respectively. Then $X_n \xrightarrow{L} X$.

The following result is easy to establish.

Theorem 3. Let $\{X_n\}$ be a sequence of rv's such that $X_n \xrightarrow{L} X$, and let c be a constant. Then

(a) $\qquad\qquad X_n + c \xrightarrow{L} X + c,$

(b) $\qquad\qquad cX_n \xrightarrow{L} cX, \qquad c \neq 0.$

A slightly stronger concept of convergence is defined by *convergence in probability*.

Definition 2. Let $\{X_n\}$ be a sequence of rv's defined on some probability space $(\Omega, \mathscr{S}, P)$. We say that the sequence $\{X_n\}$ converges in probability to the rv X if, for every $\varepsilon > 0$.

(2) $\qquad\qquad P\{|X_n - X| > \varepsilon\} \to 0 \qquad \text{as} \quad n \to \infty.$

We write $X_n \xrightarrow{P} X$.

Remark 1. We emphasize that the definition says nothing about the convergence of the rv's X_n to the rv X in the sense in which it is understood in real analysis. Thus $X_n \xrightarrow{P} X$ does not imply that, given $\varepsilon > 0$, we can find an N such that $|X_n - X| < \varepsilon$ for $n \geq N$. Definition 2 speaks only of the convergence of the sequence of probabilities $P\{|X_n - X| > \varepsilon\}$ to 0.

Example 5. Let $\{X_n\}$ be a sequence of rv's with pmf

$$P\{X_n = 1\} = \frac{1}{n}, \qquad P\{X_n = 0\} = 1 - \frac{1}{n}.$$

Then

$$P\{|X_n| > \varepsilon\} = \begin{cases} P\{X_n = 1\} = \dfrac{1}{n} & \text{if } 0 < \varepsilon < 1, \\ 0 & \text{if } \varepsilon \geq 1. \end{cases}$$

It follows that $P\{|X_n| > \varepsilon\} \to 0$ as $n \to \infty$, and we conclude that $X_n \xrightarrow{P} 0$.

The truth of the following statements can easily be verified.

1. $\qquad X_n \xrightarrow{P} X \Leftrightarrow X_n - X \xrightarrow{P} 0.$
2. $\qquad X_n \xrightarrow{P} X, \quad X_n \xrightarrow{P} Y \Rightarrow P\{X = Y\} = 1,$

for $P\{|X - Y| > c\} \le P\{|X_n - X| > \frac{c}{2}\} + P\{|X_n - Y| > \frac{c}{2}\}$,
and it follows that $P\{|X - Y| > c\} = 0$ for every $c > 0$.

3. $\qquad X_n \xrightarrow{P} X \Rightarrow X_n - X_m \xrightarrow{P} 0 \qquad$ as $n, m \to \infty$,

for

$$P\{|X_n - X_m| > \varepsilon\} \le P\{|X_n - X| > \tfrac{\varepsilon}{2}\} + P\{|X_m - X| > \tfrac{\varepsilon}{2}\}.$$

4. $\qquad X_n \xrightarrow{P} X, \quad Y_n \xrightarrow{P} Y \Rightarrow X_n \pm Y_n \xrightarrow{P} X \pm Y.$
5. $\qquad X_n \xrightarrow{P} X, k$ constant, $\Rightarrow k X_n \xrightarrow{P} kX.$
6. $\qquad X_n \xrightarrow{P} k \Rightarrow X_n^2 \xrightarrow{P} k^2.$
7. $\qquad X_n \xrightarrow{P} a, \quad Y_n \xrightarrow{P} b, a, b$ constants $\Rightarrow X_n Y_n \xrightarrow{P} ab,$

for

$$X_n Y_n = \frac{(X_n + Y_n)^2 - (X_n - Y_n)^2}{4} \xrightarrow{P} \frac{(a+b)^2 - (a-b)^2}{4} = ab.$$

8. $\qquad X_n \xrightarrow{P} 1 \Rightarrow X_n^{-1} \xrightarrow{P} 1,$

for

$$P\{\left|\tfrac{1}{X_n} - 1\right| \ge \varepsilon\} = P\{\tfrac{1}{X_n} \ge 1 + \varepsilon\} + P\{\tfrac{1}{X_n} \le 1 - \varepsilon\}$$
$$= P\{\tfrac{1}{X_n} \ge 1 + \varepsilon\} + P\{\tfrac{1}{X_n} \le 0\}$$
$$+ P\{0 < \tfrac{1}{X_n} \le 1 - \varepsilon\}$$

and each of the three terms on the right goes to 0 as $n \to \infty$.

9. $X_n \xrightarrow{P} a, \quad Y_n \xrightarrow{P} b, a, b$ constants, $b \ne 0 \Rightarrow X_n Y_n^{-1} \xrightarrow{P} ab^{-1}.$
10. $X_n \xrightarrow{P} X,$ and Y an rv $\Rightarrow X_n Y \xrightarrow{P} XY.$

Note that Y is an rv so that, given $\delta > 0$ there exists a $k > 0$ such that $P\{|Y| > k\} < \delta/2$. Thus

$$P\{|X_nY - XY| > \varepsilon\} = P\{|X_n - X||Y| > \varepsilon, |Y| > k\}$$
$$+ P\{|X_n - X||Y| > \varepsilon, |Y| \le k\}$$
$$< \frac{\delta}{2} + P\{|X_n - X| > \frac{\varepsilon}{k}\}.$$

11.
for
$$X_n \xrightarrow{P} X, \quad Y_n \xrightarrow{P} Y \Rightarrow X_nY_n \xrightarrow{P} XY,$$

$$(X_n - X)(Y_n - Y) \xrightarrow{P} 0.$$

The result now follows on multiplication, using result 10.

Theorem 4. Let $X_n \xrightarrow{P} X$, and g be a continuous function defined on $\mathscr{R}$. Then $g(X_n) \xrightarrow{P} g(X)$ as $n \to \infty$.

Proof. Since X is an rv, we can, given $\varepsilon > 0$, find a constant $k = k(\varepsilon)$ such that

$$P\{|X| > k\} < \frac{\varepsilon}{2}.$$

Also, g is continuous on $\mathscr{R}$, so that g is uniformly continuous on $[-k, k]$. It follows that there exists a $\delta = \delta(\varepsilon, k)$ such that

$$|g(x_n) - g(x)| < \varepsilon$$

whenever $|x| \le k$ and $|x_n - x| < \delta$. Let

$$A = \{|X| \le k\}, \quad B = \{|X_n - X| < \delta\}, \quad C = \{|g(X_n) - g(X)| < \varepsilon\}.$$

Then $\omega \in A \cap B \Rightarrow \omega \in C$, so that

$$A \cap B \subseteq C.$$

It follows that

$$P\{C^c\} \le P\{A^c\} + P\{B^c\},$$

that is,

$$P\{|g(X_n) - g(X)| \ge \varepsilon\} \le P\{|X_n - X| \ge \delta\} + P\{|X| > k\} < \varepsilon$$

for $n \ge N(\varepsilon, \delta, k)$, where $N(\varepsilon, \delta, k)$ is chosen so that

$$P\{|X_n - X| \ge \delta\} < \frac{\varepsilon}{2} \quad \text{for} \quad n \ge N(\varepsilon, \delta, k).$$

Corollary. $X_n \xrightarrow{P} c$, where c is a constant $\Rightarrow g(X_n) \xrightarrow{P} g(c)$, g being a continuous function.

We remark that a more general result than Theorem 4 is true and state it without proof (see Rao [97], 104): $X_n \xrightarrow{L} X$, and g continuous on $\mathscr{R} \Rightarrow g(X_n) \xrightarrow{L} g(X)$.

The following two theorems explain the relationship between weak convergence and convergence in probability.

Theorem 5. $X_n \xrightarrow{P} X \Rightarrow X_n \xrightarrow{L} X$.

Proof. Let F_n and F, respectively, be the df's of X_n and X. We have
$$\{\omega: X(\omega) \leq x'\} = \{\omega: X_n(\omega) \leq x, X(\omega) \leq x'\} \cup \{\omega: X_n(\omega) > x,$$
$$X(\omega) \leq x'\} \subseteq \{X_n \leq x\} \cup \{X_n > x, X \leq x'\}.$$
It follows that
$$F(x') \leq F_n(x) + P\{X_n > x, X \leq x'\}.$$
Since $X_n - X \xrightarrow{P} 0$, we have for $x' < x$
$$P\{X_n > x, X \leq x'\} \leq P\{|X_n - X| > x - x'\} \to 0 \quad \text{as} \quad n \to \infty.$$
Therefore
$$F(x') \leq \varliminf_{n \to \infty} F_n(x), \qquad x' < x.$$
Similarly, by interchanging X and X_n, and x and x', we get
$$\varlimsup_{n \to \infty} F_n(x) \leq F(x''), \qquad x < x''.$$
Thus, for $x' < x < x''$, we have
$$F(x') \leq \varliminf F_n(x) \leq \varlimsup F_n(x) \leq F(x'').$$
Since F has only a countable number of discontinuity points, we choose x to be a point of continuity of F, and letting $x'' \downarrow x$ and $x' \uparrow x$, we have
$$F(x) = \lim_{n \to \infty} F_n(x)$$
at all points of continuity of F.

Theorem 6. Let k be a constant. Then
$$X_n \xrightarrow{L} k \Rightarrow X_n \xrightarrow{P} k.$$

Proof. The proof is left as an exercise.

Corollary. Let k be a constant. Then
$$X_n \xrightarrow{L} k \Leftrightarrow X_n \xrightarrow{P} k.$$

Remark 2. We emphasize that we cannot improve the above result by replacing k by an rv; that is, $X_n \xrightarrow{L} X$ in general, does not imply $X_n \xrightarrow{P} X$, for let $X, X_1, X_2 \cdots$ be identically distributed rv's, and let the joint distribution of (X_n, X) be as follows:

$X \backslash X_n$	0	1	
0	0	$\frac{1}{2}$	$\frac{1}{2}$
1	$\frac{1}{2}$	0	$\frac{1}{2}$
	$\frac{1}{2}$	$\frac{1}{2}$	1

Clearly, $X_n \xrightarrow{L} X$. But

$$P\{|X_n - X| > \tfrac{1}{2}\} \geq P\{|X_n - X| = 1\}$$
$$= P\{X_n = 0, X = 1\} + P\{X_n = 1, X = 0\}$$
$$= 1 \not\to 0.$$

Hence $X_n \not\xrightarrow{P} X$, but $X_n \xrightarrow{L} X$.

Remark 3. Example 3 shows that $X_n \xrightarrow{P} X$ does not imply $EX_n^k \to EX^k$ for any $k > 0$, k integral.

Definition 3. Let $\{X_n\}$ be a sequence of rv's such that $E|X_n|^r < \infty$, for some $r > 0$. We say that X_n converges in the rth mean to an rv X if $E|X|^r < \infty$ and

(3) $$E|X_n - X|^r \to 0 \quad \text{as} \quad n \to \infty,$$

and we write $X_n \xrightarrow{r} X$.

Example 6. Let $\{X_n\}$ be a sequence of rv's defined by

$$P\{X_n = 0\} = 1 - \frac{1}{n}, \quad P\{X_n = 1\} = \frac{1}{n}, \quad n = 1, 2, \cdots.$$

Then

$$E|X_n|^2 = \frac{1}{n} \to 0 \quad \text{as} \quad n \to \infty,$$

and we see that $X_n \xrightarrow{2} X$, where the rv X is degenerate at 0.

Theorem 7. Let $X_n \xrightarrow{r} X$ for some $r > 0$. Then $X_n \xrightarrow{P} X$.

Proof. The proof is left as an exercise.

Example 7. Let $\{X_n\}$ be a sequence of rv's defined by
$$P\{X_n = 0\} = 1 - \frac{1}{n^r}, \quad P\{X_n = n\} = \frac{1}{n^r}, \quad r > 0, \quad n = 1, 2, \cdots.$$
Then $E|X_n|^r = 1$, so that $X_n \not\xrightarrow{r} 0$. We show that $X_n \xrightarrow{P} 0$.
$$P\{|X_n| > \varepsilon\} = \begin{cases} P\{X_n = n\} & \text{if } \varepsilon < n \\ 0 & \text{if } \varepsilon > n \end{cases} \to 0 \text{ as } n \to \infty.$$

Theorem 8. Let $\{X_n\}$ be a sequence of rv's such that $X_n \xrightarrow{2} X$. Then $EX_n \to EX$, and $EX_n^2 \to EX^2$ as $n \to \infty$.

Proof. We have
$$|E(X_n - X)| \le E|X_n - X| \le E^{1/2}|X_n - X|^2 \to 0 \quad \text{as } n \to \infty.$$
To see that $EX_n^2 \to EX^2$ (see also Theorem 9), we write
$$EX_n^2 = E(X_n - X)^2 + EX^2 + 2E\{X(X_n - X)\}$$
and note that
$$|E\{X(X_n - X)\}| \le \sqrt{EX^2 E(X_n - X)^2}$$
by the Cauchy-Schwarz inequality. The result follows on passing to the limits.

We get, in addition, that $X_n \xrightarrow{2} X$ implies var $(X_n) \to$ var (X).

Corollary. Let $\{X_m\}$, $\{Y_n\}$ be two sequences of rv's such that $X_m \xrightarrow{2} X$, $Y_n \xrightarrow{2} Y$. Then $E(X_m Y_n) \to E(XY)$ as $m, n \to \infty$.

Proof. The proof is left to the reader.

As a simple consequence of Theorem 8 and its corollary we see that $X_m \xrightarrow{2} X$, $Y_n \xrightarrow{2} Y$ together imply cov $(X_m, Y_n) \to$ cov (X, Y).

Theorem 9. If $X_n \xrightarrow{r} X$, then $E|X_n|^r \to E|X|^r$.

Proof. Let $0 < r \le 1$. Then
$$E|X_n|^r = E|X_n - X + X|^r$$
so that
$$E|X_n|^r - E|X|^r \le E|X_n - X|^r.$$

Interchanging X_n and X, we get
$$E|X|^r - E|X_n|^r \le E|X_n - X|^r.$$
It follows that
$$|E|X|^r - E|X_n|^r| \le E|X_n - X|^r \to 0 \quad \text{as} \quad n \to \infty.$$
For $r > 1$, we use Minkowski's inequality and obtain
$$[E|X_n|^r]^{1/r} \le [E|X_n - X|^r]^{1/r} + [E|X|^r]^{1/r}$$
and
$$[E|X|^r]^{1/r} \le [E|X_n - X|^r]^{1/r} + [E|X_n|^r]^{1/r}.$$
It follows that
$$|E^{1/r}|X_n|^r - E^{1/r}|X|^r| \le E^{1/r}|X_n - X|^r \to 0 \quad \text{as} \quad n \to \infty.$$
This completes the proof.

Theorem 10. Let $r > s$. Then $X_n \xrightarrow{r} X \Rightarrow X_n \xrightarrow{s} X$.

Proof. From Theorem 3.4.3 it follows that for $s < r$
$$E|X_n - X|^s \le [E|X_n - X|^r]^{s/r} \to 0 \quad \text{as} \quad n \to \infty$$
since $X_n \xrightarrow{r} X$.

Remark 4. Clearly the converse to Theorem 10 cannot hold, since $E|X|^s < \infty$ for $s < r$ does not imply $E|X|^r < \infty$.

Remark 5. In view of Theorem 9, it follows that $X_n \xrightarrow{r} X \Rightarrow E|X_n|^s \to E|X|^s$ for $s \le r$.

Definition 4.† Let $\{X_n\}$ be a sequence of rv's. We say that X_n converges almost surely (a.s.) to an rv X if and only if

(4) $\qquad P\{\omega : X_n(\omega) \to X(\omega) \text{ as } n \to \infty\} = 1,$

and we write $X_n \xrightarrow{a.s.} X$ or $X_n \to X$ with probability 1.

The following result elucidates Definition 4.

Theorem 11. $X_n \xrightarrow{a.s.} X$ if and only if $\lim_{n \to \infty} P\{\sup_{m \ge n}|X_m - X| > \varepsilon\} = 0$ for all $\varepsilon > 0$.

---

†May be omitted on the first reading.

Proof. Since $X_n \xrightarrow{a.s.} X$, $X_n - X \xrightarrow{a.s.} 0$, and it will be sufficient to show the equivalence of

(a) $\quad\quad\quad X_n \xrightarrow{a.s.} 0 \quad$ and $\quad$ (b) $\lim_{n \to \infty} P\{\sup_{m \geq n} |X_m| > \varepsilon\} = 0$.

Let us suppose that (a) holds. Let $\varepsilon > 0$, and write

$$A_n(\varepsilon) = \{\sup_{m \geq n} |X_m| > \varepsilon\} \quad \text{and} \quad C = \{\lim_{n \to \infty} X_n = 0\}.$$

Also write $B_n(\varepsilon) = C \cap A_n(\varepsilon)$, and note that $B_{n+1}(\varepsilon) \subset B_n(\varepsilon)$, and the limit set $\bigcap_{n=1}^{\infty} B_n(\varepsilon) = \phi$. It follows that

$$\lim_{n \to \infty} PB_n(\varepsilon) = P\{\bigcap_{n=1}^{\infty} B_n(\varepsilon)\} = 0.$$

Since $PC = 1$, $PC^c = 0$, and we have

$$PB_n(\varepsilon) = P(A_n \cap C) = 1 - P(C^c \cup A_n^c)$$
$$= 1 - PC^c - PA_n^c + P(C^c \cap A_n^c)$$
$$= PA_n + P(C^c \cap A_n^c)$$
$$= PA_n.$$

It follows that (b) holds.

Conversely, let $\lim_{n \to \infty} PA_n(\varepsilon) = 0$, and write

$$D(\varepsilon) = \{\overline{\lim_{n \to \infty}} |X_n| > \varepsilon > 0\}.$$

Since $D(\varepsilon) \subset A_n(\varepsilon)$ for $n = 1, 2, \cdots$, it follows that $PD(\varepsilon) = 0$. Also,

$$C^c = \{\lim_{n \to \infty} X_n \neq 0\} \subset \bigcup_{k=1}^{\infty} \{\overline{\lim} |X_n| > \frac{1}{k}\},$$

so that

$$1 - PC \leq \sum_{k=1}^{\infty} PD(\frac{1}{k}) = 0,$$

and (a) holds.

Remark 6. Thus $X_n \xrightarrow{a.s.} 0$ means that, for $\varepsilon > 0$, $\eta > 0$ arbitrary, we can find an n_0 such that

(5) $$P\{\sup_{n \geq n_0} |X_n| > \varepsilon\} < \eta.$$

Indeed, we can write, equivalently, that

(6) $$\lim_{n_0 \to \infty} P[\bigcup_{n \geq n_0} \{|X_n| > \varepsilon\}] = 0.$$

Theorem 12. $X_n \xrightarrow{a.s.} X \Rightarrow X_n \xrightarrow{P} X$.

MODES OF CONVERGENCE

Proof. By Remark 6, $X_n \xrightarrow{a.s.} X$ implies that, for arbitrary $\varepsilon > 0, \eta > 0$, we can choose an $n_0 = n_0(\varepsilon, \eta)$ such that

$$P[\bigcap_{n=n_0}^{\infty} \{|X_n - X| \leq \varepsilon\}] \geq 1 - \eta.$$

Clearly

$$\bigcap_{n=n_0}^{\infty} \{|X_n - X| \leq \varepsilon\} \subset \{|X_n - X| \leq \varepsilon\} \quad \text{for} \quad n \geq n_0.$$

It follows that for $n \geq n_0$

$$P\{|X_n - X| \leq \varepsilon\} \geq P[\bigcap_{n=n_0}^{\infty} \{|X_n - X| \leq \varepsilon\}] \geq 1 - \eta,$$

that is

$$P\{|X_n - X| > \varepsilon\} < \eta \quad \text{for} \quad n \geq n_0,$$

which is the same as saying $X_n \xrightarrow{P} X$.

That the converse of Theorem 12 does not hold is shown in the following example.

Example 8. For each positive integer n there exist integers m and k (uniquely determined) such that

$$n = 2^k + m, \quad 0 \leq m < 2^k, \quad k = 0, 1, 2, \cdots.$$

Thus, for $n = 1$, $k = 0$ and $m = 0$; for $n = 5$, $k = 2$ and $m = 1$; and so on. Define rv's X_n, for $n = 1, 2, \cdots$, on $\Omega = [0, 1]$ by

$$X_n(\omega) = \begin{cases} 2^k, & \dfrac{m}{2^k} \leq \omega < \dfrac{m+1}{2^k}, \\ 0, & \text{otherwise.} \end{cases}$$

Let the probability distribution of X_n be given by $P\{I\}$ = length of the interval $I \subseteq \Omega$. Thus

$$P\{X_n = 2^k\} = \frac{1}{2^k}, \quad P\{X_n = 0\} = 1 - \frac{1}{2^k}.$$

The limit $\lim_{n \to \infty} X_n(\omega)$ does not exist for any $\omega \in \Omega$, so that X_n does not converge almost surely. But

$$P\{|X_n| > \varepsilon\} = P\{X_n > \varepsilon\} = \begin{cases} 0 & \text{if } \varepsilon \geq 2^k, \\ \dfrac{1}{2^k} & \text{if } 0 < \varepsilon < 2^k, \end{cases}$$

and we see that

$$P\{|X_n| > \varepsilon\} \to 0 \quad \text{as } n \text{ (and hence } k) \to \infty.$$

Theorem 13. Let $\{X_n\}$ be a strictly decreasing sequence of positive rv's, and suppose that $X_n \xrightarrow{P} 0$. Then $X_n \xrightarrow{a.s.} 0$.

Proof. The proof is left as an exercise.

Example 9. Let $\{X_n\}$ be a sequence of independent rv's defined by

$$P\{X_n = 0\} = 1 - \frac{1}{n}, \quad P\{X_n = 1\} = \frac{1}{n}, \quad n = 1, 2, \cdots.$$

Then

$$E|X_n - 0|^2 = E|X_n|^2 = \frac{1}{n} \to 0 \quad \text{as} \quad n \to \infty,$$

so that $X_n \xrightarrow{2} 0$. Also

$$P\{X_n = 0 \quad \text{for every } m \leq n \leq n_0\}$$
$$= \prod_{n=m}^{n_0} (1 - \frac{1}{n}) = \frac{m-1}{n_0},$$

which diverges to zero as $n_0 \to \infty$ for all values of m (see P.2.9). Thus X_n does not converge to 0 with probability 1.

Example 10. Let $\{X_n\}$ be defined by

$$P\{X_n = 0\} = 1 - \frac{1}{n^r}, \quad P\{X_n = n\} = \frac{1}{n^r}, \quad r \geq 2, \quad n = 1, 2, \cdots.$$

Then

$$P\{X_n = 0 \quad \text{for } m \leq n \leq n_0\} = \prod_{n=m}^{n_0} (1 - \frac{1}{n^r}).$$

As $n_0 \to \infty$, the infinite product converges to some nonzero quantity (see P. 2.9), which itself converges to 1 as $m \to \infty$. Thus $X_n \xrightarrow{a.s.} 0$. However, $E|X_n|^r = 1$ and $X_n \xrightarrow{r}\!\!\!\!\!\!/\;\, 0$ as $n \to \infty$.

Example 11. Let $\{X_n\}$ be a sequence of rv's with $P\{X_n = \pm 1/n\} = \frac{1}{2}$. Then $E|X_n|^r = 1/n^r \to 0$ as $n \to \infty$, and $X_n \xrightarrow{r} 0$. For $j < k$, $|X_j| > |X_k|$, so that $\{|X_k| > \varepsilon\} \subset \{|X_j| > \varepsilon\}$. It follows that

$$\bigcup_{j=n}^{\infty} \{|X_j| > \varepsilon\} = \{|X_n| > \varepsilon\}.$$

Choosing $n > 1/\varepsilon$, we see that
$$P[\bigcup_{j=n}^{\infty} \{|X_j| > \varepsilon\}] = P\{|X_n| > \varepsilon\} \leq P\{|X_n| > \frac{1}{n}\} = 0,$$
and (6) implies that $X_n \xrightarrow{a.s.} 0$.

Theorem 14. Let $\{X_n, Y_n\}$, $n = 1, 2, \cdots$, be a sequence of rv's. Then
$$|X_n - Y_n| \xrightarrow{P} 0 \text{ and } Y_n \xrightarrow{L} Y \Rightarrow X_n \xrightarrow{L} Y.$$

Proof. Let x be a point of continuity of the df of Y and $\varepsilon > 0$. Then
$$\begin{aligned} P\{X_n \leq x\} &= P\{Y_n \leq x + Y_n - X_n\} \\ &= P\{Y_n \leq x + Y_n - X_n; Y_n - X_n \leq \varepsilon\} \\ &\quad + P\{Y_n \leq x + Y_n - X_n; Y_n - X_n > \varepsilon\} \\ &\leq P\{Y_n \leq x + \varepsilon\} + P\{Y_n - X_n > \varepsilon\}. \end{aligned}$$

It follows that
$$\varlimsup_{n\to\infty} P\{X_n \leq x\} \leq \varlimsup_{n\to\infty} P\{Y_n \leq x + \varepsilon\}.$$

Similarly
$$\varliminf_{n\to\infty} P\{X_n \leq x\} \geq \varliminf_{n\to\infty} P\{Y_n \leq x - \varepsilon\}.$$

Since $\varepsilon > 0$ is arbitrary and x is a continuity point of $P\{Y \leq x\}$, we get the result by letting $\varepsilon \to 0$.

Corollary. $X_n \xrightarrow{P} X \Rightarrow X_n \xrightarrow{L} X$.

Theorem 15 (Cramér [18], 254). Let $\{X_n, Y_n\}$, $n = 1, 2, \cdots$, be a sequence of pairs of rv's, and let c be a constant. Then

(a) $\quad X_n \xrightarrow{L} X, \quad Y_n \xrightarrow{P} c \Rightarrow X_n \pm Y_n \xrightarrow{L} X \pm c;$

(b) $\quad X_n \xrightarrow{L} X, \quad Y_n \xrightarrow{P} c \Rightarrow \begin{cases} X_n Y_n \xrightarrow{L} cX & \text{if } c \neq 0, \\ X_n Y_n \xrightarrow{P} 0 & \text{if } c = 0; \end{cases}$

(c) $\quad X_n \xrightarrow{L} X, \quad Y_n \xrightarrow{P} c \Rightarrow \dfrac{X_n}{Y_n} \xrightarrow{L} X/c \quad \text{if } c \neq 0.$

Proof.

(a) $X_n \xrightarrow{L} X \Rightarrow X_n \pm c \xrightarrow{L} X \pm c$ (Theorem 3). Also, $Y_n - c = (Y_n \pm X_n) \mp (X_n \pm c) \xrightarrow{P} 0$.

A simple use of Theorem 14 shows that
$$X_n \pm Y_n \xrightarrow{L} X \pm c.$$

(b) We first consider the case where $c = 0$. We have, for any fixed number $k > 0$,

$$P\{|X_n Y_n| > \varepsilon\} = P\{|X_n Y_n| > \varepsilon, |Y_n| \le \frac{\varepsilon}{k}\} + P\{|X_n Y_n| > \varepsilon, |Y_n| > \frac{\varepsilon}{k}\}$$

$$\le P\{|X_n| > k\} + P\{|Y_n| > \frac{\varepsilon}{k}\}.$$

Since $Y_n \xrightarrow{P} 0$ and $X_n \xrightarrow{L} X$, it follows that, for any fixed $k > 0$,

$$\varlimsup_{n \to \infty} P\{|X_n Y_n| > \varepsilon\} \le P\{|X| > k\}.$$

Since k is arbitrary, we can make $P\{|X| > k\}$ as small as we please by choosing k large. It follows that

$$X_n Y_n \xrightarrow{P} 0.$$

Now, let $c \ne 0$. Then

$$X_n Y_n - c X_n = X_n(Y_n - c)$$

and, since $X_n \xrightarrow{L} X$, $Y_n \xrightarrow{P} c$, $X_n(Y_n - c) \xrightarrow{P} 0$. Using Theorem 14, we get the result that

$$X_n Y_n \xrightarrow{L} cX.$$

(c) $Y_n \xrightarrow{P} c$, and $c \ne 0 \Rightarrow Y_n^{-1} \xrightarrow{P} c^{-1}$. It follows that $X_n \xrightarrow{L} X$, $Y_n \xrightarrow{P} c \Rightarrow X_n Y_n^{-1} \xrightarrow{L} c^{-1} X$, and the proof of the theorem is complete.

As an application of Theorem 15 we present the following example.

Example 12. Let $X_1, X_2, \cdots$, be iid rv's with common law $\mathcal{N}(0, 1)$. We shall determine the limiting distribution of the rv

$$W_n = \sqrt{n} \, \frac{X_1 + X_2 + \cdots + X_n}{X_1^2 + X_2^2 + \cdots + X_n^2}.$$

Let us write

$$U_n = \frac{1}{\sqrt{n}}(X_1 + X_2 + \ldots + X_n) \quad \text{and} \quad V_n = \frac{X_1^2 + X_2^2 + \ldots + X_n^2}{n}.$$

Then

$$W_n = \frac{U_n}{V_n}.$$

For the mgf of U_n we have

$$M_{U_n}(t) = \prod_{i=1}^{n} E e^{tX_i/\sqrt{n}} = \prod_{i=1}^{n} e^{t^2/2n}$$
$$= e^{t^2/2},$$

so that U_n is an $\mathcal{N}(0, 1)$ variate (see also Corollary 2 to Theorem 5.3.25). It follows that $U_n \xrightarrow{L} Z$, where Z is an $\mathcal{N}(0, 1)$ rv. As for V_n, we note that each X_i^2 is a chi-square variate with 1 d.f. Thus

$$M_{V_n}(t) = \prod_{i=1}^{n} \left(\frac{1}{1 - 2t/n}\right)^{1/2}, \qquad t < \frac{n}{2},$$
$$= \left(1 - \frac{2t}{n}\right)^{-n/2}, \qquad t < \frac{n}{2},$$

which is the mgf of a gamma variate with parameters $\alpha = n/2$ and $\beta = 2/n$. Thus the density function of V_n is given by

$$f_{V_n}(x) = \begin{cases} \dfrac{1}{\Gamma(n/2)} \dfrac{1}{(2/n)^{n/2}} x^{n/2-1} e^{-nx/2}, & 0 < x < \infty, \\ 0, & \text{otherwise.} \end{cases}$$

We will show that $V_n \xrightarrow{P} 1$. We have, for any $\varepsilon > 0$,

$$P\{|V_n - 1| > \varepsilon\} \le \frac{\text{var}(V_n)}{\varepsilon^2} = \left(\frac{n}{2}\right)\left(\frac{2}{n}\right)^2 \frac{1}{\varepsilon^2} \to 0 \qquad \text{as} \quad n \to \infty.$$

We have thus shown that

$$U_n \xrightarrow{L} Z \quad \text{and} \quad V_n \xrightarrow{P} 1.$$

It follows by Theorem 15 (c) that $W_n = U_n/V_n \xrightarrow{L} Z$, where Z is an $\mathcal{N}(0, 1)$ rv.

For another example see Section 7.3.

PROBLEMS 6.2

1. Let $X_1, X_2, \cdots$ be a sequence of rv's with corresponding df's given by $F_n(x) = 0$ if $x < -n$, $= (x + n)/2n$ if $-n \le x < n$, and $= 1$ if $x \ge n$. Does F_n converge to a df?

2. Let $X_1, X_2, \cdots$ be iid $\mathcal{N}(0, 1)$ rv's. Consider the sequence of rv's $\{\bar{X}_n\}$, where $\bar{X}_n = n^{-1} \sum_{i=1}^{n} X_i$. Let F_n be the df of $\bar{X}_n$, $n = 1, 2, \cdots$. Find $\lim_{n \to \infty} F_n(x)$. Is this limit a df?

3. Let $X_1, X_2, \cdots$ be iid $U(0, \theta)$ rv's. Let $N_n = \min(X_1, X_2, \cdots, X_n)$, and consider

the sequence $Y_n = nN_n$. Does Y_n converge in distribution to some rv Y? If so, find the df of rv Y.

4. Let $X_1, X_2, \cdots$ be iid rv's with common absolutely continuous df F. Let $M_n = \max(X_1, X_2, \cdots, X_n)$, and consider the sequence of rv's $Y_n = n[1 - F(M_n)]$. Find the limiting df of Y_n.

5. Let $X_1, X_2, \cdots$ be a sequence of iid rv's with common pdf $f(x) = e^{-x+\theta}$ if $x \geq \theta$, and $= 0$ if $x < \theta$. Write $\bar{X}_n = n^{-1}\sum_{i=1}^n X_i$.
   (a) Show that $\bar{X}_n \xrightarrow{P} 1 + \theta$.
   (b) Show that $\min\{X_1, X_2, \cdots, X_n\} \xrightarrow{P} \theta$.

6. Let $X_1, X_2, \cdots$ be iid $U[0, \theta]$ rv's. Show that $\max\{X_1, X_2, \cdots, X_n\} \xrightarrow{P} \theta$.

7. Let $\{X_n\}$ be a sequence of rv's such that $X_n \xrightarrow{L} X$. Let a_n be a sequence of positive constants such that $a_n \to \infty$ as $n \to \infty$. Show that $a_n^{-1} X_n \xrightarrow{P} 0$.

8. Let $\{X_n\}$ be a sequence of rv's such that $P\{|X_n| \leq k\} = 1$ for all n and some constant $k > 0$. Suppose that $X_n \xrightarrow{P} X$. Show that $X_n \xrightarrow{r} X$ for any $r > 0$.

9. Let $X_1, X_2, \cdots, X_{2n}$ be iid $\mathcal{N}(0,1)$ rv's. Define
$$U_n = \left\{\frac{X_1}{X_2} + \frac{X_3}{X_4} + \cdots + \frac{X_{2n-1}}{X_{2n}}\right\}, \quad V_n = X_1^2 + X_2^2 + \cdots + X_n^2, \text{ and}$$
$$Z_n = \frac{U_n}{V_n}.$$
Find the limiting distribution of Z_n.

10. Let $\{X_n\}$ be a sequence of geometric rv's with parameter λ/n, $n > \lambda > 0$. Also, let $Z_n = X_n/n$. Show that $Z_n \xrightarrow{L} G(1, 1/\lambda)$ as $n \to \infty$. (Prochaska [91])

11. Let X_n be a sequence of rv's such that $X_n \xrightarrow{a.s.} 0$, and let c_n be a sequence of real numbers such that $c_n \to 0$ as $n \to \infty$. Show that $X_n + c_n \xrightarrow{a.s.} 0$.

12. Does convergence almost surely imply convergence of moments?

13. Let $X_1, X_2, \cdots$ be a sequence of iid rv's with common df F, and write $M_n = \max\{X_1, X_2, \cdots, X_n\}$, $n = 1, 2, \cdots$.

(a) For $\alpha > 0$, $\lim_{x \to \infty} x^\alpha P\{X_1 > x\} = b > 0$. Find the limiting distribution of $(bn)^{-1/\alpha} M_n$. Also, find the pdf corresponding to the limiting df and compute its moments.

(b) If F satisfies
$$\lim_{x \to \infty} e^x[1 - F(x)] = b > 0,$$
find the limiting df of $M_n - \log(bn)$ and compute the corresponding pdf and the mgf.

(c) If X_i is bounded above by x_0 with probability 1, and for some $\alpha > 0$
$$\lim_{x \to x_0^-}(x_0 - x)^{-\alpha}[1 - F(x)] = b > 0,$$
find the limiting distribution of $(bn)^{1/\alpha}\{M_n - x_0\}$, the corresponding pdf, and the moments of the limiting distribution.

(*The above remarkable result, due to Gnedenko [38], exhausts all limiting distributions of M_n with suitable norming and centering.*)

14. Let $\{F_n\}$ be a sequence of df's that converges weakly to a df F which is continuous everywhere. Show that $F_n(x)$ converges to $F(x)$ uniformly.
15. Prove Theorem 1.
16. Prove Theorem 6.
17. Prove Theorem 13.
18. Prove the corollary to Theorem 8.
19. Let V be the class of all random variables defined on a probability space with finite expectations, and for $X \in V$ define

$$\rho(X) = E\left\{\frac{|X|}{1 + |X|}\right\}.$$

Show the following:

(a) $\rho(X + Y) \le \rho(X) + \rho(Y)$; $\rho(\sigma X) \le \max(|\sigma|, 1)\rho(X)$.
(b) $d(X, Y) = \rho(X - Y)$ is a distance function on V (assuming that we identify rv's that are a.s. equal).
(c) $\lim_{n\to\infty} d(X_n, X) = 0 \Leftrightarrow X_n \xrightarrow{P} X$.

6.3 THE WEAK LAW OF LARGE NUMBERS

Let $\{X_n\}$ be a sequence of rv's. Write $S_n = \sum_{k=1}^{n} X_k$, $n = 1, 2, \cdots$. In this section we answer the following question in the affirmative: Do there exist sequences of constants A_n and $B_n > 0$, $B_n \to \infty$ as $n \to \infty$, such that the sequence of rv's $B_n^{-1}(S_n - A_n)$ converges in probability to 0 as $n \to \infty$?

Definition 1. Let $\{X_n\}$ be a sequence of rv's, and let $S_n = \sum_{k=1}^{n} X_k$, $n = 1, 2, \cdots$. We say that $\{X_n\}$ obeys the weak law of large numbers (WLLN) with respect to the sequence of constants $\{B_n\}$, $B_n > 0$, $B_n \uparrow \infty$, if there exists a sequence of real constants A_n such that $B_n^{-1}(S_n - A_n) \xrightarrow{P} 0$ as $n \to \infty$. A_n are called centering constants, and B_n norming constants.

Theorem 1. Let $\{X_n\}$ be a sequence of pairwise uncorrelated rv's with $EX_i = \mu_i$ and $\text{var}(X_i) = \sigma_i^2$, $i = 1, 2, \cdots$. If $\sum_{i=1}^{n} \sigma_i^2 \to \infty$ as $n \to \infty$, we can choose $A_n = \sum_{k=1}^{n} \mu_k$ and $B_n = \sum_{i=1}^{n} \sigma_i^2$, that is,

$$\sum_{i=1}^{n} \frac{X_i - \mu_i}{\sum_{i=1}^{n} \sigma_i^2} \xrightarrow{P} 0 \quad \text{as} \quad n \to \infty.$$

Proof. We have, by Chebychev's inequality,

$$P\left\{\left|S_n - \sum_{k=1}^n \mu_k\right| > \varepsilon \sum_{i=1}^n \sigma_i^2\right\} \le \frac{E\{\sum_{i=1}^n (X_i - \mu_i)\}^2}{\varepsilon^2 (\sum_{i=1}^n \sigma_i^2)^2}$$

$$= \frac{1}{\varepsilon^2 \sum_{i=1}^n \sigma_i^2} \to 0 \quad \text{as} \quad n \to \infty.$$

Corollary 1. If the X_n's are identically distributed and pairwise uncorrelated with $EX_i = \mu$ and var $(X_i) = \sigma^2 < \infty$, we can choose $A_n = n\mu$ and $B_n = n\sigma^2$.

Corollary 2. In Theorem 1 we can choose $B_n = n$, provided that $n^{-2}\sum_{i=1}^n \sigma_i^2 \to 0$ as $n \to \infty$.

Corollary 3. In Corollary 1, we can take $A_n = n\mu$ and $B_n = n$, since $n\sigma^2/n^2 \to 0$ as $n \to \infty$.

Thus, if $\{X_n\}$ are pairwise-uncorrelated identically distributed rv's with finite variance, $S_n/n \xrightarrow{P} \mu$.

Example 1. Let $X_1, X_2, \cdots$ be iid rv's with common law $b(1, p)$. Then $EX_i = p$, var $(X_i) = p(1 - p)$, and we have

$$\frac{S_n}{n} \xrightarrow{P} p \quad \text{as} \quad n \to \infty.$$

Note that S_n/n is the proportion of successes in n trials.

Hereafter, we shall be interested mainly in the case where $B_n = n$. When we say that $\{X_n\}$ obeys the WLLN, this is so with respect to the sequence $\{n\}$.

Theorem 2. Let $\{X_n\}$ be any sequence of rv's. Let $EX_n = \mu_n$, $n = 1, 2, \cdots$, and write $Y_n = n^{-1} \sum_{k=1}^n (X_k - \mu_k)$. A necessary and sufficient condition for the sequence $\{X_n\}$ to satisfy the weak law of large numbers is that

(1) $$E\left\{\frac{Y_n^2}{1 + Y_n^2}\right\} \to 0 \quad \text{as} \quad n \to \infty.$$

Proof. For any two positive numbers a, b, $a \ge b > 0$, we have

(2) $$\left(\frac{a}{1+a}\right)\left(\frac{1+b}{b}\right) \ge 1.$$

Let $A = \{|Y_n| \geq \varepsilon\}$. Then $\omega \in A \Rightarrow |Y_n|^2 \geq \varepsilon^2 > 0$. Using (2), we see that $\omega \in A$ implies

$$\frac{Y_n^2}{1 + Y_n^2} \frac{1 + \varepsilon^2}{\varepsilon^2} \geq 1.$$

It follows that

$$PA \leq P\left\{\frac{Y_n^2}{1 + Y_n^2} \geq \frac{\varepsilon^2}{1 + \varepsilon^2}\right\}$$

$$\leq E\frac{|Y_n^2/(1 + Y_n^2)|}{\varepsilon^2/(1 + \varepsilon^2)} \quad \text{by Markov's inequality}$$

$$\to 0 \quad \text{as} \quad n \to \infty.$$

That is, $Y_n \xrightarrow{P} 0$ as $n \to \infty$.

Conversely, we will show that for every $\varepsilon > 0$

(3) $$P\{|Y_n| \geq \varepsilon\} \geq E\left\{\frac{Y_n^2}{1 + Y_n^2}\right\} - \varepsilon^2.$$

We will prove (3) for the case in which Y_n is of the continuous type. The discrete case being similar, we ask the reader to complete the proof. If Y_n has pdf $f(y)$, then

$$\int_{-\infty}^{\infty} \frac{y^2}{1 + y^2} f(y)\,dy = \left(\int_{|y| > \varepsilon} + \int_{|y| \leq \varepsilon}\right) \frac{y^2}{1 + y^2} f(y)\,dy$$

$$\leq P\{|Y_n| > \varepsilon\} + \int_{-\varepsilon}^{\varepsilon} \left(1 - \frac{1}{1 + y^2}\right) f(y)\,dy$$

$$\leq P\{|Y_n| > \varepsilon\} + \frac{\varepsilon^2}{1 + \varepsilon^2} \leq P\{|Y_n| > \varepsilon\} + \varepsilon^2,$$

which is (3).

Remark 1. Since condition (1) applies not to the individual variables but to their sum, Theorem 2 is of limited use. We note, however, that all the weak laws of large numbers obtained as corollaries to Theorem 1 follow easily from Theorem 2 (Problem 6).

Example 2. Let $X_1, X_2, \cdots, X_n$) be jointly normal with $EX_1 = 0$, $EX_i^2 = 1$ for all i, and $(X_i, X_j) = \rho$ if $|j - i| = 1$, and $= 0$ otherwise. Then $S_n = \sum_{k=1}^{n} X_k$ is $\mathcal{N}(0, \sigma^2)$, where

$$\sigma^2 = \text{var}(S_n) = n + 2(n - 1)\rho,$$

$$E\left\{\frac{Y_n^2}{1 + Y_n^2}\right\} = E\left\{\frac{S_n^2}{n^2 + S_n^2}\right\}$$

$$= \frac{2}{\sigma\sqrt{2\pi}} \int_0^\infty \frac{x^2}{n^2 + x^2} e^{-x^2/2\sigma^2}\, dx$$

$$= \frac{2}{\sqrt{2\pi}} \int_0^\infty \frac{y^2[n + 2(n-1)\rho]}{n^2 + y^2[n + 2(n-1)\rho]} e^{-y^2/2}\, dy$$

$$\leq \frac{n + 2(n-1)\rho}{n^2} \int_0^\infty \frac{2}{\sqrt{2\pi}} y^2 e^{-y^2/2}\, dy \to 0 \quad \text{as } n \to \infty.$$

It follows from Theorem 2 that $n^{-1} S_n \xrightarrow{P} 0$. We invite the reader to compare this result to that of Problem 6.5.6.

Example 3. Let $X_1, X_2, \cdots$ be iid $\mathscr{C}(1, 0)$ rv's. We have seen (corollary to Theorem 5.3.21) that $n^{-1}S_n \sim \mathscr{C}(1, 0)$, so that $n^{-1}S_n$ does not converge in probability to 0. It follows that the WLLN does not hold. (See also Problem 10.)

Let $X_1, X_2, \cdots$ be an arbitrary sequence of rv's, and let $S_n = \sum_{k=1}^n X_k$, $n = 1, 2, \cdots$. Let us truncate each X_i at $c > 0$, that is, let

$$X_i^c = \begin{cases} X_i & \text{if } |X_i| \leq c \\ 0 & \text{if } |X_i| > c \end{cases}, \quad i = 1, 2, \cdots, n.$$

Write $\quad S_n^c = \sum_{i=1}^n X_i^c$, and $m_n = \sum_{i=1}^n EX_i^c$.

Lemma 1. For any $\varepsilon > 0$,

(4) $\quad P\{|S_n - m_n| > \varepsilon\} \leq P\{|S_n^c - m_n| > \varepsilon\} + \sum_{k=1}^n P\{|X_k| > c\}.$

Proof. We have

$$P\{|S_n - m_n| > \varepsilon\} = P\{|S_n - m_n| > \varepsilon \text{ and } |X_k| \leq c \text{ for } k = 1, 2, \cdots, n\}$$
$$+ P\{|S_n - m_n| > \varepsilon \text{ and } |X_k| > c \text{ for at least one } k,$$
$$k = 1, 2, \cdots, n\}$$
$$\leq P\{|S_n^c - m_n| > \varepsilon\} + P\{|X_k| > c \text{ for at least one } k,$$
$$1 \leq k \leq n\}$$
$$\leq P\{|S_n^c - m_n| > \varepsilon\} + \sum_{k=1}^n P\{|X_k| > c\}.$$

Corollary. If $X_1, X_2, \cdots, X_n$ are identically distributed, then

(5) $\quad P\{|S_n - m_n| > \varepsilon\} \leq P\{|S_n^c - m_n| > \varepsilon\} + nP\{|X_1| > c\}.$

If, in addition, the rv's $X_1, X_2, \cdots, X_n$ are independent, then

(6) $\quad P\{|S_n - m_n| > \varepsilon\} \leq \dfrac{nE(X_1^c)^2}{\varepsilon^2} + nP\{|X_1| > c\}.$

Inequality (6) yields the following important theorem.

Theorem 3. Let $\{X_n\}$ be a sequence of iid rv's with common finite mean $\mu = EX_1$. Then
$$n^{-1}S_n \xrightarrow{P} \mu \qquad \text{as } n \to \infty.$$

Proof. Let us take $c=n$ in (6) and replace ε by $n\varepsilon$; then we have
$$P\{|S_n - m_n| > n\varepsilon\} \leq \frac{1}{n\varepsilon^2} E(X_1^n)^2 + nP\{|X_1| > n\}.$$

First note that $E|X_1| < \infty \Rightarrow nP\{|X_1| > n\} \to 0$ as $n \to \infty$. Now (see Lemma 3.2.1)
$$E(X_1^n)^2 = 2\int_0^n xP\{|X_1| > x\}\, dx$$
$$= 2\left(\int_0^A + \int_A^n\right) xP\{|X_1| > x\}\, dx,$$

where A is chosen sufficiently large that
$$xP\{|X_1| > x\} < \frac{\delta}{2} \qquad \text{for all } x \geq A, \delta > 0 \text{ arbitrary}.$$

Thus
$$E(X_1^n)^2 \leq c + \delta \int_A^n dx \leq c + n\delta,$$

where c is a constant. It follows that
$$\frac{1}{n\varepsilon^2} E(X_1^n)^2 \leq \frac{c}{n\varepsilon^2} + \frac{\delta}{\varepsilon^2},$$

and since δ is arbitrary, $(1/n\varepsilon^2) E(X_1^n)^2$ can be made arbitrarily small for sufficiently large n. The proof is now completed by the simple observation that, since $EX_j = \mu$,
$$\frac{m_n}{n} \to \mu \qquad \text{as } n \to \infty.$$

We emphasize that in Theorem 3 we require only that $E|X_1| < \infty$; nothing is said about the variance. Theorem 3 is due to Khintchine.

Example 4. Let $X_1, X_2, \cdots$ be iid rv's with $E|X_1|^k < \infty$ for some positive integer k. Then
$$\sum_{j=1}^n \frac{X_j^k}{n} \xrightarrow{P} EX_1^k \qquad \text{as } n \to \infty.$$

Thus, if $EX_1^2 < \infty$, then $\sum_1^n X_j^2/n \xrightarrow{P} EX_1^2$; and since $\left(\sum_{j=1}^n X_j/n\right)^2 \xrightarrow{P} (EX_1)^2$ it follows that
$$\frac{\sum X_j^2}{n} - \left(\frac{\sum X_j}{n}\right)^2 \xrightarrow{P} \text{var}(X_1).$$

Example 5. Let $X_1, X_2, \cdots$ be iid rv's with common pdf

$$f(x) = \begin{cases} \dfrac{1+\delta}{x^{2+\delta}}, & x \geq 1 \\ 0, & x < 1 \end{cases}, \quad \delta > 0.$$

Then

$$E|X| = (1+\delta) \int_1^\infty \frac{1}{x^{1+\delta}} \, dx$$

$$= \frac{1+\delta}{\delta} < \infty,$$

and the law of large numbers holds, that is,

$$n^{-1} S_n \xrightarrow{P} \frac{1+\delta}{\delta} \quad \text{as } n \to \infty.$$

PROBLEMS 6.3

1. Let $X_1, X_2, \cdots$ be a seqence of iid rv's with common uniform distribution on $[0, 1]$. Also, let $Z_n = (\prod_{i=1}^n X_i)^{1/n}$ be the geometric mean of $X_1, X_2, \cdots, X_n$, $n = 1, 2, \cdots$. Show that $Z_n \xrightarrow{P} c$, where c is some constant. Find c.

2. Let $X_1, X_2, \cdots$ be iid rv's with finite second moment. Let

$$Y_n = \frac{2}{n(n+1)} \sum_{i=1}^n i X_i.$$

Show that $Y_n \xrightarrow{P} EX_1$.

3. Let $X_1, X_2, \cdots$ be a sequence of iid rv's with $EX_i = \mu$ and var $(X_i) = \sigma^2$. Let $S_k = \sum_{j=1}^k X_j$. Does the sequence $\{S_k\}$ obey the WLLN? Does the sequence S_k obey the WLLN in the sense of Definition 1? If so, find the centering and the norming constants.

4. Let $\{X_n\}$ be a sequence of rv's for which var $(X_n) \leq C$ for all n and $\rho_{ij} = \text{cov}(X_i, X_j) \to 0$ as $|i - j| \to \infty$. Show that the WLLN holds.

5. For the following sequences of independent rv's does the WLLN hold?
  (a) $P\{X_k = \pm 2^k\} = \frac{1}{2}$.
  (b) $P\{X_k = \pm k\} = 1/2\sqrt{k}$, $P\{X_k = 0\} = 1 - (1/\sqrt{k})$.
  (c) $P\{X_k = \pm 2^k\} = 1/2^{2k+1}$, $P\{X_k = 0\} = 1 - (1/2^{2k})$.
  (d) $P\{X_k = \pm 2^k\} = 1/2^{k+1}$, $P\{X_k = \pm 1\} = \frac{1}{2}[1 - (1/2^k)]$.
  (e) $P\{X_k = \pm \sqrt{k}\} = \frac{1}{2}$.

6. Let $X_1, X_2, \cdots$ be a sequence of independent rv's such that var $(X_k) < \infty$ for $k = 1, 2, \cdots$, and $(1/n^2) \sum_{k=1}^n$ var $(X_k) \to 0$ as $n \to \infty$. Prove the WLLN, using Theorem 2.

7. Let X_n be a sequence of rv's with common finite variance σ^2. Suppose that the

correlation coefficient between X_i and X_j is < 0 for all $i \neq j$. Show that the WLLN holds for the sequence $\{X_n\}$.

8. Let $\{X_k\}$ be a sequence of rv's such that X_k is independent of X_j for $j \neq k + 1$ or $j \neq k - 1$. If var $(X_k) < C$ for all k, where C is some constant, the WLLN holds for $\{X_k\}$.

9. For any sequence of rv's $\{X_n\}$ show that
$$\max_{1 \leq k \leq n} |X_k| \xrightarrow{P} 0 \Rightarrow n^{-1} S_n \xrightarrow{P} 0.$$

10. Let $X_1, X_2, \ldots$ be iid $\mathscr{C}(1, 0)$ rv's. Use Theorem 2 to show that the weak law of large numbers does not hold. That is, show that
$$E \frac{S_n^2}{n^2 + S_n^2} \not\to 0 \text{ as } n \to \infty, \text{ where } S_n = \sum_{k=1}^{n} X_k, n = 1, 2, \cdots.$$

6.4 THE STRONG LAW OF LARGE NUMBERS†

In this section we obtain a stronger form of the law of large numbers discussed in Section 6.3. Let $X_1, X_2, \cdots$ be a sequence of rv's defined on some probability space $(\Omega, \mathscr{S}, P)$.

Definition 1. We say that the sequence $\{X_n\}$ obeys the strong law of large numbers (SLLN) with respect to the norming constants $\{B_n\}$ if there exists a sequence of (centering) constants $\{A_n\}$ such that

(1) $$B_n^{-1} (S_n - A_n) \xrightarrow{a.s.} 0 \quad \text{as} \quad n \to \infty.$$

Here $B_n > 0$ and $B_n \to \infty$ as $n \to \infty$.

We will obtain sufficient conditions for a sequence $\{X_n\}$ to obey the SLLN. In what follows, we will be interested chiefly in the case $B_n = n$. Indeed, when we speak of the SLLN we will assume that we are speaking of the norming constants $B_n = n$, unless specified otherwise.

We start with the *Borel-Cantelli lemma*. Let $\{A_j\}$ be any sequence of events in $\mathscr{S}$. We recall that

(2) $$\overline{\lim_{n \to \infty}} A_n = \lim_{n \to \infty} \bigcup_{k=n}^{\infty} A_k = \bigcap_{n=1}^{\infty} \bigcup_{k=n}^{\infty} A_k.$$

We will write $A = \overline{\lim}_{n \to \infty} A_n$. Note that A is the event that *infinitely many* of the A_n occur. We will sometimes write

$$PA = P(\overline{\lim_{n \to \infty}} A_n) = P(A_n \text{ i.o.}),$$

---
†This section may be omitted on the first reading.

where "i.o." stands for "infinitely often." In view of Theorem 6.2.11 and Remark 6.2.6 we have $X_n \xrightarrow{a.s.} 0$ if and only if $P\{|X_n| > \varepsilon \text{ i.o.}\} = 0$ for all $\varepsilon > 0$.

Theorem 1 (Borel-Cantelli Lemma).

(a) Let $\{A_n\}$ be a sequence of events such that $\sum_{n=1}^{\infty} PA_n < \infty$. Then $PA = 0$.

(b) If $\{A_n\}$ is an independent sequence of events such that $\sum_{n=1}^{\infty} PA_n = \infty$, then $PA = 1$.

Proof.

(a) $PA = P\left(\lim_{n \to \infty} \bigcup_{k=n}^{\infty} A_k\right) = \lim_{n \to \infty} P\left(\bigcup_{k=n}^{\infty} A_k\right) \leq \lim_{n \to \infty} \sum_{k=n}^{\infty} PA_k = 0$.

(b) We have $A^c = \bigcup_{n=1}^{\infty} \bigcap_{k=n}^{\infty} A_k^c$, so that

$$PA^c = P\left(\lim_{n \to \infty} \bigcap_{k=n}^{\infty} A_k^c\right) = \lim_{n \to \infty} P\left(\bigcap_{k=n}^{\infty} A_k^c\right).$$

For $n_0 > n$, we see that $\bigcap_{k=n}^{\infty} A_k^c \subset \bigcap_{k=n}^{n_0} A_k^c$, so that

$$P\left(\bigcap_{k=n}^{\infty} A_k^c\right) \leq \lim_{n_0 \to \infty} P\left(\bigcap_{k=n}^{n_0} A_k^c\right) = \lim_{n_0 \to \infty} \prod_{k=n}^{n_0} (1 - PA_k),$$

because $\{A_n\}$ is an independent sequence of events. Now we use the elementary inequality

$$1 - \exp\left(-\sum_{j=n}^{n_0} \alpha_j\right) \leq 1 - \prod_{j=n}^{n_0} (1 - \alpha_j) \leq \sum_{j=n}^{n_0} \alpha_j, \quad n_0 > n, \quad 1 \geq \alpha_j \geq 0,$$

to conclude that

$$P\left(\bigcap_{k=n}^{\infty} A_k^c\right) \leq \lim_{n_0 \to \infty} \exp\left(-\sum_{k=n}^{n_0} PA_k\right).$$

Since the series $\sum_{n=1}^{\infty} PA_n$ diverges, it follows that $PA^c = 0$ or $PA = 1$.

Corollary. Let $\{A_n\}$ be a sequence of independent events. Then PA is either 0 or 1.

The corollary follows since $\sum_{n=1}^{\infty} PA_n$ either converges or diverges.

As a simple application of the Borel-Cantelli lemma, we obtain a version of the SLLN.

Theorem 2. If $X_1, X_2, \cdots$ are iid rv's with common mean μ and finite fourth moment, then

$$P\left\{\lim_{n\to\infty}\frac{S_n}{n}=\mu\right\}=1.$$

Proof. We have

$$E\{\Sigma(X_i - \mu)\}^4 = nE(X_1 - \mu)^4 + 6\binom{n}{2}\sigma^4 \le Cn^2.$$

By Markov's inequality

$$P\{|\Sigma_1^n(X_i - \mu)| > n\varepsilon\} \le \frac{E\{\Sigma_1^n(X_i - \mu)\}^4}{(n\varepsilon)^4} \le \frac{Cn^2}{(n\varepsilon)^4} = \frac{C'}{n^2}.$$

Therefore

$$\sum_{n=1}^{\infty} P\{|S_n - \mu n| > n\varepsilon\} < \infty,$$

and it follows by the Borel-Cantelli lemma that with probability 1 only finitely many of the events $\{\omega: |(S_n/n) - \mu| > \varepsilon\}$ occur, that is, $PA_\varepsilon = 0$, where

$$A_\varepsilon = \limsup_{n\to\infty} \{|\frac{S_n}{n} - \mu| > \varepsilon\}.$$

The sets A_ε increase, as $\varepsilon \to 0$, to the ω set on which $S_n/n \not\to \mu$. Letting $\varepsilon \to 0$ through a countable set of values, we have

$$P\left\{\frac{S_n}{n} - \mu \not\to 0\right\} = P\{\bigcup_k A_{1/k}\} = 0.$$

Corollary. If $X_1, X_2, \cdots$ are iid rv's such that $P\{|X_n| < K\} = 1$ for all n, where K is a positive constant, then $n^{-1} S_n \xrightarrow{a.s.} \mu$.

Theorem 3. Let $X_1, X_2, \cdots$ be a sequence of independent rv's. Then

$$X_n \xrightarrow{a.s.} 0 \Leftrightarrow \sum_{n=1}^{\infty} P\{|X_n| > \varepsilon\} < \infty \quad \text{for all } \varepsilon > 0.$$

Proof. Writing $A_n = \{|X_n| > \varepsilon\}$, we see that $\{A_n\}$ is a sequence of independent events. Since $X_n \xrightarrow{a.s.} 0$, $X_n \to 0$ on a set E^c with $PE = 0$. A point $\omega \in E^c$ belongs only to a finite number of A_n. It follows that

$$\limsup_{n\to\infty} A_n \subset E,$$

hence $P(A_n \text{ i.o.}) = 0$. By the Borel-Cantelli lemma [Theorem 1 (b)] we must have $\sum_{n=1}^{\infty} PA_n < \infty$. [Otherwise, $\sum_{n=1}^{\infty} PA_n = \infty$, and then $P(A_n \text{ i.o.}) = 1$.]
In the other direction, let

$$A_{1/k} = \limsup_{n\to\infty} \left\{ |X_n| > \frac{1}{k} \right\},$$

and use the argument in the proof of Theorem 2.

We recall here (see remarks following Definition 4.3.5) that two rv's X and Y are said to be *equivalent* if $P\{X \neq Y\} = 0$; that is, for any $\varepsilon > 0$ there exists a $\delta > 0$ such that $P\{|X - Y| > \varepsilon\} < \delta$.

Definition 2. Two sequences of rv's X_n and X'_n are said to be tail-equivalent if they differ almost surely only by a finite number of terms; that is, if for (almost) all $\omega \in \Omega$ there exists an $n(\omega)$ such that, for $n \geq n(\omega)$, the sequences $X_n(\omega)$ and $X'_n(\omega)$ are the same. In symbols, we write $P\{X_n \neq X'_n \text{ i.o.}\} = 0$.

Definition 3. If two sequences of rv's, X_n and X'_n, converge on the same event except for a subset with zero probability, we say that they are convergence-equivalent.

Let

$$S_n = \sum_{k=1}^{n} X_k \quad \text{and} \quad S'_n = \sum_{k=1}^{n} X'_k.$$

Lemma 1 (*Equivalence Lemma*). If the series $\sum_{n=1}^{\infty} P\{X_n \neq X'_n\} < \infty$, the sequences X_n and X'_n are tail-equivalent and S_n and S'_n are convergence equivalent. The sequences $B_n^{-1} S_n$ and $B_n^{-1} S'_n$, $B_n \uparrow \infty$, converge on the same event and to the same limit, except for a null event.

Proof. By the Borel-Cantelli lemma, $P\{\overline{\lim}_{n\to\infty} [X_n \neq X'_n]\} = 0$, that is, $P\{X_n \neq X'_n \text{ i. o.}\} = 0$, so that $P\{\omega : X_n(\omega) \neq X'_n(\omega) \text{ for only finitely many } n\} = 1$. It follows that, if $S_n(\omega)/B_n \to 0$ as $n \to \infty$ for ω such that $X_n(\omega) \neq X'_n(\omega)$ only finitely many times, then $S'_n(\omega)/B_n \to 0$ as $n \to \infty$, and conversely.

Remark 1. Let $\{X_n\}$ be any sequence of rv's. We can always truncate X_n suitably to get a sequence of bounded rv's that differ from X_n only on an event of arbitrarily small probability. Let X'_n be X_n truncated at c_n, where c_n is chosen so that $P\{|X_n| > c_n\} < \varepsilon/2^n$. Then

$$P\left\{ \bigcup_{n=1}^{\infty} [X_n \neq X'_n] \right\} \leq \sum_{n=1}^{\infty} P\{X_n \neq X'_n\} = \sum_{n=1}^{\infty} P\{|X_n| > c_n\} < \varepsilon.$$

By Lemma 1 we see that X_n and X'_n are tail-equivalent.

As a typical application of the equivalence lemma we prove the following result.

Theorem 4 (Jamison, Orey, and Pruitt [52]). Let $\{X_k\}$ be a sequence of iid rv's, and $\{w_k\}$ be a sequence of positive numbers. Write $S_n = \sum_{k=1}^{n} w_k X_k$ and $W_n = \sum_{k=1}^{n} w_k$. If the sequence $\{W_k\}$ diverges such that $w_n/W_n \to 0$ as $n \to \infty$, and if

(3) $\qquad TP\{|X_1| > T\} \to 0 \quad$ as $\quad T \to \infty$,

and if

(4) $\quad \lim_{T \to \infty} EX_1^T$ exists and equals μ, where X_k^T is the rv X_k truncated at T, then $S_n/W_n \xrightarrow{P} \mu$.

Proof. Let

$$X_{nk} = \begin{cases} X_k, & |X_k| \le \dfrac{W_n}{w_k} \\ 0, & |X_k| > \dfrac{W_n}{w_k} \end{cases}$$

and write $S_{nn} = \sum_{k=1}^{n} w_k X_{nk}$. Note that $\max_{1 \le k \le n} (w_k/W_n) \to 0$ as $n \to \infty$ if and only if $\sum_{k=1}^{\infty} w_k = \infty$ and $w_n/W_n \to 0$. It follows that $W_n/w_k \to \infty$ for each $k \le n$, and

$$P\{S_{nn} \ne S_n\} \le \sum_{k=1}^{n} P\{X_{nk} \ne X_k\}$$
$$= \sum_{k=1}^{n} P\left\{|X_1| > \frac{W_n}{w_k}\right\}.$$

By (3) we have

$$\frac{W_n}{w_k} P\left\{|X_1| > \frac{W_n}{w_k}\right\} < \varepsilon$$

for sufficiently large n. It follows that for sufficiently large n

$$P\{S_{nn} \ne S_n\} \le \varepsilon \sum_{k=1}^{n} \frac{w_k}{W_n} = \varepsilon.$$

It therefore suffices to prove the result for S_{nn} instead of S_n. We have

$$E \frac{S_{nn}}{W_n} = \frac{1}{W_n} \sum_{k=1}^{n} w_k EX_{nk} \to \mu$$

by condition (4). Also,

(5) $\qquad \text{var}\left(\dfrac{S_{nn}}{W_n}\right) = \dfrac{1}{W_n^2} \sum_{k=1}^{n} w_k^2 \text{ var }(X_{nk}) \le \dfrac{1}{W_n^2} \sum_{k=1}^{n} w_k^2 EX_{nk}^2$.

But

$$\frac{1}{T}\{E(X_1^T)^2\} = \frac{1}{T}\{-T^2 P\{|X_1| > T\} + 2\int_0^T x\, P\{|X_1| > x\}dx\}$$
$$\to 0 \quad \text{as} \quad T \to \infty.$$

Thus
$$\frac{w_k}{W_n} EX_{nk}^2 \to 0 \quad \text{as} \quad n \to \infty,$$

and for sufficiently large n we get from (5)

$$\text{var}\left(\frac{S_{nn}}{W_n}\right) \le \frac{1}{W_n^2} \sum_{k=1}^n w_k^2 \cdot \varepsilon \frac{W_n}{w_k} = \varepsilon.$$

Therefore it follows that

$$P\left\{\frac{|S_{nn} - ES_{nn}|}{W_n} > \delta\right\} \le \frac{\text{var}(S_{nn}/W_n)}{\delta^2} < \frac{\varepsilon}{\delta^2}$$

for $\delta > 0$. Letting $\varepsilon \to 0$, we see that

$$P\left\{\frac{|S_{nn} - ES_{nn}|}{W_n} > \delta\right\} \to 0 \quad \text{as} \quad n \to \infty \quad \text{for every } \delta > 0,$$

and the proof is complete.

Remark 2. Note that condition (4) is weaker than $E|X| < \infty$. The converse result also holds; that is, if $S_n/W_n \xrightarrow{P} \mu$, then (3) and (4) hold for all divergent sequences $\{w_k\}$ such that $w_n/W_n \to 0$. For details we refer to Jamison, Orey, and Pruitt [52], page 41.

We next prove some important lemmas that we will need subsequently.

Lemma 2 (Kolmogorov's Inequality). Let $X_1, X_2, \cdots, X_n$ be independent rv's with common mean 0 and variances σ_k^2, $k = 1, 2, \cdots, n$, respectively. Then for any $\varepsilon > 0$

(6) $$P\{\max_{1 \le k \le n} |S_k| > \varepsilon\} \le \sum_1^n \frac{\sigma_i^2}{\varepsilon^2}.$$

Proof. Let $A_0 = \Omega$,
$$A_k = \{\max_{1 \le j \le k} |S_j| \le \varepsilon\}, \quad k = 1, 2, \cdots, n$$

and
$$B_k = A_{k-1} \cap A_k^c$$
$$= \{|S_1| \le \varepsilon, \cdots, |S_{k-1}| \le \varepsilon\} \cap \{\text{at least one of } |S_1|, \cdots, |S_k| \text{ is } > \varepsilon\}$$
$$= \{|S_1| \le \varepsilon, \cdots, |S_{k-1}| \le \varepsilon, |S_k| > \varepsilon\}.$$

It follows that
$$A_n^c = \sum_{k=1}^n B_k$$
and
$$B_k \subset \{|S_{k-1}| \le \varepsilon, |S_k| > \varepsilon\}.$$

As usual, let us write I_{B_k} for the indicator function of the event B_k. Then
$$E(S_n I_{B_k})^2 = E\{(S_n - S_k)I_{B_k} + S_k I_{B_k}\}^2,$$
$$= E\{(S_n - S_k)^2 I_{B_k} + S_k^2 I_{B_k} + 2S_k(S_n - S_k)I_{B_k}\}.$$

Since $S_n - S_k = X_{k+1} + \cdots + X_n$ and $S_k I_{B_k}$ are independent, and $EX_k = 0$ for all k, it follows that
$$E(S_n I_{B_k})^2 = E\{(S_n - S_k)I_{B_k}\}^2 + E(S_k I_{B_k})^2$$
$$\ge E(S_k I_{B_k})^2 \ge \varepsilon^2 PB_k.$$

The last inequality follows from the fact that, in B_k, $|S_k| > \varepsilon$. Moreover,
$$\sum_{k=1}^n E(S_n I_{B_k})^2 = E(S_n^2 I_{A_n^c}) \le E(S_n^2) = \sum_1^n \sigma_k^2$$

so that
$$\sum_1^n \sigma_k^2 \ge \varepsilon^2 \sum_1^n PB_k = \varepsilon^2 P(A_n^c),$$

as asserted.

Corollary. Take $n = 1$; then
$$P\{|X_1| > \varepsilon\} \le \frac{\sigma_1^2}{\varepsilon^2},$$

which is Chebychev's inequality.

Lemma 3 (Kronecker Lemma). If $\sum_{n=1}^\infty x_n$ converges to s (finite) and $b_n \uparrow \infty$, then
$$b_n^{-1} \sum_{k=1}^n b_k x_k \to 0.$$

Proof. Writing $b_0 = 0$, $a_k = b_k - b_{k-1}$, and $s_{n+1} = \sum_{k=1}^n x_k$, we have
$$\frac{1}{b_n} \sum_{k=1}^n b_k x_k = \frac{1}{b_n} \sum_{k=1}^n b_k(s_{k+1} - s_k)$$
$$= \frac{1}{b_n}\left(b_n s_{n+1} + \sum_1^n b_{k-1} s_k\right) - \frac{1}{b_n} \sum_{k=1}^n b_k s_k$$

$$= s_{n+1} - \frac{1}{b_n} \sum_{k=1}^{n} (b_k - b_{k-1})s_k$$

$$= s_{n+1} - \frac{1}{b_n} \sum_{k=1}^{n} a_k s_k.$$

It therefore suffices to show that $b_n^{-1} \sum_{k=1}^{n} a_k s_k \to s$. Since $s_n \to s$, there exists an $n_0 = n_0(\varepsilon)$ such that

$$|s_n - s| < \frac{\varepsilon}{2} \qquad \text{for } n > n_0.$$

Since $b_n \uparrow \infty$, let n_1 be an integer $> n_0$ such that

$$b_n^{-1} \Big| \sum_{1}^{n_0} (b_k - b_{k-1})(s_k - s) \Big| < \frac{\varepsilon}{2} \qquad \text{for } n > n_1.$$

Writing

$$r_n = b_n^{-1} \sum_{k=1}^{n} (b_k - b_{k-1})s_k,$$

we see that

$$|r_n - s| = \frac{1}{b_n} \Big| \sum_{k=1}^{n} (b_k - b_{k-1})(s_k - s) \Big|;$$

and, choosing $n > n_1$, we have

$$|r_n - s| \leq \Big| \frac{1}{b_n} \sum_{k=1}^{n_0} (b_k - b_{k-1})(s_k - s) \Big| + \frac{1}{b_n} \Big| \sum_{k=n_0+1}^{n} (b_k - b_{k-1}) \frac{\varepsilon}{2} \Big| < \varepsilon.$$

This completes the proof.

We need the following criterion for almost sure convergence.

Theorem 5 (Cauchy Criterion). $X_n \xrightarrow{\text{a.s.}} X$ if and only if

$$\lim_{n \to \infty} P\{\sup_m |X_{n+m} - X_n| \leq \varepsilon\} = 1 \qquad \text{for every } \varepsilon > 0.$$

Proof. Let $X_n \xrightarrow{\text{a.s.}} X$. We have for $\delta > 0$

$$\{|X_{n+m} - X| \leq \tfrac{\delta}{2}\} \cap \{|X_n - X| \leq \tfrac{\delta}{2}\} \subset \{|X_{n+m} - X_n| \leq \delta\},$$

so that

$$\bigcap_{j=n}^{\infty} \{|X_j - X| \leq \tfrac{\delta}{2}\} \subset \{|X_{n+m} - X| \leq \tfrac{\delta}{2}\} \cap \{|X_n - X| \leq \tfrac{\delta}{2}\}$$

$$\subset \{|X_{n+m} - X_n| \leq \delta\}$$

for all m. It follows that

$$\bigcap_{j=n}^{\infty} \{|X_j - X| \leq \frac{\delta}{2}\} \subset \bigcap_{m=1}^{\infty} \{|X_{n+m} - X_n| \leq \delta\}.$$

Then

$$\bigcap_{m=1}^{\infty} \{|X_{n+m} - X_n| \leq \delta\} = \{|X_{n+m} - X_n| \leq \delta \text{ for all } m\}$$
$$= \{\sup_m |X_{n+m} - X_n| \leq \delta\},$$

so that

$$P\{\sup_m |X_{n+m} - X_n| \leq \delta\} \geq P\{\bigcap_{j=n}^{\infty} [|X_j - X| \leq \frac{\delta}{2}]\}.$$

Letting $n \to \infty$, we see that

$$\lim_{n \to \infty} P\{\sup_m |X_{n+m} - X_n| \leq \delta\} \geq \lim_{n \to \infty} P\{\bigcap_{j=n}^{\infty} [|X_j - X| \leq \frac{\delta}{2}]\} = 1,$$

as asserted (Remark 6.2.6).

Conversely, let

$$\lim_{n \to \infty} P\{\sup_m |X_{n+m} - X_n| \leq \varepsilon\} = 1 \quad \text{for every} \quad \varepsilon > 0.$$

For any integer $k \geq n$, we have

$$\bigcap_{m=1}^{\infty} \{|X_{n+m} - X_n| \leq \frac{\delta}{2}\} \subset \{|X_{k+m} - X_n| \leq \frac{\delta}{2}\} \cap \{|X_k - X_n| \leq \frac{\delta}{2}\}$$
$$\subset \{|X_{k+m} - X_k| \leq \delta\},$$

so that

$$\bigcap_{m=1}^{\infty} \{|X_{n+m} - X_n| \leq \frac{\delta}{2}\} \subset \{|X_{k+m} - X_k| \leq \delta\}$$

holds for all $k \geq n$ and all m. It follows that

$$\bigcap_{m=1}^{\infty} \{|X_{n+m} - X_n| \leq \frac{\delta}{2}\} \subset \bigcap_{m=1}^{\infty} \{|X_{k+m} - X_k| \leq \delta\}$$

for all $k \geq n$, so that

$$\bigcap_{m=1}^{\infty} \{|X_{n+m} - X_n| \leq \frac{\delta}{2}\} \subset \bigcap_{k=n}^{\infty} \bigcap_{m=1}^{\infty} \{|X_{k+m} - X_k| \leq \delta\}$$

and

$$\lim_{n \to \infty} P\{\bigcap_{m=1}^{\infty} [|X_{n+m} - X_n| \leq \frac{\delta}{2}]\}$$

$$\leq \lim_{n\to\infty} P\{\bigcap_{k=n}^{\infty} \bigcap_{m=1}^{\infty} [|X_{k+m} - X_k| \leq \delta]\}$$

$$= P\{\bigcup_{n=1}^{\infty} \bigcap_{k=n}^{\infty} \bigcap_{m=1}^{\infty} [|X_{k+m} - X_k| \leq \delta]\}$$

(since $\bigcap_{k=n}^{\infty} \bigcap_{m=1}^{\infty} [|X_{k+m} - X_k| \leq \delta]$ is a nondecreasing sequence of sets). Since

$$\lim_{n\to\infty} P\{\sup_m |X_{n+m} - X_n| \leq \varepsilon\} = 1 \quad \text{for every } \varepsilon > 0,$$

we see that

$$P\{\bigcup_{n=1}^{\infty} \bigcap_{k=n}^{\infty} \bigcap_{m=1}^{\infty} [|X_{k+m} - X_k| \leq \delta]\} = 1$$

for any fixed $\delta > 0$. Let $\delta_p > 0$, $\delta_p \downarrow 0$ as $p \to \infty$, and consider the set

$$A = \bigcap_{p=1}^{\infty} \bigcup_{n=1}^{\infty} \bigcap_{k=n}^{\infty} \bigcap_{m=1}^{\infty} [|X_{k+m} - X_k| \leq \delta_p].$$

Now for any sequence A_j, by Boole's inequality

$$P\left\{\bigcap_{j=1}^{\infty} A_j\right\} \geq 1 - \sum_{j=1}^{\infty} PA_j^c.$$

Thus

$$PA \geq 1 - \sum_{p=1}^{\infty} P\left[\left\{\bigcup_{n=1}^{\infty} \bigcap_{k=n}^{\infty} \bigcap_{m=1}^{\infty} [|X_{k+m} - X_k| \leq \delta_p]\right\}^c\right],$$

and it follows that $PA = 1$.

The sequence $\{X_n\}$ satisfies the Cauchy criterion for every point $\omega \in A$. Therefore there exists a function X^* such that $\lim_{n\to\infty} X_n(\omega) = X^*(\omega)$ for all $\omega \in A$. We define

$$X(\omega) = \begin{cases} X^*(\omega), & \omega \in A, \\ 0, & \omega \in A^c. \end{cases}$$

Since $PA^c = 0$, we see that $X_n \xrightarrow{\text{a.s.}} X$. This completes the proof of the theorem.

Theorem 6. If $\sum_{n=1}^{\infty} \text{var } X_n < \infty$, then $\sum_{n=1}^{\infty} (X_n - EX_n)$ converges almost surely.

Proof. By Kolmogorov's inequality

$$P\{\max_{1\leq k\leq n} |S_{m+k} - S_m - E(S_{m+k} - S_m)| \geq \varepsilon\} \leq \frac{1}{\varepsilon^2} \sum_{k=1}^{n} \text{var}(X_{m+k}).$$

Letting $m, n \to \infty$, we see that
$$\lim_{m \to \infty} P\{\sup_k |S_{m+k} - ES_{m+k} - (S_m - ES_m)| \geq \varepsilon\} = 0,$$
and by the Cauchy criterion the sequence $\{S_n - ES_n\}$ converges with probability 1, as asserted.

Corollary 1. Let $\{X_n\}$ be independent rv's. If
$$\sum_{k=1}^{\infty} \frac{\text{var}(X_k)}{B_k^2} < \infty, \qquad B_n \uparrow \infty,$$
then
$$\frac{S_n - ES_n}{B_n} \xrightarrow{\text{a.s.}} 0.$$

The corollary follows from Theorem 6 and the Kronecker lemma.

Corollary 2. Every sequence $\{X_n\}$ of independent rv's with uniformly bounded variances obeys the SLLN.

If $\text{var}(X_k) \leq A$ for all k, and $B_k = k$, then
$$\sum_{k=1}^{\infty} \frac{\sigma_k^2}{k^2} \leq A \sum_{1}^{\infty} \frac{1}{k^2} < \infty,$$
and it follows that
$$\frac{S_n - ES_n}{n} \xrightarrow{\text{a.s.}} 0.$$

Corollary 3 (Borel's Strong Law of Large Numbers). For a sequence of Bernoulli trials with (constant) probability p of success, the SLLN holds (with $B_n = n$, and $A_n = np$).

Since $\qquad EX_k = p, \qquad \text{var}(X_k) = p(1-p) \leq \frac{1}{4}, \quad 0 < p < 1,$
the result follows from Corollary 2.

Lemma 4.
(7) $$\sum_{n=1}^{\infty} P\{|X| \geq n\} \leq E|X| \leq 1 + \sum_{n=1}^{\infty} P\{|X| \geq n\}.$$

Proof. Let X be of the continuous type with pdf f. Then
$$E|X| = \int_{-\infty}^{\infty} |x| f(x) \, dx = \sum_{k=0}^{\infty} \int_{k \leq |x| < k+1} |x| f(x) \, dx,$$

and it follows that

$$\sum_{k=0}^{\infty} k\, P\{k \le |X| < k+1\} \le E|X| \le \sum_{k=0}^{\infty} (k+1)\, P\{k \le |X| < k+1\}.$$

Now

$$\sum_{k=0}^{\infty} k\, P\{k \le |X| < k+1\} = \sum_{n=1}^{\infty} \sum_{k=n}^{\infty} P\{k \le |X| < k+1\}$$

$$= \sum_{n=1}^{\infty} P\{|X| \ge n\},$$

and

$$\sum_{k=0}^{\infty} (k+1)\, P\{k \le |X| < k+1\} = \sum_{k=0}^{\infty} k\, P\{k \le |X| < k+1\} + 1$$

$$= \sum_{n=1}^{\infty} P\{|X| \ge n\} + 1.$$

The proof for the case in which X is of the discrete type is similar. See also Lemma 3.2.2.

Theorem 7 (Kolmogorov's Strong Law of Large Numbers). Let $X_1, X_2, \cdots$ be iid rv's with common law $\mathscr{L}(X)$. Then

$$n^{-1} S_n \stackrel{\text{a.s.}}{\to} \mu \text{ finite}$$

if and only if $E|X| < \infty$, and then $\mu = EX$.

Proof. If $n^{-1} S_n \stackrel{\text{a.s.}}{\to} \mu$ finite, then

$$\frac{X_n}{n} = \frac{S_n}{n} - \frac{n-1}{n} \frac{S_{n-1}}{n-1} \stackrel{\text{a.s.}}{\to} 0.$$

By the Borel-Cantelli lemma, we must have $\sum_{n=1}^{\infty} P\{|X_n| \ge n\} < \infty$, which by Lemma 4 implies $E|X| < \infty$. It follows, for example, by the WLLN, that $\mu = EX$.

Conversely, let $E|X| < \infty$. We define a sequence of truncated rv's X'_k as follows:

$$X'_k = \begin{cases} X_k & \text{if } |X_k| \le k, \\ 0 & \text{otherwise.} \end{cases}$$

Writing $S'_n = \sum_{k=1}^{n} X'_k$, we have

$$\sum_{k=1}^{\infty} P\{X_k \ne X'_k\} = \sum_{k=1}^{\infty} P\{|X_k| \ge k\} = \sum_{k=1}^{\infty} P\{|X| \ge k\},$$

since the X_i's are identically distributed. Also, since $E|X| < \infty$,

$\sum_{k=1}^{\infty} P\{|X| \geq k\} < \infty$, and it follows by Lemma 1 that $n^{-1}S_n$ and $n^{-1}S'_n$ converge to the same limit and on the same set, except for a null set. Therefore it is sufficient to show that $n^{-1}S'_n \xrightarrow{\text{a.s.}} EX$. We have

$$\text{var}(X'_n) \leq E(X'_n)^2 = \int_{-n}^{n} x^2 f(x)\, dx$$

$$= \sum_{k=0}^{n-1} \int_{k \leq |x| < k+1} x^2 f(x)\, dx$$

$$\leq \sum_{k=0}^{n-1} (k+1)^2 P\{k \leq |X| < k+1\},$$

where f is the pdf of X. A similar argument holds for the discrete case. Thus in either case

$$\sum_{n=1}^{\infty} \frac{1}{n^2} \text{var}(X'_n) \leq \sum_{n=1}^{\infty} \sum_{k=0}^{n} \frac{(k+1)^2}{n^2} P\{k \leq |X| < k+1\}$$

$$= \sum_{n=1}^{\infty} \sum_{k=1}^{n} \frac{(k+1)^2}{n^2} P\{k \leq |X| < k+1\}$$

$$+ P\{0 \leq |X| < 1\} \sum_{n=1}^{\infty} \frac{1}{n^2}$$

$$= \sum_{k=1}^{\infty} (k+1)^2 P\{k \leq |X| < k+1\} \sum_{n=k}^{\infty} \frac{1}{n^2}$$

$$+ P\{0 \leq |X| < 1\} \sum_{1}^{\infty} \frac{1}{n^2}.$$

Then

$$\sum_{n=k}^{\infty} \frac{1}{n^2} \leq \frac{1}{k^2} + \sum_{n=k+1}^{\infty} \frac{1}{n(n-1)} = \frac{1}{k^2} + \frac{1}{k} \leq \frac{2}{k},$$

and it follows that

$$\sum_{n=1}^{\infty} \frac{1}{n^2} \text{var}(X'_n) \leq 2 \sum_{k=1}^{\infty} \frac{(k+1)^2}{k} P\{k \leq |X| < k+1\}$$

$$+ 2P\{0 \leq |X| < 1\} < \infty$$

by Lemma 4. It follows from Corollary 1 of Theorem 6 that

$$n^{-1}S'_n - n^{-1}ES'_n \xrightarrow{\text{a.s.}} 0 \quad \text{as} \quad n \to \infty.$$

Since $EX'_n \to EX$ as $n \to \infty$, it follows by Lemma 3 that $n^{-1}ES'_n \to EX$. We can therefore replace $n^{-1}ES'_n$ by EX and conclude that

$$n^{-1}S'_n \xrightarrow{\text{a.s.}} EX,$$

hence that $n^{-1}S_n \xrightarrow{\text{a.s.}} EX$.

PROBLEMS 6.4

1. For the following sequences of independent rv's does the SLLN hold?
   (a) $P\{X_k = \pm 2^k\} = \frac{1}{2}$.
   (b) $P\{X_k = \pm k\} = 1/2\sqrt{k}$, $P\{X_k = 0\} = 1 - (1/\sqrt{k})$.
   (c) $P\{X_k = \pm 2^k\} = 1/2^{2k+1}$, $P\{X_k = 0\} = 1 - (1/2^{2k})$.

2. Let $X_1, X_2, \cdots$ be a sequence of independent rv's with $\sum_{k=1}^{\infty} \text{var}(X_k)/k^2 < \infty$. Show that
$$\frac{1}{n^2} \sum_{k=1}^{n} \text{var}(X_k) \to 0 \quad \text{as } n \to \infty.$$
Does the converse also hold?

3. For what values of α does the SLLN hold for the sequence
$$P\{X_k = \pm k^\alpha\} = \frac{1}{2}?$$

4. Let $\{\sigma_k^2\}$ be a sequence of real numbers such that $\sum_{k=1}^{\infty} \sigma_k^2/k^2 = \infty$. Show that there exists a sequence of independent rv's $\{X_k\}$ with var $(X_k) = \sigma_k^2$, $k = 1, 2, \cdots$, such that $n^{-1} \sum_{k=1}^{n} (X_k - EX_k)$ does not converge to 0 almost surely.
[*Hint*: Let $P\{X_k = \pm k\} = \sigma_k^2/2k^2$, $P\{X_k = 0\} = 1 - (\sigma_k^2/k^2)$ if $\sigma_k/k \le 1$, and $P\{X_k = \pm \sigma_k\} = \frac{1}{2}$ if $\sigma_k/k > 1$. Apply the Borel-Cantelli lemma to $\{|X_n| > n\}$.]

5. Let X_n be a sequence of iid rv's with $E|X_n| = +\infty$. Show that, for every positive number A, $P\{|X_n| > nA \text{ i.o.}\} = 1$ and $P\{|S_n| > nA \text{ i.o.}\} = 1$.

6. Construct an example to show that the converse of Theorem 1(a) does not hold.

6.5 LIMITING MOMENT GENERATING FUNCTIONS

Let $X_1, X_2, \cdots$ be a sequence of rv's. Let F_n be the df of X_n, $n = 1, 2, \cdots$, and suppose that the mgf $M_n(t)$ of F_n exists. What happens to $M_n(t)$ as $n \to \infty$? Does it always converge to an mgf?

Example 1. Let $\{X_n\}$ be a sequence of rv's with pmf $P\{X_n = -n\} = 1$, $n = 1, 2, \cdots$. We have
$$M_n(t) = Ee^{tX_n} = e^{-tn} \to 0 \quad \text{as } n \to \infty \text{ for all } t > 0,$$
$$M_n(t) \to +\infty \quad \text{for all } t < 0, \text{ and } M_n(t) \to 1 \text{ at } t = 0.$$

Thus
$$M_n(t) \to M(t) = \begin{cases} 0, & t > 0 \\ 1, & t = 0 \\ \infty, & t < 0 \end{cases} \quad \text{as } n \to \infty.$$

But $M(t)$ is not an mgf. Note that if F_n is the df of X_n then

$$F_n(x) = \begin{cases} 0 & \text{if } x < -n \\ 1 & \text{if } x \geq -n \end{cases} \to F(x) = 1 \quad \text{for all } x,$$

and F is not a df.

Next suppose that X_n has mgf M_n and $X_n \xrightarrow{L} X$, where X is an rv with mgf M. Does $M_n(t) \to M(t)$ as $n \to \infty$? The answer to this question is in the negative.

Example 2 (Curtiss [20]). Consider the df

$$F_n(x) = \begin{cases} 0, & x < -n, \\ \tfrac{1}{2} + c_n \tan^{-1}(nx), & -n \leq x < n, \\ 1, & x \geq n, \end{cases}$$

where $c_n = 1/[2\tan^{-1}(n^2)]$. Clearly, as $n \to \infty$,

$$F_n(x) \to F(x) = \begin{cases} 0, & x < 0, \\ 1, & x \geq 0, \end{cases}$$

at all points of continuity of the df F. The mgf associated with F_n is

$$M_n(t) = \int_{-n}^{n} c_n e^{tx} \frac{n}{1 + n^2 x^2} \, dx,$$

which exists for all t. The mgf corresponding to F is $M(t) = 1$ for all t. But $M_n(t) \not\to M(t)$, since $M_n(t) \to \infty$ if $t \neq 0$. Indeed

$$M_n(t) > \int_0^n c_n \frac{|t|^3 x^3}{6} \frac{n}{1 + n^2 x^2} \, dx.$$

The following result is a weaker version of the continuity theorem due to Lévy and Cramér. We refer the reader to Lukacs [75], page 47, or Curtiss [20], for details of the proof.

Theorem 1 (Continuity Theorem). Let $\{F_n\}$ be a sequence of df's with corresponding mgf's $\{M_n\}$, and suppose that $M_n(t)$ exists for $|t| \leq t_0$ for every n. If there exists a df F with corresponding mgf M which exists for $|t| \leq t_1 < t_0$, such that $M_n(t) \to M(t)$ as $n \to \infty$ for every $t \in [-t_1, t_1]$, then $F_n \xrightarrow{w} F$.

Example 3. Let X_n be an rv with pmf

$$P\{X_n = 1\} = \frac{1}{n}, \quad P\{X_n = 0\} = 1 - \frac{1}{n}.$$

Then $M_n(t) = (1/n) e^t + [1 - (1/n)]$ exists for all $t \in \mathscr{R}$, and $M_n(t) \to 1$ as

$n \to \infty$ for all t. Here $M(t) = 1$ is the mgf of an rv X degenerate at 0. Thus $X_n \xrightarrow{L} X$.

The following lemma is quite useful in applications of Theorem 1.

Lemma 1. Let us write $f(x) = o(x)$, if $f(x)/x \to 0$ as $x \to 0$. We have

$$\lim_{n \to \infty} \left\{ 1 + \frac{a}{n} + o\left(\frac{1}{n}\right) \right\}^n = e^a \qquad \text{for every real } a.$$

Proof. By Taylor's expansion (Theorem P.2.7) we have

$$f(x) = f(0) + x f'(\theta x)$$
$$= f(0) + x f'(0) + \{f'(\theta x) - f'(0)\} x, \qquad 0 < \theta < 1.$$

If $f'(x)$ is continuous at $x = 0$, then

$$f(x) = f(0) + x f'(0) + o(x).$$

Taking $f(x) = \log(1 + x)$, we have $f'(x) = (1+x)^{-1}$, which is continuous at $x = 0$, so that

$$\log(1 + x) = x + o(x).$$

Then

$$n \log \left\{ 1 + \frac{a}{n} + o\left(\frac{1}{n}\right) \right\} = n \left\{ \frac{a}{n} + o\left(\frac{1}{n}\right) + o\left[\frac{a}{n} + o\left(\frac{1}{n}\right) \right] \right\}$$
$$= a + n\, o\left(\frac{1}{n}\right)$$
$$= a + o(1).$$

It follows that

$$\left\{ 1 + \frac{a}{n} + o\left(\frac{1}{n}\right) \right\}^n = e^{a + o(1)},$$

as asserted.

Example 4. Let $X_1, X_2, \cdots$ be iid $b(1, p)$ rv's. Also, let $S_n = \sum_1^n X_k$, and let $M_n(t)$ be the mgf of S_n. Then

$$M_n(t) = (q + pe^t)^n \qquad \text{for all } t,$$

where $q = 1 - p$. If we let $n \to \infty$ in such a way that np remains constant at λ, say, then, by Lemma 1,

$$M_n(t) = \left(1 - \frac{\lambda}{n} + \frac{\lambda}{n} e^t\right)^n = \left\{1 + \frac{\lambda}{n}(e^t - 1)\right\}^n \to \exp\{\lambda(e^t - 1)\}$$

for all t,

which is the mgf of a $P(\lambda)$ rv. Thus the binomial distribution function approaches the Poisson df, provided that $n \to \infty$ in such a way that $np = \lambda > 0$.

Example 5. Let $X \sim P(\lambda)$. The mgf of X is given by

$$M(t) = \exp\{\lambda(e^t - 1)\} \quad \text{for all } t.$$

Let $Y = (X - \lambda)/\sqrt{\lambda}$. Then the mgf of Y is given by

$$M_Y(t) = e^{-t\sqrt{\lambda}} M\left(\frac{t}{\sqrt{\lambda}}\right).$$

Also,

$$\log M_Y(t) = -t\sqrt{\lambda} + \log M\left(\frac{t}{\sqrt{\lambda}}\right)$$
$$= -t\sqrt{\lambda} + \lambda(e^{t/\sqrt{\lambda}} - 1)$$
$$= -t\sqrt{\lambda} + \lambda\left(\frac{t}{\sqrt{\lambda}} + \frac{t^2}{2\lambda} + \frac{t^3}{3!\lambda^{3/2}} + \cdots\right)$$
$$= \frac{t^2}{2} + \frac{t^3}{3!\lambda^{1/2}} + \cdots.$$

It follows that

$$\log M_Y(t) \to \frac{t^2}{2} \quad \text{as } \lambda \to \infty,$$

so that $M_Y(t) \to e^{t^2/2}$ as $\lambda \to \infty$, which is the mgf of an $\mathcal{N}(0, 1)$ rv.

For more examples see Section 6.6.

PROBLEMS 6.5

1. Let $X \sim NB(r; p)$. Show that
$$2pX \xrightarrow{L} Y \quad \text{as } p \to 0,$$
where $Y \sim \chi^2(2r)$.
2. Let $X_n \sim NB(r_n; 1 - p_n)$, $n = 1, 2, \cdots$. Show that $X_n \xrightarrow{L} X$ as $r_n \to \infty$, $p_n \to 0$, in such a way that $r_n p_n \to \lambda$, where $X \sim P(\lambda)$.
3. Let $X_1, X_2, \cdots$ be independent rv's with pmf given by $P\{X_n = \pm 1\} = \frac{1}{2}$, $n = 1, 2, \cdots$. Let $Z_n = \sum_{j=1}^n X_j/2^j$. Show that $Z_n \xrightarrow{L} Z$, where $Z \sim U[-1, 1]$.
4. Let $\{X_n\}$ be a sequence of rv's with $X_n \sim G(n, \beta)$ where $\beta > 0$ is a constant (independent of n). Find the limiting distribution of X_n/n.
5. Let $X_n \sim \chi^2(n)$, $n = 1, 2, \cdots$. Find the limiting distribution of X_n/n^2.

6. Let $X_1, X_2, \cdots, X_n$ be jointly normal with $EX_i = V$, $EX_i^2 = 1$ for all i and cov $(X_i, X_j) = \rho$, $i, j, = 1, 2, \cdots (i \neq j)$. What is the limiting distribution of $n^{-1} S_n$, where $S_n = \sum_{k=1}^{n} X_k$?

6.6 THE CENTRAL LIMIT THEOREM

Let $X_1, X_2, \cdots$ be a sequence of rv's, and let $S_n = \sum_{k=1}^{n} X_k$, $n = 1, 2, \cdots$. In Sections 3 and 4 we investigated the convergence of the sequence of rv's $B_n^{-1}(S_n - A_n)$ to the degenerate rv. In this section we examine the convergence of $B_n^{-1}(S_n - A_n)$ to a nondegenerate rv. Suppose that, for a suitable choice of constants A_n and $B_n > 0$, the rv's $B_n^{-1}(S_n - A_n) \xrightarrow{L} Y$. What are the properties of this limit rv Y? The question as posed is far too general and is not of much interest unless the rv's X_i are suitably restricted. For example, if we take X_1 with df F and $X_2, X_3, \cdots$ to be 0 with probability 1, choosing $A_n = 0$ and $B_n = 1$ leads to F as the limit df.

We recall (corollary to Theorem 5.3.21) that, if $X_1, X_2, \cdots, X_n$ are iid rv's with common law $\mathscr{C}(1, 0)$, then $n^{-1} S_n$ is also $\mathscr{C}(1, 0)$. Again, if $X_1, X_2, \cdots, X_n$ are iid $\mathcal{N}(0, 1)$ rv's then $n^{-1/2} S_n$ is also $\mathcal{N}(0, 1)$ (Corollary 2 to Theorem 5.3.25). We note thus that for certain sequences of rv's there exist sequences A_n and $B_n > 0$, $B_n \to \infty$, such that $B_n^{-1}(S_n - A_n) \xrightarrow{L} Y$. In the Cauchy case $B_n = n$, $A_n = 0$, and in the normal case $B_n = n^{1/2}$, $A_n = 0$. Moreover, we see that Cauchy and normal distributions appear as limiting distributions —in these two cases, because of the reproductive nature of the distributions. Cauchy and normal distributions are examples of *stable distributions*.

Definition 1. Let X_1, X_2 be iid nondegenerate rv's with common df F. Let a_1, a_2 be any positive constants. We say that F is stable if there exist constants A and B (depending on a_1, a_2) such that the rv $B^{-1}(a_1 X_1 + a_2 X_2 - A)$ also has the df F.

Let $X_1, X_2, \cdots$ be iid rv's with common df F. We remark without proof (see Loève [71], 327) that only stable distributions occur as limits. To make this statement more precise we make the following definition.

Definition 2. Let $X_1, X_2, \cdots$ be iid rv's with common df F. We say that F belongs to the domain of attraction of a distribution V if there exist norming constants $B_n > 0$ and centering constants A_n such that, as $n \to \infty$,

(1) $$P\{B_n^{-1}(S_n - A_n) \leq x\} \to V(x),$$

at all continuity points x of V.

THE CENTRAL LIMIT THEOREM

In view of the statement after Definition 1, we see that only stable distributions possess domains of attraction. From Definition 1 we also note that each stable law belongs to its own domain of attraction. The study of stable distributions is beyond the scope of this book. We shall restrict ourselves to seeking conditions under which the limit law V is the normal distribution. The importance of the normal distribution in statistics is due largely to the fact that a wide class of distributions F belongs to the domain of attraction of the normal law. Let us consider some examples.

Example 1. Let $X_1, X_2, \cdots, X_n$ be iid $b(1, p)$ rv's. Let

$$S_n = \sum_{k=1}^{n} X_k, \quad A_n = ES_n = np, \quad B_n = \sqrt{\operatorname{var}(S_n)} = \sqrt{np(1-p)}.$$

Then

$$M_n(t) = E \exp\left\{\frac{S_n - np}{\sqrt{np(1-p)}} t\right\}$$

$$= \prod_{i=1}^{n} E \exp\left\{\frac{X_i - p}{\sqrt{np(1-p)}} t\right\}$$

$$= \exp\left\{-\frac{npt}{\sqrt{np(1-p)}}\right\} \left\{q + p \exp\left[\frac{t}{\sqrt{np(1-p)}}\right]\right\}^n,$$

$$q = 1 - p,$$

$$= \left\{q \exp\left(-\frac{pt}{\sqrt{npq}}\right) + p \exp\left(\frac{qt}{\sqrt{npq}}\right)\right\}^n$$

$$= \left[1 + \frac{t^2}{2n} + o\left(\frac{1}{n}\right)\right]^n.$$

It follows from Lemma 6.5.1 that

$$M_n(t) \to e^{t^2/2} \quad \text{as } n \to \infty,$$

and since $e^{t^2/2}$ is the mgf of an $\mathcal{N}(0, 1)$ rv, we have by the continuity theorem

$$P\left\{\frac{S_n - np}{\sqrt{npq}} \le x\right\} \to \frac{1}{\sqrt{2\pi}} \int_{-\infty}^{x} e^{-t^2/2}\, dt \quad \text{for all } x \in \mathcal{R}.$$

Example 2. Let $X_1, X_2, \cdots, X_n$ be iid $\chi^2(1)$ rv's. Then $S_n \sim \chi^2(n)$, $ES_n = n$ and $\operatorname{var}(S_n) = 2n$. Also let $Z_n = (S_n - n)/\sqrt{2n}$; then

$$M_n(t) = E e^{tZ_n}$$

$$= \exp\left(-t\sqrt{\frac{n}{2}}\right)\left(1 - \frac{2t}{\sqrt{2n}}\right)^{-n/2}, \quad 2t < \sqrt{2n},$$

$$= \left[\exp\left(t\sqrt{\frac{2}{n}}\right) - t\sqrt{\frac{2}{n}} \exp\left(t\sqrt{\frac{2}{n}}\right)\right]^{-n/2}, \quad t < \sqrt{\frac{n}{2}}.$$

Using Taylor's approximation, we get

$$\exp\left(t\sqrt{\frac{2}{n}}\right) = 1 + t\sqrt{\frac{2}{n}} + \frac{t^2}{2}\left(\sqrt{\frac{2}{n}}\right)^2 + \frac{1}{6}\exp(\theta_n)\left(t\sqrt{\frac{2}{n}}\right)^3,$$

where $0 < \theta_n < t\sqrt{(2/n)}$. It follows that

$$M_n(t) = \left(1 - \frac{t^2}{n} + \frac{\zeta(n)}{n}\right)^{-n/2},$$

where

$$\zeta(n) = -\sqrt{\frac{2}{n}}\,t^3 + \left(\frac{t^3}{3}\sqrt{\frac{2}{n}} - \frac{2t^4}{3n}\right)\exp(\theta_n) \to 0 \quad \text{as } n \to \infty,$$

for every fixed t. We have from Lemma 6.5.1 that $M_n(t) \to e^{t^2/2}$ as $n \to \infty$ for all real t, and it follows that $Z_n \xrightarrow{L} Z$, where Z is $\mathcal{N}(0, 1)$.

These examples suggest that, if we take iid rv's with finite variance, and take $A_n = ES_n$, $B_n = \sqrt{\text{var}(S_n)}$, then $B_n^{-1}(S_n - A_n) \xrightarrow{L} Z$, where Z is $\mathcal{N}(0, 1)$. This is the central limit result, which we now prove. The reader should note that in both Examples 1 and 2 we used more than just the existence of $E|X|^2$. Indeed, the mgf exists and hence moments of all order exist. The existence of mgf is not a necessary condition.

Theorem 1† (Lindeberg—the Central Limit Theorem). Let $X_1, X_2, \cdots$ be independent nondegenerate rv's with distribution functions $F_1, F_2, \cdots$. Assume that $EX_k = \mu_k$, $\text{var}(X_k) = \sigma_k^2$, and write

$$s_n^2 = \sigma_1^2 + \sigma_2^2 + \cdots + \sigma_n^2.$$

If F_k are absolutely continuous with pdf F_k', assume that the relation

(2) $$\lim_{n \to \infty} \frac{1}{s_n^2} \sum_{k=1}^{n} \int_{|x - \mu_k| > \varepsilon s_n} (x - \mu_k)^2 F_k'(x)\, dx = 0$$

holds for all $\varepsilon > 0$. If X_k are of the discrete type with jump points x_{kl} and jumps p_{kl} ($l = 1, 2, \cdots$), suppose that the relation

(3) $$\lim_{n \to \infty} \frac{1}{s_n^2} \sum_{k=1}^{n} \sum_{|x_{kl} - \mu_k| > \varepsilon s_n} (x_{kl} - \mu_k)^2 p_{kl} = 0$$

holds for every $\varepsilon > 0$. Then the distribution of the normalized sum

(4) $$S_n^* = \frac{X_1 + X_2 + \cdots + X_n - \sum_{k=1}^{n} \mu_k}{s_n} = \frac{\sum_{k=1}^{n}(X_k - \mu_k)}{s_n}$$

---

† The reader may omit the proof of Theorem 1 on the first reading. The corollary to Theorem 1 will be used frequently in the rest of the book, however, and the reader should familiarize himself with the application of this result. For an alternative proof of the corollary see Problem 4.

converges to the standard normal distribution.

Corollary. Let $X_1, X_2, \cdots$ be iid nondegenerate rv's with common df F. Let both $EX_k = \mu$ and var $(X_k) = \sigma^2$ be finite. Then

(5) $$S_n^* \xrightarrow{L} Z \quad \text{as } n \to \infty,$$

where Z is $\mathcal{N}(0, 1)$ and $S_n^* = (S_n - n\mu)/(\sigma\sqrt{n})$.

We first prove that the corollary follows from Theorem 1. If F is absolutely continuous with pdf f, then

$$\mu = \int_{-\infty}^{\infty} x f(x) \, dx, \qquad \sigma^2 = \int_{-\infty}^{\infty} x^2 f(x + \mu) \, dx,$$

and condition (2) becomes

$$\frac{1}{n\sigma^2} \sum_{k=1}^{n} \int_{|x|>\varepsilon\sigma\sqrt{n}} x^2 f(x + \mu) \, dx = \frac{1}{\sigma^2} \int_{|x|>\varepsilon\sigma\sqrt{n}} x^2 f(x + \mu) \, dx \to 0$$

as $n \to \infty$. Since $E|X|^2 < \infty$, this last condition holds and the corollary follows from Theorem 1. A similar argument applies in the discrete case.

A conventional proof of Theorem 1 uses the notion of characteristic functions, which we do not study in this book. It is possible to imitate this proof using mgf's, but this requires the restrictive assumption of existence of all moments. Instead, we will prove Theorem 1 with the help of probability operators. The methods used here utilize the work of Trotter [131]. We refer the reader also to Feller [29], page 256, Lukacs [77], page 213, and Rényi [98], page 515.

Let C_3 be the class of all uniformly bounded and uniformly continuous functions defined on $\mathcal{R}$ that can be differentiated three times and whose first, second, and third derivatives are also uniformly bounded and uniformly continuous on $\mathcal{R}$. For $f \in C_3$ let us write

(6) $$\|f\| = \sup_x |f(x)|$$

and call $\|f\|$ the *norm* of f.

The following result is easy to prove.

Lemma 1. $f, g \in C_3$, and $\alpha \in \mathcal{R}$

$\Rightarrow$

(i) $f + g \in C_3$, and $\alpha f \in C_3$.
(ii) $\|f + g\| \leq \|f\| + \|g\|$.
(iii) $\|\alpha f\| \leq |\alpha| \|f\|$.

Definition 3. Let A be a transformation that assigns to each $f \in C_3$ a function $Af \in C_3$. We say that A is a linear operator if it satisfies:

(i) $A(f + g) = Af + Ag$ for all $f, g \in C_3$;
(ii) $A(\alpha f) = \alpha Af$ for $f \in C_3$ and $\alpha \in \mathscr{R}$;

and

(iii) there exists a positive number K such that $\|Af\| \leq K\|f\|$ holds for all $f \in C_3$.

A is called a contraction operator if (iii) holds for $K = 1$.

For two operators A and B we define $A + B$ and AB by the relations

$$(A + B)f = Af + Bf \quad \text{and} \quad (AB)f = A(Bf).$$

Definition 4. Let F be the df of an rv X. With F we associate an operator A_F as follows. If F has pdf F', then

(7) $$A_F f = \int_{-\infty}^{\infty} f(x + y) F'(y) \, dy \quad \text{for all } f \in C_3.$$

If F is a df with jump points x_j and jumps p_j, then

(8) $$A_F f = \sum_{j=1}^{\infty} p_j f(x + x_j) \quad \text{for all } f \in C_3.$$

The operator A_F is called the probability operator associated with F.

Lemma 2. The probability operator associated with an arbitrary df F is a linear contraction operator.

Proof. The proof is left as an exercise.

Lemma 3. Let F and G be any two df's. Then A_F and A_G commute, and $A_F A_G = A_H$, where H is the convolution df, $H = F * G$.

Proof. The proof is left as an exercise.

Lemma 4. Let $X_1, X_2, \cdots, X_n$ be independent rv's with df's $F_1, F_2, \cdots, F_n$, respectively. Also, let $S_n = \sum_{j=1}^{n} X_j$, and let H be the df of S_n. Then $H = F_1 * F_2 * \cdots * F_n$, and the associated probability operator A_H is given by

(9) $$A_H = A_{F_1} A_{F_2} \cdots A_{F_n},$$

where A_{F_j} is the probability operator associated with F_j, $j = 1, 2, \cdots, n$.

Lemma 5. Let F and G be any df's. Then

(10) $$\|A_F A_G f\| \leq \|A_G f\| \quad \text{for all} \quad f \in C_3.$$

Lemma 5 follows since A_F is a contraction operator.

Lemma 6. Let $U_1, U_2, \cdots, U_n$ and $V_1, V_2, \cdots, V_n$ be probability operators. Then

(11) $$\|U_1 U_2 \cdots U_n f - V_1 V_2 \cdots V_n f\| \leq \sum_{k=1}^{n} \|U_k f - V_k f\|.$$

In particular,

(12) $$\|U^n f - V^n f\| \leq n \|Uf - Vf\|.$$

Proof. For arbitrary linear operators we have the obvious identity

$$U_1 U_2 \cdots U_n - V_1 V_2 \cdots V_n = \sum_{k=1}^{n} U_1 U_2 \cdots U_{k-1}(U_k - V_k) V_{k+1} \cdots V_n.$$

By triangular inequality and from the fact that the U's and V's are contraction operators, result (11) follows.

Proof of Theorem 1. Without loss of generality let us assume that $\mu_k = 0$ and $s_n^2 = 1$. Let U_k be the probability operator associated with F_k, and V_k be the probability operator associated with the df of an $\mathcal{N}(0, \sigma_k^2)$ rv. Let F_n^* be the df of $S_n^* = \sum_{j=1}^{n} X_j$. Then

$$A_{F_n^*} = A_n = U_1 U_2 \cdots U_n,$$

and the operator B associated with

$$\Phi(x) = \Phi\left(\frac{x}{\sigma_1}\right) * \Phi\left(\frac{x}{\sigma_2}\right) * \cdots * \Phi\left(\frac{x}{\sigma_n}\right)$$

is given by

$$B = V_1 V_2 \cdots V_n.$$

Here and subsequently we have written

$$\Phi(x) = \frac{1}{\sqrt{2\pi}} \int_{-\infty}^{x} e^{-t^2/2} \, dt.$$

By Lemma 6, for $f \in C_3$,

(13) $$\|A_n f - Bf\| \leq \sum_{1}^{n} \|U_k f - V_k f\|.$$

We will show that

(14) $$\|A_n f - Bf\| \to 0 \quad \text{as} \quad n \to \infty.$$

By Taylor's formula

(15) $$f(x+y) = f(x) + yf'(x) + \frac{y^2}{2} f''(x+\theta_1 y),$$

(16) $$f(x+y) = f(x) + yf'(x) + \frac{y^2}{2} f''(x) + \frac{y^3}{6} f'''(x+\theta_2 y),$$

where $0 < \theta_1 < 1$ and $0 < \theta_2 < 1$, and θ_1, θ_2 depend on x and y.

In the following we will assume that F_k is absolutely continuous. Now, for $\varepsilon > 0$, we have

$$U_k f = \int_{-\varepsilon}^{\varepsilon} f(x+y) F_k'(y) \, dy + \int_{|y|>\varepsilon} f(x+y) F_k'(y) \, dy.$$

Substituting (16) into the first and (15) into the second integral, we obtain

(17) $$U_k f = f(x) + \tfrac{1}{2} f''(x) \sigma_k^2 + \tfrac{1}{6} \int_{|y| \le \varepsilon} y^3 f'''(x+\theta_2 y) F_k'(y) \, dy$$
$$+ \frac{1}{2} \int_{|y|>\varepsilon} y^2 [f''(x+\theta_1 y) - f''(y)] F_k'(y) \, dy.$$

Letting $M_2 = \sup |f''(x)|$ and $M_3 = \sup |f'''(x)|$, we get

(18) $$\left| U_k f - f(x) - \tfrac{1}{2} \sigma_k^2 f''(x) \right| \le \frac{\varepsilon M_3}{6} \sigma_k^2 + M_2 \int_{|y|>\varepsilon} y^2 F_k'(y) \, dy.$$

Also,

$$V_k f = \frac{1}{\sigma_k \sqrt{2\pi}} \int_{-\infty}^{\infty} f(x+y) e^{-(y^2/2\sigma_k^2)} \, dy,$$

so that, on using (16), we have

(19) $$\left| V_k f - f(x) - \frac{\sigma_k^2}{2} f''(x) \right| \le \frac{M_3}{3} \sigma_k^3.$$

It follows from (18) and (19) that

$$\|U_k f - V_k f\| \le \frac{\varepsilon M_3 \sigma_k^2}{6} + M_2 \int_{|y|>\varepsilon} y^2 F_k'(y) \, dy + \frac{M_3}{3} \sigma_k^3,$$

so that

(20) $$\sum_{k=1}^{n} \|U_k f - V_k f\| \le \frac{\varepsilon M_3}{6} + M_2 \sum_{k=1}^{n} \int_{|y|>\varepsilon} y^2 F_k'(y) \, dy + \frac{M_3}{3} \sum_{k=1}^{n} \sigma_k^3.$$

Now

…

$$\sigma_k^2 = \int_{|y|>\varepsilon} y^2 F_k'(y)\, dy + \int_{|y|\le\varepsilon} y^2 F_k'(y)\, dy \le \int_{|y|>\varepsilon} y^2 F_k'(y)\, dy + \varepsilon^2,$$

and, since $\sum_1^n \sigma_k^2 = 1$,

$$\sum_1^n \sigma_k^3 \le \max_{1\le k\le n} \sigma_k \le [\varepsilon^2 + \sum_1^n \int_{|y|>\varepsilon} y^2 F_k'(y)\, dy]^{1/2}.$$

It follows from (20) and condition (2) that

$$\lim_{n\to\infty} \sum_{k=1}^n \|U_k f - V_k f\| = 0$$

and hence that (14) holds. In view of the definition of the norm, this implies that

(21) $$\lim_{n\to\infty} \int_{-\infty}^{\infty} f(x+y) F_n^{*\prime}(y)\, dy = \int_{-\infty}^{\infty} f(x+y) \Phi'(y)\, dy,$$

for any x (indeed, uniformly in x), $f \in C_3$.

Let $\varepsilon > 0$ and consider a function $f_\varepsilon \in C_3$ with these properties: $f_\varepsilon(x) = 1$ for $x \le 0$, and $= 0$ for $x \ge \varepsilon$, and f_ε is monotonically decreasing in $(0, \varepsilon)$. Such a function exists. For example, take

$$f_\varepsilon(x) = \begin{cases} 1 & \text{if } x \le 0, \\ [1 - (x\varepsilon^{-1})^4]^4 & \text{if } 0 \le x \le \varepsilon, \\ 0 & \text{if } x \ge \varepsilon. \end{cases}$$

Then

$$\Phi(-x+\varepsilon) \ge \int_{-\infty}^{\infty} f_\varepsilon(x+y) \Phi'(y)\, dy \ge \Phi(-x), \qquad x \in \mathscr{R}.$$

Similarly,

$$F_n^*(-x+\varepsilon) \ge \int_{-\infty}^{\infty} f_\varepsilon(x+y) F_n^{*\prime}(y)\, dy \ge F_n^*(-x), \qquad x \in \mathscr{R}.$$

Using (21), we conclude that

(22) $$\overline{\lim_{n\to\infty}} F_n^*(-x) \le \lim_{n\to\infty} \int_{-\infty}^{\infty} f_\varepsilon(x+y) F_n^{*\prime}(y)\, dy$$
$$= \int_{-\infty}^{\infty} f_\varepsilon(x+y) \Phi'(y)\, dy \le \Phi(-x+\varepsilon)$$

and

(23) $$\underline{\lim_{n\to\infty}} F_n^*(-x+\varepsilon) \ge \lim_{n\to\infty} \int_{-\infty}^{\infty} f_\varepsilon(x+y) F_n^{*\prime}(y)\, dy$$
$$= \int_{-\infty}^{\infty} f_\varepsilon(x+y) \Phi'(y)\, dy \ge \Phi(-x)$$

for all real x and arbitrary $\varepsilon > 0$. Rewriting (23) as

(24) $$\lim_{n\to\infty} F_n^*(-x) \geq \Phi(-x-\varepsilon),$$

and combining (22) and (24), we get

$$\Phi(-x-\varepsilon) \leq \varliminf_{n\to\infty} F_n^*(-x) \leq \varlimsup_{n\to\infty} F_n^*(-x) \leq \Phi(-x+\varepsilon)$$

for all $x \in \mathscr{R}$ and $\varepsilon > 0$. Since ε is arbitrary, we see that

$$\lim_{n\to\infty} F_n^*(-x) = \Phi(-x), \qquad x \in \mathscr{R},$$

as asserted.

Exactly similar considerations apply to the discrete case, and the proof is complete.

Remark 1. Feller [27] has shown that Lindeberg's condition (2) or (3) is necessary as well, in the following sense: If $X_1, X_2 \cdots$ are independent rv's with finite standard deviations, and if F_k is the df of X_k, then

(25) $$\lim_{n\to\infty} P\left\{\frac{S_n - ES_n}{\sqrt{\operatorname{var}(S_n)}} \leq x\right\} = \Phi(x)$$

and

(26) $$\lim_{n\to\infty} P\{\max_{1\leq k\leq n} |X_k - E(X_k)| > \varepsilon\sqrt{\operatorname{var}(S_n)}\} = 0$$

hold if and only if (2) or (3) is fulfilled for every $\varepsilon > 0$.

If (26) holds, we say that the rv's $X_k - EX_k$ are *infinitesimal*. Relation (26) follows from (2) or (3), since

$$P\{\max_{1\leq k\leq n} |X_k - EX_k| > \varepsilon\sqrt{\operatorname{var}(S_n)}\}$$

$$\leq \sum_{k=1}^{n} P\{|X_k - EX_k| > \varepsilon\sqrt{\operatorname{var}(S_n)}\}$$

$$\leq \begin{cases} \dfrac{1}{\varepsilon^2 \operatorname{var}(S_n)} \sum_{k=1}^{n} \int_{|x-\mu_k|>\varepsilon s_n} (x-\mu_k)^2 F_k'(x)\,dx, \\ \dfrac{1}{\varepsilon^2 s_n^2} \sum_{k=1}^{n} \sum_{|x_{kl}-\mu_k|>\varepsilon s_n} (x_{kl}-\mu_k)^2 p_{kl}. \end{cases}$$

Thus condition (2) or (3) implies that the rv's $(X_k - EX_k)/s_n$ are infinitesimal. We will not prove here the necessity of Lindeberg's condition.

The following converse to the corollary of Theorem 1 holds.

THE CENTRAL LIMIT THEOREM

Theorem 2. Let $X_1, X_2, \cdots, X_n$ be iid rv's such that $n^{-1/2}S_n$ has the same distribution for every $n = 1, 2, \cdots$. Then, if $EX_i = 0$, var $(X_i) = 1$, the distribution of X_i must be $\mathcal{N}(0, 1)$.

Proof. Let F be the df of $n^{-1/2}S_n$. By the central limit theorem,
$$\lim_{n \to \infty} P\{n^{-1/2} S_n \leq x\} = \Phi(x).$$
Also, $P\{n^{-1/2} S_n \leq x\} = F(x)$ for each n. It follows that we must have $F(x) = \Phi(x)$.

Example 3. Let $X_1, X_2, \cdots$ be iid rv's with common pmf
$$P\{X = k\} = p(1-p)^k, \quad k = 0, 1, 2, \cdots, \quad 0 < p < 1, q = 1 - p.$$
Then $EX = q/p$, var $(X) = q/p^2$. By the corollary to Theorem 1 we see that
$$P\left\{\frac{S_n - n(q/p)}{\sqrt{nq}} p \leq x\right\} \to \Phi(x) \quad \text{as } n \to \infty \quad \text{for all } x \in \mathcal{R}.$$

Example 4. Let $X_1, X_2, \cdots$ be iid rv's with common $B(\alpha, \beta)$ distribution. Then
$$EX = \frac{\alpha}{\alpha + \beta} \quad \text{and} \quad \text{var}(X) = \frac{\alpha\beta}{(\alpha+\beta)^2(\alpha+\beta+1)}.$$
By the corollary to Theorem 1, it follows that
$$\frac{S_n - n[\alpha/(\alpha+\beta)]}{\sqrt{\alpha\beta n/[(\alpha+\beta+1)(\alpha+\beta)^2]}} \xrightarrow{L} Z,$$
where Z is $\mathcal{N}(0, 1)$.

Example 5. Let $X_1, X_2, \cdots$ be independent rv's such that X_k is $U(-a_k, a_k)$. Then $EX_k = 0$, var $(X_k) = (1/3)a_k^2$. Suppose that $|a_k| < a$ and $\sum_1^n a_k^2 \to \infty$ as $n \to \infty$. Then
$$\frac{1}{s_n^2} \sum_{k=1}^n \int_{|x|>\varepsilon s_n} x^2 F_k'(x)\, dx \leq \frac{1}{s_n^2} \sum_{k=1}^n \int_{|x|>\varepsilon s_n} a^2 \frac{1}{2a_k}\, dx$$
$$\leq \frac{a^2}{s_n^2} \sum_{k=1}^n P\{|X_k| > \varepsilon s_n\} \leq \frac{a^2}{s_n^2} \sum_{k=1}^n \frac{\text{var}(X_k)}{\varepsilon^2 s_n^2}$$
$$= \frac{a^2}{\varepsilon^2 s_n^2} \to 0 \quad \text{as } n \to \infty.$$

If $\sum_1^\infty a_k^2 < \infty$, then $s_n^2 \uparrow A^2$, say, as $n \to \infty$. For fixed k, we can find ε_k such that $\varepsilon_k A < a_k$ and then $P\{|X_k| > \varepsilon_k s_n\} \geq P\{|X_k| > \varepsilon_k A\} > 0$. For $n \geq k$, we have

$$\frac{1}{s_n^2} \sum_{j=1}^n \int_{|x|>\varepsilon_k s_n} x^2 F_j'(x)\, dx \geq \frac{s_n^2 \varepsilon_k^2}{s_n^2} \sum_{j=1}^n P\{|X_j| > \varepsilon_k s_n\}$$
$$\geq \varepsilon_k^2 P\{|X_k| > \varepsilon_k s_n\}$$
$$> 0,$$

so that the Lindeberg condition does not hold. Indeed, if X_1, X_2, $\cdots$ are independent rv's such that there exists a constant A with $P\{|X_n| \leq A\} = 1$ for all n, the Lindeberg condition (2) or (3) is satisfied if $s_n^2 \to \infty$ as $n \to \infty$. To see this, suppose that $s_n^2 \to \infty$. Since the X_k's are uniformly bounded, so are the rv's $X_k - EX_k$. It follows that for every $\varepsilon > 0$ we can find an N_ε such that, for $n \geq N_\varepsilon$, $P\{|X_k - EX_k| < \varepsilon s_n, k = 1, 2, \cdots, n\} = 1$. The Lindeberg condition follows immediately. The converse also holds, for, if $\lim_{n \to \infty} s_n^2 < \infty$ and the Lindeberg condition holds, there exists a constant $A < \infty$ such that $s_n^2 \to A^2$. For any fixed j, we can find an $\varepsilon > 0$ such that $P\{|X_j - \mu_j| > \varepsilon A\} > 0$. Then, for $n \geq j$,

$$\frac{1}{s_n^2} \sum_{k=1}^n \int_{|x-\mu_k|>\varepsilon s_n} (x - \mu_k)^2 F_k'(x)\, dx$$
$$\geq \varepsilon^2 \sum_{k=1}^n P\{|X_k - \mu_k| > \varepsilon s_n\}$$
$$\geq \varepsilon^2 P\{|X_j - \mu_j| > \varepsilon A\}$$
$$> 0,$$

and the Lindeberg condition does not hold. This contradiction shows that $s_n^2 \to \infty$ is also a necessary condition; that is, for a sequence of uniformly bounded independent rv's, a necessary and sufficient condition for the central limit theorem to hold is $s_n^2 \to \infty$ as $n \to \infty$.

Example 6. Let X_1, X_2, $\cdots$ be independent rv's such that $\alpha_k = E|X_k|^{2+\delta} < \infty$ for some $\delta > 0$ and $\alpha_1 + \alpha_2 + \cdots + \alpha_n = o(s_n^{2+\delta})$. Then the Lindeberg condition is satisfied, and the central limit theorem holds. This result is due to Lyapunov. We have

$$\frac{1}{s_n^2} \sum_{k=1}^n \int_{|x|>\varepsilon s_n} x^2 F_k'(x)\, dx$$
$$\leq \frac{1}{\varepsilon^\delta s_n^{2+\delta}} \sum_{k=1}^n \int_{-\infty}^\infty |x|^{2+\delta} F_k'(x)\, dx$$
$$= \frac{\sum_{k=1}^n \alpha_k}{\varepsilon^\delta s_n^{2+\delta}} \to 0 \quad \text{as } n \to \infty.$$

A similar argument applies in the discrete case.

Remark 2. Let X_1, X_2, $\cdots$ be a sequence of iid rv's. The condition that

$E|X|^2 < \infty$ for the central limit theorem (corollary to Theorem 1) to hold is sufficient but not necessary. It is possible, however, to obtain a necessary and sufficient condition for the central limit result to hold. We refer the reader to Feller [29], page 313, for details.

Example 7. It is known (see, for example, Rohatgi [101]) that, if $X_1, X_2, \cdots$ is a sequence of iid rv's with common law $\mathscr{L}(X)$ such that

$$P\{|X| > x\} = \frac{L(x)}{x^\rho}, \quad \rho \geq 2,$$

where L is a function of slow variation (that is, $L(cx)/L(x) \to 1$ as $x \to \infty$ for all $c > 0$), then X_i belong to the domain of attraction of the normal distribution. Thus, for example, if X is an rv with pdf

$$f(x) = 2|x|^{-3} \log|x| \quad \text{for } |x| \geq 1, \quad f(x) = 0 \quad \text{for } |x| \leq 1,$$

then $S_n/(\sqrt{2n} \log n) \xrightarrow{L} Z$, where Z is $\mathscr{N}(0, 1)$. (See Feller [29], 260.) Note that $E|X|^2 = \infty$.

Remark 3. Both the central limit theorem (CLT) and the (weak) law of large numbers (WLLN) hold for a large class of sequences of rv's $\{X_n\}$. If the $\{X_n\}$ are independent uniformly bounded rv's, that is, if $P\{|X_n| \leq M\} = 1$, the WLLN (Theorem 6.3.1) holds; the CLT holds provided that $s_n^2 \to \infty$ (Example 5).

If the rv's $\{X_n\}$ are iid, then the CLT is a stronger result than the WLLN in that the former provides an estimate of the probability $P\{|S_n - n\mu|/n \geq \varepsilon\}$. Indeed,

$$P\{|S_n - n\mu| > n\varepsilon\} = P\left\{\frac{|S_n - n\mu|}{\sigma\sqrt{n}} > \frac{\varepsilon}{\sigma}\sqrt{n}\right\}$$

$$\approx 1 - P\{|Z| \leq \frac{\varepsilon}{\sigma}\sqrt{n}\},$$

where Z is $\mathscr{N}(0, 1)$, and the law of large number follows. On the other hand, we note that the WLLN does not require the existence of a second moment.

Remark 4. If $\{X_n\}$ are independent rv's, it is quite possible that the CLT may apply to the X_n's, but not the WLLN.

Example 8 (Feller [28], 255). Let $\{X_k\}$ be independent rv's with pmf

$$P\{X_k = k^\lambda\} = P\{X_k = -k^\lambda\} = \tfrac{1}{2}, \quad k = 1, 2, \cdots.$$

Then $EX_k = 0$, var $(X_k) = k^{2\lambda}$. Also let $\lambda > 0$; then

$$s_n^2 = \sum_{k=1}^{n} k^{2\lambda} \leq \int_0^n x^{2\lambda} \, dx = \frac{n^{2\lambda+1}}{2\lambda + 1}.$$

It follows that, if $0 < \lambda < \frac{1}{2}$, $s_n/n \to 0$, and by Corollary 2 to Theorem 6.3.1 the WLLN holds. Now $k^\lambda < n^\lambda$, so that the sum $\sum_{k=1}^{n} \sum_{|x_{kl}| > \varepsilon s_n} x_{kl}^2 \, p_{kl}$ will be nonzero if $n^\lambda > \varepsilon s_n \approx \varepsilon [n^{\lambda+1/2} / \sqrt{(2\lambda + 1)}]$. It follows that, as long as $n > (2\lambda + 1)\varepsilon^{-2}$,

$$\frac{1}{s_n^2} \sum_{k=1}^{n} \sum_{|x_{kl}| > \varepsilon s_n} x_{kl}^2 \, p_{kl} = 0$$

and the Lindeberg condition holds. Thus the CLT holds for $\lambda > 0$. This means that

$$P\left\{ a < \sqrt{\frac{2\lambda + 1}{n^{2\lambda+1}}} \, S_n < b \right\} \to \int_a^b \frac{e^{-t^2/2} \, dt}{\sqrt{2\pi}}.$$

Thus

$$P\left\{ \frac{an^{\lambda+1/2-1}}{\sqrt{2\lambda + 1}} < \frac{S_n}{n} < \frac{bn^{\lambda+1/2-1}}{\sqrt{2\lambda + 1}} \right\} \to \int_a^b \frac{e^{-t^2/2}}{\sqrt{2\pi}} \, dt$$

and the WLLN cannot hold for $\lambda \geq \frac{1}{2}$.

We conclude this section with some remarks concerning the application of the CLT. Let $X_1, X_2, \cdots$ be iid rv's with common mean μ and variance σ^2. Let us write

$$Z_n = \frac{S_n - n\mu}{\sigma \sqrt{n}},$$

and let z_1, z_2 be two arbitrary real numbers with $z_1 < z_2$. If F_n is the df of Z_n, then

$$\lim_{n \to \infty} P\{z_1 < Z_n \leq z_2\} = \lim_{n \to \infty} [F_n(z_2) - F_n(z_1)]$$
$$= \frac{1}{\sqrt{2\pi}} \int_{z_1}^{z_2} e^{-t^2/2} \, dt,$$

that is,

$$\lim_{n \to \infty} P\{z_1 \sigma \sqrt{n} + n\mu < S_n \leq z_2 \sigma \sqrt{n} + n\mu\} = \frac{1}{\sqrt{2\pi}} \int_{z_1}^{z_2} e^{-t^2/2} \, dt.$$

It follows that the rv $S_n = \sum_{k=1}^{n} X_k$ is asymptotically normally distributed with mean $n\mu$ and variance $n\sigma^2$. Equivalently, the rv $n^{-1} S_n$ is asymptotically $\mathcal{N}(\mu, \sigma^2/n)$. This result is of great importance in statistics.

Example 9. Let $X_1, X_2, \cdots, X_n$ be iid $b(1, p)$ rv's. Then S_n is asymptotically $\mathcal{N}(np, np(1-p))$. Thus for large enough n we can estimate $P\{S_n \leq x\}$ by

$$P\left\{\frac{S_n - np}{\sqrt{np(1-p)}} \le \frac{x - np}{\sqrt{np(1-p)}}\right\} \approx \Phi\left[\frac{x - np}{\sqrt{np(1-p)}}\right].$$

In practice, $n \ge 20$ suffices. In particular, if $n = 25$, $p = \frac{1}{2}$, then

$$P\{S_n \le 12\} = P\left\{\frac{S_n - 12.5}{2.5} \le \frac{-.5}{2.5}\right\}$$
$$\approx P\{Z \le -.2\} = .421,$$

where Z is $\mathcal{N}(0,1)$.

A somewhat better approximation results if we apply the *continuity correction*. If X is an integer-valued rv, then $P\{x_1 \le X \le x_2\}$, where x_1 and x_2 are integers, is exactly the same as $P\{x_1 - \frac{1}{2} < X < x_2 + \frac{1}{2}\}$. This amounts to making the discrete space of values of X continuous by considering intervals of length 1 with midpoints at integers.

In Example 9, we have

$$P\{S_n \le 12\} = P\{S_n < 12.5\} \approx P\{Z < 0\} = .50,$$

which is the exact probability of at most 12 successes in 25 trials with a fair coin.

Example 10. Let $X_1, X_2, \cdots$ be iid $P(\lambda)$ rv's. Then S_n is approximately $\mathcal{N}(n\lambda, n\lambda)$ for large n. If $\lambda = 1$, $n = 64$, then

$$P\{50 < S_n \le 80\} = P\{51 \le S_n \le 80\}$$
$$= P\{50.5 < S_n < 80.5\}$$
$$\approx P\left\{\frac{-13.5}{8} < Z < \frac{16.5}{8}\right\}$$
$$= P\{-1.69 < Z < 2.06\}$$
$$= .9348.$$

PROBLEMS 6.6

1. Let $\{X_n\}$ be a sequence of independent rv's with the following distributions. In each case, does the Lindeberg condition hold?
   (a) $P\{X_n = \pm(1/2^n)\} = \frac{1}{2}$.
   (b) $P\{X_n = \pm 2^{n+1}\} = 1/2^{n+3}$, $P\{X_n = 0\} = 1 - (1/2^{n+2})$.
   (c) $P\{X_n = \pm 1\} = (1 - 2^{-n})/2$, $P\{X_n = \pm 2^{-n}\} = 1/2^{n+1}$.
   (d) $\{X_n\}$ is a sequence of independent Poisson rv's with parameter λ_n, $n = 1, 2, \cdots$, such that $\sum_{k=1}^{n} \lambda_k \to \infty$.
   (e) $P\{X_n = \pm 2^n\} = \frac{1}{2}$.
   (f) $P\{X_n = \pm 2^n\} = 2^{-n-1}$, $P\{X_n = \pm 1\} = 2^{-1}\{1 - 2^{-n}\}$.

2. Let $X_1, X_2, \cdots$ be iid rv's with mean 0, variance 1, and $EX_i^4 < \infty$. Find the limiting distribution of

$$Z_n = \sqrt{n}\, \frac{X_1 X_2 + X_3 X_4 + \cdots + X_{2n-1} X_{2n}}{X_1^2 + X_2^2 + \cdots + X_{2n}^2}.$$

3. Let $X_1, X_2 \cdots$ be iid rv's with mean α and variance σ^2, and let $Y_1, Y_2, \cdots$ be iid rv's with mean $\beta (\neq 0)$ and variance τ^2. Find the limiting distribution of $Z_n = \sqrt{n}(\bar{X}_n - \alpha)/\bar{Y}_n$, where $\bar{X}_n = n^{-1} \sum_{i=1}^n X_i$ and $\bar{Y}_n = n^{-1} \sum_{i=1}^n Y_i$.

4. Let $\{X_n\}$ be a sequence of iid rv's with mean μ and variance σ^2. Let $M(t)$ be the common mgf that is assumed to exist for $|t| < h$, $h > 0$. Also, let $Y_n = (S_n - n\mu)/(\sigma \sqrt{n})$, and write $M_n(t)$ for the mgf of Y_n. Use the continuity theorem to show that

$$M_n(t) \to e^{t^2/2} \quad \text{as} \quad n \to \infty \text{ for all real } t.$$

[Hint: The existence of $M(t)$ for $|t| < h$ implies the existence of moments of all orders. Expand $M_n(t)$ by Taylor's theorem with remainder.]

5. Let $X_1, X_2 \cdots$ be a sequence of iid rv's with common mean μ and variance σ^2. Also, let $\bar{X} = n^{-1} \sum_{k=1}^n X_k$ and $S^2 = (n-1)^{-1} \sum_{i=1}^n (X_i - \bar{X})^2$. Show that $\sqrt{n}(\bar{X} - \mu)/S \xrightarrow{L} Z$, where $Z \sim \mathcal{N}(0, 1)$.

6. Let $X_1, X_2, \cdots, X_{100}$ be iid rvs with mean 75 and variance 225. Use Chebyshev's inequality to calculate the probability that the sample mean will not differ from the population mean by more than 6. Then use the CLT to calculate the same probability, and compare your results.

7. Let $X \sim b(n, \theta)$. Use the CLT to find n such that $P_\theta\{X > n/2\} \geq 1 - \alpha$. In particular, let $\alpha = .10$ and $\theta = .45$. Calculate n, satisfying $P\{X > n/2\} \geq .90$.

8. Let $X_1, X_2, \cdots, X_{100}$ be iid $P(\lambda)$ rv's, where $\lambda = .02$. Let $S = S_{100} = \sum_{i=1}^{100} X_i$. Use the central limit result to evaluate $P\{S \geq 3\}$, and compare your result to the exact probability of the event $S \geq 3$.

9. Let $X_1, X_2, \cdots, X_{81}$ be iid rvs with mean 54 and variance 225. Use Chebychev's inequality to find the possible difference between the sample mean and the population mean with a probability of at least .75. Also use the CLT to do the same.

10. Let $X_1, X_2, \cdots$ be a sequence of iid rv's with mean μ and variance σ^2, and assume that $EX_1^4 < \infty$. Write $V_n = \sum_{k=1}^n (X_k - \mu)^2$. Find the centering and norming constants A_n and B_n such that $B_n^{-1}(V_n - A_n) \xrightarrow{L} Z$, where Z is $\mathcal{N}(0, 1)$.

11. From an urn containing 10 identical balls numbered 0 through 9, n balls are drawn with replacement.

(a) What does the law of large numbers tell you about the appearance of 0's in the n drawings?

(b) How many drawings must be made in order that, with probability at least .95, the relative frequency of the occurrence of 0's will be between .09 and .11?

(c) Use the CLT to find the probability that among the n numbers thus chosen the number 5 will appear between $(n - 3\sqrt{n})/10$ and $(n + 3\sqrt{n})/10$ times (inclusive) if (i) $n = 25$, (ii) $n = 100$.

12. Let $X_1, X_2, \cdots, X_n$ be iid rv's with $EX_1 = 0$ and $EX_1^2 = \sigma^2 < \infty$. Let $\bar{X} = \sum_{i=1}^{n} X_i/n$, and for any positive real number ε let $P_{n,\varepsilon} = P\{\bar{X} \geq \varepsilon\}$. Show that

$$P_{n,\varepsilon} \approx \frac{\sigma}{\varepsilon\sqrt{n}} \frac{1}{\sqrt{2\pi}} e^{-n\varepsilon^2/2\sigma^2} \quad \text{as} \quad n \to \infty.$$

[*Hint:* Use (5.3.68).]

13. Prove Lemmas 1 through 4.

CHAPTER 7

Sample Moments and Their Distributions

7.1 INTRODUCTION

In the preceding chapters we discussed fundamental ideas and techniques of probability theory. In this development we created a mathematical model of a random experiment by associating with it a sample space in which random events correspond to sets of a certain σ-field. The notion of probability defined on this σ-field corresponds to the notion of uncertainty in the outcome on any performance of the random experiment.

In this chapter we begin the study of some problems of mathematical statistics. The methods of probability theory learned in preceding chapters will be used extensively in this study.

Suppose that we seek information about some numerical characteristics of a collection of elements, called a *population*. For reasons of time or cost we may not wish or be able to study each individual element of the population. Our object is to draw conclusions about the unknown population characteristics on the basis of information on some characteristics of a suitably selected *sample*. Formally, let X be a random variable which describes the population under investigation, and let F be the df of X. There are two possibilities. Either X has a df F_θ with a known functional form (except perhaps for the parameter θ, which may be a vector), or X has a df F about which we know nothing (except perhaps that F is, say, absolutely continuous). In the former case let Θ be the set of possible values of the unknown parameter θ. Then the job of a statistician is to decide, on the basis of a suitably selected sample, which member or members of the family $\{F_\theta, \theta \in \Theta\}$ can represent the df of X. Problems of this type are called problems of *parametric statistical inference* and will be the subject of investigation in Chapters 8 through 12. The case in which nothing is known about the functional form of the df F of X is clearly much more difficult. Inference problems of this type fall into the domain of *nonparametric statistics* and will be discussed in Chapter 13.

To be sure, the scope of statistical methods is much wider than the statistical inference problems discussed in this book. Statisticians, for example, deal with problems of planning and designing experiments, of collecting information, and of deciding how best the collected information should be used. However, here we concern ourselves only with the best methods of making inferences about probability distributions.

In Section 2 of this chapter we introduce the notions of a (simple) *random sample* and *sample statistics*. In Section 3 we study sample moments and their distributions, and in Section 4 we consider some important distributions that arise in sampling from a normal population. Sections 5 and 6 are devoted to the study of sampling from univariate and bivariate normal distributions.

7.2 RANDOM SAMPLING

Consider a statistical experiment that culminates in outcomes x, which are the values assumed by an rv X. Let F be the df of X. In practice, F will not be completely known, that is, one or more parameters associated with F will be unknown. The job of a statistician is to estimate these unknown parameters or to test the validity of certain statements about them. He can obtain n independent observations on X. This means that he observes n values x_1, x_2, $\cdots$, x_n assumed by the rv X. Each x_i can be regarded as the value assumed by an rv X_i, $i = 1, 2, \cdots, n$, where X_1, X_2, $\cdots$, X_n are independent rv's with common df F. The observed values $(x_1, x_2, \cdots, x_n)$ are then values assumed by $(X_1, X_2, \cdots, X_n)$. The set $\{X_1, X_2, \cdots, X_n\}$ is then a *sample* of size n taken from a *population distribution* F. The set of n values $x_1, x_2, \cdots, x_n$ is called a *realization* of the sample. Note that the possible values of the random vector $(X_1, X_2, \cdots, X_n)$ can be regarded as points in $\mathscr{R}_n$, which may be called the *sample space*. In practice one observes not $x_1, x_2, \cdots, x_n$ but some function $f(x_1, x_2, \cdots, x_n)$. Then $f(x_1, x_2 \cdots, x_n)$ are values assumed by the rv $f(X_1, X_2, \cdots, X_n)$.

Let us now formalize these concepts.

Definition 1. Let X be an rv with df F, and let $X_1, X_2, \cdots, X_n$ be iid rv's with common df F. Then the collection $X_1, X_2, \cdots, X_n$ is known as a random sample of size n from the df F or simply as n independent observations on X.

If $X_1, X_2, \cdots, X_n$ is a random sample from F, the joint df is given by

$$(1) \qquad F^*(x_1, x_2, \cdots, x_n) = \prod_{i=1}^{n} F(x_i).$$

Definition 2. Let $X_1, X_2, \cdots, X_n$ be n independent observations on an rv X, and let $f: \mathscr{R}_n \to \mathscr{R}_k$ be a Borel-measurable function. Then the rv $f(X_1, X_2, \cdots, X_n)$ is called a (sample) statistic provided that it is not a function of any unknown parameter(s).

Two of the most commonly used statistics are defined as follows.

Definition 3. Let $X_1, X_2, \cdots, X_n$ be a random sample from a distribution function F. Then the statistic

$$(2) \qquad \bar{X} = n^{-1} S_n = \sum_{i=1}^{n} \frac{X_i}{n}$$

is called the sample mean, and the statistic

$$(3) \qquad S^2 = \sum_{1}^{n} \frac{(X_i - \bar{X})^2}{n-1} = \frac{\sum_{i=1}^{n} X_i^2 - n\bar{X}^2}{n-1}$$

is called the sample variance.

Remark 1. Whenever the word "sample" is used subsequently, it will mean "random sample."

Remark 2. It should be remembered that sample statistics $\bar{X}$, S^2 (and others that we will define later on) are random variables, while the population parameters μ, σ^2, and so on are fixed constants that may be unknown.

Remark 3. In (3) we divide by $n-1$ rather than n. The reason for this will become clear in the next section.

Remark 4. Other frequently occurring examples of statistics are sample order statistics $X_{(1)}, X_{(2)}, \cdots, X_{(n)}$ and their functions, as well as sample moments, which will be studied in the next section.

Example 1. Let $X \sim b(1, p)$, where p is possibly unknown. The df of X is given by

$$F(x) = p\,\varepsilon(x-1) + (1-p)\,\varepsilon(x), \qquad x \in \mathscr{R}.$$

Suppose that five independent observations on X are 0, 1, 1, 1, 0. Then 0, 1, 1, 1, 0 is a realization of the sample $X_1, X_2, \cdots, X_5$. The sample mean is

$$\bar{x} = \frac{0+1+1+1+0}{5} = .6,$$

SAMPLE CHARACTERISTICS AND THEIR DISTRIBUTIONS 299

which is the value assumed by the rv $\bar{X}$. The sample variance is

$$s^2 = \sum_{i=1}^{5} \frac{(x_i - \bar{x})^2}{5 - 1} = \frac{2(.6)^2 + 3(.4)^2}{4} = .3,$$

which is the value assumed by the rv S^2.

Example 2. Let $X \sim \mathcal{N}(\mu, \sigma^2)$, where μ is known but σ^2 is unknown. Let $X_1, X_2, \cdots, X_n$ be a sample from $\mathcal{N}(\mu, \sigma^2)$. Then, according to our definition, $\sum_{i=1}^{n} X_i / \sigma^2$ is not a statistic.

Suppose that five observations on X are $-.864, .561, 2.355, .582, -.774$. Then the sample mean is $.372$, and the sample variance is 1.648.

PROBLEMS 7.2

1. Let X be a $b(1, \frac{1}{2})$ rv, and consider all possible random samples of size 3 on X. Compute $\bar{X}$ and S^2 for each of the eight samples, and also compute the pmf's of $\bar{X}$ and S^2.

2. A die is rolled. Let X be the face value that turns up, and X_1, X_2 be two independent observations on X. Compute the pmf of $\bar{X}$.

3. Let $X_1, X_2, \cdots, X_n$ be a sample from some population. Show that

$$\max_{1 \leq i \leq n} |X_i - \bar{X}| < \frac{(n-1)S}{\sqrt{n}}$$

unless either all the n observations are equal or exactly $n-1$ of the X_j's are equal.
(Samuelson [109])

7.3 SAMPLE CHARACTERISTICS AND THEIR DISTRIBUTIONS

Let $X_1, X_2, \cdots, X_n$ be a sample from a population df F. In this section we consider some commonly used sample characteristics and their distributions.

Definition 1. Let $F_n^*(x) = n^{-1} \sum_{j=1}^{n} \varepsilon(x - X_j)$. Then $nF_n^*(x)$ is the number of X_k's $(1 \leq k \leq n)$ that are $\leq x$. $F_n^*(x)$ is called the sample (or empirical) distribution function.

If $X_{(1)}, X_{(2)}, \cdots, X_{(n)}$ is the order statistic for $X_1, X_2, \cdots, X_n$, then clearly

(1) $$F_n^*(x) = \begin{cases} 0 & \text{if } x < X_{(1)} \\ \dfrac{k}{n} & \text{if } X_{(k)} \leq x < X_{(k+1)} \quad (k = 1, 2, \cdots, n-1) \\ 1 & \text{if } x \geq X_{(n)}. \end{cases}$$

For fixed but otherwise arbitrary $x \in \mathcal{R}$, $F_n^*(x)$ itself is an rv. The following result is immediate.

Theorem 1. The rv $F_n^*(x)$ has the probability function

(2) $$P\left\{F_n^*(x) = \frac{j}{n}\right\} = \binom{n}{j}[F(x)]^j[1 - F(x)]^{n-j}, \quad j = 0, 1, \cdots, n,$$

with mean

(3) $$EF_n^*(x) = F(x)$$

and variance

(4) $$\text{var}(F_n^*(x)) = \frac{F(x)[1 - F(x)]}{n}.$$

Proof. Since $\varepsilon(x - X_j)$, $j = 1, 2, \cdots, n$, are iid rv's, each with pmf
$$P\{\varepsilon(x - X_j) = 1\} = P\{x - X_j \geq 0\} = F(x)$$
and
$$P\{\varepsilon(x - X_j) = 0\} = 1 - F(x),$$
their sum $nF_n^*(x)$ is a $b(n, p)$ rv, where $p = F(x)$. Relations (2), (3), and (4) follow immediately.

Corollary 1.
$$F_n^*(x) \xrightarrow{P} F(x) \quad \text{as } n \to \infty.$$

Corollary 2.
$$\frac{\sqrt{n}\,[F_n^*(x) - F(x)]}{\sqrt{F(x)[1 - F(x)]}} \xrightarrow{L} Z \quad \text{as } n \to \infty,$$
where Z is $\mathcal{N}(0, 1)$.

Corollary 1 follows from the WLLN, and Corollary 2 from the CLT. The convergence in Corollary 1 is for each value of x. It is possible to make a probability statement simultaneously for all x. We state the result without proof.

Theorem 2 (Glivenko-Cantelli Theorem). $F_n^*(x)$ converges uniformly to $F(x)$, that is, for $\varepsilon > 0$,
$$\lim_{n \to \infty} P\{\sup_{-\infty < x < \infty} |F_n^*(x) - F(x)| > \varepsilon\} = 0.$$

Proof. For a proof of Theorem 2 we refer to Fisz [32], page 391.

We next consider some typical values of the df $F_n^*(x)$, called *sample* statistics. Since $F_n^*(x)$ has jump points X_j, $j = 1, 2, \cdots, n$, it is clear that all moments of F_n^* exist. Let us write

(5) $$a_k = n^{-1} \sum_{j=1}^{n} X_j^k$$

for the moment of order k about 0. Here a_k will be called the *sample moment of order k*. In this notation

(6) $$a_1 = n^{-1} \sum_{j=1}^{n} X_j = \bar{X}.$$

The *sample central moments* are defined by

(7) $$b_k = n^{-1} \sum_{j=1}^{n} (X_j - a_1)^k = n^{-1} \sum_{j=1}^{n} (X_j - \bar{X})^k.$$

Clearly,

$$b_1 = 0 \quad \text{and} \quad b_2 = \left(\frac{n-1}{n}\right) S^2.$$

As mentioned earlier, we do not call b_2 the sample variance. S^2 will be referred to as the *sample variance* for reasons that will subsequently become clear. We have

(8) $$b_2 = a_2 - a_1^2.$$

For the mgf of df F_n^*, we have

(9) $$M^*(t) = n^{-1} \sum_{j=1}^{n} e^{tX_j}.$$

Similar definitions are made for sample moments of bivariate and multivariate distributions. For example, if $(X_1, Y_1), (X_2, Y_2), \cdots, (X_n, Y_n)$ is a sample from a bivariate distribution, we write

(10) $$\bar{X} = n^{-1} \sum_{j=1}^{n} X_j, \quad \bar{Y} = n^{-1} \sum_{j=1}^{n} Y_j$$

for the two sample means, and for the second-order sample central moments we write

(11) $$b_{20} = n^{-1} \sum_{j=1}^{n} (X_j - \bar{X})^2, \quad b_{02} = n^{-1} \sum_{j=1}^{n} (Y_j - \bar{Y})^2,$$
$$b_{11} = n^{-1} \sum_{j=1}^{n} (X_j - \bar{X})(Y_j - \bar{Y}).$$

Once again we write

(12) $$S_1^2 = (n-1)^{-1} \sum_{j=1}^{n} (X_j - \bar{X})^2, \quad S_2^2 = (n-1)^{-1} \sum_{j=1}^{n} (Y_j - \bar{Y})^2$$

for the two *sample variances*, and for the *sample covariance* we use the quantity

(13) $$S_{11} = (n-1)^{-1} \sum_{j=1}^{n} (X_j - \bar{X})(Y_j - \bar{Y}).$$

In particular, the *sample correlation coefficient* is defined by

(14) $$R = \frac{b_{11}}{\sqrt{b_{20}b_{02}}} = \frac{S_{11}}{S_1 S_2}.$$

It can be shown (Problem 4) that $|R| \leq 1$, the extreme values ± 1 can occur only when all sample points $(X_1, Y_1), \cdots, (X_n, Y_n)$ lie on a straight line.

One can also work out the formulas for the two lines of regression. Thus the *line of regression* of Y on X can be shown to be

(15) $$y - \bar{Y} = R \frac{S_2}{S_1} (x - \bar{X}),$$

and is called the *sample (linear) regression of Y on X*. RS_2/S_1 is called the *sample regression coefficient of Y on X*. Similar discussion applies to the sample regression line of X on Y.

The sample quantiles are defined in a similar manner. Thus, if $0 < p < 1$, the *sample quantile of order p*, denoted by Z_p, is the order statistic $X_{(r)}$, where

$$r = \begin{cases} np & \text{if } np \text{ is an integer,} \\ [np+1] & \text{if } np \text{ is not an integer.} \end{cases}$$

As usual, $[x]$ is the largest integer $\leq x$. Note that, if np is an integer, we can take any value between $X_{(np)}$ and $X_{(np)+1}$ as the pth sample quantile. Thus, if $p = \frac{1}{2}$ and n is even, we can take any value between $X_{(n/2)}$ and $X_{(n/2)+1}$, the two middle values, as the median. It is customary to take the average. Thus the sample median is defined as

(16) $$Z_{1/2} = \begin{cases} X_{((n+1)/2)} & \text{if } n \text{ is odd,} \\ \dfrac{X_{(n/2)} + X_{((n/2)+1)}}{2} & \text{if } n \text{ is even.} \end{cases}$$

Note that

$$\left[\frac{n}{2} + 1\right] = \left(\frac{n+1}{2}\right)$$

if n is odd.

Next we consider the moments of sample characteristics. In the following we write $EX^k = m_k$ and $E(X - \mu)^k = \mu_k$ for the kth-order population moments. Whenever we use m_k (or μ_k), it will be assumed to exist. Also, σ^2 represents the population variance.

SAMPLE CHARACTERISTICS AND THEIR DISTRIBUTIONS

Theorem 3. Let $X_1, X_2, \cdots, X_n$ be a sample from a population with df F. Then

(17) $$E\bar{X} = \mu,$$

(18) $$\text{var}(\bar{X}) = \frac{\sigma^2}{n},$$

(19) $$E(\bar{X})^3 = \frac{m_3 + 3(n-1)m_2\mu + (n-1)(n-2)\mu^3}{n^2},$$

and

(20) $$E(\bar{X})^4 = \frac{m_4 + 4(n-1)m_3\mu + 6(n-1)(n-2)m_2\mu^2 + 3(n-1)m_2^2}{n^3} + (n-1)(n-2)(n-3)\mu^4.$$

Proof. In view of Theorems 4.6.1 and 4.6.5, it suffices to prove (19) and (20). We have

$$\left(\sum_{j=1}^n X_j\right)^3 = \sum_{j=1}^n X_j^3 + 3 \sum_{\substack{j=1 \\ j \neq k}}^n \sum_{k=1}^n X_j^2 X_k + \sum_{\substack{j=1 \\ j \neq k, j \neq l \\ k \neq l}}^n \sum_{k=1}^n \sum_{l=1}^n X_j X_k X_l,$$

and (19) follows. Similarly

$$\left(\sum_{i=1}^n X_i\right)^4 = \left(\sum_{i=1}^n X_i\right)\left(\sum_{j=1}^n X_j^3 + 3 \sum_{j \neq k} \sum X_j^2 X_k + \sum_{j \neq k \neq l} \sum \sum X_j X_k X_l\right)$$
$$= \sum_{i=1}^n X_i^4 + 4 \sum_{j \neq k} X_j X_k^3 + 3 \sum_{j \neq k} X_j^2 X_k^2 + 6 \sum_{i \neq j \neq k} X_i^2 X_j X_k$$
$$+ \sum_{i \neq j \neq k \neq l} X_i X_j X_k X_l,$$

and (20) follows.

Theorem 4. For the third and fourth central moments of $\bar{X}$, we have

(21) $$\mu_3(\bar{X}) = \frac{\mu_3}{n^2},$$

(22) $$\mu_4(\bar{X}) = \frac{\mu_4}{n^3} + 3\frac{(n-1)\mu_2^2}{n^3}.$$

Proof. We have

$$\mu_3(\bar{X}) = E(\bar{X} - \mu)^3 = \frac{1}{n^3} E\left\{\sum_{i=1}^n (X_i - \mu)\right\}^3$$
$$= \frac{1}{n^3} \sum_{i=1}^n E(X_i - \mu)^3 = \frac{\mu_3}{n^2},$$

$$\mu_4(\bar{X}) = E(\bar{X} - \mu)^4 = \frac{1}{n^4} E\left\{\sum_{i=1}^{n}(X_i - \mu)\right\}^4$$

$$= \frac{1}{n^4} \sum_{i=1}^{n} E(X_i - \mu)^4 + \binom{4}{2}\frac{1}{n^4} \sum\sum_{i<j} E\{(X_i - \mu)^2 (X_j - \mu)^2\}$$

$$= \frac{\mu_4}{n^3} + \frac{3(n-1)}{n^3} \mu_2^2.$$

Theorem 5. For the moments of b_2, we have

(23) $$E(b_2) = \frac{(n-1)\sigma^2}{n},$$

(24) $$\text{var}(b_2) = \frac{\mu_4 - \mu_2^2}{n} - \frac{2(\mu_4 - 2\mu_2^2)}{n^2} + \frac{\mu_4 - 3\mu_2^2}{n^3},$$

(25) $$E(b_3) = \frac{(n-1)(n-2)}{n^2} \mu_3,$$

(26) $$E(b_4) = \frac{(n-1)(n^2 - 3n + 3)}{n^3} \mu_4 + \frac{3(n-1)(2n-3)}{n^3} \mu_2^2.$$

Proof. We have

$$Eb_2 = E\left\{\frac{1}{n}\sum_{1}^{n} X_i^2 - \frac{1}{n^2}\left(\sum_{i=1}^{n} X_i\right)^2\right\}$$

$$= m_2 - \frac{1}{n^2} E\left\{\sum_{i=1}^{n} X_i^2 + \sum\sum_{i\neq j} X_i X_j\right\}$$

$$= m_2 - \frac{1}{n^2} \{nm_2 + n(n-1)\mu^2\}$$

$$= \left(\frac{n-1}{n}\right)(m_2 - \mu^2).$$

Now

$$n^2 b_2^2 = \left[\sum_{i=1}^{n}(X_i - \mu)^2 - n(\bar{X} - \mu)^2\right]^2.$$

Writing $Y_i = X_i - \mu$, we see that $EY_i = 0$, $\text{var}(Y_i) = \sigma^2$, and $EY_i^4 = \mu_4$. We have

$$n^2 E b_2^2 = E\left(\sum_{1}^{n} Y_i^2 - n\bar{Y}^2\right)^2$$

$$= E\left[\sum_{i=1}^{n} Y_i^4 + \sum_{i\neq j} Y_i^2 Y_j^2 - \frac{2}{n}\left(\sum_{i\neq j} Y_i^2 Y_j^2 + \sum_{j=1}^{n} Y_j^4\right)\right.$$

$$\left. + \frac{1}{n^2}\left(3\sum_{i\neq j} Y_i^2 Y_j^2 + \sum_{1}^{n} Y_j^4\right)\right].$$

It follows that

$$n^2 Eb_2^2 = n\mu_4 + n(n-1)\sigma^4 - \frac{2}{n}[n(n-1)\sigma^4 + n\mu_4]$$
$$+ \frac{1}{n^2}[3n(n-1)\sigma^4 + n\mu_4]$$
$$= \left(n - 2 + \frac{1}{n}\right)\mu_4 + \left(n - 2 + \frac{3}{n}\right)(n-1)\mu_2^2 \quad (\mu_2 = \sigma^2).$$

Therefore

$$\text{var}(b_2) = Eb_2^2 - (Eb_2)^2$$
$$= \left(n - 2 + \frac{1}{n}\right)\frac{\mu_4}{n^2} + (n-1)\left(n - 2 + \frac{3}{n}\right)\frac{\mu_2^2}{n^2} - \left(\frac{n-1}{n}\right)^2 \mu_2^2$$
$$= \left(n - 2 + \frac{1}{n}\right)\frac{\mu_4}{n^2} + (n-1)(3-n)\frac{\mu_2^2}{n^3},$$

as asserted.

Relations (25) and (26) can be proved similarly.

Corollary 1. $ES^2 = \sigma^2$.

This is precisely the reason why we call S^2, and not b_2, the sample variance.

Corollary 2. $\text{var}(S^2) = \dfrac{\mu_4}{n} + \dfrac{3-n}{n(n-1)}\mu_2^2.$

The following result provides a justification for our definition of sample covariance.

Theorem 6. Let $(X_1, Y_1), (X_2, Y_2), \cdots, (X_n, Y_n)$ be a sample from a bivariate population with variances σ_1^2, σ_2^2 and covariance $\rho\sigma_1\sigma_2$. Then

(27) $\qquad ES_1^2 = \sigma_1^2, \qquad ES_2^2 = \sigma_2^2, \qquad \text{and} \qquad ES_{11} = \rho\sigma_1\sigma_2,$

where $S_1^2, S_2^2,$ and S_{11} are defined in (12) and (13).

Proof. It follows from Corollary 1 to Theorem 5 that $ES_1^2 = \sigma_1^2$ and $ES_2^2 = \sigma_2^2$. To prove that $ES_{11} = \rho\sigma_1\sigma_2$ we note that X_i is independent of $X_j(i \ne j)$ and $Y_j(i \ne j)$. We have

$$(n-1)ES_{11} = E\left\{\sum_{j=1}^{n}(X_j - \bar{X})(Y_j - \bar{Y})\right\}.$$

Now

$$E\{(X_j - \bar{X})(Y_j - \bar{Y})\}$$

$$= E\left\{X_j Y_j - X_j \frac{\sum_1^n Y_j}{n} - Y_j \frac{\sum_1^n X_j}{n} + \frac{\sum X_j \sum Y_j}{n^2}\right\}$$

$$= EXY - \frac{1}{n}[EXY + (n-1)EX\,EY] - \frac{1}{n}[EXY + (n-1)EX\,EY]$$

$$+ \frac{1}{n^2}[nEXY + n(n-1)EX\,EY]$$

$$= \frac{n-1}{n}(EXY - EX\,EY)$$

and it follows that

$$(n-1)ES_{11} = n\left(\frac{n-1}{n}\right)(EXY - EX\,EY),$$

that is,

$$ES_{11} = EXY - EX\,EY = \text{cov}(X, Y) = \rho\sigma_1\sigma_2,$$

as asserted.

We next turn our attention to the distributions of sample characteristics. Several possibilities exist. If the exact sampling distribution is required, the method of transformation described in Section 4.4. can be used. Sometimes the technique of mgf can be applied. Thus, if $X_1, X_2, \cdots, X_n$ is a random sample from a population distribution for which the mgf exists, the mgf of the sample mean $\bar{X}$ is given by

(28) $$M_{\bar{X}}(t) = \prod_{i=1}^n E e^{tX_i/n} = \left[M\left(\frac{t}{n}\right)\right]^n,$$

where M is the mgf of the population distribution. If $M_{\bar{X}}(t)$ has one of the known forms, it is possible to write the pdf of $\bar{X}$. Although this method has the obvious drawback that it applies only to distributions for which all moments exist, we will see in Section 7.4 its effectiveness in the important case of sampling from a normal population.

Example 1. Let $X_1, X_2, \cdots, X_n$ be a sample from a $G(\alpha, 1)$ distribution. We will compute the pdf of $\bar{X}$. We have

$$M_{\bar{X}}(t) = \left[M\left(\frac{t}{n}\right)\right]^n = \frac{1}{(1 - t/n)^{\alpha n}}, \quad \frac{t}{n} < 1,$$

so that $\bar{X}$ is a $G(\alpha n, 1/n)$ variate.

Example 2. Let $X_1, X_2, \cdots, X_n$ be a random sample from a uniform distribution on $(0, 1)$. Consider the geometric mean

SAMPLE CHARACTERISTICS AND THEIR DISTRIBUTIONS

$$Y_n = \left(\prod_{i=1}^{n} X_i\right)^{1/n}.$$

We have $\log Y_n = (1/n) \sum_{i=1}^{n} \log X_i$, so that $\log Y_n$ is the mean of $\log X_1$, $\cdots$, $\log X_n$.

The common pdf of $\log X_1, \cdots, \log X_n$ is

$$f(x) = \begin{cases} e^x & \text{if } x < 0, \\ 0 & \text{otherwise,} \end{cases}$$

which is the negative exponential distribution with parameter $\beta = 1$. We see that the mgf of $\log Y_n$ is given by

$$M(t) = \prod_{i=1}^{n} E e^{t \log X_i/n} = \frac{1}{(1 + t/n)^n},$$

and the pdf of $\log Y_n$ is given by

$$f^*(x) = \begin{cases} \dfrac{n^n}{\Gamma(n)}(-x)^{n-1} e^{nx}, & -\infty < x < 0, \\ 0, & \text{otherwise.} \end{cases}$$

It follows that Y_n has pdf

$$f_{Y_n}(y) = \begin{cases} \dfrac{n^n}{\Gamma(n)} y^{n-1}(-\log y)^{n-1}, & 0 < y < 1, \\ 0, & \text{otherwise.} \end{cases}$$

Example 3 (Hogben [49]). Let $X_1, X_2, \cdots, X_n$ be a random sample from a Bernoulli distribution with parameter p, $0 < p < 1$. Let $\bar{X}$ be the sample mean, and S^2 be the sample variance. We will find the pmf of S^2. Note that $S_n = \sum_{i=1}^{n} X_i = \sum_{i=1}^{n} X_i^2$, and that S_n is $b(n, p)$. Since

$$(n-1)S^2 = \sum_{i=1}^{n} X_i^2 - n(\bar{X})^2$$

$$= S_n - \frac{S_n^2}{n}$$

$$= \frac{S_n(n - S_n)}{n},$$

S^2 only assumes values of the form

$$t = \frac{i(n-i)}{n(n-1)}, \quad i = 0, 1, 2, \cdots, \left[\frac{n}{2}\right],$$

where $[x]$ is the largest integer $\leq x$. Thus

$$P\{S^2 = t\} = P\{nS_n - S_n^2 = i(n-i)\}$$

$$= P\left\{\left(S_n - \frac{n}{2}\right)^2 = \left(i - \frac{n}{2}\right)^2\right\}$$

$$= P\{S_n = i \text{ or } S_n = n - i\}$$
$$= \binom{n}{i} p^i (1-p)^{n-i} + \binom{n}{i} p^{n-i} (1-p)^i$$
$$= \binom{n}{i} p^i (1-p)^i \{(1-p)^{n-2i} + p^{n-2i}\}, \quad i < \left[\frac{n}{2}\right].$$

If n is even, $n = 2m$, say, where $m \geq 0$ is an interger, and $i = m$, then

$$P\left\{S^2 = \frac{m}{2(2m-1)}\right\} = 2\binom{2m}{m} p^m (1-p)^m.$$

In particular, if $n = 7$, $S^2 = 0, \frac{1}{7}, \frac{5}{21}$, and $\frac{2}{7}$ with probabilities $\{p^7 + (1-p)^7\}$, $7p(1-p)\{p^5 + (1-p)^5\}$, $21p^2(1-p)^2\{p^3 + (1-p)^3\}$, and $35p^3(1-p)^3$, respectively.

If $n = 6$, then $S^2 = 0, \frac{1}{6}, \frac{4}{15}$, and $\frac{3}{10}$ with probabilities $\{p^6 + (1-p)^6\}$, $6p(1-p)\{p^4 + (1-p)^4\}$, $15p^2(1-p)^2\{p^2 + (1-p)^2\}$, and $40p^3(1-p)^3$, respectively.

We have already considered the distribution of the sample quantiles in Section 4.5 and the distribution of the range $X_{(n)} - X_{(1)}$ in Example 4.4.10. It can be shown, without much difficulty, that the distribution of the sample median is given by

(29) $\quad f_r(y) = \dfrac{n!}{(r-1)!(n-r)!} [F(y)]^{r-1} [1 - F(y)]^{n-r} f(y) \quad \text{if } r = \dfrac{n+1}{2},$

where F and f are the population df and pdf, respectively. If $n = 2m$, and the median is taken as the average of $X_{(m)}$ and $X_{(m+1)}$, then

(30) $\quad f_r(y) = \dfrac{2(2m)!}{[(m-1)!]^2} \displaystyle\int_y^\infty [F(2y-v)]^{m-1} [1 - F(v)]^{m-1} f(2y-v) f(v) \, dv.$

Example 4. Let $X_1, X_2, \cdots, X_n$ be a random sample from $U(0, 1)$. Then the integrand in (30) is positive for the intersection of the regions $0 < 2y - v < 1$ and $0 < v < 1$. This gives $(v/2) < y < (v+1)/2$, $y < v$ and $0 < v < 1$. The shaded area in Fig. 1 gives the limits on the integral as

and
$$y < v < 2y \quad \text{if } 0 < y \leq \tfrac{1}{2},$$
$$y < v < 1 \quad \text{if } \tfrac{1}{2} < y < 1.$$

In particular, if $m = 2$, the pdf of the median, $(X_{(2)} + X_{(3)})/2$, is given by

$$f_r(y) = \begin{cases} 8y^2(3 - 4y) & \text{if } 0 < y \leq \tfrac{1}{2}, \\ 8(4y^3 - 9y^2 + 6y - 1) & \text{if } \tfrac{1}{2} < y < 1, \\ 0 & \text{otherwise.} \end{cases}$$

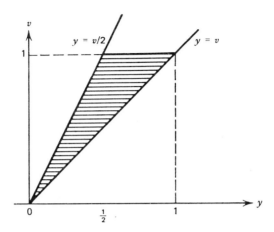

Fig. 1

If the asymptotic distribution is required, either the central limit theorem or the theorem of Cramér (Theorem 6.2.15) can be used.

Consider, for example, n independent observations $X_1, X_2, \cdots, X_n$ on an rv X. Let k be a positive integer, and assume that $E|X|^{2k} < \infty$. Then each X_i^k has mean m_k and variance $m_{2k} - (m_k)^2$. By the CLT it follows that, if a_k is the kth sample moment, then

$$(31) \qquad \frac{\sum X_i^k - nm_k}{\sqrt{n(m_{2k} - m_k)}} = \sqrt{n}\, \frac{a_k - m_k}{\sqrt{m_{2k} - m_k^2}} \xrightarrow{L} Z,$$

where Z is $\mathcal{N}(0, 1)$. Thus a_k is asymptotically normal $(m_k, (m_{2k} - m_k^2)/n)$. In particular, $\bar{X}$ is asymptotically $\mathcal{N}(\mu, \sigma^2/n)$.

In a similar manner, but with the help of a somewhat more difficult argument, we can show that the sample central moment of order k is also asymptotically normally distributed. We refer to Cramér [18], page 365, for details. Indeed, we can show under mild conditions that any sample characteristic based on (sample) moments is asymptotically normal with parameters that are identical with the corresponding population characteristic; see Cramér [18], pages 366–367.

Example 5. Let $X_1, X_2, \cdots$ be iid $\mathcal{N}(\mu, \sigma^2)$. Also, let $\bar{X}$ be the sample mean, and S^2 the sample variance. Consider the rv

$$T_n = \frac{\sqrt{n}(\bar{X} - \mu)/\sigma}{S/\sigma} = \frac{U_n}{V_n},$$

where U_n is $\mathcal{N}(0, 1)$. In Section 7.5 we will determine the distribution of T_n. Here we use Cramer's theorem to show that $T_n \xrightarrow{L} Z$, where Z is $\mathcal{N}(0, 1)$. From Example 6.3.4 we see that $(n-1)/n\, S^2 \xrightarrow{P} \sigma^2$, so that $(S/\sigma) \xrightarrow{P} 1$. It follows that $T_n \xrightarrow{L} Z$. See Problem 6.6.5 for a more general result.

The following result gives the asymptotic distribution of the rth-order statistic, $1 \leq r \leq n$, in sampling from a population with an absolutely continuous df F with pdf f.

Theorem 7. If $X_{(r)}$ denotes the rth-order statistic of a sample $X_1, X_2, \cdots, X_n$ from an absolutely continuous df F with pdf f, then

(32) $$\left\{\frac{n}{p(1-p)}\right\}^{1/2} f(\zeta_p) \{X_{(r)} - \zeta_p\} \xrightarrow{L} Z \quad \text{as } n \to \infty,$$

so that r/n remains fixed, $r/n = p$, where Z is $\mathcal{N}(0, 1)$, and ζ_p is the unique solution of $F(\zeta_p) = p$ (that is, ζ_p is the population quantile of order p assumed unique).

Proof. For the proof we refer to Problem 10.

Remark 1. The sample quantile of order p, Z_p, is asymptotically

$$\mathcal{N}\left(\zeta_p, \frac{1}{[f(\zeta_p)]^2} \frac{p(1-p)}{n}\right),$$

where ζ_p is the corresponding population quantile, and f is the pdf of the population distribution function. It also follows that $Z_p \xrightarrow{P} \zeta_p$.

PROBLEMS 7.3

1. Let $X_1, X_2, \cdots, X_n$ be random sample from a df F, and let $F_n^*(x)$ be the sample distribution function. Find cov $(F_n^*(x), F_n^*(y))$ for fixed real numbers x, y.

2. Let F_n^* be the empirical df of a random sample from df F. Show that
$$P\left\{|F_n^*(x) - F(x)| \geq \frac{\varepsilon}{2\sqrt{n}}\right\} \leq \frac{1}{\varepsilon^2} \quad \text{for all } \varepsilon > 0.$$

3. For the data of Example 7.2.2 compute the sample distribution function.

4. (a) Show that the sample correlation coefficient R satisfies $|R| \leq 1$ with equality if and only if all sample points lie on a straight line. (b) If we write $U_i = aX_i + b$ $(a \neq 0)$ and $V_i = cY_i + d$ $(c \neq 0)$, what is the sample correlation coefficient between the U's and the V's?

5. Derive (15) for the line of regression of Y on X, based on the sample $(x_1, y_1), (x_2, y_2), \cdots, (x_n, y_n)$.

6. Let $X_1, X_2, \cdots, X_n$ be a random sample from $\mathcal{N}(\mu, \sigma^2)$. Compute the first four sample moments of $\bar{X}$ about the origin and about the mean. Also compute the first four sample moments of S^2 about the mean.

7. Derive the pdf of the median as given in (29) and (30).

8. Let $U_{(1)}$, $U_{(2)}$, $\cdots$, $U_{(n)}$ be the order statistic of a sample of size n from $U(0, 1)$. Compute $E U_{(r)}^k$ for any $1 \le r \le n$ and integer $k(> 0)$. In particular, show that

$$EU_{(r)} = \frac{r}{n+1} \quad \text{and} \quad \text{var}(U_{(r)}) = \frac{r(n-r+1)}{(n+1)^2(n+2)}.$$

Show also that the correlation coefficient between $U_{(r)}$ and $U_{(s)}$ for $1 \le r < s \le n$ is given by $[r(n-s+1)/s(n-r+1)]^{1/2}$.

9. Let $X_1, X_2, \cdots, X_n$ be a sample from an absolutely continuous df F with pdf f. Show that

$$EX_{(r)} \approx F^{-1}\left(\frac{r}{n+1}\right) \quad \text{and} \quad \text{var}(X_{(r)}) \approx \frac{r(n-r+1)}{(n+1)^2(n+2)} \frac{1}{\{f[F^{-1}(r/n+1)]\}^2}.$$

[*Hint*: Let Y be an rv with mean μ, and φ be a Borel function such that $E\varphi(Y)$ exists. Expand $\varphi(Y)$ about the point μ by a Taylor series expansion, and use the fact that $F(X_{(r)}) = U_{(r)}$.]

10. Prove Theorem 7.

[*Hint*: For any real μ and $\sigma(> 0)$ compute the pdf of $(U_{(r)} - \mu)/\sigma$ and show that the standardized $U_{(r)}$, $(U_{(r)} - \mu)/\sigma$, is asymptotically $\mathcal{N}(0, 1)$ under the conditions of the theorem.]

11. Let $X_1, X_2, \cdots, X_n$ be n independent observations on X. Find the sampling distribution of $\bar{X}$, the sample mean, if (a) $X \sim P(\lambda)$, (b) $X \sim \mathscr{C}(1, 0)$, (c) $X \sim \chi^2(m)$.

12. Let $X_1, X_2, \cdots, X_n$ be a random sample from $G(\alpha, \beta)$. Let us write $Y_n = (\bar{X} - \alpha\beta)/\beta\sqrt{(\alpha/n)}$, $n = 1, 2, \cdots$.

(a) Compute the first four moments of Y_n, and compare them with the first four moments of the standard normal distribution.

(b) Compute the coefficients of skewness α_3 and of kurtosis α_4 for the rv's Y_n. (For definitions of α_3, α_4 see Problem 3.2.10.)

13. Let $X_1, X_2, \cdots, X_n$ be a random sample from $U[0, 1]$. Also, let $Z_n = (\bar{X} - .5)/\sqrt{(1/12n)}$. Repeat Problem 12 for the sequence Z_n.

14. Let $X_1, X_2, \cdots, X_n$ be a random sample from $P(\lambda)$. Find var (S^2), and compare it with var $(\bar{X})$. Note that $E\bar{X} = \lambda = ES^2$.
(*Hint*: Use Problem 3.2.9.)

15. Prove (25) and (26).

7.4. CHI-SQUARE, t-, AND F-DISTRIBUTIONS: EXACT SAMPLING DISTRIBUTIONS

In this section we investigate certain distributions that arise in sampling from a normal population. Let $X_1, X_2, \cdots, X_n$ be a sample from $\mathcal{N}(\mu, \sigma^2)$.

Then we know that $\bar{X} \sim \mathcal{N}(\mu, \sigma^2/n)$. Also, $\{\sqrt{n}(\bar{X} - \mu)/\sigma\}^2$ is $\chi^2(1)$. We will determine the distribution of S^2 in the next section. Here we mainly define chi-square, t-, and F-distributions and study their properties. Their importance will become evident in the next section and later in the testing of statistical hypotheses (Chapter 10).

The first distribution of interest is the *chi-square distribution*, defined in Chapter 5 as a special case of the gamma distribution. Let $n > 0$ be an integer. Then $G(n/2, 2)$ is a $\chi^2(n)$ rv. In view of Theorem 5.3.32 and Corollary 2 to Theorem 5.3.4, the following result holds.

Theorem 1. Let $X_1, X_2, \cdots, X_n$ be iid rv's, and let $S_n = \sum_{k=1}^{n} X_k$. Then

(a) $S_n \sim \chi^2(n) \Leftrightarrow X_1 \sim \chi^2(1)$,

and

(b) $X_1 \sim \mathcal{N}(0, 1) \Rightarrow \sum_{k=1}^{n} X_k^2 \sim \chi^2(n)$.

If X has a chi-square distribution with n d.f., we write $X \sim \chi^2(n)$. We recall that, if $X \sim \chi^2(n)$, its pdf is given by

(1) $$f(x) = \begin{cases} \dfrac{x^{n/2-1} e^{-x/2}}{2^{n/2} \Gamma(n/2)} & \text{if } x \geq 0, \\ 0 & \text{if } x < 0, \end{cases}$$

the mgf by

(2) $$M(t) = (1 - 2t)^{-n/2} \quad \text{for } t < \tfrac{1}{2},$$

and the mean and the variance by

(3) $$EX = n, \quad \text{var}(X) = 2n.$$

The $\chi^2(n)$ distribution is tabulated for values of $n = 1, 2, \cdots$. Tables usually go up to $n = 30$, since for $n > 30$ it is possible to use the normal approximation.

Theorem 2 (Fisher). If $X \sim \chi^2(n)$, then

(4) $$\lim_{n \to \infty} P\{\sqrt{2X} - \sqrt{2n - 1} \leq z\} = \int_{-\infty}^{z} \frac{1}{\sqrt{2\pi}} e^{-y^2/2} \, dy.$$

Proof. Since X is the sum of n iid $\chi^2(1)$ rv's, we apply the CLT to see that

$$Z_n = \frac{X - n}{\sqrt{2n}}$$

is asymptotically normal, that is,
$$\lim_{n\to\infty} P\{Z_n \le z\} = \lim_{n\to\infty} P\left\{\frac{X-n}{\sqrt{2n}} \le z\right\} = \int_{-\infty}^{z} \frac{e^{-t^2/2}}{\sqrt{2\pi}}\, dt.$$

Then
$$\begin{aligned}
\lim_{n\to\infty} P\left\{\frac{X-n}{\sqrt{2n}} \le z\right\} &= \lim_{n\to\infty} P\left\{\frac{\chi^2(n) - n}{\sqrt{2n}} \le z + \frac{z^2 - 1}{2\sqrt{2n-1}}\right\} \\
&= \lim_{n\to\infty} P\left\{\frac{\chi^2(n) - n}{\sqrt{2n-1}} \le z + \frac{z^2 - 1}{2\sqrt{2n-1}}\right\} \\
&= \lim_{n\to\infty} P\left\{\chi^2(n) \le n + z\sqrt{2n-1} + \frac{z^2-1}{2}\right\} \\
&= \lim_{n\to\infty} P\{2\chi^2(n) \le 2n - 1 + 2z\sqrt{2n-1} + z^2\} \\
&= \lim_{n\to\infty} P\{\sqrt{2\chi^2(n)} \le z + \sqrt{2n-1}\}.
\end{aligned}$$

Remark 1. It follows that for $z > 0$
$$\begin{aligned}
P\{\chi^2(n) \le z\} &= P\{2\chi^2(n) \le 2z\} = P\{\sqrt{2\chi^2(n)} \le \sqrt{2z}\} \\
&= P\{\sqrt{2\chi^2(n)} - \sqrt{2n-1} \le \sqrt{2z} - \sqrt{2n-1}\}
\end{aligned}$$
(5)
$$\approx \int_{-\infty}^{\sqrt{2z} - \sqrt{2n-1}} \frac{e^{-t^2/2}}{\sqrt{2\pi}}\, dt.$$

We will write $\chi^2_{n,\alpha}$ for the upper α percent point of the $\chi^2(n)$ distribution, that is,
(6)
$$P\{\chi^2(n) > \chi^2_{n,\alpha}\} = \alpha,$$
and use approximation (4) or (5) for $n \ge 30$ since this provides a better approximation than the CLT. Table 3 on page 652 gives the values of $\chi^2_{n,\alpha}$ for some selected values of n and α.

Example 1. Let $n = 25$. Then, from Table 3, page 652,
$$P\{\chi^2(25) \le 34.382\} = .90.$$
Let us approximate this probability by normal tables using Theorem 2 and compare it with the approximation obtained from the CLT. From (5) we have
$$\begin{aligned}
P\{\chi^2(25) \le 34.382\} &\approx P\{Z \le \sqrt{68.764} - 7\} \\
&= P\{Z \le 1.29\} = .9015,
\end{aligned}$$
where Z is $\mathcal{N}(0, 1)$.

To use the CLT, we see that $E\chi^2(25) = 25$, $\operatorname{var} \chi^2(25) = 50$, so that

$$P\{\chi^2(25) \le 34.382\} = P\left\{\frac{\chi^2(25) - 25}{\sqrt{50}} \le \frac{34.382 - 25}{5\sqrt{2}}\right\}$$
$$\approx P\{Z \le 1.32\}$$
$$= .9066.$$

Definition 1. Let $X_1, X_2, \cdots, X_n$ be independent normal rv's with $EX_i = \mu_i$ and var $(X_i) = \sigma^2$, $i = 1, 2, \cdots, n$. Also, let $Y = \sum_{i=1}^n X_i^2/\sigma^2$. The rv Y is said to be a noncentral chi-square rv with noncentrality parameter $\sum_{i=1}^n \mu_i^2/\sigma^2$ and n d.f. We will write $Y \sim \chi^2(n, \delta)$, where $\delta = \sum_{i=1}^n \mu_i^2/\sigma^2$.

The pdf of a $\chi^2(n, \delta)$ rv can be shown to be

(7) $$f_n(y, \delta) = \begin{cases} \dfrac{1}{\sqrt{\pi}\, 2^{n/2}} \exp\left\{-\tfrac{1}{2}(\delta + y)\right\} y^{(n-2)/2} \\ \qquad \cdot \sum_{j=0}^{\infty} \dfrac{(\delta y)^j \, \Gamma(j + 1/2)}{(2j)!\, \Gamma(j + 1/2n)}, & y > 0, \\ 0, & y \le 0, \end{cases}$$

where $\delta = \sum_{i=1}^n \mu_i^2/\sigma^2$. Letting $\delta = 0$, we see that (7) reduces to the central $\chi^2(n)$ pdf given in (1).

Although pdf (7) looks formidable, its mgf can easily be evaluated. We have

$$M(t) = E e^{t \sum_1^n X_i^2/\sigma^2} = \prod_1^n E e^{t X_i^2/\sigma^2},$$

where $X_i \sim \mathcal{N}(\mu_i, \sigma^2)$. Thus

$$E e^{t X_i^2/\sigma^2} = \int_{-\infty}^{\infty} \frac{1}{\sigma\sqrt{2\pi}} \exp\left\{\frac{t x_i^2}{\sigma^2} - \frac{(x_i - \mu_i)^2}{2\sigma^2}\right\} dx_i,$$

where the integral exists for $t < \tfrac{1}{2}$. In the integrand we complete squares, and after some simple algebra we obtain

$$E e^{t X_i^2/\sigma^2} = \frac{1}{\sqrt{1 - 2t}} \exp\left\{\frac{t\mu_i^2}{\sigma^2(1 - 2t)}\right\}, \quad t < \tfrac{1}{2}.$$

It follows that

(8) $$M(t) = (1 - 2t)^{-n/2} \exp\left(\frac{t}{1 - 2t} \frac{\sum \mu_i^2}{\sigma^2}\right), \quad t < \tfrac{1}{2},$$

and the mgf of a $\chi^2(n, \delta)$ rv is therefore

(9) $$M(t) = (1 - 2t)^{-n/2} \exp\left(\frac{t}{1 - 2t} \delta\right), \quad t < \tfrac{1}{2}.$$

It is immediate that, if $X_1, X_2, \cdots, X_k$ are independent, $X_i \sim \chi^2(n_i, \delta_i)$, $i = 1, 2, \cdots, k$, then $\sum_{i=1}^k X_i$ is $\chi^2\left(\sum_{i=1}^k n_i, \sum_{i=1}^k \delta_i\right)$.

The mean and variance of $\chi^2(n, \delta)$ are easy to calculate. We have

$$EY = \frac{\sum_{1}^{n} EX_i^2}{\sigma^2} = \frac{\sum_{1}^{n} [\text{var}(X_i) + (EX_i)^2]}{\sigma^2}$$

$$= \frac{n\sigma^2 + \sum_{1}^{n} \mu_i^2}{\sigma^2} = n + \delta,$$

and

$$\text{var}(Y) = \text{var}\left(\frac{\sum_{1}^{n} X_i^2}{\sigma^2}\right) = \frac{1}{\sigma^4}\left[\sum_{i=1}^{n} \text{var}(X_i^2)\right]$$

$$= \frac{1}{\sigma^4}\left\{\sum_{i=1}^{n} EX_i^4 - \sum_{i=1}^{n} [E(X_i^2)]^2\right\}$$

$$= \frac{1}{\sigma^4}\left\{\sum_{i=1}^{n}(3\sigma^4 + 6\sigma^2\mu_i^2 + \mu_i^4) - \sum_{i=1}^{n}(\sigma^2 + \mu_i^2)^2\right\}$$

$$= \frac{1}{\sigma^4}(2n\sigma^4 + 4\sigma^2 \sum \mu_i^2) = 2n + 4\delta.$$

We next turn our attention to *Student's t-statistic,* which arises quite naturally in sampling from a normal population.

Definition 2. Let $X \sim \mathcal{N}(0, 1)$ and $Y \sim \chi^2(n)$, and let X and Y be independent. Then the statistic

(10) $$T = \frac{X}{\sqrt{Y/n}}$$

is said to have a *t*-distribution with n d.f. and we write $T \sim t(n)$.

Theorem 3. The pdf of T defined in (10) is given by

(11) $$f_n(t) = \frac{\Gamma[(n + 1)/2]}{\Gamma(n/2)\sqrt{n\pi}}(1 + t^2/n)^{-(n+1)/2}, \qquad -\infty < t < \infty.$$

Proof. The proof is left as an exercise.

Remark 2. For $n = 1$, T is a Cauchy rv. We will therefore assume that $n > 1$. For each n, we have a different pdf. Like the normal distribution, the t-distribution is important in the theory of statistics and hence is tabulated (Table 4, page 654).

Remark 3. The pdf $f_n(t)$ is symmetric in t, and $f_n(t) \to 0$ as $t \to +\infty$. For large n, the t-distribution is close to the normal distribution. Indeed,

$(1 + t^2/n)^{-(n+1)/2} \to e^{-t^2/2}$ as $n \to \infty$. For small n, however, T deviates considerably from the normal. In fact,

$$P\{|T| \geq t_0\} \geq P\{|Z| > t_0\}, \quad Z \sim \mathcal{N}(0, 1),$$

that is, there is more probability in the tail of the t-distribution than in the tail of the standard normal. In what follows we will write $t_{n,\alpha/2}$ for the value (Fig. 1) of T for which

(12) $$P\{|T| > t_{n,\alpha/2}\} = \alpha.$$

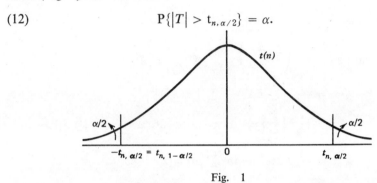

Fig. 1

In Table 4 on page 654 positive values of $t_{n,\alpha}$ are tabulated for some selected values of n and α. Negative values may be obtained from symmetry, $t_{n,1-\alpha} = -t_{n,\alpha}$.

Example 2. Let $n = 5$. Then from Table 4 on page 654, we get $t_{5,.025} = 2.571$ and $t_{5,.05} = 2.015$. The corresponding values under the $\mathcal{N}(0, 1)$ distribution are $z_{.025} = 1.96$ and $z_{.05} = 1.65$. For $n = 30$,

$$t_{30,.05} = 1.697 \quad \text{and} \quad z_{.05} = 1.65.$$

Theorem 4. Let $X \sim t(n)$, $n > 1$. Then EX^r exists for $r < n$. In particular, if $r < n$ is odd,

(13) $$EX^r = 0,$$

and if $r < n$ is even,

(14) $$EX^r = n^{r/2} \frac{\Gamma[(r+1)/2] \, \Gamma[(n-r)/2]}{\Gamma(1/2) \, \Gamma(n/2)}.$$

Corollary. If $n > 2$, $EX = 0$ and $EX^2 = \text{var}(X) = n/(n-2)$.

Remark 4. If in Definition 2 we take $X \sim \mathcal{N}(\mu, \sigma^2)$, $Y/\sigma^2 \sim \chi^2(n)$, and X and Y independent,

EXACT SAMPLING DISTRIBUTIONS

$$T = \frac{X}{\sqrt{Y/n}}$$

is said to have a *noncentral t-distribution with parameter* (also called *noncentrality parameter*) $\delta = \mu/\sigma$ and d.f. n. The pdf of a noncentral t-distribution is given by

(15) $$f_n(t, \delta) = \frac{n^{n/2} e^{-\delta^2/2}}{\sqrt{\pi}\, \Gamma(n/2)(n + t^2)^{(n+1)/2}} \sum_{s=0}^{\infty} \Gamma\left(\frac{n+s+1}{2}\right)\left(\frac{\delta^s}{s!}\right)\left(\frac{2t^2}{n+t^2}\right)^{s/2}.$$

Putting $\delta = 0$ in (15), we get the pdf $f_n(t)$ of a central t-distribution, given in (11).

It is easy to show (Problem 3) that, if T has a noncentral t-distribution with n d.f. and noncentrality parameter δ, then

(16) $$ET = \delta \frac{\Gamma[(n-1)/2]}{\Gamma(n/2)} \sqrt{\frac{n}{2}}, \quad n > 1,$$

and

(17) $$\text{var}(T) = \frac{n(1 + \delta^2)}{n - 2} - \frac{\delta^2 n}{2}\left(\frac{\Gamma[(n-1)/2]}{\Gamma(n/2)}\right)^2, \quad n > 2.$$

Definition 3. Let X and Y be independent χ^2 rv's with m and n d.f., respectively. The rv

(18) $$F = \frac{X/m}{Y/n}$$

is said to have an *F-distribution* with (m, n) d.f., and we write $F \sim F(m, n)$.

Theorem 5. The pdf of the F-statistic defined in (18) is given by

(19) $$g(f) = \begin{cases} \frac{\Gamma[(m+n)/2]}{\Gamma(m/2)\,\Gamma(n/2)}\left(\frac{m}{n}\right)\left(\frac{m}{n}f\right)^{(m/2)-1} \\ \qquad \cdot \left(1 + \frac{m}{n}f\right)^{-(m+n)/2}, & f > 0, \\ 0, & f \leq 0. \end{cases}$$

Proof. The proof is left as an exercise.

Remark 5. If $X \sim F(m, n)$, then $1/X \sim F(n, m)$. If we take $m = 1$, then $F = [t(n)]^2$, so that $F(1, n)$ and $t^2(n)$ have the same distribution. It also follows that, if Z is $\mathcal{C}(1, 0)$ [which is the same as $t(1)$], Z^2 is $F(1, 1)$.

Remark 6. As usual, we write $F_{m,n,\alpha}$ for the upper α percent point of the $F(m, n)$ distribution, that is,

(20) $$P\{F(m, n) > F_{m,n,\alpha}\} = \alpha.$$

From Remark 5, we have the following relation:

(21) $$F_{m,n,1-\alpha} = \frac{1}{F_{n,m,\alpha}}.$$

It therefore suffices to tabulate values of F that are ≥ 1. This is done in Table 5 on page 655, where values of $F_{m,n,\alpha}$ are listed for some selected values of m, n, and α.

Theorem 6. Let $X \sim F(m, n)$. Then, for $k > 0$, integral,

(22) $$EX^k = \left(\frac{n}{m}\right)^k \frac{\Gamma[k+(m/2)]\,\Gamma[(n/2)-k]}{\Gamma[(m/2)\,\Gamma(n/2)]} \quad \text{for } n > 2k.$$

In particular,

(23) $$EX = \frac{n}{n-2}, \quad n > 2,$$

and

(24) $$\operatorname{var}(X) = \frac{n^2(2m + 2n - 4)}{m(n-2)^2(n-4)}, \quad n > 4.$$

Proof. We have, for a positive integer k,

$$\int_0^\infty f^k f^{m/2-1} \left(1 + \frac{m}{n}f\right)^{-(m+n)/2} df$$

(25) $$= \left(\frac{n}{m}\right)^{k+(m/2)} \int_0^1 x^{k+(m/2)-1}(1-x)^{(n/2)-k-1}\,dx,$$

where we have changed the variable to $x = (m/n)f[1 + (m/n)f]^{-1}$. The integral in the right side of (25) converges for $(n/2) - k > 0$ and diverges for $(n/2) - k \leq 0$. We have

$$EX^k = \frac{\Gamma[(m+n)/2]}{\Gamma(m/2)\,\Gamma(n/2)} \left(\frac{m}{n}\right)^{m/2} \left(\frac{n}{m}\right)^{k+(m/2)} B\left(k+\frac{m}{2}, \frac{n}{2}-k\right),$$

as asserted.

For $k = 1$, we get

$$EX = \frac{n}{m}\,\frac{m/2}{(n/2)-1} = \frac{n}{n-2}, \quad n > 2.$$

Also,
$$EX^2 = \left(\frac{n}{m}\right)^2 \frac{(m/2)\,[(m/2)+1]}{[(n/2)-1]\,[(n/2)-2]}, \quad n > 4,$$
$$= \left(\frac{n}{m}\right)^2 \frac{m(m+2)}{(n-2)(n-4)},$$

and
$$\operatorname{var}(X) = \left(\frac{n}{m}\right)^2 \frac{m(m+2)}{(n-2)(n-4)} - \left(\frac{n}{n-2}\right)^2$$
$$= \frac{2n^2(m+n-2)}{m(n-2)^2(n-4)}, \quad n > 4.$$

Theorem 7. If $X \sim F(m, n)$, then $Y = 1/[1 + (m/n)X]$ is $B(n/2, m/2)$. Consequently, for each $x > 0$,
$$F_X(x) = 1 - F_Y\left[\frac{1}{1+(m/n)x}\right].$$

Proof. The proof is left as an exercise.

If in Definition 3 we take X to be a noncentral χ^2 rv with n d.f. and noncentrality parameter δ, we get a *noncentral F rv*.

Definition 4. Let $X \sim \chi^2(m, \delta)$ and $Y \sim \chi^2(n)$, and let X and Y be independent. Then the rv

$$(26) \qquad F = \frac{X/m}{Y/n}$$

is said to have a noncentral F-distribution with (m, n) d.f. and noncentrality parameter δ.

It can be shown that the pdf of F defined in (26) is given by

$$(27) \quad g(f; m, n, \delta) = \begin{cases} \dfrac{m^{m/2} n^{n/2}}{\Gamma(n/2)} e^{-\delta/2} f^{(m/2)-1} \\ \quad \cdot \sum_{j=0}^{\infty} \dfrac{(\delta mf/2)^j \, \Gamma[(m+n)/2+j]}{j!\,\Gamma[(m/2)+j]\,(mf+n)^{(m+n)/2+j}} \\ \qquad \qquad \qquad \qquad \text{if } f > 0, \\ 0 \qquad \qquad \qquad \quad \text{if } f \le 0. \end{cases}$$

Substituting $\delta = 0$, we get the central $F(m, n)$ pdf given in (19).

It is shown in Problem 2 that, if F has a noncentral F-distribution with (m, n) d.f. and noncentrality parameter δ,

$$EF = \frac{n(m + \delta)}{m(n - 2)}, \qquad n > 2,$$

and

$$\text{var}(F) = \frac{2n^2}{m^2(n - 4)(n - 2)^2} [(m + \delta)^2 + (n - 2)(m + 2\delta)], \qquad n > 4.$$

PROBLEMS 7.4

1. Let
$$P_x = \left\{ \Gamma\left(\frac{n}{2}\right) 2^{n/2} \right\}^{-1} \int_0^x \omega^{(n-2)/2} e^{-\omega/2} \, d\omega, \quad x > 0.$$
Show that
$$x < \frac{n}{1 - P_x}.$$

2. Let $X \sim F(m, n, \delta)$. Find EX and var(X).

3. Let T be a noncentral t-statistic with n d.f. and noncentrality parameter δ. Find ET and var(T).

4. Let $F \sim F(m, n)$. Then
$$Y = \left(1 + \frac{m}{n} F\right)^{-1} \sim B\left(\frac{n}{2}, \frac{m}{2}\right).$$
Deduce that for $x > 0$
$$P\{F \leq x\} = 1 - P\left\{Y \leq \left(1 + \frac{m}{n} x\right)^{-1}\right\}.$$

5. Derive the pdf of an F-statistic with (m, n) d.f.

6. Show that the square of a noncentral t-statistic is a noncentral F-statistic.

7. A sample of size 16 showed variance of 5.76. Find c such that $P\{|\bar{X} - \mu| < c\} = .95$, where $\bar{X}$ is the sample mean and μ is the population mean. Assume that the sample comes from a normal population.

8. A sample from a normal population produced variance 4.0. Find the size of the sample if the sample mean deviates from the population mean by tior more than 2.0 with a probability of at least .95.

9. Let X_1, X_2, X_3, X_4, X_5 be a sample from $\mathcal{N}(0,4)$. Find $P\{\sum_{i=1}^{5} X_i^2 \geq 5.75\}$.

10. Let $X \sim \chi^2(61)$. Find $P\{X > 50\}$.

11. Let $F \sim F(m, n)$. The random variable $Z = \frac{1}{2} \log F$ is known as Fisher's Z-statistic. Find the pdf of Z.

12. Prove Theorem 1.

13. Prove Theorem 3.

14. Prove Theorem 4.
15. Prove Theorem 5.

7.5 THE DISTRIBUTION OF $(\bar{X}, S^2)$ IN SAMPLING FROM A NORMAL POPULATION

Let $X_1, X_2, \cdots, X_n$ be a sample from $\mathcal{N}(\mu, \sigma^2)$, and write $\bar{X} = n^{-1} \sum_{i=1}^{n} X_i$ and $S^2 = (n-1)^{-1} \sum_{i=1}^{n} (X_i - \bar{X})^2$. In this section we show that $\bar{X}$ and S^2 are independent and derive the distribution of S^2. More precisely, we prove the following important result.

Theorem 1. Let $X_1, X_2, \cdots, X_n$ be iid $\mathcal{N}(\mu, \sigma^2)$ rv's. Then $\bar{X}$ and $(X_1 - \bar{X}, X_2 - \bar{X}, \cdots, X_n - \bar{X})$ are independent.

Proof. We compute the mgf of $\bar{X}$ and $X_1 - \bar{X}, X_2 - \bar{X}, \cdots, X_n - \bar{X}$ as follows:

$$\begin{aligned} M(t, t_1, t_2, \cdots, t_n) &= E \exp\{t\bar{X} + t_1(X_1 - \bar{X}) + t_2(X_2 - \bar{X}) + \cdots + t_n(X_n - \bar{X})\} \\ &= E \exp\left\{\sum_{i=1}^{n} t_i X_i - \left(\sum_{i=1}^{n} t_i - t\right)\bar{X}\right\} \\ &= E \exp\left\{\sum_{i=1}^{n} X_i\left(t_i - \frac{t_1 + t_2 + \cdots + t_n - t}{n}\right)\right\} \\ &= E\left[\prod_{i=1}^{n} \exp\left\{\frac{X_i(nt_i - n\bar{t} + t)}{n}\right\}\right] \quad \left(\text{where } \bar{t} = n^{-1} \sum_{i=1}^{n} t_i\right) \\ &= \prod_{i=1}^{n} E \exp\left\{\frac{X_i[t + n(t_i - \bar{t})]}{n}\right\} \\ &= \prod_{1}^{n} \exp\left\{\frac{\mu[t + n(t_i - \bar{t})]}{n} + \frac{\sigma^2}{2}\frac{1}{n^2}[t + n(t_i - \bar{t})]^2\right\} \\ &= \exp\left\{\frac{\mu}{n}\left[nt + n\sum_{i=1}^{n}(t_i - \bar{t})\right] + \frac{\sigma^2}{2n^2}\sum_{i=1}^{n}[t + n(t_i - \bar{t})]^2\right\} \\ &= \exp(\mu t) \exp\left\{\frac{\sigma^2}{2n^2}\left(nt^2 + n^2\sum_{i=1}^{n}(t_i - \bar{t})^2\right)\right\} \\ &= \exp\left(\mu t + \frac{\sigma^2}{2n}t^2\right) \exp\left\{\frac{\sigma^2}{2}\sum_{i=1}^{n}(t_i - \bar{t})^2\right\} \\ &= M_{\bar{X}}(t) M_{X_1 - \bar{X}, \cdots, X_n - \bar{X}}(t_1, t_2, \cdots, t_n) \\ &= M(t, 0, 0, \cdots, 0) M(0, t_1, t_2, \cdots, t_n). \end{aligned}$$

The result follows from Theorem 4.6.9.

Corollary 1. $\bar{X}$ and S^2 are independent.

Corollary 2. $(n - 1) S^2/\sigma^2$ is $\chi^2(n - 1)$.

Since

$$\sum_{i=1}^n \frac{(X_i - \mu)^2}{\sigma^2} \sim \chi^2(n), \quad n\left(\frac{\bar{X} - \mu}{\sigma}\right)^2 \sim \chi^2(1),$$

and $\bar{X}$ and S^2 are independent, it follows from

$$\sum_1^n \frac{(X_i - \mu)^2}{\sigma^2} = n\left(\frac{\bar{X} - \mu}{\sigma}\right)^2 + (n - 1)\frac{S^2}{\sigma^2}$$

that

$$E\left\{\exp\left[t \sum_1^n \frac{(X_i - \mu)^2}{\sigma^2}\right]\right\} = E\left\{\exp\left[tn\left(\frac{\bar{X} - \mu}{\sigma}\right)^2 + (n - 1)\frac{S^2}{\sigma^2} t\right]\right\}$$

$$= E \exp\left[n\left(\frac{\bar{X} - \mu}{\sigma}\right)^2 t\right] E \exp\left[(n - 1)\frac{S^2}{\sigma^2} t\right],$$

that is,

$$(1 - 2t)^{-n/2} = (1 - 2t)^{-1/2} E \exp\left[(n - 1)\frac{S^2}{\sigma^2} t\right], \quad t < \tfrac{1}{2},$$

and it follows that

$$E \exp\left[(n - 1)\frac{S^2}{\sigma^2} t\right] = (1 - 2t)^{-(n-1)/2}, \quad t < \tfrac{1}{2}.$$

By the uniqueness of the mgf it follows that $(n - 1) S^2/\sigma^2$ is $\chi^2(n - 1)$.

Corollary 3. The distribution of $\sqrt{n}(\bar{X} - \mu)/S$ is $t(n - 1)$.

Proof. Since $\sqrt{n}(\bar{X} - \mu)/\sigma$ is $\mathcal{N}(0, 1)$, and $(n - 1) S^2/\sigma^2 \sim \chi^2(n - 1)$ and since $\bar{X}$ and S^2 are independent,

$$\frac{\sqrt{n}(\bar{X} - \mu)/\sigma}{\sqrt{[(n - 1)S^2/\sigma^2]/(n - 1)}} = \frac{\sqrt{n}(\bar{X} - \mu)}{S} \text{ is } t(n - 1).$$

Corollary 4. If $X_1, X_2, \cdots, X_m$ are iid $\mathcal{N}(\mu_1, \sigma_1^2)$ rv's, $Y_1, Y_2, \cdots, Y_n$ are iid $\mathcal{N}(\mu_2, \sigma_2^2)$ rv's, and the two samples are independently taken, $(S_1^2/\sigma_1^2)/(S_2^2/\sigma_2^2)$ is $F(m - 1, n - 1)$. If, in particular, $\sigma_1 = \sigma_2$, then S_1^2/S_2^2 is $F(m - 1, n - 1)$.

Corollary 5. Let $X_1, X_2, \cdots, X_m$ and $Y_1, Y_2, \cdots, Y_n$, respectively, be independent samples from $\mathcal{N}(\mu_1, \sigma_1^2)$ and $\mathcal{N}(\mu_2, \sigma_2^2)$, Then

$$\frac{\bar{X} - \bar{Y} - (\mu_1 - \mu_2)}{\{[(m - 1) S_1^2/\sigma_1^2] + [(n - 1)S_2^2/\sigma_2^2]\}^{1/2}} \sqrt{\frac{m + n - 2}{\sigma_1^2/m + \sigma_2^2/n}} \sim t(m + n - 2).$$

In particular, if $\sigma_1 = \sigma_2$, then

$$\frac{\bar{X} - \bar{Y} - (\mu_1 - \mu_2)}{\sqrt{[(m-1)S_1^2 + (n-1)S_2^2]}} \sqrt{\frac{mn(m+n-2)}{m+n}} \sim t(m+n-2).$$

Corollary 5 follows since

$$\bar{X} - \bar{Y} \sim \mathcal{N}\left(\mu_1 - \mu_2, \frac{\sigma_1^2}{m} + \frac{\sigma_2^2}{n}\right) \quad \text{and}$$

$$\frac{(n-1)S_1^2}{\sigma_1^2} + \frac{(n-1)S_2^2}{\sigma_2^2} \sim \chi^2(m+n-2)$$

and the two statistics are independent.

Remark 1. The converse of Corollary 1 also holds. See Theorem 5.3.31.

Remark 2. In sampling from a symmetric distribution, $\bar{X}$ and S^2 are uncorrelated. See Problem 4.8.4.

Remark 3. Alternatively, Corollary 1 could have been derived from Corollary 2 to Theorem 5.4.6 by using the Helmert orthogonal matrix:

$$\mathbf{A} = \begin{bmatrix} 1/\sqrt{n} & 1/\sqrt{n} & 1/\sqrt{n} & \cdots & 1/\sqrt{n} \\ -1/\sqrt{2} & 1/\sqrt{2} & \cdot & \cdots & 0 \\ -1/\sqrt{6} & -1/\sqrt{6} & 2/\sqrt{6} & \cdots & 0 \\ \cdot & \cdot & \cdot & \cdots & \cdot \\ \cdot & \cdot & \cdot & \cdots & \cdot \\ \cdot & \cdot & \cdot & \cdots & \cdot \\ -1/\sqrt{n(n-1)} & -1/\sqrt{n(n-1)} & -1/\sqrt{n(n-1)} & \cdots & (n-1)/\sqrt{n(n-1)} \end{bmatrix}$$

For the case of $n = 3$ this was done in Example 4.4.8. In Problem 7 the reader is asked to work out the details in the general case.

Remark 4. An analytic approach to the development of the distribution of $\bar{X}$ and S^2 is as follows. Assuming without loss of generality that X_i is $\mathcal{N}(0, 1)$, we have as the joint pdf of $(X_1, X_2, \cdots, X_n)$

$$f(x_1, x_2, \cdots, x_n) = \frac{1}{(2\pi)^{n/2}} \exp\left\{-\tfrac{1}{2} \sum_{j=1}^n x_j^2\right\}$$

$$= \frac{1}{(2\pi)^{n/2}} \exp\left\{-\frac{(n-1)s^2 + n\bar{x}^2}{2}\right\}.$$

Changing the variables to $y_1, y_2, \cdots, y_n$ by using the transformation $y_k = (x_k - \bar{x})/s$, we see that

$$\sum_{k=1}^{n} y_k = 0 \quad \text{and} \quad \sum_{k=1}^{n} y_k^2 = n - 1.$$

It follows that two of the y_k's, say y_{n-1} and y_n, are functions of the remaining y_k. Thus either

$$y_{n-1} = \frac{\alpha + \beta}{2}, \quad y_n = \frac{\alpha - \beta}{2},$$

or

$$y_{n-1} = \frac{\alpha - \beta}{2}, \quad y_n = \frac{\alpha + \beta}{2},$$

where

$$\alpha = -\sum_{k=1}^{n-2} y_k \quad \text{and} \quad \beta = \sqrt{2(n-1) - 2\sum_{k=1}^{n-2} y_k^2 - \left(\sum_{k=1}^{n-2} y_k\right)^2}.$$

We leave the reader to derive the joint pdf of $(Y_1, Y_2, \cdots, Y_{n-2}, \bar{X}, S^2)$, using the result described in Remark 4.4.2, and to show that the rv's $\bar{X}, S^2$, and $(Y_1, Y_2, \cdots, Y_{n-2})$ are independent.

PROBLEMS 7.5

1. Let $X_1, X_2, \cdots, X_n$ be a random sample from $\mathcal{N}(\mu, \sigma^2)$, and $\bar{X}$ and S^2, respectively, be the sample mean and the sample variance. Let $X_{n+1} \sim \mathcal{N}(\mu, \sigma^2)$, and assume that $X_1, X_2, \cdots, X_n, X_{n+1}$ are independent. Find the sampling distribution of $[(X_{n+1} - \bar{X})/S]\sqrt{n/(n+1)}$

2. Let $X_1, X_2, \cdots, X_m$ and $Y_1, Y_2, \cdots, Y_n$ be independent random samples from $\mathcal{N}(\mu_1, \sigma^2)$ and $\mathcal{N}(\mu_2, \sigma^2)$, respectively. Also, let α, β be two fixed real numbers. If $\bar{X}, \bar{Y}$ denote the corresponding sample means, what is the sampling distribution of

$$\frac{\alpha(\bar{X} - \mu_1) + \beta(\bar{Y} - \mu_2)}{\sqrt{\frac{(m-1)S_1^2 + (n-1)S_2^2}{m+n-2}} \sqrt{\frac{\alpha^2}{m} + \frac{\beta^2}{n}}},$$

where S_1^2 and S_2^2, respectively, denote the sample variances of the X's and the Y's?

3. Let $X_1, X_2, \cdots, X_n$ be a random sample from $\mathcal{N}(\mu, \sigma^2)$, and k be a positive integer. Find $E(S^{2k})$. In particular, find $E(S^2)$ and var (S^2).

4. A random sample of 5 is taken from a normal population with mean 2.5 and variance $\sigma^2 = 36$.
(a) Find the probability that the sample variance lies between 30 and 44.
(b) Find the probability that the sample mean lies between 1.3 and 3.5, while the sample variance lies between 30 and 44.

5. The mean life of a sample of 10 light bulbs was observed to be 1327 hours with a standard deviation of 425 hours. A second sample of 6 bulbs chosen from a different batch showed a mean life of 1215 hours with a standard deviation of 375 hours. If the means of the two batches are assumed to be same, how probable is the observed difference between the two sample means?

6. Let S_1^2 and S_2^2 be the sample variances from two independent samples of sizes $n_1 = 5$ and $n_2 = 4$ from two populations having the same unknown variance σ^2. Find (approximately) the probability that $S_1^2/S_2^2 < 1/5.2$ or > 6.25.

7. Let $X_1, X_2, \cdots, X_n$ be a sample from $\mathcal{N}(\mu, \sigma^2)$. By using the Helmert orthogonal transformation defined in Remark 3, show that $\bar{X}$ and S^2 are independent.

8. Derive the joint pdf of $\bar{X}$ and S^2 by using the transformation described in Remark 4.

7.6 SAMPLING FROM A BIVARIATE NORMAL DISTRIBUTION

Let $(X_1, Y_1), (X_2, Y_2), \cdots, (X_n, Y_n)$ be a sample from a bivariate normal population with parameters $\mu_1, \mu_2, \rho, \sigma_1^2, \sigma_2^2$. Let us write

$$\bar{X} = n^{-1} \sum_{i=1}^{n} X_i, \qquad \bar{Y} = n^{-1} \sum_{i=1}^{n} Y_i,$$

$$S_1^2 = (n-1)^{-1} \sum_{i=1}^{n} (X_i - \bar{X})^2, \qquad S_2^2 = (n-1)^{-1} \sum_{i=1}^{n} (Y_i - \bar{Y})^2,$$

and

$$S_{11} = (n-1)^{-1} \sum_{i=1}^{n} (X_i - \bar{X})(Y_i - \bar{Y}).$$

In this section we show that $(\bar{X}, \bar{Y})$ is independent of (S_1^2, S_{11}, S_2^2) and obtain the distribution of the sample correlation coefficient and regression coefficients (at least in the special case where $\rho = 0$).

Theorem 1. The random vectors $(\bar{X}, \bar{Y})$ and $(X_1 - \bar{X}, X_2 - \bar{X}, \cdots, X_n - \bar{X}, Y_1 - \bar{Y}, Y_2 - \bar{Y}, \cdots, Y_n - \bar{Y})$ are independent. The joint distribution of $(\bar{X}, \bar{Y})$ is bivariate normal with parameters $\mu_1, \mu_2, \rho, \sigma_1^2/n, \sigma_2^2/n$.

Proof. The proof follows along the lines of the proof of Theorem 7.5.1. The mgf of $(\bar{X}, \bar{Y}, X_1 - \bar{X}, \cdots, X_n - \bar{X}, Y_1 - \bar{Y}, \cdots, Y_n - \bar{Y})$ is given by

$$M^* = M(u, v, t_1, t_2, \cdots, t_n, s_1, s_2, \cdots, s_n)$$
$$= E \exp\left\{u\bar{X} + v\bar{Y} + \sum_{i=1}^{n} t_i(X_i - \bar{X}) + \sum_{i=1}^{n} s_i(Y_i - \bar{Y})\right\}$$
$$= E \exp\left\{\sum_{i=1}^{n} X_i\left(\frac{u}{n} + t_i - \bar{t}\right) + \sum_{i=1}^{n} Y_i\left(\frac{v}{n} + s_i - \bar{s}\right)\right\},$$

where $\bar{t} = n^{-1} \sum_{i=1}^{n} t_i$, $\bar{s} = n^{-1} \sum_{i=1}^{n} s_i$. Therefore

$$M^* = \prod_{i=1}^{n} E \exp\left\{\left(\frac{u}{n} + t_i - \bar{t}\right) X_i + \left(\frac{v}{n} + s_i - \bar{s}\right) Y_i\right\}$$

$$= \prod_{i=1}^{n} \exp\left\{\left(\frac{u}{n} + t_i - \bar{t}\right)\mu_1 + \left(\frac{v}{n} + s_i - \bar{s}\right)\mu_2 \right.$$

$$+ \frac{\sigma_1^2 [(u/n) + t_i - \bar{t}]^2 + 2\rho\sigma_1\sigma_2 [(u/n) + t_i - \bar{t}][(v/n) + s_i - \bar{s}]}{2}$$

$$\left. + \frac{\sigma_2^2 [(v/n) + s_i - \bar{s}]^2}{2} \right\}$$

$$= \exp\left(\mu_1 u + \mu_2 v + \frac{u^2\sigma_1^2 + 2\rho\sigma_1\sigma_2 uv + v^2\sigma_2^2}{2n}\right)$$

$$\cdot \exp\left\{\tfrac{1}{2}\sigma_1^2 \sum_{i=1}^{n}(t_i - \bar{t})^2 + \rho\sigma_1\sigma_2 \sum_{i=1}^{n}(t_i - \bar{t})(s_i - \bar{s}) \right.$$

$$\left. + \tfrac{1}{2}\sigma_2^2 \sum_{i=1}^{n}(s_i - \bar{s})^2\right\}$$

$$= M_1(u, v, 0, 0, \cdots, 0) M_2(0, 0, t_1, t_2, \cdots, t_n, s_1, s_2, \cdots, s_n)$$

for all real $u, v, t_1, \cdots, t_n, s_1, \cdots s_n$ where M_1 is the mgf of $(\bar{X}, \bar{Y})$ and M_2 is the mgf of $(X_1 - \bar{X}, \cdots, X_n - \bar{X}, Y_1 - \bar{Y}, \cdots, Y_n - \bar{Y})$. Also, M_1 is the mgf of a bivariate normal distribution. This completes the proof.

Corollary. The sample mean vector $(\bar{X}, \bar{Y})$ is independent of the sample variance-covariance matrix $\begin{pmatrix} S_1^2 & S_{11} \\ S_{11} & S_2^2 \end{pmatrix}$ in sampling from a bivariate normal population.

Remark 1. The result of Theorem 1 can be generalized to the case of sampling from a k-variate normal population. We do not propose to do so here.

Remark 2. Unfortunately the method of proof of Theorem 1 does not lead to the distribution of the variance-covariance martrix. The distribution of $(\bar{X}, \bar{Y}, S_1^2, S_{11}, S_2^2)$ was found by Fisher [31] and Romanovsky [102]. The general case is due to Wishart [142], who determined the distribution of the sample variance-covariance matrix in sampling from a k-dimensional normal distribution. The distribution is named after him.

We will next compute the distribution of the sample correlation coefficient:

$$(1) \quad R = \frac{\sum_{i=1}^{n}(X_i - \bar{X})(Y_i - \bar{Y})}{\left\{\sum_{i=1}^{n}(X_i - \bar{X})^2 \sum_{i=1}^{n}(Y_i - \bar{Y})^2\right\}^{1/2}} = \frac{S_{11}}{S_1 S_2},$$

and that of the sample regression coefficient:

$$(2) \quad B_{Y|X} = \frac{\sum_{i=1}^{n}(X_i - \bar{X})(Y_i - \bar{Y})}{\sum_{i=1}^{n}(X_i - \bar{X})^2} = \frac{S_{11}}{S_1^2} = R\frac{S_2}{S_1}$$

of Y on X. The distribution of the regression coefficient of X on Y is similarly determined. Since we will need only the distribution of R and $B_{Y|X}$ whenever $\rho = 0$, we will make this simplfying assumption in what follows. The general case is computationally quite complicated. We refer the reader to Cramér [18] for details.

We note that

$$(3) \quad R = \frac{\sum_{i=1}^{n} Y_i(X_i - \bar{X})}{(n-1)S_1 S_2}$$

and

$$(4) \quad B_{Y|X} = \frac{\sum_{i=1}^{n} Y_i(X_i - \bar{X})}{(n-1)S_1^2}.$$

Moreover,

$$(5) \quad R^2 = \frac{B_{Y|X}^2 S_1^2}{S_2^2}.$$

In the following we write $B = B_{Y|X}$.

Theorem 2. Let $(X_1, Y_1), \cdots, (X_n, Y_n), n \geq 2$, be a sample from a bivariate normal population with parameters $EX = \mu_1$, $EY = \mu_2$, $\text{var}(X) = \sigma_1^2$, $\text{var}(Y) = \sigma_2^2$, and $\text{cov}(X, Y) = 0$. In other words, let $X_1, X_2, \cdots, X_n$ be iid $\mathcal{N}(\mu_1, \sigma_1^2)$ rv's, and $Y_1, Y_2, \cdots, Y_n$ be iid $\mathcal{N}(\mu_2, \sigma_2^2)$ rv's, and suppose that the X's and Y's are independent. Then the pdf of R is given by

$$(6) \quad f_1(r) = \begin{cases} \dfrac{\Gamma[(n-1)/2]}{\Gamma(\frac{1}{2})\,\Gamma[(n-2)/2]} (1 - r^2)^{(n-4)/2}, & -1 \leq r \leq 1, \\ 0, & \text{otherwise}; \end{cases}$$

and the pdf of B is given by

$$(7) \quad h_1(b) = \frac{\Gamma(n/2)}{\Gamma(\frac{1}{2})\,\Gamma[(n-1)/2]} \frac{\sigma_1 \sigma_2^{n-1}}{(\sigma_2^2 + \sigma_1^2 b^2)^{n/2}}, \quad -\infty < b < \infty.$$

Proof. Without any loss of generality, we assume that $\mu_1 = \mu_2 = 0$ and $\sigma_1^2 = \sigma_2^2 = 1$, for we can always define

(8) $$X_i^* = \frac{X_i - \mu_1}{\sigma_1} \quad \text{and} \quad Y_i^* = \frac{Y_i - \mu_2}{\sigma_2}.$$

Now note that the conditional distribution of Y_i, given $X_1, X_2, \cdots, X_n$, is $\mathcal{N}(0, 1)$, and $Y_1, Y_2, \cdots, Y_n$, given $X_1, X_2, \cdots, X_n$, are mutually independent. Let us define the following orthogonal transformation:

(9) $$u_i = \sum_{j=1}^n c_{ij} y_j, \quad i = 1, 2, \cdots, n,$$

where $((c_{ij}))_{i,j=1,2,\cdots,n}$ is an orthogonal matrix with the first two rows

(10) $$c_{1j} = \frac{1}{\sqrt{n}}, \quad j = 1, 2, \cdots, n,$$

(11) $$c_{2j} = \frac{x_j - \bar{x}}{\left\{\sum_{i=1}^n (x_i - \bar{x})^2\right\}^{1/2}}, \quad j = 1, 2, \cdots, n.$$

It follows from orthogonality that for any $i \geq 2$

(12) $$\sum_{j=1}^n c_{ij} = \sqrt{n} \sum_{j=1}^n c_{ij} \frac{1}{\sqrt{n}} = \sqrt{n} \sum_{j=1}^n c_{ij} c_{1j} = 0,$$

and

$$\sum_{i=1}^n u_i^2 = \sum_{i=1}^n \left(\sum_{j=1}^n c_{ij} y_j \sum_{j'=1}^n c_{ij'} y_{j'}\right)$$

$$= \sum_{j=1}^n \sum_{j'=1}^n \left(\sum_{i=1}^n c_{ij} c_{ij'}\right) y_j y_{j'}$$

(13) $$= \sum_{j=1}^n y_j^2.$$

Moreover,

(14) $$u_1 = \sqrt{n}\, \bar{y}$$

and

(15) $$u_2 = b \sqrt{\sum (x_i - \bar{x})^2},$$

where b is a value assumed by rv B. Also $U_1, U_2, \cdots, U_n$, given $X_1, X_2, \cdots, X_n$, are normal rv's (being linear combinations of the Y's). Thus

(16) $$E\{U_i | X_1, X_2, \cdots, X_n\} = \sum_{j=1}^n c_{ij} E\{Y_j | X_1, X_2, \cdots, X_n\} = 0$$

and

$$\text{cov}\{U_i, U_k | X_1, X_2, \cdots, X_n\} = \text{cov}\left\{\sum_{j=1}^n c_{ij} Y_j, \sum_{p=1}^n c_{kp} Y_p \Big| X_1, X_2, \cdots, X_n\right\}$$

$$= \sum_{j=1}^{n} \sum_{p=1}^{n} c_{ij} c_{kp} \operatorname{cov}\{Y_j, Y_p | X_1, X_2, \cdots, X_n\}$$

$$= \sum_{j=1}^{n} c_{ij} c_{kj}.$$

This last equality follows since

$$\operatorname{cov}\{Y_j, Y_p | X_1, X_2, \cdots, X_n\} = \begin{cases} 0, & j \neq p, \\ 1, & j = p. \end{cases}$$

From orthogonality, we have

(17) $$\operatorname{cov}\{U_i, U_k | X_1, X_2, \cdots, X_n\} = \begin{cases} 0, & i \neq k, \\ 1, & i = k; \end{cases}$$

and it follows that the rv's $U_1, U_2, \cdots, U_n$, given $X_1, X_2, \cdots, X_n$, are mutually independent $\mathcal{N}(0, 1)$. Now

$$\sum_{j=1}^{n}(y_j - \bar{y}^2) = \sum_{i=1}^{n} y_i^2 - n\bar{y}^2$$

$$= \sum_{j=1}^{n} u_j^2 - u_1^2$$

(18) $$= \sum_{j=2}^{n} u_j^2.$$

Thus

(19) $$R^2 = \frac{U_2^2}{\sum_{i=2}^{n} U_i^2} = \frac{U_2^2}{U_2^2 + \sum_{i=3}^{n} U_i^2}.$$

Writing $U = U_2^2$ and $W = \sum_{i=3}^{n} U_i^2$, we see that the conditional distribution of U, given $X_1, X_2, \cdots, X_n$, is $\chi^2(1)$, and that of W, given $X_1, X_2, \cdots, X_n$, is $\chi^2(n-2)$. Moreover U and W are independent. Since these conditional distributions do not involve the X's, we see that U and W are unconditionally independent with $\chi^2(1)$ and $\chi^2(n-2)$ distributions, respectively. The joint pdf of U and W is

$$f(u, w) = \frac{1}{\Gamma(\frac{1}{2})\sqrt{2}} u^{1/2-1} e^{-u/2} \frac{1}{\Gamma[(n-2)/2] \, 2^{(n-2)/2}} w^{(n-2)/2-1} e^{-w/2}.$$

Let $u + w = z$; then $u = r^2 z$ and $w = z(1 - r^2)$. The Jacobian of this transformation is z, so that the joint pdf of R^2 and Z is given by

$$f^*(r^2, z) = \frac{1}{\Gamma(\frac{1}{2}) \Gamma[(n-2)/2] \, 2^{(n-1)/2}} z^{n/2-3/2} e^{-z/2} (r^2)^{-1/2} (1 - r^2)^{n/2-2}.$$

The marginal pdf of R^2 is easily computed as

(20) $$f_1^*(r^2) = \frac{\Gamma[(n-1)/2]}{\Gamma(\frac{1}{2}) \Gamma[(n-2)/2]} (r^2)^{-1/2} (1 - r^2)^{n/2-2}, \quad 0 \leq r^2 \leq 1.$$

Finally, using Theorem 2.5.4, we get the pdf of R as

$$f_1(r) = \frac{\Gamma[(n-1)/2]}{\Gamma(\frac{1}{2})\,\Gamma[(n-2)/2]} (1-r^2)^{n/2-2}, \quad -1 \leq r \leq 1.$$

As for the distribution of B, note that the conditional pdf of $U_2 = \sqrt{n-1}\,BS_1$, given $X_1, X_2, \ldots, X_n$, is $\mathcal{N}(0, 1)$, so that the conditional pdf of B, given $X_1, X_2, \ldots, X_n$, is $\mathcal{N}(0, 1/\sum(x_i - \bar{x})^2)$. Let us write $\Lambda = (n-1)S_1^2$. Then the pdf of rv Λ is that of a $\chi^2(n-1)$ rv. Thus the joint pdf of B and Λ is given by

(21) $$h(b, \lambda) = g(b|\lambda)h_2(\lambda),$$

where $g(b|\lambda)$ is $\mathcal{N}(0, 1/\lambda)$, and $h_2(\lambda)$ is $\chi^2(n-1)$. We have

$$h_1(b) = \int_0^\infty h(b, \lambda)\,d\lambda$$

$$= \frac{1}{2^{n/2}\,\Gamma(\frac{1}{2})\,\Gamma[(n-1)/2]} \int_0^\infty \lambda^{n/2-1} e^{-\lambda/2(1+b^2)}\,d\lambda$$

(22) $$= \frac{\Gamma(n/2)}{\Gamma(\frac{1}{2})\,\Gamma[(n-1)/2]} \frac{1}{(1+b^2)^{n/2}}, \quad -\infty < b < \infty.$$

To complete the proof let us write

$$X_i = \mu_1 + X_i^* \sigma_1 \quad \text{and} \quad Y_i = \mu_2 + Y_i^* \sigma_2,$$

where $X_i^* \sim \mathcal{N}(0, 1)$ and $Y_i^* \sim \mathcal{N}(0, 1)$. Then $X_i \sim \mathcal{N}(\mu_1, \sigma_1^2)$, $Y_i \sim \mathcal{N}(\mu_2, \sigma_2^2)$, and

$$R = \frac{\sum_{i=1}^n (X_i^* - \bar{X}^*)(Y_i^* - \bar{Y}^*)}{\sqrt{\sum_{i=1}^n (X_i^* - \bar{X}^*)^2 \sum_{i=1}^n (Y_i^* - \bar{Y}^*)^2}}$$

(23) $$= R^*,$$

so that pdf of R is the same as derived above. Also

$$B = \frac{\sigma_1\sigma_2 \sum_{i=1}^n (X_i^* - \bar{X}^*)(Y_i^* - \bar{Y}^*)}{\sigma_1^2 \sum_{i=1}^n (X_i^* - \bar{X}^*)^2}$$

(24) $$= \frac{\sigma_2}{\sigma_1} B^*,$$

where the pdf of B^* is given by (22). Relations (22) and (24) are used to find the pdf of B. We leave the reader to carry out these simple details.

Remark 3. It is remarkable that for fixed n the sampling distribution of R is independent of $\mu_1, \mu_2, \sigma_1, \sigma_2$. In the general case where $\rho \neq 0$, one can show that for fixed n the distribution of R depends only on ρ. See, for example, Cramér [18], page 398.

Remark 4. Let us change the variable to

(25) $$T = \frac{R}{\sqrt{1-R^2}} \sqrt{n-2}.$$

Then

$$1 - R^2 = \left(1 + \frac{T^2}{n-2}\right)^{-1},$$

and the pdf of T is given by

(26) $$p(t) = \frac{1}{\sqrt{n-2}} \frac{1}{B[(n-2)/2, \frac{1}{2}]} \frac{1}{[1 + t^2/(n-2)]^{(n-1)/2}},$$

which is the pdf of a t-statistic with $n-2$ d.f. Thus T defined by (25) has a $t(n-2)$ distribution, provided that $\rho = 0$. This result facilitates the computation of probabilities under the pdf of R when $\rho = 0$.

Remark 5. To compute the pdf of $B_{X|Y} = R(S_1/S_2)$, the sample regression coefficient of X on Y, all we need to do is to interchange σ_1 and σ_2 in (7).

Remark 6. In the line of regression of Y on X, namely,

$$y - \bar{Y} = R \frac{S_2}{S_1} (x - \bar{X}),$$

we have already computed the pdf of the slope. We leave the reader to compute the distribution of the intercept

$$A = A_{Y|X} = \bar{Y} - B\bar{X}.$$

This is easily done if we note that $(\bar{X}, \bar{Y})$ has a bivariate normal distribution, B has the distribution given by (7), and $(\bar{X}, \bar{Y})$ and B are independent (by Theorem 1).

Remark 7. From (7) we can compute the mean and variance of B. For $n > 2$, clearly

$$EB = 0,$$

and for $n > 3$, we can show that

$$EB^2 = \text{var}(B) = \frac{\sigma_2^2}{\sigma_1^2} \frac{1}{n-3}.$$

Moreover,
$$EA = E\bar{Y} - EB\,E\bar{X} = \mu_2$$
and
$$\text{var}(A) = \frac{\sigma_2^2}{n}\frac{n-2}{n-3} + \frac{\sigma_2^2}{\sigma_1^2}\frac{\mu_1^2}{n-3}.$$

Similarly, we can use (6) to compute the mean and variance of R. We have, for $n > 4$,
$$ER = 0$$
and
$$ER^2 = \text{var}(R) = \frac{1}{n-1}.$$

PROBLEMS 7.6

1 Let $(X_1, Y_1), (X_2, Y_2), \cdots, (X_n, Y_n)$ be a random sample from a bivariate normal population with $EX = \mu_1$, $EY = \mu_2$, $\text{var}(X) = \text{var}(Y) = \sigma^2$, and $\text{cov}(X, Y) = \rho\sigma^2$. Let $\bar{X}, \bar{Y}$ denote the corresponding sample means, S_1^2, S_2^2, the corresponding sample variances, and S_{11}, the sample covariance. Write $R = 2S_{11}/(S_1^2 + S_2^2)$. Find the pdf of R.
[*Hint:* Let $U = (X + Y)/2$, and $V = (X - Y)/2$, and observe that the random vector (U, V) is also bivariate normal. In fact, U and V are independent.]

2. Compute the sampling distribution of the intercept
$$A = A_{Y|X} = \bar{Y} - B\bar{X},$$
assuming that $EX_i = 0$.

3. Let X and Y be independent normal rv's. A sample of $n = 11$ observations on (X, Y) produces sample correlation coefficient $r = .40$. Find the probability of obtaining a value of R that exceeds the observed value.

4. Let X_1, X_2 be jointly normally distributed with zero means, unit variances, and correlation coefficient ρ. Let S be a $\chi^2(n)$ rv that is independent of (X_1, X_2). Then the joint distribution of $Y_1 = X_1/\sqrt{S/n}$ and $Y_2 = X_2/\sqrt{S/n}$ is known as a *central bivariate t-distribution*. Find the joint pdf of (Y_1, Y_2) and the marginal pdf's of Y_1 and Y_2, respectively.

5. Let $(X_1, Y_1), \cdots, (X_n, Y_n)$ be a sample from a bivariate normal distribution with parameters $EX_i = \mu_1$, $EY_i = \mu_2$, $\text{var}(X_i) = \text{var}(Y_i) = \sigma^2$, and $\text{cov}(X_i, Y_i) = \rho\sigma^2$, $i = 1, 2, \cdots, n$. Find the distribution of the statistic
$$T(\mathbf{X}, \mathbf{Y}) = \sqrt{n}\,\frac{(\bar{X} - \mu_1) - (\bar{Y} - \mu_2)}{\sqrt{\sum_{i=1}^n (X_i - Y_i - \bar{X} + \bar{Y})^2}}.$$

CHAPTER 8

The Theory of Point Estimation

8.1 INTRODUCTION

In this chapter we study the theory of point estimation. Suppose, for example, that a random variable X is known to have a normal distribution $\mathcal{N}(\mu, \sigma^2)$, but we do not know one of the parameters, say μ. Suppose further that a sample $X_1, X_2, \cdots, X_n$ is taken on X. The problem of point estimation is to pick a (one-dimensional) statistic $T(X_1, X_2, \cdots, X_n)$ that best estimates the parameter μ. The numerical value of T when the realization is $x_1, x_2, \cdots, x_n$ is frequently called an *estimate* of μ, while the statistic T is called an *estimator* of μ. However, we will not observe this distinction and will use the word "estimate" for both the function T and its numerical value. If both μ and σ^2 are unknown, we seek a joint statistic $T = (U, V)$ as an estimate of (μ, σ^2).

In Sections 2 and 3 we do the preliminary work of formally describing the problem and investigate the desirable properties of an estimate that we may seek. Sections 4 and 5 deal with the theory of unbiased estimation. In Sections 6, 7, and 8 we discuss some commonly used methods of estimation, while in Section 9 we consider in detail the construction of a minimal sufficient statistic. The material of Section 9 properly belongs in Section 3, but we do not need it in Sections 4 through 8. Moreover, it is an important concept that merits special attention—hence the separation.

8.2 THE PROBLEM OF POINT ESTIMATION

Let $\mathbf{X}$ be an rv defined on a probability space $(\Omega, \mathscr{S}, P)$. Suppose that the df F of $\mathbf{X}$ depends on a certain number of parameters, and suppose further that the functional form of F is known except perhaps for a finite number of these parameters. Let $\boldsymbol{\theta}$ be the vector of (unknown) parameters associated with F.

Definition 1. The set of all admissible values of the parameters of a df F is called the parameter space.

We will write $F_{\boldsymbol{\theta}}$ for the df of $\mathbf{X}$ if $\boldsymbol{\theta}$ is the vector of parameters associated with the df of $\mathbf{X}$. Let $\boldsymbol{\theta}$ take values in the parameter set Θ. Then the set $\{F_{\boldsymbol{\theta}}: \boldsymbol{\theta} \in \Theta\}$ is called *the family of* df's of $\mathbf{X}$. Similarly we speak of *the family of pdf's* of $\mathbf{X}$ if $\mathbf{X}$ is continuous, and *the family of pmf's* if $\mathbf{X}$ is discrete.

Example 1. Let $X \sim b(n, p)$, and p be unknown. Then $\Theta = \{p: 0 < p < 1\}$ and $\{b(n, p): 0 < p < 1\}$ is the family of possible pmf's of X.

Example 2. Let $X \sim \mathcal{N}(\mu, \sigma^2)$. If both μ and σ^2 are unknown, $\Theta = \{(\mu, \sigma^2): -\infty < \mu < \infty, \sigma^2 > 0\}$. If $\mu = \mu_0$, say, and σ^2 is not known, $\Theta = \{(\mu_0, \sigma^2): \sigma^2 > 0\}$ or, simply, $\Theta = \{\sigma > 0\} = (0, \infty)$.

Let X be an rv with df $F_{\boldsymbol{\theta}}$, where $\boldsymbol{\theta} = (\theta_1, \theta_2, \cdots, \theta_k)$ is the vector of unknown parameters. Suppose that $\boldsymbol{\theta} \in \Theta$. Let $X_1, X_2, \cdots, X_n$ be a random sample of n iid observations on X. In this chapter we investigate the problem of approximating $\boldsymbol{\theta}$ on the basis of the available sample. In the following we restrict ourselves to the case where $\Theta \subseteq \mathcal{R}$ or to the case where we have to approximate some function d of Θ into $\mathcal{R}$.

Definition 2. Let $X_1, X_2, \cdots, X_n$ be a sample from F_{θ}, where $\theta \in \Theta \subseteq \mathcal{R}$. A statistic $T(X_1, X_2, \cdots, X_n)$ is said to be a (point) estimate of θ if T maps $\mathcal{R}_n$ into Θ.

The problem of parametric point estimation is to find an estimate T, for the unknown parmeter θ, that has some nice properties. We remark that we have already encountered a problem of point estimation in Section 4.8.

Example 3. Let $X_1, X_2, \cdots, X_n$ be a sample from $P(\lambda)$, where λ is not known. Then $\bar{X} = n^{-1} \sum_{k=1}^{n} X_k$ is an estimate of λ, and so also is $[2/n(n+1)] \sum_{i=1}^{n} iX_i$. Indeed, any X_i is an estimate of λ.

Example 4. Let $X_1, X_2, \cdots, X_n$ be a sample from $b(1, p)$, where p is unknown. Then $\bar{X} = n^{-1} \sum_{i=1}^{n} X_i$ is an estimate of p. Some other estimates are $T = \frac{1}{2}$, $T = X_1$, and $T = (X_1 + X_n)/2$.

From these examples it is clear that we need some criterion to choose among possible estimates. In the following sections we will consider some properties that we may require our estimates to possess and discuss some commonly used methods of estimation. We conclude this section by emphasizing that we demand in Definition 2 that T be a statistic (independent of any unknown parameters). We also remark that we do not consider here the much harder problem of (nonparametric) estimation when not even the functional form of F is known. (See Chapter 13.)

8.3 PROPERTIES OF ESTIMATES

We have seen that many estimates will be available to us in any given situation. It is therefore desirable to investigate some properties of point estimates. This will help us to decide which estimate to choose. We first consider the concept of *consistency*.

Definition 1. Let $X_1, X_2, \cdots$ be a sequence of iid rv's with common df $F_\theta, \theta \in \Theta$. A sequence of point estimates $T_n(X_1, X_2, \cdots, X_n) = T_n$ will be called consistent for θ if

$$T_n \xrightarrow{P} \theta \quad \text{as } n \to \infty$$

for each fixed $\theta \in \Theta$.

Remark 1. Recall that $T_n \xrightarrow{P} \theta$ if and only if $P\{|T_n - \theta| > \varepsilon\} \to 0$ as $n \to \infty$ for every $\varepsilon > 0$. One can similarly define *strong consistency* of a sequence of estimates T_n if $T_n \xrightarrow{a.s.} \theta$. Sometimes one speaks of *consistency in the rth mean* when $T_n \xrightarrow{r} \theta$. In what follows, "consistency" will mean *weak consistency* of T_n for θ, that is, $T_n \xrightarrow{P} \theta$.

Remark 2. It is important to remember that consistency is essentially a large-sample property. Moreover, we speak of consistency of a sequence of estimates rather than one point estimate.

Example 1. Let $X_1, X_2, \cdots$ be iid $b(1, p)$ rv's. Then $EX_i = p$, and it follows by the WLLN that

$$\frac{\sum_{1}^{n} X_i}{n} \xrightarrow{P} p.$$

Thus $\bar{X}$ is consistent for p. Also $(\sum_{1}^{n} X_i + 1)/(n + 2) \xrightarrow{P} p$, so that a consistent estimate need not be unique. Indeed, if $T_n \xrightarrow{P} p$, and $c_n \to 0$ as $n \to \infty$, then $T_n + c_n \xrightarrow{P} p$.

Theorem 1. If $X_1, X_2, \cdots$ are iid rv's with common law $\mathscr{L}(X)$, and $E|X|^p < \infty$ for some positive integer p, then

$$\frac{\sum_1^n X_i^k}{n} \xrightarrow{P} EX^k \quad \text{for } 1 \leq k \leq p,$$

and $n^{-1} \sum_1^n X_i^k$ is consistent for EX^k, $1 \leq k \leq p$. Moreover, if c_n is any sequence of constants such that $c_n \to 0$ as $n \to \infty$, then $\{n^{-1} \sum X_i^k + c_n\}$ is also consistent for EX^k, $1 \leq k \leq p$.

This is simply a restatement of the WLLN for iid rv's.

Example 2. Let $X_1, X_2, \cdots$ be iid $\mathscr{N}(\mu, \sigma^2)$ rv's. If S^2 is the sample variance, we know that $(n-1)S^2/\sigma^2 \sim \chi^2(n-1)$. Thus $E(S^2/\sigma^2) = 1$ and $\text{var}(S^2/\sigma^2) = 2/(n-1)$. It follows that

$$P\{|S^2 - \sigma^2| > \varepsilon\} \leq \frac{\text{var}(S^2)}{\varepsilon^2} = \frac{2\sigma^4}{(n-1)\varepsilon^2} \to 0 \quad \text{as } n \to \infty.$$

Thus $S^2 \xrightarrow{P} \sigma^2$. Actually, this result holds for any sequence of iid rv's with $E|X|^2 < \infty$ and can be obtained from Theorem 1. See Example 6.3.4.

Example 2 is a particular case of the following theorem.

Theorem 2. If T_n is a sequence of estimates such that $ET_n \to \theta$ and $\text{var}(T_n) \to 0$ as $n \to \infty$, then T_n is consistent for θ.

Proof. We have

$$P\{|T_n - \theta| > \varepsilon\} \leq \varepsilon^{-2} E\{T_n - ET_n + ET_n - \theta\}^2$$
$$= \varepsilon^{-2} \{\text{var}(T_n) + (ET_n - \theta)^2\} \to 0 \quad \text{as } n \to \infty.$$

We next consider the concept of *invariance*. Consider, for example, an experiment in which the length of life of a piece of equipment is measured. Then an estimate obtained from the measurements expressed in hours and minutes must agree with an estimate obtained from the measurements expressed in minutes. Similarly, if the relative speed of two cars is to be estimated, the estimates based on speed measured in miles per hour and in kilometers per hour must correspond.

We now formalize these notions. Let $\mathscr{G}$ be a group of Borel-measurable functions of $\mathscr{R}_n$ onto itself. The group operation is composition, that is, if g_1 and g_2 are mappings from $\mathscr{R}_n$ onto itself, $g_2 g_1$ is defined by $g_2 g_1(\mathbf{x}) = g_2(g_1(\mathbf{x}))$. Also, $\mathscr{G}$ is closed under composition and inverse, so that all maps in $\mathscr{G}$ are one-to-one and onto.

Definition 2. A family of probability distributions $\{P_\theta: \theta \in \Theta\}$ is said to be invariant under a group $\mathscr{G}$ if, for each $g \in \mathscr{G}$ and every $\theta \in \Theta$, we can find a unique $\theta' \in \Theta$ such that the distribution of $g(\mathbf{X})$ (which is an rv) is given by $P_{\theta'}$ whenever $\mathbf{X}$ has the distribution P_θ. We write $\theta' = \bar{g}\theta$.

Remark 3. Here $\mathbf{X} = (X_1, X_2, \cdots, X_n)$ is an n-dimensional rv so that $\mathscr{R}_n$ is the space of values of $\mathbf{X}$. P_θ is the distribution of $\mathbf{X}$ which determines the df of $g(\mathbf{X})$. (See Section 4.4.)

Remark 4. The condition that $\{P_\theta: \theta \in \Theta\}$ be invariant under $\mathscr{G}$ is the same as saying that

(1) $$P_\theta\{g(\mathbf{X}) \in A\} = P_{\bar{g}\theta}\{\mathbf{X} \in A\}$$

for all Borel subsets in $\mathscr{R}_n$. Equivalently,

(2) $$F_\theta^*(x_1, x_2, \cdots, x_n) = F_{\bar{g}\theta}(x_1, x_2, \cdots, x_n),$$

where F^* is the df of $g(X_1, X_2, \cdots, X_n)$ and F is the df of $(X_1, X_2, \cdots, X_n)$.

Example 3. Let $X \sim b(n, p)$, $0 \leq p \leq 1$. Let $\mathscr{G}$ be the group consisting of the identity map e and mappings g, where $g(x) = n - x$. Then $g(X)$ is $b(n, 1 - p)$, and we see that $\bar{g}p = 1 - p$. The group $\mathscr{G}$ leaves the family $\{b(n, p): 0 \leq p \leq 1\}$ invariant.

Example 4. Let $X_1, X_2, \cdots, X_n$ be a sample from $\mathscr{N}(\mu, \sigma^2)$. The joint pdf of $(X_1, X_2, \cdots, X_n)$ is given by

$$f(x_1, x_2, \cdots, x_n) = \frac{1}{(\sqrt{2\pi}\sigma)^n} \exp\left\{-\frac{1}{2\sigma^2} \sum_{i=1}^{n} (x_i - \mu)^2\right\}.$$

Consider the group of transformations $\mathscr{G}$, which contains elements g:

$$g(x_1, \cdots, x_n) = (ax_1 + b, \cdots, ax_n + b), \quad a > 0, \quad -\infty < b < \infty.$$

The pdf of $g(\mathbf{X})$ is given by

$$f^*(x_1, x_2, \cdots, x_n) = \frac{1}{(\sqrt{2\pi}\,a\sigma)^n} \exp\left\{-\frac{1}{2a^2\sigma^2} \sum_{i=1}^{n} (x_i - a\mu - b)^2\right\}.$$

We see that

$$\bar{g}(\mu, \sigma^2) = (a\mu + b, a^2\sigma^2), \quad -\infty < a\mu + b < \infty, \quad a^2\sigma^2 > 0.$$

Thus $\mathscr{G}$ leaves the family of pdf's $\{f: -\infty < \mu < +\infty, \sigma^2 > 0\}$ invariant.

Definition 3. Let $\mathscr{G}$ be a group of transformations that leaves the family $\{F_\theta: \theta \in \Theta\}$ of df's invariant. An estimate T is said to be invariant under $\mathscr{G}$ if

(3) $$T(g(X_1), g(X_2), \cdots, g(X_n)) = T(X_1, X_2, \cdots, X_n)$$
for all $g \in \mathcal{G}$.

The *principle of invariance* says that we should restrict attention to invariant estimates. A comprehensive investigation of the invariance principle and invariant estimates is beyond the scope of this book. We will consider here only some commonly used invariant estimates.

Definition 4. An estimate is said to be location invariant if
(4) $$T(X_1 + a, X_2 + a, \cdots, X_n + a) = T(X_1, X_2, \cdots, X_n), \quad a \in \mathcal{R};$$
scale invariant if
(5) $$T(cX_1, cX_2, \cdots, cX_n) = T(X_1, X_2, \cdots, X_n), \quad c \neq 0, \quad c \in \mathcal{R};$$
location and scale invariant if
(6) $$T(cX_1 + a, cX_2 + a, \cdots, cX_n + a) = T(X_1, X_2, \cdots, X_n),$$
$$c \neq 0, \quad a, c \in \mathcal{R};$$
and permutation invariant if
(7) $$T(X_{i_1}, X_{i_2}, \cdots, X_{i_n}) = T(X_1, X_2, \cdots, X_n)$$
for all permutations $(i_1, i_2, \cdots, i_n)$ of $1, 2, \cdots, n$.

Remark 5. In Definition 4 it is assumed that the family of distributions is left invariant by the group under consideration in the sense of Definition 2.

Example 5. Let $X_1, X_2, \cdots, X_n$ be iid $\mathcal{N}(\mu, \sigma^2)$ rv's. Then the family $\{\mathcal{N}(\mu, \sigma^2): -\infty < \mu < \infty, \sigma^2 > 0\}$ is invariant under translations. The estimate S^2 for σ^2 is invariant under this group since
$$T(g(X_1), g(X_2), \cdots, g(X_n)) = T(X_1 + a, \cdots, X_n + a)$$
$$= \frac{1}{n-1} \sum_1^n (X_i + a - \bar{X} - a)^2 = S^2$$
$$= T(X_1, X_2, \cdots, X_n).$$
But S^2 is not invariant under scale changes since
$$T(cX_1, cX_2, \cdots, cX_n) = \frac{c^2}{n-1} \sum_1^n (X_i - \bar{X})^2 = c^2 S^2.$$
Clearly the family of joint densities of $(X_1, X_2, \cdots, X_n)$ is also invariant under permutations. The estimate $\bar{X}$ is a permutation-invariant estimate of μ and S^2 is a permutation-invariant estimate of σ^2.

We next consider the important concept of a *sufficient statistic*. After the completion of any experiment, the job of a statistician is to interpret the data he has collected and to draw some statistically valid conclusions about the population under investigation. The raw data by themselves, besides being costly to store, are not suitable for this purpose. Therefore the statistician would like to condense the data by computing some statistics from them and to base his analysis on these statistics, provided that there is no loss of information in doing so. In many problems of statistical inference a function of the observations contains as much information about the unknown parameter as do all the observed values. The following example illustrates this point.

Example 6. Let $X_1, X_2, \cdots, X_n$ be a sample from $\mathcal{N}(\mu, 1)$, where μ is unknown. Suppose that we transform variables $X_1, X_2, \cdots, X_n$ to $Y_1, Y_2, \cdots, Y_n$ with the help of an orthogonal transformation so that Y_1 is $\mathcal{N}(\sqrt{n}\,\mu, 1)$, $Y_2, \cdots, Y_n$ are iid $\mathcal{N}(0, 1)$, and $Y_1, Y_2, \cdots, Y_n$ are independent. (Take $y_1 = \sqrt{n}\,\bar{x}$, and, for $k = 2, \cdots, n$, $y_k = [(k-1)x_k - (x_1 + \cdots + x_{k-1})] / \sqrt{k(k-1)}$.) To estimate μ we can use either the observed values of $X_1, X_2, \cdots, X_n$ or simply the observed value of $Y_1 = \sqrt{n}\,\bar{X}$. The rv's $Y_2, Y_3, \cdots, Y_n$ provide no information about μ. Clearly, Y_1 is preferable since one need not keep a record of all the observations; it suffices to cumulate the observations and compute y_1. Any analysis of the data based on y_1 is just as effective as any analysis that could be based on x_i's.

A rigorous definition of the concept involved in the above discussion requires the notion of a conditional distribution and is beyond the scope of this book. In view of the discussion of conditional probability distributions in Section 4.2, the following definition will suffice for our purposes.

Definition 5. Let $\mathbf{X} = (X_1, X_2, \cdots, X_n)$ be a sample from $\{F_\theta: \theta \in \Theta\}$. A statistic $T = T(\mathbf{X})$ is sufficient for θ or for the family of distributions $\{F_\theta: \theta \in \Theta\}$ if and only if the conditional distribution of X, given $T = t$, does not depend on θ (except perhaps for a null set A, $P_\theta\{T \in A\} = 0$ for all θ).

Remark 6. The outcome $X_1, X_2, \cdots, X_n$ is always sufficient, but we will exclude this trivial statistic from consideration. According to Definition 5, if T is sufficient for θ, we need only concentrate on T since it exhausts all the information that the sample has about θ. In practice, there will be several sufficient statistics for a family of distributions, and the question arises as to which of these should be used in a given problem. Fortunately we do not need to make this choice in the problems of elementary statistical inference considered in this book. We will return to this topic in more detail in Section 9.

Example 7. We show that the statistic Y_1 in Example 6 is sufficient for μ. By construction $Y_2, \cdots, Y_n$ are iid $\mathcal{N}(0, 1)$ rv's that are independent of Y_1. Hence the conditional distribution of $Y_2, \cdots, Y_n$, given $Y_1 = \sqrt{n}\,\bar{X}$, is the same as the unconditional distribution of $(Y_2, \cdots, Y_n)$, which is multivariate normal with mean $(0, 0, \cdots, 0)$ and dispersion matrix $\mathbf{I}_{n-1}$. Since this distribution is independent of μ, the conditional distribution of $(Y_1, Y_2, \cdots, Y_n)$, and hence $(X_1, X_2, \cdots, X_n)$, given $Y_1 = y_1$, is also independent of μ and Y_1 is sufficient. (See Problem 5.4.1.)

Example 8. Let $X_1, X_2, \cdots, X_n$ be iid $b(1, p)$ rv's. Intuitively, if a loaded coin is tossed with probability p of heads n times, it seems unnecessary to know which toss resulted in a head. To estimate p, it should be sufficient to know the number of heads in n trials. We show that this is consistent with our definition. Let $T(X_1, X_2, \cdots, X_n) = \sum_{i=1}^{n} X_i$. Then

$$P\left\{X_1 = x_1, \cdots, X_n = x_n \bigg| \sum_{i=1}^{n} X_i = t\right\} = \frac{P\{X_1 = x_1, \cdots, X_n = x_n, T = t\}}{\binom{n}{t} p^t (1-p)^{n-t}},$$

if $\sum_{1}^{n} x_i = t$, and $= 0$ otherwise. Thus, for $\sum_{1}^{n} x_i = t$, we have

$$P\{X_1 = x_1, \cdots, X_n = x_n | T = t\} = \frac{p^{\sum_{1}^{n} x_i}(1-p)^{n - \Sigma x_i}}{\binom{n}{t} p^t (1-p)^{n-t}}$$

$$= \frac{1}{\binom{n}{t}},$$

which is independent of p. It is therefore sufficient to concentrate on $\sum_{1}^{n} X_i$.

Example 9. Let X_1, X_2 be iid $P(\lambda)$ rv's. Then $X_1 + X_2$ is sufficient for λ, for

$$P\{X_1 = x_1, X_2 = x_2 | X_1 + X_2 = t\} = \begin{cases} \dfrac{P\{X_1 = x_1, X_2 = t - x_1\}}{P\{X_1 + X_2 = t\}} \\ \qquad \text{if } t = x_1 + x_2, x_i = 0, 1, 2, \cdots, \\ 0 \qquad \text{otherwise.} \end{cases}$$

Thus, for $x_i = 0, 1, 2, \cdots, i = 1, 2, x_1 + x_2 = t$, we have

$$P\{X_1 = x_1, X_2 = x_2 | X_1 + X_2 = t\} = \binom{t}{x_1}\left(\frac{1}{2}\right)^t,$$

which is independent of λ.

Not every statistic is sufficient.

Example 10. Let X_1, X_2 be iid $P(\lambda)$ rv's, and consider the statistic $T = X_1 + 2X_2$. We have

PROPERTIES OF ESTIMATES

$$P\{X_1 = 0, X_2 = 1 | X_1 + 2X_2 = 2\} = \frac{P\{X_1 = 0, X_2 = 1\}}{P\{X_1 + 2X_2 = 2\}}$$
$$= \frac{e^{-\lambda}(\lambda e^{-\lambda})}{P\{X_1 = 0, X_2 = 1\} + P\{X_1 = 2, X_2 = 0\}}$$
$$= \frac{\lambda e^{-2\lambda}}{\lambda e^{-2\lambda} + (\lambda^2/2) e^{-2\lambda}} = \frac{1}{1 + (\lambda/2)},$$

and we see that $X_1 + 2X_2$ is not sufficient for λ.

Definition 5 is not a constructive definition since it requires that we first guess a statistic T and then check to see whether T is sufficient. Moreover, the procedure for checking that T is sufficient is quite time-consuming. We now give a criterion for determining sufficient statistics.

Theorem 3 (**The Factorization Criterion**). Let $X_1, X_2, \cdots, X_n$ be discrete rv's with pmf $p_\theta(x_1, x_2, \cdots, x_n)$, $\theta \in \Theta$. Then $T(X_1, X_2, \cdots, X_n)$ is sufficient for θ if and only if we can write

(8) $p_\theta(x_1, x_2, \cdots, x_n) = h(x_1, x_2, \cdots, x_n) g_\theta(T(x_1, x_2, \cdots, x_n)),$

where h is a nonnegative function of $x_1, x_2, \cdots, x_n$ only and does not depend on θ, and g is a nonnegative function of θ and $T(x_1, x_2, \cdots, x_n)$ only. The statistic $T(X_1, \cdots, X_n)$ and parameter θ may be vectors.

Proof. Let T be sufficient for θ. Then $P\{\mathbf{X} = \mathbf{x} | T = t\}$ is independent of θ, and we may write

$$P_\theta\{\mathbf{X} = \mathbf{x}\} = P_\theta\{\mathbf{X} = \mathbf{x}, T(X_1, X_2, \cdots, X_n) = t\}$$
$$= P_\theta\{T = t\} P\{\mathbf{X} = \mathbf{x} | T = t\},$$

provided that $P\{\mathbf{X} = \mathbf{x} | T = t\}$ is well defined.

For values of $\mathbf{x}$ for which $P_\theta\{\mathbf{X} = \mathbf{x}\} = 0$ for all θ, let us define $h(x_1, x_2, \cdots, x_n) = 0$, and for $\mathbf{x}$ for which $P_\theta\{\mathbf{X} = \mathbf{x}\} > 0$ for some θ, we define

$$h(x_1, x_2, \cdots, x_n) = P\{X_1 = x_1, \cdots, X_n = x_n | T = t\}$$

and define

$$g_\theta(T(x_1, x_2, \cdots, x_n)) = P_\theta\{T(x_1, \cdots, x_n) = t\}.$$

Thus we see that (8) holds.

Conversely, suppose that (8) holds. Then for fixed t_0 we have

$$P_\theta\{T = t_0\} = \sum_{\{\mathbf{x}: T(\mathbf{x}) = t_0\}} P_\theta\{\mathbf{X} = \mathbf{x}\}$$
$$= \sum_{\{\mathbf{x}: T(\mathbf{x}) = t_0\}} g_\theta(T(\mathbf{x})) h(\mathbf{x})$$
$$= g_\theta(t_0) \sum_{T(\mathbf{x}) = t_0} h(\mathbf{x}).$$

Suppose that $P_\theta\{T = t_0\} > 0$ for some $\theta > 0$. Then

$$P_\theta\{X=x|T=t_0\} = \frac{P_\theta\{X=x, T(x)=t_0\}}{P_\theta\{T(x)=t_0\}} = \begin{cases} 0 & \text{if } T(x) \neq t_0, \\ \frac{P_\theta\{X=x\}}{P_\theta\{T(x)=t_0\}} & \text{if } T(x) = t_0. \end{cases}$$

Thus, if $T(x) = t_0$, then

$$\frac{P_\theta\{X = x\}}{P_\theta\{T(x) = t_0\}} = \frac{g_\theta(t_0) h(x)}{g_\theta(t_0) \sum_{T(x)=t_0} h(x)},$$

which is independent of θ, as asserted. This completes the proof.

Remark 7. Theorem 3 holds also for the continuous case and, indeed, for quite arbitrary families of distributions. The general proof is beyond the scope of this book, and we refer the reader to Halmos and Savage [44] or to Lehmann [70], pages 47–49. We will assume that the result holds for the absolutely continuous case. We leave the reader to write the analogue of (8) and to prove it, at least under the regularity conditions assumed in Theorem 4.4.6.

Remark 8. Theorem 3 (and its analogue for the continuous case) holds if θ is a vector of parameters and T is a random vector, and we say that T is *jointly sufficient* for θ. We emphasize that, even if θ is scalar, T may be a vector (Example 14). If θ and T are vectors of the same dimension, and if T is sufficient for θ, it does not follow that the *j*th component of T is sufficient for the *j*th component of θ (Example 13). The converse is true under mild conditions (see Fraser [33], 21).

Remark 9. If T is sufficient for θ, any one-to-one function of T is also sufficient. This follows from Theorem 3 since, if $U = k(T)$ is a one-to-one function of T, then $t = k^{-1}(u)$; and we can write

$$f_\theta(x) = g_\theta(t) h(x) = g_\theta(k^{-1}(u)) h(x) = g_\theta^*(u) h(x).$$

If T_1, T_2 are two distinct sufficient statistics, then

$$f_\theta(x) = g_\theta(t_1) h_1(x) = g_\theta(t_2) h_2(x),$$

and it follows that T_1 is a function of T_2. It does not follow, however, that every function of a sufficient statistic is itself sufficient. For example, in sampling from a normal population, $\bar{X}$ is sufficient for the mean μ, but $\bar{X}^2$ is not sufficient for μ. Note that $\bar{X}$ is sufficient for μ^2.

Remark 10. As a rule, Theorem 3 cannot be used to show that a given statistic T is not sufficient. To do this, one would normally have to use the definition of sufficiency. In most cases Theorem 3 will lead to a sufficient statistic if it exists. However, it does not answer the question of whether a family of densities (pmf's) admits a sufficient statistic (other than the trivial sufficient statistic).

Remark 11. If $T(\mathbf{X})$ is sufficient for $\{F_\theta: \theta \in \Theta\}$, then T is sufficient for $\{F_\theta: \theta \in \omega\}$, where $\omega \subseteq \Theta$. This follows trivially from the definition.

Example 11. Let $X_1, X_2, \cdots, X_n$ be iid $b(1, p)$ rv's. Then $T = \sum_{i=1}^{n} X_i$ is sufficient. We have

$$P_p\{X_1 = x_1, X_2 = x_2, \cdots, X_n = x_n\} = p^{\sum_1^n x_i}(1 - p)^{n - \sum_1^n x_i},$$

and, taking

$$h(x_1, x_2, \cdots, x_n) = 1 \quad \text{and} \quad g_p(x_1, x_2, \cdots, x_n) = (1 - p)^n \left(\frac{p}{1 - p}\right)^{\sum_{i=1}^n x_i},$$

we see that T is sufficient.

Example 12. Let $X_1, X_2, \cdots, X_n$ be iid rv's with common pmf

$$P\{X_i = k\} = \frac{1}{N}, \quad k = 1, 2, \cdots, N; \quad i = 1, 2, \cdots, n.$$

Then

$$P_N\{X_1 = k_1, X_2 = k_2, \cdots, X_n = k_n\} = \frac{1}{N^n} \quad \text{if } 1 \le k_1, \cdots, k_n \le N,$$

$$= \frac{1}{N^n} \varphi(1, \min_{1 \le i \le n} k_i) \varphi(\max_{1 \le i \le n} k_i, N),$$

where $\varphi(a, b) = 1$ if $b \ge a$, and $= 0$ if $b < a$. It follows, by taking $g_N[\max (k_1, \cdots, k_n)] = (1/N^n) \varphi(\max_{1 \le i \le n} k_i, N)$ and $h = \varphi(1, \min k_i)$, that $\max (X_1, X_2, \cdots, X_n)$ is sufficient for the family of joint pmf's P_N.

Example 13. Let $X_1, X_2, \cdots, X_n$ be a sample from $\mathcal{N}(\mu, \sigma^2)$, where both μ and σ^2 are unknown. The pdf of $(X_1, X_2, \cdots, X_n)$ is

$$f_{\mu,\sigma^2}(\mathbf{x}) = \frac{1}{(\sigma\sqrt{2\pi})^n} \exp\left\{-\frac{\sum(x_i - \mu)^2}{2\sigma^2}\right\}$$

$$= \frac{1}{(\sigma\sqrt{2\pi})^n} \exp\left(-\frac{\sum_1^n x_i^2}{2\sigma^2} + \frac{\mu \sum_1^n x_i}{\sigma^2} - \frac{n\mu^2}{2\sigma^2}\right).$$

It follows that the statistic

$$T(X_1, \cdots, X_n) = \left(\sum_1^n X_i, \sum_1^n X_i^2\right)$$

is jointly sufficient for the parameter (μ, σ^2). An equivalent sufficient statistic that is frequently used is $T_1(X_1, \cdots, X_n) = (\bar{X}, S^2)$. Note that $\bar{X}$ is not sufficient for μ if σ^2 is unknown, and S^2 is not sufficient for σ^2 if μ is unknown. If, however, σ^2 is known, $\bar{X}$ is sufficient for μ. If $\mu = \mu_0$ is known, $\sum_1^n (X_i - \mu_0)^2$ is sufficient for σ^2.

Example 14. Let $X_1, X_2, \cdots, X_n$ be a sample from pdf

$$f_\theta(x) = \begin{cases} \dfrac{1}{\theta}, & x \in \left[-\dfrac{\theta}{2}, \dfrac{\theta}{2}\right], \quad \theta > 0, \\ 0, & \text{otherwise.} \end{cases}$$

The joint pdf of $X_1, X_2, \cdots, X_n$ is given by

$$f_\theta(x_1, x_2, \cdots, x_n) = \frac{1}{\theta^n} I_A(x_1, \cdots, x_n),$$

where

$$A = \left\{(x_1, x_2, \cdots, x_n): -\frac{\theta}{2} \leq \min x_i \leq \max x_i \leq \frac{\theta}{2}\right\}.$$

It follows that $(\min_{1 \leq i \leq n} X_i, \max_{1 \leq i \leq n} X_i)$ is sufficient for θ.

The concept of sufficiency is frequently used with another concept, called *completeness*, which we now define.

Definition 6. Let $\{f_\theta(x), \theta \in \Theta\}$ be a family of pdf's (or pmf's). We say that this family is complete if

$$E_\theta g(X) = 0 \qquad \text{for all } \theta \in \Theta$$

implies

$$P_\theta\{g(X) = 0\} = 1 \qquad \text{for all } \theta \in \Theta.$$

Definition 7. A statistic $T(X)$ is said to be complete if the family of distributions of T is complete.

In Definition 7 X will usually be a random vector. The family of distributions of T is obtained from the family of distributions of $X_1, X_2, \cdots, X_n$ by the usual transformation technique discussed in Section 4.4.

Example 15. Let $X_1, X_2, \cdots, X_n$ be iid $b(1, p)$ rv's. Then $T = \sum_1^n X_i$ is a sufficient statistic. We show that T is also complete; that is, the family of distributions of T, $\{b(n, p), 0 < p < 1\}$, is complete.

$$E_p g(T) = \sum_{t=0}^{n} g(t) \binom{n}{t} p^t (1-p)^{n-t} = 0 \quad \text{for all } p \in (0, 1)$$

may be rewritten as

$$(1-p)^n \sum_{t=0}^{n} g(t) \binom{n}{t} \left(\frac{p}{1-p}\right)^t = 0 \quad \text{for all } p \in (0, 1).$$

This is a polynomial in $p/(1-p)$. Hence the coefficients must vanish, and it follows that $g(t) = 0$ for $t = 0, 1, 2, \cdots, n$, as required.

Example 16. Let X be $\mathcal{N}(0, \theta)$. Then the family of pdf's $\{\mathcal{N}(0, \theta), \theta > 0\}$ is not complete since $EX = 0$ and $g(x) = x$ is not identically zero. Note that $T(X) = X^2$ is complete, for the pdf of $X^2 \sim \theta \chi^2(1)$ is given by

$$f(t) = \begin{cases} \dfrac{e^{-t/2\theta}}{\sqrt{2\pi t \theta}}, & t > 0, \\ 0, & \text{otherwise.} \end{cases}$$

$$E_\theta g(T) = \frac{1}{\sqrt{2\pi \theta}} \int_0^\infty g(t) \, t^{-1/2} e^{-t/2\theta} \, dt = 0 \quad \text{for all } \theta > 0,$$

which holds if and only if $\int_0^\infty g(t) t^{-1/2} e^{-t/2\theta} \, dt = 0$, and using the uniqueness property of Laplace transforms (Theorem P. 2.27), it follows that

$$g(t) \, t^{-1/2} = 0 \quad \text{for } t > 0,$$

that is, $g(t) = 0$.

Example 17. Let $X_1, X_2, \cdots, X_n$ be a sample from $\mathcal{N}(\theta, \theta^2)$. Then $T = (\sum_1^n X_i, \sum_1^n X_i^2)$ is sufficient for θ. However, T is not complete since

$$E_\theta \left\{ 2\left(\sum_1^n X_i\right)^2 - (n+1) \sum_1^n X_i^2 \right\} = 0 \quad \text{for all } \theta,$$

and the function $g(x_1, \cdots, x_n) = 2(\sum_1^n x_i)^2 - (n+1) \sum_1^n x_j^2$ is not identically zero.

Example 18. Let $X \sim U(0, \theta)$, $\theta \in (0, \infty)$. We show that the family of pdf's of X is complete. We need to show that

$$E_\theta g(X) = \int_0^\theta \frac{1}{\theta} g(x) \, dx = 0 \quad \text{for all } \theta > 0$$

if and only if $g(x) = 0$ for all x. In general, this result follows from Lebesgue integration theory. If g is continuous, we differentiate both sides in

$$\int_0^\theta g(x) \, dx = 0$$

to get $g(\theta) = 0$ for all $\theta > 0$.

Now let $X_1, X_2, \cdots, X_n$ be iid $U(0, \theta)$ rv's. Also, let $M_n = \max(X_1, X_2, \cdots, X_n)$. Then the pdf of M_n is given by

$$f_n(x|\theta) = \begin{cases} n\theta^{-n} x^{n-1}, & 0 < x < \theta, \\ 0, & \text{otherwise.} \end{cases}$$

We see by a similar argument that M_n is complete, which is the same as saying that $\{f_n(x|\theta); \theta > 0\}$ is a complete family of densities. Clearly M_n is sufficient.

Example 19. Let $X_1, X_2, \cdots, X_n$ be a sample from pmf

$$P_N(x) = \begin{cases} \dfrac{1}{N}, & x = 1, 2, \cdots, N, \\ 0, & \text{otherwise.} \end{cases}$$

We first show that the family of pmf's $\{P_N, N \geq 1\}$ is complete. We have

$$E_N g(X) = \frac{1}{N} \sum_{k=1}^{N} g(k) = 0 \quad \text{for all } N \geq 1,$$

and this happens if and only if $g(k) = 0$, $k = 1, 2, \cdots, N$. Next we consider the family of pmf's of $M_n = \max(X_1, \cdots, X_n)$. The pmf of M_n is given by

$$P_N^{(n)}(x) = \frac{x^n}{N^n} - \frac{(x-1)^n}{N^n}, \quad x = 1, 2, \cdots, N.$$

Also,

$$E_N g(M_n) = \sum_{k=1}^{N} g(k) \left[\frac{k^n}{N^n} - \frac{(k-1)^n}{N^n} \right] = 0 \quad \text{for all } N \geq 1.$$

$$E_1 g(M_n) = g(1) = 0$$

implies $g(1) = 0$. Again,

$$E_2 g(M_n) = \frac{g(1)}{2^n} + g(2)\left(1 - \frac{1}{2^n}\right) = 0$$

so that $g(2) = 0$.

Using an induction argument, we conclude that $g(1) = g(2) = \cdots = g(N) = 0$ and hence $g(x) = 0$. It follows that $P_N^{(n)}$ is a complete family of distributions, and M_n is a complete sufficient statistic.

Now suppose that we exclude the value $N = n_0$ for some fixed $n_0 \geq 1$ from the family $\{P_N: N \geq 1\}$. Let us write $\mathscr{P} = \{P_N: N \geq 1, N \neq n_0\}$. Then $\mathscr{P}$ is not complete. We ask the reader to show that the class of all functions g such that $E_P g(X) = 0$ for all $P \in \mathscr{P}$ consists of functions of the form

$$g(k) = \begin{cases} 0, & k = 1, 2, \cdots, n_0 - 1, n_0 + 2, n_0 + 3, \cdots, \\ c, & k = n_0, \\ -c, & k = n_0 + 1, \end{cases}$$

where c is a constant, $c \neq 0$.

Remark 12. Completeness is a property of a family of distributions. In Remark 11 we saw that if a statistic is sufficient for a class of distributions it is sufficient for any subclass of those distributions. Completeness works in the opposite direction. Example 19 shows that the exclusion of even one member from the family $\{P_N: N \geq 1\}$ destroys completeness. If a family $\mathscr{P}$ is complete, it is sometimes possible to conclude completeness for a larger class, but we will not go into details here. See Fraser [33], page 25.

The following result covers a large class of probability distributions for which a complete sufficient statistic exists.

Theorem 4. Let $\{f_\theta: \theta \in \Theta\}$ be a k-parameter exponential family given by

(9) $$f_\theta(\mathbf{x}) = \exp\left\{\sum_{j=1}^{k} Q_j(\theta) T_j(\mathbf{x}) + D(\theta) + S(\mathbf{x})\right\},$$

where $\theta = (\theta_1, \theta_2, \cdots, \theta_k) \in \Theta$, an interval in $\mathscr{R}_k$, $T_1, T_2, \cdots, T_k$, and S are defined on $\mathscr{R}_n$, $\mathbf{T} = (T_1, T_2, \cdots, T_k)$, and $\mathbf{x} = (x_1, x_2, \cdots, x_n)$, $k \leq n$. Let $\mathbf{Q} = (Q_1, Q_2, \cdots, Q_k)$, and suppose that the range of $\mathbf{Q}$ contains an open set in $\mathscr{R}_k$. Then

$$\mathbf{T} = (T_1(\mathbf{X}), T_2(\mathbf{X}), \cdots, T_k(\mathbf{X}))$$

is a complete sufficient statistic.

Proof. For a complete proof in a general setting we refer the reader to Lehmann [70], pages 132–133. Essentially, the unicity of the Laplace transform is used on the probability distribution induced by $\mathbf{T}$. We will content ourselves here by proving the result for the $k = 1$ case when f_θ is a pmf.

Let us write $Q(\theta) = \theta$ in (9), and let $(\alpha, \beta) \subseteq \Theta$. We wish to show that

(10) $$\begin{aligned} E_\theta g(T(\mathbf{X})) &= \sum_t g(t) P_\theta\{T(\mathbf{X}) = t\} \\ &= \sum_t g(t) \exp\{\theta t + D(\theta) + S^*(t)\} = 0 \quad \text{for all } \theta \end{aligned}$$

implies that $g(t) = 0$. (Note that we have used Theorem 5.5.1 in next to the last equality.)

Let us write $x^+ = x$ if $x \geq 0$, $= 0$ if $x < 0$, and $x^- = -x$ if $x < 0$, $= 0$ if $x \geq 0$. Then $g(t) = g^+(t) - g^-(t)$, and both g^+ and g^- are nonnegative functions. In terms of g^+ and g^- (10) is the same as

(11) $$\sum_t g^+(t) e^{\theta t + S^*(t)} = \sum_t g^-(t) e^{\theta t + S^*(t)}$$

for all θ.

Let $\theta_0 \in (\alpha, \beta)$ be fixed, and write

(12) $$p^+(t) = \frac{g^+(t) e^{\theta_0 t + S^*(t)}}{\sum_t g^+(t) e^{\theta_0 t + S^*(t)}} \quad \text{and} \quad p^-(t) = \frac{g^-(t) e^{\theta_0 t + S^*(t)}}{\sum_t g^-(t) e^{\theta_0 t + S^*(t)}}.$$

Then both p^+ and p^- are pmf's, and it follows from (11) that

$$\text{(13)} \qquad \sum_t e^{\delta t} p^+(t) = \sum_t e^{\delta t} p^-(t)$$

for all $\delta \in (\alpha - \theta_0, \beta - \theta_0)$. By the uniqueness of mgf's (13) implies that

$$p^+(t) = p^-(t) \qquad \text{for all } t$$

and hence that $g^+(t) = g^-(t)$ for all t, which is equivalent to $g(t) = 0$ for all t. Since T is clearly sufficient (by the factorization criterion), it is proved that T is a complete sufficient statistic.

Example 20. Let $X_1, X_2, \cdots, X_n$ be iid $\mathcal{N}(\mu, \sigma^2)$ rv's where both μ and σ^2 are unknown. We know that the family of distributions of $\mathbf{X} = (X_1, \cdots, X_n)$ is a two-parameter exponential family with $T(X_1, \cdots, X_n) = (\sum_1^n X_i, \sum_1^n X_i^2)$. From Theorem 4 it follows that T is a complete sufficient statistic. Examples 15 and 16 fall in the domain of Theorem 4.

PROBLEMS 8.3

1. Suppose that T_n is a sequence of estimates for parameter θ that satisfies the conditions of Theorem 2. Then $T_n \xrightarrow{2} \theta$, that is, T_n is squared error consistent for θ.

   If T_n is consistent for θ and $|T_n - \theta| \le A < \infty$ for all θ and all $(x_1, x_2, \cdots, x_n) \in \mathcal{R}_n$, show that $T_n \xrightarrow{2} \theta$. If, however, $|T_n - \theta| \le A_n < \infty$, then show that T_n may not be squared error consistent for θ.

2. Let $X_1, X_2, \cdots, X_n$ be a sample from $U[0, \theta]$, $\theta \in \Theta = (0, \infty)$. Let $M_n = \max\{X_1, X_2, \cdots, X_n\}$. Show that $M_n \xrightarrow{P} \theta$. Write $Y_n = 2\bar{X}$. Is Y_n consistent for θ?

3. Let $X_1, X_2, \cdots, X_n$ be iid rv's with $EX_i = \mu$ and $E|X_i|^2 < \infty$. Show that $T(X_1, X_2, \cdots, X_n) = 2[n(n+1)]^{-1} \sum_{i=1}^n iX_i$ is a consistent estimate for μ.

4. Let $X_1, X_2, \cdots X_n$ be a sample fom $U[0, \theta]$. Show that $T(X_1, X_2, \cdots, X_n) = (\prod_{i=1}^n X_i)^{1/n}$ is a consistent estimate for θe^{-1}.

5. Find a sufficient statistic in each of the following cases based on a random sample of size n.
   (a) $X \sim B(\alpha, \beta)$ when (i) α is unknown, β known; (ii) β is unknown, α known, and (iii) α, β are both unknown.
   (b) $X \sim G(\alpha, \beta)$ when (i) α is unknown, β known, (ii) β is unknown, α known, and (iii) α, β are both unknown.
   (c) $X \sim P_{N_1, N_2}(x)$, where

   $$P_{N_1, N_2}(x) = \frac{1}{N_2 - N_1}, \qquad x = N_1 + 1, N_1 + 2, \cdots, N_2,$$

and N_1, $N_2 (N_1 < N_2)$ are integers, when (i) N_1 is known, N_2 unknown; (ii) N_2 known, N_1 unknown, and (iii) N_1, N_2 are both unknown.

(d) $X \sim f_\theta(x)$, where

$$f_\theta(x) = \begin{cases} e^{-x+\theta} & \text{if } \theta < x < \infty, \\ 0 & \text{otherwise.} \end{cases}$$

(e) $X \sim f(x; \mu, \sigma)$, where

$$f(x; \mu, \sigma) = \frac{1}{x\sigma\sqrt{2\pi}} \exp\left\{-\frac{1}{2\sigma^2}(\log x - \mu)^2\right\}, \quad x > 0.$$

(f) $X \sim f_\theta(x)$, where

$$f_\theta(x) = P_\theta\{X = x\} = 2^{-x/\theta}, \quad x = \theta, \theta + 1, \cdots, \theta > 0.$$

(g) $X \sim P_{\theta,p}(x)$, where

$$P_{\theta,p}(x) = (1-p)p^{x-\theta}, \quad x = \theta, \theta + 1, \cdots, 0 < p < 1,$$

when (i) p is known, θ unknown; (ii) p is unknown, θ known, and (iii) p, θ are both unknown.

6. Let $\mathbf{X} = (X_1, X_2, \cdots, X_n)$ be a sample from $\mathcal{N}(\alpha\sigma, \sigma^2)$, where α is a known real number. Show that the statistic $T(\mathbf{X}) = (\sum_{i=1}^n X_i, \sum_{i=1}^n X_i^2)$ is sufficient for σ but that the family of distributions of $T(\mathbf{X})$ is not complete.

7. Let $X_1, X_2, \cdots, X_n$ be a sample from $\mathcal{N}(\mu, \sigma^2)$. Then $\mathbf{X} = (X_1, X_2, \cdots, X_n)$ is clearly sufficient for the family $\mathcal{N}(\mu, \sigma^2)$, $\mu \in \mathcal{R}$, $\sigma > 0$. Is the family of distributions of $\mathbf{X}$ complete?

8. Let $X_1, X_2, \cdots, X_n$ be a sample from $U(\theta - \frac{1}{2}, \theta + \frac{1}{2})$, $\theta \in \mathcal{R}$. Show that the statistic $T(X_1, \cdots, X_n) = (\min X_i, \max X_i)$ is sufficient for θ but not complete.

9. Let X be an rv with pmf

$$P_\theta\{X = -1\} = \theta \quad \text{and} \quad P_\theta\{X = x\} = (1-\theta)^2 \theta^x, \quad x = 0, 1, 2, \cdots.$$

Is $\{P_\theta : \theta \in (0, 1)\}$ a complete family?

10. If $T = g(U)$ and T is sufficient, then so also is U.

11. Let $\Theta \subseteq \mathcal{R}$. We say that a parameter $\theta \in \Theta$ is a location parameter for the df $F_\theta(x)$ of an rv X if $F_\theta(x) = F(x - \theta)$, where F is a df. We say that a positive parameter $\theta \in \Theta$ is a scale parameter for the df $F_\theta(x)$ of an rv X if $F_\theta(x) = F(x/\theta)$, where F is a df.

(a) θ is a location parameter if and only if the distribution of $X - \theta$ computed under θ, that is, $P_\theta\{X - \theta \leq x\}$, is independent of θ. Find a group of transformations that leaves a location parameter family of distributions invariant.

(b) θ is a scale parameter if and only if the distribution of X/θ computed under θ, that is, $P_\theta\{X/\theta \leq x\}$, is independent of θ. Find a group of transformations that leaves a scale parameter family invariant.

12. In Example 19 show that the class of all functions g for which $E_P g(X) = 0$ for all $P \in \mathcal{P}$ consists of functions of the form

$$g(k) = \begin{cases} 0, & k = 1, 2, \cdots, n_0 - 1, n_0 + 2, n_0 + 3, \cdots, \\ c, & k = n_0, \\ -c, & k = n_0 + 1, \end{cases}$$

where c is a constant.

13. For the class $\{F_{\theta_1}, F_{\theta_2}\}$ of two df's, where F_{θ_1} is $\mathcal{N}(0,1)$ and F_{θ_2} is $\mathscr{C}(1,0)$, find a sufficient statistic.

14. Consider the class of hypergeometric probability distributions $\{P_D: D = 0, 1, 2, \cdots, N\}$, where

$$P_D\{X = x\} = \binom{N}{n}^{-1}\binom{D}{x}\binom{N-D}{n-x}, \quad x = 0, 1, \cdots, \min\{n, D\}.$$

Show that it is a complete class. If $\mathscr{P} = \{P_D: D = 0, 1, 2, \cdots, N, D \neq d, d \text{ integral } 0 \leq d \leq N\}$, is $\mathscr{P}$ complete?

15. Is the family of distributions of the order statistic in sampling from a Poisson distribution complete?

16. Let $(X_1, X_2, \cdots, X_n)$ be a random vector of the discrete type. Is the statistic $T(X_1, \cdots, X_n) = (X_1, \cdots, X_{n-1})$ sufficient?

8.4 UNBIASED ESTIMATION

In this section we describe yet another property of estimates and study some procedures for finding estimates possessing this property.

Definition 1. Let $\{F_\theta, \theta \in \Theta\}$, $\Theta \subseteq \mathscr{R}$, be a nonempty set of probability distribution functions. A Borel-measurable function T of $\mathscr{R}_n \to \Theta$ is said to be an unbiased estimate of θ if

(1) $$E_\theta(T) = \theta \quad \text{for all } \theta \in \Theta.$$

Any function $d(\theta)$ for which there exists a T satisfying (1) is usually referred to as an *estimable function*. An estimate that is not unbiased is called *biased*, and the function $b(\theta, T)$, defined by

(2) $$b(\theta, T) = E_\theta T - \theta,$$

is called the *bias* of T.

Remark 1. Definition 1, in particular, requires that T be integrable (summable), that is, $E_\theta T$ exist for every $\theta \in \Theta$. This definition can easily be extended to the case where θ is a vector of parameters and T is a random vector.

Example 1. Let $X_1, X_2, \cdots, X_n$ be a random sample from some population with finite mean. Then $\bar{X}$ is unbiased for the population mean. If the population variance is finite, the sample variance S^2 is unbiased for the population variance. In general, if the kth population moment m_k exists, the kth sample moment is unbiased for m_k.

Note that S is not, in general, unbiased for σ. If $X_1, X_2, \cdots, X_n$ are iid $\mathcal{N}(\mu, \sigma^2)$ rv's we know that $(n-1)S^2/\sigma^2$ is $\chi^2(n-1)$. Therefore

$$E(S\sqrt{n-1}/\sigma) = \int_0^\infty \sqrt{x}\, \frac{1}{2^{(n-1)/2}\Gamma[(n-1)/2]} x^{(n-1)/2-1} e^{-x/2}\, dx$$

$$= \sqrt{2}\, \Gamma\left(\frac{n}{2}\right)\left[\Gamma\left(\frac{n-1}{2}\right)\right]^{-1},$$

$$E_\sigma(S) = \sigma\left\{\sqrt{\frac{2}{n-1}}\, \Gamma\left(\frac{n}{2}\right)\left[\Gamma\left(\frac{n-1}{2}\right)\right]^{-1}\right\}.$$

The bias of S is given by

$$b(\sigma,S) = \sigma\left\{\sqrt{\frac{2}{n-1}}\, \Gamma\left(\frac{n}{2}\right)\left[\Gamma\left(\frac{n-1}{2}\right)\right]^{-1} - 1\right\}.$$

If T is unbiased for θ, $g(T)$ is not, in general, an unbiased estimate of $g(\theta)$ unless g is a linear function.

Example 2. Unbiased estimates do not always exist. Consider an rv with pmf $b(1, p)$. Suppose that we wish to estimate $d(p) = p^2$. Then, in order that T be unbiased for p^2, we must have

$$p^2 = E_p T = pT(1) + (1-p)T(0), \quad 0 \le p \le 1,$$

that is,

$$p^2 = p\{T(1) - T(0)\} + T(0)$$

must hold for all p in the interval $[0,1]$, which is impossible. (If a convergent power series vanishes in an open interval, each of the coefficients must be 0. See theorem P. 2.10.) See also Problem 1.

Example 3. Sometimes an unbiased estimate may be absurd.

Let X be $P(\lambda)$, and $d(\lambda) = e^{-3\lambda}$. We show that $T(X) = (-2)^X$ is unbiased for $d(\lambda)$. We have

$$E_\lambda T(X) = e^{-\lambda}\sum_{x=0}^\infty (-2)^x \frac{\lambda^x}{x!} = e^{-\lambda}\sum_{x=0}^\infty \frac{(-2\lambda)^x}{x!} = e^{-\lambda}e^{-2\lambda} = d(\lambda).$$

However, $T(x) = (-2)^x > 0$ if x is even, and < 0 if x is odd, which is absurd since $d(\lambda) > 0$.

Example 4. Let $X_1, X_2, \cdots, X_n$ be a sample from $P(\lambda)$. Then $\bar{X}$ is unbiased for λ and so also is S^2, since both the mean and the variance are equal to λ. Indeed, $\alpha \bar{X} + (1-\alpha)S^2$, $0 \le \alpha \le 1$, is unbiased for λ.

Let θ be estimable, and let T be an unbiased estimate of θ. Let T_1 be another unbiased estimate of θ, different from T. This means that there exists at least one θ such that $P_\theta\{T \ne T_1\} > 0$. In this case there exist infinitely many unbiased estimates of θ of the form $\alpha T + (1-\alpha)T_1$, $0 < \alpha < 1$. It is therefore desirable to find a procedure to differentiate between these estimates.

Definition 2. Let $\theta_0 \in \Theta$ and $U(\theta_0)$ be the class of all unbiased estimates T of θ_0 such that $E_{\theta_0} T^2 < \infty$. Then $T_0 \in U(\theta_0)$ is called a locally minimum variance unbiased estimate (LMVUE) at θ_0 if

(3) $$E_{\theta_0}(T_0 - \theta_0)^2 \le E_{\theta_0}(T - \theta_0)^2$$

holds for all $T \in U(\theta_0)$.

Definition 3. Let U be the set of all unbiased estimates T of $\theta \in \Theta$ such that $E_\theta T^2 < \infty$ for all $\theta \in \Theta$. An estimate $T_0 \in U$ is called a uniformly minimum variance unbiased estimate (UMVUE) of θ if

(4) $$E_\theta(T_0 - \theta)^2 \le E_\theta(T - \theta)^2$$

for all $\theta \in \Theta$ and every $T \in U$.

Remark 2. Let $a_1, a_2, \cdots, a_n$ be any set of real numbers with $\sum_{i=1}^n a_i = 1$. Let $X_1, X_2, \cdots, X_n$ be independent rv's with common mean μ and variances σ_k^2, $k = 1, 2, \cdots, n$. Then $T = \sum_{i=1}^n a_i X_i$ is an unbiased estimate of μ with variance $\sum_{i=1}^n a_i^2 \sigma_i^2$ (see the corollary to Theorem 4.6.5). T is called a *linear unbiased estimate* of μ. Linear unbiased estimates of μ that have minimum variance (among all linear unbiased estimates) are called *best linear unbiased estimates* (BLUE's). In Theorem 4.6.6 we have shown that, if X_i are iid rv's with common variance σ^2, the BLUE of μ is $\bar X = n^{-1} \sum_{i=1}^n X_i$. If X_i are independent with common mean μ but different variances σ_i^2, the BLUE of μ is obtained if we choose a_i proportional to $1/\sigma_i^2$; then the minimum variance is H/n, where H is the harmonic mean of $\sigma_1^2, \cdots, \sigma_n^2$. (See Example 4.6.4.)

Remark 3. Sometimes the precision of an estimate T of parameter θ is measured by the so-called *mean square error* (MSE). We say that an estimate T_0 is at least as good as any other estimate T in the sense of the MSE if

(5) $$E_\theta(T_0 - \theta)^2 \le E_\theta(T - \theta)^2 \quad \text{for all } \theta \in \Theta.$$

In general, a particular estimate will be better than another for some values of θ and worse for others. Definitions 2 and 3 are special cases of this concept if we restrict attention only to unbiased estimates.

The following result gives a necessary and sufficient condition for an unbiased estimate to be a UMVUE.

Theorem 1. Let U be the class of all unbiased estimates T of a parameter $\theta \in \Theta$ with $E_\theta T^2 < \infty$ for all θ, and suppose that U is nonempty. Let U_0 be the set of all unbiased estimates v of 0, that is,

$$U_0 = \{v: E_\theta v = 0, E_\theta v^2 < \infty \text{ for all } \theta \in \Theta\}.$$

Then $T_0 \in U$ is a UMVUE if and only if

(6) $\qquad E_\theta(vT_0) = 0 \qquad$ for all θ and all $v \in U_0$.

Proof. The conditions of the theorem guarantee the existence of $E_\theta(vT_0)$ for all θ and $v \in U_0$. Suppose that $T_0 \in U$ is a UMVUE and $E_{\theta_0}(v_0 T_0) \neq 0$ for some θ_0 and some $v_0 \in U_0$. Then $T_0 + \lambda v_0 \in U$ for all real λ. If $E_{\theta_0} v_0^2 = 0$, then $E_{\theta_0}(v_0 T_0) = 0$ must hold since $P_{\theta_0}\{v_0 = 0\} = 1$. Let $E_{\theta_0} v_0^2 > 0$. Choose $\lambda_0 = -E_{\theta_0}(T_0 v_0)/E_{\theta_0} v_0^2$. Then

(7) $\qquad E_{\theta_0}(T_0 + \lambda_0 v_0)^2 = E_{\theta_0} T_0^2 - \dfrac{E_{\theta_0}^2(v_0 T_0)}{E_{\theta_0} v_0^2} < E_{\theta_0} T_0^2.$

Since $T_0 + \lambda_0 v_0 \in U$ and $T_0 \in U$, it follows from (7) that

(8) $\qquad \mathrm{var}_{\theta_0}(T_0 + \lambda_0 v_0) < \mathrm{var}_{\theta_0}(T_0),$

which is a contradiction. It follows that (6) holds.

Conversely, let (6) hold for some $T_0 \in U$, all $\theta \in \Theta$ and all $v \in U_0$, and let $T \in U$. Then $T_0 - T \in U_0$, and for every θ

$$E_\theta\{T_0(T_0 - T)\} = 0.$$

We have

$$E_\theta T_0^2 = E_\theta(TT_0) \leq (E_\theta T_0^2)^{1/2} (E_\theta T^2)^{1/2}$$

by the Cauchy-Schwarz inequality. If $E_\theta T_0^2 = 0$, then $P(T_0 = 0) = 1$ and there is nothing to prove. Otherwise

$$(E_\theta T_0^2)^{1/2} \leq (E_\theta T^2)^{1/2}$$

or $\mathrm{var}_\theta(T_0) \leq \mathrm{var}_\theta(T)$. Since T is arbitrary, the proof is complete.

Theorem 2. Let U be the nonempty class of unbiased estimates as defined in Theorem 1. Then there exists at most one UMVUE for θ.

Proof. If T and $T_0 \in U$ are both UMVUE's, then $T - T_0 \in U_0$ and

$$E_\theta\{T_0(T - T_0)\} = 0 \qquad \text{for all } \theta \in \Theta,$$

that is, $E_\theta T_0^2 = E_\theta(TT_0)$, and it follows that

$$\text{cov }(T, T_0) = \text{var}_\theta (T_0) \quad \text{for all } \theta.$$

Since T_0 and T are both UMVUE's $\text{var}_\theta(T) = \text{var}_\theta(T_0)$, and it follows that the correlation coefficient between T and T_0 is 1. This implies that $P_\theta\{aT + bT_0 = 0\} = 1$ for some a, b and all $\theta \in \Theta$. Since T and T_0 are both unbiased for θ, we must have $P_\theta\{T = T_0\} = 1$ for all θ.

Remark 4. Both Theorems 1 and 2 have analogues for LMVUE's at $\theta_0 \in \Theta$, θ_0 fixed.

Theorem 3. If UMVUE's T_i exist for real functions d_i, $i = 1, 2$, of θ, they also exist for λd_i (λ real), as well as for $d_1 + d_2$, and are given by λT_i and $T_1 + T_2$, respectively.

Proof. The proof is left as an exercise.

Theorem 4. Let $\{T_n\}$ be a sequence of UMVUE's and T be a statistic with $E_\theta T^2 < \infty$ and such that $E_\theta\{T_n - T\}^2 \to 0$ as $n \to \infty$ for all $\theta \in \Theta$. Then T is also the UMVUE.

Proof. That T is unbiased follows from $|E_\theta T - \theta| \leq E_\theta|T - T_n| \leq E_\theta^{1/2}\{T_n - T\}^2$. For all $v \in U_0$, all θ, and every $n = 1, 2, \cdots$,

$$E_\theta(T_n v) = 0$$

by Theorem 1. Therefore

$$E_\theta(vT) = E_\theta(vT) - E_\theta(vT_n)$$
$$= E_\theta[v(T - T_n)],$$

and

$$|E_\theta(vT)| \leq (E_\theta v^2)^{1/2}[E_\theta(T - T_n)^2]^{1/2} \to 0 \quad \text{as } n \to \infty$$

for all θ and all $v \in U$. Thus

$$E_\theta(vT) = 0 \quad \text{for all } v \in U_0, \text{ all } \theta \in \Theta,$$

and, by Theorem 1, T must be the UMVUE.

Example 5. Let $X \sim P(\lambda)$. Then X is the UMVUE of λ. Surely X is unbiased. Let g be an unbiased estimate of 0. Then $T(X) = X + g(X)$ is unbiased for θ. But the family of pmf's $\{P(\lambda): \lambda > 0\}$ is complete. It follows that

$$E_\lambda g(X) = 0 \quad \text{for all } \lambda > 0 \Rightarrow g(x) = 0 \quad \text{for } x = 0, 1, 2, \cdots.$$

Hence X must be the UMVUE of λ.

Example 6. Sometimes an estimate with larger variance may be preferable.

Let X be a $G(1, 1/\beta)$ rv. X is usually taken as a good model to describe the time to failure of a piece of equipment. Let $X_1, X_2, \cdots, X_n$ be a sample of n observations on X. Then $\bar{X}$ is unbiased for $EX = 1/\beta$ with variance $1/n\beta^2$. ($\bar{X}$ is actually the UMVUE for $1/\beta$.) Now consider $M_n = \min(X_1, X_2, \cdots, X_n)$. Then nM_n is unbiased for $1/\beta$ with variance $1/\beta^2$, and it has a larger variance than $\bar{X}$. However, if the length of time is of importance, nM_n may be preferable to $\bar{X}$, since to observe nM_n one needs to wait only until the first piece of equipment fails, whereas to compute $\bar{X}$ one would have to wait until all the n observations $X_1, X_2, \cdots, X_n$ are available.

Theorem 5. If a sample consists of n independent observations $X_1, X_2, \cdots, X_n$ from the same distribution, the UMVUE, if it exists, is a symmetric function of the X_i's.

Proof. The proof is left as an exercise.

The converse of Theorem 5 is not true. If $X_1, X_2, \cdots, X_n$ are iid $P(\lambda)$ rv's, $\lambda > 0$, both $\bar{X}$ and S^2 are unbiased for θ. But $\bar{X}$ is the UMVUE, whereas S^2 is not.

We now turn our attention to some methods for finding UMVUE's.

Theorem 6 (Rao [96], Blackwell [9]). Let $\{F_\theta : \theta \in \Theta\}$ be a family of probability df's, and h be any statistic in U, where U is the (nonempty) class of all unbiased estimates of θ with $E_\theta h^2 < \infty$. Let T be a sufficient statistic for $\{F_\theta, \theta \in \Theta\}$. Then the conditional expectation $E_\theta\{h|T\}$ is independent of θ and is an unbiased estimate of θ. Moreover,

$$(9) \quad E_\theta(E\{h|T\} - \theta)^2 \leq E_\theta(h - \theta)^2 \quad \text{for all } \theta \in \Theta.$$

The equality in (9) holds if and only if $h = E\{h|T\}$ (that is, $P_\theta\{h = E\{h|T\}\} = 1$ for all θ).

Proof. We have

$$E_\theta\{E\{h|T\}\} = E_\theta h = \theta.$$

It is therefore sufficient to show that

$$(10) \quad E_\theta\{E\{h|T\}\}^2 \leq E_\theta h^2 \quad \text{for all } \theta \in \Theta.$$

But $E_\theta h^2 = E_\theta\{E\{h^2|T\}\}$, so that it will be sufficient to show that

$$(11) \quad [E\{h|T\}]^2 \leq E\{h^2|T\}.$$

By the Cauchy-Schwarz inequality
$$E^2\{h|T\} \le E\{h^2|T\}E\{1|T\},$$
and (11) follows. The equality holds in (9) if and only if
(12) $$E_\theta[E\{h|T\}]^2 = E_\theta h^2,$$
that is,
$$E_\theta[E\{h^2|T\} - E^2\{h|T\}] = 0,$$
which is the same as
$$E_\theta\{\text{var }\{h|T\}\} = 0.$$
This happens if and only if var $\{h|T\} = 0$, that is, if and only if
$$E\{h^2|T\} = E^2\{h|T\},$$
as will be the case if and only if h is a function of T. Thus $h = E\{h|T\}$.

Theorem 6 is applied along with completeness to yield the following powerful result.

Theorem 7 (Lehmann-Scheffé [69]). If T is a complete sufficient statistic and there exists an unbiased estimate h of θ, there exists a unique UMVUE of θ, which is given by $E\{h|T\}$.

Proof. If $h_1, h_2 \in U$, then $E\{h_1|T\}$ and $E\{h_2|T\}$ are both unbiased and
$$E_\theta[E\{h_1|T\} - E\{h_2|T\}] = 0 \quad \text{for all } \theta \in \Theta.$$
Since T is a complete sufficient statistic, it follows that $E\{h_1|T\} = E\{h_2|T\}$. By Theorem 6 $E\{h|T\}$ is the UMVUE.

Remark 5. According to Theorem 6, we should restrict our search to Borel-measurable functions of a sufficient statistic (whenever it exists). According to Theorem 7, if a complete sufficient statistic T exists, all we need to do is to find a Borel-measurable function of T that is unbiased. If a complete sufficient statistic does not exist, an UMVUE may still exist (see Example 10).

Example 7. Let $X_1, X_2, \cdots, X_n$ be $\mathcal{N}(\theta, 1)$. X_1 is unbiased for θ. However, $\bar{X} = n^{-1} \sum_1^n X_i$ is a complete sufficient statistic, so that $E\{X_1|\bar{X}\}$ is the UMVUE.

We will show that $E\{X_1|\bar{X}\} = \bar{X}$. Let $Y = n\bar{X}$. Then Y is $\mathcal{N}(n\theta, n)$, X_1 is $\mathcal{N}(\theta, 1)$, and (X_1, Y) is a bivariate normal rv with variance covariance matrix $\begin{pmatrix} 1 & 1 \\ 1 & n \end{pmatrix}$. Therefore

$$E\{X_1|y\} = EX_1 + \frac{\text{cov}(X_1, Y)}{\text{var}(Y)}(y - EY)$$
$$= \theta + \frac{1}{n}(y - n\theta) = \frac{y}{n},$$

as asserted.

If we let $d(\theta) = \theta^2$, we can show similarly that $\bar{X}^2 - 1/n$ is the UMVUE for $d(\theta)$. Note that $\bar{X}^2 - 1/n$ may occasionally be negative, so that an UMVUE for θ^2 is not very sensible in this case.

Example 8. Let $X_1, X_2, \cdots, X_n$ be iid $b(1, p)$ rv's. Then $T = \sum_1^n X_i$ is a complete sufficient statistic. The UMVUE for p is clearly $\bar{X}$. To find the UMVUE for $d(p) = p(1 - p)$, we have $E(nT) = n^2 p$, $ET^2 = np + n(n-1)p^2$, so that $E\{nT - T^2\} = n(n-1)p(1-p)$, and it follows that $(nT - T^2)/n(n-1)$ is the UMVUE for $d(p) = p(1 - p)$.

Example 9. Let $X_1, X_2, \cdots, X_n$ be a sample from $\mathcal{N}(\mu, \sigma^2)$. Then $(\bar{X}, S^2)$ is a complete sufficient statistic for (μ, σ^2). $\bar{X}$ is the UMVUE for μ, and S^2 is the UMVUE for σ^2. Also $k(n)S$ is the UMVUE for σ, where $k(n) = \sqrt{[(n-1)/2]}\, \Gamma[(n-1)/2]/\Gamma(n/2)$. We wish to find the UMVUE for the pth quantile ζ_p. We have

$$p = P\{X \leq \zeta_p\} = P\left\{Z \leq \frac{\zeta_p - \mu}{\sigma}\right\},$$

where Z is $\mathcal{N}(0, 1)$. Thus $\zeta_p = \sigma z_{1-p} + \mu$, and the UMVUE is

$$T(X_1, X_2, \cdots, X_n) = z_{1-p} k(n) S + \bar{X}.$$

Example 10 (Stigler [127]). We return to Example 8.3.19. We have seen that the family $\{P_N^{(n)}: N \geq 1\}$ of pmf's of $M_n = \max_{1 \leq i \leq n} X_i$ is complete and M_n is sufficient for N. Now $EX_1 = (N+1)/2$, so that $T(X_1) = 2X_1 - 1$ is unbiased for N. It follows from Theorem 7 that $E\{T(X_1)|M_n\}$ is the UMVUE of N. We have

$$P\{X_1 = x_1 | M_n = y\} = \begin{cases} \dfrac{y^{n-1} - (y-1)^{n-1}}{y^n - (y-1)^n} & \text{if } x_1 = 1, 2, \cdots, y-1, \\ \dfrac{y^{n-1}}{y^n - (y-1)^n} & \text{if } x_1 = y. \end{cases}$$

Thus

$$E\{T(X_1)|M_n = y\} = \frac{y^{n-1} - (y-1)^{n-1}}{y^n - (y-1)^n} \sum_{x_1=1}^{y-1}(2x_1 - 1) + (2y - 1)\frac{y^{n-1}}{y^n - (y-1)^n}$$
$$= \frac{y^{n+1} - (y-1)^{n+1}}{y^n - (y-1)^n}$$

is the UMVUE of N.

If we consider the family $\mathcal{P}$ instead, we have seen (Example 8.3.19 and

Problem 8.3.12) that $\mathscr{P}$ is not complete. The UMVUE for the family $\{P_N: N \geq 1\}$ is $T(X_1) = 2X_1 - 1$, which is not the UMVUE for $\mathscr{P}$. The UMVUE for $\mathscr{P}$ is in fact, given by

$$T_1(k) = \begin{cases} 2k - 1, & k \neq n_0, \ k \neq n_0 + 1, \\ 2n_0, & k = n_0, \ k = n_0 + 1. \end{cases}$$

The reader is asked to check that T_1 has covariance 0 with all unbiased estimates g of 0 that are of the form described in Example 8.3.19 and Problem 8.3.12, and hence Theorem 1 implies that T_1 is the UMVUE. Since $E_{n_0} T_1(X_1) = n_0 + 1/n_0$, T_1 is not even unbiased for the family $\{P_N: N \geq 1\}$. The minimum variance is given by

$$\mathrm{var}_N(T_1(X_1)) = \begin{cases} \mathrm{var}_N(T(X_1)) & \text{if } N < n_0, \\ \mathrm{var}_N(T(X_1)) - \dfrac{2}{N} & \text{if } N > n_0. \end{cases}$$

PROBLEMS 8.4

1. Let $X_1, X_2, \cdots, X_n$ ($n \geq 2$) be a sample from $b(1,p)$. Find an unbiased estimate for $d(p) = p^2$.

2. Let $X_1, X_2, \cdots, X_n$ ($n \geq 2$) be a sample from $\mathcal{N}(\mu, \sigma^2)$. Find an unbiased estimate for σ^p, where $p + n > 1$. Find a minimum MSE estimate of σ^p.

3. Let $X_1, X_2, \cdots, X_n$ be iid $\mathcal{N}(\mu, \sigma^2)$ rv's. Find a minimum MSE estimate of the form αS^2 for the parameter σ^2. Compare the variances of the minimum MSE estimate and the obvious estimate S^2.

4. Let $X \sim b(1, \theta^2)$. Does there exist an unbiased estimate of θ?

5. Let $X \sim P(\lambda)$. Does there exist an unbiased estimate of $d(\lambda) = \lambda^{-1}$?

6. Let $X_1, X_2, \cdots, X_n$ be a sample from $b(1, p)$, $0 < p < 1$, and $0 < s < n$ be an integer. Find the UMVUE for (a) $d(p) = p^s$, (b) $d(p) = p^s + (1 - p)^{n-s}$.

7. Let $X_1, X_2, \cdots, X_n$ be a sample from a population with mean θ and finite variance, and T be an estimate of θ of the form $T(X_1, X_2, \cdots, X_n) = \sum_{i=1}^{n} \alpha_i X_i$. If T is an unbiased estimate of θ that has minimum variance and T' is another linear unbiased estimate of θ, then

$$\mathrm{cov}_\theta(T, T') = \mathrm{var}_\theta(T).$$

8. Let T_1, T_2 be two unbiased estimates having common variance $\alpha \sigma^2 (\alpha > 1)$, where σ^2 is the variance of the UMVUE. Show that the correlation coefficient between T_1 and T_2 is $\geq (2 - \alpha)/\alpha$.

9. Let $X \sim NB(1; \theta)$ and $d(\theta) = P_\theta\{X = 0\}$. Let $X_1, X_2, \cdots, X_n$ be a sample on X. Find the UMVUE of $d(\theta)$.

10. This example covers most discrete distributions. Let $X_1, X_2, \cdots, X_n$ be a sample from pmf

$$P_\theta\{X = x\} = \frac{\alpha(x)\theta^x}{f(\theta)}, \quad x = 0, 1, 2, \cdots,$$

where $\theta > 0$, $\alpha(x) > 0$, $f(\theta) = \sum_{x=0}^{\infty} \alpha(x)\theta^x$, $\alpha(0) = 1$, and let $T = X_1 + X_2 + \cdots + X_n$. Write

$$c(t, n) = \sum_{x_1, x_2, \cdots, x_n} \prod_{i=1}^{n} \alpha(x_i).$$

with $\sum_{i=1}^{n} x_i = t$

Show that T is a complete sufficient statistic for θ, and that the UMVUE for $d(\theta) = \theta^r (r > 0$ is an integer) is given by

$$Y_r(t) = \begin{cases} 0 & \text{if } t < r. \\ \dfrac{c(t-r, n)}{c(t, n)} & \text{if } t \geq r. \end{cases} \quad \text{(Roy and Mitra [105])}$$

11. Let X be a hypergeometric rv with pmf

$$P_M\{X = x\} = \binom{N}{n}^{-1} \binom{M}{x} \binom{N-M}{n-x},$$

where $\max(0, M + n - N) \leq x \leq \min(M, n)$.
(a) Find the UMVUE for M when N is assumed to be known.
(b) Does there exist an unbiased estimate of N (M known)?

12. Let $X_1, X_2, \cdots, X_n$ be iid $G(1, 1/\lambda)$ rv's $\lambda > 0$. Find the UMVUE of $P_\lambda\{X_1 \leq t_0\}$, where $t_0 > 0$ is a fixed real number.

13. Let $X_1, X_2, \cdots, X_n$ be a random sample from $P(\lambda)$. Let $\psi(\lambda) = \sum_{k=0}^{\infty} c_k \lambda^k$ be a parametric function. Find the UMVUE for $\psi(\lambda)$. In particular, find the UMVUE for (a) $\psi(\lambda) = 1/(1-\lambda)$, (b) $\psi(\lambda) = \lambda^s$ for some fixed integer $s > 0$, (c) $\psi(\lambda) = P_\lambda\{X = 0\}$, and (d) $\psi(\lambda) = P_\lambda\{X = 0 \text{ or } 1\}$.

14. Let $X_1, X_2, \cdots, X_n$ be a sample from pmf

$$P_N(x) = \frac{1}{N}, \quad x = 1, 2, \cdots, N.$$

Let $\psi(N)$ be some function of N. Find the UMVUE of $\psi(N)$.

15. Let $X_1, X_2, \cdots, X_n$ be a random sample from $P(\lambda)$. Find the UMVUE of $\psi(\lambda) = P_\lambda\{X = k\}$, where k is a fixed positive integer.

16. Let $(X_1, Y_1), (X_2, Y_2), \cdots, (X_n, Y_n)$ be a sample from a bivariate normal population with parameters $\mu_1, \mu_2, \sigma_1^2, \sigma_2^2$, and ρ. Assume that $\mu_1 = \mu_2 = \mu$, and it is required to find an unbiased estimate of μ. Since a complete sufficient statistic does not exist, consider the class of all linear unbiased estimates

$$\hat{\mu}(\alpha) = \alpha \bar{X} + (1 - \alpha) \bar{Y}.$$

(a) Find the variance of $\hat{\mu}$.
(b) Choose $\alpha = \alpha_0$ to minimize var $(\hat{\mu})$, and consider the estimate
$$\hat{\mu}_0 = \alpha_0 \bar{X} + (1 - \alpha_0) \bar{Y}.$$
Compute var $(\hat{\mu}_0)$. If $\sigma_1 = \sigma_2$, the BLUE of μ (in the sense of minimum variance) is
$$\hat{\mu}_1 = \frac{\bar{X} + \bar{Y}}{2}$$
irrespective of whether σ_1 and ρ are known or unknown.
(c) If $\sigma_1 \neq \sigma_2$ and ρ, σ_1, σ_2 are unknown, replace these values in α_0 by their corresponding estimates. Let
$$\hat{\alpha} = \frac{S_2^2 - S_{11}}{S_1^2 + S_2^2 - 2S_{11}}.$$
Show that
$$\hat{\mu}_2 = \bar{Y} + (\bar{X} - \bar{Y})\hat{\alpha}$$
is an unbiased estimate of μ.

17. Prove Theorem 3.

18. Prove Theorem 5.

19. In Example 10 show that T_1 is the UMVUE for N (restricted to the family $\mathscr{P}$), and compute the minimum variance.

20. Let $(X_1, Y_1), \cdots, (X_n, Y_n)$ be a sample from a bivariate population with finite variances σ_1^2 and σ_2^2, respectively, and covariance γ. Show that
$$\text{var}(S_{11}) = \frac{1}{n}\left(\mu_{22} - \frac{n-2}{n-1}\gamma^2 + \frac{\sigma_1^2 \sigma_2^2}{n-1}\right),$$
where $\mu_{22} = E[(X - EX)^2(Y - EY)^2]$. It is assumed that appropriate order moments exist.

21. Suppose that a random sample is taken on (X, Y) and it is desired to estimate γ, the unknown covariance between X and Y. Suppose that for some reason a set S of n observations is available on both X and Y, an additional $n_1 - n$ observations are available on X but the corresponding Y values are missing, and an additional $n_2 - n$ observations of Y are available for which the X values are missing. Let S_1 be the set of all $n_1 (\geq n)$ X values, and S_2, the set of all $n_2 (\geq n)$ Y values, and write
$$\hat{X} = \frac{\sum_{j \in S_1} X_j}{n_1}, \quad \hat{Y} = \frac{\sum_{j \in S_2} Y_j}{n_2}, \quad \bar{X} = \frac{\sum_{i \in S} X_i}{n}, \quad \bar{Y} = \frac{\sum_{i \in S} Y_i}{n}.$$
Show that
$$\hat{\gamma} = \frac{n_1 n_2}{n(n_1 n_2 - n_1 - n_2 + n)} \sum_{i \in S}(X_i - \hat{X})(Y_i - \hat{Y})$$
is an unbiased estimate of γ. Find the variance of $\hat{\gamma}$, and show that var $(\hat{\gamma}) \leq$ var (S_{11}), where S_{11} is the usual unbiased estimate of γ based on the n observations in S. (Boas [10])

8.5 UNBIASED ESTIMATION (*CONTINUED*): A LOWER BOUND FOR THE VARIANCE OF AN ESTIMATE

In this section we consider two inequalities, each of which provides a lower bound for the variance of an estimate. These inequalities can sometimes be used to show that an unbiased estimate is the UMVUE. We first consider an inequality due to Fréchet, Cramér, and Rao (the FCR inequality).

Theorem 1 (Fréchet [34], Cramér [19], Rao [95]). Let Θ be an open interval of the real line, and $\{f_\theta : \theta \in \Theta\}$ be a family of pdf's or pmf's. Assume that the set $\{f_\theta(\mathbf{x}) = 0 \text{ for every } \theta \in \Theta\}$ is independent of θ. For every θ let $\partial f_\theta(\mathbf{x})/\partial \theta$ be defined. Suppose that

(1)
$$\frac{\partial}{\partial \theta} \int f_\theta(\mathbf{x}) \, d\mathbf{x} = \int \frac{\partial}{\partial \theta} f_\theta(\mathbf{x}) \, d\mathbf{x} = 0 \quad \text{if } f_\theta \text{ is a pdf,}$$
$$\frac{\partial}{\partial \theta} \sum_\mathbf{x} f_\theta(\mathbf{x}) = \sum_\mathbf{x} \frac{\partial}{\partial \theta} f_\theta(\mathbf{x}) = 0 \quad \text{if } f_\theta \text{ is a pmf,}$$

for every $\theta \in \Theta$. [Here, as usual, $\mathbf{x} = (x_1, x_2, \cdots, x_n)$.]

Let ψ be defined on Θ and be differentiable there, and let T be an unbiased estimate of ψ such that $E_\theta T^2 < \infty$ for all θ. Assume that

(2)
$$\frac{\partial}{\partial \theta} \int T(\mathbf{x}) f_\theta(\mathbf{x}) \, d\mathbf{x} = \int T(\mathbf{x}) \frac{\partial}{\partial \theta} f_\theta(\mathbf{x}) \, d\mathbf{x} \quad \text{if } f_\theta \text{ is a pdf,}$$
$$\frac{\partial}{\partial \theta} \sum_\mathbf{x} T(\mathbf{x}) f_\theta(\mathbf{x}) = \sum T(\mathbf{x}) \frac{\partial}{\partial \theta} f_\theta(\mathbf{x}) \quad \text{if } f_\theta \text{ is a pmf,}$$

for every $\theta \in \Theta$. Let φ be any function of $\Theta \to \mathscr{R}$. Then

(3) $$[\psi'(\theta)]^2 \leq E_\theta \{T - \varphi(\theta)\}^2 E_\theta \left\{ \frac{\partial \log f_\theta(\mathbf{X})}{\partial \theta} \right\}^2$$

for every $\theta \in \Theta$.

For any $\theta_0 \in \Theta$, either $\psi'(\theta_0) = 0$ and equality holds in (3) for $\theta = \theta_0$, or we have

(4) $$E_{\theta_0}\{T - \varphi(\theta)\}^2 \geq \frac{[\psi'(\theta_0)]^2}{E_{\theta_0}\{\partial \log f_\theta(\mathbf{X})/\partial \theta\}^2}.$$

If, in the latter case, equality holds in (4), then there exists a real number $K(\theta_0) \neq 0$ such that

(5) $$T(\mathbf{X}) - \varphi(\theta_0) = K(\theta_0) \frac{\partial \log f_{\theta_0}(\mathbf{X})}{\partial \theta} \quad \text{with probability 1,}$$

provided that T is not a constant.

Remark 1. Conditions (1) and (2) are frequently known as *regularity conditions*. It is clear that $E_\theta \{\partial \log f_\theta(\mathbf{X})/\partial \theta\}^2$ is well defined and satisfies

$$0 \leq E_\theta \left\{ \frac{\partial \log f_\theta(\mathbf{X})}{\partial \theta} \right\}^2 \leq \infty.$$

Remark 2. Sufficient conditions on $\{f_\theta, \theta \in \Theta\}$ for (1) and (2) to hold may be found in P.2.12 and P.2.13.

Proof of Theorem 1. It follows from (1) that $E_\theta \{\partial \log f_\theta(\mathbf{X})/\partial \theta\} = 0$ and from (2) that

$$E_\theta \left\{ T(\mathbf{X}) \frac{\partial \log f_\theta(\mathbf{X})}{\partial \theta} \right\} = \psi'(\theta),$$

so that

$$E_\theta \left\{ [T - \varphi(\theta)] \frac{\partial \log f_\theta(\mathbf{X})}{\partial \theta} \right\} = \psi'(\theta).$$

By the Cauchy-Schwarz inequality, (3) follows immediately.

To prove (4) it suffices to consider either the case where $\psi'(\theta_0) \neq 0$ or the one where in (3) the equality sign does not hold for $\theta = \theta_0$. In either case it follows from (3) that $E_{\theta_0} \{\partial \log f_\theta(\mathbf{X})/\partial \theta\}^2 > 0$, and (4) follows.

If the equality holds in (4), then necessarily $\psi'(\theta_0) \neq 0$. Therefore from the Cauchy-Schwarz inequality there exists a real number $K(\theta_0)$ such that

$$T(\mathbf{x}) - \varphi(\theta_0) = K(\theta_0) \frac{\partial \log f_{\theta_0}(\mathbf{x})}{\partial \theta},$$

and (5) holds. Since T is not a constant, it also follows that $K(\theta_0) \neq 0$.

Remark 3. If we take $\varphi = \psi$ in (4), we get

(6) $$\mathrm{var}_\theta \, (T(\mathbf{X})) \geq \frac{[\psi'(\theta)]^2}{E_\theta \{\partial \log f_\theta(\mathbf{X})/\partial \theta\}^2}.$$

In particular, if $\psi(\theta) = \theta$, then (6) reduces to

(7) $$\mathrm{var}_\theta \, (T(\mathbf{X})) \geq \left[E_\theta \left\{ \frac{\partial \log f_\theta(\mathbf{X})}{\partial \theta} \right\}^2 \right]^{-1}.$$

Remark 4. If $\mathbf{X} = (X_1, X_2, \cdots, X_n)$ is a sample from a pdf (pmf) $f_\theta(x)$, $\theta \in \Theta$, then

$$E_\theta \left\{ \frac{\partial \log f_\theta(\mathbf{X})}{\partial \theta} \right\}^2 = \sum_{i=1}^n E_\theta \left\{ \frac{\partial \log f_\theta(X_i)}{\partial \theta} \right\}^2 = n E_\theta \left\{ \frac{\partial \log f_\theta(X_1)}{\partial \theta} \right\}^2$$

and (6) reduces to

$$\text{(8)} \qquad \text{var}_\theta\,(T(\mathbf{X})) \geq \frac{[\psi'(\theta)]^2}{nE_\theta\{\partial \log f_\theta(X_1)/\partial\theta\}^2}.$$

Example 1. Let $X \sim b(n, p)$; $\Theta = [0, 1] \subset \mathscr{R}$. The only condition that need be checked is differentiability under the summation sign. We have

$$\psi(p) = ET(X) = \sum_{x=0}^{n} \binom{n}{x} T(x)\, p^x(1-p)^{n-x},$$

which is a polynomial in p and hence can be differentiated with respect to p. For any unbiased estimate $T(X)$ of p we have

$$\text{var}_p\,(T(X)) \geq \frac{1}{n} p(1-p),$$

and since

$$\text{var}\left(\frac{X}{n}\right) = \frac{np(1-p)}{n^2} = \frac{p(1-p)}{n},$$

it follows that the variance of the estimate X/n attains the lower bound of the FCR inequality, and hence the estimate is the UMVUE.

Example 2. Let X be $U[0, \theta]$. Then

$$f_\theta(x) = \begin{cases} \dfrac{1}{\theta} & \text{if } 0 \leq x \leq \theta, \\ 0 & \text{otherwise}, \end{cases}$$

and $\Theta = (0, \infty)$.

Here $\partial f_\theta(x)/\partial\theta$ exists for $\theta \neq x$, but not if $\theta = x$, so that the regularity conditions do not hold. Let us compute $nE_\theta\{\partial \log f_\theta(X)/\partial\theta\}^2$. We have $E_\theta\{\partial \log f_\theta(X)/\partial\theta\}^2 = 1/\theta^2$, so that the lower bound of the FCR inequality is θ^2/n. We leave the reader to check that $[(n+1)/n] \max(X_1, X_2, \cdots, X_n)$ is unbiased for θ and has variance $\theta^2/[n(n+2)]$, which is much smaller than the FCR lower bound. This is not surprising since the regularity conditions do not hold. Note that $[(n+1)/n] \max(X_1,\cdots, X_n)$ is the UMVUE since $\max(X_1, X_2,\cdots,X_n)$ is a complete sufficient statistic.

Example 3. Let $X \sim P(\lambda)$. We leave the reader to check that the regularity conditions are satisfied and

$$\text{var}_\lambda\,(T(X)) \geq \lambda.$$

Since $T(X) = X$ has variance λ, X is the UMVUE of λ. Similarly, if we take a sample of size n from $P(\lambda)$, we can show that

$$\text{var}_\lambda\,(T(X_1, \cdots, X_n)) \geq \frac{\lambda}{n}$$

and $\bar{X}$ is the UMVUE.

Let us next consider the problem of unbiased estimation of $d(\lambda) = e^{-\lambda}$, based on a sample of size 1. The estimate

$$\partial(X) = \begin{cases} 1 & \text{if } X = 0, \\ 0 & \text{if } X \geq 1, \end{cases}$$

is unbiased for $d(\lambda)$ since

$$E_\lambda \, \partial(X) = E_\lambda \, [\partial(X)]^2 = P_\lambda \{X = 0\} = e^{-\lambda}.$$

Also,

$$\text{var}_\lambda \, (\partial(X)) = e^{-\lambda}(1 - e^{-\lambda}).$$

To compute the FCR lower bound we have

$$\log f_\lambda(x) = x \log \lambda - \lambda - \log x!.$$

This has to be differentiated with respect to $e^{-\lambda}$, since we want a lower bound for an estimate of the parameter $e^{-\lambda}$. Let $\theta = e^{-\lambda}$. Then

$$\log f_\theta(x) = x \log \log \frac{1}{\theta} + \log \theta - \log x!,$$

$$\frac{\partial}{\partial \theta} \log f_\theta(x) = x \frac{1}{\theta \log \theta} + \frac{1}{\theta},$$

and

$$E_\theta \left\{ \frac{\partial}{\partial \theta} \log f_\theta(X) \right\}^2 = \frac{1}{\theta^2}\left\{ 1 + \frac{2}{\log \theta} \log \frac{1}{\theta} \right.$$
$$\left. + \frac{1}{(\log \theta)^2}\left(\log \frac{1}{\theta} + \left(\log \frac{1}{\theta}\right)^2 \right) \right\}$$
$$= e^{2\lambda} \left\{ 1 - 2 + \frac{1}{\lambda^2}(\lambda + \lambda^2) \right\}$$
$$= \frac{e^{2\lambda}}{\lambda},$$

so that

$$\text{var}_\theta \, T(X) \geq \frac{\lambda}{e^{2\lambda}},$$

where $\theta = e^{-\lambda}$.

Since $e^{-\lambda}(1 - e^{-\lambda}) > \lambda e^{-2\lambda}$ for $\lambda > 0$, we see that $\text{var}(\delta(X))$ is greater than the lower bound obtained from the FCR inequality. We show next that $\delta(X)$ is the only unbiased estimate of θ, and hence is the UMVUE.

If h is any unbiased estimate of θ, it must satisfy $E_\theta h(X) = \theta$. Now the pmf of X in terms of θ is

$$f_\theta(x) = \frac{\theta(\log 1/\theta)^x}{x!}, \quad x = 0, 1, 2, \cdots,$$

so that

$$\theta = E_\theta h(X) = \theta \sum_{k=0}^{\infty} h(k) \frac{(\log 1/\theta)^k}{k!} \quad \text{for all} \quad 1 > \theta > 0.$$

It follows that $h(0) = 1$ and $h(k) = 0$ for $k = 1, 2, \cdots$, that is, that $h(X) = \partial(X)$, as asserted.

We next consider an inequality due to Chapman, Robbins, and Kiefer (the CRK inequality) that gives a lower bound for the variance of an estimate but does not require regularity conditions of the Fréchet-Cramer-Rao type.

Theorem 2 (Chapman and Robbins [11], Kiefer [58]). Let $\Theta \subset \mathcal{R}$ and $\{f_\theta(\mathbf{x}): \theta \in \Theta\}$ be a class of pdf's (pmf's). Let ψ be defined on Θ, and let T be an unbiased estimate of $\psi(\theta)$ with $E_\theta T^2 < \infty$ for all $\theta \in \Theta$. If $\theta \neq \varphi$, assume that f_θ and f_φ are different and assume further that there exists a $\varphi \in \Theta$ such that $\theta \neq \varphi$ and

(9) $$S(\theta) = \{f_\theta(\mathbf{x}) > 0\} \supset S(\varphi) = \{f_\varphi(\mathbf{x}) > 0\}.$$

Then

(10) $$\operatorname{var}_\theta (T(\mathbf{X})) \geq \sup_{\{\varphi: S(\varphi) \subset S(\theta), \varphi \neq \theta\}} \frac{[\psi(\varphi) - \psi(\theta)]^2}{\operatorname{var}_\theta \{f_\varphi(\mathbf{X})/f_\theta(\mathbf{X})\}}$$

for all $\theta \in \Omega$.

Proof. Since T is unbiased for ψ, $E_\varphi T(\mathbf{X}) = \psi(\varphi)$ for all $\varphi \in \Theta$. Hence, for $\varphi \neq \theta$,

(11) $$\int_{S(\theta)} T(\mathbf{x}) \frac{f_\varphi(\mathbf{x}) - f_\theta(\mathbf{x})}{f_\theta(\mathbf{x})} f_\theta(\mathbf{x}) \, d\mathbf{x} = \psi(\varphi) - \psi(\theta),$$

which yields

$$\operatorname{cov}_\theta \left\{ T(\mathbf{X}), \frac{f_\varphi(\mathbf{X})}{f_\theta(\mathbf{X})} - 1 \right\} = \psi(\varphi) - \psi(\theta).$$

Using the Cauchy-Schwarz inequality, we get

$$\operatorname{cov}_\theta^2 \left\{ T(\mathbf{X}), \frac{f_\varphi(\mathbf{X})}{f_\theta(\mathbf{X})} - 1 \right\} \leq \operatorname{var}_\theta (T(\mathbf{X})) \operatorname{var}_\theta \left\{ \frac{f_\varphi(\mathbf{X})}{f_\theta(\mathbf{X})} - 1 \right\}$$

$$= \operatorname{var}_\theta (T(\mathbf{X})) \operatorname{var}_\theta \left(\frac{f_\varphi(\mathbf{X})}{f_\theta(\mathbf{X})} \right).$$

Thus

$$\operatorname{var}_\theta (T(\mathbf{X})) \geq \frac{[\psi(\varphi) - \psi(\theta)]^2}{\operatorname{var}_\theta \{f_\varphi(\mathbf{X})/f_\theta(\mathbf{X})\}},$$

and the result follows. In the discrete case it is necessary only to replace the integral in the left side of (11) by a sum. The rest of the proof needs no change.

Remark 5. Inequality (10) holds without any regularity conditions on f_θ or $\psi(\theta)$. We will show that it covers some nonregular cases of the FCR inequality. Sometimes (10) is available in an alternative form. Let θ and $\theta + \delta$ ($\delta \neq 0$) be any two distinct values in Θ such that $S(\theta + \delta) \subset S(\theta)$, and take $\psi(\theta) = \theta$. Write

$$J = J(\theta, \delta) = \frac{1}{\delta^2}\left\{\left(\frac{f_{\theta+\delta}(\mathbf{X})}{f_\theta(\mathbf{X})}\right)^2 - 1\right\}.$$

Then (10) can be written as

(12) $$\operatorname{var}_\theta(T(\mathbf{X})) \geq \frac{1}{\inf_\delta E_\theta J},$$

where the infimum is taken over all $\delta \neq 0$ such that $S(\theta + \delta) \subset S(\theta)$.

Remark 6. Inequality (10) applies if the parameter space is discrete, but the Fréchet-Cramér-Rao regularity conditions do not hold in this case.

Example 4. Let X be $U[0, \theta]$. In Example 2 we showed that the regularity conditions of FCR inequality do not hold in this case. Let $\psi(\theta) = \theta$. If $\varphi < \theta$, then $S(\varphi) \subset S(\theta)$. Also,

$$E_\theta\left\{\frac{f_\varphi(X)}{f_\theta(X)}\right\}^2 = \int_0^\varphi \left(\frac{\theta}{\varphi}\right)^2 \frac{1}{\theta}\, dx = \frac{\theta}{\varphi}.$$

Thus

$$\operatorname{var}_\theta(T(X)) \geq \sup_{\{\varphi:\, \varphi < \theta\}} \frac{(\varphi - \theta)^2}{(\theta/\varphi) - 1} = \sup_{\{\varphi:\, \varphi < \theta\}} \{\varphi(\theta - \varphi)\} = \frac{\theta^2}{4}$$

for any unbiased estimate $T(X)$ of θ. X is a complete sufficient statistic, and $2X$ is unbiased for θ so that $T(X) = 2X$ is the UMVUE. Also

$$\operatorname{var}_\theta(2X) = 4 \operatorname{var} X = \frac{\theta^2}{3} > \frac{\theta^2}{4}.$$

Thus the lower bound $\theta^2/4$ of the CRK inequality is not achieved by any unbiased estimate of θ.

Example 5. Let X have pmf

$$P_N\{X = k\} = \begin{cases} \dfrac{1}{N}, & k = 1, 2, \cdots, N \\ 0, & \text{otherwise.} \end{cases}$$

Let $\Theta = \{N: N \geq M, M > 1 \text{ given}\}$. Take $\psi(N) = N$. Although the regularity conditions do not hold, (10) is applicable since, for $N \neq N' \in \Theta \subset \mathcal{R}$,

$$S(N) = \{1, 2, \cdots, N\} \supset S(N') = \{1, 2, \cdots, N'\} \quad \text{if } N' < N.$$

Also, P_N and $P_{N'}$ are different for $N \neq N'$. Thus

$$\text{var}_N(T) \geq \sup_{N' < N} \frac{(N - N')^2}{\text{var}_N \{P_{N'}/P_N\}}.$$

Now

$$\frac{P_{N'}}{P_N}(x) = \frac{P_{N'}(x)}{P_N(x)} = \begin{cases} \frac{N}{N'}, & x = 1, 2, \cdots, N', \ N' < N, \\ 0, & \text{otherwise,} \end{cases}$$

$$E_N \left\{ \frac{P_{N'}(X)}{P_N(X)} \right\}^2 = \frac{1}{N} \sum_1^{N'} \left(\frac{N}{N'} \right)^2 = \frac{N}{N'},$$

and

$$\text{var}_N \left\{ \frac{P_{N'}(X)}{P_N(X)} \right\} = \frac{N}{N'} - 1 > 0 \quad \text{for } N > N'.$$

It follows that

$$\text{var}_N(T(X)) \geq \sup_{N' < N} \frac{(N - N')^2}{(N - N')/N'} = \sup_{N' < N} N'(N - N').$$

Now

$$\frac{k(N - k)}{(k - 1)(N - k + 1)} > 1 \quad \text{if and only if } k < \frac{N + 1}{2},$$

so that $N'(N - N')$ increases as long as $N' < (N + 1)/2$ and decreases if $N' > (N + 1)/2$. The maximum is achieved at $[(N + 1)/2]$. Therefore

$$\text{var}_N(T(X)) \geq \left[\frac{N + 1}{2} \right] \left\{ N - \left[\frac{N + 1}{2} \right] \right\},$$

where $[x]$ is the largest integer $\leq x$.

In this case X is a complete sufficient statistic, and $2X - 1$ is unbiased for N and hence is the UMVUE for N. The variance of the UMVUE is $(N^2 - 1)/3$, which is $> [(N + 1)/2] \{N - [(N + 1)/2]\}$ for $N > 2$.

Example 6. Let $X \sim \mathcal{N}(0, \sigma^2)$. Let us compute J for $\delta \neq 0$.

$$J = \frac{1}{\delta^2} \left\{ \left(\frac{f_{\sigma+\delta}(\mathbf{X})}{f_\sigma(\mathbf{X})} \right)^2 - 1 \right\} = \frac{1}{\delta^2} \left[\frac{\sigma^{2n}}{(\sigma + \delta)^{2n}} \exp \left\{ -\frac{\sum X_i^2}{(\sigma+\delta)^2} + \frac{\sum X_i^2}{\sigma^2} \right\} - 1 \right]$$

$$= \frac{1}{\delta^2} \left[\left(\frac{\sigma}{\sigma + \delta} \right)^{2n} \exp \left\{ \frac{\sum X_i^2(\delta^2 + 2\sigma\delta)}{\sigma^2(\sigma + \delta)^2} \right\} - 1 \right],$$

$$E_\sigma J = \frac{1}{\delta^2} \left(\frac{\sigma}{\sigma + \delta} \right)^{2n} E_\sigma \left\{ \exp \left(c \frac{\sum X_i^2}{\sigma^2} \right) \right\} - \frac{1}{\delta^2},$$

where $c = (\delta^2 + 2\sigma\delta)/(\sigma + \delta)^2$.

Since $\sum X_i^2/\sigma^2 \sim \chi^2(n)$

$$E_\sigma J = \frac{1}{\delta^2}\left\{\left(\frac{\sigma}{\sigma + \delta}\right)^{2n}\frac{1}{(1-2c)^{n/2}} - 1\right\} \quad \text{for } c < \tfrac{1}{2}.$$

Let $k = \delta/\sigma$; then

$$c = \frac{2k + k^2}{(1 + k)^2} \quad \text{and} \quad 1 - 2c = \frac{1 - 2k - k^2}{(1 + k)^2},$$

$$E_\sigma J = \frac{1}{k^2\sigma^2}[(1 + k)^{-n}(1 - 2k - k^2)^{-n/2} - 1].$$

Here $1 + k > 0$ and $1 - 2c > 0$, so that $1 - 2k - k^2 > 0$, implying $-\sqrt{2} < k + 1 < \sqrt{2}$ and also $k > -1$. Thus $-1 < k < \sqrt{2} - 1$ and $k \neq 0$. Also,

$$\lim_{k \to 0} E_\sigma J = \lim_{k \to 0} \frac{(1 + k)^{-n}(1 - 2k - k^2)^{-n/2} - 1}{k^2\sigma^2}$$
$$= \frac{2n}{\sigma^2}$$

by L'Hospital's rule. We leave the reader to check that this is the FCR lower bound for $\text{var}_\sigma (T(X))$. But the minimum value of $E_\sigma J$ is not achieved in the neighborhood of $k = 0$ so that the CRK inequality is sharper than the FCR inequality. Next, we show that for $n = 2$ we can do better with the CRK inequality. We have

$$E_\sigma J = \frac{1}{k^2\sigma^2}\left\{\frac{1}{(1 - 2k - k^2)(1 + k)^2} - 1\right\}$$
$$= \frac{(k + 2)^2}{\sigma^2(1 + k)^2(1 - 2k - k^2)}, \quad -1 < k < \sqrt{2} - 1, \quad k \neq 0.$$

For $k = -.1607$ we achieve the lower bound as $(E_\sigma J)^{-1} = .2698\sigma^2$, so that $\text{var}_\sigma (T(X)) \geq .2698\sigma^2 > \sigma^2/4$. Finally, we show that this bound is by no means the best available; it is possible to improve on the Chapman-Robbins-Kiefer bounds too in some cases. Take

$$T(X_1, X_2, \cdots, X_n) = \frac{\Gamma(n/2)}{\Gamma[(n + 1)/2]}\frac{\sigma}{\sqrt{2}}\sqrt{\sum_1^n \frac{X_i^2}{\sigma^2}}$$

to be an estimate of σ. Now $E_\sigma T = \sigma$ and

$$E_\sigma T^2 = \frac{\sigma^2}{2}\left(\frac{\Gamma(n/2)}{\Gamma[(n + 1)/2]}\right)^2 E\left\{\sum_1^n \frac{X_i^2}{\sigma^2}\right\}$$
$$= \frac{n\sigma^2}{2}\left\{\frac{\Gamma(n/2)}{\Gamma[(n + 1)/2]}\right\}^2,$$

so that

$$\text{var}_\sigma(T) = \sigma^2\left\{\frac{n}{2}\left(\frac{\Gamma(n/2)}{\Gamma[(n+1)/2]}\right)^2 - 1\right\}.$$

For $n = 2$,

$$\text{var}_\sigma(T) = \sigma^2\left[\frac{4}{\pi} - 1\right] = .2732\sigma^2,$$

which is $> .2698\sigma^2$, the CRK bound. Note that T is the UMVUE.

Remark 7. In general the CRK inequality is as sharp as the FCR inequality. See Chapman and Robbins [11], pages 584–585, for details.

We next introduce the concept of *efficiency*.

Definition 1. Let T_1, T_2 be two unbiased estimates for a parameter θ. Suppose that $E_\theta T_1^2 < \infty$, $E_\theta T_2^2 < \infty$. We define the efficiency of T_1 relative to T_2 by

(13) $$\text{eff}_\theta(T_1|T_2) = \frac{\text{var}_\theta(T_1)}{\text{var}_\theta(T_2)}$$

and say that T_1 is more efficient than T_2 if

(14) $$\text{eff}_\theta(T_1|T_2) < 1.$$

It is usual to consider the performance of an unbiased estimate by comparing its variance with the lower bound given by the FCR inequality.

Definition 2. Assume that the regularity conditions of the FCR inequality are satisfied by the family of df's $\{F_\theta, \theta \in \Theta\}$, $\Theta \subseteq \mathcal{R}$. We say that an unbiased estimate T for parameter θ is most efficient for the family $\{F_\theta\}$ if

(15) $$\text{var}_\theta(T) = \left[E_\theta\left\{\frac{\partial \log f_\theta(\mathbf{X})}{\partial \theta}\right\}^2\right]^{-1}.$$

Definition 3. Let T be the most efficient estimate for the regular family of df's $\{F_\theta, \theta \in \Theta\}$. Then the efficiency of any unbiased estimate T_1 of θ is defined as

(16) $$\text{eff}_\theta(T_1) = \text{eff}_\theta(T_1|T) = \frac{\text{var}_\theta(T_1)}{\text{var}_\theta(T)}.$$

Clearly the efficiency of the most efficient estimate is 1, and the efficiency of any unbiased estimate T_1 is > 1.

Definition 4. We say that an estimate T_1 is asymptotically (most) efficient if
$$\lim_{n \to \infty} \text{eff}_\theta (T_1) = 1 \tag{17}$$
and T_1 is at least asymptotically unbiased in the sense that $\lim_{n \to \infty} E_\theta T_1 = \theta$. Here n is the sample size.

Remark 8. Definition 2, although in common usage, has many drawbacks. We have already seen cases in which the regularity conditions are not satisfied and yet UMVUE's exist. The definition does not cover such cases. Moreover, in many cases where the regularity conditions are satisfied and UMVUE's exist, the UMVUE is not most efficient since the variance of the best estimate (the UMVUE) does not achieve the lower bound of the FCR inequality.

Example 7. Let $X \sim b(n, p)$. Then we have seen in Example 1 that X/n is the UMVUE since its variance achieves the lower bound of the FCR inequality. It follows that X/n is most efficient.

Example 8. Let $X_1, X_2, \cdots, X_n$ be a sample from $\mathcal{N}(\mu, \sigma^2)$, where both μ and σ^2 are unknown. Then $(\bar{X}, S^2)$ is jointly sufficient for (μ, σ^2). Consider the estimate S^2 for σ^2. Since $(n-1)S^2/\sigma^2 \sim \chi^2(n-1)$, we have $\text{var}(S^2) = 2\sigma^4/(n-1)$. The FCR inequality lower bound for an unbiased estimate of σ^2 can be shown to be $2\sigma^4/n$, so that

$$\text{eff}(S^2) = \frac{2\sigma^4/(n-1)}{2\sigma^4/n} = \frac{n}{n-1} > 1.$$

However, since $\text{eff}(S^2) \to 1$ as $n \to \infty$, S^2 is asymptotically efficient.

Finally we explore the relationship between most efficient unbiased estimates and UMVUE's.

In Theorem 1 we showed that, if $\psi'(\theta) \neq 0$ and T is most efficient for θ, then

$$\frac{1}{K(\theta)} [T(\mathbf{x}) - \psi(\theta)] = \frac{\partial \log f_\theta(\mathbf{x})}{\partial \theta} \quad \text{for all} \quad \theta \in \Theta. \tag{18}$$

Let us assume for simplicity of notation that $\Theta = \mathcal{R}$ and that

$$C(\theta) = \int_{-\infty}^{\theta} \frac{d\theta}{K(\theta)}, \quad \Psi(\theta) = \int_{-\infty}^{\theta} \frac{\psi(\theta)}{K(\theta)} d\theta \tag{19}$$

are defined, and let $\lim_{\theta \to -\infty} f_\theta(\mathbf{x}) = \lambda(\mathbf{x})$. Then integrating (18) with respect to θ yields

$$f_\theta(\mathbf{x}) = \exp\{T(\mathbf{x}) C(\theta) - \Psi(\theta) + \lambda(\mathbf{x})\} \tag{20}$$

for all $\mathbf{x} \in \mathcal{R}_n$ and all $\theta \in \mathcal{R}$. This is a one-parameter exponential family, and we see that T is sufficient.

We have thus shown that the most efficient estimate T of a parameter $\psi(\theta)$ is sufficient and satisfies (18). In the converse direction we can state that, if T is sufficient for $\psi(\theta)$ and satisfies (18), it is most efficient. To see this, first note that sufficiency of T implies that the conditional distribution, given $T = t$, is independent of θ. Using (18), we have

$$(21) \qquad E_\theta \left\{ \frac{\partial \log f_\theta(\mathbf{X})}{\partial \theta} \right\}^2 = \left[\frac{1}{K(\theta)} \right]^2 \mathrm{var}_\theta\, (T(\mathbf{X})).$$

Using (18) in

$$E_\theta \left\{ [T(\mathbf{X}) - \psi(\theta)] \frac{\partial \log f_\theta(\mathbf{X})}{\partial \theta} \right\} = \psi'(\theta),$$

we get

$$(22) \qquad K(\theta)\, E_\theta \left\{ \frac{\partial \log f_\theta(\mathbf{X})}{\partial \theta} \right\}^2 = \psi'(\theta).$$

From (21) and (22), it follows that

$$\mathrm{var}_\theta\, (T(\mathbf{X})) = [\psi'(\theta)]^2 \left[E_\theta \left\{ \frac{\partial \log f_\theta(\mathbf{X})}{\partial \theta} \right\}^2 \right]^{-1},$$

as asserted. We have thus proved the following theorem.

Theorem 3. A necessary and sufficient condition for an estimate T to be most efficient is that T be sufficient and (18) hold.

Clearly, an estimate T satisfying the conditions of Theorem 3 will be the UMVUE, and two estimates coincide. We emphasize that we have assumed the regularity conditions of FCR inequality in making this statement.

Corollary. If $\{f_\theta\}$ is a one-parameter exponential family given by (5.5.1) and $Q(\theta)$ has a continuous nonvanishing derivative (on Θ), $T(\mathbf{X})$ is the most efficient estimate and the UMVUE of $E_\theta(T(\mathbf{X}))$.

Remark 9. For a rigorous proof of the fact that, if the variance of an unbiased estimate attains the FCR lower bound, the family of distributions must be a one-parameter exponential family, the reader is referred to Wijsman [140].

PROBLEMS 8.5

1. Are the following families of distributions regular in the sense of Fréchet, Cramér, and Rao? If so, find the lower bound for the variance of an unbiased estimate based on a sample of size n.

(a) $f_\theta(x) = \theta^{-1} e^{-x/\theta}$ if $x > 0$, and $= 0$ otherwise; $\theta > 0$.
(b) $f_\theta(x) = e^{-(x-\theta)}$ if $\theta < x < \infty$, and $= 0$ otherwise.
(c) $f_\theta(x) = \theta(1-\theta)^x$, $x = 0, 1, 2, \cdots; 0 < \theta < 1$.
(d) $f(x; \sigma^2) = (1/\sigma\sqrt{2\pi}) e^{-x^2/2\sigma^2}$, $-\infty < x < \infty; \sigma^2 > 0$.
(e) In (d) find the lower bound for the variance of an unbiased estimate for σ.

2. Find the CRK lower bound for the variance of an unbiased estimate of θ, based on a sample of size n from the pdf of Problem 1(b).

3. Find the CRK bound for the variance of an unbiased estimate of θ in sampling from $\mathcal{N}(\theta, 1)$.

4. In Problem 1 check to see whether there exists a most efficient estimate in each case.

5. Let $X_1, X_2, \cdots, X_n$ be a sample from a three-point distribution:

$$P\{X = y_1\} = \frac{1-\theta}{2}, \quad P\{X = y_2\} = \frac{1}{2}, \quad P\{X = y_3\} = \frac{\theta}{2},$$

where $0 < \theta < 1$. Does the FCR inequality apply in this case? If so, what is the lower bound for the variance of an unbiased estimate of θ?

6. Let $X_1, X_2, \cdots, X_n$ be iid rv's with mean μ and finite variance. What is the efficiency of the unbiased (and consistent) estimate $[2/n(n+1)] \sum_{i=1}^n iX_i$ relative to $\bar{X}$?

7. When does the equality hold in the CRK inequality?

8. Let $X_1, X_2, \cdots, X_n$ be a sample from $\mathcal{N}(\mu, 1)$, and let $d(\mu) = \mu^2$.

(a) Show that the minimum variance of any estimate of μ^2 from the FCR inequality is $4\mu^2/n$.

(b) Show that $T(X_1, X_2, \cdots, X_n) = \bar{X}^2 - (1/n)$ is the UMVUE of μ^2 with variance $(4\mu^2/n + 2/n^2)$.

9. Let $X_1, X_2, \cdots, X_n$ be iid $G(1, 1/\alpha)$ rv's.

(a) Show that the estimate $T(X_1, X_2, \cdots, X_n) = (n-1)/n \bar{X}$ is the UMVUE for α with variance $\alpha^2/(n-2)$.

(b) Show that the minimum variance from FCR inequality is α^2/n.

10. In Problem 8.4.16 compute the relative efficiency of $\hat{\mu}_0$ with respect to $\hat{\mu}_1$.

11. Let $X_1, X_2, \cdots, X_n$ and $Y_1, Y_2, \cdots, Y_m$ be independent samples from $\mathcal{N}(\mu, \sigma_1^2)$ and $\mathcal{N}(\mu, \sigma_2^2)$, respectively, where $\mu, \sigma_1^2, \sigma_2^2$ are unknown. Let $\rho = \sigma_2^2/\sigma_1^2$ and $\theta = m/n$, and consider the problem of unbiased estimation of μ.

(a) If ρ is known, show that

$$\hat{\mu}_0 = \alpha \bar{X} + (1-\alpha) \bar{Y}$$

where $\alpha = \rho/(\rho + \theta)$ is the BLUE of μ. Compute var $(\hat{\mu}_0)$.

(b) If ρ is unknown, the unbiased estimate

$$\tilde{\mu} = \frac{\bar{X} + \theta \bar{Y}}{1 + \theta}$$

is optimum in the neighborhood of $\rho = 1$. Find the variance of $\tilde{\mu}$.

(c) Compute the efficiency of $\tilde{\mu}$ relative to $\hat{\mu}_0$.
(d) Another unbiased estimate of μ is

$$\hat{\mu} = \frac{\rho F \bar{X} + \theta \bar{Y}}{\theta + \rho F},$$

where $F = S_2^2 / \rho S_1^2$ is an $F(m - 1, n - 1)$ rv.

8.6 THE METHOD OF MOMENTS

One of the simplest methods of estimation is the *method of moments*, which we study in this section. In a wide variety of problems the parameter to be estimated is some known function of a given (finite) number of moments. Let us suppose that we are interested in estimating

(1) $$\theta = h(m_1, m_2, \cdots, m_k),$$

where h is some known numerical function, and m_j is the jth-order moment of the population distribution that is known to exist for $1 \leq j \leq k$.

Definition 1. The method of moments consists in estimating θ by the statistic

(2) $$T(X_1, \cdots, X_n) = h\left(n^{-1} \sum_1^n X_i, n^{-1} \sum_1^n X_i^2, \cdots, n^{-1} \sum_1^n X_i^k\right).$$

To make sure that T is a statistic, we will assume that $h \colon \mathcal{R}_k \to \mathcal{R}$ is a Borel-measurable function.

Remark 1. It is easy to extend the method to the estimation of joint moments. Thus we use $n^{-1} \sum_1^n X_i Y_i$ to estimate $E(XY)$, and so on,

Remark 2. From the WLLN, $n^{-1} \sum_{i=1}^n X_i^j \xrightarrow{P} EX^j$. Thus, if one is interested in estimating the population moments, the method of moments leads to consistent and unbiased estimates. Moreover, the method of moments estimates in this case are asymptotically normally distributed (see Section 7.3).

Again, if one estimates parameters of the type θ defined in (1) and h is a continuous function, the estimates $T(X_1, X_2, \cdots, X_n)$ defined in (2) are consistent for θ (see Problem 1). Under some mild conditions on h, the estimate T is also asymptotically normal (see Cramér [18], 386–387).

Example 1. Let $X_1, X_2, \cdots, X_n$ be iid rv's with common mean μ and variance σ^2. Then $\sigma = \sqrt{(m_2 - m_1^2)}$, and the method of moments estimate for σ is given by

$$T(X_1, \cdots, X_n) = \sqrt{\frac{1}{n}\sum_1^n X_i^2 - \left(\frac{\sum X_i}{n}\right)^2}.$$

Although T is consistent and asymptotically normal for σ, it is not unbiased.

In particular, if $X_1, X_2, \cdots, X_n$ are iid $P(\lambda)$ rv's, we know that $EX_1 = \lambda$ and var $(X_1) = \lambda$. The method of moments leads to using either $\bar{X}$ or $\sum_1^n (X_i - \bar{X})^2/n$ as an estimate of λ. [Note that $\bar{X}$ and $\sum_1^n (X_i - \bar{X})^2/n$ have different units of measurement.] To avoid this kind of ambiguity we take the estimate involving the lowest-order sample moment.

Example 2. Let $X_1, X_2, \cdots, X_n$ be a sample from

$$f(x) = \begin{cases} \dfrac{1}{b-a}, & a \leq x \leq b, \\ 0, & \text{otherwise.} \end{cases}$$

Then

$$EX = \frac{a+b}{2} \quad \text{and} \quad \text{var}(X) = \frac{(b-a)^2}{12}.$$

The method of moments leads to estimating EX by $\bar{X}$ and var (X) by $\sum_1^n (X_1 - \bar{X})^2/n$, so that the estimates for a and b, respectively, are

$$T_1(X_1, \cdots, X_n) = \bar{X} - \sqrt{\frac{3\sum_1^n (X_i - \bar{X})^2}{n}}$$

and

$$T_2(X_1, \cdots, X_n) = \bar{X} + \sqrt{\frac{3\sum_1^n (X_i - \bar{X})^2}{n}}.$$

Example 3. Let $X_1, X_2, \cdots, X_N$ be iid $b(n, p)$ rv's, where both n and p are unknown. The method of moments estimates of p and n are given by

$$\bar{X} = EX = np$$

and

$$\frac{1}{N}\sum_1^N X_i^2 = EX^2 = np(1-p) + n^2p^2.$$

Solving for n and p, we get the estimate for p as

$$T_1(X_1, \cdots, X_N) = \frac{\bar{X}}{T_2(X_1, \cdots, X_N)},$$

where $T_2(X_1, \cdots, X_N)$ is the estimate for n, given by

$$T_2(X_1, X_2, \cdots, X_N) = \frac{(\bar{X})^2}{\bar{X} + \bar{X}^2 - \left(\sum_{1}^{N} X_i^2/N\right)}.$$

Note that $\bar{X} \xrightarrow{P} np$, $\sum_{1}^{N} X_i^2/N \xrightarrow{P} np(1-p) + n^2p^2$, but T_2 does not converge in probability to n (see Remark 2 and Problem 1).

PROBLEMS 8.6

1. Let $X_n \xrightarrow{P} a$ and $Y_n \xrightarrow{P} b$, where a and b are constants Let $h: \mathcal{R}_2 \to \mathcal{R}$ be a continuous function. Show that $h(X_n, Y_n) \xrightarrow{P} h(a, b)$.

2. Let $X_1, X_2, \cdots, X_n$ be a sample from $G(\alpha, \beta)$. Find the method of moments estimate for (α, β).

3. Let $X_1, X_2, \cdots, X_n$ be a sample from $\mathcal{N}(\mu, \sigma^2)$. Find the method of moments estimate for (μ, σ^2).

4. Let $X_1, X_2, \cdots, X_n$ be a sample from $B(\alpha, \beta)$. Find the method of moments estimate for (α, β).

5. A random sample of size n is taken from the lognormal pdf

$$f(x; \mu, \sigma) = (\sigma\sqrt{2\pi})^{-1} x^{-1} \exp\left\{-\frac{1}{2\sigma^2}(\log x - \mu)^2\right\}, \quad x > 0.$$

Find the method of moments estimates for μ and σ^2.

8.7 MAXIMUM LIKELIHOOD ESTIMATES

In this section we study a frequently used method of estimation, namely, the *method of maximum likelihood estimation*. Consider the following example.

Example 1. Let $X \sim b(n, p)$. One observation on X is available, and it is known that n is either 2 or 3 and $p = \frac{1}{2}$ or $\frac{1}{3}$. Our object is to find an estimate of the pair (n, p). The following table gives the probability that $X = x$ for each possible pair (n, p):

x	$(2, \frac{1}{2})$	$(2, \frac{1}{3})$	$(3, \frac{1}{2})$	$(3, \frac{1}{3})$	Maximum Probability
0	$\frac{1}{4}$	$\frac{4}{9}$	$\frac{1}{8}$	$\frac{8}{27}$	$\frac{4}{9}$
1	$\frac{1}{2}$	$\frac{4}{9}$	$\frac{3}{8}$	$\frac{12}{27}$	$\frac{1}{2}$
2	$\frac{1}{4}$	$\frac{1}{9}$	$\frac{3}{8}$	$\frac{6}{27}$	$\frac{3}{8}$
3	0	0	$\frac{1}{8}$	$\frac{1}{27}$	$\frac{1}{8}$

The last column gives the maximum probability in each row, that is, for each value that X assumes. If the value $x = 1$, say, is observed, it is more probable that it came from the distribution $b(2, \frac{1}{2})$ than from any of the other distributions, and so on. The following estimate is, therefore, reasonable in that it maximizes the probability of the observed value:

$$(\hat{n}, \hat{p})(x) = \begin{cases} (2, \frac{1}{3}) & \text{if } x = 0, \\ (2, \frac{1}{2}) & \text{if } x = 1, \\ (3, \frac{1}{2}) & \text{if } x = 2, \\ (3, \frac{1}{2}) & \text{if } x = 3. \end{cases}$$

The *principle of maximum likelihood* essentially assumes that the sample is representative of the population and chooses as the estimate that value of the parameter that maximizes the pdf (pmf) $f_\theta(x)$.

Definition 1. Let $(X_1, X_2, \cdots, X_n)$ be a random vector with pdf (pmf) $f_\theta(x_1, x_2, \cdots, x_n)$, $\theta \in \Theta$. The function

(1) $$L(\theta; x_1, x_2, \cdots, x_n) = f_\theta(x_1, x_2, \cdots, x_n),$$

considered as a function of θ, is called the likelihood function.

Usually θ will be a vector of parameters. If $X_1, X_2, \cdots, X_n$ are iid with pdf (pmf) $f_\theta(x)$, the likelihood function is

(2) $$L(\theta; x_1, x_2, \cdots, x_n) = \prod_{i=1}^{n} f_\theta(x_i).$$

Let $\Theta \subseteq \mathcal{R}_k$ and $\mathbf{X} = (X_1, X_2, \cdots, X_n)$.

Definition 2. The principle of maximum likelihood estimation consists of choosing as an estimate of θ a $\hat{\theta}(\mathbf{X})$ that maximizes $L(\theta; x_1, x_2, \cdots, x_n)$, that is, to find a mapping $\hat{\theta}$ of $\mathcal{R}_n \to \mathcal{R}_k$ that satisfies

(3) $$L(\hat{\theta}; x_1, x_2, \cdots, x_n) = \sup_{\theta \in \Theta} L(\theta; x_1, x_2, \cdots, x_n).$$

(Constants are not admissible as estimates.)

If a $\hat{\theta}$ satisfying (3) exists, we call it a *maximum likelihood estimate* (MLE). The method of maximum likelihood estimation attempts to find the *mode* [the value that maximizes the pdf (pmf)] of the distribution. The fact that the mode is usually a poorer estimate than the mean or the median explains why the small-sample properties of the method are, in general, poor. For large samples, however, the mode tends to approach the mean (if it exists) and the median, and the method has many large-sample properties, as we shall see. [It is not difficult to construct examples in which θ is the mean of the distribution but the MLE is not the sample mean. See Problem 1 (f)].

MAXIMUM LIKELIHOOD ESTIMATES

It is convenient to work with the logarithm of the likelihood function. Since log is a monotone function,

(4) $$\log L(\hat{\boldsymbol{\theta}}; x_1, \cdots, x_n) = \sup_{\boldsymbol{\theta} \in \Theta} \log L(\boldsymbol{\theta}; x_1, \cdots, x_n).$$

Let Θ be an open subset of $\mathscr{R}_k$, and suppose that $f_{\boldsymbol{\theta}}(\mathbf{x})$ is a positive, differentiable function of $\boldsymbol{\theta}$ (that is, the first-order partial derivatives exist in the components of $\boldsymbol{\theta}$). If a supremum $\hat{\boldsymbol{\theta}}$ exists, it must satisfy (see P.2.3) the *likelihood equations*

(5) $$\frac{\partial \log L(\hat{\boldsymbol{\theta}}; x_1, \cdots, x_n)}{\partial \theta_j} = 0, \qquad j = 1, 2, \cdots, k, \quad \boldsymbol{\theta} = (\theta_1, \cdots, \theta_k).$$

Any nontrivial root of the likelihood equations (5) is called an MLE in the *loose sense*. A parameter value that provides the absolute maximum of the likelihood function is called an MLE in the *strict* sense or, simply, an MLE.

Remark 1. If $\Theta \subseteq \mathscr{R}$, there may still be many problems. Often the likelihood equation $\partial L/\partial \theta = 0$ has more than one root, or the likelihood function is not differentiable everywhere in Θ, or $\hat{\theta}$ may be a terminal value. Sometimes the likelihood equation may be quite complicated and difficult to solve explicitly. In that case one may have to resort to some numerical procedure to obtain the estimate. Similar remarks apply to the multiparameter case.

Example 2. Let $X_1, X_2, \cdots, X_n$ be a sample from $\mathscr{N}(\mu, \sigma^2)$, where both μ and σ^2 are unknown. Here $\Theta = \{(\mu, \sigma^2), -\infty < \mu < \infty, \sigma^2 > 0\}$. The likelihood function is

$$L(\mu, \sigma^2; x_1, \cdots, x_n) = \frac{1}{\sigma^n (2\pi)^{n/2}} \exp\left\{-\sum_{i=1}^n \frac{(x_i - \mu)^2}{2\sigma^2}\right\},$$

and

$$\log L(\mu, \sigma^2; \mathbf{x}) = -\frac{n}{2} \log \sigma^2 - \frac{n}{2} \log (2\pi) - \frac{\sum_{1}^n (x_i - \mu)^2}{2\sigma^2}.$$

The likelihood equations are

$$\frac{1}{\sigma^2} \sum_{i=1}^n (x_i - \mu) = 0$$

and

$$-\frac{n}{2}\frac{1}{\sigma^2} + \frac{1}{2\sigma^4} \sum_{i=1}^n (x_i - \mu)^2 = 0.$$

Solving the first of these equations for μ, we get $\hat{\mu} = \bar{X}$ and, substituting in the second, $\hat{\sigma}^2 = \sum_{i=1}^n [(X_i - \bar{X})^2/n]$. We see that $(\hat{\mu}, \hat{\sigma}^2) \in \Theta$ with probability 1.

We show that $(\hat{\mu}, \hat{\sigma}^2)$ maximizes the likelihood function. First note that $\bar{X}$ maximizes $L(\mu, \sigma^2; \mathbf{x})$ whatever σ^2 is, since $L(\mu, \sigma^2; \mathbf{x}) \to 0$ as $|\mu| \to \infty$, and in that case $L(\hat{\mu}, \sigma^2; \mathbf{x}) \to 0$ as $\sigma^2 \to 0$ or ∞ whenever $\hat{\boldsymbol{\theta}} \in \Theta$, $\hat{\boldsymbol{\theta}} = (\hat{\mu}, \hat{\sigma}^2)$.

Note that $\hat{\sigma}^2$ is not unbiased for σ^2. Indeed, $E\hat{\sigma}^2 = [(n-1)/n]\sigma^2$. But $n\hat{\sigma}^2/(n-1) = S^2$ is unbiased, as we already know. Also, $\hat{\mu}$ is unbiased, and both $\hat{\mu}$ and $\hat{\sigma}^2$ are consistent. In addition, $\hat{\mu}$ and $\hat{\sigma}^2$ are method of moments estimates for μ and σ^2, and $(\hat{\mu}, \hat{\sigma}^2)$ is jointly sufficient.

Finally, note that $\hat{\mu}$ is the MLE of μ if σ^2 is known; but if μ is known, the MLE of σ^2 is not $\hat{\sigma}^2$ but $\sum_1^n (X_i - \mu)^2/n$.

Example 3. Let $X_1, X_2, \cdots, X_n$ be a sample from pmf

$$P_N(k) = \begin{cases} \dfrac{1}{N}, & k = 1, 2, \cdots, N, \\ 0 & \text{otherwise.} \end{cases}$$

The likelihood function is

$$L(N; k_1, k_2, \cdots, k_n) = \begin{cases} \dfrac{1}{N^n}, & 1 \leq \max(k_1, \cdots, k_n) \leq N, \\ 0, & \text{otherwise.} \end{cases}$$

Clearly the MLE of N is given by

$$\hat{N}(X_1, X_2, \cdots, X_n) = \max(X_1, X_2, \cdots, X_n),$$

for, if we take any $\hat{\alpha} < \hat{N}$ as the MLE, then $P_{\hat{\alpha}}(k_1, k_2, \cdots, k_n) = 0$; and if we take any $\hat{\beta} > \hat{N}$ as the MLE, then $P_{\hat{\beta}}(k_1, k_2, \cdots, k_n) = 1/(\hat{\beta})^n < 1/(\hat{N})^n = P_{\hat{N}}(k_1, k_2, \cdots, k_n)$.

We see that the MLE $\hat{N}$ is consistent, sufficient, and complete, but not unbiased.

Example 4. Consider the hypergeometric pmf

$$P_N(x) = \begin{cases} \dfrac{\binom{M}{x}\binom{N-M}{n-x}}{\binom{N}{n}}, & \max(0, n-N+M) \leq x \leq \min(n, M), \\ 0, & \text{otherwise.} \end{cases}$$

To find the MLE $\hat{N} = \hat{N}(X)$ of N consider the ratio

$$R(N) = \frac{P_N(x)}{P_{N-1}(x)} = \frac{N-n}{N} \cdot \frac{N-M}{N-M-n+x}.$$

For values of N for which $R(N) > 1$, $P_N(x)$ increases with N, and for values of N for which $R(N) < 1$, $P_N(x)$ is a decreasing function of N:

$$R(N) > 1 \quad \text{if and only if} \quad N < \frac{nM}{x},$$

and

$$R(N) < 1 \quad \text{if and only if} \quad N > \frac{nM}{x}.$$

It follows that $P_N(x)$ reaches its maximum value where $N \approx nM/x$. Thus $\hat{N}(X) = [nM/X]$, where $[x]$ denotes the largest integer $\leq x$.

Example 5. Let $X_1, X_2, \cdots, X_n$ be a sample from $U[\theta - \frac{1}{2}, \theta + \frac{1}{2}]$. The likelihood function is

$$L(\theta; x_1, x_2, \cdots, x_n) = \begin{cases} 1 & \text{if } \theta - \frac{1}{2} \leq \min(x_1, \cdots, x_n) \\ & \leq \max(x_1, \cdots, x_n) \leq \theta + \frac{1}{2}, \\ 0 & \text{otherwise.} \end{cases}$$

Thus $L(\theta; \mathbf{x})$ attains its maximum provided that

$$\theta - \tfrac{1}{2} \leq \min(x_1, \cdots, x_n) \quad \text{and} \quad \theta + \tfrac{1}{2} \geq \max(x_1, \cdots, x_n),$$

or when

$$\theta \leq \min(x_1, \cdots, x_n) + \tfrac{1}{2} \quad \text{and} \quad \theta \geq \max(x_1, \cdots, x_n) - \tfrac{1}{2}.$$

It follows that every statistic $T(X_1, X_2, \cdots, X_n)$ such that

(6) $$\max_{1 \leq i \leq n} x_i - \tfrac{1}{2} \leq T(X_1, X_2, \cdots, X_n) \leq \min_{1 \leq i \leq n} X_i + \tfrac{1}{2}$$

is an MLE of θ. Indeed, for $0 < \alpha < 1$,

$$T_\alpha(X_1, \cdots, X_n) = \max_{1 \leq i \leq n} X_i - \tfrac{1}{2} + \alpha\left(1 + \min_{1 \leq i \leq n} X_i - \max_{1 \leq i \leq n} X_i\right)$$

lies in interval (6), and hence for each α, $0 < \alpha < 1$, $T_\alpha(X_1, \cdots, X_n)$ is an MLE of θ. In particular, if $\alpha = \frac{1}{2}$,

$$T_{1/2}(X_1, \cdots, X_n) = \frac{\min X_i + \max X_i}{2}$$

is an MLE of θ.

Example 6. Let $X \sim b(1, p)$, $p \in [\frac{1}{4}, \frac{3}{4}]$. In this case $L(p; x) = p^x(1-p)^{1-x}$, $x = 0, 1$, and we cannot differentiate $L(p; x)$ to get the MLE of p, since that would lead to $\hat{p} = x$, a value that does not lie in $\Theta = [\frac{1}{4}, \frac{3}{4}]$. We have

$$L(p; x) = \begin{cases} p, & x = 1, \\ 1 - p, & x = 0, \end{cases}$$

which is maximized if we choose $\hat{p}(x) = \frac{1}{4}$ if $x = 0$, and $= \frac{3}{4}$ if $x = 1$. Thus the MLE of p is given by

$$\hat{p}(X) = \frac{2X + 1}{4}.$$

Note that $E_p \hat{p}(X) = (2p + 1)/4$, so that $\hat{p}$ is biased. Also, the mean square error for $\hat{p}$ is

$$E_p(\hat{p}(X) - p)^2 = \tfrac{1}{16} E_p(2X + 1 - 4p)^2 = \tfrac{1}{16}.$$

In the sense of the MSE, the MLE is worse than the trivial estimate $\delta(X) = \tfrac{1}{2}$, for $E_p(\tfrac{1}{2} - p)^2 = (\tfrac{1}{2} - p)^2 \leq \tfrac{1}{16}$ for $p \in [\tfrac{1}{4}, \tfrac{3}{4}]$.

Example 7. Let $X_1, X_2, \cdots, X_n$ be iid $b(1, p)$ rv's, and suppose that $p \in (0,1)$. If $(0, 0, \cdots, 0)$ $((1, 1, \cdots, 1))$ is observed, $\bar{x} = 0$ ($\bar{x} = 1$) is the MLE, which is not an admissible value of p. Hence an MLE does not exist.

Example 8 (Oliver [87]). This example illustrates a distribution for which an MLE is necessarily an actual observation, but not necessarily any particular observation. Let $X_1, X_2, \cdots, X_n$ be a sample from pdf

$$f_\theta(x) = \begin{cases} \dfrac{2}{\alpha} \dfrac{x}{\theta}, & 0 \leq x \leq \theta, \\ \dfrac{2}{\alpha} \dfrac{\alpha - x}{\alpha - \theta}, & \theta \leq x \leq \alpha, \\ 0, & \text{otherwise,} \end{cases}$$

where $\alpha > 0$ is a (known) constant. The likelihood function is

$$L(\theta; x_1, x_2, \cdots, x_n) = \left(\frac{2}{\alpha}\right)^n \prod_{x_i \leq \theta} \left(\frac{x_i}{\theta}\right) \prod_{x_i > \theta} \left(\frac{\alpha - x_i}{\alpha - \theta}\right),$$

where we have assumed that observations are arranged in increasing order of magnitude, $0 \leq x_1 < x_2 < \cdots < x_n \leq \alpha$. Clearly L is continuous in θ (even for $\theta =$ some x_i) and differentiable for values of θ between any two x_i's. Thus, for $x_j < \theta < x_{j+1}$, we have

$$L(\theta) = \left(\frac{2}{\alpha}\right)^n \theta^{-j} (\alpha - \theta)^{-(n-j)} \prod_{i=1}^{j} x_i \prod_{i=j+1}^{n} (\alpha - x_i),$$

$$\frac{\partial \log L}{\partial \theta} = -\frac{j}{\theta} + \frac{n - j}{\alpha - \theta}, \quad \text{and} \quad \frac{\partial^2 \log L}{\partial \theta^2} = \frac{j}{\theta^2} + \frac{n - j}{(\alpha - \theta)^2} > 0.$$

It follows that any stationary value that exists must be a minimum, so that there can be no maximum in any range $x_j < \theta < x_{j+1}$. Moreover, there can be no maximum in $0 \leq \theta < x_1$ or $x_n < \theta \leq \alpha$. This follows since, for $0 \leq \theta < x_1$,

$$L(\theta) = \left(\frac{2}{\alpha}\right)^n (\alpha - \theta)^{-n} \prod_{i=1}^{n} (\alpha - x_i)$$

is a strictly increasing function of θ. By symmetry, $L(\theta)$ is a strictly decreasing function of θ in $x_n < \theta \leq \alpha$. We conclude that an MLE has to be one of the observations.

In particular, let $\alpha = 5$ and $n = 3$, and suppose that the observations, arranged in increasing order of magnitude, are 1, 2, 4. In this case the MLE can be shown to be $\hat{\theta} = 1$, which corresponds to the first-order statistic. If the sample values are 2, 3, 4, the third-order statistic is the MLE.

Example 9. Let $X_1, X_2, \cdots, X_n$ be a sample from $G(r, 1/\beta)$; $\beta > 0$ and $r > 0$ are both unknown. The likelihood function is

$$L(\beta, r; x_1, x_2, \cdots, x_n) = \begin{cases} \dfrac{\beta^{nr}}{\{\Gamma(r)\}^n} \prod_{i=1}^{n} x_i^{r-1} \exp\left(-\beta \sum_{i=1}^{n} x_i\right), & x_i \geq 0, \\ 0, & \text{otherwise.} \end{cases}$$

Then

$$\log L(\beta, r) = nr \log \beta - n \log \Gamma(r) + (r-1) \sum_{i=1}^{n} \log x_i - \beta \sum_{i=1}^{n} x_i,$$

$$\frac{\partial \log L(\beta, r)}{\partial \beta} = \frac{nr}{\beta} - \sum_{i=1}^{n} x_i = 0,$$

$$\frac{\partial \log L(\beta, r)}{\partial r} = n \log \beta - n \frac{\Gamma'(r)}{\Gamma(r)} + \sum_{i=1}^{n} \log x_i = 0.$$

The first of the likelihood equations yields $\hat{\beta}(x_1, x_2, \ldots, x_n) = \hat{r}/\bar{x}$, while the second gives

$$n \log \frac{r}{\bar{x}} + \sum_{i=1}^{n} \log x_i - n \frac{\Gamma'(r)}{\Gamma(r)} = 0,$$

that is,

$$\log r - \frac{\Gamma'(r)}{\Gamma(r)} = \log \bar{x} - \frac{1}{n} \sum_{i=1}^{n} \log x_i,$$

which is to be solved for $\hat{r}$. In this case, the likelihood equation is not easily solvable and it is necessary to resort to numerical methods, using tables for $\Gamma'(r)/\Gamma(r)$.

Remark 2. We have seen that MLE's may not be unique, although frequently they are. Also, they are not necessarily unbiased even if a unique MLE exists. In terms of MSE, an MLE may be worthless. Moreover, MLE's may not even exist. We have also seen that MLE's are functions of sufficient statistics. This is a general result, which we now prove.

Theorem 1. Let T be a sufficient statistic for the family of pdf's (pmf's) $f_\theta(\mathbf{x})$, $\theta \in \Theta$. If an MLE of θ exists, it is a function of T.

Proof. Since T is sufficient, we can write

$$f_\theta(\mathbf{x}) = h(\mathbf{x}) g_\theta(T(\mathbf{x})),$$

using the factorization criterion. Maximization of the likelihood function with respect to θ is therefore equivalent to maximization of $g_\theta(T(\mathbf{x}))$, which is a function of T alone.

Remark 3. If the likelihood equations exist and T is sufficient, the MLE's are given by

$$\frac{\partial \log g_\theta(T)}{\partial \theta_j} = 0, \quad j = 1, 2, \cdots, k,$$

and every nonconstant solution of these equations is a function that depends on T alone.

Remark 4. Theorem 1 does not say that an MLE is itself a sufficient statistic, although this will usually be the case. Consider, for example, the case of sampling from $U[\theta, \theta + 1]$, $\theta \in \mathscr{R}$. Then

$$f_\theta(\mathbf{x}) = \begin{cases} 1, & \theta \leq \min x_i \leq \max x_i \leq \theta + 1, \\ 0, & \text{otherwise,} \end{cases}$$

and it follows that $(\min X_i, \max X_i)$ is jointly sufficient for θ. Any value of θ satisfying

$$\max x_i - 1 \leq \theta \leq \min x_i$$

is an MLE. In particular, $\min_{1 \leq i \leq n} X_i$ is an MLE that is not sufficient.

Theorem 2. Suppose that the regularity conditions of the FCR inequality are satisfied and θ belongs to an open interval on the real line. If an estimate $\hat{\theta}$ of θ attains the FCR lower bound for the variance, the likelihood equation has a unique solution $\hat{\theta}$ that maximizes the likelihood.

Proof. If $\hat{\theta}$ attains the FCR lower bound, we have (see Theorem 8.5.1)

$$\frac{\partial \log f_\theta(\mathbf{X})}{\partial \theta} = [K(\theta)]^{-1} [\hat{\theta}(\mathbf{X}) - \theta]$$

with probability 1, and the likelihood equation has a unique solution $\theta = \hat{\theta}$. Let us write $A(\theta) = [K(\theta)]^{-1}$. Then

$$\frac{\partial^2 \log f_\theta(\mathbf{X})}{\partial \theta^2} = A'(\theta) (\hat{\theta} - \theta) - A(\theta),$$

so that

$$\left.\frac{\partial^2 \log f_\theta(\mathbf{X})}{\partial \theta^2}\right|_{\theta = \hat{\theta}} = - A(\theta).$$

We need only to show that $A(\theta) > 0$. But by (8.5.22)

$$A(\theta) = E_\theta \left\{ \frac{\partial \log f_\theta(\mathbf{X})}{\partial \theta} \right\}^2,$$

and the proof is complete.

Remark 5. In Theorem 2 we assumed the differentiability of $A(\theta)$ and the existence of the second-order partial derivative $\partial^2 \log f_\theta / \partial \theta^2$. If the conditions of Theorem 2 are satisfied, the most efficient estimate is necessarily the MLE. It does not follow, however, that every MLE is most efficient. For example, in sampling from a normal population, $\hat{\sigma}^2 = \sum_1^n (X_i - \bar{X})^2 / n$ is the MLE of σ^2, but it is not most efficient. Since $\sum (X_i - \bar{X})^2 / \sigma^2$ is $\chi^2(n-1)$, we see that var $(\hat{\sigma}^2) = 2(n-1) \sigma^4 / n^2$, which is not equal to the FCR lower bound, $2\sigma^4/n$. Note that $\hat{\sigma}^2$ is not even an unbiased estimate.

We next consider an important property of MLE's that is not a characteristic of unbiased estimates.

Theorem 3 (Zehna [147]). Let $\{f_\theta : \theta \in \Theta\}$ be a family of pdf's (pmf's), and let $L(\theta)$ be the likelihood function. Suppose that $\Theta \subseteq \mathcal{R}_k, k \geq 1$. Let $h: \Theta \to \Lambda$ be a mapping of Θ onto Λ, where Λ is an interval in $\mathcal{R}_p$ ($1 \leq p \leq k$). If $\hat{\theta}$ is an MLE of θ, then $h(\hat{\theta})$ is an MLE of $h(\theta)$.

Proof. For each $\lambda \in \Lambda$, let us define

$$\Theta_\lambda = \{\theta : \theta \in \Theta, h(\theta) = \lambda\}$$

and

$$M(\lambda; \mathbf{x}) = \sup_{\theta \in \Theta_\lambda} L(\theta; \mathbf{x}).$$

Then M defined on Λ is called the likelihood function induced by h. If $\hat{\theta}$ is any MLE of θ, then $\hat{\theta}$ belongs to one and only one set, $\Theta_{\hat{\lambda}}$ say. Since $\hat{\theta} \in \Theta_{\hat{\lambda}}$, $\hat{\lambda} = h(\hat{\theta})$. Now

$$M(\hat{\lambda}; \mathbf{x}) = \sup_{\theta \in \Theta_{\hat{\lambda}}} L(\theta; \mathbf{x}) \geq L(\hat{\theta}; \mathbf{x})$$

and $\hat{\lambda}$ maximizes M, since

$$M(\hat{\lambda}; \mathbf{x}) \leq \sup_{\lambda \in \Lambda} M(\lambda; \mathbf{x}) = \sup_{\theta \in \Theta} L(\theta; \mathbf{x}) = L(\hat{\theta}; \mathbf{x}),$$

so that $M(\hat{\lambda}; \mathbf{x}) = \sup_{\lambda \in \Lambda} M(\lambda; \mathbf{x})$. It follows that $\hat{\lambda}$ is an MLE of $h(\theta)$, where $\hat{\lambda} = h(\hat{\theta})$.

Example 10. Let $X \sim b(1, p)$, $0 \leq p \leq 1$, and let $h(p) = p(1-p)$. We wish to find the MLE of $h(p)$. Note that $\Lambda = [0, \frac{1}{4}]$. The function h is not one-to-one. The MLE of p based on a sample of size n is $\hat{p}(X_1, \ldots, X_n) = \bar{X}$. Hence the MLE of parameter $h(p)$ is $h(\bar{X}) = \bar{X}(1 - \bar{X})$.

Example 11. Consider a random sample from $G(1, \beta)$. It is required to find the MLE of β in the following manner. A sample of size n is taken, and it is known only that k, $0 \leq k \leq n$, of these observations are $\leq M$, where M is a fixed positive number.

Let $p = P\{X_i \leq M\} = 1 - e^{-M/\beta}$, so that $- M/\beta = \log(1 - p)$ and $\beta = M/\log[1/(1 - p)]$. Therefore the MLE of β is $M/\log[1/(1 - \hat{p})]$, where $\hat{p}$ is the MLE of p. To compute the MLE of p we have

$$L(p; x_1, x_2, \ldots, x_n) = p^k(1 - p)^{n-k},$$

so that the MLE of p is $\hat{p} = k/n$. Thus the MLE of β is

$$\hat{\beta} = \frac{M}{\log[n/(n-k)]}.$$

Finally we consider some important large-sample properties of MLE's. In the following we assume that $\{f_\theta, \theta \in \Theta\}$ is a family of pdf's (pmf's), where Θ is an open interval on $\mathscr{R}$. The conditions listed below are stated when f_θ is a pdf. Modifications for the case where f_θ is a pmf are obvious and will be left to the reader.

(i) $\partial \log f_\theta/\partial \theta$, $\partial^2 \log f_\theta/\partial \theta^2$, $\partial^3 \log f_\theta/\partial \theta^3$ exist for all $\theta \in \Theta$ and every x. Also,

$$\int_{-\infty}^{\infty} \frac{\partial f_\theta(x)}{\partial \theta} dx = E_\theta \frac{\partial \log f_\theta(X)}{\partial \theta} = 0 \quad \text{for all } \theta \in \Theta.$$

(ii) $$\int_{-\infty}^{\infty} \frac{\partial^2 f_\theta(x)}{\partial \theta^2} dx = 0 \quad \text{for all } \theta \in \Theta.$$

(iii) $$-\infty < \int_{-\infty}^{\infty} \frac{\partial^2 \log f_\theta(x)}{\partial \theta^2} f_\theta(x) dx < 0 \quad \text{for all } \theta.$$

(iv) There exists a function $H(x)$ such that for all $\theta \in \Theta$

$$\left|\frac{\partial^3 \log f_\theta(x)}{\partial \theta^3}\right| < H(x) \quad \text{and} \quad \int_{-\infty}^{\infty} H(x) f_\theta(x) dx = M(\theta) < \infty.$$

(v) There exists a function $g(\theta)$ that is positive and twice differentiable for every $\theta \in \Theta$ and a function $H(x)$ such that for all θ

$$\left|\frac{\partial^2}{\partial \theta^2}\left[g(\theta)\frac{\partial \log f_\theta}{\partial \theta}\right]\right| < H(x) \quad \text{and} \quad \int_{-\infty}^{\infty} H(x) f_\theta(x) dx < \infty.$$

Note that condition (v) is equivalent to condition (iv) with the added qualification that $g(\theta) = 1$.

We state the following results without proof.

Theorem 4 (Cramér [18]).

(a) Conditions (i), (iii) and (iv) imply that, with probability approaching 1, as $n \to \infty$, the likelihood equation has a consistent solution.

(b) Conditions (i) through (iv) imply that a consistent solution $\hat{\theta}_n$ of the liklihood equation is asymptotically normal, that is,

$$\sigma^{-1}\sqrt{n}(\hat{\theta}_n - \theta) \xrightarrow{L} Z$$

where Z is $\mathcal{N}(0, 1)$, and

$$\sigma^2 = \left[E_\theta\left\{\frac{\partial \log f_\theta(X)}{\partial \theta}\right\}^2\right]^{-1}.$$

On occasions one encounters examples where the conditions of Theorem 4 are not satisfied and yet a solution of the likelihood equation is consistent and asymptotically normal.

Example 12 (Kulldorf [65]). Let $X \sim \mathcal{N}(0, \theta)$, $\theta > 0$. Let $X_1, X_2, \ldots, X_n$ be n independent observations on X. The solution of the likelihood equation is $\hat{\theta}_n = \sum_{i=1}^{n} X_i^2/n$. Also, $EX^2 = \theta$, var $(X^2) = 2\theta^2$, and

$$E_\theta\left\{\frac{\partial \log f_\theta(X)}{\partial \theta}\right\}^2 = \frac{1}{2\theta^2}.$$

We note that

$$\hat{\theta}_n \xrightarrow{a.s.} \theta$$

and

$$\sqrt{n}(\hat{\theta}_n - \theta) = \sqrt{2\theta}\,\frac{\sum_{1}^{n} X_i^2 - n\theta}{\sqrt{2n\theta}} \xrightarrow{L} \mathcal{N}(0, 2\theta^2).$$

However,

$$\frac{\partial^3 \log f_\theta}{\partial^3 \theta} = -\frac{1}{\theta^3} + \frac{3x^2}{\theta^4} \to \infty \quad \text{as } \theta \to 0$$

and is not bounded in $0 < \theta < \infty$. Thus condition (iv) does not hold.

The following theorem covers such cases also.

Theorem 5 (Kulldorf [65]).

(a) Conditions (i), (iii), and (v) imply that, with probability appoaching 1 as $n \to \infty$, the likelihood equation has a solution.
(b) Conditions (i), (ii), (iii), and (v) imply that a consistent solution of the likelihood equation is asymptotically normal.

Proof of Theorems 4 and 5. For proofs we refer to Cramér [18], page 500, and Kulldorf [65].

Remark 6. It is important to note that the results in Theorems 4 and 5 establish the consistency of some root of the likelihood equation but not necessarily that of the MLE when the likelihood equation has several roots. Huzurbazar [51] has shown that under certain conditions the likelihood equation has at most one consistent solution and that the likelihood function has a relative maximum for such a solution. Since there may be several solutions for which the likelihood function has relative maxima, Cramér's and Huzurbazar's results still do not imply that a solution of the likelihood equation that makes the likelihood function an absolute maximum is necessarily consistent.

Wald [136] has shown that under certain conditions the MLE is strongly consistent. It is important to note that Wald does not make any differentiability assumptions.

In any event, if the MLE is a unique solution of the likelihood equation, we can use Theorems 4 and 5 to conclude that it is consistent and asymptotically normal. Note that the asymptotic variance is the same as the lower bound of the FCR inequality.

Example 13. Consider $X_1, X_2, \ldots, X_n$ iid $P(\lambda)$ rv's, $\lambda \in \Theta = (0, \infty)$. The likelihood equation has a unique solution, $\hat{\lambda}(x_1, \ldots, x_n) = \bar{x}$, which maximizes the likelihood function. We leave the reader to check that the conditions of Theorem 4 hold and that MLE $\bar{X}$ is consistent and asymptotically normal with mean λ and variance λ/n, a result that is immediate otherwise.

We leave the reader to check that in Example 12 conditions of Theorem 5 are satisfied.

PROBLEMS 8.7

1. Let $X_1, X_2, \ldots, X_n$ be iid rv's with common pmf (pdf) $f_\theta(x)$. Find an MLE for θ in each of the following cases.

(a) $f(x) = 1/\theta$, $x = 1, 2, \ldots, \theta$, $1 \leq \theta \leq \theta_0$ an integer.
(b) $f_\theta(x) = \frac{1}{2} e^{-|x-\theta|}$, $-\infty < x < \infty$.
(c) $f_\theta(x) = e^{-x+\theta}$, $\theta \leq x < \infty$.
(d) $f_\theta(x) = (\theta\alpha) x^{\alpha-1} e^{-\theta x^\alpha}$, $x > 0$ and α known.
(e) $f_\theta(x) = \theta(1-x)^{\theta-1}$, $0 \leq x \leq 1$, $\theta > 1$.
(f) $f_\theta(x) = \theta(1-\theta)^{-1} x^{(2\theta-1)/(1-\theta)}$, $0 < x \leq 1$, $\frac{1}{2} < \theta < 1$.

2. Find an MLE, if it exists, in each of the following cases.

(a) $X \sim b(n, \theta)$: both n and θ are unknown, and one observation is available.

(b) $X_1, X_2, \cdots, X_k$ are iid $b(n, \theta)$ rv's, and both n and θ are unknown.
(c) $X_1, X_2, \cdots, X_n \sim b(1, \theta)$, $\theta \in [\frac{1}{2}, \frac{3}{4}]$.
(d) $X_1, X_2, \cdots, X_n \sim \mathcal{N}(\theta, \theta^2)$, $\theta \in \mathcal{R}$.
(e) $X_1, X_2, \cdots, X_n$ is a sample from

$$P\{X = y_1\} = \frac{1-\theta}{2}, \quad P\{X = y_2\} = \frac{1}{2}, \quad P\{X = y_3\} = \frac{\theta}{2} (0 < \theta < 1).$$

(f) $X_1, X_2, \cdots, X_n \sim \mathcal{N}(\theta, \theta)$, $0 < \theta < \infty$.
(g) $X \sim \mathcal{C}(\theta, 0)$.

3. Suppose that n observations are taken on an rv X with distribution $\mathcal{N}(\mu, 1)$, but instead of recording all the observations one notes only whether or not the observation is less than 0. If $\{X < 0\}$ occurs m ($< n$) times, find the MLE of μ.

4. Let $X_1, X_2, \cdots, X_n$ be a random sample from pdf

$$f(x; \alpha, \beta) = \beta^{-1} e^{-\beta^{-1}(x-\alpha)}, \quad \alpha < x < \infty, \quad -\infty < \alpha < \infty, \quad \beta > 0.$$

(a) Find the MLE of (α, β).
(b) Find the MLE of $P_{\alpha,\beta}\{X_1 \geq 1\}$.

5. Let $X_1, X_2, \cdots, X_n$ be a sample from exponential density $f_\theta(x) = \theta e^{-\theta x}$, $x \geq 0$, $\theta > 0$. Find the MLE of θ, and show that it is consistent and asymptotically normal.

6. For Problem 8.6.5 find the MLE for (μ, σ^2).

7. For a sample of size 1 taken from $\mathcal{N}(\mu, \sigma^2)$, show that no MLE of (μ, σ^2) exists.

8. For Problem 5.2.5 suppose that we wish to estimate N on the basis of observations $X_1, X_2, \cdots, X_M$.
(a) Find the UMVUE of N.
(b) Find the MLE of N.
(c) Compare the MSE's of the UMVUE and the MLE.

9. Let X_{ij} ($i = 1, 2, \cdots, s; j = 1, 2, \cdots, n$) be independent rv's where $X_{ij} \sim \mathcal{N}(\mu_i, \sigma^2)$, $i = 1, 2, \cdots, s$. Find MLE's for $\mu_1, \mu_2, \cdots, \mu_s$, and σ^2. Show that the MLE for σ^2 is not consistent as $s \to \infty$ (n fixed).

(Neyman and Scott [86])

10. Let (X, Y) have a bivariate normal distribution with parameters $\mu_1, \mu_2, \sigma_1^2, \sigma_2^2$, and ρ. Suppose that n observations are made on the pair (X, Y), and $N - n$ observations on X; that is, $N - n$ observations on Y are missing. Find the MLE's of $\mu_1, \mu_2, \sigma_1^2, \sigma_2^2$, and ρ. (Anderson [1])

[*Hint*: If $f(x, y; \mu_1, \mu_2, \sigma_1^2, \sigma_2^2, \rho)$ is the joint pdf of (X, Y) write

$$f(x, y; \mu_1, \mu_2, \sigma_1^2, \sigma_2^2, \rho) = f_1(x; \mu_1, \sigma_1^2) f_{Y|X}(y|\beta_x, \sigma_2^2(1-\rho^2)),$$

where f_1 is the marginal (normal) pdf of X, and $f_{Y|X}$ is the conditional (normal) pdf of Y, given x with mean

$$\beta_x = \left(\mu_2 - \rho \frac{\sigma_2}{\sigma_1} \mu_1\right) + \rho \frac{\sigma_2}{\sigma_1} x$$

and variance $\sigma_2^2(1 - \rho^2)$. Maximize the likelihood function first with respect to μ_1 and σ_1^2 and then with respect to $\mu_2 - \rho(\sigma_2/\sigma_1)\mu_1$, $\rho\sigma_2/\sigma_1$, and $\sigma_2^2(1 - \rho^2)$.]

8.8 BAYES AND MINIMAX ESTIMATION

In this section we consider the problem of point estimation in a general setting. This discussion falls under the heading of *decision theory*.

Let $\{f_\theta: \theta \in \Theta\}$ be a family of pdf's (pmf's), and $X_1, X_2, \ldots, X_n$ be a sample of size n from this distribution. Once the sample point $(x_1, x_2, \ldots, x_n)$ is observed, the statistician takes an *action* on the basis of these data. Let us denote by $\mathscr{A}$ the set of all *actions* or *decisions* open to the statistician.

Definition 1. A decision function d is a statistic that takes values in $\mathscr{A}$, that is, d is a Borel-measurable function that maps $\mathscr{R}_n$ into $\mathscr{A}$.

If $\mathbf{X} = \mathbf{x}$ is observed, the statistician takes action $d(\mathbf{x}) \in \mathscr{A}$.

Example 1. Let $\mathscr{A} = \{a_1, a_2\}$. Then any decision function d divides the space of values of $(X_1, \ldots, X_n)$, namely, $\mathscr{R}_n$, into a set C and its complement C^c, such that if $\mathbf{x} \in C$ we take action a_1, and if $\mathbf{x} \in C^c$ action a_2 is taken. This is the problem of testing of hypotheses, which we will discuss in Chapter 9.

Example 2. Let $\mathscr{A} = \Theta$. In this case we face the problem of estimation.

Another element of decision theory is the specification of a *loss function*, which measures the loss incurred when we make a decision.

Definition 2. Let $\mathscr{A}$ be an arbitrary space of actions. A nonnegative function L that maps $\Theta \times \mathscr{A}$ into $\mathscr{R}$ is called a loss function.

The value $L(\theta, a)$ is the loss to the statistician if he takes action a when θ is the true parameter value. If we use the decision function $d(\mathbf{X})$, and L is the loss function and θ the true parameter value, the loss is the rv $L(\theta, d(\mathbf{X}))$. (As always, we will assume that L is a Borel-measurable function for this to be true.)

Definition 3. Let $\mathscr{D}$ be a class of decision functions that map $\mathscr{R}_n$ into $\mathscr{A}$, and let L be a loss function on $\Theta \times \mathscr{A}$. The function R defined on $\Theta \times \mathscr{D}$ by

$$(1) \qquad R(\theta, d) = E_\theta L(\theta, d(\mathbf{X}))$$

is known as the risk function associated with d at θ.

Example 3. Let $\mathscr{A} = \Theta \subseteq \mathscr{R}$, $L(\theta, a) = |\theta - a|^2$. Then
$$R(\theta, d) = E_\theta L(\theta, d(X)) = E_\theta \{d(X) - \theta\}^2,$$
which is just the MSE. If we restrict attention to estimates that are unbiased, the risk is just the variance of the estimate.

The basic problem of decision theory is the following: Given a loss function $L(\theta, a)$, a decision function $d \in \mathscr{D}$, and the risk function $R(\theta, d)$, what criteria should we use to choose one decision function in $\mathscr{D}$ over another? An ideal solution would be to choose d so that $R(\theta, d)$ is minimum for all $\theta \in \Theta$. Unfortunately this is usually not possible.

Definition 4. The principle of minimax is to choose $d^* \in \mathscr{D}$ so that
(2)
$$\max_\theta R(\theta, d^*) \leq \max_\theta R(\theta, d)$$
for all d in $\mathscr{D}$. Such a rule d^*, if it exists, is called a minimax rule (decision function).

If the problem is one of estimation, that is, if $\mathscr{A} = \Theta$, we call d^* satisfying (2) a *minimax estimate* of θ.

Example 4. Let $X \sim b(1, p)$, $p \in \Theta = \{\frac{1}{4}, \frac{1}{2}\}$ and $\mathscr{A} = \{a_1, a_2\}$. Let the loss function be defined as follows.

	a_1	a_2
$p_1 = \frac{1}{4}$	1	4
$p_2 = \frac{1}{2}$	3	2

The set of decision rules includes four functions: d_1, d_2, d_3, d_4, defined by $d_1(0) = d_1(1) = a_1$; $d_2(0) = a_1$, $d_2(1) = a_2$; $d_3(0) = a_2$, $d_3(1) = a_1$; and $d_4(0) = d_4(1) = a_2$. The risk function takes the following values

i	$R(p_1, d_i)$	$R(p_2, d_i)$	$\max_{p_1, p_2} R(p, d_i)$	$\min_i \max_{p_1, p_2} R(p, d_i)$
1	1	3	3	
2	$\frac{7}{4}$	$\frac{5}{2}$	$\frac{5}{2}$	$\frac{5}{2}$
3	$\frac{13}{4}$	$\frac{5}{2}$	$\frac{13}{4}$	
4	4	2	4	

Thus the minimax solution is $d_2(x) = a_1$ if $x = 0$, and $= a_2$ if $x = 1$.

The computation of minimax estimates is facilitated if we use the *Bayes estimation method*. So far, we have considered θ as a fixed constant and $f_\theta(\mathbf{x})$

has represented the pdf (pmf) of the rv **X**. In Bayesian estimation we treat θ as a random variable distributed according to pdf (pmf) $\pi(\theta)$ on Θ. Also, π is called the *a priori distribution*. Now $f(\mathbf{x}|\theta)$ represents the conditional probability density (or mass) function of rv **X**, given that $\theta \in \Theta$ is held fixed. Since π is the distribution of θ, it follows that the joint density (pmf) of θ and **X** is given by

(3) $$f(\mathbf{x}, \theta) = \pi(\theta) f(\mathbf{x}|\theta).$$

In this framework $R(\theta, d)$ is the conditional average loss, $E\{L(\theta, d(\mathbf{X})) | \theta\}$, given that θ is held fixed. (Note that we are using the same symbol to denote the rv θ and a value assumed by it.)

Definition 5. The Bayes risk of a decision function d is defined by

(4) $$R(\pi, d) = E_\pi R(\theta, d).$$

If θ is a continuous rv and **X** is of the continuous type, then

$$R(\pi, d) = \int R(\theta, d)\pi(\theta)\, d\theta$$
$$= \iint L(\theta, d(\mathbf{x})) f(\mathbf{x}|\theta)\, \pi(\theta)\, d\mathbf{x}\, d\theta$$
(5) $$= \iint L(\theta, d(\mathbf{x})) f(\mathbf{x}, \theta)\, d\mathbf{x}\, d\theta.$$

If θ is discrete with pmf π and **X** is of the discrete type, then

(6) $$R(\pi, d) = \sum_\theta \sum_\mathbf{x} L(\theta, d(\mathbf{x})) f(\mathbf{x}, \theta).$$

Similar expressions may be written in the other two cases.

Definition 6. A decision function d^* is known as a Bayes rule (procedure) if it minimizes the Bayes risk, that is, if

(7) $$R(\pi, d^*) = \inf_d R(\pi, d).$$

Definition 7. The conditional distribution of rv θ, given $\mathbf{X} = \mathbf{x}$, is called the a posteriori probability distribution of θ, given the sample.

Let the joint pdf (pmf) be expressed in the form

(8) $$f(\mathbf{x}, \theta) = g(\mathbf{x})\, h(\theta|\mathbf{x}),$$

where g denotes the joint marginal density (pmf) of **X**. The a priori pdf (pmf) $\pi(\theta)$ gives the distribution of θ before the sample is taken, and the a posteriori pdf (pmf) $h(\theta|\mathbf{x})$ gives the distribution of θ after sampling. In terms of $h(\theta|\mathbf{x})$ we may write

(9) $$R(\pi, d) = \int g(\mathbf{x}) \left\{ \int L(\theta, d(\mathbf{x})) h(\theta|\mathbf{x}) d\theta \right\} d\mathbf{x}$$

or

(10) $$R(\pi, d) = \sum_{\mathbf{x}} g(\mathbf{x}) \left\{ \sum_{\theta} L(\theta, d(\mathbf{x})) h(\theta|\mathbf{x}) \right\},$$

depending on whether f and π are both continuous or both discrete. Similar expressions may be written if only one of f and π is discrete.

Theorem 1. Consider the problem of estimation of a parameter $\theta \in \Theta \subseteq \mathcal{R}$ with respect to a quadratic loss function $L(\theta, d) = (\theta - d)^2$. A Bayes solution is given by

(11) $$d(\mathbf{x}) = E\{\theta | \mathbf{X} = \mathbf{x}\}$$

($d(\mathbf{x})$ defined by (11) is called the *Bayes estimate*).

Proof. In the continuous case, if π is the prior pdf of θ, then

$$R(\pi, d) = \int g(\mathbf{x}) \left\{ \int [\theta - d(\mathbf{x})]^2 h(\theta|\mathbf{x}) d\theta \right\} d\mathbf{x},$$

where g is the marginal pdf of $\mathbf{X}$, and h is the conditional pdf of θ, given $\mathbf{x}$. The Bayes rule is a function d that minimizes $R(\pi, d)$. Minimization of $R(\pi, d)$ is the same as minimization of

$$\int [\theta - d(\mathbf{x})]^2 h(\theta|\mathbf{x}) d\theta,$$

which is minimum if and only if

$$d(\mathbf{x}) = E\{\theta|\mathbf{x}\}.$$

The proof for the remaining cases is similar.

Remark 1. The argument used in Theorem 1 shows that a Bayes estimate is one which minimizes $E\{L(\theta, d(\mathbf{X}))|\mathbf{X}\}$. Theorem 1 is a special case which says that if $L(\theta, d(\mathbf{X})) = [\theta - d(\mathbf{X})]^2$ the function

$$d(\mathbf{x}) = \int \theta \, h(\theta|\mathbf{x}) d\theta$$

is the Bayes estimate for θ with respect to π, the a priori distribution on Θ.

Example 5. Let $X \sim b(n, p)$ and $L(p, d(x)) = [p - d(x)]^2$. Let $\pi(p) = 1$ for $0 < p < 1$ be the a priori pdf of p. Then

$$h(p|x) = \frac{\binom{n}{x} p^x (1-p)^{n-x}}{\int_0^1 \binom{n}{x} p^x (1-p)^{n-x} dp}.$$

It follows that
$$E\{p|x\} = \int_0^1 p\, h\{p|x\}\, dp$$
$$= \frac{x+1}{n+2}.$$

Hence the Bayes estimate is
$$d^*(X) = \frac{X+1}{n+2}.$$

The Bayes risk is
$$R(\pi, d^*) = \int \pi(p) \sum_{x=0}^n [d^*(x) - p]^2 f(x|p)\, dp$$
$$= \int_0^1 E\left\{\left(\frac{X+1}{n+2} - p\right)^2 \Big| p\right\} dp$$
$$= \frac{1}{(n+2)^2} \int_0^1 [np(1-p) + (1-2p)^2]\, dp$$
$$= \frac{1}{6(n+2)}.$$

Example 6. Let $X \sim \mathcal{N}(\mu, 1)$, and let the a priori pdf of μ be $\mathcal{N}(0, 1)$. Also, let $L(\mu, d) = [\mu - d(X)]^2$. Then
$$h(\mu|\mathbf{x}) = \frac{f(\mathbf{x}, \mu)}{g(\mathbf{x})} = \frac{\pi(\mu) f(\mathbf{x}|\mu)}{g(\mathbf{x})},$$

where
$$g(\mathbf{x}) = \int f(\mathbf{x}, \mu)\, d\mu$$
$$= \frac{1}{(2\pi)^{(n+1)/2}} \exp\left(-\tfrac{1}{2} \sum_1^n x_i^2\right)$$
$$\cdot \int_{-\infty}^{\infty} \exp\left\{-\frac{n+1}{2}\left(\mu^2 - 2\mu \frac{n\bar{x}}{n+1}\right)\right\} d\mu$$
$$= \frac{(n+1)^{-1/2}}{(2\pi)^{n/2}} \exp\left\{-\tfrac{1}{2} \sum x_i^2 + \frac{n^2 \bar{x}^2}{2(n+1)}\right\}.$$

It follows that
$$h(\mu|\mathbf{x}) = \frac{1}{\sqrt{2\pi/(n+1)}} \exp\left\{-\frac{n+1}{2}\left(\mu - \frac{n\bar{x}}{n+1}\right)^2\right\},$$

and the Bayes estimate is
$$d^*(\mathbf{x}) = E\{\mu|\mathbf{x}\} = \frac{n\bar{x}}{n+1} = \frac{\sum_1^n x_i}{n+1}.$$

The Bayes risk is

$$R(\pi, d^*) = \int \pi(\mu) \int [d^*(\mathbf{x}) - \mu]^2 f(\mathbf{x}|\mu) \, d\mathbf{x} \, d\mu$$

$$= \int_{-\infty}^{\infty} E_\mu \left\{ \frac{n\bar{X}}{n+1} - \mu \right\}^2 \pi(\mu) \, d\mu$$

$$= \int_{-\infty}^{\infty} (n+1)^{-2} (n + \mu^2) \, \pi(\mu) \, d\mu$$

$$= \frac{1}{n+1}.$$

The quadratic loss function used in Theorem 1 is but one example of a loss function in frequent use. Some other loss functions that may be used are

$$|\theta - d(\mathbf{X})|, \quad \frac{|\theta - d(\mathbf{X})|^2}{|\theta|}, \quad |\theta - d|^4, \quad \text{and} \left(\frac{|\theta - d(\mathbf{X})|}{|\theta| + 1} \right)^{1/2}.$$

Example 7. Let $X_1, X_2, \ldots, X_n$ be iid $\mathcal{N}(\mu, \sigma^2)$ rv's. It is required to find a Bayes estimate of μ of the form $d(x_1, \ldots, x_n) = d(\bar{x})$, where $\bar{x} = \sum_1^n x_i/n$, using the loss function $L(\mu, d) = |\mu - d(\bar{x})|$. From the argument used in the proof of Theorem 1 (or by Remark 1), the Bayes estimate is one that minimizes the integral $\int |\mu - d(\bar{x})| h(\mu|\bar{x}) \, d\mu$. This will be the case if we choose d to be the median of the conditional distribution (see Problem 3.2.5).

Let the a priori distribution of μ be $\mathcal{N}(0, \tau^2)$. Since $\bar{X} \sim \mathcal{N}(\mu, \sigma^2/n)$, we have

$$f(\bar{x}, \mu) = \frac{\sqrt{n}}{2\pi\sigma\tau} \exp\left\{ -\frac{(\mu - \theta)^2}{2\tau^2} - \frac{n(\bar{x} - \mu)^2}{2\sigma^2} \right\}.$$

Writing

$$(\bar{x} - \mu)^2 = (\bar{x} - \theta + \theta - \mu)^2 = (\bar{x} - \theta)^2 - 2(\bar{x} - \theta)(\mu - \theta) + (\mu - \theta)^2,$$

we see that the exponent in $f(\bar{x}, \mu)$ is

$$-\frac{1}{2} \left\{ (\mu - \theta)^2 \left(\frac{1}{\tau^2} + \frac{n}{\sigma^2} \right) - \frac{2n(\bar{x} - \theta)(\mu - \theta)}{\sigma^2} + \frac{n}{\sigma^2} (\bar{x} - \theta)^2 \right\}.$$

It follows that the joint pdf of μ and $\bar{X}$ is bivariate normal with means θ, θ, variances τ^2, $\tau^2 + (\sigma^2/n)$, and correlation coefficient $\tau/\sqrt{[\tau^2 + (\sigma^2/n)]}$. The marginal of $\bar{X}$ is $\mathcal{N}(0, \tau^2 + (\sigma^2/n))$, and the conditional distribution of μ, given $\bar{X}$, is normal with mean

$$\theta + \frac{\tau}{\sqrt{\tau^2 + (\sigma^2/n)}} \frac{\tau}{\sqrt{\tau^2 + (\sigma^2/n)}} (\bar{x} - \theta) = \frac{\theta(\sigma^2/n) + \bar{x}\tau^2}{\tau^2 + (\sigma^2/n)}$$

and variance

$$\tau^2\left[1 - \frac{\tau^2}{\tau^2 + (\sigma^2/n)}\right] = \frac{\tau^2\sigma^2/n}{\tau^2 + (\sigma^2/n)}$$

(see the proof of Theorem 5.4.1). The Bayes estimate is therefore the median of this conditional distribution, and since the distribution is symmetric about the mean,

$$d^*(\bar{x}) = \frac{\theta(\sigma^2/n) + \bar{x}\tau^2}{\tau^2 + (\sigma^2/n)}$$

is the Bayes estimate.

The following theorem provides a method for determining minimax estimates.

Theorem 2. Let $\{f_\theta : \theta \in \Theta\}$ be a family of pdf's (pmf's), and suppose that an estimate d^* of θ is a Bayes estimate corresponding to an a priori distribution π on Θ. If the risk function $R(\theta, d^*)$ is constant on Θ, then d^* is a minimax estimate for θ.

Proof. The proof is left as an exercise.

Example 8 (Hodges and Lehmann [48]). Let $X \sim b(n, p)$, $0 \le p \le 1$. We seek a minimax estimate of p of the form $\alpha X + \beta$, using the squared error loss function. We have

$$R(p, d) = E_p\{\alpha X + \beta - p\}^2 = E_p\{\alpha(X - np) + \beta + (\alpha n - 1)p\}^2$$
$$= [(\alpha n - 1)^2 - \alpha^2 n]p^2 + [\alpha^2 n + 2\beta(\alpha n - 1)]p + \beta^2,$$

which is a quadratic equation in p. To find α and β such that $R(p, d)$ is constant for all $p \in \Theta$, we set the coefficients of p^2 and p equal to 0 to get

$$(\alpha n - 1)^2 - \alpha^2 n = 0 \quad \text{and} \quad \alpha^2 n + 2\beta(\alpha n - 1) = 0.$$

It follows that

$$\alpha = \frac{1}{\sqrt{n}(1 + \sqrt{n})} \quad \text{or} \quad \frac{1}{\sqrt{n}(\sqrt{n} - 1)}$$

and

$$\beta = \frac{1}{2(1 + \sqrt{n})} \quad \text{or} \quad -\frac{1}{2(\sqrt{n} - 1)}.$$

Since $0 \le p \le 1$, we discard the second set of roots for both α and β, and then the estimate is of the form

BAYES AND MINIMAX ESTIMATION

$$d^*(X) = \frac{X}{\sqrt{n}(1+\sqrt{n})} + \frac{1}{2(1+\sqrt{n})}$$

It remains to show that d^* is Bayes against some a priori pdf π.

Consider the a priori pdf

$$\pi(p) = [B(\alpha', \beta')]^{-1} p^{\alpha'-1}(1-p)^{\beta'-1}, \quad 0 \le p \le 1, \quad \alpha', \beta' > 0.$$

The a posteriori pdf of p, given x, is expressed by

$$h(p|x) = \frac{p^{x+\alpha'-1}(1-p)^{n-x+\beta'-1}}{\int_0^1 p^{x+\alpha'-1}(1-p)^{n-x+\beta'-1} dp}.$$

It follows that

$$E\{p|x\} = \frac{\int_0^1 p^{x+\alpha'}(1-p)^{n-x+\beta'-1} dp}{\int_0^1 p^{x+\alpha'-1}(1-p)^{n-x+\beta'-1} dp} = \frac{B(x+\alpha'+1, n-x+\beta')}{B(x+\alpha', n-x+\beta')}$$

$$= \frac{x+\alpha'}{n+\alpha'+\beta'},$$

which is the Bayes estimate for a squared error loss. For this to be of the form d^*, we must have

$$\frac{1}{\sqrt{n}(1+\sqrt{n})} = \frac{1}{n+\alpha'+\beta'} \quad \text{and} \quad \frac{1}{2(1+\sqrt{n})} = \frac{\alpha'}{n+\alpha'+\beta'},$$

giving $\alpha' = \beta' = \sqrt{n}/2$. It follows that the estimate $d^*(X)$ is minimax with constant risk

$$R(p, d^*) = \frac{1}{4(1+\sqrt{n})^2} \quad \text{for all} \quad p \in [0, 1].$$

Note that the UMVUE (which is also the MLE) is $d(X) = X/n$ with risk $R(p, d) = p(1-p)/n$. Comparing the two risks (Figs. 1 and 2), we see that

$$\frac{p(1-p)}{n} \le \frac{1}{4(1+\sqrt{n})^2} \quad \text{if and only if} \quad |p - \tfrac{1}{2}| \ge \frac{\sqrt{1+2\sqrt{n}}}{2(1+\sqrt{n})},$$

so that

$$R(p, d^*) < R(p, d)$$

in the interval $(\tfrac{1}{2} - a_n, \tfrac{1}{2} + a_n)$, where $a_n \to 0$ as $n \to \infty$. Moreover,

$$\frac{\sup_p R(p, d)}{\sup_p R(p, d^*)} = \frac{1/4n}{1/[4(1+\sqrt{n})^2]} = \frac{n+2\sqrt{n}+1}{n} \to 1 \quad \text{as } n \to \infty.$$

Clearly we would prefer the minimax estimate if n is small, and would prefer the UMVUE because of its simplicity if n is large.

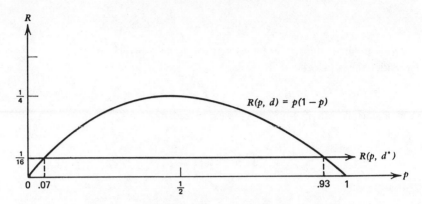

Fig. 1 Comparison of $R(p, d)$ and $R(p, d^*)$ for $n = 1$.

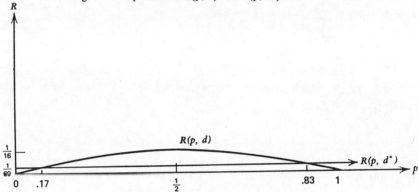

Fig. 2 Comparison of $R(p, d)$ $R(p, d^*)$ for $n = 9$.

Example 9 (Hodges and Lehmann [48]). A lot contains N elements, of which D are defective. A random sample of size n produces X defectives. We wish to estimate D. Clearly

$$P_D\{X = k\} = \binom{D}{k}\binom{N-D}{n-k}\binom{N}{n}^{-1},$$

$$E_D X = n\frac{D}{N}, \quad \text{and} \quad \sigma_D^2 = \frac{nD(N-n)(N-D)}{N^2(N-1)}.$$

Proceeding as in Example 8, we find a linear function of X with constant risk. Indeed, $E_D(\alpha X + \beta - D)^2 = \beta^2$ when

$$\alpha = \frac{N}{n + \sqrt{n(N-n)/(N-1)}}, \quad \beta = \frac{N}{2}\left(1 - \frac{\alpha n}{N}\right).$$

We show that $\alpha X + \beta$ is the Bayes estimate corresponding to a priori pmf

$$P\{D = d\} = c\int_0^1 \binom{N}{d} p^d(1-p)^{N-d} p^{a-1}(1-p)^{b-1} \, dp,$$

where $a, b > 0$ and $c = \Gamma(a+b)/\Gamma(a)\Gamma(b)$. First note that $\sum_{d=0}^N P\{D = d\} = 1$ so that

$$\sum_{d=0}^N \binom{N}{d} \frac{\Gamma(a+b)}{\Gamma(a)\Gamma(b)} \frac{\Gamma(a+d)\Gamma(N+b-d)}{\Gamma(N+a+b)} = 1.$$

The Bayes estimate is given by

$$d^*(k) = \frac{\sum_{d=k}^{N-n+k} d\binom{d}{k}\binom{N-d}{n-k}\binom{N}{d}\Gamma(a+d)\Gamma(N+b-d)}{\sum_{d=k}^{N-n+k} \binom{d}{k}\binom{N-d}{n-k}\binom{N}{d}\Gamma(a+d)\Gamma(N+b-d)}.$$

A little simplification, writing $d = (d-a) + a$ and using

$$\binom{d}{k}\binom{N-d}{n-k}\binom{N}{d} = \binom{N-n}{d-k}\binom{N}{n}\binom{n}{k},$$

yields

$$d^*(k) = \frac{\sum_{i=0}^{N-n}\binom{N-n}{i}\Gamma(d+a+1)\Gamma(N+b-d)}{\sum_0^{N-n}\binom{N-n}{i}\Gamma(d+a)\Gamma(N+b-d)} - a$$

$$= k\frac{a+b+N}{a+b+n} + \frac{a(N-n)}{a+b+n}.$$

Now putting

$$\alpha = \frac{a+b+N}{a+b+n} \quad \text{and} \quad \beta = \frac{a(N-n)}{a+b+n}$$

and solving for a and b, we get

$$a = \frac{\beta}{\alpha - 1}, \quad b = \frac{N - \alpha n - \beta}{\alpha - 1}.$$

Since $a > 0$, $\beta > 0$, and since $b > 0$, $N > \alpha n + \beta$. Moreover, $\alpha > 1$ if $N > n + 1$. If $N = n + 1$, the result is obtained if we give D a binomial distribution with parameter $p = \frac{1}{2}$. If $N = n$, the result is immediate.

PROBLEMS 8.8

1. It rains quite often in Bowling Green, Ohio. On a rainy day a teacher has essentially three choices: (1) to take an umbrella and face the possible prospect of

carrying it around in the sunshine; (2) to leave the umbrella at home and perhaps get drenched; or (3) to just give up the lecture and stay at home. Let $\Theta = \{\theta_1, \theta_2\}$, where θ_1 corresponds to rain, and θ_2, to no rain. Let $\mathscr{A} = \{a_1, a_2, a_3\}$, where a_i corresponds to the choice i, $i = 1, 2, 3$. Suppose that the following table gives the losses for the decision problem:

	θ_1	θ_2
a_1	1	2
a_2	4	0
a_3	5	5

The teacher has to make a decision on the basis of a weather report that depends on θ as follows.

	θ_1	θ_2
W_1(rain)	.7	.2
W_2(no rain)	.3	.8

Find the minimax rule to help the teacher reach a decision.

2. Let $X_1, X_2, \cdots, X_n$ be a random sample from $P(\lambda)$. For estimating λ, using the quadratic error loss function, an a priori distribution over Θ, given by pdf

$$\pi(\lambda) = e^{-\lambda} \quad \text{if} \quad \lambda > 0,$$
$$= 0 \quad \text{otherwise,}$$

is used.

(a) Find the Bayes estimate for λ.
(b) If it is required to estimate $\varphi(\lambda) = e^{-\lambda}$ with the same loss function and same a priori pdf, find the Bayes estimate for $\varphi(\lambda)$.

3. Let $X_1, X_2, \cdots, X_n$ be a sample from $b(1, \theta)$. Consider the class of decision rules d of the form $d(x_1, x_2, \cdots, x_n) = n^{-1}\sum_{i=1}^{n} x_i + \alpha$, where α is a constant to be determined. Find α according to the minimax principle, using the loss function $(\theta - d)^2$, where d is an estimate for θ.

4. Let d^* be a minimax estimate for $\psi(\theta)$ with respect to the squared error loss function. Show that $ad^* + b$ (a, b constants) is a minimax estimate for $a\psi(\theta) + b$.

5. Let $X \sim b(n, \theta)$, and suppose that the a priori pdf of θ is $U(0,1)$. Find the Bayes estimate of θ, using loss function $L(\theta, d) = (\theta - d)^2/[\theta(1 - \theta)]$. Find a minimax estimate for θ.

6. In Example 5 find the Bayes estimate for p^2.

7. Let $X_1, X_2, \cdots, X_n$ be a random sample from $G(1, 1/\lambda)$. To estimate λ, let the a priori pdf on λ be $\pi(\lambda) = e^{-\lambda}$, $\lambda > 0$, and let the loss function be squared error. Find the Bayes estimate of λ.

8.9 MINIMAL SUFFICIENT STATISTIC

In Section 3 we alluded to the fact that a given family of probability distributions that admits a nontrivial sufficient statistic usually admits several sufficient statistics. Clearly we would like to be able to choose the sufficient statistic that results in the greatest reduction of data collection. In this section we study the notion of a *minimal sufficient statistic*. For this purpose it is convenient to introduce the notion of a *sufficient partition*. The reader will recall that a *partition* of a space $\mathcal{X}$ is just a collection of disjoint sets E_α such that $\sum_\alpha E_\alpha = \mathcal{X}$. Any statistic $T(X_1, X_2, \cdots, X_n)$ induces a partition of the space of values of $(X_1, X_2, \cdots, X_n)$, that is, T induces a covering of $\mathcal{X}$ by a family $\mathfrak{A}$ of disjoint sets $A_t = \{(x_1, x_2, \cdots, x_n) \in \mathcal{X} : T(x_1, x_2, \cdots, x_n) = t\}$, where t belongs to the range of T. The sets A_t are called *partition sets*. Conversely, given a partition, any assignment of a number to each set so that no two partition sets have the same number assigned defines a statistic. Clearly this function is not, in general, unique.

Definition 1. Let $\{F_\theta : \theta \in \Theta\}$ be a family of df's, and $\mathbf{X} = (X_1, X_2, \cdots, X_n)$ be a sample from F_θ. Let $\mathfrak{A}$ be a partition of the sample space induced by a statistic $T = T(X_1, X_2, \cdots, X_n)$. We say that $\mathfrak{A} = \{A_t : t \text{ is in the range of } T\}$ is a sufficient partition for θ (or the family $\{F_\theta : \theta \in \Theta\}$) if the conditional distribution of $\mathbf{X}$, given $T = t$, does not depend on θ for any A_t, provided that the conditional probability is well defined.

Example 1. Let $X_1, X_2, \cdots, X_n$ be iid $b(1, p)$ rv's. The sample space of values of $(X_1, X_2, \cdots, X_n)$ is the set of n-tuples $(x_1, x_2, \cdots, x_n)$, where each $x_i = 0$ or $= 1$ and consists of 2^n points. Let $T(X_1, X_2, \cdots, X_n) = \sum_1^n X_i$, and consider the partition $\mathfrak{A} = \{A_0, A_1, \cdots, A_n\}$, where $\mathbf{x} \in A_j$ if and only if $\sum_1^n x_i = j$, $0 \leq j \leq n$. Each A_j contains $\binom{n}{j}$ sample points. The conditional probability

$$P_p\{\mathbf{x} \mid A_j\} = \frac{P_p\{\mathbf{x}\}}{P_p(A_j)} = \binom{n}{j}^{-1} \quad \text{if} \quad \mathbf{x} \in A_j,$$

and we see that $\mathfrak{A}$ is a sufficient partition.

Example 2. Let $X_1, X_2, \cdots, X_n$ be iid $U[0, \theta]$ rv's. Consider the statistic $T(\mathbf{X}) = \max_{1 \leq i \leq n} X_i$. The space of values of $X_1, X_2, \cdots, X_n$ is the set of points $\{\mathbf{x} : 0 \leq x_i \leq \theta, i = 1, 2, \cdots, n\}$. T induces a partition $\mathfrak{A}$ on this set. The sets of this partition are $A_t = \{(x_1, x_2, \cdots, x_n) : \max(x_1, \cdots, x_n) = t\}$, $t \in [0, \theta]$.

We have

$$f_\theta(\mathbf{x} \mid t) = \frac{f_\theta(\mathbf{x})}{f_\theta^T(t)} \quad \text{if} \quad \mathbf{x} \in A_t,$$

where $f_\theta^T(t)$ is the pdf of T. We have

$$f_\theta(\mathbf{x}|t) = \frac{1/\theta^n}{nt^{n-1}/\theta^n} = \frac{1}{nt^{n-1}} \quad \text{if} \quad \mathbf{x} \in A_t.$$

It follows that $\mathfrak{A} = \{A_t\}$ defines a sufficient partition.

Remark 1. Clearly a sufficient statistic T for a family of df's $\{F_\theta : \theta \in \Theta\}$ induces a sufficient partition; and, conversely, given a sufficient partition, we can define a sufficient statistic (not necessarily uniquely) for the family.

Remark 2. Two statistics T_1, T_2 that define the same partition must be in one-to-one correspondence, that is, there exists a function h such that $T_1 = h(T_2)$ with a unique inverse, $T_2 = h^{-1}(T_1)$. It follows that if T_1 is sufficient every one-to-one function of T_1 is also sufficient.

Let $\mathfrak{A}_1$, $\mathfrak{A}_2$ be two partitions of a space $\mathfrak{X}$. We say that $\mathfrak{A}_1$ is a *subpartition* of $\mathfrak{A}_2$ if every partition set in $\mathfrak{A}_2$ is a union of sets of $\mathfrak{A}_1$. We sometimes say also that $\mathfrak{A}_1$ is *finer* than $\mathfrak{A}_2$ ($\mathfrak{A}_2$ is *coarser* than $\mathfrak{A}_1$) or that $\mathfrak{A}_2$ is a *reduction* of $\mathfrak{A}_1$. In this case, a statistic T_2 that defines $\mathfrak{A}_2$ must be a function of any statistic T_1 that defines $\mathfrak{A}_1$. Clearly, this function need not have a unique inverse unless the two partitions have exactly the same partition sets.

Given a family of distributions $\{F_\theta : \theta \in \Theta\}$ for which a sufficient partition exists, we seek to find a sufficient partition $\mathfrak{A}$ that is as coarse as possible, that is, any reduction of $\mathfrak{A}$ leads to a partition that is not sufficient.

Definition 2. A partition $\mathfrak{A}$ is said to be minimal sufficient if

(i) $\mathfrak{A}$ is a sufficient partition, and
(ii) if $\mathscr{C}$ is any sufficient partition, $\mathscr{C}$ is a subpartition of $\mathfrak{A}$.

The question of the existence of the minimal partition was settled by Lehmann and Scheffé [69] and, in general, involves measure-theoretic considerations. However, in the cases that we consider where the sample space is either discrete or a finite-dimensional Euclidean space, and the family of distributions of $\mathbf{X}$ is defined by a family of pdf's (pmf's) $\{f_\theta, \theta \in \Theta\}$, such difficulties do not arise. The construction may be described as follows.

Two points $\mathbf{x}$ and $\mathbf{y}$ in the sample space are said to be equivalent, and we write $\mathbf{x} \sim \mathbf{y}$, if and only if the ratio $f_\theta(\mathbf{x})/f_\theta(\mathbf{y})$ does not depend on θ (whenever the ratio is defined). We leave the reader to check that "$\sim$" is an equivalence relation (that is, it is reflexive, symmetric, and transitive) and hence "$\sim$" defines a partition of the sample space. This partition defines the minimal sufficient partition.

Example 3. Consider again Example 1. Then

$$\frac{f_p(\mathbf{x})}{f_p(\mathbf{y})} = p^{\Sigma x_i - \Sigma y_i}(1-p)^{-\Sigma x_i + \Sigma y_i},$$

and this ratio is independent of p if and only if

$$\sum_1^n x_i = \sum_1^n y_i,$$

so that $\mathbf{x} \sim \mathbf{y}$ if and only if $\sum_1^n x_i = \sum_1^n y_i$. It follows that the partition $\mathfrak{A} = \{A_0, A_1, \cdots, A_n\}$, where $\mathbf{x} \in A_j$ if and only if $\sum_1^n x_i = j$, introduced in Example 1, is minimal sufficient.

A rigorous proof of the above assertion is beyond the scope of this book. The basic ideas are outlined in the following theorem.

Theorem 1. The relation " $\sim$ " defined above induces a minimal sufficient partition.

Proof. If T is a sufficient statistic, we have to show that $\mathbf{x} \sim \mathbf{y}$ whenever $T(\mathbf{x}) = T(\mathbf{y})$. This will imply that every set of the minimal sufficient partition is a union of sets of the form $A_t = \{T = t\}$, proving condition (ii) of Definition 2.

Sufficiency of T means that, whenever $\mathbf{x} \in A_t$, then

$$f_\theta\{\mathbf{x} \mid T = t\} = \frac{f_\theta(\mathbf{x})}{f_\theta^T(t)} \qquad \text{if } \mathbf{x} \in A_t$$

is independent of θ. It follows that, if both $\mathbf{x}$ and $\mathbf{y} \in A_t$, then

$$\frac{f_\theta(\mathbf{x} \mid t)}{f_\theta(\mathbf{y} \mid t)} = \frac{f_\theta(\mathbf{x})}{f_\theta(\mathbf{y})}$$

is independent of θ, and hence $\mathbf{x} \sim \mathbf{y}$.

To prove the sufficiency of the minimal sufficient partition $\mathfrak{A}$, let T_1 be an rv that induces $\mathfrak{A}$. Then T_1 takes on distinct values over distinct sets of $\mathfrak{A}$, but remains constant on the same set. If $\mathbf{x} \in \{T_1 = t_1\}$, then

(1) $$f_\theta(\mathbf{x} \mid T_1 = t_1) = \frac{f_\theta(\mathbf{x})}{P_\theta\{T_1 = t_1\}}.$$

Now

$$P_\theta\{T_1 = t_1\} = \int_{\{y : T_1(y) = t_1\}} f_\theta(\mathbf{y}) \, d\mathbf{y} \quad \text{or} \quad \sum_{\{y : T_1(y) = t_1\}} f_\theta(\mathbf{y}),$$

depending on whether the joint distribution of $\mathbf{X}$ is absolutely continuous or discrete. Since $f_\theta(\mathbf{x})/f_\theta(\mathbf{y})$ is independent of θ whenever $\mathbf{x} \sim \mathbf{y}$, it follows that the ratio on the right-hand side of (1) does not depend on θ. Thus T_1 is sufficient.

Definition 3. A statistic that induces the minimal sufficient partition is called a minimal sufficient statistic.

In view of Remark 2, a minimal sufficient statistic is not unique; every one-to-one function of a minimal sufficient statistic is itself minimal sufficient. In view of Theorem 1 a minimal sufficient statistic is a function of every sufficient statistic. We caution once again that not every function of a sufficient statistic is sufficient. It was shown by Lehmann and Scheffé [69] that, if $T_1(\mathbf{X})$ is sufficient for θ, $\theta \in \Theta$, then $T_2(\mathbf{X})$ is sufficient if and only if there exist a Borel set B with $P_\theta B = 1$ for all $\theta \in \Theta$ and a Borel-measurable function g such that $T_1(\mathbf{x}) = g(T_2(\mathbf{x}))$ for all $\mathbf{x} \in B$. This result can sometimes be used to test the sufficiency of a statistic. The following example is motivated by Harris and Soms [47].

Example 4. Let $X_1, X_2, \cdots, X_n$ be a sample from $\mathcal{N}(\mu, 1)$. Then the statistic $T(\mathbf{X}) = \sum_{i=1}^{n} X_i$ is easily shown to be sufficient—indeed, minimal sufficient—for μ. We show that $T_1(\mathbf{X}) = \sum_{i=1}^{n} X_i^2$ is not sufficient for μ. To establish the sufficiency of T_1 we have to show the existence of a Borel set B with $P_\mu(B) = 1$ and a Borel-measurable function g such that $T(\mathbf{x}) = g(T_1(\mathbf{x}))$ for $\mathbf{x} \in B$, that is,

$$\sum_{i=1}^{n} x_i = g(\sum_{i=1}^{n} x_i^2), \qquad \mathbf{x} \in B.$$

Differentiating with respect to x_i ($i = 1, 2, \cdots, n$), we get

$$\frac{\partial T(\mathbf{x})}{\partial x_i} = 1 = \frac{\partial g}{\partial T_1} \frac{\partial T_1}{\partial x_i}, \qquad i = 1, 2, \cdots, n, \quad \mathbf{x} \in B.$$

It follows that g must satisfy

$$n = 2 \frac{\partial g}{\partial T_1} \sum_{i=1}^{n} x_i = 2 \frac{\partial g}{\partial T_1} g(T_1),$$

hence

$$n T_1(\mathbf{x}) = [g(T_1(\mathbf{x}))]^2, \qquad \mathbf{x} \in B,$$

where we have used the fact that $T_1(0) = T(0) = 0$. Thus we have that such a g exists if and only if

$$n \sum_{i=1}^{n} x_i^2 = (\sum_{i=1}^{n} x_i)^2, \qquad \mathbf{x} \in B,$$

and therefore if and only if all the components of **x** are equal, or in the trivial case when $n = 1$. It follows that T_1 is not sufficient.

PROBLEMS 8.9

1. Let $X_1, X_2, \cdots, X_n$ be a random sample from a population with law $\mathscr{L}(X)$. Find a minimal sufficient statistic in each of the following cases.

   (a) $X \sim P(\lambda)$.
   (b) $X \sim U[0, \theta]$.
   (c) $X \sim NB(1; p)$.
   (d) $X \sim P_N$, where $P_N\{X = k\} = 1/N$ if $k = 1, 2, \cdots, N$, and $= 0$ otherwise.
   (e) $X \sim \mathcal{N}(\mu, \sigma^2)$.
   (f) $X \sim G(\alpha, \beta)$.
   (g) $X \sim B(\alpha, \beta)$.
   (h) $X \sim f_\theta(x)$, where $f_\theta(x) = (2/\theta^2)(\theta - x)$, $0 < x < \theta$.

2. Let X_1, X_2 be a sample of size 2 from $P(\lambda)$. Show that the statistic $X_1 + \alpha X_2$, where $\alpha > 1$ is an integer, is not sufficient for λ.

3. Let $X_1, X_2, \cdots, X_n$ be a sample from the pdf

$$f_\theta(x) = \begin{cases} \dfrac{x}{\theta} e^{-x^2/2\theta} & \text{if } x > 0 \\ 0 & \text{if } x \le 0 \end{cases} \quad \theta > 0.$$

Show that $\sum_{i=1}^n X_i^2$ is a minimal sufficient statistic for θ, but $\sum_{i=1}^n X_i$ is not sufficient.

4. Let $X_1, X_2, \cdots, X_n$ be a sample from $\mathcal{N}(0, \sigma^2)$. Show that $\sum_{i=1}^n X_i^2$ is a minimal sufficient statistic but $\sum_{i=1}^n X_i$ is not sufficient for σ^2.

5. Let $X_1, X_2, \cdots, X_n$ be a sample from pdf $f_{\alpha,\beta}(x) = \beta e^{-\beta(x-\alpha)}$ if $x > \alpha$, and $= 0$ if $x \le \alpha$. Find a minimal sufficient statistic for (α, β).

CHAPTER 9

Neyman-Pearson Theory of Testing of Hypotheses

9.1 INTRODUCTION

Let $X_1, X_2, \cdots, X_n$ be a random sample from a population distribution F_θ, $\theta \in \Theta$, where the functional form of F_θ is known except, perhaps, for the parameter θ. Thus, for example, the X_i's may be a random sample from $\mathcal{N}(\theta, 1)$, where $\theta \in \mathcal{R}$ is not known. In many practical problems the experimenter is interested in testing the validity of an assertion about the unknown parameter θ. For example, in a coin-tossing experiment it is of interest to test, in some sense, whether the (unknown) probability of heads p equals a given number p_0, $0 < p_0 < 1$. Similarly, it is of interest to check the claim of a car manufacturer about the average mileage per gallon of gasoline achieved by a particular model. A problem of this type is usually referred to as a problem of *testing of hypotheses* and is the subject of discussion in this chapter. We will develop the fundamentals of Neyman-Pearson theory. In Section 2 we introduce the various concepts involved. In Section 3 the fundamental Neyman-Pearson lemma is proved, and Sections 4 and 5 deal with some basic results in the testing of composite hypotheses.

9.2 SOME FUNDAMENTAL NOTIONS OF HYPOTHESES TESTING

In Chapter 8 we discussed the problem of point estimation in sampling from a population whose distribution is known except for a finite number of unknown parameters. Here we consider another important problem in statistical inference, the testing of statistical hypotheses. We begin by considering the following examples.

Example 1. In coin-tossing experiments one frequently assumes that the coin is fair, that is, the probability of getting heads or tails is the same: $\frac{1}{2}$. How does one test whether the coin is fair (unbiased) or loaded (biased)? If one is guided by intuition, a reasonable procedure would be to toss the coin n times say, and count the number of heads. If the proportion of heads observed does not deviate "too much" from $p = \frac{1}{2}$, one would tend to conclude that the coin is fair.

Example 2. It is usual for manufactures to make quantitative assertions about their products. For example, a manufacturer of 12 volt batteries may claim that a certain brand of his batteries lasts for N hours. How does one go about checking the truth of this assertion? A reasonable procedure suggests itself: Take a random sample of n batteries of the brand in question, and note their length of life under more or less identical conditions. If the average length of life is "much smaller" than N, one would tend to doubt the manufacturer's claim.

To fix ideas, let us define formally the concepts involved. As usual, $\mathbf{X} = (X_1, X_2, \cdots, X_n)$ and let $\mathbf{X} \sim f_\theta$, $\theta \in \Theta \subseteq \mathcal{R}_k$. It will be assumed that the functional form of f_θ is known except for the parameter θ. Also, we assume that Θ contains at least two points.

Definition 1. A parametric hypothesis is an assertion about the unknown parameter θ. It is usually referred to as the null hypothesis, $H_0: \theta \in \Theta_0 \subset \Theta$. The statement $H_1: \theta \in \Theta_1 = \Theta - \Theta_0$ is usually referred to as the alternative hypothesis.

Definition 2. If Θ_0 (Θ_1) contains only one point, we say that $\Theta_0(\Theta_1)$ is simple; otherwise, composite. Thus, if a hypothesis is simple, the probability distribution of $\mathbf{X}$ is completely specified under the hypothesis.

Example 3. Let $X \sim \mathcal{N}(\mu, \sigma^2)$. If both μ and σ^2 are unknown, $\Theta = \{(\mu, \sigma^2): -\infty < \mu < \infty, \sigma^2 > 0\}$. The hypothesis $H_0: \mu \leq \mu_0, \sigma^2 > 0$, where μ_0 is a known constant, is a composite null hypothesis. The alternative hypothesis is $H_1: \mu > \mu_0, \sigma^2 > 0$, which is also composite. Similarly, the null hypothesis $\mu = \mu_0, \sigma^2 > 0$ is also composite.

If $\sigma^2 = \sigma_0^2$ is known, the hypothesis $H_0: \mu = \mu_0$ is a simple hypothesis.

Example 4. Let $X_1, X_2, \cdots, X_n$ be iid $b(1, p)$ rv's. Some hypotheses of interest are $p = \frac{1}{2}, p \leq \frac{1}{2}, p \geq \frac{1}{2}$ or, quite generally, $p = p_0, p \leq p_0, p \geq p_0$, where p_0 is a known number, $0 < p_0 < 1$.

The problem of testing of hypotheses may be described as follows: Given the sample point $\mathbf{x} = (x_1, x_2, \cdots, x_n)$, find a decision rule (function) that will lead to a decision to accept or reject the null hypothesis. In other words, partition the space $\mathscr{R}_n$ into two disjoint sets C and C^c such that, if $\mathbf{x} \in C$, we reject $H_0 : \boldsymbol{\theta} \in \Theta_0$ (and accept H_1), and if $\mathbf{x} \in C^c$, we accept H_0 that $\mathbf{X} \sim f_{\boldsymbol{\theta}}, \boldsymbol{\theta} \in \Theta_0$.

Definition 3. Let $\mathbf{X} \sim f_{\boldsymbol{\theta}}, \boldsymbol{\theta} \in \Theta$. A subset C of $\mathscr{R}_n$ such that, if $\mathbf{x} \in C$, then H_0 is rejected (with probability 1) is called the critical region (set):

$$C = \{\mathbf{x} \in \mathscr{R}_n : H_0 \text{ is rejected if } \mathbf{x} \in C\}.$$

There are two types of errors that can be made if one uses such a procedure. One may reject H_0 when in fact it is true, called a *type I error*, or accept H_0 when it is false, called a *type II error*.

		True	
		H_0	H_1
Accept	H_0	Correct	Type II error
	H_1	Type I error	Correct

If C is the critical region of a rule, $P_{\boldsymbol{\theta}} C$, $\boldsymbol{\theta} \in \Theta_0$, is a *probability of type I error*, and $P_{\boldsymbol{\theta}} C^c$, $\boldsymbol{\theta} \in \Theta_1$, is a *probability of type II error*. Ideally one would like to find a critical region for which both these probabilities are 0. This will be the case if we can find a subset $S \subseteq \mathscr{R}_n$ such that $P_{\boldsymbol{\theta}} S = 1$ for every $\boldsymbol{\theta} \in \Theta_0$ and $P_{\boldsymbol{\theta}} S = 0$ for every $\boldsymbol{\theta} \in \Theta_1$. Unfortunately situations such as this do not arise in practice, although they are conceivable. For example, let $X \sim \mathscr{C}(1, \theta)$ under H_0 and $X \sim P(\theta)$ under H_1. Usually, if a critical region is such that the probability of type I error is 0, it will be of the form "always accept H_0" and the probability of type II error will then be 1.

The procedure used in practice is to limit the probability of type I error to some preassigned level α (usually .01 or .05) that is small and to minimize the probability of type II error. To reformulate our problem in terms of this requirement, let us formalize these notions.

Definition 4. Every Borel-measurable mapping φ of $\mathscr{R}_n \to [0, 1]$ is known as a test function.

Some simple examples of test functions are $\varphi(\mathbf{x}) = 1$ for all $\mathbf{x} \in \mathscr{R}_n$, $\varphi(\mathbf{x}) = 0$ for all $\mathbf{x} \in \mathscr{R}_n$, or $\varphi(\mathbf{x}) = \alpha$, $0 \leq \alpha \leq 1$, for all $\mathbf{x} \in \mathscr{R}_n$. In fact, Definition 4 includes Definition 3 in the sense that, whenever φ is the indicator function of some Borel subset A of $\mathscr{R}_n$, A is called the *critical region* (of the test φ).

Definition 5. The mapping φ is said to be a test of hypothesis $H_0: \theta \in \Theta_0$ against the alternatives $H_1: \theta \in \Theta_1$ with error probability α (also called level of significance or, simply, level) if

(1) $\qquad E_\theta \varphi(X) \le \alpha \qquad$ for all $\theta \in \Theta_0$.

We shall say, in short, that φ is a test for the problem $(\alpha, \Theta_0, \Theta_1)$.

Let us write $\beta_\varphi(\theta) = E_\theta \varphi(X)$. Our object, in practice, will be to seek a test φ for a given α, $0 \le \alpha \le 1$, such that

(2) $\qquad \sup_{\theta \in \Theta_0} \beta_\varphi(\theta) \le \alpha.$

The left-hand side of (2) is usually known as the *size* of the test φ. Condition (1) therefore restricts attention to tests whose size does not exceed a given level of significance α.

The following interpretation may be given to all tests φ satisfying $\beta_\varphi(\theta) \le \alpha$ for all $\theta \in \Theta_0$. To every $x \in \mathcal{R}_n$ we assign a number $\varphi(x)$, $0 \le \varphi(x) \le 1$, which is the probability of rejecting H_0 that $X \sim f_\theta$, $\theta \in \Theta_0$, if x is observed. The restriction $\beta_\varphi(\theta) \le \alpha$ for $\theta \in \Theta_0$ then says that, if H_0 were true, φ rejects it with a probability $\le \alpha$. We will call such a test a *randomized test function*. If $\varphi(x) = I_A(x)$, φ will be called a *nonrandomized* test. If $x \in A$, we reject H_0 with probability 1; and if $x \notin A$, this probability is 0. Needless to say, $A \in \mathfrak{B}_n$.

We next turn our attention to the type II error.

Definition 6. Let φ be a test function for the problem $(\alpha, \Theta_0, \Theta_1)$. For every $\theta \in \Theta$ define

(3) $\qquad \beta_\varphi(\theta) = E_\theta \varphi(X) = P_\theta \{\text{Reject } H_0\}.$

As a function of θ, $\beta_\varphi(\theta)$ is called the power function of the test φ. For any $\theta \in \Theta_1$, $\beta_\varphi(\theta)$ is called the power of φ against the alternative θ.

In view of Definitions 5 and 6 the problem of testing of hypotheses may now be reformulated. Let $X \sim f_\theta$, $\theta \in \Theta \subseteq \mathcal{R}_k$, $\Theta = \Theta_0 + \Theta_1$. Also, let $0 \le \alpha \le 1$ be given. Given a sample point x, find a test $\varphi(x)$ such that $\beta_\varphi(\theta) \le \alpha$ for $\theta \in \Theta_0$, and $\beta_\varphi(\theta)$ is maximum for $\theta \in \Theta_1$.

Definition 7. Let Φ_α be the class of all tests for the problem $(\alpha, \Theta_0, \Theta_1)$. A test $\varphi_0 \in \Phi_\alpha$ is said to be a most powerful (MP) test against an alternative $\theta \in \Theta_1$ if

(4) $\qquad \beta_{\varphi_0}(\theta) \ge \beta_\varphi(\theta) \qquad$ for all $\qquad \varphi \in \Phi_\alpha.$

If Θ_1 contains only one point, this definition suffices. If, on the other hand, Θ_1 contains at least two points, as will usually be the case, we will have an MP test corresponding to each $\theta \in \Theta_1$.

Definition 8. A test $\varphi_0 \in \Phi_\alpha$ for the problem $(\alpha, \Theta_0, \Theta_1)$ is said to be a uniformly most powerful (UMP) test if

(5) $\qquad \beta_{\varphi_0}(\theta) \geq \beta_\varphi(\theta) \qquad$ for all $\varphi \in \Phi_\alpha$, uniformly in $\theta \in \Theta_1$.

Thus, if Θ_0 and Θ_1 are both composite, the problem is to find a UMP test φ for the problem $(\alpha, \Theta_0, \Theta_1)$. We will see that UMP tests very frequently do not exist, and we will have to place further restrictions on the class of all tests, Φ_α.

Note that, if φ_1, φ_2 are two tests and λ is a real number, $0 < \lambda < 1$, then $\lambda \varphi_1 + (1 - \lambda) \varphi_2$ is also a test function, and it follows that the class of all test functions Φ_α is convex.

Example 5. Let $X_1, X_2, \cdots, X_n$ be iid $\mathcal{N}(\mu, 1)$ rv's, where μ is unknown but it is known that $\mu \in \Theta = \{\mu_0, \mu_1\}$, $\mu_0 < \mu_1$. Let $H_0: X_i \sim \mathcal{N}(\mu_0, 1)$, $H_1: X_i \sim \mathcal{N}(\mu_1, 1)$. Both H_0 and H_1 are simple hypotheses. Intuitively, one would accept H_0 if the sample mean $\bar{X}$ is "closer" to μ_0 than to μ_1; that is to say, one would reject H_0 if $\bar{X} > k$, and accept H_0 otherwise. The constant k is determined from the level requirements. Note that, under H_0, $\bar{X} \sim \mathcal{N}(\mu_0, 1/n)$, and, under H_1, $\bar{X} \sim \mathcal{N}(\mu_1, 1/n)$. Given $0 < \alpha < 1$, we have

$$P_{\mu_0}\{\bar{X} > k\} = P\left\{\frac{\bar{X} - \mu_0}{1/\sqrt{n}} > \frac{k - \mu_0}{1/\sqrt{n}}\right\}$$
$$= P\{\text{Type I error}\} = \alpha,$$

so that $k = \mu_0 + z_\alpha/\sqrt{n}$. The test, therefore, is (Fig. 1)

$$\varphi(\mathbf{x}) = \begin{cases} 1 & \text{if } \bar{x} > \mu_0 + z_\alpha/\sqrt{n}, \\ 0 & \text{otherwise.} \end{cases}$$

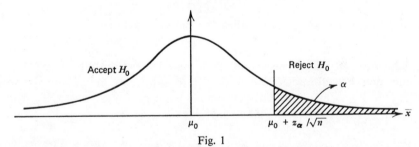

Fig. 1

$\bar{X}$ is known as a *test statistic*, and the test φ is nonrandomized with critical region $C = \{\mathbf{x}: \bar{x} > \mu_0 + z_\alpha/\sqrt{n}\}$. Note that in this case the continuity of $\mathbf{X}$ (that is, the absolute continuity of the df of $\mathbf{X}$) allows us to achieve any size α, $0 < \alpha < 1$.

The power of the test at μ_1 is given by

$$E_{\mu_1}\varphi(\mathbf{X}) = P_{\mu_1}\left\{\bar{X} > \mu_0 + \frac{z_\alpha}{\sqrt{n}}\right\}$$

$$= P\left\{\frac{\bar{X} - \mu_1}{1/\sqrt{n}} > (\mu_0 - \mu_1)\sqrt{n} + z_\alpha\right\}$$

$$= P\{Z > z_\alpha - \sqrt{n}\,(\mu_1 - \mu_0)\},$$

where $Z \sim \mathcal{N}(0, 1)$. In particular, $E_{\mu_1}\varphi(\mathbf{X}) > \alpha$ since $\mu_1 > \mu_0$. The probability of type II error is given by

$$P\{\text{Type II error}\} = 1 - E_{\mu_1}\varphi(X)$$
$$= P\{Z \leq z_\alpha - \sqrt{n}\,(\mu_1 - \mu_0)\}.$$

Example 6. Let X_1, X_2, X_3, X_4, X_5, be a sample from $b(1, p)$, where p is unknown and $0 \leq p \leq 1$. Consider the simple null hypothesis $H_0: X_i \sim b(1, \frac{1}{2})$, that is, under $H_0, p = \frac{1}{2}$. Then $H_1: X_i \sim b(1, p), p \neq 1/2$. A reasonable procedure would be to compute the average number of 1's, namely, $\bar{X} = \sum_1^5 X_i/5$, and to accept H_0 if $|\bar{X} - \frac{1}{2}| \leq c$, where c is to be determined. Let $\alpha = .10$. Then we would like to choose c such that the size of our test is α, that is,

$$.10 = P_{p=1/2}\{|\bar{X} - \tfrac{1}{2}| > c\},$$

or

$$.90 = P_{p=1/2}\{-5c \leq \sum_1^5 X_i - \tfrac{5}{2} \leq 5c\}$$

(6)
$$= P_{p=1/2}\{-k \leq \sum_1^5 X_i - \tfrac{5}{2} \leq k\},$$

where $k = 5c$. Now $\sum_1^5 X_i \sim b(5, \frac{1}{2})$ under H_0, so that the pmf of $\sum_1^5 X_i - \frac{5}{2}$ is given in the following table.

$\sum_1^5 x_i$	$\sum_1^5 x_i - \tfrac{5}{2}$	$P_{p=1/2}\{\sum_1^5 X_i = \sum_1^5 x_i\}$
0	-2.5	.03125
1	-1.5	.15625
2	$-.5$	.31250
3	$.5$	.31250
4	1.5	.15625
5	2.5	.03125

Note that we cannot choose any k to satisfy (6) exactly. It is clear that we have to reject H_0 when $k = \pm 2.5$, that is, when we observe $\sum X_i = 0$ or 5. The resulting size if we use this test is $\alpha = .03125 + .03125 = .0625 < .10$. A second procedure would be to reject H_0 if $k = \pm 1.5$ or ± 2.5 ($\sum X_i = 0, 1, 4, 5$), in which case the resulting size is $\alpha = .0625 + 2(.15625) = .375$, which is considerably larger than .10. A third alternative, if we insist on achieving $\alpha = .10$, is to randomize on the boundary. Instead of accepting or rejecting H_0 with probability 1 when $\sum X_i = 1$ or 4, we reject H_0 with probability γ where

$$.10 = P_{p=1/2}\{\sum_1^5 X_i = 0 \text{ or } 5\} + \gamma P_{p=1/2}\{\sum_1^5 X_i = 1 \text{ or } 4\}.$$

Thus

$$\gamma = \frac{.0375}{.3125} = .114.$$

A randomized test of size $\alpha = .10$ is therefore given by

$$\varphi(\mathbf{x}) = \begin{cases} 1 & \text{if } \sum_1^5 x_i = 0 \text{ or } 5, \\ .114 & \text{if } \sum_1^5 x_i = 1 \text{ or } 4, \\ 0 & \text{otherwise.} \end{cases}$$

The power of this test is

$$E_p\varphi(\mathbf{X}) = P_p\{\sum_1^5 X_i = 0 \text{ or } 5\} + .114\, P_p\{\sum_1^5 X_i = 1 \text{ or } 4\},$$

where $p \neq \frac{1}{2}$ and can be computed for any value of p.

We conclude this section with the following remarks.

Remark 1. So far we have been somewhat vague about the specification of the null hypothesis. The convention used is to choose H_0 such that it represents the status quo or some important situation. If H_0 is then rejected when in fact it is true, there is a significant change or loss. Thus the type I error is the one most important to the experimenter, and he would like to limit its probability at a certain fixed level α, usually small.

Remark 2. The problem of testing of hypotheses may be considered as a special case of the general decision problem described in Section 8.8. Let $\mathscr{A} = \{a_0, a_1\}$, where a_0 represents the decision to accept $H_0: \theta \in \Theta_0$, and a_1 represents the decision to reject H_0. A decision function δ is a mapping of $\mathscr{R}_n$ into $\mathscr{A}$. Let us introduce the following loss functions:

$$L_1(\theta, a_1) = \begin{cases} 1 & \text{if } \theta \in \Theta_0 \\ 0 & \text{if } \theta \in \Theta_1, \end{cases} \text{ and } L_1(\theta, a_0) = 0 \text{ for all } \theta,$$

and

$$L_2(\theta, a_0) = \begin{cases} 0 & \text{if } \theta \in \Theta_0 \\ 1 & \text{if } \theta \in \Theta_1, \end{cases} \text{ and } L_2(\theta, a_1) = 0 \text{ for all } \theta.$$

FUNDAMENTAL NOTIONS OF HYPOTHESES TESTING   411

Then the minimization of $E_\theta L_2(\theta, \delta(\mathbf{X}))$ subject to $E_\theta L_1(\theta, \delta(\mathbf{X})) \leq \alpha$ is the hypotheses testing problem discussed above. We have

$$E_\theta L_2(\theta, \delta(\mathbf{X})) = P_\theta\{\mathbf{x}: \delta(\mathbf{x}) = a_0\}, \qquad \theta \in \Theta_1,$$
$$= P_\theta\{\text{Accept } H_0 | H_1 \text{ true}\},$$

and

$$E_\theta L_1(\theta, \delta(\mathbf{X})) = P_\theta\{\mathbf{x}: \delta(\mathbf{x}) = a_1\}, \qquad \theta \in \Theta_0,$$
$$= P_\theta\{\text{Reject } H_0 | \theta \in \Theta_0 \text{ true}\}.$$

PROBLEMS 9.2

1. A sample of size 1 is taken from a population distribution $P(\lambda)$. To test $H_0: \lambda = 1$ against $H_1: \lambda = 2$, consider the nonrandomized test $\varphi(x) = 1$ if $x > 3$, and $= 0$ if $x \leq 3$. Find the probabilities of type I and type II errors and the power of the test against $\lambda = 2$. If it is required to achieve a size equal to .05, how should one modify the test φ?

2. Let $X_1, X_2, \cdots, X_n$ be a sample from a population with finite mean μ and finite variance σ^2. Suppose that μ is not known, but σ is known, and it is required to test $\mu = \mu_0$ against $\mu = \mu_1 (\mu_1 > \mu_0)$. Let n be sufficiently large so that the central limit theorem holds, and consider the test

$$\varphi(x_1, x_2, \cdots, x_n) = 1 \quad \text{if } \bar{x} > k,$$
$$= 0 \quad \text{if } \bar{x} \leq k,$$

where $\bar{x} = n^{-1} \sum_{i=1}^n x_i$. Find k such that the test has (approximately) size α. What is the power of this test at $\mu = \mu_1$? If the probabilities of type I and type II errors are fixed at α and β, respectively, find the smallest sample size needed.

3. In Problem 2, if σ is not known, find k such that the test φ has size α.

4. Let $X_1, X_2, \cdots, X_n$ be a sample from $\mathcal{N}(\mu, 1)$. For testing $\mu \leq \mu_0$ against $\mu > \mu_0$ consider the test function

$$\varphi(x_1, x_2, \cdots, x_n) = \begin{cases} 1 & \text{if } \bar{x} > \mu_0 + \dfrac{z_\alpha}{\sqrt{n}}, \\ 0 & \text{if } \bar{x} \leq \mu_0 + \dfrac{z_\alpha}{\sqrt{n}}. \end{cases}$$

Show that the power function of φ is a nondecreasing function of μ. What is the size of the test?

5. A sample of size 1 is taken from an exponential pdf with parameter θ, that is, $X \sim G(1, \theta)$. To test $H_0: \theta = 1$ against $H_1: \theta > 1$, the test to be used is the nonrandomized test

$$\varphi(x) = 1 \quad \text{if } x > 2,$$
$$= 0 \quad \text{if } x \leq 2.$$

Find the size of the test. What is the power function?

6. Let $X_1, X_2, \cdots, X_n$ be a sample from $\mathcal{N}(0, \sigma^2)$. To test $H_0: \sigma = \sigma_0$ against $H_1: \sigma \ne \sigma_0$, it is suggested that the test

$$\varphi(x_1, x_2, \cdots, x_n) = \begin{cases} 1 & \text{if } \sum x_i^2 > c_1 \text{ or } \sum x_i^2 < c_2, \\ 0 & \text{if } c_2 \le \sum x_i^2 \le c_1, \end{cases}$$

be used. How will you find c_1 and c_2 such that the size of φ is a preassigned number $\alpha, 0 < \alpha < 1$? What is the power function of this test?

7. An urn contains 10 marbles, of which M are white and $10 - M$ are black. To test that $M = 5$ against the alternative hypothesis that $M = 6$, one draws 3 marbles from the urn without replacement. The null hypothesis is rejected if the sample contains 2 or 3 white marbles; otherwise it is accepted. Find the size of the test and its power.

9.3 THE NEYMAN-PEARSON LEMMA

In this section we prove the fundamental lemma due to Neyman and Pearson [85], which gives a general method for finding a best (most powerful) test of a simple hypothesis against a simple alternative. Let $\{f_\theta, \theta \in \Theta\}$, where $\Theta = \{\theta_0, \theta_1\}$, be a family of possible distributions of $\mathbf{X}$. Also, f_θ represents the pdf of $\mathbf{X}$ if $\mathbf{X}$ is a continuous type rv, and the pmf of $\mathbf{X}$ if $\mathbf{X}$ is of the discrete type. Let us write $f_0(\mathbf{x}) = f_{\theta_0}(\mathbf{x})$ and $f_1(\mathbf{x}) = f_{\theta_1}(\mathbf{x})$ for convenience.

Theorem 1 (The Neyman-Pearson Fundamental Lemma).

(a) Any test φ of the form

(1) $$\varphi(\mathbf{x}) = \begin{cases} 1 & \text{if } f_1(\mathbf{x}) > k f_0(\mathbf{x}), \\ \gamma(\mathbf{x}) & \text{if } f_1(\mathbf{x}) = k f_0(\mathbf{x}), \\ 0 & \text{if } f_1(\mathbf{x}) < k f_0(\mathbf{x}), \end{cases}$$

for some $k \ge 0$ and $0 \le \gamma(\mathbf{x}) \le 1$, is most powerful of its size for testing $H_0: \theta = \theta_0$ against $H_1: \theta = \theta_1$. If $k = \infty$, the test

(2) $$\varphi(\mathbf{x}) = 1 \quad \text{if } f_0(\mathbf{x}) = 0,$$
$$= 0 \quad \text{if } f_0(\mathbf{x}) > 0,$$

is most powerful of size 0 for testing H_0 against H_1.

(b) Given $\alpha, 0 \le \alpha \le 1$, there exists a test of form (1) or (2) with $\gamma(\mathbf{x}) = \gamma$ (a constant), for which $E_{\theta_0} \varphi(\mathbf{X}) = \alpha$.

Proof. Let φ be a test satisfying (1), and φ^* be any test with $E_{\theta_0} \varphi^*(\mathbf{X}) \le E_{\theta_0} \varphi(\mathbf{X})$. In the continuous case

$$\int (\varphi(\mathbf{x}) - \varphi^*(\mathbf{x})) (f_1(\mathbf{x}) - k f_0(\mathbf{x})) d\mathbf{x}$$
$$= \left(\int_{f_1 > k f_0} + \int_{f_1 < k f_0} \right) (\varphi(\mathbf{x}) - \varphi^*(\mathbf{x})) (f_1(\mathbf{x}) - k f_0(\mathbf{x})) d\mathbf{x}.$$

For any $\mathbf{x} \in \{f_1(\mathbf{x}) > k f_0(\mathbf{x})\}$, $\varphi(\mathbf{x}) - \varphi^*(\mathbf{x}) = 1 - \varphi^*(\mathbf{x}) \geq 0$, so that the integrand is ≥ 0. For $\mathbf{x} \in \{f_1(\mathbf{x}) < k f_0(\mathbf{x})\}$, $\varphi(\mathbf{x}) - \varphi^*(\mathbf{x}) = -\varphi^*(\mathbf{x}) \leq 0$, so that the integrand is again ≥ 0. It follows that

$$\int (\varphi(\mathbf{x}) - \varphi^*(\mathbf{x})) (f_1(\mathbf{x}) - k f_0(\mathbf{x})) d\mathbf{x}$$
$$= E_{\theta_1}\varphi(\mathbf{X}) - E_{\theta_1}\varphi^*(\mathbf{X}) - k(E_{\theta_0}\varphi(\mathbf{X}) - E_{\theta_0}\varphi^*(\mathbf{X})) \geq 0,$$

which implies

$$E_{\theta_1}\varphi(\mathbf{X}) - E_{\theta_1}\varphi^*(\mathbf{X}) \geq k(E_{\theta_0}\varphi(\mathbf{X}) - E_{\theta_0}\varphi^*(\mathbf{X})) \geq 0$$

since $E_{\theta_0}\varphi^*(\mathbf{X}) \leq E_{\theta_0}\varphi(\mathbf{X})$.

If $k = \infty$, any test φ^* of size 0 must vanish on the set $\{f_0(\mathbf{x}) > 0\}$. We have

$$E_{\theta_1}\varphi(\mathbf{X}) - E_{\theta_1}\varphi^*(\mathbf{X}) = \int_{\{f_0(\mathbf{x})=0\}} (1 - \varphi^*(\mathbf{x})) f_1(\mathbf{x}) d\mathbf{x} \geq 0.$$

The proof for the discrete case requires the usual change of integral by a sum throughout.

To prove (b) we need to restrict ourselves to the case where $0 < \alpha \leq 1$, since the MP size 0 test is given by (2). Let $\gamma(\mathbf{x}) = \gamma$, and let us compute the size of a test of form (1). We have

$$E_{\theta_0}\varphi(\mathbf{X}) = P_{\theta_0}\{f_1(\mathbf{X}) > k f_0(\mathbf{X})\} + \gamma P_{\theta_0}\{f_1(\mathbf{X}) = k f_0(\mathbf{X})\}$$
$$= 1 - P_{\theta_0}\{f_1(\mathbf{X}) \leq k f_0(\mathbf{X})\} + \gamma P_{\theta_0}\{f_1(\mathbf{X}) = k f_0(\mathbf{X})\}.$$

Since $P_{\theta_0}\{f_0(\mathbf{X}) = 0\} = 0$, we may rewrite $E_{\theta_0}\varphi(\mathbf{X})$ as

(3) $$E_{\theta_0}\varphi(\mathbf{X}) = 1 - P_{\theta_0}\left\{\frac{f_1(\mathbf{X})}{f_0(\mathbf{X})} \leq k\right\} + \gamma P_{\theta_0}\left\{\frac{f_1(\mathbf{X})}{f_0(\mathbf{X})} = k\right\}.$$

Given $0 < \alpha \leq 1$, we wish to find k and γ such that $E_{\theta_0}\varphi(\mathbf{X}) = \alpha$, that is,

(4) $$P_{\theta_0}\left\{\frac{f_1(\mathbf{X})}{f_0(\mathbf{X})} \leq k\right\} - \gamma P_{\theta_0}\left\{\frac{f_1(\mathbf{X})}{f_0(\mathbf{X})} = k\right\} = 1 - \alpha.$$

Note that

$$P_{\theta_0}\left\{\frac{f_1(\mathbf{X})}{f_0(\mathbf{X})} \leq k\right\}$$

is a df so that it is a nondecreasing and right continuous function of k. If there exists a k_0 such that

$$P_{\theta_0}\left\{\frac{f_1(\mathbf{X})}{f_0(\mathbf{X})} \le k_0\right\} = 1 - \alpha,$$

we choose $\gamma = 0$ and $k = k_0$. Otherwise there exists a k_0 such that

(5) $$P_{\theta_0}\left\{\frac{f_1(\mathbf{X})}{f_0(\mathbf{X})} < k_0\right\} \le 1 - \alpha < P_{\theta_0}\left\{\frac{f_1(\mathbf{X})}{f_0(\mathbf{X})} \le k_0\right\},$$

that is, there is a jump at k_0 (see Fig. 1). In this case we choose $k = k_0$ and

(6) $$\gamma = \frac{P_{\theta_0}\{f_1(\mathbf{X})/f_0(\mathbf{X}) \le k_0\} - (1-\alpha)}{P_{\theta_0}\{f_1(\mathbf{X})/f_0(\mathbf{X}) = k_0\}}.$$

Since γ given by (6) satisfies (4), and $0 \le \gamma \le 1$, the proof is complete.

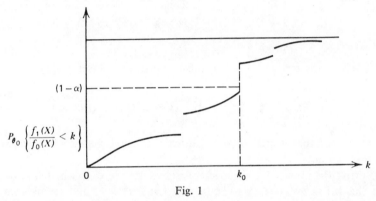

Fig. 1

Remark 1. It is possible to show (see Problem 6) that the test given by (1) or (2) is unique (except on a null set), that is, if φ is an MP test of size α of H_0 against H_1, it must have form (1) or (2), except perhaps for a set A with $P_{\theta_0}(A) = P_{\theta_1}(A) = 0$.

Remark 2. For generalizations of Theorem 1 we refer to Lehmann [70], page 83, or Ferguson [30], page 204.

Theorem 2. If a sufficient statistic T exists for the family $\{f_\theta : \theta \in \Theta\}$, $\Theta = \{\theta_0, \theta_1\}$, the Neyman-Pearson MP test is a function of T.

Proof. The proof of this result is left as an exercise.

Remark 3. If the family $\{f_\theta; \theta \in \Theta\}$ admits a sufficient statistic, one can restrict attention to tests based on the sufficient statistic, that is, to tests that are functions of the sufficient statistic. If φ is a test function and T is a sufficient statistic, $E\{\varphi(\mathbf{X})|T\}$ is itself a test function, $0 \le E\{\varphi(\mathbf{X})|T\} \le 1$, and

$$E_\theta\{E\{\varphi(\mathbf{X})|T\}\} = E_\theta\varphi(\mathbf{X}),$$

so that φ and $E\{\varphi|T\}$ have the same power function.

Example 1. Let $X \sim \mathcal{N}(0, 1)$ under H_0 and $X \sim \mathscr{C}(1, 0)$ under H_1. To find an MP size α test of H_0 against H_1,

$$\frac{f_1(x)}{f_0(x)} = \frac{(1/\pi)[1/(1+x^2)]}{(1/\sqrt{2\pi})\,e^{-x^2/2}}$$

$$= \sqrt{\frac{2}{\pi}}\,\frac{e^{x^2/2}}{1+x^2}.$$

Thus the MP test is of the form

$$\varphi(x) = \begin{cases} 1 & \text{if } \sqrt{\dfrac{2}{\pi}}\,\dfrac{e^{x^2/2}}{1+x^2} > k, \\ 0 & \text{otherwise,} \end{cases}$$

where k is determined so that $E_0\,\varphi(X) = \alpha$. Note that the boundary set has probability 0 under both H_0 and H_1. Since $f_1(x)/f_0(x)$ is a nondecreasing function of $|x|$, it follows that

$$\varphi(x) = \begin{cases} 1 & \text{if } |x| > k_1, \\ 0 & \text{if } |x| < k_1, \end{cases}$$

where k_1 is determined from

$$\int_{-k_1}^{k_1} \frac{1}{\sqrt{2\pi}} e^{-x^2/2}\,dx = 1 - \alpha.$$

It follows that $k_1 = z_{\alpha/2}$. The power of the test is given by

$$E_1\varphi(X) = 1 - \int_{-k_1}^{k_1} \frac{1}{\pi}\,\frac{1}{1+x^2}\,dx = 1 - \frac{2}{\pi}\tan^{-1} k_1$$

$$= 1 - \frac{2}{\pi}\tan^{-1} z_{\alpha/2}.$$

Example 2. Let $X_1, X_2, \cdots, X_n$ be iid $b(1, p)$ rv's, and let $H_0: p = p_0$, $H_1: p = p_1$, $p_1 > p_0$. The MP size α test of H_0 against H_1 is of the form

$$\varphi(x_1, \cdots, x_n) = \begin{cases} 1, & \lambda(\mathbf{x}) = \dfrac{p_1^{\Sigma x_i}(1-p_1)^{n-\Sigma x_i}}{p_0^{\Sigma x_i}(1-p_0)^{n-\Sigma x_i}} > k, \\ \gamma, & \lambda(\mathbf{x}) = k, \\ 0, & \lambda(\mathbf{x}) < k, \end{cases}$$

where k and γ are determined from

$$E_{p_0}\varphi(\mathbf{X}) = \alpha.$$

Now

$$\lambda(\mathbf{x}) = \left(\frac{p_1}{p_0}\right)^{\Sigma x_i}\left(\frac{1-p_1}{1-p_0}\right)^{n-\Sigma x_i},$$

and since $p_1 > p_0$, $\lambda(\mathbf{x})$ is an increasing function of $\sum x_i$. It follows that $\lambda(\mathbf{x}) > k$ if and only if $\sum x_i > k_1$, where k_1 is some constant. Thus the MP size α test is of the form

$$\varphi(\mathbf{x}) = \begin{cases} 1 & \text{if } \sum x_i > k_1, \\ \gamma & \text{if } \sum x_i = k_1, \\ 0 & \text{otherwise.} \end{cases}$$

Also, k_1 and γ are determined from

$$\alpha = E_{p_0}\varphi(\mathbf{X}) = P_{p_0}\left\{\sum_1^n X_i > k_1\right\} + \gamma\, P_{p_0}\left\{\sum_1^n X_i = k_1\right\}$$

$$= \sum_{r=k_1+1}^{n}\binom{n}{r} p_0^r (1-p_0)^{n-r} + \gamma\binom{n}{k_1}p_0^{k_1}(1-p_0)^{n-k_1}.$$

Note that the MP size α test is independent of p_1 as long as $p_1 > p_0$, that is, it remains an MP size α test against any $p > p_0$ and is therefore a UMP test of $p = p_0$ against $p > p_0$.

In particular, let $n = 5$, $p_0 = \frac{1}{2}$, $p_1 = \frac{3}{4}$, and $\alpha = .05$. Then the MP test is given by

$$\varphi(\mathbf{x}) = \begin{cases} 1 & \sum x_i > k, \\ \gamma & \sum x_i = k, \\ 0 & \sum x_i < k, \end{cases}$$

where k and γ are determined from

$$.05 = \alpha = \sum_{k+1}^{5}\binom{5}{r}(\tfrac{1}{2})^5 + \gamma\binom{5}{k}(\tfrac{1}{2})^5.$$

It follows that $k = 4$ and $\gamma = .122$. Thus the MP size $\alpha = .05$ test is to reject $p = \frac{1}{2}$ in favor of $p = \frac{3}{4}$ if $\sum_1^n X_i = 5$ and reject $p = \frac{1}{2}$ with probability .122, if $\sum_1^n X_i = 4$.

It is simply a matter of reversing inequalities to see that the MP size α test of $H_0: p = p_0$ against $H_1: p = p_1 (p_1 < p_0)$ is given by

$$\varphi(\mathbf{x}) = \begin{cases} 1 & \text{if } \sum x_i < k, \\ \gamma & \text{if } \sum x_i = k, \\ 0 & \text{if } \sum x_i > k, \end{cases}$$

where γ and k are determined from $E_{p_0}\varphi(\mathbf{X}) = \alpha$.

We remark that in both cases ($p_1 > p_0$, $p_1 < p_0$) the MP test is quite intuitive. We would tend to accept the larger probability if a larger number

Example 3. Let $X_1, X_2, \cdots, X_n$ be iid $\mathcal{N}(\mu, \sigma^2)$ rv's where both μ and σ^2 are unknown. We wish to test the null hypothesis $H_0: \mu = \mu_0, \sigma^2 = \sigma_0^2$ against the alternative $H_1: \mu = \mu_1, \sigma^2 = \sigma_0^2$. The fundamental lemma leads to the following MP test:

$$\varphi(\mathbf{x}) = \begin{cases} 1 & \text{if } \lambda(\mathbf{x}) > k, \\ 0 & \text{if } \lambda(\mathbf{x}) < k, \end{cases}$$

where

$$\lambda(\mathbf{x}) = \frac{(1/\sigma_0 \sqrt{2\pi})^n \exp\{-[\sum(x_i - \mu_1)^2/2\sigma_0^2]\}}{(1/\sigma_0 \sqrt{2\pi})^n \exp\{-[\sum(x_i - \mu_0)^2/2\sigma_0^2]\}},$$

and k is determined from $E_{\mu_0, \sigma_0^2} \varphi(\mathbf{X}) = \alpha$. We have

$$\lambda(\mathbf{x}) = \exp\left\{\sum x_i \left(\frac{\mu_1}{\sigma_0^2} - \frac{\mu_0}{\sigma_0^2}\right) + n\left(\frac{\mu_0^2}{2\sigma_0^2} - \frac{\mu_1^2}{2\sigma_1^2}\right)\right\}.$$

If $\mu_1 > \mu_0$, then

$$\lambda(\mathbf{x}) > k \quad \text{if and only if } \sum_{i=1}^n x_i > k',$$

where k' is determined from

$$\alpha = P_{\mu_0, \sigma_0}\left\{\sum_{i=1}^n X_i > k'\right\} = P\left\{\frac{\sum X_i - n\mu_0}{\sqrt{n}\, \sigma_0} > \frac{k' - n\mu_0}{\sqrt{n}\, \sigma_0}\right\},$$

giving $k' = z_\alpha \sqrt{n}\, \sigma_0 + n\mu_0$. The case $\mu_1 < \mu_0$ is treated similarly. If σ_0 is known, the test determined above is independent of μ_1 as long as $\mu_1 > \mu_0$, and it follows that the test is UMP against $H_1': \mu > \mu_1, \sigma^2 = \sigma_0^2$. If, however, σ_0 is not known, that is, the null hypothesis is a composite hypothesis $H_0'': \mu = \mu_0, \sigma^2 > 0$ to be tested against the alternatives $H_1'': \mu = \mu_1, \sigma^2 > 0$ ($\mu_1 > \mu_0$), then the MP test determined above depends on σ^2. In other words, an MP test against the alternative μ_1, σ_0^2 will not be MP against μ_1, σ_1^2, where $\sigma_1^2 \neq \sigma_0^2$.

PROBLEMS 9.3

1. A sample of size 1 is taken from pdf

$$f_\theta(x) = \begin{cases} \dfrac{2}{\theta^2}(\theta - x) & \text{if } 0 < x < \theta, \\ 0 & \text{otherwise.} \end{cases}$$

Find an MP test of $H_0: \theta = \theta_0$ against $H_1: \theta = \theta_1 (\theta_1 < \theta_0)$.

2. Find the Neyman-Pearson size α test of $H_0: \theta = \theta_0$ against $H_1: \theta = \theta_1 (\theta_1 < \theta_0)$, based on a sample of size 1 from the pdf

$$f_\theta(x) = 2\theta x + 2(1 - \theta)(1 - x), \qquad 0 < x < 1, \quad \theta \in [0, 1].$$

3. Find the Neyman-Pearson size α test of $H_0: \beta = 1$ against $H_1: \beta = \beta_1 (> 1)$, based on a sample of size 1 from

$$f(x; \beta) = \begin{cases} \beta x^{\beta-1}, & 0 < x < 1, \\ 0, & \text{otherwise.} \end{cases}$$

4. Find an MP size α test of $H_0: X \sim f_0(x)$, where $f_0(x) = (2\pi)^{-1/2} e^{-x^2/2}$, $-\infty < x < \infty$, against $H_1: X \sim f_1(x)$ where $f_1(x) = 2^{-1} e^{-|x|}$, $-\infty < x < \infty$, based on a sample of size 1.

5. For the pdf $f_\theta(x) = e^{-(x-\theta)}$, $x \geq \theta$, find an MP size α test of $\theta = \theta_0$ against $\theta = \theta_1 (> \theta_0)$, based on a sample of size n.

6. If φ^* is an MP size α test of $H_0: \mathbf{X} \sim f_0(\mathbf{x})$ against $H_1: \mathbf{X} \sim f_1(\mathbf{x})$, show that it has to be either of form (1) or form (2) (except for a set of $\mathbf{x}$ that has probability 0 under H_0 and H_1).

7. Let φ^* be an MP size α $(0 < \alpha \leq 1)$ test of H_0 against H_1, and let $k(\alpha)$ denote the value of k in (1). Show that if $\alpha_1 < \alpha_2$, then $k(\alpha_2) \leq k(\alpha_1)$.

8. For the family of Neyman-Pearson tests show that the larger the α, the smaller the β ($= P\{\text{Type II error}\}$).

9. Let $1 - \beta$ be the power of an MP size α test, where $0 < \alpha < 1$. Show that $\alpha < 1 - \beta$ unless $P_{\theta_0} = P_{\theta_1}$.

10. Let α be a real number, $0 < \alpha < 1$, and φ^* be an MP size α test of H_0 against H_1. Also, let $\beta = E_{H_1} \varphi^*(\mathbf{X}) < 1$. Show that $1 - \varphi^*$ is an MP test for testing H_1 against H_0 at level $1 - \beta$.

11. Let $X_1, X_2, \cdots, X_n$ be a random sample from pdf

$$f_\theta(x) = \frac{\theta}{x^2} \quad \text{if} \quad 0 < \theta \leq x < \infty.$$

Find an MP test of $\theta = \theta_0$ against $\theta = \theta_1$ ($\neq \theta_0$).

12. Let X be an observation in $(0, 1)$. Find an MP size α test of $H_0: X \sim f(x) = 2x$ if $0 < x < \frac{1}{2}$, and $= 2 - 2x$ if $\frac{1}{2} \leq x < 1$, against $H_1: X \sim f(x) = 1$ if $0 < x < 1$. Assume that $0 < \alpha < \frac{1}{2}$. Find the power of your test.

9.4 FAMILIES WITH MONOTONE LIKELIHOOD RATIO

In this section we consider the problem of testing one-sided hypotheses on a single real-valued parameter. Let $\{f_\theta, \theta \in \Theta\}$ be a family of pdf's (pmf's), $\Theta \subseteq \mathcal{R}$, and suppose that we wish to test $H_0: \theta \leq \theta_0$ against the alternatives $H_1: \theta > \theta_0$ or its dual, $H_0': \theta \geq \theta_0$, against $H_1': \theta < \theta_0$. In general, it is not possible to find a UMP test for this problem. The MP test of $H_0: \theta \leq \theta_0$, say, against the alternative $\theta = \theta_1(> \theta_0)$ depends on θ_1 and cannot be UMP. Here we consider a special class of distributions that is large enough to include the one-parameter exponential family, for which a UMP test of a one-sided hypothesis exists.

Definition 1. Let $\{f_\theta, \theta \in \Theta\}$ be a family of pdf's (pmf's), $\Theta \subseteq \mathcal{R}$. We say that $\{f_\theta\}$ has a monotone likelihood ratio (MLR) in the statistic $T(\mathbf{x})$ if for $\theta_1 < \theta_2$, whenever $f_{\theta_1}, f_{\theta_2}$ are distinct, the ratio $f_{\theta_2}(\mathbf{x})/f_{\theta_1}(\mathbf{x})$ is a nondecreasing function of $T(\mathbf{x})$ for the set of values $\mathbf{x}$ for which at least one of f_{θ_1} and f_{θ_2} is > 0.

It is also possible to define families of densities with nonincreasing MLR in $T(\mathbf{x})$, but such families can be treated by symmetry.

Example 1. Let $X_1, X_2, \cdots, X_n \sim U[0, \theta], \theta > 0$. The joint pdf of $X_1, \cdots, X_n$ is

$$f_\theta(\mathbf{x}) = \begin{cases} \dfrac{1}{\theta^n}, & 0 \leq \max x_i \leq \theta, \\ 0, & \text{otherwise.} \end{cases}$$

Let $\theta_1 > \theta_2$ and consider the ratio

$$\frac{f_{\theta_1}(\mathbf{x})}{f_{\theta_2}(\mathbf{x})} = \frac{(1/\theta_1^n) I_{[\max x_i \leq \theta_1]}}{(1/\theta_2^n) I_{[\max x_i \leq \theta_2]}}$$

$$= \left(\frac{\theta_2}{\theta_1}\right)^n I_{[\max x_i \leq \theta_1]} / I_{[\max x_i \leq \theta_2]}.$$

Let $R(\mathbf{x}) = I_{[\max x_i \leq \theta_1]} / I_{[\max x_i \leq \theta_2]}$

$$= \begin{cases} 1, & \max x_i \in [0, \theta_2], \\ \infty, & \max x_i \in [\theta_2, \theta_1]. \end{cases}$$

Define $R(\mathbf{x}) = \infty$ if $\max x_i > \theta_1$. It follows that $f_{\theta_1}/f_{\theta_2}$ is a nondecreasing function of $\max_{1 \leq i \leq n} x_i$, and the family of uniform densities on $[0, \theta]$ has an MLR in $\max_{1 \leq i \leq n} x_i$.

Theorem 1. The one-parameter exponential family

(1) $$f_\theta(\mathbf{x}) = \exp\{Q(\theta) T(\mathbf{x}) + S(\mathbf{x}) + D(\theta)\},$$

where $Q(\theta)$ is nondecreasing, has an MLR in $T(\mathbf{x})$.

Proof. The proof is left as an exercise.

Remark 1. The nondecreasingness of $Q(\theta)$ can be obtained by a reparametrization, putting $\vartheta = Q(\theta)$, if necessary.

Theorem 1 includes normal, binomial, Poisson, gamma (one parameter fixed), beta (one parameter fixed), and so on. In Example 1 we have already seen that $U[0, \theta]$, which is not an exponential family, has an MLR.

Example 2. Let $X \sim \mathscr{C}(1, \theta)$. Then

$$\frac{f_{\theta_2}(x)}{f_{\theta_1}(x)} = \frac{1 + (x - \theta_1)^2}{1 + (x - \theta_2)^2} \to 1 \qquad \text{as } x \to \pm \infty,$$

and we see that $\mathscr{C}(1, \theta)$ does not have an MLR.

Theorem 2. Let $\mathbf{X} \sim f_\theta$, $\theta \in \Theta$, where $\{f_\theta\}$ has an MLR in $T(\mathbf{x})$. For testing $H_0: \theta \leq \theta_0$ against $H_1: \theta > \theta_0$, $\theta_0 \in \Theta$, any test of the form

(2) $$\varphi(\mathbf{x}) = \begin{cases} 1 & \text{if } T(\mathbf{x}) > t_0, \\ \gamma & \text{if } T(\mathbf{x}) = t_0, \\ 0 & \text{if } T(\mathbf{x}) < t_0, \end{cases}$$

has a nondecreasing power function and is UMP of its size $E_{\theta_0} \varphi(\mathbf{X}) = \alpha$ (provided that the size is not 0).

Moreover, for every $0 \leq \alpha \leq 1$ and every $\theta_0 \in \Theta$, there exists a t_0, $-\infty \leq t_0 \leq \infty$, and $0 \leq \gamma \leq 1$ such that the test described in (2) is the UMP size α test of H_0 against H_1.

Proof. Let $\theta_1, \theta_2 \in \Theta$, $\theta_1 < \theta_2$. By the fundamental lemma any test of the form

(3) $$\varphi(\mathbf{x}) = \begin{cases} 1, & \lambda(\mathbf{x}) > k, \\ \gamma(\mathbf{x}), & \lambda(\mathbf{x}) = k, \\ 0, & \lambda(\mathbf{x}) < k, \end{cases}$$

where $\lambda(\mathbf{x}) = f_{\theta_2}(\mathbf{x})/f_{\theta_1}(\mathbf{x})$, is MP of its size for testing $\theta = \theta_1$ against $\theta = \theta_2$, provided that $0 \leq k < \infty$; and if $k = \infty$, the test

(4) $$\varphi(\mathbf{x}) = \begin{cases} 1 & \text{if } f_{\theta_1}(\mathbf{x}) = 0, \\ 0 & \text{if } f_{\theta_1}(\mathbf{x}) > 0, \end{cases}$$

is MP of size 0. Since f_θ has an MLR in T, it follows that any test of form (2) is also of form (3), provided that $E_{\theta_1} \varphi(\mathbf{X}) > 0$, that is, provided that its size is > 0. The trivial test $\varphi'(\mathbf{x}) \equiv \alpha$ has size α and power α, so that the power of any test (2) is at least α, that is,

$$E_{\theta_2} \varphi(\mathbf{X}) \geq E_{\theta_2} \varphi'(\mathbf{X}) = \alpha = E_{\theta_1} \varphi(\mathbf{X}).$$

It follows that, if $\theta_1 < \theta_2$ and $E_{\theta_1} \varphi(\mathbf{X}) > 0$, then $E_{\theta_1} \varphi(\mathbf{X}) \leq E_{\theta_2} \varphi(\mathbf{X})$, as asserted.

Let $\theta_1 = \theta_0$ and $\theta_2 > \theta_0$, as above. We know that (2) is an MP test of its size $E_{\theta_0}\varphi(\mathbf{X})$ for testing $\theta = \theta_0$ against $\theta = \theta_2 (\theta_2 > \theta_0)$, provided that $E_{\theta_0}\varphi(\mathbf{X}) > 0$. Since the power function of φ is nondecreasing,

(5) $$E_\theta\varphi(\mathbf{X}) \le E_{\theta_0}\varphi(\mathbf{X}) = \alpha_0 \quad \text{for all } \theta \le \theta_0.$$

Since, however, φ does not depend on θ_2 (it depends only on constants k and r), it follows that φ is the UMP size α_0 test for testing $\theta = \theta_0$ against $\theta > \theta_0$. Thus φ is UMP among the class of tests φ'' for which

(6) $$E_{\theta_0}\varphi''(\mathbf{X}) \le E_{\theta_0}\varphi(\mathbf{X}) = \alpha_0.$$

Now the class of tests satisfying (5) is contained in the class of tests satisfying (6) [there are more restrictions in (5)]. It follows that φ, which is UMP in the larger class satisfying (6), must also be UMP in the smaller class satisfying (5). Thus, provided that $\alpha_0 > 0$, φ is the UMP size α_0 test for $\theta \le \theta_0$ against $\theta > \theta_0$.

We ask the reader to complete the proof of the final part of the theorem, using the fundamental lemma.

Remark 2. By interchanging inequalities throughout in Theorem 2, we see that this theorem also provides a solution of the dual problem $H_0': \theta \ge \theta_0$ against $H_1': \theta < \theta_0$.

Example 3. Let X have the pmf

$$P_M\{X = x\} = \frac{\binom{M}{x}\binom{N-M}{n-x}}{\binom{N}{n}}, \quad x = 0, 1, 2, \cdots, M.$$

Since

$$\frac{P_{M+1}\{X = x\}}{P_M\{X = x\}} = \frac{M+1}{N-M} \cdot \frac{N-M-n+x}{M+1-x},$$

we see that $\{P_M\}$ has an MLR in x (P_{M_2}/P_{M_1} where $M_2 > M_1$ is just a product of such ratios). It follows that there exists a UMP test of $H_0: M \le M_0$ against $H_1: M > M_0$, which rejects H_0 when X is too large, that is, the UMP size α test is given by

$$\varphi(x) = \begin{cases} 1, & x > k, \\ r, & x = k, \\ 0, & x < k, \end{cases}$$

where (integer) k and γ are determined from

$$E_{M_0}\varphi(X) = \alpha.$$

For the one-parameter exponential family UMP tests exist also for some two-sided hypotheses of the form

(7) $$H_0 : \theta \leq \theta_1 \text{ or } \theta \geq \theta_2 \ (\theta_1 < \theta_2).$$

We state the following result without proof.

Theorem 3. For the one-parameter exponential family (1), there exists a UMP test of the hypothesis $H_0 : \theta \leq \theta_1$ or $\theta \geq \theta_2$ $(\theta_1 < \theta_2)$ against $H_1 : \theta_1 < \theta < \theta_2$ that is of the form

(8) $$\varphi(\mathbf{x}) = \begin{cases} 1 & \text{if } c_1 < T(\mathbf{x}) < c_2, \\ \gamma_i & \text{if } T(\mathbf{x}) = c_i, \ i = 1, 2 \quad (c_1 < c_2), \\ 0 & \text{if } T(\mathbf{x}) < c_1 \text{ or } > c_2, \end{cases}$$

where the c's and the γ's are given by

(9) $$E_{\theta_1}\varphi(\mathbf{X}) = E_{\theta_2}\varphi(\mathbf{X}) = \alpha.$$

See Lehmann [70], page 88, for proof.

Example 4. Let $X_1, X_2, \cdots, X_n$ be iid $\mathcal{N}(\mu, 1)$ rv's. To test $H_0: \mu \leq \mu_0$ or $\mu \geq \mu_1$ $(\mu_1 > \mu_0)$ against $H_1: \mu_0 < \mu < \mu_1$, the UMP test is given by

$$\varphi(\mathbf{x}) = \begin{cases} 1 & \text{if } c_1 < \sum_{1}^{n} x_i < c_2, \\ \gamma_i & \text{if } \sum x_i = c_1 \text{ or } c_2, \\ 0 & \text{if } \sum x_i < c_1 \text{ or } > c_2, \end{cases}$$

where we determine c_1, c_2 from

$$\alpha = P_{\mu_0}\{c_1 < \sum X_i < c_2\} = P_{\mu_1}\{c_1 < \sum X_i < c_2\}$$

and $\gamma_1 = \gamma_2 = 0$. Thus

$$\alpha = P\left\{\frac{c_1 - n\mu_0}{\sqrt{n}} < \frac{\sum X_i - n\mu_0}{\sqrt{n}} < \frac{c_2 - n\mu_0}{\sqrt{n}}\right\}$$
$$= P\left\{\frac{c_1 - n\mu_1}{\sqrt{n}} < \frac{\sum X_i - n\mu_1}{\sqrt{n}} < \frac{c_2 - n\mu_1}{\sqrt{n}}\right\}$$

$$= P\left\{\frac{c_1 - n\mu_0}{\sqrt{n}} < Z < \frac{c_2 - n\mu_0}{\sqrt{n}}\right\}$$
$$= P\left\{\frac{c_1 - n\mu_1}{\sqrt{n}} < Z < \frac{c_2 - n\mu_1}{\sqrt{n}}\right\},$$

where Z is $\mathcal{N}(0, 1)$. Given α, n, μ_0, and μ_1, we can solve for c_1 and c_2 from the simultaneous equations

$$\Phi\left(\frac{c_2 - n\mu_0}{\sqrt{n}}\right) - \Phi\left(\frac{c_1 - n\mu_0}{\sqrt{n}}\right) = \alpha,$$

$$\Phi\left(\frac{c_2 - n\mu_1}{\sqrt{n}}\right) - \Phi\left(\frac{c_1 - n\mu_1}{\sqrt{n}}\right) = \alpha,$$

where Φ is the df of Z.

Remark 3. We caution the reader that UMP tests for testing $H_0: \theta_1 \leq \theta \leq \theta_2$ and $H_0': \theta = \theta_0$ for the one-parameter exponential family do not exist. An example will suffice.

Example 5. Let $X_1, X_2, \cdots, X_n$ be a sample from $\mathcal{N}(0, \sigma^2)$. Since the family of joint pdf's of $\mathbf{X} = (X_1, \cdots, X_n)$ has an MLR in $\sum_1^n X_i^2$, it follows that UMP tests exist for one-sided hypotheses $\sigma \geq \sigma_0$ and $\sigma \leq \sigma_0$.

Consider now the null hypothesis $H_0: \sigma = \sigma_0$ against the alternative $H_1: \sigma \neq \sigma_0$. We will show that a UMP test of H_0 does not exist. For testing $\sigma = \sigma_0$ against $\sigma > \sigma_0$, a test of the form

$$\varphi_1(\mathbf{x}) = \begin{cases} 1, & \sum_i x_i^2 > c_1, \\ 0, & \text{otherwise}, \end{cases}$$

is UMP, and for testing $\sigma = \sigma_0$ against $\sigma < \sigma_0$, a test of the form

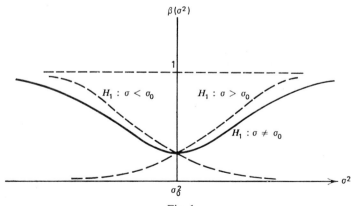

Fig. 1

$$\varphi_2(\mathbf{x}) = \begin{cases} 1, & \sum x_i^2 < c_2, \\ 0, & \text{otherwise,} \end{cases}$$

is UMP. If the size is chosen as α, then $c_1 = \sigma_0^2 \chi_{n,\alpha}^2$ and $c_2 = \sigma_0^2 \chi_{n,1-\alpha}^2$. Clearly, neither φ_1 nor φ_2 is UMP for H_0 against $H_1: \sigma \neq \sigma_0$. The power of any test of H_0 for values of $\sigma > \sigma_0$ cannot exceed that of φ_1, and for values of $\sigma < \sigma_0$ it cannot exceed the power of test φ_2. Hence no test of H_0 can be UMP (see Fig. 1).

PROBLEMS 9.4

1. For the following families of pmf's (pdf's) $f_\theta(x)$, $\theta \in \Theta \subseteq \mathcal{R}$, find a UMP size α test of $H_0: \theta \leq \theta_0$ against $H_1: \theta > \theta_0$, based on a sample of n observations.
   (a) $f_\theta(x) = \theta^x(1-\theta)^{1-x}$, $x = 0, 1; 0 < \theta < 1$.
   (b) $f_\theta(x) = (1/\sqrt{2\pi}) \exp\{-(x-\theta)^2/2\}$, $-\infty < x < \infty$, $-\infty < \theta < \infty$.
   (c) $f_\theta(x) = e^{-\theta}(\theta^x/x!)$, $x = 0, 1, 2, \cdots$; $\theta > 0$.
   (d) $f_\theta(x) = (1/\theta)e^{-x/\theta}$, $x > 0$, $\theta > 0$.
   (e) $f_\theta(x) = [1/\Gamma(\theta)]x^{\theta-1}e^{-x}$, $x > 0$, $\theta > 0$.
   (f) $f_\theta(x) = (1/\theta)x^{\theta-1}$, $0 < x < 1$, $\theta > 0$.

2. Let $X_1, X_2, \cdots, X_n$ be a sample of size n from the pmf
$$P_N(x) = \frac{1}{N}, \quad x = 1, 2, \cdots, N; N \in \{1, 2, \cdots\}.$$

   (a) Show that the test
   $$\varphi(x_1, x_2, \cdots, x_n) = \begin{cases} 1 & \text{if } \max(x_1, x_2, \cdots, x_n) > N_0, \\ \alpha & \text{if } \max(x_1, x_2, \cdots, x_n) \leq N_0, \end{cases}$$
   is UMP size α for testing $H_0: N \leq N_0$ against $H_1: N > N_0$.
   (b) Show that
   $$\varphi(x_1, x_2, \cdots, x_n) = \begin{cases} 1 & \text{if } \max(x_1, x_2, \cdots, x_n) > N_0 \text{ or } \\ & \max(x_1, x_2, \cdots, x_n) \leq \alpha^{1/n} N_0, \\ 0 & \text{otherwise,} \end{cases}$$
   is a UMP size α test of $H_0': N = N_0$ aganst $H_1': N \neq N_0$.

3. Let $X_1, X_2 \cdots, X_n$ be a sample of size n from $U(0, \theta)$, $\theta > 0$. Show that the test
$$\varphi_1(x_1, x_2, \cdots x_n) = \begin{cases} 1 & \text{if } \max(x_1, \cdots, x_n) > \theta_0, \\ \alpha & \text{if } \max(x_1, x_2, \cdots x_n) \leq \theta_0. \end{cases}$$
is UMP size α for testing $H_0: \theta \leq \theta_0$ against $H_1: \theta > \theta_0$ and that the test
$$\varphi_2(x_1, x_2, \cdots, x_n) = \begin{cases} 1 & \text{if } \max(x_1, \cdots, x_n) > \theta_0 \text{ or} \\ & \max(x_1, x_2 \cdots, x_n) \leq \theta_0 \alpha^{1/n}, \\ 0 & \text{otherwise,} \end{cases}$$
is UMP size α for $H_0': \theta = \theta_0$ against $H_1': \theta \neq \theta_0$.

4. Does the Laplace family of pdf's
$$f_\theta(x) = \tfrac{1}{2} \exp\{-|x-\theta|\}, \quad -\infty < x < \infty, \quad \theta \in \mathcal{R},$$
possess an MLR?

9.5 UNBIASED AND INVARIANT TESTS

We have seen that, if we restrict ourselves to the class Φ_α of all size α tests, there do not exist UMP tests for many important hypotheses. This suggests that we reduce the class of tests under consideration by imposing certain restrictions.

Definition 1. A size α test φ of $H_0: \theta \in \Theta_0$ against the alternatives $H_1: \theta \in \Theta_1$ is said to be unbiased if

(1) $\qquad E_\theta \varphi(\mathbf{X}) \geq \alpha \qquad$ for all $\theta \in \Theta_1$.

It follows that a test φ is unbiased if and only if its power function $\beta_\varphi(\theta)$ satisfies

(2) $\qquad \beta_\varphi(\theta) \leq \alpha \qquad$ for $\theta \in \Theta_0$

and

(3) $\qquad \beta_\varphi(\theta) \geq \alpha \qquad$ for $\theta \in \Theta_1$.

This seems to be a reasonable requirement to place on a test. The probability of rejecting H_0 when false is larger than the probability of rejecting H_0 when true.

Definition 2. Let U_α be the class of all unbiased size α tests of H_0. If there exists a test $\varphi \in U_\alpha$ that has maximum power at each $\theta \in \Theta_1$, we call φ a UMP unbiased size α test.

Clearly $U_\alpha \subset \Phi_\alpha$. If a UMP test exists in Φ_α, it is UMP in U_α. This follows by comparing the power of the UMP test with that of the trivial test $\varphi(\mathbf{x}) = \alpha$. It is convenient to introduce another class of tests.

Definition 3. A test φ is said to be α-similar on a subset Θ^* of Θ if

(4) $\qquad \beta_\varphi(\theta) = E_\theta \varphi(\mathbf{X}) = \alpha \qquad$ for $\theta \in \Theta^*$.

A test is said to be similar on a set $\Theta^* \subseteq \Theta$ if it is α-similar on Θ^* for some α, $0 \leq \alpha \leq 1$.

It is clear that there exists at least one similar test on every Θ^*, namely, $\varphi(\mathbf{x}) \equiv \alpha$, $0 \leq \alpha \leq 1$.

Theorem 1. Let $\beta_\varphi(\theta)$ be continuous in θ for any φ. If φ is an unbiased size α test of $H_0\colon \theta \in \Theta_0$ against $H_1\colon \theta \in \Theta_1$, it is α-similar on the boundary $\Lambda = \bar{\Theta}_0 \cap \bar{\Theta}_1$. (Here $\bar{A}$ is the closure of set A.)

Proof. Let $\theta \in \Lambda$. Then there exists a sequence $\{\theta_n\}$, $\theta_n \in \Theta_0$, such that $\theta_n \to \theta$. Since $\beta_\varphi(\theta)$ is continuous, $\beta_\varphi(\theta_n) \to \beta_\varphi(\theta)$; and since $\beta_\varphi(\theta_n) \leq \alpha$, for $\theta_n \in \Theta_0$, $\beta_\varphi(\theta) \leq \alpha$. Similarly there exists a sequence $\{\theta'_n\}$, $\theta'_n \in \Theta_1$, such that $\beta_\varphi(\theta'_n) \geq \alpha$ (φ is unbiased) and $\theta'_n \to \theta$. Thus $\beta_\varphi(\theta'_n) \to \beta_\varphi(\theta)$, and it follows that $\beta_\varphi(\theta) \geq \alpha$. Hence $\beta_\varphi(\theta) = \alpha$ for $\theta \in \Lambda$, and φ is α-similar on Λ.

Remark 1. Thus, if $\beta_\varphi(\theta)$ is continuous in θ for any φ, an unbiased size α test of H_0 against H_1 is also α-similar for the pdf's (pmf's) of Λ, that is, for $\{f_\theta, \theta \in \Lambda\}$. If we can find an MP similar test of $H_0\colon \theta \in \Lambda$ against H_1, and if this test is unbiased size α, then necessarily it is MP in the smaller class.

Definition 4. A test φ that is UMP among all α-similar tests on the boundary $\Lambda = \bar{\Theta}_0 \cap \bar{\Theta}_1$ is said to be a UMP α-similar test.

It is frequently easier to find a UMP α-similar test. Moreover, tests that are UMP similar on the boundary are often UMP unbiased.

Theorem 2. Let the power function of every test φ of $H_0\colon \theta \in \Theta_0$ against $H_1\colon \theta \in \Theta_1$ be continuous in θ. Then a UMP α-similar test is UMP unbiased, provided that its size is α for testing H_0 against H_1.

Proof. Let φ_0 be UMP α-similar. Then $E_\theta \varphi_0(\mathbf{X}) \leq \alpha$ for $\theta \in \Theta_0$. Comparing its power with that of the trivial similar test $\varphi(\mathbf{x}) \equiv \alpha$, we see that φ_0 is unbiased also. By the continuity of $\beta_\varphi(\theta)$ we see that the class of all unbiased size α tests is a subclass of the class of all α-similar tests. It follows that φ_0 is a UMP unbiased size α test.

Remark 2. The continuity of power function $\beta_\varphi(\theta)$ is not always easy to check. For some sufficient conditions see P.2.13.

Example 1. Let $X_1, X_2, \cdots, X_n$ be a sample from $\mathcal{N}(\mu, 1)$. We wish to test $H_0\colon \mu \leq 0$ against $H_1\colon \mu > 0$. Since the family of densities has an MLR in $\sum_1^n X_i$, we can use Theorem 9.4.2 to conclude that a UMP test rejects H_0 if $\sum_1^n X_i > c$. This test is also UMP unbiased. Nevertheless we use this example to illustrate the concepts introduced above.

Here $\Theta_0 = \{\mu\colon \mu \leq 0\}$, $\Theta_1 = \{\mu\colon \mu > 0\}$, $\Lambda = \bar{\Theta}_0 \cap \bar{\Theta}_1 = \{\mu = 0\}$. Also the power function of any test φ is

$$\beta_\varphi(\mu) = \int \cdots \int \varphi(x_1, \cdots, x_n) \frac{1}{(\sqrt{2\pi})^n} \exp\left\{-\frac{\sum_{1}^{n}(x_i - \mu)^2}{2}\right\} d\mathbf{x},$$

which is continuous in μ. It follows that any unbiased test of size α of H_0 has the property $\beta_\varphi(0) = \alpha$ of similarity over $\Lambda = \{\mu = 0\}$. To use Theorem 2, we find a UMP test of H_0': $\mu \in \Lambda$ against H_1. Let $\mu_1 > 0$. Then the fundamental lemma gives an MP test of $\mu = 0$ against $\mu = \mu_1$ as

$$\varphi(\mathbf{x}) = \begin{cases} 1 & \text{if } \exp\left\{\frac{\sum_{1}^{n} x_i^2}{2} - \frac{\sum_{1}^{n}(x_i - \mu_1)^2}{2}\right\} > k', \\ 0 & \text{otherwise,} \end{cases}$$

that is,

$$\varphi(\mathbf{x}) = \begin{cases} 1 & \text{if } \sum_{1}^{n} x_i > k, \\ 0 & \text{if } \sum_{1}^{n} x_i \le k, \end{cases}$$

and k is determined from

$$\alpha = P_0\{\sum X_i > k\} = P\left\{\frac{\sum X_i}{\sqrt{n}} > \frac{k}{\sqrt{n}}\right\}.$$

Thus $k = \sqrt{n}\, z_\alpha$. Since φ is independent of μ_1 as long as $\mu_1 > 0$, we see that the test

$$\varphi(\mathbf{x}) = \begin{cases} 1, & \sum x_i > \sqrt{n}\, z_\alpha \\ 0, & \text{otherwise,} \end{cases}$$

is UMP α-similar. We need only to check that φ is of the right size for testing H_0 against H_1. We have, for $\mu \le 0$,

$$E_\mu \varphi(\mathbf{X}) = P_\mu\{\sum X_i > \sqrt{n}\, z_\alpha\}$$
$$= P\left\{\frac{\sum X_i - n\mu}{\sqrt{n}} > z_\alpha - \sqrt{n}\, \mu\right\}$$
$$\le P\{Z > z_\alpha\},$$

since $-\sqrt{n}\, \mu \ge 0$. Here Z is $\mathcal{N}(0, 1)$. It follows that

$$E_\mu \varphi(\mathbf{X}) \le \alpha \quad \text{for } \mu \le 0,$$

hence φ is UMP unbiased.

Theorem 2 can be used only if it is possible to find a UMP α-similar test. Unfortunately this requires heavy use of conditional expection, and we will not pursue the subject here. We refer to Lehmann [70], Chapters 4 and 5, and

Ferguson [30], pages 224–233, for further details. Here we content ourselves with the following important results, which we state without proof.

Theorem 3. Let $X_1, X_2, \cdots, X_n$ be a sample from $\mathcal{N}(\mu, \sigma^2)$, where both μ and σ^2 are unknown.
(a) For testing $H_{01}: \mu \leq \mu_0, \sigma^2 > 0$ against $H_{11}: \mu > \mu_0, \sigma^2 > 0$, the test

(5) $$\varphi(\mathbf{x}) = \begin{cases} 1, & \dfrac{\bar{x} - \mu_0}{\sqrt{[1/(n-1)] \sum(x_i - \bar{x})^2}} \sqrt{n} \geq c_1, \\ 0, & \text{otherwise,} \end{cases}$$

is UMP unbiased. The UMP unbiased test of $H_{02}: \mu \geq \mu_0, \sigma^2 > 0$ against $H_{12}: \mu < \mu_0, \sigma^2 > 0$ is obtained by reversing the inequalities in (5).
(b) For testing $H_{03}: \sigma \leq \sigma_0, -\infty < \mu < \infty$ against $H_{13}: \sigma > \sigma_0, -\infty < \mu < \infty$, the test

(6) $$\varphi(\mathbf{x}) = \begin{cases} 1, & \sum(x_i - \bar{x})^2 \geq c_2, \\ 0, & \text{otherwise,} \end{cases}$$

is UMP unbiased. The UMP unbiased test of $H_{04}: \sigma \geq \sigma_0, -\infty < \mu < \infty$, against $H_{14}: \sigma < \sigma_0, -\infty < \mu < \infty$, is obtained by reversing the inequalities in (6).

For proof we refer to Lehmann [70], pages 94–97, 163–168.

Remark 3. The test defined in (5) and its dual are called *one-tailed t-tests*, since the test statistic has a $t(n-1)$ distribution. The test defined in (6) and its dual are called *one-tailed χ^2-tests* since the test statistic has a χ^2-distribution. The constants c_1 and c_2 are determined from

$$P_{\mu_0}\left\{\frac{\bar{X} - \mu_0}{S} \sqrt{n} \geq c_1\right\} = \alpha \quad \text{and} \quad P_{\sigma_0}\left\{\sum(X_i - \bar{X})^2 \geq c_2\right\} = \alpha.$$

Theorem 4. Let $X_1, X_2, \cdots, X_n$ be a sample from $\mathcal{N}(\mu, \sigma^2)$, where both μ and σ^2 are unknown.
(a) For testing $H_0: \mu = \mu_0, \sigma^2 > 0$ against $H_1: \mu \neq \mu_0, \sigma^2 > 0$, the test

(7) $$\varphi(\mathbf{x}) = \begin{cases} 1, & \left|\dfrac{\sqrt{n}(\bar{x} - \mu_0)}{\sqrt{[1/(n-1)] \sum(x_i - \bar{x})^2}}\right| > c, \\ 0, & \text{otherwise,} \end{cases}$$

is UMP unbiased.
(b) For testing $H_0: \sigma = \sigma_0, -\infty < \mu < \infty$ against $H_1: \sigma \neq \sigma_0, -\infty < \mu < \infty$, the test

(8) $$\varphi(\mathbf{x}) = \begin{cases} 1, & \dfrac{\Sigma(x_i - \bar{x})^2}{\sigma_0^2} < c_1 \text{ or } > c_2, \\ 0, & \text{otherwise,} \end{cases}$$

is UMP unbiased.

For proof see Lehmann [70], pages 164–166.

Remark 4. The test defined in (7) is known as a *two-tailed t-test*, and that defined in (8), as a *two-tailed χ^2-test*. The constant c is determined from

$$P_{\mu=\mu_0}\left\{\frac{\sqrt{n}\,|\bar{X} - \mu_0|}{S} > c\right\} = \alpha,$$

so that $c = t_{n-1,\alpha/2}$. The constants c_1 and c_2 are determined from

$$\int_{c_1}^{c_2} \chi^2_{n-1}(y)\,dy = \frac{1}{n-1}\int_{c_1}^{c_2} y\,\chi^2_{n-1}(y)\,dy = 1 - \alpha,$$

where $\chi^2_{n-1}(y)$ is the pdf of a χ^2 rv with $n-1$ d.f. Since

$$y\,\chi^2_{n-1}(y) = (n-1)\,\chi^2_{n+1}(y),$$

the second of these conditions can be written as

$$\int_{c_1}^{c_2} \chi^2_{n+1}(y)\,dy = 1 - \alpha.$$

In practice, one uses the equal tails test, given by

$$\int_0^{c_1} \chi^2_{n-1}(y)\,dy = \int_{c_2}^{\infty} \chi^2_{n-1}(y)\,dy = \frac{\alpha}{2},$$

so that $c_1 = \chi^2_{n-1,1-\alpha/2}$, $c_2 = \chi^2_{n-1,\alpha/2}$. The equal tails test is a good approximation to the unbiased test, at least for moderately large n.

Theorem 5. Let $X_1, \cdots, X_m$ and $Y_1, \cdots, Y_n$ be independent samples from normal distributions $\mathcal{N}(\mu, \sigma^2)$ and $\mathcal{N}(\theta, \sigma^2)$, respectively. For testing hypotheses $\mu \leq \theta$, $\mu \geq \theta$, and $\mu = \theta$, the tests with rejection regions $t(\mathbf{x}, \mathbf{y}) > c_1$, $t(\mathbf{x}, \mathbf{y}) < c_2$, and $|t(\mathbf{x}, \mathbf{y})| > c_3$, respectively, where

(9) $$t(\mathbf{x}, \mathbf{y}) = \frac{(\bar{x} - \bar{y})/\sqrt{1/m + 1/n}}{\sqrt{\left[\sum_1^m (x_i - \bar{x})^2 + \sum_1^n (y_i - \bar{y})^2\right]/(m + n - 2)}},$$

are UMP unbiased.

Proof. For proof see Lehmann [70], pages 168–173, or Ferguson [30], page 234.

Remark 5. By Corollary 5 to Theorem 7.5.1 we see that $t(X, Y) \sim t(m + n - 2)$ when $\mu = \theta$. The constants c_1, c_2, and c_3 are determined, using the central t-distribution for $m + n - 2$ d.f., and are given by $t_{m+n-2,\alpha}$, $t_{m+n-2,1-\alpha}$, and $t_{m+n-2,\alpha/2}$, respectively. The tests defined in Theorem 5 are called *two-sample t-tests*.

Theorem 6. Let $X_1, X_2, \cdots, X_m$ and $Y_1, Y_2, \cdots, Y_n$ be independent samples from normal distributions $\mathcal{N}(\mu, \sigma^2)$ and $\mathcal{N}(\theta, \tau^2)$, respectively. For testing $\sigma^2/\tau^2 \leq 1$, $\sigma^2/\tau^2 \geq 1$, and $\sigma^2/\tau^2 = 1$, the tests with rejection regions $F(\mathbf{x}, \mathbf{y}) \geq c_1$, $F(\mathbf{x}, \mathbf{y}) \leq c_2$, and $c_3 \leq F(\mathbf{x}, \mathbf{y}) \leq c_4$, respectively, where

(10) $$F(\mathbf{x}, \mathbf{y}) = \frac{\sum_{1}^{m}(x_j - \bar{x})^2/(m-1)}{\sum_{1}^{n}(y_j - \bar{y})^2/(n-1)},$$

are UMP unbiased.

For proof see Lehmann [70], pages 169–170.

Remark 6. By Corollary 4 to Theorem 7.5.1, $F(\mathbf{X}, \mathbf{Y})$, defined in (10), has an $F(m - 1, n - 1)$ distribution when $\tau^2 = \sigma^2$. The constants c_1 and c_2 are given by $c_1 = F_{m-1,n-1,\alpha}$ and $c_2 = F_{m-1,n-1,1-\alpha}$. The constants c_3 and c_4 are determined from

$$\int_{c_3}^{c_4} B_{(1/2)(m-1),(1/2)(n-1)}(y)\,dy = \int_{c_3}^{c_4} B_{(1/2)(m+1),(1/2)(n-1)}(y)\,dy = 1 - \alpha,$$

where $B_{\alpha,\beta}(y)$ is the pdf of a $B(\alpha, \beta)$ rv. In practice, one uses the equal tails test:

$$\int_{0}^{c_3} F_{m-1,n-1}(y)\,dy = \int_{c_4}^{\infty} F_{m-1,n-1}(y)\,dy = \frac{\alpha}{2},$$

where $F_{\alpha,\beta}(y)$ is the pdf of an $F(\alpha, \beta)$ rv. The tests described in Theorem 6 are called *F-tests*.

Yet another reduction is obtained if we apply the principle of invariance to hypothesis testing problems. We recall that a class of distributions is invariant under a group of transformations $\mathscr{G}$ if for every $g \in \mathscr{G}$ and every $\theta \in \Theta$ there exists a unique $\theta' \in \Theta$ such that $g(\mathbf{X})$ has distribution $P_{\theta'}$ whenever $\mathbf{X} \sim P_\theta$. We write $\theta' = \bar{g}\theta$.

Definition 5. A group $\mathscr{G}$ of transformations on the space of values of $\mathbf{X}$

leaves a hypothesis testing problem invariant if $\mathscr{G}$ leaves both $\{P_\theta: \theta \in \Theta_0\}$ and $\{P_\theta: \theta \in \Theta_1\}$ invariant.

Definition 6. We say that φ is invariant under $\mathscr{G}$ if

$$\varphi(g(\mathbf{x})) = \varphi(\mathbf{x}) \quad \text{for all } \mathbf{x} \text{ and all } g \in \mathscr{G}.$$

Definition 7. Let $\mathscr{G}$ be a group of transformations on the space of values of the rv $\mathbf{X}$. We say that a statistic $T(\mathbf{x})$ is maximal invariant under $\mathscr{G}$ if (a) T is invariant; (b) T is maximal, that is, $T(\mathbf{x}_1) = T(\mathbf{x}_2) \Rightarrow \mathbf{x}_1 = g(\mathbf{x}_2)$ for some $g \in \mathscr{G}$.

Example 2. Let $\mathbf{x} = (x_1, x_2, \cdots, x_n)$, and $\mathscr{G}$ be the group of translations

$$g_c(\mathbf{x}) = (x_1 + c, \cdots, x_n + c), \quad -\infty < c < \infty.$$

Here the space of values of $\mathbf{X}$ is $\mathscr{R}_n$. Consider the statistic

$$T(\mathbf{x}) = (x_n - x_1, \cdots, x_n - x_{n-1}).$$

Clearly,

$$T(g_c(\mathbf{x})) = (x_n - x_1, \cdots, x_n - x_{n-1}) = T(\mathbf{x}).$$

If $T(\mathbf{x}) = T(\mathbf{x}')$, then $x_n - x_i = x_n' - x_i'$, $i = 1, 2, \cdots, n-1$, and we have $x_i - x_i' = x_n - x_n' = c$ ($i = 1, 2, \cdots, n-1$), that is, $g_c(\mathbf{x}') = (x_1' + c, \cdots, x_n' + c) = \mathbf{x}$ and T is maximal invariant.

Next consider the group of scale changes

$$g_c(\mathbf{x}) = (cx_1, \cdots, cx_n), \quad c > 0.$$

Then

$$T(\mathbf{x}) = \begin{cases} 0 & \text{if all } x_i = 0, \\ \left(\dfrac{x_1}{z}, \cdots, \dfrac{x_n}{z}\right) & \text{if at least one } x_i \neq 0, \end{cases} \quad z = \left(\sum_1^n x_i^2\right)^{1/2},$$

is maximal invariant; for

$$T(g_c(\mathbf{x})) = T(cx_1, \cdots, cx_n) = T(\mathbf{x}),$$

and if $T(\mathbf{x}) = T(\mathbf{x}')$, then either $T(\mathbf{x}) = T(\mathbf{x}') = 0$, in which case $x_i = x_i' = 0$, or $T(\mathbf{x}) = T(\mathbf{x}') \neq 0$, in which case $x_i/z = x_i'/z'$, implying $x_i' = (z'/z)x_i = cx_i$, and T is maximal.

Finally, if we consider the group of translation and scale changes,

$$g(\mathbf{x}) = (ax_1 + b, \cdots, ax_n + b), \quad a > 0, \; -\infty < b < \infty,$$

a maximal invariant is

$$T(\mathbf{x}) = \begin{cases} 0 & \text{if } \beta = 0, \\ \left(\dfrac{x_1 - \bar{x}}{\beta}, \dfrac{x_2 - \bar{x}}{\beta}, \cdots, \dfrac{x_n - \bar{x}}{\beta} \right) & \text{if } \beta \neq 0, \end{cases}$$

where $\bar{x} = n^{-1} \sum_1^n x_i$ and $\beta = n^{-1} \sum_1^n (x_i - \bar{x})^2$.

Definition 8. Let I_α denote the class of all invariant size α tests of $H_0: \theta \in \Theta_0$ against $H_1: \theta \in \Theta_1$. If there exists a UMP member in I_α, we call the test a UMP invariant test of H_0 against H_1.

The search for UMP invariant tests is greatly facilitated by the use of the following result.

Theorem 7. Let $T(\mathbf{x})$ be a maximal invariant with respect to $\mathscr{G}$. Then φ is invariant under $\mathscr{G}$ if and only if φ is a function of T.

Proof. Let φ be invariant. We have to show that $T(\mathbf{x}_1) = T(\mathbf{x}_2) \Rightarrow \varphi(\mathbf{x}_1) = \varphi(\mathbf{x}_2)$. If $T(\mathbf{x}_1) = T(\mathbf{x}_2)$, there is a $g \in \mathscr{G}$ such that $\mathbf{x}_1 = g(\mathbf{x}_2)$, so that $\varphi(\mathbf{x}_1) = \varphi(g(\mathbf{x}_2)) = \varphi(\mathbf{x}_2)$.

Conversely, if φ is a function of T, $\varphi(\mathbf{x}) = h[T(\mathbf{x})]$, then

$$\varphi(g(\mathbf{x})) = h[T(g(\mathbf{x}))] = h[T(\mathbf{x})] = \varphi(\mathbf{x}),$$

and φ is invariant.

Remark 7. The use of Theorem 7 is obvious. If a hypothesis testing problem is invariant under a group $\mathscr{G}$, it suffices to restrict attention to test functions that are functions of maximal invariant T.

Example 3. Let $X_1, X_2, \cdots, X_n$ be a sample from $\mathcal{N}(\mu, \sigma^2)$, where both μ and σ^2 are unknown. We wish to test $H_0: \sigma \geq \sigma_0$, $-\infty < \mu < \infty$, against $H_1: \sigma < \sigma_0$, $-\infty < \mu < \infty$. The family $\{\mathcal{N}(\mu, \sigma^2)\}$ remains invariant under translations $x_i' = x_i + c$, $-\infty < c < \infty$. Moreover, since $\text{var}(X + c) = \text{var}(X)$, the hypothesis testing problem remains invariant under the group of translations, that is, both $\{\mathcal{N}(\mu, \sigma^2): \sigma^2 \geq \sigma_0^2\}$ and $\{\mathcal{N}(\mu, \sigma^2): \sigma^2 < \sigma_0^2\}$ remain invariant. The joint sufficient statistic is $(\bar{X}, \sum(X_i - \bar{X})^2)$, which is transformed to $(\bar{X} + c, \sum(X_i - \bar{X})^2)$ under translations. A maximal invariant is $\sum(X_i - \bar{X})^2$. It follows that the class of invariant tests consists of tests that are functions of $\sum(X_i - \bar{X})^2$.

Now $\sum(X_i - \bar{X})^2/\sigma^2 \sim \chi^2(n - 1)$, so that the pdf of $Z = \sum(X_i - \bar{X})^2$ is given by

$$f_{\sigma^2}(z) = \frac{\sigma^{-(n-1)}}{\Gamma[(n-1)/2] \, 2^{(n-1)/2}} z^{(n-3)/2} e^{-z/2\sigma^2}, \qquad z > 0.$$

The family of densities $\{f_{\sigma^2}: \sigma^2 > 0\}$ has an MLR in z, and it follows that a UMP test is to reject $H_0: \sigma^2 \geq \sigma_0^2$ if $z \leq k$, that is, a UMP invariant test is given by

$$\varphi(\mathbf{x}) = \begin{cases} 1 & \text{if } \sum(x_i - \bar{x})^2 \leq k, \\ 0 & \text{if } \sum(x_i - \bar{x})^2 > k, \end{cases}$$

where k is determined from the size restriction

$$\alpha = P_{\sigma_0}\{\sum(X_i - \bar{X})^2 \leq k\} = P\left\{\frac{\sum(X_i - \bar{X})^2}{\sigma_0^2} \leq \frac{k}{\sigma_0^2}\right\},$$

that is,

$$k = \sigma_0^2 \chi_{n-1,1-\alpha}^2.$$

This was the UMP unbiased test described in Theorem 3.

Example 4. Let $\mathbf{X}$ have pdf $f_i(x_1 - \theta, \cdots, x_n - \theta)$ under H_i ($i = 0, 1$), $-\infty < \theta < \infty$. Let $\mathscr{G}$ be the group of translations

$$g_c(\mathbf{x}) = (x_1 + c, \cdots, x_n + c), \quad -\infty < c < \infty.$$

Clearly, g induces $\bar{g}$ on Θ, where $\bar{g}\theta = \theta + c$. The hypothesis testing problem remains invariant under $\mathscr{G}$. A maximal invariant under $\mathscr{G}$ is $T(\mathbf{X}) = (X_1 - X_n, \cdots, X_{n-1} - X_n) = (T_1, T_2, \cdots, T_{n-1})$. The class of invariant tests coincides with the class of tests that are functions of T. The pdf of T under H_i is independent of θ and is given by $\int_{-\infty}^{\infty} f_i(t_1 + z, \cdots, t_{n-1} + z, z)\,dz$. The problem is thus reduced to testing a simple hypothesis against a simple alternative. By the fundamental lemma the MP test

$$\varphi(t_1, t_2, \cdots, t_{n-1}) = \begin{cases} 1 & \text{if } \lambda(\mathbf{t}) > c, \\ 0 & \text{if } \lambda(\mathbf{t}) < c, \end{cases}$$

where $\mathbf{t} = (t_1, t_2, \cdots, t_{n-1})$ and

$$\lambda(\mathbf{t}) = \frac{\int_{-\infty}^{\infty} f_1(t_1 + z, \cdots, t_{n-1} + z, z)\,dz}{\int_{-\infty}^{\infty} f_0(t_1 + z, \cdots, t_{n-1} + z, z)\,dz}$$

is UMP invariant.

A particular case of Example 4 will be, for instance, to test $H_0: X \sim \mathcal{N}(\theta, 1)$ against $H_1: X \sim \mathscr{C}(1, \theta)$, $\theta \in \mathscr{R}$.

We conclude this section by remarking, without proof, that the tests defined in Theorem 3 are all UMP invariant. Test (7) is also UMP invariant, but invariance considerations do not reduce the problem of testing $\sigma = \sigma_0$ to lead to a UMP invariant test. The two-sample t-tests in Theorem 5 are also UMP invariant, as is the F-test of $\sigma^2/\tau^2 \geq 1$ (or $\sigma^2/\tau^2 \leq 1$) considered in Theorem 6. We refer the reader to Lehmann [70], pages 228–229.

PROBLEMS 9.5

1. To test $H_0: X \sim \mathcal{N}(\theta, 1)$, against $H_1 : X \sim \mathscr{C}(1, \theta)$, a sample of size 2 is available on X. Find a UMP invariant test of H_0 against H_1.

2. Let $\mathbf{X} = (X_1, X_2, \cdots, X_n)$ have pdf

$$\frac{1}{\theta^n} f_i\left(\frac{x_1 - \mu}{\theta}, \frac{x_2 - \mu}{\theta}, \cdots, \frac{x_n - \mu}{\theta}\right)$$

under H_i ($i = 0, 1$), $-\infty < \mu < \infty$, $\theta > 0$, where both μ and θ are unknown and f is even. Show that the problem of testing H_0 against H_1 remains invariant under the group of transformations $x_i' = ax_i + b$ ($i = 1, 2, \cdots, n$), $a \neq 0$, $b \in \mathscr{R}$. Find a UMP invariant test.

(Lehmann [70], 248–249)

3. Let $X_1, X_2, \cdots, X_n$ be a sample from $P(\lambda)$. Find a UMP unbiased size α test for the null hypothesis $H_0 : \lambda \leq \lambda_0$ against alternatives $\lambda > \lambda_0$ by the methods of this section.

4. Let $X \sim NB(1; \theta)$. By the methods of this section find a UMP unbiased size α test of $H_0 : \theta \geq \theta_0$ against $H_1 : \theta < \theta_0$.

CHAPTER 10

Some Further Results on Hypotheses Testing

10.1 INTRODUCTION

In this chapter we study some procedures commonly used in the theory of hypotheses testing. In Section 2 we describe the classical procedure for constructing tests based on likelihood ratios, which will be used in Chapter 12. Sections 3, 4, and 5 deal with some most frequently used tests of hypotheses, and in Section 6 we look at the problem of testing hypotheses from a decision-theoretic viewpoint.

10.2 THE LIKELIHOOD RATIO TESTS

In Chapter 9 we saw that UMP tests do not exist for a wide variety of problems of hypotheses testing. In cases where UMP tests do exist, the methods apply only to special families of distributions. Moreover, some of the reductions suggested, such as invariance, do not apply to all families of distributions.

In this section we consider a procedure for constructing tests that has some intuitive appeal and that frequently, though not necessarily, leads to UMP tests or UMP unbiased or invariant tests. Also, the procedure leads to tests that have some desirable large-sample properties.

Let $\theta \in \Theta \subseteq \mathcal{R}_k$ be a vector of parameters, and let $\mathbf{X}$ be a random vector with pdf (pmf) f_θ. Consider the problem of testing the null hypothesis $H_0: \mathbf{X} \sim f_\theta$, $\theta \in \Theta_0$ against the alternatives $H_1: \mathbf{X} \sim f_\theta$, $\theta \in \Theta_1$.

Definition 1. For testing H_0 against H_1, a test of the form: reject H_0 if and only if $\lambda(\mathbf{x}) < c$, where c is some constant, and

$$\lambda(\mathbf{x}) = \frac{\sup_{\theta \in \Theta_0} f_\theta(x_1, x_2, \cdots, x_n)}{\sup_{\theta \in \Theta} f_\theta(x_1, x_2, \cdots, x_n)},$$

is called a likelihood ratio test.

The numerator of the likelihood ratio λ is the best *explanation* of $\mathbf{X}$ (in the sense of maximum likelihood) that the null hypothesis H_0 can provide, and the denominator is the best possible explanation of $\mathbf{X}$. H_0 is rejected if there is a much better explanation of $\mathbf{X}$ than the best one provided by H_0.

It is clear that $0 \leq \lambda \leq 1$. The constant c is determined from the size restriction

$$\sup_{\theta \in \Theta_0} P_\theta\{\mathbf{x} : \lambda(\mathbf{x}) < c\} = \alpha.$$

If the distribution of λ is continuous (that is, the df is absolutely continuous), any size α is attainable. If, however, $\lambda(\mathbf{X})$ is a discrete rv, it may not be possible to find a likelihood ratio test whose size exactly equals α. This problem arises because of the nonrandomized nature of the likelihood ratio test and can be handled by randomization. The following result holds.

Theorem 1. If for given α, $0 \leq \alpha \leq 1$, nonrandomized Neyman-Pearson and likelihood ratio tests of a simple hypothesis against a simple alternative exist, they are equivalent.

Proof. The proof is left as an exercise.

Theorem 2. For testing $\theta \in \Theta_0$ against $\theta \in \Theta_1$, the likelihood ratio test is a function of every sufficient statistic for θ.

Theorem 2 follows from the factorization theorem for sufficient statistics.

Example 1. Let $X \sim b(n, p)$, and we seek a level α likelihood ratio test of $H_0: p \leq p_0$ against $H_1: p > p_0$:

$$\lambda(x) = \frac{\sup_{p \leq p_0} \binom{n}{x} p^x (1-p)^{n-x}}{\sup_{0 \leq p \leq 1} \binom{n}{x} p^x (1-p)^{n-x}}.$$

Now

$$\sup_{0 \leq p \leq 1} p^x (1-p)^{n-x} = \left(\frac{x}{n}\right)^x \left(1 - \frac{x}{n}\right)^{n-x}.$$

THE LIKELIHOOD RATIO TESTS

The function $p^x(1-p)^{n-x}$ first increases, then achieves its maximum at $p = x/n$, and finally decreases, so that

$$\sup_{p \le p_0} p^x(1-p)^{n-x} = \begin{cases} p_0^x(1-p_0)^{n-x} & \text{if } p_0 < \dfrac{x}{n}, \\ \left(\dfrac{x}{n}\right)^x \left(1 - \dfrac{x}{n}\right)^{n-x} & \text{if } \dfrac{x}{n} \le p_0. \end{cases}$$

It follows that

$$\lambda(x) = \begin{cases} \dfrac{p_0^x(1-p_0)^{n-x}}{(x/n)^x [1-(x/n)]^{n-x}} & \text{if } p_0 < \dfrac{x}{n}, \\ 1 & \text{if } \dfrac{x}{n} \le p_0. \end{cases}$$

Note that $\lambda(x) \le 1$ for $np_0 < x$ and $\lambda(x) = 1$ if $x \le np_0$, and it follows that $\lambda(x)$ is a decreasing function of x. Thus $\lambda(x) < c$ if and only if $x > c'$, and the likelihood ratio test rejects H_0 if $x > c'$.

The likelihood ratio test is of the type obtained in Section 9.4 for families with an MLR except for the boundary $\lambda(x) = c$. In other words, if the size of the test happens to be exactly α, the likelihood ratio test is a UMP level α test. Since X is a discrete rv, however, to obtain size α may not be possible. We have

$$\alpha = \sup_{p \le p_0} P_p\{X > c'\} = P_{p_0}\{X > c'\}.$$

If such a c' does not exist, we choose an integer c' such that

$$P_{p_0}\{X > c'\} \le \alpha \quad \text{and} \quad P_{p_0}\{X > c' - 1\} > \alpha.$$

Example 2. Consider the problem of testing $\mu = \mu_0$ against $\mu \ne \mu_0$ in sampling from $\mathcal{N}(\mu, \sigma^2)$, where both μ and σ^2 are unknown. In this case $\Theta_0 = \{(\mu_0, \sigma^2): \sigma^2 > 0\}$ and $\Theta = \{(\mu, \sigma^2): -\infty < \mu < \infty, \sigma^2 > 0\}$. We write $\boldsymbol{\theta} = (\mu, \sigma^2)$:

$$\sup_{\boldsymbol{\theta} \in \Theta_0} f_{\boldsymbol{\theta}}(\mathbf{x}) = \sup_{\sigma^2 > 0} \left[\frac{1}{(\sigma\sqrt{2\pi})^n} \exp\left\{ -\frac{\sum_{1}^{n}(x_i - \mu_0)^2}{2\sigma^2} \right\} \right]$$
$$= f_{\hat{\sigma}_0^2}(\mathbf{x}),$$

where $\hat{\sigma}_0^2$ is the MLE, $\hat{\sigma}_0^2 = (1/n) \sum_{i=1}^{n}(x_i - \mu_0)^2$. Thus

$$\sup_{\boldsymbol{\theta} \in \Theta_0} f_{\boldsymbol{\theta}}(\mathbf{x}) = \frac{1}{(2\pi/n)^{n/2} \left\{\sum_{1}^{n}(x_i - \mu_0)^2\right\}^{n/2}} e^{-n/2}.$$

The MLE of $\boldsymbol{\theta} = (\mu, \sigma^2)$ when both μ and σ^2 are unknown is $(\sum_{1}^{n} x_i/n, \sum_{1}^{n}(x_i - \bar{x})^2/n)$. It follows that

$$\sup_{\theta \in \Theta} f_\theta(\mathbf{x}) = \sup_{\mu, \sigma^2} \left[\frac{1}{(\sigma\sqrt{2\pi})^n} \exp\left\{ -\frac{\sum_1^n (x_i - \mu)^2}{2\sigma^2} \right\} \right]$$

$$= \frac{1}{(2\pi/n)^{n/2} \left\{ \sum_1^n (x_i - \bar{x})^2 \right\}^{n/2}} e^{-n/2}.$$

Thus

$$\lambda(\mathbf{x}) = \left\{ \frac{\sum_1^n (x_i - \bar{x})^2}{\sum_1^n (x_i - \mu_0)^2} \right\}^{n/2}$$

$$= \left\{ \frac{1}{1 + [n(\bar{x} - \mu_0)^2 / \sum_1^n (x_i - \bar{x})^2]} \right\}^{n/2}.$$

The likelihood ratio test rejects H_0 if

$$\lambda(\mathbf{x}) < c,$$

and since $\lambda(\mathbf{x})$ is a decreasing function of $n(\bar{x} - \mu_0)^2 / \sum_1^n (x_i - \bar{x})^2$, we reject H_0 if

$$\left| \frac{\bar{x} - \mu_0}{\sqrt{\sum_1^n (x_i - \bar{x})^2}} \right| > c',$$

that is, if

$$\left| \frac{\sqrt{n}(\bar{x} - \mu_0)}{s} \right| > c'',$$

where $s^2 = (n-1)^{-1} \sum_1^n (x_i - \bar{x})^2$. The statistic

$$t(\mathbf{X}) = \frac{\sqrt{n}(\bar{X} - \mu_0)}{S}$$

has a t-distribution with $n-1$ d.f. Under H_0: $\mu = \mu_0$, $t(\mathbf{X})$ has a central $t(n-1)$ distribution, but under H_1: $\mu \neq \mu_0$, $t(\mathbf{X})$ has a noncentral t-distribution with $n-1$ d.f. and noncentrality parameter $\delta = (\mu - \mu_0)/\sigma$. We choose $c'' = t_{n-1, \alpha/2}$ in accordance with the distribution of $t(\mathbf{X})$ under H_0. Note that the two-sided t-test obtained here is UMP unbiased (Theorem 9.5.4). Similarly one can obtain one-sided t-tests also as likelihood ratio tests.

Example 3. Let $X_1, X_2, \cdots, X_m$ and $Y_1, Y_2, \cdots, Y_n$ be independent random samples from $\mathcal{N}(\mu_1, \sigma_1^2)$ and $\mathcal{N}(\mu_2, \sigma_2^2)$, respectively. We wish to test the null hypothesis H_0: $\sigma_1^2 = \sigma_2^2$ against H_1: $\sigma_1^2 \neq \sigma_2^2$. Here

$$\Theta = \{(\mu_1, \sigma_1^2, \mu_2, \sigma_2^2) : -\infty < \mu_i < \infty, \sigma_i^2 > 0, i = 1, 2\}$$

and
$$\Theta_0 = \{(\mu_1, \sigma_1^2, \mu_2, \sigma_2^2): -\infty < \mu_i < \infty, i = 1, 2, \sigma_1^2 = \sigma_2^2 > 0\}.$$

Let $\theta = (\mu_1, \sigma_1^2, \mu_2, \sigma_2^2)$. Then the joint pdf is

$$f_\theta(\mathbf{x}, \mathbf{y}) = \frac{1}{(2\pi)^{(m+n)/2} \sigma_1^m \sigma_2^n} \exp\left\{-\frac{1}{2\sigma_1^2} \sum_1^m (x_i - \mu_1)^2 - \frac{1}{2\sigma_2^2} \sum_1^n (y_i - \mu_2)^2\right\}.$$

Also,

$$\log f_\theta(\mathbf{x}, \mathbf{y}) = -\frac{m+n}{2} \log 2\pi - \frac{m}{2} \log \sigma_1^2 - \frac{n}{2} \log \sigma_2^2 - \frac{\sum_1^m (x_i - \mu_1)^2}{2\sigma_1^2}$$
$$- \frac{1}{2\sigma_2^2} \sum_1^n (y_i - \mu_2)^2.$$

Differentiating with respect to μ_1 and μ_2, we obtain the MLE's

$$\hat{\mu}_1 = \bar{x}, \qquad \hat{\mu}_2 = \bar{y}.$$

Differentiating with respect to σ_1^2 and σ_2^2, we obtain the MLE's

$$\hat{\sigma}_1^2 = \frac{1}{m} \sum_1^m (x_i - \bar{x})^2, \qquad \hat{\sigma}_2^2 = \frac{1}{n} \sum_1^n (y_i - \bar{y})^2.$$

If, however, $\sigma_1^2 = \sigma_2^2 = \sigma^2$, the MLE of σ^2 is

$$\hat{\sigma}^2 = \frac{\sum_1^m (x_i - \bar{x})^2 + \sum_1^n (y_i - \bar{y})^2}{m+n}.$$

Thus

$$\sup_{\theta \in \Theta_0} f_\theta(\mathbf{x}, \mathbf{y}) = \frac{e^{-(m+n)/2}}{[2\pi/(m+n)]^{(m+n)/2} \left\{\sum_1^m (x_i - \bar{x})^2 + \sum_1^n (y_i - \bar{y})^2\right\}^{(m+n)/2}}$$

and

$$\sup_{\theta \in \Theta} f_\theta(\mathbf{x}, \mathbf{y}) = \frac{e^{-(m+n)/2}}{(2\pi/m)^{m/2} (2\pi/n)^{n/2} \left\{\sum_1^m (x_i - \bar{x})^2\right\}^{m/2} \left\{\sum_1^n (y_i - \bar{y})^2\right\}^{n/2}},$$

so that

$$\lambda(\mathbf{x}, \mathbf{y}) = \left(\frac{m}{m+n}\right)^{m/2} \left(\frac{n}{m+n}\right)^{n/2} \frac{\left\{\sum_1^m (x_i - \bar{x})^2\right\}^{m/2} \left\{\sum_1^n (y_i - \bar{y})^2\right\}^{n/2}}{\left\{\sum_1^m (x_i - \bar{x})^2 + \sum_1^n (y_i - \bar{y})^2\right\}^{(m+n)/2}}.$$

Now

$$\frac{\left\{\sum_{1}^{m}(x_i-\bar{x})^2\right\}^{m/2}\left\{\sum_{1}^{n}(y_i-\bar{y})^2\right\}^{n/2}}{\left\{\sum_{1}^{m}(x_i-\bar{x})^2+\sum_{1}^{n}(y_i-\bar{y})^2\right\}^{(m+n)/2}}$$

$$=\frac{1}{\left\{1+\sum_{1}^{m}(x_i-\bar{x})^2/\sum_{1}^{n}(y_i-\bar{y})^2\right\}^{n/2}\left\{1+\sum_{1}^{n}(y_i-\bar{y})^2/\sum_{1}^{m}(x_i-\bar{x})^2\right\}^{m/2}}.$$

Writing

$$f=\frac{\sum_{1}^{m}(x_i-\bar{x})^2/(m-1)}{\sum_{1}^{n}(y_i-\bar{y})^2/(n-1)},$$

we have

$$\lambda(\mathbf{x},\mathbf{y})=\left(\frac{m}{m+n}\right)^{m/2}\left(\frac{n}{m+n}\right)^{n/2}$$
$$\cdot\frac{1}{\{1+[(m-1)/(n-1)]f\}^{n/2}\{1+[(n-1)/(m-1)](1/f)\}^{m/2}}.$$

We leave the reader to check that $\lambda(\mathbf{x},\mathbf{y})<c$ is equivalent to $f<c_1$ or $f>c_2$. (Take logarithms, and use properties of convex functions. Alternatively, differentiate log λ.)

Under H_0, the statistic

$$F=\frac{\sum_{1}^{m}(X_i-\bar{X})^2/(m-1)}{\sum_{1}^{n}(Y_i-\bar{Y})^2/(n-1)}$$

has an $F(m-1, n-1)$ distribution, so that c_1, c_2 can be selected. It is usual to take

$$P\{F\leq c_1\}=P\{F\geq c_2\}=\frac{\alpha}{2}.$$

Under H_1, $(\sigma_2^2/\sigma_1^2)F$ has an $F(m-1, n-1)$ distribution.

We remark that the likelihood ratio test in this case coincides with the UMP unbiased test (Theorem 9.5.6).

In certain situations the likelihood ratio test does not perform well. We reproduce here an example due to Stein and Rubin.

Example 4. Let X be a discrete rv with pmf

$$P_{p=0}\{X=x\}=\begin{cases}\dfrac{\alpha}{2} & \text{if } x=\pm 2,\\[4pt] \dfrac{1-2\alpha}{2} & \text{if } x=\pm 1,\\[4pt] \alpha & \text{if } x=0,\end{cases}$$

under the null hypothesis $H_0: p = 0$, and

$$P_p\{X = x\} = \begin{cases} pc & \text{if } x = -2, \\ \dfrac{1-c}{1-\alpha}(\tfrac{1}{2} - \alpha) & \text{if } x = \pm 1, \\ \alpha\left(\dfrac{1-c}{1-\alpha}\right) & \text{if } x = 0, \\ (1-p)c & \text{if } x = 2, \end{cases}$$

under the alternative $H_1: p \in (0, 1)$, where α and c are constants with

$$0 < \alpha < \tfrac{1}{2} \quad \text{and} \quad \dfrac{\alpha}{2-\alpha} < c < \alpha.$$

To test the simple null hypothesis against the composite alternative at the level of significance α, let us compute the likelihood ratio λ. We have

$$\lambda(2) = \frac{P_0\{X = 2\}}{\sup_{0 \le p < 1} P_p\{X = 2\}} = \frac{\alpha/2}{c} = \frac{\alpha}{2c}$$

since $\alpha/2 < c$. Similarly $\lambda(-2) = \alpha/(2c)$. Also

$$\lambda(1) = \lambda(-1) = \frac{\tfrac{1}{2} - \alpha}{[(1-c)/(1-\alpha)](\tfrac{1}{2} - \alpha)} = \frac{1-\alpha}{1-c}, \qquad \alpha < \tfrac{1}{2},$$

and

$$\lambda(0) = \frac{1-\alpha}{1-c}.$$

The test rejects H_0 if $\lambda(x) < k$, where k is to be determined so that the level is α. We see that

$$P_0\left\{\lambda(X) < \frac{1-\alpha}{1-c}\right\} = P_0\{X = \pm 2\} = \alpha,$$

provided that $\alpha/2c < [(1-\alpha)/(1-c)]$. But $\alpha/(2-\alpha) < c < \alpha$ implies $\alpha < 2c - c\alpha$, so that $\alpha - c\alpha < 2c - 2c\alpha$, or $\alpha(1-c) < 2c(1-\alpha)$, as required. Thus the likelihood ratio size α test is to reject H_0 if $X = \pm 2$. The power of the likelihood ratio test is

$$P_p\left\{\lambda(X) < \frac{1-\alpha}{1-c}\right\} = P_p\{X = \pm 2\} = pc + (1-p)c = c < \alpha$$

for all $p \in (0, 1)$. The test is not unbiased and is even worse than the trivial test $\varphi(x) \equiv \alpha$.

Another test that is better than the trivial test is to reject H_0 whenever $x = 0$ (this is opposite to what the likelihood ratio test says). Then

$$P_0\{X = 0\} = \alpha,$$

$$P_p\{X = 0\} = \alpha\,\frac{1-c}{1-\alpha} > \alpha \qquad (\text{since } c < \alpha),$$

for all $p \in (0, 1)$, and the test is unbiased.

We will use the likelihood ratio procedure quite frequently hereafter. We conclude this discussion with the following large-sample property of the likelihood ratio, which solves the problem of distribution.

Theorem 3. Under some regularity conditions on $f_\theta(\mathbf{x})$, the random variable $-2 \log \lambda(\mathbf{X})$ is asymptotically distributed as a chi-square rv with degrees of freedom equal to the difference between the number of independent parameters in Θ and the number in Θ_0.

We will not prove this result here; the reader is referred to Wilks [141], page 419. The regularity condtions are essentially the ones associated with Theorem 8.7.4. In Example 2 the number of parameters unspecified under H_0 is one (namely, σ^2), and under H_1 two parameters are unspecified (μ and σ^2), so that the asymptotic chi-square distribution will have 1 d.f. Similarly, in Example 3, the d.f. $= 4 - 3 = 1$.

Example 5. In Example 2 we showed that, in sampling from a normal population with unknown mean μ and unknown variance σ^2, the likelihood ratio for testing $H_0: \mu = \mu_0$ against $H_1: \mu \neq \mu_0$ is

$$\lambda(\mathbf{x}) = \left\{ 1 + \frac{n(\bar{x} - \mu_0)^2}{\sum_{i=1}^{n}(x_i - \bar{x})^2} \right\}^{-n/2}.$$

Thus

$$-2 \log \lambda(\mathbf{X}) = n \log \left\{ 1 + n \frac{(\bar{X} - \mu_0)^2}{\sum_{1}^{n}(X_i - \bar{X})^2} \right\}.$$

Under H_0, $\sqrt{n}(\bar{X} - \mu_0)/\sigma \sim \mathcal{N}(0, 1)$ and $\sum_{1}^{n}(X_i - \bar{X})^2/\sigma^2 \sim \chi^2(n - 1)$. Moreover, $\bar{X}$ and $\sum_{1}^{n}(X_i - \bar{X})^2$ are independent. It follows that $\sqrt{n}(\bar{X} - \mu_0)/\sqrt{[\sum(X_i - \bar{X})^2/(n - 1)]}$ has a (central) t-distribution with $(n - 1)$ d.f.; hence $[n(n - 1)(\bar{X} - \mu_0)^2/\sum_{1}^{n}(X - \bar{X})^2]$ is an F-statistic with $(1, n - 1)$ d.f. Therefore we can write for the mgf of $-2 \log \lambda(\mathbf{X})$

$$M_n(t) = E_{H_0} \exp\{-2t \log \lambda(\mathbf{X})\}$$
$$= E_{H_0} \exp\left\{\log\left(1 + \frac{F}{n-1}\right)^{nt}\right\},$$

where F has an $F(1, n - 1)$ distribution. Thus

$$M_n(t) = E_{H_0}\left(1 + \frac{F}{n-1}\right)^{nt}.$$

$$= \int_0^\infty \frac{\Gamma(n/2)}{\Gamma[(n-1)/2]\,\Gamma(\tfrac{1}{2})\,(n-1)^{1/2}}$$
$$\cdot f^{-1/2}\left(1 + \frac{f}{n-1}\right)^{-n/2}\left(1 + \frac{f}{n-1}\right)^{nt} df.$$

Let us substitute $y = 1/\{1 + [f/(n-1)]\}$. Then

$$\frac{f}{n-1} = \frac{1-y}{y} \quad \text{and} \quad df = -\frac{n-1}{y^2}\,dy.$$

We have

$$\int_0^\infty f^{-1/2}\left(1 + \frac{f}{n-1}\right)^{nt-(n/2)} df$$
$$= (n-1)^{1/2} \int_0^1 y^{(n-3)/2-nt}(1-y)^{1/2-1}\,dy$$
$$= (n-1)^{1/2}\, B\!\left(\frac{n-1}{2} - nt,\, \frac{1}{2}\right) \quad \text{for } t < \frac{1}{2} - \frac{1}{2n}.$$

Thus

$$E_{H_0}\!\left(1 + \frac{F}{n-1}\right)^{nt} = \frac{\Gamma(n/2)}{\Gamma[(n-1)/2]}\,\frac{\Gamma[(n-1)/2 - nt]}{\Gamma[n/2 - nt]} \quad \text{for } t < \frac{1}{2} - \frac{1}{2n}.$$

As $n \to \infty$, we treat $\Gamma(\alpha(n) + 1)$ as $(\alpha(n))!$, and applying Stirling's approximation (see P.2.8) to $(\alpha(n))!$, we see that

$$M_n(t) \to \frac{1}{(1-2t)^{1/2}} \quad \text{as } n \to \infty.$$

It follows by the continuity theorem that $-2\log\lambda(\mathbf{X}) \xrightarrow{L} Y$, where $Y \sim \chi^2(1)$. This result is consistent with Theorem 3.

PROBLEMS 10.2

1. Prove Theorem 1.
2. Prove Theorem 2.
3. Find the likelihood ratio test of $H_0: p = p_0$ against $H_1: p \neq p_0$, based on a sample of size 1 from $b(n, p)$.
4. Let $X_1, X_2, \cdots, X_n$ be a sample from $\mathcal{N}(\mu, \sigma^2)$, where both μ and σ^2 are unknown. Find the likelihood ratio test of $H_0: \sigma = \sigma_0$ against $H_1: \sigma \neq \sigma_0$.
5. Let $X_1, X_2, \cdots, X_k$ be a sample from pmf

$$P_N\{X = j\} = \frac{1}{N}, \quad j = 1, 2, \cdots, N,\ N \geq 1 \text{ is an integer}.$$

(a) Find the likelihood ratio test of $H_0: N \leq N_0$ against $H_1: N > N_0$.

(b) Find the likelihood ratio test of $H_0: N = N_0$ against $H_1: N \neq N_0$.

6. For a sample of size 1 from pdf

$$f_\theta(x) = \frac{2}{\theta^2}(\theta - x), \qquad 0 < x < \theta,$$

find the likelihood ratio test of $\theta = \theta_0$ against $\theta \neq \theta_0$.

7. Let $X_1, X_2, \cdots, X_n$ be a sample from $G(1, \beta)$.
   (a) Find the likelihood ratio test of $\beta = \beta_0$ against $\beta \neq \beta_0$.
   (b) Find the likelihood ratio test of $\beta \leq \beta_0$ against $\beta > \beta_0$.

8. Let $(X_1, Y_1), (X_2, Y_2), \cdots, (X_n, Y_n)$ be a random sample from a bivariate normal population with $EX_i = \mu_1$, $EY_i = \mu_2$, var $(X_i) = \sigma^2$, var $(Y_i) = \sigma^2$, and cov$(X_i, Y_i) = \rho\sigma^2$. Show that the likelihood ratio test of the null hypothesis $H_0: \rho = 0$ against $H_1: \rho \neq 0$ reduces to rejecting H_0 if $|R| > c$, where $R = 2S_{11}/(S_1^2 + S_2^2)$, S_{11}, S_1^2, and S_2^2 being the sample covariance and the sample variances, respectively. (For the pdf of the test statistic R, see Problem 7.6.1.)

10.3 THE CHI-SQUARE TESTS

In this section we consider some tests based on the chi-square statistic for some parametric hypotheses. The types of problems discussed here arise quite often in practice. Chi-square tests are also used for testing some nonparametric hypotheses and will be taken up again in Chapter 13.

We first consider some tests concerning variances. Given a sample of size n from a normal population $\mathcal{N}(\mu, \sigma^2)$, where σ^2 is unknown, we need to test a hypothesis of the type $\sigma^2 \geq \sigma_0^2$, $\sigma^2 \leq \sigma_0^2$, or $\sigma^2 = \sigma_0^2$, where σ_0 is some given positive number. In the following table we summarize the tests.

			Reject H_0 at level α if	
H_0	H_1	μ Known		μ Unknown
I. $\sigma \geq \sigma_0$	$\sigma < \sigma_0$	$\sum_1^n (x_i - \mu)^2 \leq \chi_{n,1-\alpha}^2 \sigma_0^2$		$s^2 \leq \dfrac{\sigma_0^2}{n-1} \chi_{n-1, 1-\alpha}^2$
II. $\sigma \leq \sigma_0$	$\sigma > \sigma_0$	$\sum_1^n (x_i - \mu)^2 \geq \chi_{n,\alpha}^2 \sigma_0^2$		$s^2 \geq \dfrac{\sigma_0^2}{n-1} \chi_{n-1,\alpha}^2$
III. $\sigma = \sigma_0$	$\sigma \neq \sigma_0$	$\sum_1^n (x_i - \mu)^2 \leq \chi_{n, 1-\alpha/2}^2 \sigma_0^2$ or $\sum_1^n (x_i - \mu)^2 \geq \chi_{n, \alpha/2}^2 \sigma_0^2$		$s^2 \leq \dfrac{\sigma_0^2}{n-1} \chi_{n-1, 1-\alpha/2}^2$ or $s^2 \geq \dfrac{\sigma_0^2}{n-1} \chi_{n-1, \alpha/2}^2$

Remark 1. All these tests can be derived by the standard likelihood ratio procedure. From Theorem 9.5.3 we see that, if μ is unknown, tests I and

II are UMP unbiased (and UMP invariant). In fact, test II is UMP (see Lehmann [70], 95). If μ is known, tests I and II are UMP (see Example 9.4.5). For tests III we have chosen constants c_1, c_2 so that each tail has probability $\alpha/2$. This is the customary procedure, even though it destroys the unbiasedness property of the tests, at least for small samples.

Example 1. A manufacturer claims that the lifetime of a certain brand of batteries produced by his factory has a variance of 5000 (hours)^2. A sample of size 26 has a variance of 7200 (hours)^2. Assuming that it is reasonable to treat these data as a random sample from a normal population, let us test the manufacturer's claim at the $\alpha = .02$ level. Here $H_0: \sigma^2 = 5000$ is to be tested against $H_1: \sigma^2 \neq 5000$. We reject H_0 if either

$$s^2 = 7200 \leq \frac{\sigma_0^2}{n-1} \chi^2_{n-1, 1-\alpha/2} \quad \text{or} \quad s^2 > \frac{\sigma_0^2}{n-1} \chi^2_{n-1, \alpha/2}.$$

We have

$$\frac{\sigma_0^2}{n-1} \chi^2_{n-1, 1-\alpha/2} = \frac{5000}{25} \times 11.524 = 2304.8$$

$$\frac{\sigma_0^2}{n-1} \chi^2_{n-1, \alpha/2} = \frac{5000}{25} \times 44.314 = 8862.8$$

Since s^2 is neither ≤ 2304.8 nor ≥ 8862.8, we cannot reject H_0 at level .02.

A test based on a chi-square statistic is also used for testing the equality of several proportions. Let $X_1, X_2, \cdots, X_k$ be independent rv's with $X_i \sim b(n_i, p_i)$, $i = 1, 2, \cdots, k$, $k \geq 2$.

Theorem 1. The rv $\sum_{i=1}^{k} \{(X_i - n_i p_i)/\sqrt{[n_i p_i (1 - p_i)]}\}^2$ converges in distribution to the $\chi^2(k)$ rv as $n_1, n_2, \cdots, n_k \to \infty$.

Proof. The proof is left as an exercise.

If $n_1, n_2, \cdots, n_k$ are large, we can use Theorem 1 to test $H_0: p_1 = p_2 = \cdots = p_k = p$ against all alternatives. If p is known, we compute

$$y = \sum_{1}^{k} \left\{\frac{x_i - n_i p}{\sqrt{n_i p(1-p)}}\right\}^2,$$

and if $y \geq \chi^2_{k,\alpha}$, we reject H_0. In practice p will be unknown. Let $\mathbf{p} = (p_1, p_2, \cdots, p_k)$. Then the likelihood function is

$$L(\mathbf{p}; x_1, \cdots, x_k) = \prod_{1}^{k} \left\{\binom{n_i}{x_i} p_i^{x_i}(1 - p_i)^{n_i - x_i}\right\}$$

so that

$$\log L(\mathbf{p}; \mathbf{x}) = \sum_{i=1}^{k} \log \binom{n_i}{x_i} + \sum_{i=1}^{k} x_i \log p_i + \sum_{i=1}^{k} (n_i - x_i) \log (1 - p_i).$$

The MLE $\hat{p}$ of p, under H_0, is therefore given by

$$\frac{\sum_{1}^{k} x_i}{p} - \frac{\sum_{1}^{k} (n_i - x_i)}{1 - p} = 0,$$

that is,

$$\hat{p} = \frac{x_1 + x_2 + \cdots + x_k}{n_1 + n_2 + \cdots + n_k}.$$

Under certain regularity assumptions (see Cramér [18], 426–427) it can be shown that the statistic

(1) $$Y_1 = \sum_{1}^{k} \frac{(X_i - n_i \hat{p})^2}{n_i \hat{p}(1 - \hat{p})}$$

is asymptotically $\chi^2(k - 1)$. Thus the test rejects $H_0: p_1 = p_2 = \cdots = p_k = p$, p unknown, at level α if $y_1 \geq \chi^2_{k-1, \alpha}$.

It should be remembered that the tests based on Theorem 1 are all large-sample tests and hence not exact, in contrast to the tests concerning the variance discussed above, which are all exact tests. In the case $k = 1$, UMP tests of $p \geq p_0$ and $p \leq p_0$ exist and can be obtained by the MLR method described in Section 9.4. For testing $p = p_0$, the usual test is UMP unbiased.

In the case $k = 2$, it is possible to find exact UMP unbiased tests of $p_1 \leq p_2$, $p_1 \geq p_2$, and $p_1 \neq p_2$. But the tests are based on the conditional distribution of X_1, given $X_1 + X_2$, and are too complicated to describe here. We refer the reader to Lehmann [70], page 143. If n_1 and n_2 are large, a test based on the normal distribution can be used instead of Theorem 1. In this case the statistic

(2) $$Z = \frac{X_1/n_1 - X_2/n_2}{\sqrt{\hat{p}(1 - \hat{p})(1/n_1 + 1/n_2)}},$$

where $\hat{p} = (X_1 + X_2)/(n_1 + n_2)$, is asymptotically $\mathcal{N}(0, 1)$ under $H_0: p_1 = p_2 = p$. If p is known, one uses p instead of $\hat{p}$. It is not too difficult to show that Z^2 is equal to Y_1, so that the two tests are equivalent.

In applications a wide variety of problems can be reduced to the multinomial distribution model. We therefore consider the problem of testing the parameters of a multinomial distribution. Let $(X_1, X_2, \cdots, X_{k-1})$ be a sample from a multinomial distribution with parameters $n, p_1, p_2, \cdots, p_{k-1}$, and let us write $X_k = n - X_1 - \cdots - X_{k-1}$, and $p_k = 1 - p_1 - \cdots - p_{k-1}$. The difference between the model of Theorem 1 and the multinomial model is the independence of the X_i's.

Theorem 2. Let $(X_1, X_2, \cdots, X_{k-1})$ be a multinomial rv with parameters $n, p_1, p_2, \cdots, p_{k-1}$. Then the rv

$$(3) \quad U_k = \sum_{i=1}^{k} \left\{ \frac{(X_i - np_i)^2}{np_i} \right\}$$

is asymptotically distributed as a $\chi^2(k-1)$ rv (as $n \to \infty$).

Proof. For the general proof we refer the reader to Cramér [18], pages 417–419. We will consider here the $k = 2$ case to make the result a little more plausible. We have

$$U_2 = \frac{(X_1 - np_1)^2}{np_1} + \frac{(X_2 - np_2)^2}{np_2} = \frac{(X_1 - np_1)^2}{np_1} + \frac{[n - X_1 - n(1 - p_1)]^2}{n(1 - p_1)}$$

$$= (X_1 - np_1)^2 \left[\frac{1}{np_1} + \frac{1}{n(1 - p_1)} \right]$$

$$= \frac{(X_1 - np_1)^2}{np_1(1 - p_1)}.$$

It follows from Theorem 1 that $U_2 \xrightarrow{L} Y$ as $n \to \infty$, where $Y \sim \chi^2(1)$.

To use Theorem 2 to test $H_0: p_1 = p_1', \cdots, p_k = p_k'$, we need only to compute the quantity

$$u = \sum_{1}^{k} \left\{ \frac{(x_i - np_i')^2}{np_i'} \right\}$$

from the sample; if n is large, we reject H_0 if $u > \chi^2_{k-1, \alpha}$.

Example 2. A die is rolled 120 times with the following results:

	1	2	3	4	5	6
Frequency:	20	30	20	25	15	10

Let us test the hypothesis that the die is fair at level $\alpha = .05$. The null hypothesis is $H_0: p_i = \frac{1}{6}$, $i = 1, 2, \cdots, 6$, where p_i is the probability that the face value is i, $1 \leq i \leq 6$. By Theorem 2 we reject H_0 if

$$u = \sum_{1}^{6} \frac{[x_i - 120(\frac{1}{6})]^2}{120(\frac{1}{6})} > \chi^2_{5, .05}.$$

We have

$$u = 0 + \frac{10^2}{20} + 0 + \frac{5^2}{20} + \frac{5^2}{20} + \frac{10^2}{20} = 12.5.$$

Since $\chi^2_{5, .05} = 11.07$, we reject H_0. Note that, if we choose $\alpha = .025$, then $\chi^2_{5, .025} = 12.8$, and we cannot reject at this level.

Theorem 2 has much wider applicability, and we will later study its application to contingency tables. Here we consider the application of Theorem 2 to testing the null hypothesis that the df of an rv X has a specified form.

Theorem 3. Let $X_1, X_2, \cdots, X_n$ be a random sample on X. Also, let $H_0: X \sim F$, where the functional form of the df F is known completely. Consider a collection of disjoint Borel sets $A_1, A_2, \cdots, A_k$ that form a partition of the real line. Let $P\{X \in A_i\} = p_i$, $i = 1, 2, \cdots, k$, and assume that $p_i > 0$ for each i. Let $Y_j =$ number of X_i's in A_j, $j = 1, 2, \cdots, k$, $i = 1, 2, \cdots, n$. Then the joint distribution of $(Y_1, Y_2, \cdots, Y_{k-1})$ is multinomial with parameters $n, p_1, p_2, \cdots, p_{k-1}$. Clearly $Y_k = n - Y_1 - \cdots - Y_{k-1}$ and $p_k = 1 - p_1 - \cdots - p_{k-1}$.

The proof of Theorem 3 is obvious. One frequently selects $A_1, A_2, \cdots, A_k$ as disjoint intervals. Theorem 3 is especially useful when one or more of the parameters associated with the df F are unknown. In that case the following result is useful.

Theorem 4. Let $H_0: X \sim F_\theta$, where $\theta = (\theta_1, \theta_2, \cdots, \theta_r)$ is unknown. Let $X_1, X_2, \cdots, X_n$ be independent observations on X, and suppose that the MLE's of $\theta_1, \theta_2, \cdots, \theta_r$ exist and are, respectively, $\hat{\theta}_1, \hat{\theta}_2, \cdots, \hat{\theta}_r$. Let $A_1, A_2, \cdots, A_k$ be a collection of disjoint Borel sets that cover the real line, and let

$$\hat{p}_i = P_{\hat{\theta}}\{X \in A_i\} > 0, \quad i = 1, 2, \cdots, k,$$

where $\hat{\theta} = (\hat{\theta}_1, \cdots, \hat{\theta}_r)$, and P_θ is the probability distribution associated with F_θ. Let $Y_1, Y_2, \cdots, Y_k$ be the rv's, defined as follows: $Y_i =$ number of $X_1, X_2, \cdots, X_n$ in A_i, $i = 1, 2, \cdots, k$.

Then the rv

$$V_k = \sum_{i=1}^{k} \left\{ \frac{(Y_i - n\hat{p}_i)^2}{n\hat{p}_i} \right\}$$

is asymptotically distributed as a $\chi^2(k - r - 1)$ rv (as $n \to \infty$).

The proof of Theorem 4 and some regularity conditions required on F_θ are given in Cramér [18], page 426.

To test $H_0: X \sim F$, where F is completely specified, we reject H_0 if

$$u = \sum_{1}^{k} \left\{ \frac{(y_i - np_i)^2}{np_i} \right\} > \chi^2_{k-1,\alpha},$$

provided that n is sufficiently large. If the null hypothesis is $H_0: X \sim F_\theta$, where F_θ is known except for the parameter θ, we use Theorem 4 and reject H_0 if

THE CHI-SQUARE TESTS

$$v = \sum_{i=1}^{k} \left\{ \frac{(y_i - n\hat{p}_i)^2}{n\hat{p}_i} \right\} > \chi^2_{k-r-1, \alpha},$$

where r is the number of parameters estimated.

Example 3. The following data were obtained from a table of random numbers of normal distribution with mean 0 and variance 1.

0.464	0.137	2.455	−0.323	−0.068
0.906	−0.513	−0.525	0.595	0.881
−0.482	1.678	−0.057	−1.229	−0.486
−1.787	−0.261	1.237	1.046	−0.508

We want to test the null hypothesis that the df F from which the data came is normal with mean 0 and variance 1. Here F is completely specified. Let us choose three intervals $(-\infty, -.5], (-.5, .5]$ and $(.5, \infty)$. We see that $Y_1 = 5$, $Y_2 = 8$, and $Y_3 = 7$.

Also, if Z is $\mathcal{N}(0, 1)$, then $p_1 = .3085$, $p_2 = .3830$, and $p_3 = .3085$. Thus

$$u = \sum_{i=1}^{3} \left\{ \frac{(y_i - np_i)^2}{np_i} \right\}$$

$$= \frac{(5 - 20 \times .3085)^2}{6.17} + \frac{(8 - 20 \times .383)^2}{7.66} + \frac{(7 - 20 \times .3085)^2}{6.17}$$

$$< 1.$$

Also, $\chi^2_{2,.05} = 5.99$, so that we accept H_0 at level .05.

Example 4. In a 72-hour period on a long holiday weekend there was a total of 306 fatal automobile accidents. The data are as follows:

Number of Fatal Accidents per Hour	Numbers of Hours
0 or 1	4
2	10
3	15
4	12
5	12
6	6
7	5
8 or more	7

Let us test the hypothesis that the number of accidents per hour is a Poisson rv.

Since the mean of the Poisson rv is not given, we estimate it by

$$\hat{\lambda} = \bar{x} = \frac{306}{72} = 4.25.$$

Let us now estimate $\hat{p}_i = P_{\hat{\lambda}}\{X = i\}$, $i = 0, 1, 2, \cdots$, $\hat{p}_0 = e^{-\hat{\lambda}} = .0143$. Note that

$$\frac{P_{\hat{\lambda}}\{X = x + 1\}}{P_{\hat{\lambda}}\{X = x\}} = \frac{\hat{\lambda}}{x + 1},$$

so that $\hat{p}_{i+1} = [\hat{\lambda}/(i + 1)] \hat{p}_i$. Thus

$\hat{p}_1 = .0606$, $\hat{p}_2 = .1288$, $\hat{p}_3 = .1825$, $\hat{p}_4 = .1939$,
$\hat{p}_5 = .1648$, $\hat{p}_6 = .1167$, $\hat{p}_7 = .0709$, $\hat{p}_8 = 1 - .9325 = .0675$.

The observed and expected frequencies are as follows:

i:	0 or 1	2	3	4	5	6	7	8 or more
Observed frequency, 0_i:	4	10	15	12	12	6	5	7
Expected frequency $= 72\hat{p}_i = e_i$:	5.38	9.28	13.14	13.96	11.87	8.41	5.10	4.86

Thus

$$u = \sum_{i=1}^{8} \frac{(0_i - e_i)^2}{e_i}$$
$$= 2.58.$$

Since we estimated one parameter, the number of degrees of freedom is $k - r - 1 = 8 - 1 - 1 = 6$. From Table 3 on page 652, $\chi^2_{6,\,.05} = 12.6$, and since $2.58 < 12.6$ we accept the null hypothesis.

Remark 2. Any application of Theorem 3 or 4 requires that we choose sets $A_1, A_2, \cdots, A_k$, and frequently these are chosen to be disjoint intervals. As a rule of thumb, we choose the length of each interval in such a way that the probability $P\{X \in A_i\}$ under H_0 is approximately $1/k$. Moreover, it is desirable to have $n/k \geq 5$ or, rather, $e_i \geq 5$ for each i. If any of the e_i's is < 5, the corresponding interval is pooled with one or more adjoining intervals to make the cell frequency at least 5. The number of degrees of freedom, if any pooling is done, is the number of classes after pooling, minus 1, minus the number of parameters estimated.

PROBLEMS 10.3

1. The standard deviation of capacity for batteries of a standard type is known to be 1.66 ampere-hours. The following capacities (ampere-hours) were recorded for 10 batteries of a new type: 146, 141, 135, 142, 140, 143, 138, 137, 142, 136. Does the new battery differ from the standard type with respect to variability of capacity? (Natrella [84], 4-1)

2. A manufacturer recorded the cut-off bias (volts) of a sample of 10 tubes as follows: 12.1, 12.3, 11.8, 12.0, 12.4, 12.0, 12.1, 11.9, 12.2, 12.2. The variability of cut-off bias for tubes of a standard type as measured by the standand deviation is .208 volts. Is the variability of the new tube with respect to cut-off bias less than that of the standard type? (Natrella [84], 4-5)

3. (Approximately) equal numbers of four different types of meters are in service and all types are believed to be equally likely to break down. The actual numbers of breakdowns reported are as follows:

Type of meter:	1	2	3	4
Number of breakdowns reported:	30	40	33	47

Is there evidence to conclude that the chances of failure of the four types are not equal? (Natrella [84], 9-4)

4. Every clinical thermometer is classified into one of four categories, A, B, C, D, on the basis of inspection and test. From past experience it is known that thermeters produced by a certain manufacturer are distributed among the four categories in the following proportions:

Category:	A	B	C	D
Proportion:	.87	.09	.03	.01

A new lot of 1336 thermometers is submitted by the manufacturer for inspection and test, and the following distribution into the four categories results:

Category:	A	B	C	D
Number of thermometers reported:	1188	91	47	10

Does this new lot of thermometers differ from the previous experience with regard to proportion of thermometers in each category? (Natrella [84], 9-2)

5. A computer program is written to generate random numbers, X, uniformly in the interval $0 \leq X < 10$. From 250 consecutive values the following data are obtained:

X value:	0-1.99	2-3.99	4-5.99	6-7.99	8-9.99
Frequency:	38	55	54	41	62

Do these data offer any evidence that the program is not written properly?

6. A machine working correctly cuts pieces of wire to a mean length of 10.5cm with a standard deviation of 0.15cm. Sixteen samples of wire were drawn at random from a production batch and measured with the following results (centimeters): 10.4, 10.6, 10.1, 10.3, 10.2, 10.9, 10.5, 10.8, 10.6, 10.5, 10.7, 10.2, 10.7, 10.3, 10.4, 10.5. Test the hypothesis that the machine is working correctly.

7. An experiment consists in tossing a coin until the first head shows up. One hundred repetitions of this experiment are performed. The frequency distribution of the number of trials required for the first head is as follows:

Number of trials:	1	2	3	4	5 or more
Frequency:	40	32	15	7	6

Can we conclude that the coin is fair?

8. Fit a binomial distribution to the following data:

x:	0	1	2	3	4
Frequency:	8	46	55	40	11

9. Prove Theorem 1.

10.4 THE t-TESTS

In this section we investigate one of the most frequently used types of tests in statistics, the tests based on a t-statistic. Let $X_1, X_2, \cdots, X_n$ be a random sample from $\mathcal{N}(\mu, \sigma^2)$, and, as usual, let us write

$$\bar{X} = n^{-1} \sum_1^n X_i, \qquad S^2 = (n-1)^{-1} \sum_1^n (X_i - \bar{X})^2.$$

In the following table we summarize some of the results we have already encountered.

		σ^2 Known	σ^2 Unknown
H_0	H_1	\multicolumn{2}{c}{Reject H_0 at level α if}
I. $\mu \leq \mu_0$	$\mu > \mu_0$	$\bar{x} \geq \mu_0 + \dfrac{\sigma}{\sqrt{n}} z_\alpha$	$\bar{x} \geq \mu_0 + \dfrac{s}{\sqrt{n}} t_{n-1,\alpha}$
II. $\mu \geq \mu_0$	$\mu < \mu_0$	$\bar{x} \leq \mu_0 + \dfrac{\sigma}{\sqrt{n}} z_{1-\alpha}$	$\bar{x} \leq \mu_0 + \dfrac{s}{\sqrt{n}} t_{n-1,1-\alpha}$
III. $\mu = \mu_0$	$\mu \neq \mu_0$	$\left|\bar{x} - \mu_0\right| \geq \dfrac{\sigma}{\sqrt{n}} z_{\alpha/2}$	$\left|\bar{x} - \mu_0\right| \geq \dfrac{s}{\sqrt{n}} t_{n-1,\alpha/2}$

Remark 1. We recall that a test based on a t-statistic is called a t-test. The t-tests in I and II are called *one-tailed tests;* the t-test in III, a *two-tailed test*.

Remark 2. If σ^2 is known, tests I and II are UMP and test III is UMP unbiased. If σ^2 is unknown, the t-tests are UMP unbiased and UMP invariant. All these tests can be derived by the usual likelihood ratio procedure.

Remark 3. If n is large (≥ 30), we may use normal tables instead of t-tables. The assumption of normality may also be dropped because of the central limit theorem. For small samples care is required in applying the proper test, since the tail probabilities under normal distribution and t-distribution differ significantly for small n. (See Remark 7.4.3).

Example 1. Nine determinations of copper in a certain solution yielded a sample mean of 8.3 percent with a standard deviation of .025 percent. Let μ be the mean of the population of such determinations. Let us test $H_0: \mu = 8.42$ against $H_1: \mu < 8.42$ at level $\alpha = .05$.

Here $n = 9$, $\bar{x} = 8.3$, $s = .025$, $\mu_0 = 8.42$, and $t_{n-1, 1-\alpha} = -t_{8, .05} = -1.860$. Thus

$$\mu_0 + \frac{s}{\sqrt{n}} t_{n-1, 1-\alpha} = 8.42 - \frac{.025}{3} 1.86$$
$$= 8.4045.$$

We reject H_0 since $8.3 < 8.4045$.

We next consider the two-sample case. Let $X_1, X_2, \cdots, X_m$ and $Y_1, Y_2, \cdots, Y_n$ be independent random samples from $\mathcal{N}(\mu_1, \sigma_1^2)$ and $\mathcal{N}(\mu_2, \sigma_2^2)$, respectively. Let us write

$$\bar{X} = m^{-1} \sum_1^m X_i, \qquad \bar{Y} = n^{-1} \sum_1^n Y_i,$$

$$S_1^2 = (m-1)^{-1} \sum_1^m (X_i - \bar{X})^2, \qquad S_2^2 = (n-1)^{-1} \sum_1^n (Y_i - \bar{Y})^2,$$

and

$$S_p^2 = \frac{(m-1) S_1^2 + (n-1) S_2^2}{m+n-2}.$$

S_p^2 is sometimes called the *pooled sample variance*. The following table summarizes the two sample tests comparing μ_1 and μ_2:

H_0 (δ = known constant)	H_1	σ_1^2, σ_2^2 Known	σ_1^2, σ_2^2 Unknown, $\sigma_1 = \sigma_2$
		Reject H_0 at level α if	
I. $\mu_1 - \mu_2 \leq \delta$	$\mu_1 - \mu_2 > \delta$	$\bar{x} - \bar{y} \geq \delta + z_\alpha \sqrt{\frac{\sigma_1^2}{m} + \frac{\sigma_2^2}{n}}$	$\bar{x} - \bar{y} \geq \delta + t_{m+n-2, \alpha} \cdot S_p \sqrt{\frac{1}{m} + \frac{1}{n}}$
II. $\mu_1 - \mu_2 \geq \delta$	$\mu_1 - \mu_2 < \delta$	$\bar{x} - \bar{y} \leq \delta - z_\alpha \sqrt{\frac{\sigma_1^2}{m} + \frac{\sigma_2^2}{n}}$	$\bar{x} - \bar{y} \leq \delta - t_{m+n-2, \alpha} \cdot S_p \sqrt{\frac{1}{m} + \frac{1}{n}}$
III. $\mu_1 - \mu_2 = \delta$	$\mu_1 - \mu_2 \neq \delta$	$\|\bar{x} - \bar{y} - \delta\| \geq z_{\alpha/2} \sqrt{\frac{\sigma_1^2}{m} + \frac{\sigma_2^2}{n}}$	$\|\bar{x} - \bar{y} - \delta\| \geq t_{m+n-2, \alpha/2} \cdot S_p \sqrt{\frac{1}{m} + \frac{1}{n}}$

Remark 4. The case of most interest is that in which $\delta = 0$. If σ_1^2, σ_2^2 are unknown and $\sigma_1^2 = \sigma_2^2 = \sigma^2$, σ^2 unknown, then S_p^2 is an unbiased estimate of σ^2. In this case all the two-sample t-tests are UMP unbiased and UMP invariant. Before applying the t-test, one should first make sure that $\sigma_1^2 = \sigma_2^2 = \sigma^2$, σ^2 unknown. This means applying another test on the data. We will consider this test in the next section.

Remark 5. If $m + n$ is large, we use normal tables; if both m and n are large, we can drop the assumption of normality, using the CLT.

Remark 6. The problem of equality of means in sampling from several populations will be considered in Chapter 12.

Example 2. The mean life of a sample of 9 light bulbs was observed to be 1309 hours with a standard deviation of 420 hours. A second sample of 16 bulbs chosen from a different batch showed a mean life of 1205 hours with a standard deviation of 390 hours. Let us test to see whether there is a significant difference between the means of the two batches, assuming that the population variances are the same (see also Example 10.5.1).

Here $H_0: \mu_1 = \mu_2$, $H_1: \mu_1 \neq \mu_2$, $m = 9$, $n = 16$, $\bar{x} = 1309$, $s_1 = 420$, $\bar{y} = 1205$, $s_2 = 390$, and let us take $\alpha = .05$. We have

$$s_p = \sqrt{\frac{8(420)^2 + 15(390)^2}{23}}$$

so that

$$t_{m+n-2,\alpha/2}\, s_p \sqrt{\frac{1}{m} + \frac{1}{n}} = t_{23,.025} \sqrt{\frac{8(420)^2 + 15(390)^2}{23}} \sqrt{\frac{1}{9} + \frac{1}{16}}$$
$$= 2.069 \times \sqrt{160552.17} \times \frac{5}{12} = 345.44.$$

Since $|\bar{x} - \bar{y}| = |1309 - 1205| = 104 \not> 345.44$, we cannot reject H_0 at level $\alpha = .05$.

Quite frequently one samples from a bivariate normal population with means μ_1, μ_2, variances σ_1^2, σ_2^2, and correlation coefficient ρ, the hypothesis of interest being $\mu_1 = \mu_2$. Let $(X_1, Y_1), (X_2, Y_2), \cdots, (X_n, Y_n)$ be a sample from a bivariate normal distribution with parameters μ_1, μ_2, σ_1^2, σ_2^2, and ρ. Then $X_j - Y_j$ is $\mathcal{N}(\mu_1 - \mu_2, \sigma^2)$, where $\sigma^2 = \sigma_1^2 + \sigma_2^2 - 2\rho\sigma_1\sigma_2$. We can therefore treat $D_j = (X_j - Y_j)$, $j = 1, 2, \cdots, n$, as a sample from a normal population (see Problem 7.6.5). Let us write

$$\bar{d} = \frac{\sum_{1}^{n} d_i}{n} \quad \text{and} \quad s_d^2 = \frac{\sum_{1}^{n}(d_i - \bar{d})^2}{n-1}.$$

The following table summarizes the resulting tests:

H_0 ($d_0 =$ known constant)	H_1	Reject H_0 at level α if
I. $\mu_1 - \mu_2 \geq d_0$	$\mu_1 - \mu_2 < d_0$	$\bar{d} \leq d_0 + \dfrac{s_d}{\sqrt{n}} \cdot t_{n-1, 1-\alpha}$
II. $\mu_1 - \mu_2 \leq d_0$	$\mu_1 - \mu_2 > d_0$	$\bar{d} \geq d_0 + \dfrac{s_d}{\sqrt{n}} \cdot t_{n-1, \alpha}$
III. $\mu_1 - \mu_2 = d_0$	$\mu_1 - \mu_2 \neq d_0$	$|\bar{d} - d_0| \geq \dfrac{s_d}{\sqrt{n}} \cdot t_{n-1, \alpha/2}$

Remark 7. The case of most importance is that in which $d_0 = 0$. Clearly all the t-tests are UMP unbiased and UMP invariant (see Remark 2). If σ is known, one can base the test on a standardized normal rv, but in practice such an assumption is quite unrealistic. If n is large (≥ 30), one can replace t-values by the corresponding critical values under the normal distribution.

Remark 8. Clearly, it is not necessary to assume that $(X_1, Y_1), \cdots, (X_n, Y_n)$ is a sample from a bivariate normal population. It suffices to assume that the differences D_i form a sample from a normal population.

Example 3. Nine adults agreed to test the efficacy of a new diet program. Their weights (pounds) were measured before and after the program and found to be as follows:

	1	2	3	4	5	6	7	8	9
Before:	132	139	126	114	122	132	142	119	126
After:	124	141	118	116	114	132	145	123	121

Let us test the null hypothesis that the diet is not effective, $H_0: \mu_1 - \mu_2 = 0$, against the alternative, $H_1: \mu_1 - \mu_2 > 0$, that it is effective at level $\alpha = .01$. We compute

$$\bar{d} = \frac{8 - 2 + 8 - 2 + 8 + 0 - 3 - 4 + 5}{9} = \frac{18}{9} = 2,$$

$$s_d^2 = \{(8-2)^2 + (-2-2)^2 + (8-2)^2 + (-2-2)^2 + (8-2)^2$$
$$+ (-2)^2 + (-3-2)^2 + (-4-2)^2 + (5-2)^2\}/8$$
$$= 26.75,$$

$$s_d = 5.17.$$

Thus

$$d_0 + \frac{s_d}{\sqrt{n}} t_{n-1,\alpha} = 0 + \frac{5.17}{\sqrt{9}} t_{8,.01} = \frac{5.17}{3} \times 2.896$$
$$= 4.99.$$

Since $\bar{d} \not> 4.99$, we accept hypothesis H_0 that the diet is not very effective.

PROBLEMS 10.4

1. The manufacturer of a certain subcompact car claims that the average mileage of this model is 30 miles per gallon of regular gasoline. For nine cars of this model driven in an identical manner, using 1 gallon of regular gasoline, the mean distance traveled was 26 miles with a standard deviation of 2.8 miles. Test the manufacturer's claim if you are willing to reject a true claim no more than twice in 100.

2. The nicotine contents of five cigarettes of a certain brand showed a mean of

21.2 milligrams with a standard deviation of 2.05 milligram. Test the hypothesis that the average nicotine content of this brand of cigarettes does not exceed 19.7 milligrams. Use $\alpha = .05$.

3. The additional hours of sleep gained by eight patients in an experiment with a certain drug were recorded as follows:

Patient:	1	2	3	4	5	6	7	8
Hours gained:	.7	−1.1	3.4	.8	2.0	.1	−.2	3.0

Assuming that these patients form a random sample from a population of such patients and that the number of additional hours gained from the drug is a normal random variable, test the hypothesis that the drug has no effect at level $\alpha = .10$.

4. The mean life of a sample of 8 light bulbs was found to be 1432 hours with a standard deviation of 436 hours. A second sample of 19 bulbs chosen from a different batch produced a mean life of 1310 hours with a standard deviation of 382 hours. Making appropriate assumptions, test the hypothesis that the two samples came from the same population of light bulbs at level $\alpha = .05$.

5. A sample of 25 observations has a mean of 57.6 and a variance of 1.8. A further sample of 20 values has a mean of 55.4 and a variance of 2.5. Test the hypothesis that the two samples came from the same normal population.

6. Two methods were used in a study of the latent heat of fusion of ice. Both method A and method B were conducted with the specimens cooled to $-0.72°C$. The following data represent the change in total heat from $-0.72°C$ to water, $0°C$, in calories per gram of mass:

Method A: 79.98, 80.04, 80.02, 80.04, 80.03, 80.03, 80.04, 79.97, 80.05, 80.03, 80.02, 80.00, 80.02

Method B: 80.02, 79.94, 79.98, 79.97, 79.97, 80.03, 79.95, 79.97

Perform a test at level .05 to see whether the two methods differ with regard to their avarage performance. (Natrella [84], 3-23)

7. In Problem 6, if it is known from past experience that the standard deviations of the two methods are $\sigma_A = .024$ and $\sigma_B = .033$, test the hypothesis that the methods are same with regard to their avarage performance at level $\alpha = .05$.

8. During World War II bacterial polysaccharides were investigated as blood plasma extenders. Sixteen samples of hydrolized polysaccharides supplied by various manufacturers in order to assess two chemical methods for determining the average molecular weight yielded the following results:

Method A: 62,700; 29,100; 44,400; 47,800; 36,300; 40,000; 43,400; 35,800; 33,900; 44,200; 34,300; 31,300; 38,400; 47,100; 42,100; 42,200

Method B: 56,400; 27,500; 42,200; 46,800; 33,300; 37,100; 37,300; 36,200; 35,200; 38,000; 32,200; 27,300; 36,100; 43,100; 38,400; 39,900

Perform an appropriate test of the hypothesis that the two averages are the same against a one-sided alternative that the average of Method A exceeds that of Method B. Use $\alpha = .05$. (Natrella [84], 3-38)

9. The following grade-point averages were collected over a period of 7 years to determine whether membership in a fraternity is beneficial or detrimental to grades:

Year:	1	2	3	4	5	6	7
Fraternity:	2.4	2.0	2.3	2.1	2.1	2.0	2.0
Nonfraternity:	2.4	2.2	2.5	2.4	2.3	1.8	1.9

Assuming that the populations were normal, test at the .025 level of significance whether membership in a fraternity is detrimental to grades.

10.5 THE F-TESTS

The term *F-tests* refers to tests based on an F-statistic. Let $X_1, X_2, \cdots, X_m$ and $Y_1, Y_2, \cdots, Y_n$ be independent samples from $\mathcal{N}(\mu_1, \sigma_1^2)$ and $\mathcal{N}(\mu_2, \sigma_2^2)$, respectively. We recall that $\sum_1^m (X_i - \bar{X})^2/\sigma^2 \sim \chi^2(m-1)$ and $\sum_1^n (Y_i - \bar{Y})^2/\sigma_2^2 \sim \chi^2(n-1)$ are independent rv's, so that the rv

$$F(X, Y) = \frac{\sum_1^m (X_i - \bar{X})^2}{\sum_1^n (Y_i - \bar{Y})^2} \cdot \frac{\sigma_2^2(n-1)}{\sigma_1^2(m-1)} = \frac{\sigma_2^2}{\sigma_1^2} \frac{S_1^2}{S_2^2}$$

is distributed as $F(m-1, n-1)$.

The following table summarizes the F-tests:

H_0	H_1	μ_1, μ_2 Known	μ_1, μ_2 Unknown
		Reject H_0 at level α if	
I. $\sigma_1^2 \leq \sigma_2^2$	$\sigma_1^2 > \sigma_2^2$	$\dfrac{\sum_1^m (x_i - \mu_1)^2}{\sum_1^n (y_i - \mu_2)^2} \geq \dfrac{m}{n} F_{m, n, \alpha}$	$\dfrac{s_1^2}{s_2^2} \geq F_{m-1,\, n-1, \alpha}$
II. $\sigma_1^2 \geq \sigma_2^2$	$\sigma_1^2 < \sigma_2^2$	$\dfrac{\sum_1^n (y_i - \mu_2)^2}{\sum_1^m (x_i - \mu_1)^2} \geq \dfrac{n}{m} F_{n, m, \alpha}$	$\dfrac{s_2^2}{s_1^2} \geq F_{n-1, m-1, \alpha}$
III. $\sigma_1^2 = \sigma_2^2$	$\sigma_1^2 \neq \sigma_2^2$	$\dfrac{\sum_1^m (x_i - \mu_1)^2}{\sum_1^n (y_i - \mu_2)^2} \geq \dfrac{m}{n} F_{m, n, \alpha/2}$ or $\dfrac{\sum_1^n (y_i - \mu_2)^2}{\sum_1^m (x_i - \mu_1)^2} \geq \dfrac{n}{m} F_{n, m, \alpha/2}$	$\dfrac{s_1^2}{s_2^2} \geq F_{m-1, n-1, \alpha/2}$ if $s_1^2 \geq s_2^2$ or $\dfrac{s_2^2}{s_1^2} \geq F_{n-1, m-1, \alpha/2}$ if $s_1^2 < s_2^2$

Remark 1. The tests are based on the right-hand tail of the F-distribution. To do so one uses the result (see Remark 7.4.6)

$$F_{m,n,1-\alpha} = \{F_{n,m,\alpha}\}^{-1}.$$

Remark 2. We have seen that the tests described can be easily obtained from the likelihood ratio procedure. Moreover, in the important case where μ_1, μ_2 are unknown, tests I and II are UMP unbiased and UMP invariant. For test III we have chosen equal tails, as is customarily done for convenience even though the unbiasedness property of the test is thereby destroyed.

Example 1 (Example 10.4.2 continued). In Example 10.4.2 let us test the validity of the assumption on which the t-test was based, namely, that the two populations have the same variance at level .05. Since $s_2 < s_1$, we compute $s_1^2/s_2^2 = (420/390)^2 = 196/169 = 1.16$. Since $F_{m-1,n-1,\alpha/2} = F_{8,15,.025} = 3.20$, we cannot reject $H_0: \sigma_1 = \sigma_2$.

An important application of the F-test involves the case where one is testing the equality of means of two normal populations under the assumption that the variances are the same, that is, testing whether the two samples come from the same population. Let $X_1, X_2, \cdots, X_m$ and $Y_1, Y_2, \cdots, Y_n$ be independent samples from $\mathcal{N}(\mu_1, \sigma_1^2)$ and $\mathcal{N}(\mu_2, \sigma_2^2)$, respectively. If $\sigma_1^2 = \sigma_2^2$ but is unknown, the t-test rejects $H_0: \mu_1 = \mu_2$ if $|T| > c$, where c is selected so that $\alpha_2 = P\{|T| > c \mid \mu_1 = \mu_2, \sigma_1 = \sigma_2\}$, that is, $c = t_{m+n-2,\alpha_2/2} s_p \sqrt{(1/m + 1/n)}$, where

$$s_p^2 = \frac{(m-1)s_1^2 + (n-1)s_2^2}{m+n-2},$$

s_1, s_2 being the sample variances. If first an F-test is performed to test $\sigma_1 = \sigma_2$, and then a t-test to test $\mu_1 = \mu_2$ at levels α_1 and α_2, respectively, the probability of accepting both hypotheses when they are true is

$$P\{|T| \leq c, c_1 < F < c_2 \mid \mu_1 = \mu_2, \sigma_1 = \sigma_2\};$$

and if F is independent of T, this probability is $(1 - \alpha_1)(1 - \alpha_2)$. It follows that the combined test has a significance level $\alpha = 1 - (1 - \alpha_1)(1 - \alpha_2)$. We see that

$$\alpha = \alpha_1 + \alpha_2 - \alpha_1\alpha_2 \leq \alpha_1 + \alpha_2$$

and $\alpha \geq \max(\alpha_1, \alpha_2)$. In fact, α will be closer to $\alpha_1 + \alpha_2$, since for small α_1 and α_2, $\alpha_1\alpha_2$ will be closer to 0.

We show that F is independent of T whenever $\sigma_1 = \sigma_2$. The statistic $V = (\bar{X}, \bar{Y}, \sum_1^m (X_i - \bar{X})^2 + \sum_1^n (Y_i - \bar{Y})^2)$ is a complete sufficient statistic for the parameter $(\mu_1, \mu_2, \sigma_1 = \sigma_2)$ (see Theorem 8.3.4). Since the distribution of F does not depend on μ_1, μ_2 and $\sigma_1 = \sigma_2$, it follows (Problem 5) that F is independent of V whenever $\sigma_1 = \sigma_2$. But T is a function of V alone, so that F must be independent of T also.

In Example 1, the combined test has a significance level of

$$\alpha = 1 - (.95)(.95) = 1 - .9025 = .0975.$$

PROBLEMS 10.5

1. For the data of Problem 10.4.4 is the assumption of equality of variances, on which the t-test is based, valid?
2. Answer the same question for Problems 10.4.5 and 10.4.6.
3. The performance of each of two different dive bombing methods is measured a dozen times. The sample variances for the two methods are computed to be 5545 and 4073, respectively. Do the two methods differ in variability?
4. In Problem 3 does the variability of the first method exceed that of the second method?
5. Let $\mathbf{X} = (X_1, X_2, \cdots, X_n)$ be a random sample from a distribution with pdf (pmf) $f(x, \boldsymbol{\theta})$, $\boldsymbol{\theta} \in \Theta$, where Θ is an interval in $\mathcal{R}_k$. Let $T(\mathbf{X})$ be a complete sufficient statistic for the family $\{f(\mathbf{x};\boldsymbol{\theta}): \boldsymbol{\theta} \in \Theta\}$. If $U(\mathbf{X})$ is a statistic (not a function of T alone) whose distribution does not depend on $\boldsymbol{\theta}$, show that U is independent of T.

10.6 BAYES AND MINIMAX PROCEDURES

Let $X_1, X_2, \cdots, X_n$ be a sample from a probability distribution with pdf (pmf) f_θ, $\theta \in \Theta$. In Section 8.8 we described the general decision problem, namely, once the statistician observes $\mathbf{x}$, he has a set $\mathcal{A}$ of options available. The problem is to find a decision function d that minimizes the risk $R(\theta, d) = E_\theta L(\theta, d)$ in some sense. Thus a minimax solution requires the minimization of max $R(\theta, d)$, while a Bayes solution requires the minimization of $R(\pi, d) = ER(\theta, d)$, where π is the a priori distribution on Θ. In Remark 9.2.2 we considered the problem of hypothesis testing as a special case of the general decision problem. The set $\mathcal{A}$ contains two points, a_0 and a_1; a_0 corresponds to the acceptance of $H_0: \theta \in \Theta_0$, and a_1 corresponds to the rejection of H_0. Suppose that the loss function is defined by

(1)
$$\begin{cases} L(\theta, a_0) = a(\theta) & \text{if } \theta \in \Theta_1, \quad a(\theta) > 0, \\ L(\theta, a_1) = b(\theta) & \text{if } \theta \in \Theta_0, \quad b(\theta) > 0, \\ L(\theta, a_0) = 0 & \text{if } \theta \in \Theta_0, \\ L(\theta, a_1) = 0 & \text{if } \theta \in \Theta_1. \end{cases}$$

Then

(2) $\quad R(\theta, d(\mathbf{X})) = L(\theta, a_0) P_\theta\{d(\mathbf{X}) = a_0\} + L(\theta, a_1) P_\theta\{d(\mathbf{X}) = a_1\}$

(3) $\quad = \begin{cases} a(\theta) P_\theta\{d(\mathbf{X}) = a_0\} & \text{if } \theta \in \Theta_1 \\ b(\theta) P_\theta\{d(\mathbf{X}) = a_1\} & \text{if } \theta \in \Theta_0. \end{cases}$

A *minimax solution* to the problem of testing $H_0: \theta \in \Theta_0$ against $H_1: \theta \in \Theta_1$, where $\Theta = \Theta_0 + \Theta_1$, is to find a rule that minimizes

$$\max_\theta [a(\theta) P_\theta\{d(\mathbf{X}) = a_0\}, \; b(\theta) P_\theta\{d(\mathbf{X}) = a_1\}].$$

We will consider here only the special case of testing $H_0: \theta = \theta_0$ against $H_1: \theta = \theta_1$. In that case we want to find a rule d which minimizes

(4) $\quad \max [a \, P_{\theta_1}\{d(\mathbf{X}) = a_0\}, \quad b \, P_{\theta_0}\{d(\mathbf{X}) = a_1\}].$

We will show that the solution is to reject H_0 if

(5) $\quad \dfrac{f_{\theta_1}(\mathbf{x})}{f_{\theta_0}(\mathbf{x})} \geq k,$

provided that the constant k is chosen so that

(6) $\quad R(\theta_0, d(\mathbf{X})) = R(\theta_1, d(\mathbf{X})),$

where d is the rule defined in (5); that is, the minimax rule d is obtained if we choose k in (5) so that

(7) $\quad a \, P_{\theta_1}\{d(\mathbf{X}) = a_0\} = b \, P_{\theta_0}\{d(\mathbf{X}) = a_1\},$

or, equivalently, we choose k so that

(8) $\quad a P_{\theta_1}\left\{\dfrac{f_{\theta_1}(\mathbf{X})}{f_{\theta_0}(\mathbf{X})} < k\right\} = b P_{\theta_0}\left\{\dfrac{f_{\theta_1}(\mathbf{X})}{f_{\theta_0}(\mathbf{X})} \geq k\right\}.$

Let d^* be any other rule. If $R(\theta_0, d) < R(\theta_0, d^*)$, then $R(\theta_0, d) = R(\theta_1, d) < \max [R(\theta_0, d^*), R(\theta_1, d^*)]$ and d^* cannot be minimax. Thus $R(\theta_0, d) \geq R(\theta_0, d^*)$, which means that

(9) $\quad P_{\theta_0}\{d^*(\mathbf{X}) = a_1\} \leq P_{\theta_0}\{d(\mathbf{X}) = a_1\} = P\{\text{Reject } H_0 | H_0 \text{ true}\}.$

By the Neyman-Pearson lemma, rule d is the most powerful of its size, so that its power must be at least that of d^*, that is,

$$P_{\theta_1}\{d(\mathbf{X}) = a_1\} \geq P_{\theta_1}\{d^*(\mathbf{X}) = a_1\}$$

so that

$$P_{\theta_1}\{d(\mathbf{X}) = a_0\} \leq P_{\theta_1}\{d^*(\mathbf{X}) = a_0\}.$$

It follows that

$$a P_{\theta_1}\{d(\mathbf{X}) = a_0\} \leq a P_{\theta_1}\{d^*(\mathbf{X}) = a_0\}$$

and hence that

(10) $$R(\theta_1, d) \leq R(\theta_1, d^*).$$

This means that

$$\max [R(\theta_0, d), R(\theta_1, d)] = R(\theta_1, d) \leq R(\theta_1, d^*)$$

and thus

$$\max [R(\theta_0, d), R(\theta_1, d)] \leq \max [R(\theta_0, d^*), R(\theta_1, d^*)].$$

Note that in the discrete case one may need some randomization procedure in order to achieve equality in (8).

Example 1. Let $X_1, X_2, \cdots, X_n$ be iid $\mathcal{N}(\mu, 1)$ rv's. To test $H_0: \mu = \mu_0$ against $H_1: \mu = \mu_1 \ (> \mu_0)$, we should choose k so that (8) is satisfied. This is the same as choosing c, and thus k, so that

$$a P_{\mu_1}\{\bar{X} < c\} = b P_{\mu_0}\{\bar{X} \geq c\}$$

or

$$a P_{\mu_1}\left\{\frac{\bar{X} - \mu_1}{1/\sqrt{n}} < \frac{c - \mu_1}{1/\sqrt{n}}\right\} = b P\left\{\frac{\bar{X} - \mu_0}{1/\sqrt{n}} \geq \frac{c - \mu_0}{1/\sqrt{n}}\right\}.$$

Thus

$$a\Phi[\sqrt{n}(c - \mu_1)] = b\{1 - \Phi[\sqrt{n}(c - \mu_0)]\},$$

where Φ is the df of an $\mathcal{N}(0, 1)$ rv. This can easily be accomplished with the help of normal tables once we know a, b, μ_0, μ_1, and n.

We next consider the problem of testing $H_0: \theta \in \Theta_0$ against $\mu_1: \theta \in \Theta_1$ from a Bayesian point of view. Let $\pi(\theta)$ be the a priori probability distribution on Θ. Then

$$R(\pi, d) = E_\theta R(\theta, d(\mathbf{X}))$$

$$= \begin{cases} \int_\Theta R(\theta, d) \pi(\theta) \, d\theta & \text{if } \pi \text{ is a pdf,} \\ \sum_\Theta R(\theta, d) \pi(\theta) & \text{if } \pi \text{ is a pmf,} \end{cases}$$

(11)
$$= \begin{cases} \int_{\Theta_0} b(\theta) \pi(\theta) P_\theta\{d(\mathbf{X}) = a_1\} \, d\theta + \int_{\Theta_1} a(\theta) \pi(\theta) P_\theta\{d(\mathbf{X}) = a_0\} d\theta \\ \qquad \text{if } \pi \text{ is a pdf,} \\ \sum_{\Theta_0} b(\theta) \pi(\theta) P_\theta\{d(\mathbf{X}) = a_1\} + \sum_{\Theta_1} a(\theta) \pi(\theta) P_\theta\{d(\mathbf{X}) = a_0\} \\ \qquad \text{if } \pi \text{ is a pmf.} \end{cases}$$

The Bayes solution is a decision rule that minimizes $R(\pi, d)$. In what follows we restrict our attention to the case where both H_0 and H_1 have exactly one point each, that is, $\Theta_0 = \{\theta_0\}$, $\Theta_1 = \{\theta_1\}$. Let $\pi(\theta_0) = \pi_0$ and $\pi(\theta_1) = 1 - \pi_0 = \pi_1$. Then

(12) $\qquad R(\pi, d) = b\,\pi_0 P_{\theta_0}\{d(\mathbf{X}) = a_1\} + a\,\pi_1 P_{\theta_1}\{d(\mathbf{X}) = a_0\},$

where $b(\theta_0) = b$, $a(\theta_1) = a$; $(a, b > 0)$.

Theorem 1. Let $\mathbf{X} = (X_1, X_2, \cdots, X_n)$ be an rv of the discrete (continuous) type with pmf (pdf) f_θ, $\theta \in \Theta = \{\theta_0, \theta_1\}$. Let $\pi(\theta_0) = \pi_0$, $\pi(\theta_1) = 1 - \pi_0 = \pi_1$ be the a priori probability mass function on Θ. A Bayes solution for testing $H_0: \mathbf{X} \sim f_{\theta_0}$ against $H_1: \mathbf{X} \sim f_{\theta_1}$, using the loss function (1), is to reject H_0 if

(13) $$\frac{f_{\theta_1}(\mathbf{x})}{f_{\theta_0}(\mathbf{x})} \geq \frac{b\pi_0}{a\pi_1}.$$

Proof. We wish to find d which minimizes

$$R(\pi, d) = b\pi_0 P_{\theta_0}\{d(\mathbf{X}) = a_1\} + a\,\pi_1 P_{\theta_1}\{d(\mathbf{X}) = a_0\}.$$

Now

$$R(\pi, d) = E_\theta R(\theta, d)$$
$$= E\{E_\theta\{L(\theta, d) | \mathbf{X}\}\}$$

so it suffices to minimize $E_\theta\{L(\theta, d) | \mathbf{X}\}$.

The a posteriori distribution of θ is given by

$$h(\theta | \mathbf{x}) = \frac{\pi(\theta) f_\theta(\mathbf{x})}{\sum_\theta f_\theta(\mathbf{x}) \pi(\theta)}$$

$$= \frac{\pi(\theta) f_\theta(\mathbf{x})}{\pi_0 f_{\theta_0}(\mathbf{x}) + \pi_1 f_{\theta_1}(\mathbf{x})}$$

(14) $$= \begin{cases} \dfrac{\pi_0 f_{\theta_0}(\mathbf{x})}{\pi_0 f_{\theta_0}(\mathbf{x}) + \pi_1 f_{\theta_1}(\mathbf{x})} & \text{if } \theta = \theta_0, \\ \dfrac{\pi_1 f_{\theta_1}(\mathbf{x})}{\pi_0 f_{\theta_0}(\mathbf{x}) + \pi_1 f_{\theta_1}(\mathbf{x})} & \text{if } \theta = \theta_1. \end{cases}$$

Thus

$$E_\theta\{L(\theta, d(\mathbf{X})) | \mathbf{X} = \mathbf{x}\} = \begin{cases} b\,h(\theta_0|\mathbf{x}), & \theta = \theta_0,\ d(\mathbf{X}) = a_1, \\ a\,h(\theta_1|\mathbf{x}), & \theta = \theta_1,\ d(\mathbf{X}) = a_0. \end{cases}$$

It follows that we reject H_0, that is, $d(\mathbf{X}) = a_1$ if

$$b\,h(\theta_0|\mathbf{x}) \leq a\,h(\theta_1|\mathbf{x}),$$

which is the case if and only if

$$b\,\pi_0 f_{\theta_0}(\mathbf{x}) \leq a\,\pi_1 f_{\theta_1}(\mathbf{x}),$$

as asserted.

Remark 1. In the Neyman-Pearson lemma we fixed $P_{\theta_0}\{d(\mathbf{X}) = a_1\}$, the probability of rejecting H_0 when it is true, and minimized $P_{\theta_1}\{d(\mathbf{X}) = a_0\}$, the probability of accepting H_0 when it is false. Here we no longer have a fixed level α for $P_{\theta_0}\{d(\mathbf{X}) = a_1\}$. Instead we allow it to assume any value as long as $R(\pi, d)$, defined in (12), is minimum.

Remark 2. It is easy to generalize Theorem 1 to the case of multiple decisions. Let $\mathbf{X}$ be an rv with pdf (pmf) f_θ, where θ can take any of the k values $\theta_1, \theta_2, \cdots, \theta_k$. The problem is to observe $\mathbf{x}$ and decide which of the θ_i's is the correct value of θ. Let us write $H_i: \theta = \theta_i$, $i = 1, 2, \cdots, k$, and assume that $\pi(\theta_i) = \pi_i$, $i = 1, 2, \cdots, k$, $\sum_1^k \pi_i = 1$, is the prior probability distribution on $\Theta = \{\theta_1, \theta_2, \cdots, \theta_k\}$. Let

$$L(\theta_i, d) = \begin{cases} 1 & \text{if } d \text{ chooses } \theta_j, j \neq i, \\ 0 & \text{if } d \text{ chooses } \theta_i. \end{cases}$$

The problem is to find a rule d that minimizes $R(\pi, d)$. We leave the reader to show that a Bayes solution is to accept $H_i: \theta = \theta_i$ ($i = 1, 2, \cdots, k$) if

$$\pi_i f_{\theta_i}(\mathbf{x}) \geq \pi_j f_{\theta_j}(\mathbf{x}) \quad \text{for all } j \neq i, j = 1, 2, \cdots, k, \tag{15}$$

where any point lying in more than one such region is assigned to any one of them.

Example 2. Let $X_1, X_2, \cdots, X_n$ be iid $\mathcal{N}(\mu, 1)$ rv's. To test $H_0: \mu = \mu_0$ against $H_1: \mu = \mu_1 (> \mu_0)$, let us take $a = b$ in the loss function (1). Then Theorem 1 says that the Bayes rule is one that rejects H_0 if

$$\frac{f_{\theta_1}(\mathbf{x})}{f_{\theta_0}(\mathbf{x})} \geq \frac{\pi_0}{1 - \pi_0},$$

that is,

$$\exp\left\{-\frac{\sum_1^n (x_i - \mu_1)^2}{2} + \frac{\sum_1^n (x_i - \mu_0)^2}{2}\right\} \geq \frac{\pi_0}{1 - \pi_0},$$

$$\exp\left\{(\mu_1 - \mu_0)\sum_1^n x_i + \frac{n(\mu_0^2 - \mu_1^2)}{2}\right\} \geq \frac{\pi_0}{1 - \pi_0}.$$

This happens if and only if

$$\frac{1}{n}\sum_1^n x_i \geq \frac{1}{n}\frac{\log[\pi_0/(1 - \pi_0)]}{\mu_1 - \mu_0} + \frac{\mu_0 + \mu_1}{2},$$

where the logarithm is to the base e. It follows that, if $\pi_0 = \frac{1}{2}$, the rejection region consists of

$$\bar{x} \geq \frac{\mu_0 + \mu_1}{2}.$$

Example 3. This example illustrates the result described in Remark 2. Let $X_1, X_2, \cdots, X_n$ be a sample from $\mathcal{N}(\mu, 1)$, and suppose that μ can take any one of the three values $\mu_1, \mu_2,$ or μ_3. Let $\mu_1 < \mu_2 < \mu_3$. Assume, for simplicity, that $\pi_1 = \pi_2 = \pi_3$. Then we accept $H_i: \mu = \mu_i$, $i = 1, 2, 3$, if

$$\pi_i \exp\left\{-\sum_{k=1}^{n} \frac{(x_k - \mu_i)^2}{2}\right\} \geq \pi_j \exp\left\{-\sum_{k=1}^{n} \frac{(x_k - \mu_j)^2}{2}\right\}$$

for each $j \neq i$, $j = 1, 2, 3$.

It follows that we accept H_i if

$$(\mu_i - \mu_j)\bar{x} + \frac{\mu_j^2 - \mu_i^2}{2} \geq 0, \qquad j = 1, 2, 3 \ (j \neq i),$$

that is,

$$\bar{x}(\mu_i - \mu_j) \geq \frac{(\mu_i - \mu_j)(\mu_i + \mu_j)}{2}, \qquad j = 1, 2, 3 \ (j \neq i).$$

Thus the acceptance region of H_1 is given by

$$\bar{x} \leq \frac{\mu_1 + \mu_2}{2} \quad \text{and} \quad \bar{x} \leq \frac{\mu_1 + \mu_3}{2}.$$

Also, the acceptance region of H_2 is given by

$$\bar{x} \geq \frac{\mu_1 + \mu_2}{2} \quad \text{and} \quad \bar{x} \leq \frac{\mu_2 + \mu_3}{2},$$

and that of H_3 by

$$\bar{x} \geq \frac{\mu_1 + \mu_3}{2} \quad \text{and} \quad \bar{x} \geq \frac{\mu_2 + \mu_3}{2}.$$

In particular, if $\mu_1 = 0$, $\mu_2 = 2$, $\mu_3 = 4$, we accept H_1 if $\bar{x} \leq 1$, H_2 if $1 \leq \bar{x} \leq 3$, and H_3 if $\bar{x} \geq 3$. In this case, boundary points 1 and 3 have zero probability, and it does not matter where we include them.

PROBLEMS 10.6

1. In Example 1 let $n = 15$, $\mu_0 = 4.7$, and $\mu_1 = 5.2$, and choose $a = b > 0$. Find the minimax test, and compute its power at $\mu = 4.7$ and $\mu = 5.2$.

2. A sample of five observations is taken on a $b(1, \theta)$ rv to test $H_0: \theta = \frac{1}{2}$ against $H_1: \theta = \frac{3}{4}$.
   (a) Find the most powerful test of size $\alpha = .05$.
   (b) If $L(\frac{1}{2}, \frac{1}{2}) = L(\frac{3}{4}, \frac{3}{4}) = 0$, $L(\frac{1}{2}, \frac{3}{4}) = 1$, and $L(\frac{3}{4}, \frac{1}{2}) = 2$, find the minimax rule.
   (c) If the prior probabilities of $\theta = \frac{1}{2}$ and $\theta = \frac{3}{4}$ are $\pi_0 = \frac{1}{3}$ and $\pi_1 = \frac{2}{3}$, respectively, find the Bayes rule.

3. A sample of size n is to be used from the pdf

$$f_\theta(x) = \theta\, e^{-\theta x}, \qquad x > 0,$$

to test $H_0: \theta = 1$ against $H_1: \theta = 2$. If the a priori distribution on θ is $\pi_0 = \tfrac{2}{3}$, $\pi_1 = \tfrac{1}{3}$, and $a = b$, find the Bayes solution. Find the power of the test at $\theta = 1$ and $\theta = 2$.

4. Given two normal densities with variances 1 and with means -1 and 1, respectively, find the Bayes solution based on a single observation when $a = b$ and (a) $\pi_0 = \pi_1 = \tfrac{1}{2}$, and (b) $\pi_0 = \tfrac{1}{4}$, $\pi_1 = \tfrac{3}{4}$.

5. Given three normal densities with variances 1 and with means $-1, 0, 1$, respectively, find the Bayes solution to the multiple decision problem based on a single observation when $\pi_1 = \tfrac{2}{5}$, $\pi_2 = \tfrac{2}{5}$, $\pi_3 = \tfrac{1}{5}$.

6. For the multiple decision problem described in Remark 2 show that a Bayes solution is to accept $H_i: \theta = \theta_i (i = 1, 2, \cdots k)$ if (15) holds.

CHAPTER 11

Confidence Estimation

11.1 INTRODUCTION

In many problems of statistical inference the experimenter is interested in constructing a family of sets that contain the true (unknown) parameter value with a specified (high) probability. If X, for example, represents the length of life of a piece of equipment, the experimenter is interested in a lower bound $\underline{\theta}$ for the mean θ of X. Since $\underline{\theta} = \underline{\theta}(X)$ will be a function of the observations, one cannot ensure with probability 1 that $\underline{\theta}(X) \leq \theta$. All that one can do is to choose a number $1 - \alpha$ that is close to 1 so that $P_\theta\{\underline{\theta}(X) \leq \theta\} \geq 1 - \alpha$ for all θ. Problems of this type are called problems of *confidence estimation*. In this chapter we restrict ourselves largely to the case where $\Theta \subseteq \mathcal{R}$ and consider the problem of setting confidence limits for the parameter θ.

In Section 2 we introduce the basic ideas and study some simple methods of finding confidence intervals. Section 3 deals with shortest-length confidence intervals, while Section 4 explores the relationship between tests of hypotheses and confidence intervals. In Section 5 we study the theory of unbiased confidence intervals, and in Section 6 the method of construction of Bayes confidence intervals is described.

11.2 SOME FUNDAMENTAL NOTIONS OF CONFIDENCE ESTIMATION

So far we have considered a random variable or some function of it as the basic observable quantity. Let X be an rv, and a, b, be two given positive real numbers. Then

$$P\{a < X < b\} = P\{a < X \text{ and } X < b\}$$
$$= P\left\{\frac{bX}{a} > b \text{ and } X < b\right\}$$

$$= P\left\{X < b < \frac{bX}{a}\right\},$$

and if we know the distribution of X and a, b, we can determine the probability $P\{a < X < b\}$. Consider the interval $I(X) = (X, bX/a)$. This is an interval with end points that are functions of the rv X, and hence it takes the value $(x, bx/a)$ when X takes the value x. In other words, $I(X)$ assumes the value $I(x)$ whenever X assumes the value x. Thus $I(X)$ is a random quantity and is an example of a *random interval*. Note that $I(X)$ includes the value b with a certain fixed probability. For example, if $b = 1$, $a = \frac{1}{2}$, and X is $U(0, 1)$, the interval $(X, 2X)$ includes point 1 with probability $\frac{1}{2}$.

Definition 1. Let $\mathscr{P}_\theta$, $\theta \in \Theta \subseteq \mathscr{R}_k$, be the set of probability distributions of an rv $\mathbf{X}$. A family of subsets $S(\mathbf{x})$ of Θ, where $S(\mathbf{x})$ depends on the observation $\mathbf{x}$ but not on θ, is called a family of random sets. If, in particular, $\Theta \subseteq \mathscr{R}$ and $S(\mathbf{x})$ is an interval $(\underline{\theta}(\mathbf{x}), \bar{\theta}(\mathbf{x}))$, where $\underline{\theta}(\mathbf{x})$ and $\bar{\theta}(\mathbf{x})$ are functions of $\mathbf{x}$ alone (and not θ), we call $S(\mathbf{X})$ a random interval with $\underline{\theta}(\mathbf{X})$ and $\bar{\theta}(\mathbf{X})$ as lower and upper bounds, respectively. $\underline{\theta}(\mathbf{X})$ may be $-\infty$, and $\bar{\theta}(\mathbf{X})$ may be $+\infty$.

In a wide variety of inference problems one is not interested in estimating the parameter or testing some hypothesis concerning it. Rather, one wishes to establish a lower or an upper bound, or both, for the real-valued parameter. For example, if X is the time to failure of a piece of equipment, one is interested in a lower bound for the mean of X. If the rv X measures the toxicity of a drug, the concern is to find an upper bound for the mean. Similarly, if the rv X measures the nicotine content of a certain brand of cigarettes, one is interested in determining an upper and a lower bound for the average nicotine content of these cigarettes.

In this chapter we are interested in the *problem of confidence estimation*, namely, that of finding a family of random sets $S(\mathbf{x})$ for a parameter θ such that, for a given α, $0 < \alpha < 1$ (usually small),

(1) $\qquad P_\theta\{S(\mathbf{X}) \ni \theta\} \geq 1 - \alpha \qquad$ for all $\theta \in \Theta$.

We restrict our attention mainly to the case where $\theta \in \Theta \subseteq \mathscr{R}$.

Definition 2. Let $\theta \in \Theta \subseteq \mathscr{R}$ and $0 < \alpha < 1$. A function $\underline{\theta}(\mathbf{X})$ satisfying

(2) $\qquad P_\theta\{\underline{\theta}(\mathbf{X}) \leq \theta\} \geq 1 - \alpha \qquad$ for all θ

is called a lower confidence bound for θ at confidence level $1 - \alpha$. The quantity

(3) $$\inf_{\theta \in \Theta} P_\theta\{\underline{\theta}(\mathbf{X}) \leq \theta\}$$

is called the confidence coefficient.

Definition 3. A function $\underline{\theta}$ that minimizes

$$P_\theta\{\underline{\theta}(X) \leq \theta'\} \quad \text{for all } \theta' < \theta \tag{4}$$

subject to (2) is known as a uniformly most accurate (UMA) lower confidence bound for θ at confidence level $1 - \alpha$.

Similar definitions are given for an upper confidence bound for θ and a UMA upper confidence bound.

Definition 4. A family of subsets $S(\mathbf{x})$ of $\Theta \subseteq \mathscr{R}_k$ is said to constitute a family of confidence sets at confidence level $1 - \alpha$ if

$$P_\mathbf{\theta}\{S(\mathbf{X}) \ni \mathbf{\theta}\} \geq 1 - \alpha \quad \text{for all } \mathbf{\theta} \in \Theta, \tag{5}$$

that is, the random set $S(\mathbf{X})$ covers the true parameter value $\mathbf{\theta}$ with probability $\geq 1 - \alpha$. A lower confidence bound corresponds to the special case where $k = 1$ and

$$S(\mathbf{x}) = \{\theta : \underline{\theta}(\mathbf{x}) \leq \theta < \infty\}; \tag{6}$$

and an upper confidence bound, to the case where

$$S(\mathbf{x}) = \{\theta : \bar{\theta}(\mathbf{x}) \geq \theta > -\infty\}. \tag{7}$$

If $S(\mathbf{x})$ is of the form

$$S(\mathbf{x}) = (\underline{\theta}(\mathbf{x}), \bar{\theta}(\mathbf{x})) \tag{8}$$

we will call it a confidence interval at confidence level $1 - \alpha$, provided that

$$P_\theta\{\underline{\theta}(\mathbf{X}) < \theta < \bar{\theta}(\mathbf{X})\} \geq 1 - \alpha \quad \text{for all } \theta, \tag{9}$$

and the quantity

$$\inf_\theta P_\theta\{\underline{\theta}(\mathbf{X}) < \theta < \bar{\theta}(\mathbf{X})\} \tag{10}$$

will be referred to as the confidence coefficient associated with the random interval.

Remark 1. It is not quite correct to read (5) or (9) as follows: The probability that θ lies in the set $S(\mathbf{X})$ is $\geq 1 - \alpha$. θ is fixed; it is $S(\mathbf{X})$ that is random here. One can give the following interpretation to (5) or (9).

Choose and fix α, $0 < \alpha < 1$ (usually small, say .01 or .05). Consider a sequence of independent experiments in which each experiment consists of taking a sample of size n from a population distribution with unknown parameter $\mathbf{\theta}$, where n and $\mathbf{\theta}$ may vary from experiment to experiment. For each sample point $\mathbf{x}$, compute $S(\mathbf{x})$. Then the parameter $\mathbf{\theta}$ of the corresponding population may or may not be covered by $S(\mathbf{x})$. Although the set $S(\mathbf{x})$ will vary from sample to sample, the probability that the statement "$S(\mathbf{x})$ includes $\boldsymbol{\theta}$" will be true is roughly (at least) $1 - \alpha$. Consequently the probability of making a false statement is roughly (at most) α. In the long run we therefore expect to be correct in at least $(1 - \alpha) 100$ percent of all cases. Note that a given confidence set either includes the true parameter point $\boldsymbol{\theta}$ or does not. One would never know this unless $\boldsymbol{\theta}$ was known. But the point is that one would be successful in trapping this true value $\boldsymbol{\theta}$ in sets of this type at least $(1 - \alpha) 100$ percent of the time.

Definition 5. A family of $1 - \alpha$ level confidence sets $\{S(\mathbf{x})\}$ is said to be a UMA family of confidence sets at level $1 - \alpha$ if

$$P_{\boldsymbol{\theta}'}\{S(\mathbf{X}) \text{ contains } \boldsymbol{\theta}\} \leq P_{\boldsymbol{\theta}'}\{S'(\mathbf{X}) \text{ contains } \boldsymbol{\theta}\}$$

for all $\boldsymbol{\theta}, \boldsymbol{\theta}'$ and any $1 - \alpha$ level family of confidence sets $S'(\mathbf{X})$.

Example 1. Let $X_1, X_2, \cdots, X_n$ be iid rv's, $X_i \sim \mathcal{N}(\mu, \sigma^2)$. Consider the interval $(\bar{X} - c_1, \bar{X} + c_2)$. In order for this to be a $1 - \alpha$ level confidence interval, we must have

$$P\{\bar{X} - c_1 < \mu < \bar{X} + c_2\} \geq 1 - \alpha,$$

which is the same as

$$P\{\mu - c_2 < \bar{X} < \mu + c_1\} \geq 1 - \alpha.$$

Thus

$$P\left\{-\frac{c_2}{\sigma}\sqrt{n} < \frac{\bar{X} - \mu}{\sigma}\sqrt{n} < \frac{c_1}{\sigma}\sqrt{n}\right\} \geq 1 - \alpha.$$

Since $\sqrt{n}(\bar{X} - \mu)/\sigma \sim \mathcal{N}(0, 1)$, we can choose c_1 and c_2 to satisfy

$$P\left\{-\frac{c_2}{\sigma}\sqrt{n} < \frac{\bar{X} - \mu}{\sigma}\sqrt{n} < \frac{c_1}{\sigma}\sqrt{n}\right\} = 1 - \alpha,$$

provided that σ is known. There are infinitely many such pairs of values (c_1, c_2). In particular, an intuitively reasonable choice is $c_1 = -c_2 = c$, say. In that case

$$\frac{c\sqrt{n}}{\sigma} = z_{\alpha/2},$$

and the confidence interval is $(\bar{X} - (\sigma/\sqrt{n}) z_{\alpha/2}, \bar{X} + (\sigma/\sqrt{n}) z_{\alpha/2})$. The length of this interval is $(2\sigma/\sqrt{n}) z_{\alpha/2}$. Given σ and α, we can choose n to get a confidence interval of a fixed length.

If σ is not known, we have from

$$P\{-c_2 < \bar{X} - \mu < c_1\} \geq 1 - \alpha$$

that

$$P\left\{-\frac{c_2}{S}\sqrt{n} < \frac{\bar{X}-\mu}{S/\sqrt{n}} < \frac{c_1}{S}\sqrt{n}\right\} \geq 1-\alpha,$$

and once again we can choose pairs of values (c_1, c_2) using a t-distribution with $n-1$ d.f. such that

$$P\left\{-\frac{c_2\sqrt{n}}{S} < \frac{\bar{X}-\mu}{S}\sqrt{n} < \frac{c_1\sqrt{n}}{S}\right\} = 1-\alpha.$$

In particular, if we take $c_1 = -c_2 = c$, say, then

$$c\frac{\sqrt{n}}{S} = t_{n-1,\alpha/2},$$

and $(\bar{X} - (S/\sqrt{n})t_{n-1,\alpha/2}, \bar{X} + (S/\sqrt{n})t_{n-1,\alpha/2})$, is a $1-\alpha$ level confidence interval for μ. The length of this interval is $(2S/\sqrt{n})t_{n-1,\alpha/2}$, which is no longer constant. Therefore we cannot choose n to get a fixed-width confidence interval of level $1-\alpha$. Indeed, the length of this interval can be quite large if σ is large. Its expected length is

$$\frac{2}{\sqrt{n}}t_{n-1,\alpha/2}E_\sigma S = \frac{2}{\sqrt{n}}t_{n-1,\alpha/2}\sqrt{\frac{2}{n-1}}\frac{\Gamma(n/2)}{\Gamma[(n-1)/2]}\sigma,$$

which can be made as small as we please by choosing n large enough.

Example 2. In Example 1, suppose that we wish to find a confidence interval for σ^2 instead when μ is unknown. Consider the interval $(c_1 S^2, c_2 S^2)$, $c_1, c_2 > 0$. We have

$$P\{c_1 S^2 < \sigma^2 < c_2 S^2\} \geq 1-\alpha,$$

so that

$$P\left\{c_2^{-1} < \frac{S^2}{\sigma^2} < c_1^{-1}\right\} \geq 1-\alpha.$$

Since $(n-1)S^2/\sigma^2$ is $\chi^2(n-1)$, we can choose pairs of values (c_1, c_2) from the tables of the chi-square distribution. In particular, we can choose c_1, c_2 so that

$$P\left\{\frac{S^2}{\sigma^2} \geq \frac{1}{c_1}\right\} = \frac{\alpha}{2} = P\left\{\frac{S^2}{\sigma^2} \leq \frac{1}{c_2}\right\}.$$

Then

$$\frac{n-1}{c_1} = \chi^2_{n-1,\,\alpha/2} \quad \text{and} \quad \frac{n-1}{c_2} = \chi^2_{n-1,\,1-\alpha/2}.$$

Thus

$$\left(\frac{(n-1)S^2}{\chi^2_{n-1,\alpha/2}}, \frac{(n-1)S^2}{\chi^2_{n-1,1-\alpha/2}} \right)$$

is a $1 - \alpha$ level confidence interval for σ^2 whenever μ is unknown. If μ is known, then

$$\sum_1^n \frac{(X_i - \mu)^2}{\sigma^2} \sim \chi^2(n).$$

Thus we can base the confidence interval on $\sum_1^n (X_i - \mu)^2$. Proceeding similarly, we get a $1 - \alpha$ level confidence interval as

$$\left(\frac{\sum_1^n (X_i - \mu)^2}{\chi^2_{n,\alpha/2}}, \frac{\sum_1^n (X_i - \mu)^2}{\chi^2_{n,1-\alpha/2}} \right).$$

Next suppose that both μ and σ^2 are unknown and that we want a confidence set for (μ, σ^2). We have from Boole's inequality

$$P\left\{ \bar{X} - \frac{S}{\sqrt{n}} t_{n-1,\alpha_1/2} < \mu < \bar{X} + \frac{S}{\sqrt{n}} t_{n-1,\alpha_1/2}, \right.$$
$$\left. \frac{(n-1)S^2}{\chi^2_{n-1,\alpha_2/2}} < \sigma^2 < \frac{(n-1)S^2}{\chi^2_{n-1,1-\alpha_2/2}} \right\}$$
$$\geq 1 - P\left\{ \bar{X} + \frac{S}{\sqrt{n}} t_{n-1,\alpha_1/2} \leq \mu \text{ or } \bar{X} - \frac{S}{\sqrt{n}} t_{n-1,\alpha_1/2} \geq \mu \right\}$$
$$- P\left\{ \frac{(n-1)S^2}{\chi^2_{n-1,1-\alpha_2/2}} \leq \sigma^2 \text{ or } \frac{(n-1)S^2}{\chi^2_{n-1,\alpha_2/2}} \geq \sigma^2 \right\}$$
$$= 1 - \alpha_1 - \alpha_2,$$

so that the Cartesian product,

$$S(\mathbf{X}) = \left(\bar{X} - \frac{S}{\sqrt{n}} t_{n-1,\alpha_1/2}, \bar{X} + \frac{S}{\sqrt{n}} t_{n-1,\alpha_1/2} \right) \times \left(\frac{(n-1)S^2}{\chi^2_{n-1,\alpha_2/2}}, \frac{(n-1)S^2}{\chi^2_{n-1,1-\alpha_2/2}} \right)$$

is a $1 - \alpha_1 - \alpha_2$ level confidence set for (μ, σ^2).

The following result provides a general method of finding confidence intervals and covers most cases in practice.

Theorem 1. Let $X_1, X_2, \cdots, X_n$ be a random sample from a df F_θ, $\theta \in \Theta$, where Θ is an interval of $\mathcal{R}$. Let $T(\mathbf{X}, \theta) = T(X_1, X_2, \cdots, X_n, \theta)$ be a real-valued function defined on $\mathcal{R}_n \times \Theta$, such that, for each θ, $T(\mathbf{X}, \theta)$ is a statistic, and as a function of θ, T is strictly increasing or decreasing at every $\mathbf{x} = (x_1, x_2, \cdots, x_n) \in \mathcal{R}_n$. Let $\Lambda \subseteq \mathcal{R}$ be the range of T, and for every $\lambda \in \Lambda$ and $\mathbf{x} \in \mathcal{R}_n$ let the equation $\lambda = T(\mathbf{x}, \theta)$ be solvable. If the distribution of $T(\mathbf{X}, \theta)$ is independent of θ, one can construct a confidence interval for θ at any level.

Proof. Let $0 < \alpha < 1$. Then we can choose a pair of numbers $\lambda_1(\alpha)$ and $\lambda_2(\alpha)$ in Λ not necessarily unique, such that

(11) $\qquad P_\theta\{\lambda_1(\alpha) < T(\mathbf{X}, \theta) < \lambda_2(\alpha)\} \geq 1 - \alpha \qquad$ for all θ.

Since the distribution of T is independent of θ, it is clear that λ_1 and λ_2 are independent of θ. Since, moreover, T is monotone in θ, we can solve the equations

(12) $\qquad T(\mathbf{x}, \theta) = \lambda_1(\alpha) \qquad$ and $\qquad T(\mathbf{x}, \theta) = \lambda_2(\alpha)$

for every $\mathbf{x}$ uniquely for θ. We have

(13) $\qquad P_\theta\{\underline{\theta}(\mathbf{X}) < \theta < \bar{\theta}(\mathbf{X})\} \geq 1 - \alpha \qquad$ for all θ,

where $\underline{\theta}(\mathbf{X}) < \bar{\theta}(\mathbf{X})$ are rv's. This completes the proof.

Remark 2. The condition that $\lambda = T(\mathbf{x}, \theta)$ be solvable will be satisfied if, for example, T is continuous and strictly increasing or decreasing as a function of θ in Θ.

Remark 3. It is usually possible to find a function T that is monotone and has a distribution independent of θ. For example, if F_θ is continuous and monotone, in θ, we can take

$$T(\mathbf{X}, \theta) = \prod_{i=1}^{n} F_\theta(X_i).$$

Since F_θ is continuous, $F_\theta(X_i)$ are iid with $U[0, 1]$ distribution. Then

$$-\log T(\mathbf{X}, \theta) = - \sum_{i=1}^{n} \log F_\theta(X_i),$$

where $-\log F_\theta(X_i)$ are iid rv's, each with common $G(1, 1)$ distribution. It follows that $-\log T(\mathbf{X}, \theta) \sim G(n, 1)$, and we can find λ_1, λ_2 such that

$$P_\theta\{-\log \lambda_2 < -\log T(\mathbf{X}, \theta) < -\log \lambda_1\} = \frac{1}{\Gamma(n)} \int_{-\log \lambda_2}^{-\log \lambda_1} x^{n-1} e^{-x} \, dx$$
$$= 1 - \alpha.$$

Thus

$$P_\theta\{\lambda_1 < T(\mathbf{X}, \theta) < \lambda_2\} = 1 - \alpha.$$

This last statement is equivalent to

$$P_\theta\{\underline{\theta}(\mathbf{X}) < \theta < \bar{\theta}(\mathbf{X})\} = 1 - \alpha.$$

Note that in the continuous case we can find a confidence interval with equality on the right side of (11). In the discrete case, however, this is usually not possible.

Remark 4. Relation (11) is valid even when the assumption of monotonicity of T in the theorem is dropped. In that case inversion of the inequalities may yield a set of intervals (random set) $S(\mathbf{X})$ in Θ instead of a confidence interval.

Remark 5. The argument used in Theorem 1 can be extended to cover the multiparameter case, and the method will determine a confidence set for all the parameters of a distribution.

Example 3. Let $X_1, X_2, \cdots, X_n \sim \mathcal{N}(\mu, \sigma^2)$, where σ is unknown and we seek a $1 - \alpha$ level confidence interval for μ. Let us choose

$$T(\mathbf{X}, \mu) = \frac{\bar{X} - \mu}{S} \sqrt{n},$$

where $\bar{X}$, S^2 are the usual sample statistics. The conditions of Theorem 1 are clearly fulfilled. The rv $T(\mathbf{X}, \mu)$ has Student's t-distribution with $n - 1$ d.f., which is independent of μ, and $T(\mathbf{X}, \mu)$, as a function of μ is monotone. We can clearly choose $\lambda_1(\alpha)$, $\lambda_2(\alpha)$ (not necessarily uniquely) so that

$$P\{\lambda_1(\alpha) < T(\mathbf{X}, \mu) < \lambda_2(\alpha)\} = 1 - \alpha \quad \text{for all } \mu.$$

Solving

$$\lambda_i(\alpha) = \frac{\bar{X} - \mu}{S} \sqrt{n},$$

we get

$$\underline{\mu}(\mathbf{X}) = \bar{X} - \frac{S}{\sqrt{n}} \lambda_2(\alpha), \quad \bar{\mu}(\mathbf{X}) = \bar{X} - \frac{S}{\sqrt{n}} \lambda_1(\alpha),$$

and a $1 - \alpha$ level confidence interval is

$$\left(\bar{X} - \frac{S}{\sqrt{n}} \lambda_2(\alpha), \bar{X} - \frac{S}{\sqrt{n}} \lambda_1(\alpha) \right).$$

In practice, one chooses $\lambda_2(\alpha) = -\lambda_1(\alpha) = t_{n-1, \alpha/2}$.

Remark 6. If a sufficiently large sample is available and the conditions of the central limit theorem hold, it is frequently possible to construct approximate confidence intervals of any desired level. We illustrate the method with the following example.

Example 4. Let $X_1, X_2, \cdots, X_n$ be iid rv's with finite variance. Also, let $EX_i = \mu$ and $EX_i^2 = \sigma^2 + \mu^2$. From the CLT it follows that

$$\frac{\bar{X} - \mu}{\sigma/\sqrt{n}} \xrightarrow{L} Z,$$

where $Z \sim \mathcal{N}(0, 1)$. Suppose that we want a $1 - \alpha$ level confidence interval for μ when σ is not known. We know from Example 7.3.5 that for large n the quantity $[\sqrt{n}(\bar{X} - \mu)/S]$ is approximately normally distributed with mean 0 and variance 1. Hence, for large n, we can find constants c_1, c_2 such that

$$P\left\{c_1 < \frac{\bar{X} - \mu}{S} \sqrt{n} < c_2\right\} = 1 - \alpha.$$

In particular, we can choose $-c_1 = c_2 = z_{\alpha/2}$ to give

$$\left(\bar{x} - \frac{s}{\sqrt{n}} z_{\alpha/2}, \bar{x} + \frac{s}{\sqrt{n}} z_{\alpha/2}\right)$$

as an approximate $1 - \alpha$ level confidence inverval for μ.

For large samples there is yet another possibility: using the maximum likelihood estimate for the parameter θ, provided that it exists. If $\hat{\theta}$ is the MLE of θ and the conditions of Theorem 8.7.4 or 8.7.5 are satisfied (*caution:* see Remark 8.7.6), then

$$\frac{\sqrt{n}(\hat{\theta} - \theta)}{\sigma} \xrightarrow{L} \mathcal{N}(0, 1) \qquad \text{as } n \to \infty,$$

where

$$\sigma^2 = \left[E_\theta\left\{\frac{\partial \log f_\theta(X)}{\partial \theta}\right\}^2\right]^{-1}.$$

In such cases, we can invert the statement

$$P_\theta\left\{-z_{\alpha/2} < \frac{\hat{\theta} - \theta}{\sigma} \sqrt{n} < z_{\alpha/2}\right\} \geq 1 - \alpha$$

to give an approximate $1 - \alpha$ level confidence interval for θ.

Yet another possible procedure has universal applicability and hence can be used for large or small samples. Unfortunately, however, this procedure usually yields confidence intervals that are much too large. The method employs the well-known Chebychev inequality (see Section 3.4):

$$P\{|X - EX| < \varepsilon \sqrt{\text{var}(X)}\} > 1 - \frac{1}{\varepsilon^2}.$$

If $\hat{\theta}$ is an estimate of θ (not necessarily unbiased) with finite variance $\sigma^2(\theta)$, then by Chebychev's inequality

$$P\{|\hat{\theta} - \theta| < \sqrt{E(\hat{\theta} - \theta)^2}\, \varepsilon\} > 1 - \frac{1}{\varepsilon^2}.$$

It follows that

$$(\hat{\theta} - \varepsilon\sqrt{E(\hat{\theta} - \theta)^2}, \hat{\theta} + \varepsilon\sqrt{E(\hat{\theta} - \theta)^2})$$

is a $1 - (1/\varepsilon^2)$ level confidence interval for θ. Under some mild consistency conditions one can replace the normalizing constant $\sqrt{[E(\hat{\theta} - \theta)^2]}$, which will be some function $\lambda(\theta)$ of θ, by $\lambda(\hat{\theta})$.

Note that the estimate $\hat{\theta}$ need not have a limiting normal law.

Example 5. Let $X_1, X_2, \cdots, X_n$ be iid $b(1, p)$ rv's, and it is required to find a confidence interval for p. We know that $E\bar{X} = p$, and

$$\text{var}(\bar{X}) = \frac{\text{var}(X)}{n} = \frac{p(1-p)}{n}.$$

It follows that

$$P\left\{|\bar{X} - p| < \varepsilon\sqrt{\frac{p(1-p)}{n}}\right\} > 1 - \frac{1}{\varepsilon^2}.$$

Since $p(1-p) \leq \frac{1}{4}$, we have

$$P\left\{\bar{X} - \frac{1}{2\sqrt{n}}\varepsilon < p < \bar{X} + \frac{1}{2\sqrt{n}}\varepsilon\right\} > 1 - \frac{1}{\varepsilon^2}.$$

One can now choose ε and n or, if n is kept constant at a given number, ε to get the desired level.

Actually the confidence interval obtained above can be improved somewhat. We note that

$$P\left\{|\bar{X} - p| < \varepsilon\sqrt{\frac{p(1-p)}{n}}\right\} > 1 - \frac{1}{\varepsilon^2},$$

so that

$$P\left\{|\bar{X} - p|^2 < \frac{\varepsilon^2 p(1-p)}{n}\right\} > 1 - \frac{1}{\varepsilon^2}.$$

Now

$$|\bar{X} - p|^2 < \frac{\varepsilon^2}{n} p(1-p)$$

if and only if

$$\left(1 + \frac{\varepsilon^2}{n}\right)p^2 - \left(2\bar{X} + \frac{\varepsilon^2}{n}\right)p + \bar{X}^2 < 0.$$

This last inequality holds if and only if p lies between the two roots of the quadratic equation

$$\left(1 + \frac{\varepsilon^2}{n}\right)p^2 - \left(2\bar{X} + \frac{\varepsilon^2}{n}\right)p + \bar{X}^2 = 0.$$

The two roots are

$$p_1 = \frac{2\bar{X} + (\varepsilon^2/n) - \sqrt{[2\bar{X} + (\varepsilon^2/n)]^2 - 4[1 + (\varepsilon^2/n)]\bar{X}^2}}{2[1 + (\varepsilon^2/n)]}$$

$$= \frac{\bar{X}}{1 + (\varepsilon^2/n)} + \frac{(\varepsilon^2/n) - \sqrt{4(\varepsilon^2/n)\bar{X}(1 - \bar{X}) + (\varepsilon^4/n^2)}}{2[1 + (\varepsilon^2/n)]}$$

and

$$p_2 = \frac{2\bar{X} + (\varepsilon^2/n) + \sqrt{[2\bar{X} + (\varepsilon^2/n)]^2 - 4[1 + (\varepsilon^2/n)]\bar{X}^2}}{2[1 + (\varepsilon^2/n)]}$$

$$= \frac{\bar{X}}{1 + (\varepsilon^2/n)} + \frac{(\varepsilon^2/n) + \sqrt{4(\varepsilon^2/n)\bar{X}(1 - \bar{X}) + (\varepsilon^4/n^2)}}{2[1 + (\varepsilon^2/n)]}.$$

It follows that

$$P\{p_1 < p < p_2\} > 1 - \frac{1}{\varepsilon^2}.$$

Note that when n is large

$$p_1 \approx \bar{X} - \varepsilon\sqrt{\frac{\bar{X}(1 - \bar{X})}{n}}, \quad p_2 \approx \bar{X} + \varepsilon\sqrt{\frac{\bar{X}(1 - \bar{X})}{n}},$$

as one should expect in view of the fact that $\bar{X} \to p$ with probability 1 and $\sqrt{[\bar{X}(1 - \bar{X})/n]}$ estimates $\sqrt{[p(1 - p)/n]}$. Alternatively, we could have used the CLT (or large-sample property of the MLE) to arrive at the same result but with ε replaced by $z_{\alpha/2}$.

Example 6. Let $X_1, X_2, \cdots, X_n$ be a sample from $U(0, \theta)$. We seek a confidence interval for the parameter θ. The estimate

$$\hat{\theta} = \max(X_1, \cdots, X_n) = M_n$$

is the MLE of θ, which is also sufficient for θ. The pdf of M_n is given by

$$f_n(y) = \begin{cases} \dfrac{ny^{n-1}}{\theta^n}, & 0 < y < \theta, \\ 0, & \text{otherwise.} \end{cases}$$

The rv $T_n = M_n/\theta$ has pdf

$$h(t) = \begin{cases} nt^{n-1}, & 0 < t < 1, \\ 0, & \text{otherwise,} \end{cases}$$

which is independent of θ. Applying Theorem 1, we find numbers $\lambda_1(\alpha)$, $\lambda_2(\alpha)$ such that

$$P\{\lambda_1 < T_n < \lambda_2\} = 1 - \alpha \quad \text{for all } \theta,$$

that is
$$n \int_{\lambda_1}^{\lambda_2} t^{n-1} \, dt = 1 - \alpha,$$
so that
$$\lambda_2^n - \lambda_1^n = 1 - \alpha.$$
This equation has infinitely many solutions. If we choose $\lambda_2 = 1$, then $\lambda_1 = (\alpha)^{1/n}$ and a $1 - \alpha$ level confidence interval is given by $(M_n, \alpha^{-1/n} M_n)$. In Example 11.3.4 we will show that this confidence interval has the smallest length among all confidence intervals for θ based on M_n.

Let us now apply the method of Chebychev's inequality to the same problem. We have
$$E_\theta M_n = \frac{n}{n+1} \theta$$
and
$$E_\theta (M_n - \theta)^2 = \theta^2 \frac{2}{(n+1)(n+2)}.$$
Thus
$$P\left\{ \frac{|M_n - \theta|}{\theta} \sqrt{\frac{(n+1)(n+2)}{2}} < \varepsilon \right\} > 1 - \frac{1}{\varepsilon^2}.$$
Since $M_n \xrightarrow{P} \theta$, we replace θ by M_n, and, for moderately large n,
$$P\left\{ \frac{|M_n - \theta|}{M_n} \sqrt{\frac{(n+1)(n+2)}{2}} < \varepsilon \right\} > 1 - \frac{1}{\varepsilon^2}.$$
It follows that
$$\left(M_n - \varepsilon M_n \frac{\sqrt{2}}{\sqrt{(n+1)(n+2)}}, M_n + \varepsilon M_n \frac{\sqrt{2}}{\sqrt{(n+1)(n+2)}} \right)$$
is a $1 - (1/\varepsilon^2)$ confidence interval for θ. Choosing $1 - (1/\varepsilon^2) = 1 - \alpha$, or $\varepsilon = 1/\sqrt{\alpha}$, and noting that $1/\sqrt{[(n+1)(n+2)]} \approx 1/n$ for large n, and the fact that $M_n \leq \theta$, we can use the approximate confidence interval
$$\left(M_n, M_n \left(1 + \frac{1}{n} \sqrt{\frac{2}{\alpha}} \right) \right)$$
for θ.

In the examples given above we see that, for a given confidence level $1 - \alpha$, a wide choice of confidence intervals is available. Clearly, the larger the interval, the better is the chance of trapping a true parameter value. Thus the interval $(-\infty, +\infty)$, which ignores the data completely, will include the real-valued parameter θ with confidence level 1. However, the larger the confidence interval, the less meaningful it is. Therefore, for a given confidence level $1 - \alpha$, it is desirable to choose the shortest possible confidence interval. Since the length $\bar{\theta} - \underline{\theta}$, in general, is a random variable, one can show that a confidence interval of level $1 - \alpha$ with uniformly minimum length among all such intervals does not exist in most cases. The alternative, to minimize $E_\theta(\bar{\theta} - \underline{\theta})$, is also quite unsatisfactory. In the next section we consider the problem of finding shortest-length confidence interval based on some suitable statistic.

PROBLEMS 11.2

1. A sample of size 25 from a normal population with variance 81 produced a mean of 81.2. Find a .95 level confidence interval for the mean μ.

2. Let $\bar{X}$ be the mean of a random sample of size n from $\mathcal{N}(\mu, 16)$. Find the smallest sample size n such that $(\bar{X} - 1, \bar{X} + 1)$ is a .90 level confidence interval for μ.

3. Let $X_1, X_2, \cdots, X_m$ and $Y_1, Y_2, \cdots, Y_n$ be independent random samples from $\mathcal{N}(\mu_1, \sigma^2)$ and $\mathcal{N}(\mu_2, \sigma^2)$, respectively. Find a confidence interval for $\mu_1 - \mu_2$ at confidence level $1 - \alpha$ when (a) σ is known, and (b) σ is unknown.

4. Two independent samples, each of size 7, from normal populations with common unknown variance σ^2 produced sample means 4.8 and 5.4 and sample variances 8.38 and 7.62, respectively. Find a .95 level confidence interval for $\mu_1 - \mu_2$, the difference between the means of samples 1 and 2.

5. In Problem 3 suppose that the first population has variance σ_1^2 and the second population has variance σ_2^2, where both σ_1^2 and σ_2^2 are known. Find a $1 - \alpha$ level confidence interval for $\mu_1 - \mu_2$. What happens if both σ_1^2 and σ_2^2 are unknown and unequal?

6. In Problem 5 find a confidence interval for the ratio σ_2^2/σ_1^2, both when μ_1, μ_2 are known and when μ_1, μ_2 are unknown. What happens if either μ_1 or μ_2 is unknown but the other is known?

7. Let $X_1, X_2, \cdots, X_n$ be a sample from a $G(1, \beta)$ distribution. Find a confidence interval for the parameter β with confidence level $1 - \alpha$.

8. (a) Use the large-sample properties of the MLE to construct a $1 - \alpha$ level confidence interval for the parameter θ in each of the following cases: (i) $X_1, X_2, \cdots, X_n$ is a sample from $G(1, 1/\theta)$, and (ii) $X_1, X_2, \cdots, X_n$ is a sample from $P(\theta)$.
   (b) In part (a) use Chebychev's inequality to do the same.

9. For a sample of size 1 from the population

$$f_\theta(x) = \frac{2}{\theta^2}(\theta - x), \quad 0 < x < \theta,$$

find a $1 - \alpha$ level confidence interval for θ.

10. Let $X_1, X_2, \cdots, X_n$ be a sample from the uniform distribution on N points. Find an upper $1 - \alpha$ level confidence bound for N, based on max $(X_1, X_2, \cdots, X_n)$.

11. In Example 6 find the smallest n such that the length of the $1 - \alpha$ level confidence interval $(M_n, \alpha^{-1/n} M_n) < d$, provided it is known that $\theta \leq a$, where a is a known constant.

12. Let X and Y be independent rv's with pdf's $\lambda e^{-\lambda x}$ $(x > 0)$ and $\mu e^{-\mu y}$ $(y > 0)$, respectively. Find a $1 - \alpha$ level confidence region for (λ, μ) of the form $\{(\lambda, \mu): \lambda X + \mu Y \leq k\}$.

11.3 SHORTEST-LENGTH CONFIDENCE INTERVALS

We have already remarked that we can increase the confidence level by simply taking a larger-length confidence interval. Indeed, the worthless interval $-\infty < \theta < \infty$, which simply says that θ lies somewhere on the real line, has confidence level 1. In practice, one would like to set the level at a given fixed number $1 - \alpha$ $(0 < \alpha < 1)$ and, if possible, construct an interval as short as possible among all confidence intervals with the same level. Such an interval is desirable since it is more informative. We have already remarked that shortest-length confidence intervals do not always exist. In this section we will investigate the possibility of constructing shortest-length confidence intervals based on some simple rv's. The discussion here is based on Guenther [40]. Theorem 11.2.1 is really the key to the following discussion.

Let $X_1, X_2, \cdots, X_n$ be a sample from a pdf $f_\theta(x)$, and $T(X_1, X_2, \cdots, X_n, \theta) = T_\theta$ be an rv with distribution independent of θ. Also, let $\lambda_1 = \lambda_1(\alpha)$, $\lambda_2 = \lambda_2(\alpha)$ be chosen so that

(1) $$P\{\lambda_1 < T_\theta < \lambda_2\} = 1 - \alpha,$$

and suppose that (1) can be rewritten as

(2) $$P\{\underline{\theta}(\mathbf{X}) < \theta < \bar{\theta}(\mathbf{X})\} = 1 - \alpha.$$

(See Theorem 11.2.1 for a set of sufficient conditions.)

For every T_θ, λ_1 and λ_2 can be chosen in many ways. We would like to choose λ_1 and λ_2 so that $\bar{\theta} - \underline{\theta}$ is minimum. Such an interval is a $1 - \alpha$ level shortest-length confidence interval based on T_θ. It may be possible, however, to find another rv T_θ^* that may yield an even shorter interval. Therefore we are not asserting that the procedure, if it succeeds, will lead to a $1 - \alpha$ level confidence interval that has shortest length among all intervals of this level.

For T_θ we use the simplest rv that is a function of a sufficient statistic and θ. Let us first give a suitable name to $T(\mathbf{X}, \theta)$.

Definition 1. An rv $T(\mathbf{X}, \theta)$, which is a function of $\mathbf{X} = (X_1, X_2, \cdots, X_n)$ and θ, and whose distribution is independent of θ, is called a pivot.

For example, in sampling from a normal population, if the variance is known, $(\bar{X} - \mu)\sqrt{n}/\sigma$ is a natural choice for a pivot. If σ is unknown, $(\bar{X} - \mu)\sqrt{n}/S$ is a natural choice for a pivot. If one desires a confidence interval for the variance σ^2, $\sum_1^n (X_i - \mu)^2/\sigma^2$ is a pivot; and, if μ is unknown, S^2/σ^2 is a pivot.

Remark 1. An alternative to minimizing the length of the confidence interval is to minimize the expected length $E_\theta\{\bar{\theta}(\mathbf{X}) - \underline{\theta}(\mathbf{X})\}$. Unfortunately, this also is quite unsatisfactory since, in general, there does not exist a member in the class of all $1 - \alpha$ level confidence intervals that minimizes $E_\theta\{\bar{\theta}(\mathbf{X}) - \underline{\theta}(\mathbf{X})\}$ for all θ. The procedures applied in finding the shortest-length confidence interval based on a pivot are also applicable in finding an interval that minimizes the expected length. We remark here that the restriction to unbiased confidence intervals is natural if we wish to minimize $E_\theta\{\bar{\theta}(\mathbf{X}) - \underline{\theta}(\mathbf{X})\}$. See Sections 11.4 and 11.5 for definitions and further details.

Example 1. Let $X_1, X_2, \cdots, X_n$ be a sample from $\mathcal{N}(\mu, \sigma^2)$, where σ^2 is known. Consider the pivot

$$T_\mu(\mathbf{X}) = \frac{\bar{X} - \mu}{\sigma/\sqrt{n}}.$$

Then

$$1 - \alpha = P\left\{a < \frac{\bar{X} - \mu}{\sigma}\sqrt{n} < b\right\}$$
$$= P\left\{\bar{X} - b\frac{\sigma}{\sqrt{n}} < \mu < \bar{X} - a\frac{\sigma}{\sqrt{n}}\right\}.$$

The length of this confidence interval is $(\sigma/\sqrt{n})(b - a)$. We wish to minimize $L = (\sigma/\sqrt{n})(b - a)$ such that

$$\Phi(b) - \Phi(a) = \frac{1}{\sqrt{2\pi}} \int_a^b e^{-x^2/2} dx = \int_a^b \varphi(x)\, dx = 1 - \alpha.$$

Here φ and Φ, respectively, are the pdf and df of an $\mathcal{N}(0, 1)$ rv. Thus

$$\frac{dL}{da} = \frac{\sigma}{\sqrt{n}}\left(\frac{db}{da} - 1\right),$$

and

$$\varphi(b)\frac{db}{da} - \varphi(a) = 0,$$

giving

$$\frac{dL}{da} = \frac{\sigma}{\sqrt{n}}\left[\frac{\varphi(a)}{\varphi(b)} - 1\right].$$

The minimum occurs when $\varphi(a) = \varphi(b)$, that is, when $a = b$ or $a = -b$. Since $a = b$ does not satisfy

$$\int_a^b \varphi(t)\, dt = 1 - \alpha,$$

we choose $a = -b$. The shortest confidence interval based on T_μ is therefore the equal tails interval,

$$\left(\bar{X} + z_{1-\alpha/2}\frac{\sigma}{\sqrt{n}},\ \bar{X} + z_{\alpha/2}\frac{\sigma}{\sqrt{n}}\right) \text{ or } \left(\bar{X} - z_{\alpha/2}\frac{\sigma}{\sqrt{n}},\ \bar{X} + z_{\alpha/2}\frac{\sigma}{\sqrt{n}}\right).$$

The length of this interval is $2z_{\alpha/2}(\sigma/\sqrt{n})$. In this case we can plan our experiment to give a prescribed confidence level and a prescribed length for the interval. To have level $1 - \alpha$ and length $\leq 2d$, we choose the smallest n such that

$$d \geq z_{\alpha/2}\frac{\sigma}{\sqrt{n}} \quad \text{or} \quad n \geq z_{\alpha/2}^2\frac{\sigma^2}{d^2}.$$

This can also be interpreted as follows. If we estimate μ by $\bar{X}$, taking a sample of size $n \geq z_{\alpha/2}^2(\sigma^2/d^2)$, we are $1 - \alpha$ percent confident that the error in our estimate is at most d.

Example 2. In Example 1 suppose that σ is unknown. In that case we use

$$T_\mu(\mathbf{X}) = \frac{\bar{X} - \mu}{S}\sqrt{n}$$

as a pivot. T_μ has Student's t-distribution with $n - 1$ d.f. Thus

$$1 - \alpha = P\left\{a < \frac{\bar{X} - \mu}{S}\sqrt{n} < b\right\} = P\left\{\bar{X} - b\frac{S}{\sqrt{n}} < \mu < \bar{X} - a\frac{S}{\sqrt{n}}\right\}.$$

We wish to minimize

$$L = (b - a)\frac{S}{\sqrt{n}}$$

subject to

$$\int_a^b f_{n-1}(t)\, dt = 1 - \alpha,$$

where $f_{n-1}(t)$ is the pdf of T_μ. We have

$$\frac{dL}{da} = \left(\frac{db}{da} - 1\right)\frac{S}{\sqrt{n}} \quad \text{and} \quad f_{n-1}(b)\frac{db}{da} - f_{n-1}(a) = 0,$$

giving

$$\frac{dL}{da} = \left[\frac{f_{n-1}(a)}{f_{n-1}(b)} - 1\right]\frac{S}{\sqrt{n}}.$$

It follows that the minimum occurs at $a = -b$ (the other solution, $a = b$, is not admissible). The shortest-length confidence interval based on T_μ is the equal tails interval,

$$\left(\bar{X} - t_{n-1,\alpha/2}\frac{S}{\sqrt{n}}, \bar{X} + t_{n-1,\alpha/2}\frac{S}{\sqrt{n}}\right).$$

The length of this interval is $2t_{n-1,\alpha/2}(S/\sqrt{n})$, which, being random, may be arbitrarily large. Note that the same confidence interval minimizes the expected length of the interval, namely, $EL = (b-a)c_n(\sigma/\sqrt{n})$, where c_n is a constant determined from $ES = c_n\sigma$ and the minimum expected length is $2t_{n-1,\alpha/2}\,c_n(\sigma/\sqrt{n})$. In this case Stein suggested a two-stage procedure that is discussed in Section 14.4.

Example 3. Let $X_1, X_2, \cdots, X_n$ be iid $\mathcal{N}(\mu, \sigma^2)$ rv's. Suppose that μ is known and we want a confidence interval for σ^2. The obvious choice for a pivot T_{σ^2} is given by

$$T_{\sigma^2}(\mathbf{X}) = \frac{\sum_{1}^{n}(X_i - \mu)^2}{\sigma^2},$$

which has a chi-square distribution with n d.f. Now

$$P\left\{a < \frac{\sum_{1}^{n}(X_i - \mu)^2}{\sigma^2} < b\right\} = 1 - \alpha,$$

so that

$$P\left\{\frac{\sum_{1}^{n}(X_i - \mu)^2}{b} < \sigma^2 < \frac{\sum_{1}^{n}(X_i - \mu)^2}{a}\right\} = 1 - \alpha.$$

We wish to minimize

$$L = \left(\frac{1}{a} - \frac{1}{b}\right)\sum_{1}^{n}(X_i - \mu)^2$$

subject to

$$\int_a^b f_n(t)\,dt = 1 - \alpha,$$

where f_n is the pdf of a chi-square rv with n d.f. We have

$$\frac{dL}{da} = \left(\frac{1}{a^2} - \frac{1}{b^2}\frac{db}{da}\right)\sum_1^n (X_i - \mu)^2$$

and

$$\frac{db}{da} = \frac{f_n(a)}{f_n(b)},$$

so that

$$\frac{dL}{da} = \left[\frac{1}{a^2} - \frac{1}{b^2}\frac{f_n(a)}{f_n(b)}\right]\sum_1^n (X_i - \mu)^2,$$

which vanishes if

$$\frac{1}{a^2} = \frac{1}{b^2}\frac{f_n(a)}{f_n(b)}.$$

Numerical results giving values of a and b to four significant places of decimals are available; see Tate and Klett [129]. In practice, the equal tails interval,

$$\left(\frac{\sum_{i=1}^n (X_i - \mu)^2}{\chi^2_{n,\,\alpha/2}}, \frac{\sum_{i=1}^n (X_i - \mu)^2}{\chi^2_{n,\,1-\alpha/2}}\right),$$

is used.

If μ is unknown, we use

$$T_{\sigma^2}(\mathbf{X}) = \frac{\sum_1^n (X_i - \bar{X})^2}{\sigma^2} = (n-1)\frac{S^2}{\sigma^2}$$

as a pivot. T_{σ^2} has a $\chi^2(n-1)$ distribution. Proceeding as above, we can show that the shortest-length confidence interval based on T_{σ^2} is $((n-1)(S^2/b), (n-1)(S^2/a))$; here a and b are a solution of

$$P\{a < \chi^2(n-1) < b\} = 1 - \alpha$$

and

$$a^2 f_{n-1}(a) = b^2 f_{n-1}(b),$$

where f_{n-1} is the pdf of a $\chi^2(n-1)$ rv. Numerical solutions due to Tate and Klett [129] may be used, but, in practice, the simpler equal tails confidence interval,

$$\left(\frac{(n-1)S^2}{\chi^2_{n-1,\,\alpha/2}}, \frac{(n-1)S^2}{\chi^2_{n-1,\,1-\alpha/2}}\right)$$

is employed.

Example 4. Let $X_1, X_2, \cdots, X_n$ be a sample from $U(0, \theta)$. Then $M_n = \max(X_1, X_2, \cdots, X_n)$ is sufficient for θ with density

$$f_n(y) = n\frac{y^{n-1}}{\theta^n}, \quad 0 < y < \theta.$$

The rv $T_n = M_n/\theta$ has pdf

$$h(t) = nt^{n-1}, \quad 0 < t < 1.$$

Using T_n as pivot, we see that the confidence interval is $(M_n/b, M_n/a)$ with length $L = M_n(1/a - 1/b)$. We minimize L subject to

$$\int_a^b nt^{n-1}\,dt = b^n - a^n = 1 - \alpha.$$

Now

$$(1 - \alpha)^{1/n} < b \leq 1$$

and

$$\frac{dL}{db} = M_n\left(-\frac{1}{a^2}\frac{da}{db} + \frac{1}{b^2}\right) = M_n\left(\frac{a^{n+1} - b^{n+1}}{b^2 a^{n+1}}\right) < 0,$$

so that the minimum occurs at $b = 1$. The shortest interval is therefore $(M_n, M_n/\alpha^{1/n})$. Note that

$$EL = \left(\frac{1}{a} - \frac{1}{b}\right)EM_n = \frac{n\theta}{n+1}\left(\frac{1}{a} - \frac{1}{b}\right),$$

which is minimized subject to

$$b^n - a^n = 1 - \alpha,$$

where $b = 1$ and $a = \alpha^{1/n}$. The expected length of the interval that minimizes EL is $[(1/\alpha^{1/n}) - 1][n\theta/(n + 1)]$, which is also the expected length of the shortest confidence interval based on M_n.

Note that the length of the interval $(M_n, \alpha^{-1/n}M_n)$ goes to 0 as n becomes large.

For some results on asymptotically shortest-length confidence intervals, we refer the reader to Wilks [141], pages 374–376.

PROBLEMS 11.3

1. Let $X_1, X_2, \cdots, X_n$ be a sample from

$$f_\theta(x) = \begin{cases} e^{-(x-\theta)} & \text{if } x > \theta, \\ 0 & \text{otherwise.} \end{cases}$$

Find the shortest-length confidence interval for θ at level $1 - \alpha$, based on a sufficient statistic for θ.

2. Let $X_1, X_2, \cdots, X_n$ be a sample from $G(1, \theta)$. Find the shortest-length confidence interval for θ at level $1 - \alpha$, based on a sufficient statistic for θ.

3. Let $X_1, X_2, \cdots, X_n$ be a sample from $B(\theta, 1)$. Find the shortest-length confidence interval for θ at level $1 - \alpha$, based on a sufficient statistic for θ.

4. In Problem 11.2.9 how will you find the shortest-length confidence interval for θ at level $1 - \alpha$, based on the statistic X/θ?

5. Let $T(\mathbf{X}, \theta)$ be a pivot of the form $T(\mathbf{X}, \theta) = T_1(\mathbf{X}) - \theta$. Show how one can construct a confidence interval for θ with fixed width d and maximum possible confidence coefficient. In particular, construct a confidence interval that has fixed width d and maximum possible confidence coefficient for the mean μ of a normal population with variance 1. Find the smallest size n for which this confidence interval has a confidence coefficient $\geq 1 - \alpha$. Repeat the above in sampling from an exponential pdf

$$f_\mu(x) = e^{\mu-x} \text{ for } x > \mu \text{ and } f_\mu(x) = 0 \quad \text{for } x \leq \mu. \quad \text{(Desu [24])}$$

11.4 RELATION BETWEEN CONFIDENCE ESTIMATION AND HYPOTHESES TESTING

In this section we investigate the relationship between confidence estimation and hypotheses testing. More precisely, given a level α test of $H_0: \theta = \theta_0$, say, is it possible to construct a $1 - \alpha$ level confidence interval for θ_0, and conversely?

Example 1. Let $X_1, X_2, \cdots, X_n$ be a sample from $\mathcal{N}(\mu, \sigma_0^2)$. In Example 11.2.1 we showed that

$$\left(\bar{X} - \frac{1}{\sqrt{n}} z_{\alpha/2}\sigma_0, \bar{X} + \frac{1}{\sqrt{n}} z_{\alpha/2}\sigma_0\right)$$

is a $1 - \alpha$ level confidence interval for μ. If we define a test φ that rejects a value of $\mu = \mu_0$ if and only if μ_0 lies outside this interval, that is, if and only if

$$\frac{\sqrt{n}|\bar{X} - \mu_0|}{\sigma_0} \geq z_{\alpha/2},$$

then

$$P_{\mu_0}\left\{\sqrt{n}\frac{|\bar{X} - \mu_0|}{\sigma_0} \geq z_{\alpha/2}\right\} = \alpha,$$

and the test φ is a size α test of $\mu = \mu_0$ against the alternatives $\mu \neq \mu_0$.

Conversely, a family of α level tests for the hypothesis $\mu = \mu_0$ generates a family of confidence intervals for μ by simply taking, as the confidence interval for μ_0, the region of acceptance of the test.

Similarly, we can generate a family of α level tests from a $1 - \alpha$ level lower (or upper) confidence bound. Suppose that we start with the $1 - \alpha$ level lower confidence bound $\bar{X} - z_\alpha(\sigma_0/\sqrt{n})$ Then, by defining a test $\varphi(\mathbf{X})$ that rejects $\mu \leq \mu_0$ if and only if $\mu_0 < \bar{X} - z_\alpha(\sigma_0/\sqrt{n})$, we get an α level test for a hypothesis of the form $\mu \leq \mu_0$.

Example 1 is a special case of the duality principle proved in Theorem 1 below. In the following we restrict our attention to the case in which the rejection (acceptance) region of the test is the indicator function of a (Borel-measurable) set, that is, we consider only nonrandomized tests (and confidence intervals). For notational convenience we write $H_0(\theta_0)$ for the hypothesis $H_0: \theta = \theta_0$ and $H_1(\theta_0)$ for the alternative hypothesis, which may be one or two-sided.

Theorem 1. Let $A(\theta_0)$, $\theta_0 \in \Theta$, denote the region of acceptance of an α level test of $H_0(\theta_0)$. For each observation $\mathbf{x} = (x_1, x_2, \cdots, x_n)$ let $S(\mathbf{x})$ denote the set

(1) $$S(\mathbf{x}) = \{\theta : \mathbf{x} \in A(\theta), \theta \in \Theta\}.$$

Then $S(\mathbf{x})$ is a family of confidence sets for θ at confidence level $1 - \alpha$. If, moreover, $A(\theta_0)$ is UMP for the problem $(\alpha, H_0(\theta_0), H_1(\theta_0))$, then $S(\mathbf{X})$ minimizes

(2) $$P_\theta\{S(\mathbf{X}) \ni \theta'\} \quad \text{for all } \theta \in H_1(\theta')$$

among all $1 - \alpha$ level families of confidence sets.

Proof. We have

(3) $$S(\mathbf{x}) \ni \theta \quad \text{if and only } \mathbf{x} \in A(\theta),$$

so that

$$P_\theta\{S(\mathbf{X}) \ni \theta\} = P_\theta\{\mathbf{X} \in A(\theta)\} \geq 1 - \alpha,$$

as asserted.

If $S^*(\mathbf{X})$ is any other family of $1 - \alpha$ level confidence sets, let $A^*(\theta) = \{\mathbf{x} : S^*(\mathbf{x}) \ni \theta\}$. Then

$$P_\theta\{\mathbf{X} \in A^*(\theta)\} = P_\theta\{S^*(\mathbf{X}) \ni \theta\} \geq 1 - \alpha;$$

and, since $A(\theta_0)$ is UMP for $(\alpha, H_0(\theta_0), H_1(\theta_0))$, it follows that

$$P_\theta\{\mathbf{X} \in A^*(\theta_0)\} \geq P_\theta\{\mathbf{X} \in A(\theta_0)\} \quad \text{for any } \theta \in H_1(\theta_0).$$

Hence

$$P_\theta\{S^*(\mathbf{X}) \ni \theta_0\} \geq P_\theta\{\mathbf{X} \in A(\theta_0)\} = P_\theta\{S(\mathbf{X}) \ni \theta_0\}$$

for all $\theta \in H_1(\theta_0)$. This completes the proof.

Remark 1. In view of Definition 11.2.5, the family of confidence sets associated with a UMP acceptance region is a UMA family at level $1 - \alpha$.

Example 2. Let X be an rv of the continuous type with a one-parameter exponential pdf, given by

$$f_\theta(x) = \exp\{Q(\theta) T(x) + S'(x) + D(\theta)\},$$

where $Q(\theta)$ is a nondecreasing function of θ. Let $H_0: \theta = \theta_0$ and $H_1: \theta < \theta_0$. Then the acceptance region of a UMP size α test of H_0 is of the form

$$A(\theta_0) = \{x: T(x) > c(\theta_0)\}.$$

Since, for $\theta \geq \theta'$,

$$P_{\theta'}\{T(X) \leq c(\theta')\} = \alpha = P_\theta\{T(X) \leq c(\theta)\} \leq P_{\theta'}\{T(X) \leq c(\theta)\},$$

$c(\theta)$ may be chosen to be nondecreasing. (The last inequality follows because the power of the UMP test is at least α, the size.) We have

$$S(x) = \{\theta: x \in A(\theta)\},$$

so that $S(x)$ is of the form $(-\infty, c^{-1}(T(x)))$, or $(-\infty, c^{-1}(T(x))]$, where c^{-1} is defined by

$$c^{-1}(T(x)) = \sup_\theta \{\theta: c(\theta) \leq T(x)\}.$$

In particular, if

$$f_\theta(x) = \begin{cases} \dfrac{1}{\theta} e^{-x/\theta}, & x > 0, \\ 0, & \text{otherwise}, \end{cases}$$

then $T(x) = x$; and, for testing $H_0: \theta = \theta_0$ against $H_1: \theta < \theta_0$, the UMP acceptance region is of the form

$$A(\theta_0) = \{x: x \geq c(\theta_0)\},$$

where

$$c(\theta_0) = \theta_0 \log \frac{1}{1 - \alpha}, \quad 0 < \alpha < 1.$$

The UMA family of $1 - \alpha$ level confidence sets is of the form

$$S(x) = \{\theta: x \in A(\theta)\}$$
$$= \left\{\theta: \theta \leq \frac{x}{\log[1/(1-\alpha)]}\right\}$$
$$= \left(-\infty, \frac{x}{\log[1/(1-\alpha)]}\right].$$

Since UMP tests generally do not exist, we restrict consideration to smaller subclasses of tests and look for UMP tests in these classes (see Chapter 9). We will show that UMP unbiased tests lead to *UMA unbiased confidence sets*. This analogy carries over to UMP invariant tests also, but we will not show that here.

Definition 1. A family $\{S(\mathbf{x})\}$ of confidence sets for a parameter θ is said to be unbiased at confidence level $1 - \alpha$ if

(4) $$P_\theta\{S(\mathbf{X}) \text{ contains } \theta\} \geq 1 - \alpha$$

and

(5) $$P_{\theta'}\{S(\mathbf{X}) \text{ contains } \theta\} \leq 1 - \alpha \quad \text{for all } \theta, \theta' \in \Theta.$$

If $S(\mathbf{X})$ is an interval satisfying (4) and (5), we call it a $1 - \alpha$ level unbiased confidence interval. If a family of unbiased confidence sets at level $1 - \alpha$ is UMA in the class of all $1 - \alpha$ level unbiased confidence sets, we call it a UMA unbiased (UMAU) family of confidence sets at level $1 - \alpha$. In other words if $S^*(\mathbf{x})$ satisfies (4) and (5) and minimizes

$$P_\theta\{S(\mathbf{X}) \text{ contains } \theta'\} \quad \text{for } \theta, \theta' \in \Theta$$

among all unbiased families of confidence sets $S(\mathbf{X})$ at level $1 - \alpha$, then $S^*(\mathbf{X})$ is a UMAU family of confidence sets at level $1 - \alpha$.

Remark 2. Definition 1 says that a family $S(\mathbf{X})$ of confidence sets for a parameter θ is unbiased at level $1 - \alpha$ if $S(\mathbf{X})$ traps the true parameter value with probability at least $1 - \alpha$ and $S(\mathbf{X})$ traps a false parameter value with a probability at most $1 - \alpha$. In other words, $S(\mathbf{X})$ traps a true parameter value more often than it does a false one.

Theorem 2. Let $A(\theta_0)$ be the acceptance region of a UMP unbiased size α test of $H_0(\theta_0): \theta = \theta_0$ against $H_1(\theta_0): \theta \neq \theta_0$ for each θ_0. Then $S(\mathbf{x}) = \{\theta: \mathbf{x} \in A(\theta)\}$ is a UMA unbiased family of confidence sets at level $1 - \alpha$.

Proof. To see that $S(\mathbf{x})$ is unbiased we note that, since $A(\theta)$ is the acceptance region of an unbiased test,

$$P_{\theta'}\{S(\mathbf{X}) \text{ contains } \theta\} = P_{\theta'}\{\mathbf{X} \in A(\theta)\} \leq 1 - \alpha.$$

We next show that $S(\mathbf{X})$ is UMA. Let $S^*(\mathbf{x})$ be any other unbiased $1 - \alpha$ level family of confidence sets, and write $A^*(\theta) = \{\mathbf{x}: S^*(\mathbf{x}) \text{ contains } \theta\}$. Then $P_{\theta'}\{\mathbf{X} \in A^*(\theta)\} = P_{\theta'}\{S^*(\mathbf{X}) \text{ contains } \theta\} \leq 1 - \alpha$, and it follows that $A^*(\theta)$ is the acceptance region of an unbiased size α test. Hence

$$P_{\theta'}\{S^*(\mathbf{X}) \text{ contains } \theta\} = P_{\theta'}\{\mathbf{X} \in A^*(\theta)\}$$
$$\geq P_{\theta'}\{\mathbf{X} \in A(\theta)\}$$
$$= P_{\theta'}\{S(\mathbf{X}) \text{ contains } \theta\}.$$

The inequality follows since $A(\theta)$ is the acceptance region of a UMP unbiased test. This completes the proof.

Example 3. Let $X_1, X_2, \cdots, X_n$ be a sample from $\mathcal{N}(\mu, \sigma^2)$, where both μ and σ^2 are unknown. For testing $H_0: \mu = \mu_0$ against $H_1: \mu \neq \mu_0$, we know (Theorem 9.5.4) that the t-test

$$\varphi(\mathbf{x}) = \begin{cases} 1, & \dfrac{|\sqrt{n}(\bar{x} - \mu_0)|}{s} > c, \\ 0, & \text{otherwise,} \end{cases}$$

where $\bar{x} = \sum x_i/n$ and $s^2 = (n-1)^{-1} \sum (x_i - \bar{x})^2$ is UMP unbiased. We choose c from the size requirement

$$\alpha = P_{\mu = \mu_0}\left\{\left|\frac{\sqrt{n}(\bar{X} - \mu_0)}{S}\right| > c\right\},$$

so that $c = t_{n-1, \alpha/2}$. Thus

$$A(\mu_0) = \left\{\mathbf{x}: \left|\frac{\sqrt{n}(\bar{x} - \mu_0)}{s}\right| \leq t_{n-1, \alpha/2}\right\}$$

is the acceptance region of a UMP unbiased size α test of $H_0: \mu = \mu_0$ against $H_1: \mu \neq \mu_0$. By Theorem 2, it follows that

$$S(\mathbf{x}) = \{\mu: \mathbf{x} \in A(\mu)\}$$
$$= \left\{\bar{x} - \frac{s}{\sqrt{n}} t_{n-1, \alpha/2} \leq \mu \leq \bar{x} + \frac{s}{\sqrt{n}} t_{n-1, \alpha/2}\right\}$$

is a UMA unbiased family of confidence sets at level $1 - \alpha$.

Remark 3. Example 3 and the results of Section 9.5 show that the concept of unbiasedness is most suitable where the UMP tests do not exist. This is the case when the parameter set Θ consists of points (θ, μ), say, where both θ and μ are unknown, and one is interested in obtaining confidence sets for the parameter θ alone. The parameter μ is usually referred to as a *nuisance parameter*.

Remark 4. In Section 11.3 we remarked that shortest-length confidence intervals, although very desirable, do not exist for most commonly used distributions. Pratt [90] has shown that the restriction to the unbiased family of confidence intervals makes it possible, at least in many commonly encountered problems, to obtain $1 - \alpha$ level confidence intervals that have uniformly minimum expected length among all $1 - \alpha$ level unbiased confidence intervals. We will explore this topic in somewhat greater detail in the next section.

PROBLEMS 11.4

1. Let $X_1, X_2, \cdots, X_n$ be a sample from $\mathcal{N}(\mu, \sigma^2)$, where σ^2 is known. Find a UMA $1-\alpha$ level upper confidence bound for μ.

2. Let $X_1, X_2, \cdots, X_n$ be a sample from $U(0, \theta)$. Find a UMA family of confidence intervals for θ at level $1 - \alpha$.

3. Let $X_1, X_2, \cdots, X_n$ be a sample from $\mathcal{N}(\mu_0, \sigma^2)$, where σ^2 is unknown and μ_0 is a known constant. Find a UMA unbiased family of confidence intervals for σ^2 at level $1 - \alpha$.

4. Let $X_1, X_2, \cdots, X_m$ and $Y_1, Y_2, \cdots, Y_n$ be independent samples from normal distributions $\mathcal{N}(\mu, \sigma^2)$ and $\mathcal{N}(\theta, \sigma^2)$, respectively. Find a UMA family of unbiased confidence intervals for $\mu - \theta$ at level $1 - \alpha$.

5. In Problem 4, if the variances of the two normal populations are different, say σ^2 and τ^2, respectively, find a UMA family of unbiased confidence intervals for the ratio τ^2/σ^2 at level $1 - \alpha$.

11.5 UNBIASED CONFIDENCE INTERVALS

In the earlier sections we emphasized that, if the shortest length is used as a measure of precision of the confidence interval, it is not possible to find intervals that have uniformly shortest length among all $1 - \alpha$ level confidence intervals, even for some of the most commonly used distributions. In view of the duality principle (Theorem 11.4.1), since the UMP tests do not exist in general, one frequently restricts attention to the class of unbiased confidence intervals (Theorem 11.4.2). This is specially true if a nuisance parameter is present. Indeed, Pratt [90] has shown that, if the measure of precision is the expected length of the confidence interval, one is naturally led to a consideration of the unbiased confidence intervals. The following result is due to Pratt [90].

Theorem 1. Let Θ be an interval on the real line, and f_θ be the pdf of **X**. Let $S(\mathbf{X})$ be a family of $1 - \alpha$ level confidence intervals of finite length, that is, let $S(\mathbf{X}) = (\underline{\theta}(\mathbf{X}), \bar{\theta}(\mathbf{X}))$, and suppose that $\bar{\theta}(\mathbf{X}) - \underline{\theta}(\mathbf{X})$ is (random) finite. Then

(1) $$\int (\bar{\theta}(\mathbf{x}) - \underline{\theta}(\mathbf{x})) f_\theta(\mathbf{x}) \, d\mathbf{x} = \int_{\theta' \neq \theta} P_\theta\{S(\mathbf{X}) \text{ contains } \theta'\} \, d\theta'$$

for all $\theta \in \Theta$.

Proof. We have

$$\bar{\theta} - \underline{\theta} = \int_{\underline{\theta}}^{\bar{\theta}} d\theta'.$$

Thus for all $\theta \in \Theta$

$$\begin{aligned}
E_\theta\{\bar{\theta}(\mathbf{X}) - \underline{\theta}(\mathbf{X})\} &= E_\theta\left\{\int_{\underline{\theta}}^{\bar{\theta}} d\theta'\right\} \\
&= \int f_\theta(\mathbf{x}) \left\{\int_{\underline{\theta}}^{\bar{\theta}} d\theta'\right\} d\mathbf{x} \\
&= \int \left\{\int_{\underline{\theta}}^{\bar{\theta}} f_\theta(\mathbf{x}) \, d\mathbf{x}\right\} d\theta' \\
&= \int P_\theta\{S(\mathbf{X}) \text{ contains } \theta'\} \, d\theta' \\
&= \int_{\theta' \neq \theta} P_\theta\{S(\mathbf{X}) \text{ contains } \theta'\} \, d\theta'.
\end{aligned}$$

Remark 1. Theorem 1 says that the expected length of the confidence interval is the probability that $S(\mathbf{X})$ includes θ averaged over all false values of θ.

Remark 2. If $S(\mathbf{X})$ is a family of UMAU $1 - \alpha$ level confidence intervals, the expected length of $S(\mathbf{X})$ is minimal. This follows since the left-hand side of (1) is the expected length, if θ is the true value, of $S(\mathbf{X})$ and $P_\theta\{S(\mathbf{X}) \text{ contains } \theta'\}$ is minimal [because $S(\mathbf{X})$ is UMAU], by Theorem 11.4.2, with respect to all families of $1 - \alpha$ level unbiased confidence intervals uniformly in θ ($\theta \neq \theta'$).

Example 1. Let $X_1, X_2, \cdots, X_n$ be a sample from $\mathcal{N}(\mu, \sigma^2)$, where both μ and σ^2 are unknown. To find a $1 - \alpha$ level unbiased confidence interval for μ we proceed as in Example 11.3.2. The statistic

$$T_\mu(\mathbf{X}) = \frac{\bar{X} - \mu}{S/\sqrt{n}}$$

has Student's t-distribution with $n - 1$ d.f. and leads to the shortest-length confidence interval

$$S(\mathbf{X}) = \left(\bar{X} - t_{n-1, \alpha/2} \frac{S}{\sqrt{n}}, \bar{X} + t_{n-1, \alpha/2} \frac{S}{\sqrt{n}}\right).$$

Moreover, by Theorems 9.5.4 and 11.4.2 (or by Example 11.4.3), $S(\mathbf{X})$ is a UMAU family of confidence sets at level $1 - \alpha$. It follows by Remark 2 that $ES(\mathbf{X})$ is minimal.

Since a reasonably complete discussion of UMP unbiased tests (see Section 9.5) is beyond the scope of this text, the following procedure for determining unbiased confidence intervals is sometimes quite useful. (See Guenther [41]). Let $X_1, X_2, \cdots, X_n$ be a sample from an absolutely continuous df with pdf $f_\theta(x)$, and suppose that we seek an unbiased confidence interval for θ. Following the discussion in Section 11.3, suppose that

$$T(X_1, X_2, \cdots, X_n, \theta) = T(\mathbf{X}, \theta) = T_\theta$$

is a pivot, and suppose that the statement

$$P\{\lambda_1(\alpha) < T_\theta < \lambda_2(\alpha)\} = 1 - \alpha$$

can be converted to

$$P_\theta\{\underline{\theta}(\mathbf{X}) < \theta < \bar{\theta}(\mathbf{X})\} = 1 - \alpha.$$

In order for $(\underline{\theta}, \bar{\theta})$ to be unbiased, we must have

(2) $\quad P(\theta, \theta') = P_\theta\{\underline{\theta}(\mathbf{X}) < \theta' < \bar{\theta}(\mathbf{X})\} = 1 - \alpha \quad \text{if } \theta' = \theta$

and

(3) $\quad\quad\quad\quad P(\theta, \theta') < 1 - \alpha \quad \text{if } \theta' \neq \theta.$

If $P(\theta, \theta')$ depends only on a function γ of θ, θ', we may write

(4)
(5) $\quad\quad\quad P(\gamma) \begin{cases} = 1 - \alpha & \text{if } \theta' = \theta, \\ < 1 - \alpha & \text{if } \theta' \neq \theta, \end{cases}$

and it follows that $P(\gamma)$ has a maximum at $\theta' = \theta$.

Example 2. Let $X_1, X_2, \cdots, X_n$ be iid $\mathcal{N}(\mu, \sigma^2)$ rv's, and suppose that we desire an unbiased confidence interval for σ^2. Then

$$T(\mathbf{X}, \sigma^2) = \frac{(n-1)S^2}{\sigma^2} = T_\sigma$$

has a $\chi^2(n-1)$ distribution, and we have

$$P\left\{\lambda_1 < (n-1)\frac{S^2}{\sigma^2} < \lambda_2\right\} = 1 - \alpha,$$

so that

$$P\left\{(n-1)\frac{S^2}{\lambda_2} < \sigma^2 < (n-1)\frac{S^2}{\lambda_1}\right\} = 1 - \alpha.$$

Then
$$P(\sigma^2, \sigma'^2) = P_{\sigma^2}\left\{(n-1)\frac{S^2}{\lambda_2} < \sigma'^2 < (n-1)\frac{S^2}{\lambda_1}\right\}$$
$$= P\left\{\frac{T_\sigma}{\lambda_2} < \gamma < \frac{T_\sigma}{\lambda_1}\right\},$$

where $\gamma = \sigma'^2/\sigma^2$ and $T_\sigma \sim \chi^2(n-1)$. Thus
$$P(\gamma) = P\{\lambda_1 \gamma < T_\sigma < \lambda_2 \gamma\}.$$

Equations (2) and (3) become
$$P(1) = 1 - \alpha$$
and
$$P(\gamma) < 1 - \alpha.$$

Thus we need λ_1, λ_2 such that

(4) $$P(1) = 1 - \alpha$$

and

(5) $$\left.\frac{dP(\gamma)}{d\gamma}\right|_{\gamma=1} = \lambda_2 f_{n-1}(\lambda_2) - \lambda_1 f_{n-1}(\lambda_1) = 0,$$

where f_{n-1} is the pdf of T_σ. Equations (4) and (5) have been solved numerically for λ_1, λ_2 by several authors; see, for example, Tate and Klett [129]. Having obtained λ_1, λ_2 from (4) and (5), we have as the unbiased $1-\alpha$ level confidence interval

(6) $$\left((n-1)\frac{S^2}{\lambda_2}, (n-1)\frac{S^2}{\lambda_1}\right),$$

which, in view of Theorem 9.5.4, is also the UMAU confidence interval at level $1-\alpha$. To see this, one simply has to note that, in the notation of Theorem 9.5.4, the power function of a UMP unbiased test of $\sigma^2 = \sigma_0^2$, namely,
$$P_{\sigma^2}\left\{c_1 \sigma_0^2 < (n-1)S^2 < c_2 \sigma_0^2\right\}$$

has a minimum at $\sigma^2 = \sigma_0^2$, which leads to conditions (4) and (5). (Compare with the conditions on c_1, c_2 in Remark 9.5.4.) By Remark 2 it follows that (6) minimizes the expected length with respect to all families of $1-\alpha$ level unbiased confidence intervals uniformly in σ^2 ($\neq \sigma'^2$).

Note that in this case the shortest-length confidence interval (based on T_σ) derived in Example 11.3.3, the usual equal tails confidence interval, and (6) are all different. The length of the confidence interval (6), however, can be considerably greater than that of the shortest interval of Example 11.3.3. For large n all three sets of intervals are approximately the same.

PROBLEMS 11.5

1. Let $X_1, X_2, \cdots, X_n$ be a sample from $U(0, \theta)$. Show that the unbiased confidence interval for θ based on the pivot max X_i/θ, coincides with the shortest-length confidence interval based on the same pivot.

2. Let $X_1, X_2, \cdots, X_n$ be a sample from $G(1, \theta)$. Find the unbiased confidence interval for θ, based on the pivot $2 \sum_{i=1}^{n} X_i/\theta$.

3. Let $X_1, X_2, \cdots X_n$ be a sample from pdf
$$f_\theta(x) = \begin{cases} e^{-(x-\theta)} & \text{if } x > \theta \\ 0 & \text{otherwise,} \end{cases}$$
Find the unbiased confidence interval based on the pivot $2n[\min X_i - \theta]$.

11.6 BAYES CONFIDENCE INTERVALS

So far we have considered the problem of confidence estimation by regarding the parameter θ as a fixed unknown quantity. Another approach to the problem takes into account any prior knowledge that the experimenter has about the parameter. Let Θ be the parameter set, and let the observable rv $\mathbf{X}$ have pdf (pmf) $f_\theta(\mathbf{x})$. Suppose that we consider θ as an rv with distribution $\pi(\theta)$ on Θ. Then $f_\theta(\mathbf{x})$ can be considered as the conditional pdf (pmf) of $\mathbf{X}$, given that the rv θ takes the value θ. Note that we are using the same symbol for the rv θ and the values that it assumes. We can determine the joint distribution of $\mathbf{X}$ and θ, the marginal distribution of $\mathbf{X}$, and also the conditional distribution of θ, given $\mathbf{X} = \mathbf{x}$ as usual. Thus the joint distribution is given by

(1) $$f(\mathbf{x}, \theta) = \pi(\theta) f_\theta(\mathbf{x}),$$

and the marginal distribution of $\mathbf{X}$ by

(2) $$g(\mathbf{x}) = \begin{cases} \sum \pi(\theta) f_\theta(\mathbf{x}) & \text{if } \pi \text{ is a pmf,} \\ \int \pi(\theta) f_\theta(\mathbf{x}) \, d\theta & \text{if } \pi \text{ is a pdf.} \end{cases}$$

The conditional distribution of θ, given that $\mathbf{x}$ is observed, is given by

(3) $$h(\theta|\mathbf{x}) = \frac{\pi(\theta) f_\theta(\mathbf{x})}{g(\mathbf{x})}, \qquad g(\mathbf{x}) > 0.$$

Given $h(\theta|\mathbf{x})$, it is easy to find functions $l(\mathbf{x})$, $u(\mathbf{x})$ such that

BAYES CONFIDENCE INTERVALS 495

$$P\{l(\mathbf{X}) < \theta < u(\mathbf{X})\} \geq 1 - \alpha,$$

where

(4) $$P\{l(\mathbf{X}) < \theta < u(\mathbf{X}) | \mathbf{X} = \mathbf{x}\} = \begin{cases} \int_l^u h(\theta|\mathbf{x}) \, d\theta, \\ \sum_l^u h(\theta|\mathbf{x}), \end{cases}$$

depending on whether h is a pdf or a pmf.

Definition 1. An interval $(l(\mathbf{x}), u(\mathbf{x}))$ that has probability at least $1 - \alpha$ of including θ is called a $1 - \alpha$ level Bayes interval for θ. Also, $l(\mathbf{x})$ and $u(\mathbf{x})$ are called the lower and upper limits of the interval.

One can similarly define one-sided Bayes intervals or $1 - \alpha$ level lower and upper Bayes limits. It clear is that one chooses the shortest-length Bayesian confidence interval among all $1 - \alpha$ level confidence intervals.

Example 1. Let $X_1, X_2, \cdots, X_n$ be iid $\mathcal{N}(\mu, 1)$, $\mu \in \mathcal{R}$, and let the a priori distribution of μ be $\mathcal{N}(0, 1)$. Then from Example 8.8.6 we know that $h(\mu|\mathbf{x})$ is

$$\mathcal{N}\left(\frac{\sum_1^n x_i}{n+1}, \frac{1}{n+1}\right).$$

Thus a $1 - \alpha$ level Bayesian confidence interval is

$$\left(\frac{n\bar{x}}{n+1} - \frac{z_{\alpha/2}}{\sqrt{n+1}}, \frac{n\bar{x}}{n+1} + \frac{z_{\alpha/2}}{\sqrt{n+1}}\right)$$

A $1 - \alpha$ level confidence interval for μ (treating μ as fixed) is a random interval with value

$$\left(\bar{x} - \frac{z_{\alpha/2}}{\sqrt{n}}, \bar{x} + \frac{z_{\alpha/2}}{\sqrt{n}}\right).$$

Thus the Bayesian interval is somewhat shorter in length. This is to be expected since we assumed more in the Bayesian case.

Example 2. Let $X_1, X_2, \cdots, X_n$ be iid $b(1, p)$ rv's, and let the a priori pdf on $\Theta = (0, 1)$ be $U(0, 1)$. In Example 8.8.8. we saw that the a posteriori pdf of p, given $X_1 + X_2 + \cdots + X_n$, is given by

$$h(p | \sum_1^n x_i = t) = \begin{cases} [B(t+1, n-t+1)]^{-1} p^t (1-p)^{n-t}, & 0 < p < 1, \\ 0, & \text{otherwise.} \end{cases}$$

Given suitable tables of incomplete beta integrals and the observed value of t, one can easily construct a Bayesian interval estimate of p.

PROBLEMS 11.6

1. Let $X_1, X_2, \cdots, X_n$ be a sample from a Poisson distribution with unknown parameter λ. Assuming that λ is a value assumed by a $G(\alpha, \beta)$ rv, find a Bayesian confidence interval for λ.

2. Let $X_1, X_2, \cdots, X_n$ be a sample from a geometric distribution with parameter θ. Assuming that θ has a priori pdf that is given by the density of a $B(\alpha, \beta)$ rv, find a Bayesian confidence interval for θ.

3. Let $X_1, X_2, \cdots, X_n$ be a sample from $\mathcal{N}(\mu, 1)$, and suppose that the a priori pdf for μ is $U(-1, 1)$. Find a Bayesian confidence interval for μ.

CHAPTER 12

The General Linear Hypothesis

12.1 INTRODUCTION

The last three chapters of this book will be devoted to the study of some special topics in mathematical statistics. This chapter deals with the first of these, namely, the general linear hypothesis. In a wide variety of problems the experimenter is interested in making inferences about a vector parameter. For example, he may wish to estimate the mean of a multivariate normal or to test some hypotheses concerning the mean vector. The problem of estimation can be solved, for example, by resorting to the method of maximum likelihood estimation, discussed in Section 8.7, or the method of least squares, discussed in Section 4.8. In this chapter we restrict ourselves to the so-called linear model problems and concern ourselves mainly with problems of hypotheses testing.

In Section 2 we formally describe the general model and derive a test in complete generality. In the next four sections we demonstrate the power of this test by solving four important testing problems. We will need a considerable amount of linear algebra in Section 2, and the reader may want to refresh his memory by glancing through Section P.3.

12.2 THE GENERAL LINEAR HYPOTHESIS

A wide variety of problems of hypotheses testing can be treated under a general setup. In this section we state the general problem, and derive the test statistic and its distribution. Consider the following examples.

Example 1. Let $X_1, X_2, \cdots, X_k$ be independent rv's with $EX_i = \mu_i$, $i = 1, 2, \cdots, k$, and common variance σ^2. Also, n_i observations are taken on X_i, $i = 1, 2, \cdots, k$, and $\sum_{i=1}^{k} n_i = n$. It is required to test $H_0: \mu_1 = \mu_2 = \cdots$

$= \mu_k$. The case $k = 2$ has already been treated in Theorem 9.5.5 and Section 10.4. Problems of this nature arise quite naturally, for example, in agricultural experiments where one is interested in comparing the average yield when k fertilizers are available.

Example 2. An experimenter observes the velocity of a particle moving along a line. He takes observations at given times $t_1, t_2, \cdots, t_n$. Let β_1 be the initial velocity of the particle, and β_2 the acceleration; then the velocity at time t is given by $x = \beta_1 + \beta_2 t + \varepsilon$, where ε is an rv that is nonobservable (like an error in measurement). In practice, the experimenter does not know β_1 and β_2 and has to use the random observations $X_1, X_2, \cdots, X_n$ made at times $t_1, t_2, \cdots, t_n$, respectively, to obtain some information about the unknown parameters β_1, β_2.

A similar example is the case where the relation between x and t is governed by

$$x = \beta_0 + \beta_1 t + \beta_2 t^2 + \varepsilon,$$

where t is a mathematical variable, $\beta_0, \beta_1, \beta_2$ are unknown parameters, and ε is a nonobservable rv. The experimenter takes observations $X_1, X_2, \cdots, X_n$ at predetermined values $t_1, t_2, \cdots, t_n$, respectively, and is interested in testing the hypothesis that the relation is in fact linear, that is, $\beta_2 = 0$.

Examples of the type discussed above and their much more complicated variants can all be treated under a general setup. To fix ideas, let us first make the following definition.

Definition 1. Let $\mathbf{X} = (X_1, X_2, \cdots, X_n)$ be a random vector, and $\mathbf{A}$ be an $n \times k$ matrix, $k < n$, of known constants $a_{ij}, i = 1, 2, \cdots, n; j = 1, 2, \cdots, k$. We say that the distribution of $\mathbf{X}$ satisfies a linear model if

(1) $$E\mathbf{X} = \boldsymbol{\beta}\mathbf{A}',$$

where $\boldsymbol{\beta} = (\beta_1, \beta_2, \cdots, \beta_k)$ is a vector of unknown (scalar) parameters $\beta_1, \beta_2, \cdots, \beta_k$. It is convenient to write

(2) $$\mathbf{X} = \boldsymbol{\beta}\mathbf{A}' + \boldsymbol{\varepsilon},$$

where $\boldsymbol{\varepsilon} = (\varepsilon_1, \varepsilon_2, \cdots, \varepsilon_n)$ is a vector of nonobservable rv's with $E\varepsilon_j = 0$, $j = 1, 2, \cdots, n$. Relation (2) is known as a linear model. The general linear hypothesis concerns $\boldsymbol{\beta}$, namely, that $\boldsymbol{\beta}$ satisfies $H_0: \boldsymbol{\beta}\mathbf{H}' = \mathbf{0}$, where $\mathbf{H}$ is a known $r \times k$ matrix with $r \leq k$.

In what follows we will assume that $\varepsilon_1, \varepsilon_2, \cdots, \varepsilon_n$ are independent, normal rv's with common variance σ^2 and $E\varepsilon_j = 0, j = 1, 2, \cdots, n$. In view of (2), it follows that $X_1, X_2, \cdots, X_n$ are independent normal rv's with

(3) $$EX_i = \sum_{j=1}^{k} a_{ij}\beta_j \quad \text{and} \quad \text{var}(X_i) = \sigma^2, \quad i = 1, 2, \cdots, n.$$

We will assume that $\mathbf{H}$ is a matrix of full rank r, $r \leq k$, and $\mathbf{A}$ is a matrix of full rank $k < n$. Some remarks are in order.

Remark 1. Clearly $\mathbf{X}$ satisfies a linear model if the vector of means $E\mathbf{X} = (EX_1, EX_2, \cdots, EX_n)$ lies in a k-dimensional subspace generated by the linearly independent column vectors $\mathbf{a}'_1, \mathbf{a}'_2, \cdots, \mathbf{a}'_k$ of the matrix $\mathbf{A}$. Indeed, (1) states that $E\mathbf{X}$ is a linear combination of the known vectors $\mathbf{a}'_1, \cdots, \mathbf{a}'_k$. The general linear hypothesis $H_0: \boldsymbol{\beta}\mathbf{H}' = \mathbf{0}$ states that the parameters $\beta_1, \beta_2, \cdots, \beta_k$ satisfy r independent homogeneous linear restrictions. It follows that, under H_0, $E\mathbf{X}$ lies in a $(k - r)$-dimensional subspace of the k-space generated by $\mathbf{a}'_1, \cdots, \mathbf{a}'_k$ (see Section P.3).

Remark 2. The assumption of normality, which is conventional, is made to compute the likelihood ratio test statistic of H_0 and its distribution. If the problem is to estimate $\boldsymbol{\beta}$, no such assumption is needed. One can use the principle of least squares and estimate $\boldsymbol{\beta}$ by minimizing the sum of squares,

(4) $$\sum_{i=1}^{n} \varepsilon_i^2 = \boldsymbol{\varepsilon}\boldsymbol{\varepsilon}' = (\mathbf{X} - \boldsymbol{\beta}\mathbf{A}')(\mathbf{X} - \boldsymbol{\beta}\mathbf{A}')'.$$

The minimizing value $\hat{\boldsymbol{\beta}}(\mathbf{x})$ is known as a *least square estimate of $\boldsymbol{\beta}$*. This is not a difficult problem, and we will not discuss it here in any detail but will mention only that any solution of the so-called *normal equations*

(5) $$\mathbf{A}'\mathbf{A}\hat{\boldsymbol{\beta}}' = \mathbf{A}'\mathbf{X}'$$

is a least square estimate. If the rank of $\mathbf{A}$ is k $(< n)$, then $\mathbf{A}'\mathbf{A}$, which has the same rank as $\mathbf{A}$, is a nonsingular matrix that can be inverted to give a unique least square estimate

(6) $$\hat{\boldsymbol{\beta}}' = (\mathbf{A}'\mathbf{A})^{-1}\mathbf{A}'\mathbf{X}'.$$

If the rank of $\mathbf{A}$ is $< k$, then $\mathbf{A}'\mathbf{A}$ is singular and the normal equations do not have a unique solution. One can show, for example, that $\hat{\boldsymbol{\beta}}$ is unbiased for $\boldsymbol{\beta}$, and if the X_i's are uncorrelated with common variance σ^2, the variance-covariance matrix of the $\hat{\beta}_i$'s is given by

(7) $$E\{(\hat{\boldsymbol{\beta}} - \boldsymbol{\beta})'(\hat{\boldsymbol{\beta}} - \boldsymbol{\beta})\} = \sigma^2(\mathbf{A}'\mathbf{A})^{-1}.$$

Remark 3. One can similarly compute the so-called *restricted least square estimate* of $\boldsymbol{\beta}$ by the usual method of Lagrange multipliers. For example, under $H_0: \boldsymbol{\beta}\mathbf{H}' = \mathbf{0}$ one simply minimizes $(\mathbf{X} - \boldsymbol{\beta}\mathbf{A}')(\mathbf{X} - \boldsymbol{\beta}\mathbf{A}')'$ subject to $\boldsymbol{\beta}\mathbf{H}' = \mathbf{0}$ to get the restricted least square estimate $\hat{\boldsymbol{\beta}}$. The important point is that, if $\boldsymbol{\varepsilon}$ is assumed to be a multivariate normal rv with mean vector $\mathbf{0}$ and dispersion matrix $\sigma^2\,\mathbf{I}_n$, the MLE of $\boldsymbol{\beta}$ is the same as the least square estimate. In fact, one can show that $\hat{\beta}_i$ is the UMVUE of β_i, $i = 1, 2, \cdots, k$, by the usual methods.

Example 3. Suppose that a random variable X is linearly related to a mathematical variable a that is not random. (See Example 2.) Let $X_1, X_2, \cdots, X_n$ be observations made at different known values $a_1, a_2, \cdots, a_n$ of a. For example, $a_1, a_2, \cdots, a_n$ may represent different levels of a fertilizer, and $X_1, X_2, \cdots, X_n$, respectively, the corresponding yields of a crop. Also, $\varepsilon_1, \varepsilon_2, \cdots, \varepsilon_n$ represent unobservable rv's that may be errors of measurements. Then

$$X_i = \beta_0 + \beta_1 a_i + \varepsilon_i, \quad i = 1, 2, \cdots, n,$$

and we wish to test whether $\beta_1 = 0$, that the fertilizer levels do not affect the yield. Here

$$\mathbf{A} = \begin{pmatrix} 1 & a_1 \\ 1 & a_2 \\ \vdots & \vdots \\ 1 & a_n \end{pmatrix},$$

$$\boldsymbol{\beta} = (\beta_0, \beta_1), \quad \text{and} \quad \boldsymbol{\varepsilon} = (\varepsilon_1, \varepsilon_2, \cdots, \varepsilon_n).$$

The hypothesis to be tested is $H_0: \beta_1 = 0$ so that, with $\mathbf{H} = (0, 1)$, the null hypothesis can be written as $H_0: \mathbf{H}\boldsymbol{\beta}' = 0$. This is a problem of *linear regression*.

Similarly, we may assume that the regression of X on a is *quadratic*:

$$X = \beta_0 + \beta_1 a + \beta_2 a^2 + \varepsilon,$$

and we may wish to test that a linear function will be sufficient to describe the relationship, that is, $\beta_2 = 0$. Here $\mathbf{A}$ is the $n \times 3$ matrix

$$\mathbf{A} = \begin{pmatrix} 1 & a_1 & a_1^2 \\ 1 & a_2 & a_2^2 \\ \vdots & \vdots & \vdots \\ 1 & a_n & a_n^2 \end{pmatrix},$$

$$\boldsymbol{\beta} = (\beta_0, \beta_1, \beta_2), \quad \boldsymbol{\varepsilon} = (\varepsilon_1, \varepsilon_2, \cdots, \varepsilon_n),$$

and $\mathbf{H}$ is the 1×3 matrix $(0, 0, 1)$.

In another example of regression, the X's can be written as

$$X = \beta_1 a + \beta_2 b + \beta_3 c + \varepsilon,$$

and we wish to test the hypothesis that $\beta_1 = \beta_2 = \beta_3$. In this case, $\mathbf{A}$ is the matrix

$$\mathbf{A} = \begin{pmatrix} a_1 & b_1 & c_1 \\ a_2 & b_2 & c_2 \\ \vdots & \vdots & \vdots \\ a_n & b_n & c_n \end{pmatrix},$$

and $\mathbf{H}$ may be chosen to be the 2×3 matrix

$$\mathbf{H} = \begin{pmatrix} 1 & 0 & -1 \\ 1 & -1 & 0 \end{pmatrix}.$$

Example 4. Another important example of the general linear hypothesis involves the *analysis of variance*. We have already derived tests of hypotheses regarding the equality of the means of two normal populations when the variances are equal. In practice one is frequently interested in the equality of several means when the variances are the same, that is, one has k samples from $\mathcal{N}(\mu_1, \sigma^2), \cdots, \mathcal{N}(\mu_k, \sigma^2)$, where σ^2 is unknown and one wants to test $H_0: \mu_1 = \mu_2 = \cdots = \mu_k$. (See Example 1.) Such a situation is of common occurrence in agricultural experiments. Suppose that k treatments are applied to experimental units (plots), the ith treatment is applied to n_i randomly chosen units, $i = 1, 2, \cdots, k$, $\sum_{i=1}^{k} n_i = n$, and the observation x_{ij} represents some numerical characteristic (yield) of the jth experimental unit under the ith treatment. Suppose also that

$$X_{ij} = \mu_i + \varepsilon_{ij}, \quad j = 1, 2, \cdots, n_i; \quad i = 1, 2, \cdots, k,$$

where ε_{ij} are iid $\mathcal{N}(0, \sigma^2)$ rv's. We are interested in testing $H_0: \mu_1 = \mu_2 = \cdots = \mu_k$. We write

$$\mathbf{X} = (X_{11}, X_{12}, \cdots, X_{1n_1}, X_{21}, X_{22}, \cdots, X_{2n_2}, \cdots, X_{k_1}, X_{k_2}, \cdots, X_{kn_k}),$$

$$\boldsymbol{\beta} = (\mu_1, \mu_2, \cdots, \mu_k),$$

$$\mathbf{A} = \begin{pmatrix} \mathbf{1}'_{n_1} & \mathbf{0}' & \cdots & \mathbf{0}' \\ \mathbf{0}' & \mathbf{1}'_{n_2} & \cdots & \mathbf{0}' \\ \vdots & \vdots & \vdots & \vdots \\ \mathbf{0}' & \mathbf{0}' & \cdots & \mathbf{1}'_{n_k} \end{pmatrix},$$

were $\mathbf{1}_{n_i} = (1, 1, \cdots, 1)$ is the n_i-vector ($i = 1, 2, \cdots, k$), each of whose elements is unity. Thus $\mathbf{A}$ is $n \times k$. We can choose

$$\mathbf{H} = \begin{pmatrix} 1 & -1 & 0 & \cdots & 0 \\ 1 & 0 & -1 & \cdots & 0 \\ \vdots & \vdots & \vdots & \vdots & \vdots \\ 1 & 0 & 0 & \cdots & -1 \end{pmatrix}$$

so that $H_0: \mu_1 = \mu_2 = \cdots = \mu_k$ is of the form $\mathbf{H}\boldsymbol{\beta}' = \mathbf{0}'$. Here $\mathbf{H}$ is a $(k-1) \times k$ matrix.

The model described in this example is frequently referred to as a *one–way analysis of variance* model. This is a very simple example of an analysis of variance model. Note that the matrix **A** is of a very special type, namely, the elements of **A** are either 0 or 1. **A** is known as a *design matrix*.

Returning to our general model

$$\mathbf{X} = \boldsymbol{\beta}\mathbf{A}' + \boldsymbol{\varepsilon},$$

we wish to test the null hypothesis $H_0: \boldsymbol{\beta}\mathbf{H}' = \mathbf{0}$. We will compute the likelihood ratio test and the distribution of the test statistic. In order to do so, we assume that $\boldsymbol{\varepsilon}$ has a multivariate normal distribution with mean vector **0** and variance-covariance matrix $\sigma^2 \mathbf{I}_n$, where σ^2 is unknown and $\mathbf{I}_n$ is the $n \times n$ identity matrix. This means that **X** has an n-variate normal distribution with mean $\boldsymbol{\beta}\mathbf{A}'$ and dispersion matrix $\sigma^2 \mathbf{I}_n$ for some $\boldsymbol{\beta}$ and some σ^2, both unknown. Here the parameter space Θ is the set of $(k+1)$-tuples $(\boldsymbol{\beta}, \sigma^2)$ = $(\beta_1, \beta_2, \cdots, \beta_k, \sigma^2)$, and the joint pdf of the X's is given by

$$(8) \quad f_{\boldsymbol{\beta},\sigma^2}(x_1, x_2, \cdots, x_n) = \frac{1}{(2\pi)^{n/2}\sigma^n} \exp\left\{-\frac{1}{2\sigma^2} \sum_{i=1}^{n}(x_i - \beta_1 a_{i1} - \cdots - \beta_k a_{ik})^2\right\}$$

$$= \frac{1}{(2\pi)^{n/2}\sigma^n} \exp\left\{-\frac{1}{2\sigma^2}(\mathbf{x} - \boldsymbol{\beta}\mathbf{A}')(\mathbf{x} - \boldsymbol{\beta}\mathbf{A}')'\right\}.$$

Theorem 1. Consider the linear model

$$\mathbf{X} = \boldsymbol{\beta}\mathbf{A}' + \boldsymbol{\varepsilon},$$

where **A** is an $n \times k$ matrix $((a_{ij}))$, $i = 1, 2, \cdots, n$, $j = 1, 2, \cdots, k$, of known constants and full rank $k < n$, $\boldsymbol{\beta}$ is a vector of unknown parameters β_1, $\beta_2, \cdots, \beta_k$, and $\boldsymbol{\varepsilon} = (\varepsilon_1, \varepsilon_2, \cdots, \varepsilon_n)$ is a vector of nonobservable independent normal rv's with common variance σ^2 and mean $E\boldsymbol{\varepsilon} = \mathbf{0}$. The likelihood ratio test for testing the linear hypothesis $H_0: \boldsymbol{\beta}\mathbf{H}' = \mathbf{0}$, where **H** is an $r \times k$ matrix of full rank $r \leq k$, is to reject H_0 at level α if $F \geq F_0$, where $P_{H_0}\{F \geq F_0\} = \alpha$, and F is the rv given by

$$(9) \quad F = \frac{(\mathbf{X} - \hat{\hat{\boldsymbol{\beta}}}\mathbf{A}')(\mathbf{X} - \hat{\hat{\boldsymbol{\beta}}}\mathbf{A}')' - (\mathbf{X} - \hat{\boldsymbol{\beta}}\mathbf{A}')(\mathbf{X} - \hat{\boldsymbol{\beta}}\mathbf{A}')'}{(\mathbf{X} - \hat{\boldsymbol{\beta}}\mathbf{A}')(\mathbf{X} - \hat{\boldsymbol{\beta}}\mathbf{A}')'}.$$

In (9), $\hat{\boldsymbol{\beta}}, \hat{\hat{\boldsymbol{\beta}}}$ are the MLE's of $\boldsymbol{\beta}$ under Θ and Θ_0, respectively. Moreover, the rv $[(n-k)/r] F$ has the F-distribution with $(r, n-k)$ d.f. under H_0.

Proof. The likelihood ratio test of $H_0: \mathbf{H}\boldsymbol{\beta}' = \mathbf{0}'$ is to reject H_0 if and only if $\lambda(\mathbf{x}) < c$, where

$$(10) \quad \lambda(\mathbf{x}) = \frac{\sup_{\theta \in \Theta_0} f_{\boldsymbol{\beta},\sigma^2}(\mathbf{x})}{\sup_{\theta \in \Theta} f_{\boldsymbol{\beta},\sigma^2}(\mathbf{x})},$$

$\theta = (\beta, \sigma^2)$, and $\Theta_0 = \{(\beta, \sigma^2): \mathbf{H}\beta' = \mathbf{0}'\}$. Let $\hat{\theta} = (\hat{\beta}, \hat{\sigma}^2)$ be the MLE of $\theta \in \Theta$, and $\hat{\hat{\theta}} = (\hat{\hat{\beta}}, \hat{\hat{\sigma}}^2)$ be the MLE of θ under H_0, that is, when $\beta\mathbf{H}' = \mathbf{0}$. It is easily seen that $\hat{\beta}$ is the value of β that minimizes $(\mathbf{x} - \beta\mathbf{A}')(\mathbf{x} - \beta\mathbf{A}')'$, and

(11) $$\hat{\sigma}^2 = n^{-1}(\mathbf{x} - \hat{\beta}\mathbf{A}')(\mathbf{x} - \hat{\beta}\mathbf{A}')'.$$

Similarly, $\hat{\hat{\beta}}$ is the value of β that minimizes $(\mathbf{x} - \beta\mathbf{A}')(\mathbf{x} - \beta\mathbf{A}')'$ subject to $\beta\mathbf{H}' = \mathbf{0}$, and

(12) $$\hat{\hat{\sigma}}^2 = n^{-1}(\mathbf{x} - \hat{\hat{\beta}}\mathbf{A}')(\mathbf{x} - \hat{\hat{\beta}}\mathbf{A}')'.$$

It follows that

(13) $$\lambda(\mathbf{x}) = \left(\frac{\hat{\sigma}^2}{\hat{\hat{\sigma}}^2}\right)^{n/2}.$$

The critical region $\lambda(\mathbf{x}) < c$ is equivalent to the region $\{\lambda(\mathbf{x})\}^{-2/n} < \{c\}^{-2/n}$, which is of the form

(14) $$\frac{\hat{\hat{\sigma}}^2}{\hat{\sigma}^2} > c_1.$$

This may be written as

(15) $$\frac{(\mathbf{x} - \hat{\hat{\beta}}\mathbf{A}')(\mathbf{x} - \hat{\hat{\beta}}\mathbf{A}')'}{(\mathbf{x} - \hat{\beta}\mathbf{A}')(\mathbf{x} - \hat{\beta}\mathbf{A}')'} > c_1$$

or, equivalently, as

(16) $$\frac{(\mathbf{x} - \hat{\hat{\beta}}\mathbf{A}')(\mathbf{x} - \hat{\hat{\beta}}\mathbf{A}')' - (\mathbf{x} - \hat{\beta}\mathbf{A}')(\mathbf{x} - \hat{\beta}\mathbf{A}')'}{(\mathbf{x} - \hat{\beta}\mathbf{A}')(\mathbf{x} - \hat{\beta}\mathbf{A}')'} > c_1 - 1.$$

It remains to determine the distribution of the test statistic. For this purpose it is convenient to reduce the problem to the *canonical form*. Let V_n be the vector space of the observation vector $\mathbf{X}$, V_k be the subspace of V_n generated by the column vectors $\mathbf{a}'_1, \mathbf{a}'_2, \cdots, \mathbf{a}'_k$ of $\mathbf{A}$, and V_{k-r} be the subspace of V_k in which $E\mathbf{X}$ is postulated to lie under H_0. We change variables from $X_1, X_2, \cdots, X_n$ to $Z_1, Z_2, \cdots, Z_n$, where $Z_1, Z_2, \cdots, Z_n$ are independent normal rv's with common variance σ^2 and means $EZ_i = \theta_i$, $i = 1, 2, \cdots, k$, $EZ_i = 0$, $i = k+1, \cdots, n$. This is done as follows. Let us choose an orthonormal basis of $k - r$ column vectors $\{\alpha'_i\}$ for V_{k-r}, say $\{\alpha'_{r+1}, \alpha'_{r+2}, \cdots, \alpha'_k\}$. We extend this to an orthonormal basis $\{\alpha'_1, \alpha'_2, \cdots, \alpha'_r, \alpha'_{r+1}, \cdots, \alpha'_k\}$ for V_k, and then extend once again to an orthonormal basis $\{\alpha'_1, \alpha'_2, \cdots, \alpha'_k, \alpha'_{k+1}, \cdots, \alpha'_n\}$ for V_n. This is always possible. (See Theorems P.3.2 and P.3.3.)

Let $z_1, z_2, \cdots, z_n$ be the coordinates of $\mathbf{x}$ relative to the basis

$\{\boldsymbol{\alpha}_1', \boldsymbol{\alpha}_2', \cdots, \boldsymbol{\alpha}_n'\}$. Then $z_i = \mathbf{x}\boldsymbol{\alpha}_i'$ and $\mathbf{z} = \mathbf{x}\mathbf{P}'$, where $\mathbf{P}$ is an orthogonal matrix with ith row $\boldsymbol{\alpha}_i$. Thus $EZ_i = E\mathbf{X}\,\boldsymbol{\alpha}_i' = \boldsymbol{\beta}\mathbf{A}'\,\boldsymbol{\alpha}_i'$, and $E\mathbf{Z} = \boldsymbol{\beta}\mathbf{A}'\mathbf{P}'$. Since $\boldsymbol{\beta}\mathbf{A}' \in V_k$ (Remark 1), it follows that $\boldsymbol{\beta}\mathbf{A}'\boldsymbol{\alpha}_i' = 0$ for $i > k$. Similarly, under H_0, $\boldsymbol{\beta}\mathbf{A}' \in V_{k-r} \subset V_k$, so that $\boldsymbol{\beta}\mathbf{A}'\boldsymbol{\alpha}_i' = 0$ for $i \le r$. Let us write $\boldsymbol{\omega} = \boldsymbol{\beta}\mathbf{A}'\mathbf{P}'$. Then $\omega_{k+1} = \omega_{k+2} = \cdots = \omega_n = 0$, and, under H_0, $\omega_1 = \omega_2 = \cdots = \omega_r = 0$. Finally, from Corollary 2 of Theorem 5.4.6 it follows that $Z_1, Z_2, \cdots, Z_n$ are independent normal rv's with the same variance σ^2 and $EZ_i = \omega_i$, $i = 1, 2, \cdots, n$. We have thus transformed the problem to the following simpler canonical form:

(17) $$\begin{cases} \Omega: & Z_i \text{ are independent } \mathcal{N}(\omega_i, \sigma^2),\ i = 1, 2, \cdots, n, \\ & \omega_{k+1} = \omega_{k+2} = \cdots = \omega_n = 0, \\ H_0: & \omega_1 = \omega_2 = \cdots = \omega_r = 0. \end{cases}$$

Now

(18) $$\begin{aligned}(\mathbf{x} - \boldsymbol{\beta}\mathbf{A}')(\mathbf{x} - \boldsymbol{\beta}\mathbf{A}')' &= (\mathbf{z}\mathbf{P} - \boldsymbol{\omega}\mathbf{P})(\mathbf{z}\mathbf{P} - \boldsymbol{\omega}\mathbf{P})' \\ &= (\mathbf{z} - \boldsymbol{\omega})(\mathbf{z} - \boldsymbol{\omega})' \\ &= \sum_{i=1}^{k} (z_i - \omega_i)^2 + \sum_{i=k+1}^{n} z_i^2. \end{aligned}$$

The quantity $(\mathbf{x} - \boldsymbol{\beta}\mathbf{A}')(\mathbf{x} - \boldsymbol{\beta}\mathbf{A}')'$ is minimized if we choose $\hat{\omega}_i = z_i$, $i = 1, 2, \cdots, k$, so that

(19) $$(\mathbf{x} - \hat{\boldsymbol{\beta}}\mathbf{A}')(\mathbf{x} - \hat{\boldsymbol{\beta}}\mathbf{A}')' = \sum_{i=k+1}^{n} z_i^2.$$

Under H_0, $\omega_1 = \omega_2 = \cdots = \omega_r = 0$, so that $(\mathbf{x} - \boldsymbol{\beta}\mathbf{A}')(\mathbf{x} - \boldsymbol{\beta}\mathbf{A}')'$ will be minimized if we choose $\hat{\hat{\omega}}_i = z_i$, $i = r + 1, \cdots, k$. Thus

(20) $$(\mathbf{x} - \hat{\hat{\boldsymbol{\beta}}}\mathbf{A}')(\mathbf{x} - \hat{\hat{\boldsymbol{\beta}}}\mathbf{A}')' = \sum_{i=1}^{r} z_i^2 + \sum_{i=k+1}^{n} z_i^2.$$

It follows that

$$F = \frac{\sum_{i=1}^{r} Z_i^2}{\sum_{i=k+1}^{n} Z_i^2}.$$

Now $\sum_{i=k+1}^{n} Z_i^2/\sigma^2$ has a $\chi^2(n - k)$ distribution, and, under H_0, $\sum_{i=1}^{r} Z_i^2/\sigma^2$ has a $\chi^2(r)$ distribution. Since $\sum_{i=1}^{r} Z_i^2$ and $\sum_{i=k+1}^{n} Z_i^2$ are independent, we see that $[(n - k)/r]\,F$ is distributed as $F(r, n - k)$ under H_0, as asserted. This completes the proof of the theorem.

Remark 4. In practice, one does not need to find a transformation that reduces the problem to the canonical form. As will be done in the following sections, one simply computes the estimates $\hat{\boldsymbol{\theta}}$ and $\hat{\hat{\boldsymbol{\theta}}}$ and then computes the test statistic in any of the equivalent forms (14), (15), or (16) to apply the F-test.

Remark 5. The computation of $\hat{\beta}$, $\hat{\hat{\beta}}$ is greatly facilitated, in view of Remark 3, by using the principle of least squares. Indeed, this was done in the proof of Theorem 1 when we reduced the problem of maximum likelihood estimation to that of minimization of sum of squares $(\mathbf{x} - \beta \mathbf{A}')(\mathbf{x} - \beta \mathbf{A}')'$.

Remark 6. The distribution of the test statistic under H_1 is easily determined. We note that $Z_i/\sigma \sim \mathcal{N}(\omega_i/\sigma, 1)$ for $i = 1, 2, \cdots, r$, so that $\sum_{i=1}^{r} Z_i^2/\sigma^2$ has a noncentral chi-square distribution with r d.f. and noncentrality parameter $\delta = \sum_{i=1}^{r} \omega_i^2/\sigma^2$. It follows that $[(n - k)/r] F$ has a noncentral F-distribution with d.f. $(r, n - k)$ and noncentrality parameter δ. Under H_0, $\delta = 0$, so that $[(n - k)/r] F$ has a central $F(r, n - k)$ distribution. Since $\sum_{i=1}^{r} \omega_i^2 = \sum_{i=1}^{r} (EZ_i)^2$, it follows from (19) and (20) that, if we replace each observation X_i by its expected value in the numerator of (16), we get $\sigma^2 \delta$.

Remark 7. The general linear hypothesis makes use of the assumption of common variance. For instance, in Example 4, $X_{ij} \sim \mathcal{N}(\mu_i, \sigma^2)$, $j = 1, 2, \cdots, k$. Let us suppose that $X_{ij} \sim \mathcal{N}(\mu_i, \sigma_i^2)$, $i = 1, 2, \cdots, k$. Then we need to test that $\sigma_1 = \sigma_2 = \cdots = \sigma_k$ before we can apply Theorem 1. The case $k = 2$ has already been considered in Theorem 9.5.6. For the case where $k > 2$ one can show that a UMP unbiased test does not exist. A large-sample approximation is described in Lehmann [70], pages 273-275. It is beyond the scope of this book to consider the effects of departures from the underlying assumptions. We refer the reader to Scheffé [111], Chapter 10, for a discussion of this topic.

PROBLEMS 12.2

1. Show that any solution of the normal equations (5) minimizes the sum of squares $(\mathbf{X} - \beta \mathbf{A}')(\mathbf{X} - \beta \mathbf{A}')'$.

2. Show that the least square estimate given in (6) is an unbiased estimate of β. If the rv's X_i are uncorrelated with common variance σ^2, show that the covariance matrix of the $\hat{\beta}_i$'s is given by (7).

3. Under the assumption that ε [in model (2)] has a multivariate normal distribution with mean 0 and dispersion matrix $\sigma^2 I_n$ show that the least square estimates and the MLE's of β coincide.

4. Prove statements (11) and (12).

12.3 THE REGRESSION MODEL

In this section we consider a simple *linear regression model* as a special case of the general linear hypothesis and show how some inferential questions about the parameters of the regression equation can be answered. Let $t_1, t_2, \cdots, t_n$ be n given numbers, and suppose that

(1) $$X_i = \alpha_0 + \alpha_1 t_i + \varepsilon_i, \qquad i = 1, 2, \cdots, n,$$

where α_0, α_1 are unknown parameters, and ε_i are independent normal rv's with $E\varepsilon_i = 0$ and $\text{var}(\varepsilon_i) = \sigma^2$, $i = 1, 2, \cdots, n$. Also, σ^2 is assumed to be unknown. Our object is to test hypotheses concerning α_0 and α_1 and to construct confidence intervals for α_0 and α_1. Rewriting (1) in the usual fashion, we have

(2) $$\mathbf{X} = \boldsymbol{\beta}\mathbf{A}' + \boldsymbol{\varepsilon},$$

where

$$\boldsymbol{\beta} = (\alpha_0, \alpha_1) \quad \text{and} \quad \mathbf{A} = \begin{pmatrix} 1 & t_1 \\ 1 & t_2 \\ \vdots & \vdots \\ 1 & t_n \end{pmatrix}.$$

Clearly, $X_1, X_2, \cdots, X_n$ are independent normal rv's with $EX_i = \alpha_0 + \alpha_1 t_i$ and $\text{var}(X_i) = \sigma^2$, $i = 1, 2, \cdots, n$, and $\mathbf{X}$ is an n-variate normal random vector with mean $\boldsymbol{\beta}\mathbf{A}'$ and variance $\sigma^2 \mathbf{I}_n$. The joint pdf of $\mathbf{X}$ is given by

(3) $$f(\mathbf{x}; \alpha_0, \alpha_1, \sigma^2) = \frac{1}{(2\pi)^{n/2}} \frac{1}{\sigma^n} \exp\left\{-\frac{1}{2\sigma^2} \sum_{i=1}^{n} (x_i - \alpha_0 - \alpha_1 t_i)^2\right\}.$$

It easily follows that the MLE's for α_0, α_1, and σ^2 are given by

(4) $$\hat{\alpha}_0 = \frac{\sum_{i=1}^{n} X_i}{n} - \hat{\alpha}_1 \bar{t},$$

(5) $$\hat{\alpha}_1 = \frac{\sum_{i=1}^{n} (t_i - \bar{t})(X_i - \bar{X})}{\sum_{i=1}^{n} (t_i - \bar{t})^2},$$

and

(6) $$\hat{\sigma}^2 = \frac{1}{n} \sum_{i=1}^{n} (X_i - \hat{\alpha}_0 - \hat{\alpha}_1 t_i)^2,$$

where $\bar{t} = n^{-1} \sum_{i=1}^{n} t_i$.

If we wish to test $H_0: \alpha_1 = 0$, we take $\mathbf{H} = (0, 1)$, so that the model is a special case of the general linear hypothesis with $k = 2$, $r = 1$. Under H_0 the MLE's are

(7)
$$\hat{\hat{\alpha}}_0 = \bar{X} = \frac{\sum_{i=1}^{n} X_i}{n}$$

and

(8)
$$\hat{\hat{\sigma}}^2 = \frac{1}{n} \sum_{i=1}^{n} (X_i - \bar{X})^2.$$

Thus

(9)
$$F = \frac{\sum_{i=1}^{n} (X_i - \bar{X})^2 - \sum_{i=1}^{n} (X_i - \bar{X} + \hat{\alpha}_1 \bar{t} - \hat{\alpha}_1 t_i)^2}{\sum_{i=1}^{n} (X_i - \bar{X} + \hat{\alpha}_1 \bar{t} - \hat{\alpha}_1 t_i)^2}$$

$$= \frac{\hat{\alpha}_1^2 \sum_{i=1}^{n} (t_i - \bar{t})^2}{\sum_{i=1}^{n} (X_i - \bar{X} + \hat{\alpha}_1 \bar{t} - \hat{\alpha}_1 t_i)^2}.$$

From Theorem 12.2.1, the statistic $[(n - 2)/1] F$ has a central $F(1, n - 2)$ distribution under H_0. Since $F(1, n - 2)$ is the square of a $t(n-2)$ rv, the likelihood ratio test rejects H_0 if

(10)
$$|\hat{\alpha}_1| \left\{ \frac{(n - 2) \sum_{i=1}^{n} (t_i - \bar{t})^2}{\sum_{i=1}^{n} (X_i - \bar{X} + \hat{\alpha}_1 \bar{t} - \hat{\alpha}_1 t_i)^2} \right\}^{1/2} > c_0,$$

where c_0 is computed from t-tables for $n - 2$ d.f.

For testing $H_0: \alpha_0 = 0$, we choose $\mathbf{H} = (1, 0)$ so that the model is again a special case of the general linear hypothesis. In this case

$$\hat{\hat{\alpha}}_1 = \frac{\sum_{i=1}^{n} t_i X_i}{\sum_{i=1}^{n} t_i^2}$$

and

(11)
$$\hat{\hat{\sigma}}^2 = \frac{1}{n} \sum_{i=1}^{n} (X_i - \hat{\hat{\alpha}}_1 t_i)^2.$$

It follows that

(12)
$$F = \frac{\sum_{i=1}^{n} (X_i - \hat{\hat{\alpha}}_1 t_i)^2 - \sum_{i=1}^{n} (X_i - \bar{X} + \hat{\alpha}_1 \bar{t} - \hat{\alpha}_1 t_i)^2}{\sum_{i=1}^{n} (X_i - \bar{X} + \hat{\alpha}_1 \bar{t} - \hat{\alpha}_1 t_i)^2},$$

and since

$$\hat{\hat{\alpha}}_1 = \frac{\sum_{i=1}^{n} t_i X_i}{\sum_{i=1}^{n} t_i^2} = \frac{\sum_{i=1}^{n} (t_i - \bar{t})(X_i - \bar{X}) + n\bar{t}\bar{X}}{\sum_{i=1}^{n} t_i^2}$$

$$= \frac{\hat{\alpha}_1 \sum_{i=1}^{n} (t_i - \bar{t})^2 + n\bar{t}(\hat{\alpha}_0 + \hat{\alpha}_1 \bar{t})}{\sum_{i=1}^{n} t_i^2}$$

(13)
$$= \hat{\alpha}_1 + \frac{n\hat{\alpha}_0 \bar{t}}{\sum_{i=1}^{n} t_i^2},$$

we can write the numerator of F as

$$\sum_{i=1}^{n} (X_i - \hat{\hat{\alpha}}_1 t_i)^2 - \sum_{i=1}^{n} (X_i - \bar{X} + \hat{\alpha}_1 \bar{t} - \hat{\alpha}_1 t_i)^2$$

$$= \sum_{i=1}^{n} \left(X_i - \hat{\alpha}_1 t_i + \hat{\alpha}_1 \bar{t} - \bar{X} + \bar{X} - \hat{\alpha}_1 \bar{t} - \frac{n\hat{\alpha}_0 \bar{t} t_i}{\sum_{i=1}^{n} t_i^2} \right)^2$$

$$- \sum_{i=1}^{n} (X_i - \bar{X} + \hat{\alpha}_1 \bar{t} - \hat{\alpha}_1 t_i)^2$$

$$= \sum_{i=1}^{n} \left(\bar{X} - \hat{\alpha}_1 \bar{t} - \frac{n\hat{\alpha}_0 \bar{t} t_i}{\sum_{i=1}^{n} t_i^2} \right)^2 + 2 \sum_{i=1}^{n} (X_i - \hat{\alpha}_1 t_i + \hat{\alpha}_1 \bar{t} - \bar{X})$$

$$\cdot \left(\bar{X} - \hat{\alpha}_1 \bar{t} - \frac{n\hat{\alpha}_0 \bar{t} t_i}{\sum_{i=1}^{n} t_i^2} \right)$$

(14)
$$= \frac{\hat{\alpha}_0^2 n \sum_{i=1}^{n} (t_i - \bar{t})^2}{\sum_{i=1}^{n} t_i^2}.$$

It follows from Theorem 12.2.1 that the statistic

(15)
$$\frac{\hat{\alpha}_0 \sqrt{n \sum_{i=1}^{n}(t_i - \bar{t})^2 / \sum_{i=1}^{n} t_i^2}}{\sqrt{\sum_{i=1}^{n}(X_i - \bar{X} + \hat{\alpha}_1 \bar{t} - \hat{\alpha}_1 t_i)^2/(n-2)}}$$

has a central t-distribution with $n-2$ d.f. under $H_0: \alpha_0 = 0$. The rejection region is therefore given by

(16)
$$\frac{|\hat{\alpha}_0| \sqrt{n \sum_{i=1}^{n}(t_i - \bar{t})^2 / \sum_{i=1}^{n} t_i^2}}{\sqrt{\sum_{i=1}^{n}(X_i - \hat{\alpha}_0 - \hat{\alpha}_1 t_i)^2/(n-2)}} > c_0,$$

where c_0 is determined from the tables of $t(n-2)$ distribution for a given level of significance α.

For testing $H_0: \alpha_0 = \alpha_1 = 0$, we choose $\mathbf{H} = \begin{pmatrix} 1 & 0 \\ 0 & 1 \end{pmatrix}$, so that the model is again a special case of the general linear hypothesis with $r = 2$. In this case

(17) $$\hat{\sigma}^2 = \frac{1}{n}\sum_{i=1}^{n} X_i^2$$

and

(18) $$F = \frac{\sum_{i=1}^{n} X_i^2 - \sum_{i=1}^{n}(X_i - \bar{X} + \hat{\alpha}_1 \bar{t} - \hat{\alpha}_1 t_i)^2}{\sum_{i=1}^{n}(X_i - \bar{X} + \hat{\alpha}_1 \bar{t} - \hat{\alpha}_1 t_i)^2}$$

$$= \frac{n\bar{X}^2 + \hat{\alpha}_1^2 \sum_{i=1}^{n}(t_i - \bar{t})^2}{\sum_{i=1}^{n}(X_i - \hat{\alpha}_0 - \hat{\alpha}_1 t_i)^2}$$

$$= \frac{n(\hat{\alpha}_0 + \hat{\alpha}_1 \bar{t})^2 + \hat{\alpha}_1^2 \sum_{i=1}^{n}(t_i - \bar{t})^2}{\sum_{i=1}^{n}(X_i - \hat{\alpha}_0 - \hat{\alpha}_1 t_i)^2}.$$

From Theorem 12.2.1, the statistic $[(n - 2)/2] F$ has a central $F(2, n-2)$ distribution under $H_0: \alpha_0 = \alpha_1 = 0$. It follows that the level α rejection region for H_0 is given by

(19) $$\frac{n-2}{2} F > c_0,$$

where F is given by (18), and c_0 is the upper α percent point under the $F(2, n-2)$ distribution.

Remark 1. It is quite easy to modify the analysis above to obtain tests of null hypotheses $\alpha_0 = \alpha_0'$, $\alpha_1 = \alpha_1'$, and $(\alpha_0, \alpha_1) = (\alpha_0', \alpha_1')$, where α_0', α_1' are given real numbers (Problem 4).

Remark 2. The confidence intervals for α_0, α_1 are also easily obtained. One can show that a $1 - \alpha$ level confidence interval for α_0 is given by

(20) $$\left(\hat{\alpha}_0 - t_{n-2, \alpha/2} \sqrt{\frac{\sum_{i=1}^{n} t_i^2 \sum_{i=1}^{n}(X_i - \hat{\alpha}_0 - \hat{\alpha}_1 t_i)^2}{n(n-2)\sum_{i=1}^{n}(t_i - \bar{t})^2}},\right.$$

$$\left.\hat{\alpha}_0 + t_{n-2, \alpha/2} \sqrt{\frac{\sum_{i=1}^{n} t_i^2 \sum_{i=1}^{n}(X_i - \hat{\alpha}_0 - \hat{\alpha}_1 t_i)^2}{n(n-2)\sum_{i=1}^{n}(t_i - \bar{t})^2}}\right),$$

and that for α_1 is given by

$$\left(\hat{\alpha}_1 - t_{n-2,\alpha/2}\sqrt{\frac{\sum_{i=1}^{n}(X_i - \hat{\alpha}_0 - \hat{\alpha}_1 t_i)^2}{(n-2)\sum_{i=1}^{n}(t_i - \bar{t})^2}},\right.$$

(21)

$$\left. \hat{\alpha}_1 + t_{n-2,\alpha/2}\sqrt{\frac{\sum_{i=1}^{n}(X_i - \hat{\alpha}_0 - \hat{\alpha}_1 t_i)^2}{(n-2)\sum_{i=1}^{n}(t_i - \bar{t})^2}}\right).$$

Similarly, one can obtain confidence sets for (α_0, α_1) from the likelihood ratio test of $(\alpha_0, \alpha_1) = (\alpha_0', \alpha_1')$. It can be shown that the collection of sets of points (α_0, α_1) satisfying

(22)
$$\frac{(n-2)[n(\hat{\alpha}_0 - \alpha_0)^2 + 2n\bar{t}(\hat{\alpha}_0 - \alpha_0)(\hat{\alpha}_1 - \alpha_1) + \sum_{i=1}^{n} t_i^2(\hat{\alpha}_1 - \alpha_1)^2]}{2\sum_{i=1}^{n}(X_i - \hat{\alpha}_0 - \hat{\alpha}_1 t_i)^2}$$

$$\leq F_{2,n-2,\alpha}$$

is a $1 - \alpha$ level collection of confidence sets (ellipsoids) for (α_0, α_1) centered at $(\hat{\alpha}_0, \hat{\alpha}_1)$.

Remark 3. Sometimes interest lies in constructing a confidence interval on the unknown linear regression function $E\{X|t_0\} = \alpha_0 + \alpha_1 t_0$ for a given value of t, or on a value of X, given $t = t_0$. We assume that t_0 is a value of t distinct from $t_1, t_2, \cdots, t_n$. Clearly $\hat{\alpha}_0 + \hat{\alpha}_1 t_0$ is the maximum likelihood estimate of $\alpha_0 + \alpha_1 t_0$. This is also the best linear unbiased estimate. Let us write $\hat{E}\{X|t_0\} = \hat{\alpha}_0 + \hat{\alpha}_1 t_0$. Then

$$\hat{E}\{X|t_0\} = \bar{X} - \hat{\alpha}_1 \bar{t} + \hat{\alpha}_1 t_0$$

$$= \bar{X} + (t_0 - \bar{t})\frac{\sum_{i=1}^{n}(t_i - \bar{t})(X_i - \bar{X})}{\sum_{i=1}^{n}(t_i - \bar{t})^2},$$

which is clearly a linear function of normal rv's X_i. It follows that $\hat{E}\{X|t_0\}$ is also normally distributed with mean $E(\hat{\alpha}_0 + \hat{\alpha}_1 t_0) = \alpha_0 + \alpha_1 t_0$ and variance

(23)
$$\begin{aligned}\text{var}\{\hat{E}\{X|t_0\}\} &= E\{\hat{\alpha}_0 - \alpha_0 + \hat{\alpha}_1 t_0 - \alpha_1 t_0\}^2 \\ &= \text{var}(\hat{\alpha}_0) + t_0^2 \text{var}(\hat{\alpha}_1) + 2t_0 \text{cov}(\hat{\alpha}_0, \hat{\alpha}_1) \\ &= \sigma^2\left[\frac{1}{n} + \frac{(\bar{t} - t_0)^2}{\sum_{i=1}^{n}(t_i - \bar{t})^2}\right].\end{aligned}$$

(See Problem 6.) It follows that

(24)
$$\frac{\hat{\alpha}_0 + \hat{\alpha}_1 t_0 - \alpha_0 - \alpha_1 t_0}{\sigma\{(1/n) + [(\bar{t} - t_0)^2/\sum_{i=1}^{n}(t_i - \bar{t})^2]\}^{1/2}}$$

is $\mathcal{N}(0, 1)$. But σ is not known, so that we cannot use (24) to construct a confidence interval for $E\{X|t_0\}$. Since $n\hat{\sigma}^2/\sigma^2$ is a $\chi^2(n-2)$ rv and $n\hat{\sigma}^2/\sigma^2$ is independent of $\hat{\alpha}_0 + \hat{\alpha}_1 t_0$ (why?), it follows that

(25) $$\sqrt{n-2}\,\frac{\hat{\alpha}_0 + \hat{\alpha}_1 t_0 - \alpha_0 - \alpha_1 t_0}{\hat{\sigma}\{1 + n[(\bar{t} - t_0)^2/\sum_{i=1}^{n}(t_i - \bar{t})^2]\}^{1/2}}$$

has a $t(n-2)$ distribution. Thus a $1 - \alpha$ level confidence interval for $\alpha_0 + \alpha_1 t_0$ is given by

(26) $$\left(\hat{\alpha}_0 + \hat{\alpha}_1 t_0 - t_{n-2,\alpha/2}\,\hat{\sigma}\sqrt{\frac{n}{n-2}\left[\frac{1}{n} + \frac{(\bar{t} - t_0)^2}{\sum_{i=1}^{n}(t_i - \bar{t})^2}\right]},\right.$$
$$\left.\hat{\alpha}_0 + \hat{\alpha}_1 t_0 + t_{n-2,\alpha/2}\,\hat{\sigma}\sqrt{\frac{n}{n-2}\left[\frac{1}{n} + \frac{(\bar{t} - t_0)^2}{\sum_{i=1}^{n}(t_i - \bar{t})^2}\right]}\right).$$

In a similar manner one can show (Problem 7) that

(27) $$\left(\hat{\alpha}_0 + \hat{\alpha}_1 t_0 - t_{n-2,\alpha/2}\,\hat{\sigma}\sqrt{\frac{n}{n-2}\left[\frac{n+1}{n} + \frac{(\bar{t} - t_0)^2}{\sum_{i=1}^{n}(t_i - \bar{t})^2}\right]},\right.$$
$$\left.\hat{\alpha}_0 + \hat{\alpha}_1 t_0 + t_{n-2,\alpha/2}\,\hat{\sigma}\sqrt{\frac{n}{n-2}\left[\frac{n+1}{n} + \frac{(\bar{t} - t_0)^2}{\sum_{i=1}^{n}(t_i - \bar{t})^2}\right]}\right)$$

is a $1 - \alpha$ level confidence interval for $x_0 = \alpha_0 + \alpha_1 t_0 + \varepsilon$, that is, for the estimated value x_0 of X at t_0.

Remark 4. The simple regression model (2) considered above can be generalized in many directions. Thus we may consider EX as a polynomial in t_i of a degree higher than 1, or we may regard EX as a function of several variables. Some of these generalizations will be taken up in the problems.

Remark 5. Let $(X_1, Y_1), (X_2, Y_2), \cdots, (X_n, Y_n)$ be a sample from a bivariate normal population with parameters $EX = \mu_1$, $EY = \mu_2$, var $(X) = \sigma_1^2$, var $(Y) = \sigma_2^2$, and cov $(X, Y) = \rho$. In Section 7.6 we computed the pdf of the sample correlation coefficient R and showed (Remark 7.6.4) that the statistic

(28) $$T = R\sqrt{\frac{n-2}{1-R^2}}$$

has a $t(n-2)$ distribution, provided that $\rho = 0$. If we wish to test $\rho = 0$, that is, the independence of two normal rv's, we can base a test on the statistic T. Essentially, we are testing that the population covariance is 0, which implies that the population regression coefficients are 0. Thus we are testing, in particular, that $\alpha_1 = 0$. It is therefore not surprising that (28) is identical with (10). We emphasize that we derived (28) for a bivariate normal population, but (10) was derived by taking the t's (X's) as fixed and the distribution of X's (Y's) as normal. Note that for a bivariate normal populaion $E\{Y|x\} = \mu_2 + \rho(\sigma_2/\sigma_1)(x - \mu_1)$ is linear, in consistency with our model (1) or (2).

Example 1. Let us assume that the following data satisfy a linear regression model:

$$X_i = \alpha_0 + \alpha_1 t_i + \varepsilon_i.$$

t:	0	1	2	3	4	5
x:	.475	1.007	.838	−.618	1.378	.943

Let us test the null hypothesis that $\alpha_1 = 0$. We have

$$\bar{t} = 2.5, \quad \sum_{i=0}^{5}(t_i - \bar{t})^2 = 17.5, \quad \bar{x} = .671,$$

$$\sum_{i=0}^{5}(t_i - \bar{t})(x_i - \bar{x}) = .9985,$$

$$\hat{\alpha}_1 = .0571, \quad \hat{\alpha}_0 = \bar{x} - \hat{\alpha}_1 \bar{t} = .5279,$$

$$\sum_{i=0}^{5}(x_i - \hat{\alpha}_0 - \hat{\alpha}_1 t_i)^2 = 2.3571,$$

and

$$|\hat{\alpha}_1|\sqrt{\frac{(n-2)\sum(t_i - \bar{t})^2}{\sum(x_i - \hat{\alpha}_0 - \hat{\alpha}_1 t_i)^2}} = .3106.$$

Since $t_{n-2,\alpha/2} = t_{4,.025} = 2.776 > .3106$, we accept H_0 at level $\alpha = .05$.

Let us next find a 95 percent confidence interval for $E\{X|t = 7\}$. This is given by (26). We have

$$t_{n-2,\alpha/2}\,\hat{\sigma}\sqrt{\frac{n}{n-2}\left[\frac{1}{n} + \frac{(\bar{t} - t_0)^2}{\sum(t_i - \bar{t})^2}\right]} = 2.776\sqrt{\frac{2.3571}{6}}\sqrt{\frac{6}{4}\left(\frac{1}{6} + \frac{20.25}{17.5}\right)}$$

$$= 2.3707,$$

$$\hat{\alpha}_0 + \hat{\alpha}_1 t_0 = .5279 + .0571 \times 7$$
$$= .9276,$$

so that the 95 percent confidence interval is $(-1.4431, 3.2983)$.

(The data were produced from Table 6, page 659, of random numbers with $\mu = 0$, $\sigma = 1$, by letting $\alpha_0 = 1$ and $\alpha_1 = 0$ so that $E\{X|t\} = \alpha_0 + \alpha_1 t = 1$, which surely lies in the interval.)

PROBLEMS 12.3

1. Prove statements (4), (5), and (6).
2. Prove statements (7) and (8).
3. Prove statement (11).
4. Obtain tests of null hypotheses $\alpha_0 = \alpha_0'$, $\alpha_1 = \alpha_1'$, and $(\alpha_0, \alpha_1) = (\alpha_0', \alpha_1')$, where α_0', α_1' are given real numbers.
5. Obtain the confidence intervals for α_0 and α_1 as given in (20) and (21), respectively.

ONE-WAY ANALYSIS OF VARIANCE

6. Derive the expression for var $\{\hat{E}\{X|t_0\}\}$ as given in (23).
7. Show that the interval given in (27) is a $1-\alpha$ level confidence interval for $x_0 = \alpha_0 + \alpha_1 t_0 + \varepsilon$, the estimated value of X at t_0.
8. Suppose that the regression of X on the (mathematical) variable a is a quadratic

$$X_i = \beta_0 + \beta_1 a_i + \beta_2 a_i^2 + \varepsilon_i,$$

where $\beta_0, \beta_1, \beta_2$ are unknown paramaters, $a_1, a_2, \cdots, a_n$ are known values of a, and $\varepsilon_1, \varepsilon_2, \cdots, \varepsilon_n$ are unobservable rv's that are assumed to be independently normally distributed with common mean 0 and common variance σ^2. (See Example 12.2.3.) Assume that the coefficient vectors $(a_1^k, a_2^k, \cdots, a_n^k)$, $k = 0, 1, 2$, are linearly independent. Write the normal equations for estimating the β's, and derive the likelihood ratio test of $\beta_2 = 0$.

9. Suppose that the X's can be written as

$$X_i = \beta_1 a_i + \beta_2 b_i + \beta_3 c_i + \varepsilon_i,$$

where a, b, c are three mathematical variables, and ε_i are iid $\mathcal{N}(0, 1)$ rv's. Assuming that the matrix $\mathbf{A}$ (see Example 12.2.3) is of full rank, write the normal equations and derive the likelihood ratio test of the null hypothesis $H_0: \beta_1 = \beta_2 = \beta_3$.

10. The following table gives the weight X (grams) of a crystal suspended in a saturated solution against the time suspended T (days).

Time T:	0	1	2	3	4	5	6
Weight X:	.4	.7	1.1	1.6	1.9	2.3	2.6

(a) Find the linear regression line of X on T.
(b) Test the hypothesis that $\alpha_0 = 0$ in the linear regression model $X_i = \alpha_0 + \alpha_1 T_i + \varepsilon_i$.
(c) Obtain a .95 level confidence interval for α_0.

12.4 ONE-WAY ANALYSIS OF VARIANCE

In this section we return to the problem of one-way analysis of variance considered in Examples 12.2.1 and 12.2.4. Consider the model

(1) $\qquad X_{ij} = \mu_i + \varepsilon_{ij}, \qquad j = 1, 2, \cdots, n_i; \quad i = 1, 2, \cdots, k,$

as described in Example 12.2.4. In matrix notation we write

(2) $\qquad \mathbf{X} = \boldsymbol{\beta} \mathbf{A}' + \boldsymbol{\varepsilon},$

where

$$\mathbf{X} = (X_{11}, X_{12}, \cdots, X_{1n_1}, X_{21}, X_{22}, \cdots, X_{2n_2}, \cdots, X_{k1}, X_{k2}, \cdots, X_{kn_k}),$$
$$\boldsymbol{\beta} = (\mu_1, \mu_2, \cdots, \mu_k),$$
$$\mathbf{A} = \begin{pmatrix} \mathbf{1}'_{n_1} & \mathbf{0}' & \cdots & \mathbf{0}' \\ \vdots & \vdots & \vdots\vdots\vdots & \vdots \\ \mathbf{0}' & \mathbf{0}' & \cdots & \mathbf{1}'_{n_k} \end{pmatrix},$$

$$\boldsymbol{\varepsilon} = (\varepsilon_{11}, \varepsilon_{12}, \cdots, \varepsilon_{1n_1}, \varepsilon_{21}, \varepsilon_{22}, \cdots, \varepsilon_{2n_2}, \cdots, \varepsilon_{k1}, \varepsilon_{k2}, \cdots, \varepsilon_{kn_k}).$$

As in Example 12.2.4, **X** is a vector of n-observations ($n = \sum_{i=1}^{k} n_i$), whose components X_{ij} are subject to random error $\varepsilon_{ij} \sim \mathcal{N}(0, \sigma^2)$, $\boldsymbol{\beta}$ is a vector of k unknown parameters, and **A** is a design matrix. We wish to find a test of $H_0: \mu_1 = \mu_2 = \cdots = \mu_k$ against all alternatives. We may write H_0 in the form $\mathbf{H}\boldsymbol{\beta}' = \mathbf{0}'$, where **H** is a $(k-1) \times k$ matrix of rank $(k-1)$, which can be chosen to be

$$\mathbf{H} = \begin{pmatrix} 1 & -1 & 0 & \cdots & 0 \\ 1 & 0 & -1 & \cdots & 0 \\ \vdots & \vdots & \vdots & \vdots & \vdots \\ 1 & 0 & 0 & \cdots & -1 \end{pmatrix}.$$

Let us write $\mu_1 = \mu_2 = \cdots \mu_k = \mu$ under H_0. The joint pdf of **X** is given by

$$(3) \quad f(\mathbf{x}; \mu_1, \mu_2, \cdots, \mu_k, \sigma^2) = \left(\frac{1}{2\pi\sigma^2}\right)^{n/2} \exp\left\{-\frac{1}{2\sigma^2} \sum_{i=1}^{k} \sum_{j=1}^{n_i} (x_{ij} - \mu_i)^2\right\},$$

and, under H_0, by

$$(4) \quad f(\mathbf{x}; \mu, \sigma^2) = \frac{1}{(2\pi\sigma^2)^{n/2}} \exp\left\{-\frac{1}{2\sigma^2} \sum_{i=1}^{k} \sum_{j=1}^{n_i} (x_{ij} - \mu)^2\right\}.$$

It is easy to check that the MLE's are

$$(5) \quad \hat{\mu}_i = \frac{\sum_{j=1}^{n_i} x_{ij}}{n_i} = \bar{x}_{i\cdot}, \quad i = 1, 2, \cdots, k,$$

$$(6) \quad \hat{\sigma}^2 = \frac{\sum_{i=1}^{k} \sum_{j=1}^{n_i} (x_{ij} - \bar{x}_{i\cdot})^2}{n},$$

$$(7) \quad \hat{\hat{\mu}} = \frac{\sum_{i=1}^{k} \sum_{j=1}^{n_i} x_{ij}}{n} = \bar{x},$$

and

$$(8) \quad \hat{\hat{\sigma}}^2 = \frac{\sum_{i=1}^{k} \sum_{j=1}^{n_i} (x_{ij} - \bar{x})^2}{n}.$$

By Theorem 12.2.1, the likelihood ratio test is to reject H_0 if

$$(9) \quad \frac{\sum_{i=1}^{k} \sum_{j=1}^{n_i} (X_{ij} - \bar{X})^2 - \sum_{i=1}^{k} \sum_{j=1}^{n_i} (X_{ij} - \bar{X}_{i\cdot})^2}{\sum_{i=1}^{k} \sum_{j=1}^{n_i} (X_{ij} - \bar{X}_{i\cdot})^2} \left(\frac{n-k}{k-1}\right) \geq F_0,$$

where F_0 is the upper α percent point in the $F(k-1, n-k)$ distribution. Since

ONE-WAY ANALYSIS OF VARIANCE

(10) $$\sum_{i=1}^{k}\sum_{j=1}^{n_i}(X_{ij} - \bar{X})^2 = \sum_{i=1}^{k}\sum_{j=1}^{n_i}(X_{ij} - \bar{X}_{i\cdot} + \bar{X}_{i\cdot} - \bar{X})^2$$
$$= \sum_{i=1}^{k}\sum_{j=1}^{n_i}(X_{ij} - \bar{X}_{i\cdot})^2 + \sum_{i=1}^{k}n_i(\bar{X}_{i\cdot} - \bar{X})^2,$$

we may rewrite (9) as

(11) $$\frac{\sum_{i=1}^{k}n_i(\bar{X}_{i\cdot} - \bar{X})^2/(k-1)}{\sum_{i=1}^{k}\sum_{j=1}^{n_i}(X_{ij} - \bar{X}_{i\cdot})^2/(n-k)} \geq F_0.$$

It is usual to call the sum of squares in the numerator of (11) the *between sum of squares* (BSS), and the sum of squares in the denominator of (11) *the within sum of squares* (WSS). The results are conveniently displayed in a so-called *analysis of variance table* in the following form.

One-Way Analysis of Variance

Source of Variation	Sum of Squares	Degrees of Freedom	Mean sum of Squares	F-Ratio
Between	BSS $= \sum_{i=1}^{k} n_i(\bar{X}_i - \bar{X})^2$	$k - 1$	BSS$/(k-1)$	$\dfrac{\text{BSS}/(k-1)}{\text{WSS}/(n-k)}$
Within	WSS $= \sum_{i=1}^{k}\sum_{j=1}^{n_i}(X_{ij} - \bar{X}_{i\cdot})^2$	$n - k$	WSS$/(n-k)$	
Mean	$n\bar{X}^2$	1		
Total	TSS $= \sum_{i=1}^{k}\sum_{j=1}^{n_i} X_{ij}^2$	n		

The third row, designated "Mean," has been included to make the total of the second column add up to the total sum of squares (TSS), $\sum_{i=1}^{k}\sum_{j=1}^{n_i} X_{ij}^2$.

Example 1. The lifetimes (in hours) of samples from three different brands of batteries were recorded with the following results:

	Brand	
X	Y	Z
40	60	60
30	40	50
50	55	70
50	65	65
30		75
		40

We wish to test whether the three brands have different average lifetimes. We will assume that the three samples come from normal populations with common (unknown) standard deviation σ.

From the data $n_1 = 5$, $n_2 = 4$, $n_3 = 6$, $n = 15$, and

$$\bar{x} = \frac{200}{5} = 40, \qquad \bar{y} = \frac{220}{4} = 55, \qquad \bar{z} = \frac{360}{6} = 60,$$

$$\sum_{i=1}^{5}(x_i - \bar{x})^2 = 400, \qquad \sum_{i=1}^{4}(y_i - \bar{y})^2 = 350, \qquad \sum_{i=1}^{6}(z_i - \bar{z})^2 = 850.$$

Also, the grand mean is

$$\bar{w} = \frac{200 + 220 + 360}{15} = \frac{780}{15} = 52.$$

Thus

$$\text{BSS} = 5(40 - 52)^2 + 4(55 - 52)^2 + 6(60 - 52)^2$$
$$= 1140,$$
$$\text{WSS} = 400 + 350 + 850 = 1600.$$

Analysis of Variance

Source	SS	d.f.	MSS	F-Ratio
Between	1140	2	570	570/133.33=4.28
Within	1600	12	133.33	

Choosing $\alpha = .05$, we see that $F_0 = F_{2,12,.05} = 3.89$. Thus we reject H_0: $\mu_1 = \mu_2 = \mu_3$ at level $\alpha = .05$.

Example 2. Three sections of the same elementary statistics course were taught by three instructors. The final grades of students were recorded as follows:

	Instructor		
	I	II	III
	95	88	68
	33	78	79
	48	91	91
	76	51	71
	89	85	87
	82	77	68
	60	31	79
	77	62	16
		96	35
		81	

Let us test the hypothesis that the average grades given by the three instructors are the same at level $\alpha = .05$.

From the data $n_1 = 8$, $n_2 = 10$, $n_3 = 9$, $n = 27$, $\bar{x} = 70$, $\bar{y} = 74$, $\bar{z} = 66$, $\sum_{i=1}^{8}(x_i - \bar{x})^2 = 3168$, $\sum_{i=1}^{10}(y_i - \bar{y})^2 = 3686$, $\sum_{i=1}^{9}(z_i - \bar{z})^2 = 4898$. Also, the grand mean is

$$\bar{w} = \frac{560 + 740 + 594}{27} = \frac{1894}{27} = 70.15.$$

Thus

$$\text{BSS} = 8(.15)^2 + 10(3.85)^2 + 9(4.15)^2$$
$$= 303.4075,$$
$$\text{WSS} = 3168 + 3686 + 4898$$
$$= 11{,}752.$$

Analysis of Variance

Source	SS	d.f.	MSS	F-Ratio
Between	303.41	2	151.70	151.70/489.67
Within	11,752.00	24	489.67	

We therefore cannot reject the null hypothesis that the average grades given by the three instructors are the same.

PROBLEMS 12.4

1. Prove statements (5), (6), (7), and (8).

2. The following are the coded values of the amounts of corn (in bushels per acre) obtained from four varieties, using unequal numbers of plots for the different varieties:

   A: 2, 1, 3, 2
   B: 3, 4, 2, 3, 4, 2
   C: 6, 4, 8
   D: 7, 6, 7, 4

   Test whether there is a significant difference between the yields of the varieties.

3. A consumer interested in buying a new car has reduced his search to six different brands: D, F, G, P, V, T. He would like to buy the brand that gives the highest mileage per gallon of regular gasoline. One of his friends advises him that he should use some other method of selection, since the average mileages of the six brands are the same, and offers the following data in support of his assertion.

Distance Traveled (miles) per Gallon of Gasoline

Car	Brand					
	D	F	G	P	V	T
1	42	38	28	32	30	25
2	35	33	32	36	35	32
3	37	28	35	27	25	24
4		37	37	26	30	
5				28	30	
6				19		

Should the consumer accept his friend's advice?

4. The following data give the ages of entering freshman in independent random samples from three different universities.

\multicolumn{3}{c}{University}		
A	B	C
17	16	21
19	16	23
20	19	22
21		20
18		19

Test the hypothesis that the average ages of entering freshman at these universities are the same.

12.5 TWO-WAY ANALYSIS OF VARIANCE WITH ONE OBSERVATION PER CELL

In many practical problems one is interested in investigating the effects of two factors that influence an outcome. For example, the variety of grain and the type of fertilizer used both affect the yield of a plot; or the score on a standard examination is influenced by the size of the class and the instructor.

Let us suppose that two factors affect the outcome of an experiment. Suppose also that one observation is available at each of a number of levels of these two factors. Let X_{ij} ($i = 1, 2, \cdots, a$; $j = 1, 2, \cdots, b$) be the observation when the first factor is at the ith level, and the second factor at the jth level. Assume that

(1) $\quad X_{ij} = \mu + \alpha_i + \beta_j + \varepsilon_{ij}, \quad i = 1, 2, \cdots, a; \quad j = 1, 2, \cdots, b,$

where α_i is the effect of the ith level of the first factor, β_j is the effect of the jth level of the second factor, and ε_{ij} is the random error, which is assumed to be normally distributed with mean 0 and variance σ^2. We will assume that the ε_{ij}'s are independent. It follows that X_{ij} are independent normal rv's with means $\mu + \alpha_i + \beta_j$ and variance σ^2. There is no loss of generality in assuming that $\sum_{i=1}^{a} \alpha_i = \sum_{j=1}^{b} \beta_j = 0$, for, if $\mu_{ij} = \mu' + \alpha'_i + \beta'_j$, we can write

$$\mu_{ij} = (\mu' + \bar{\alpha}' + \bar{\beta}') + (\alpha'_i - \bar{\alpha}') + (\beta'_j - \bar{\beta}')$$
$$= \mu + \alpha_i + \beta_j$$

and $\sum_{i=1}^{a} \alpha_i = 0$, $\sum_{j=1}^{b} \beta_j = 0$. Here we have written $\bar{\alpha}'$ and $\bar{\beta}'$ for the means of α'_i's and β'_j's, respectively. Thus X_{ij} may denote the yield from the use of the ith variety of some grain and the jth type of some fertilizer. The two hypotheses of interest are

$$\alpha_1 = \alpha_2 = \cdots = \alpha_a = 0 \quad \text{and} \quad \beta_1 = \beta_2 = \cdots = \beta_b = 0.$$

The first of these, for example, says that the first factor has no effect on the outcome of the experiment.

In view of the fact that $\sum_{i=1}^{a} \alpha_i = 0$ and $\sum_{j=1}^{b} \beta_j = 0$, $\alpha_a = - \sum_{i=1}^{a-1} \alpha_i$, $\beta_b = - \sum_{j=1}^{b-1} \beta_j$, and we can write our model in matrix notation as

(2) $$\mathbf{X} = \boldsymbol{\beta} \mathbf{A}' + \boldsymbol{\varepsilon},$$

where

$$\mathbf{X} = (X_{11}, X_{12}, \cdots, X_{1b}, X_{21}, X_{22}, \cdots, X_{2b}, \cdots, X_{a1}, X_{a2}, \cdots, X_{ab}),$$
$$\boldsymbol{\beta} = (\mu, \alpha_1, \alpha_2, \cdots, \alpha_{a-1}, \beta_1, \beta_2, \cdots, \beta_{b-1}),$$
$$\boldsymbol{\varepsilon} = (\varepsilon_{11}, \varepsilon_{12}, \cdots, \varepsilon_{1b}, \varepsilon_{21}, \varepsilon_{22}, \cdots, \varepsilon_{2b}, \cdots, \varepsilon_{a1}, \varepsilon_{a2}, \cdots, \varepsilon_{ab}),$$

and

$$\mathbf{A} = \begin{pmatrix}
\mu & \alpha_1 & \alpha_2 & \cdots & \alpha_{a-1} & \beta_1 & \beta_2 & \cdots & \beta_{b-1} \\
1 & 1 & 0 & \cdots & 0 & 1 & 0 & \cdots & 0 \\
1 & 1 & 0 & \cdots & 0 & 0 & 1 & \cdots & 0 \\
\cdot & \cdot & \cdot & \cdots & \cdot & \cdot & \cdot & \cdots & \cdot \\
\cdot & \cdot & \cdot & \cdots & \cdot & \cdot & \cdot & \cdots & \cdot \\
\cdot & \cdot & \cdot & \cdots & \cdot & \cdot & \cdot & \cdots & \cdot \\
1 & 1 & 0 & \cdots & 0 & 0 & 0 & \cdots & 1 \\
1 & 1 & 0 & \cdots & 0 & -1 & -1 & \cdots & -1 \\
1 & 0 & 1 & \cdots & 0 & 1 & 0 & \cdots & 0 \\
1 & 0 & 1 & \cdots & 0 & 0 & 1 & \cdots & 0 \\
\cdot & \cdot & \cdot & \cdots & \cdot & \cdot & \cdot & \cdots & \cdot \\
\cdot & \cdot & \cdot & \cdots & \cdot & \cdot & \cdot & \cdots & \cdot \\
\cdot & \cdot & \cdot & \cdots & \cdot & \cdot & \cdot & \cdots & \cdot \\
1 & 0 & 1 & \cdots & 0 & 0 & 0 & \cdots & 1 \\
1 & 0 & 1 & \cdots & 0 & -1 & -1 & \cdots & -1 \\
\cdot & \cdot & \cdot & \cdots & \cdot & \cdot & \cdot & \cdots & \cdot \\
\cdot & \cdot & \cdot & \cdots & \cdot & \cdot & \cdot & \cdots & \cdot \\
\cdot & \cdot & \cdot & \cdots & \cdot & \cdot & \cdot & \cdots & \cdot \\
1 & -1 & -1 & \cdots & -1 & 1 & 0 & \cdots & 0 \\
1 & -1 & -1 & \cdots & -1 & 0 & 1 & \cdots & 0 \\
\cdot & \cdot & \cdot & \cdots & \cdot & \cdot & \cdot & \cdots & \cdot \\
\cdot & \cdot & \cdot & \cdots & \cdot & \cdot & \cdot & \cdots & \cdot \\
\cdot & \cdot & \cdot & \cdots & \cdot & \cdot & \cdot & \cdots & \cdot \\
1 & -1 & -1 & \cdots & -1 & 0 & 0 & \cdots & 1 \\
1 & -1 & -1 & \cdots & -1 & -1 & -1 & \cdots & -1
\end{pmatrix}$$

The vector of unknown parameters $\boldsymbol{\beta}$ is $1 \times (a + b - 1)$, and the matrix $\mathbf{A}$ is $ab \times (a + b - 1)$ (b blocks of a rows each). We leave the reader to check that **A** is of full rank, $a + b - 1$. The hypothesis $H_\alpha: \alpha_1 = \alpha_2 = \cdots = \alpha_a = 0$ or $H_\beta: \beta_1 = \beta_2 = \cdots = \beta_b = 0$ can easily be put into the form $\beta\mathbf{H}' = \mathbf{0}$. For example, for H_β we can choose **H** to be the $(b-1) \times (a+b-1)$ matrix of full rank $b - 1$, given by

$$\mathbf{H} = \begin{pmatrix} \mu & \alpha_1 & \alpha_2 & \cdots & \alpha_{a-1} & \beta_1 & \beta_2 & \cdots & \beta_{b-1} \\ 0 & 0 & 0 & \cdots & 0 & 1 & 0 & \cdots & 0 \\ 0 & 0 & 0 & \cdots & 0 & 0 & 1 & \cdots & 0 \\ \cdot & \cdot & \cdot & \cdots & \cdot & \cdot & \cdot & \cdots & \cdot \\ \cdot & \cdot & \cdot & \cdots & \cdot & \cdot & \cdot & \cdots & \cdot \\ 0 & 0 & 0 & \cdots & 0 & 0 & 0 & \cdots & 1 \end{pmatrix}$$

Clearly, the model described above is a special case of the general linear hypothesis, and we can use Theorem 12.2.1 to test H_β.

To apply Theorem 12.2.1 we need the estimates $\hat{\mu}_{ij}$ and $\hat{\hat{\mu}}_{ij}$. It is easily checked that

$$\hat{\mu} = \frac{\sum_{i=1}^{a}\sum_{j=1}^{b} x_{ij}}{ab} = \bar{x} \tag{3}$$

and

$$\hat{\alpha}_i = \bar{x}_{i\cdot} - \bar{x}, \qquad \hat{\beta}_j = \bar{x}_{\cdot j} - \bar{x}, \tag{4}$$

where $\bar{x}_{i\cdot} = \sum_{j=1}^{b} x_{ij}/b$, $\bar{x}_{\cdot j} = \sum_{i=1}^{a} x_{ij}/a$. Also, under H_β, for example,

$$\hat{\hat{\mu}} = \bar{x}, \quad \text{and} \quad \hat{\hat{\alpha}}_i = \bar{x}_{i\cdot} - \bar{x}. \tag{5}$$

In the notation of Theorem 12.2.1, $n = ab$, $k = a + b - 1$, $r = b - 1$, so that $n - k = ab - a - b + 1 = (a-1)(b-1)$, and

$$F = \frac{\sum_{i=1}^{a}\sum_{j=1}^{b}(X_{ij} - \bar{X}_{i\cdot})^2 - \sum_{i=1}^{a}\sum_{j=1}^{b}(X_{ij} - \bar{X}_{i\cdot} - \bar{X}_{\cdot j} + \bar{X})^2}{\sum_{i=1}^{a}\sum_{j=1}^{b}(X_{ij} - \bar{X}_{i\cdot} - \bar{X}_{\cdot j} + \bar{X})^2}. \tag{6}$$

Since

$$\sum_{i=1}^{a}\sum_{j=1}^{b}(X_{ij} - \bar{X}_{i\cdot})^2 = \sum_{i=1}^{a}\sum_{j=1}^{b}\{(X_{ij} - \bar{X}_{i\cdot} - \bar{X}_{\cdot j} + \bar{X}) + (\bar{X}_{\cdot j} - \bar{X})\}^2 \tag{7}$$

$$= \sum_{i=1}^{a}\sum_{j=1}^{b}(X_{ij} - \bar{X}_{i\cdot} - \bar{X}_{\cdot j} + \bar{X})^2 + a\sum_{j=1}^{b}(\bar{X}_{\cdot j} - \bar{X})^2,$$

we may write

$$F = \frac{a\sum_{j=1}^{b}(\bar{X}_{\cdot j} - \bar{X})^2}{\sum_{i=1}^{a}\sum_{j=1}^{b}(X_{ij} - \bar{X}_{i\cdot} - \bar{X}_{\cdot j} + \bar{X})^2}. \tag{8}$$

It follows that, under H_β, $(a-1)F$ has a central $F(b-1, (a-1)(b-1))$ distribution.

The numerator of F in (8) measures the variability between the means $\bar{X}_{\cdot j}$, and the denominator measures the variability that exists once the effects due to the two factors have been subtracted.

If H_α is the null hypothesis to be tested, one can show that under H_α the MLE's are

(9) $$\hat{\mu} = \bar{x}, \quad \text{and} \quad \hat{\bar{\beta}}_j = \bar{x}_{\cdot j} - \bar{x}.$$

As before, $n = ab$, $k = a + b - 1$, but $r = a - 1$. Also,

(10) $$F = \frac{\sum_{i=1}^{a}\sum_{j=1}^{b}(X_{ij} - \bar{X}_{\cdot j})^2 - \sum_{i=1}^{a}\sum_{j=1}^{b}(X_{ij} - \bar{X}_{i\cdot} - \bar{X}_{\cdot j} + \bar{X})^2}{\sum_{i=1}^{a}\sum_{j=1}^{b}(X_{ij} - \bar{X}_{i\cdot} - \bar{X}_{\cdot j} + \bar{X})^2},$$

which may be rewritten as

(11) $$F = \frac{b\sum_{i=1}^{a}(\bar{X}_{i\cdot} - \bar{X})^2}{\sum_{i=1}^{a}\sum_{j=1}^{b}(X_{ij} - \bar{X}_{i\cdot} - \bar{X}_{\cdot j} + \bar{X})^2}.$$

It follows that, under H_α, $(b-1)F$ has a central $F(a-1, (a-1)(b-1))$ distribution. The numerator of F in (11) measures the variability between the means $\bar{X}_{i\cdot}$.

If the data are put into the following form:

Levels of Factor 2

$\alpha \backslash \beta$	1	2		b	Row Means
1	X_{11},	X_{12},	$\cdots$,	X_{1b}	$\bar{X}_{1\cdot}$
2	X_{21},	X_{22},	$\cdots$,	X_{2b}	$\bar{X}_{2\cdot}$
.	.	.	$\cdots$	.	.
.	.	.	$\cdots$	.	.
.	.	.	$\cdots$	.	.
a	X_{a1},	X_{a2},	$\cdots$,	X_{ab}	$\bar{X}_{a\cdot}$
Column Means	$\bar{X}_{\cdot 1}$,	$\bar{X}_{\cdot 2}$,	$\cdots$,	$\bar{X}_{\cdot b}$	$\bar{X}$

(Rows labeled: Levels of Factor 1)

so that the rows represent various levels of factor 1, and the columns, the levels of factor 2, one can write

$$\text{Between sum of squares for rows} = b\sum_{i=1}^{a}(\bar{X}_{i\cdot} - \bar{X})^2$$
$$= \text{sum of squares for factor 1}$$
$$= SS_1.$$

Similarly

Between sum of squares for columns $= a \sum_{j=1}^{b} (\bar{X}_{\cdot j} - \bar{X})^2$
$\qquad\qquad\qquad\qquad\qquad\qquad = $ sum of squares for factor 2
$\qquad\qquad\qquad\qquad\qquad\qquad = SS_2.$

It is usual to write error or residual sum of squares (SSE) for the denominator of (8) or (11). These results are conveniently presented in an analysis of variance table as follows.

Two-Way Analysis of Variance Table with One Observation per Cell

Source of Variation	Sum of Squares	Degrees of Freedom	Mean Square	F-Ratio
Rows	SS_1	$a-1$	$MS_1 = SS_1/(a-1)$	MS_1/MSE
Columns	SS_2	$b-1$	$MS_2 = SS_2/(b-1)$	MS_2/MSE
Error	SSE	$(a-1)(b-1)$	$MSE = SSE/(a-1)(b-1)$	
Mean	$ab\,\bar{X}^2$	1	$ab\,\bar{X}^2$	
Total	$\sum_{i=1}^{a}\sum_{j=1}^{b} X_{ij}^2$	ab	$\sum_{i=1}^{a}\sum_{j=1}^{b} X_{ij}^2/ab$	

Example 1. The following table gives the yield (pounds per plot) of three varieties of wheat, obtained with four different kinds of fertilizers.

| | Variety of Wheat | | |
Fertilizer	A	B	C
α	8	3	7
β	10	4	8
γ	6	5	6
δ	8	4	7

Let us test the hypothesis of equality in the average yields of the three varieties of wheat and the null hypothesis that the four fertilizers are equally effective.

In our notation, $b = 3$, $a = 4$, $\bar{x}_{1\cdot} = 6$, $\bar{x}_{2\cdot} = 7.33$, $\bar{x}_{3\cdot} = 5.67$, $\bar{x}_{4\cdot} = 6.33$, $\bar{x}_{\cdot 1} = 8$, $\bar{x}_{\cdot 2} = 4$, $\bar{x}_{\cdot 3} = 7$, $\bar{x} = 6.33$.

Also,

$$SS_1 = \text{sum of squares due to fertilizer}$$
$$\qquad = 3[(.33)^2 + 1^2 + (.66)^2 + 0^2]$$
$$\qquad = 4.67;$$
$$SS_2 = \text{sum of squares due to variety of wheat}$$
$$\qquad = 4[(1.67)^2 + (2.33)^2 + (.67)^2]$$
$$\qquad = 34.67;$$
$$SSE = \sum_{i=1}^{4}\sum_{j=1}^{3}(x_{ij} - \bar{x}_{i\cdot} - \bar{x}_{\cdot j} + \bar{x})^2$$
$$\qquad = 7.33.$$

The results are shown in the following table.

Analysis of Variance

Source	SS	d.f.	MS	F-Ratio
Variety of wheat	34.67	2	17.33	14.2
Fertilizer	4.67	3	1.56	1.28
Error	7.33	6	1.22	
Mean	481.33	1	481.33	
Total	528.00	12	44.00	

Now $F_{2,6,.05} = 5.14$ and $F_{3,6,.05} = 4.76$. Since $14.2 > 5.14$, we reject H_β, that there is equality in the average yield of the three varieties; but, since $1.27 \not> 4.76$, we accept H_α, that the four fertilizers are equally effective.

PROBLEMS 12.5

1. Show that the matrix A for the model defined in (2) is of full rank, $a + b - 1$.
2. Prove statements (3), (4), (5) and (9).
3. The following data represent the units of production per day turned out by four different brands of machines used by four machinists:

Machine	Machinist			
	A_1	A_2	A_3	A_4
B_1	15	14	19	18
B_2	17	12	20	16
B_3	16	18	16	17
B_4	16	16	15	15

Test whether the differences in the performances of the machinists are significant and also whether the differences in the performances of the four brands of machines are significant. Use $\alpha = .05$.

4. Students were classified into four ability groups, and three different teaching methods were employed. The following table gives the means for the four groups:

Ability Group	Teaching Method		
	A	B	C
1	15	19	14
2	18	17	12
3	22	25	17
4	17	21	19

Test the hypothesis that the teaching methods yield the same results. That is, that the teaching methods are equally effective.

12.6 TWO-WAY ANALYSIS OF VARIANCE WITH INTERACTION

The model described in Section 12.5 assumes that the two factors act independently, that is, are *additive*. In practice this is an assumption that needs testing. In this section we allow for the possibility that the two factors might jointly affect the outcome, that is, there might be so-called *interactions*. More precisely, if X_{ij} is the observation in the (i, j)th cell, we will consider the model

(1) $$X_{ij} = \mu + \alpha_i + \beta_j + \gamma_{ij} + \varepsilon_{ij},$$

where α_i $(i = 1, 2, \cdots, a)$ represent row effects (or effects due to factor 1), β_j $(j = 1, 2, \cdots, b)$ represent column effects (or effects due to factor 2), and γ_{ij} represent interactions or joint effects. We will assume that ε_{ij} are independently $\mathcal{N}(0, \sigma^2)$. We will further assume that

(2) $$\sum_{i=1}^{a} \alpha_i = 0 = \sum_{j=1}^{b} \beta_j, \quad \text{and} \quad \sum_{j=1}^{b} \gamma_{ij} = 0 \quad \text{for all } i, \sum_{i=1}^{a} \gamma_{ij} = 0$$
$$\text{for all } j.$$

The hypothesis of interest is

(3) $$H_0: \gamma_{ij} = 0 \quad \text{for all } i, j.$$

One may also be interested in testing that all α's are 0 or that all β's are 0 in the presence of interactions γ_{ij}.

We first note that (2) is not restrictive since we can write

$$X_{ij} = \mu' + \alpha'_i + \beta'_j + \gamma'_{ij} + \varepsilon_{ij},$$

where α'_i, β'_j and γ'_{ij} do not satisfy (2), as

$$X_{ij} = \mu' + \bar{\alpha}' + \bar{\beta}' + \bar{\gamma}' + (\alpha'_i - \bar{\alpha}' + \bar{\gamma}'_{i\cdot} - \bar{\gamma}') + (\beta'_j - \bar{\beta}' + \bar{\gamma}'_{\cdot j} - \bar{\gamma}')$$
$$+ (\gamma'_{ij} - \bar{\gamma}'_{i\cdot} - \bar{\gamma}'_{\cdot j} + \bar{\gamma}') + \varepsilon_{ij},$$

and then (2) is satisfied by choosing

$$\mu = \mu' + \bar{\alpha}' + \bar{\beta}' + \bar{\gamma}',$$
$$\alpha_i = \alpha'_i - \bar{\alpha}' + \bar{\gamma}'_{i\cdot} - \bar{\gamma}',$$
$$\beta_j = \beta'_j - \bar{\beta}' + \bar{\gamma}'_{\cdot j} - \bar{\gamma}',$$
$$\gamma_{ij} = \gamma'_{ij} - \bar{\gamma}'_{i\cdot} - \bar{\gamma}'_{\cdot j} + \bar{\gamma}'.$$

Here

$$\bar{\alpha}' = a^{-1} \sum_{i=1}^{a} \alpha'_i, \quad \bar{\beta}' = b^{-1} \sum_{j=1}^{b} \beta'_j, \quad \bar{\gamma}'_{i\cdot} = b^{-1} \sum_{j=1}^{b} \gamma'_{ij},$$
$$\bar{\gamma}'_{\cdot j} = a^{-1} \sum_{i=1}^{a} \gamma'_{ij}, \quad \text{and} \quad \bar{\gamma}' = (ab)^{-1} \sum_{i=1}^{a} \sum_{j=1}^{b} \gamma'_{ij}.$$

Next note that, unless we replicate, that is, take more than one observation per cell, there are no degrees of freedom left to estimate the error SS. (See Remark 1.)

Let X_{ijs} be the sth observation when the first factor is at the ith level, and the second factor at the jth level, $i = 1, 2, \cdots, a, j = 1, 2, \cdots, b, s = 1, 2, \cdots, m\ (> 1)$. Then the model becomes as follows:

	Levels of Factor 2			
Levels of Factor 1	1	2	$\cdots$	b
1	x_{111}	x_{121}	$\cdots$	x_{1b1}
	.	.	$\cdots$	.
	.	.	$\cdots$	.
	.	.	$\cdots$	.
	x_{11m}	x_{12m}	$\cdots$	x_{1bm}
2	x_{211}	x_{221}	$\cdots$	x_{2b1}
	.	.	$\cdots$	.
	.	.	$\cdots$	.
	.	.	$\cdots$	.
	x_{21m}	x_{22m}	$\cdots$	x_{2bm}
	.	.	$\cdots$	.
	.	.	$\cdots$	.
	.	.	$\cdots$	.
a	x_{a11}	x_{a21}	$\cdots$	x_{ab1}
	.	.	$\cdots$	.
	.	.	$\cdots$	.
	.	.	$\cdots$	.
	x_{a1m}	x_{a2m}	$\cdots$	x_{abm}

(4) $$X_{ijs} = \mu + \alpha_i + \beta_j + \gamma_{ij} + \varepsilon_{ijs},$$

$i = 1, 2, \cdots, a, j = 1, 2, \cdots, b$, and $s = 1, 2, \cdots, m$, where ε_{ijs}'s are independent $\mathcal{N}(0, \sigma^2)$. We assume that $\sum_{i=1}^{a} \alpha_i = \sum_{j=1}^{b} \beta_j = \sum_{i=1}^{a} \gamma_{ij} = 0 = \sum_{j=1}^{b} \gamma_{ij}$. Suppose that we wish to test $H_\alpha: \alpha_1 = \alpha_2 = \cdots = \alpha_a = 0$. We leave the reader to check that model (4) is then a special case of the general linear hypothesis with $n = abm$, $k = ab$, $r = a - 1$, and $n - k = ab(m - 1)$.

Let us write

(5) $$\bar{X} = \frac{\sum_{i=1}^{a} \sum_{j=1}^{b} \sum_{s=1}^{m} X_{ijs}}{n}, \quad \bar{X}_{ij\cdot} = \frac{\sum_{s=1}^{m} X_{ijs}}{m},$$

$$\bar{X}_{i\cdot\cdot} = \frac{\sum_{j=1}^{b} \sum_{s=1}^{m} X_{ijs}}{mb}, \quad \bar{X}_{\cdot j\cdot} = \frac{\sum_{i=1}^{a} \sum_{s=1}^{m} X_{ijs}}{am}.$$

Then it can be easily checked that

(6) $\begin{cases} \hat{\mu} = \hat{\hat{\mu}} = \bar{X}, & \hat{\alpha}_i = \bar{X}_{i..} - \bar{X}, & \hat{\hat{\beta}}_j = \hat{\hat{\beta}}_j = \bar{X}_{.j.} - \bar{X}, \\ \hat{\gamma}_{ij} = \hat{\hat{\gamma}}_{ij} = \bar{X}_{ij.} - \bar{X}_{i..} - \bar{X}_{.j.} + \bar{X}. \end{cases}$

It follows from Theorem 12.2.1 that

(7) $F = \dfrac{\sum_i \sum_j \sum_s (X_{ijs} - \bar{X}_{ij.} + \bar{X}_{i..} - \bar{X})^2 - \sum_i \sum_j \sum_s (X_{ijs} - \bar{X}_{ij.})^2}{\sum_i \sum_j \sum_s (X_{ijs} - \bar{X}_{ij.})^2}.$

Since

$$\sum_i \sum_j \sum_s (X_{ijs} - \bar{X}_{ij.} + \bar{X}_{i..} - \bar{X})^2$$
$$= \sum_i \sum_j \sum_s (X_{ijs} - \bar{X}_{ij.})^2 + \sum_i \sum_j \sum_s (\bar{X}_{i..} - \bar{X})^2,$$

we can write (7) as

(8) $F = \dfrac{bm \sum_i (\bar{X}_{i..} - \bar{X})^2}{\sum_i \sum_j \sum_s (X_{ijs} - \bar{X}_{ij.})^2}.$

Under H_α the statistic $[ab(m-1)/(a-1)] F$ has the central $F(a-1, ab(m-1))$ distribution, so that the likelihood ratio test rejects H_α if

(9) $\dfrac{ab(m-1)}{a-1} \dfrac{mb \sum_i (\bar{X}_{i..} - \bar{X})^2}{\sum_i \sum_j \sum_s (X_{ijs} - \bar{X}_{ij.})^2} > c.$

A similar analysis holds for testing $H_\beta: \beta_1 = \beta_2 = \cdots = \beta_b$.

Next consider the test of hypothesis $H_\gamma: \gamma_{ij} = 0$ for all i, j, that is, that the two factors are independent and the effects are additive. In this case $n = abm$, $k = ab$, $r = (a-1)(b-1)$, and $n - k = ab(m-1)$. It can be shown that

(10) $\hat{\hat{\mu}} = \bar{X}, \quad \hat{\hat{\alpha}}_i = \bar{X}_{i..} - \bar{X}, \quad \text{and } \hat{\hat{\beta}}_j = \bar{X}_{.j.} - \bar{X}.$

Thus

(11) $F = \dfrac{\sum_i \sum_j \sum_s (X_{ijs} - \bar{X}_{i..} - \bar{X}_{.j.} + \bar{X})^2 - \sum_i \sum_j \sum_s (X_{ijs} - \bar{X}_{ij.})^2}{\sum_i \sum_j \sum_s (X_{ijs} - \bar{X}_{ij.})^2}.$

Now

$$\sum_i \sum_j \sum_s (X_{ijs} - \bar{X}_{i..} - \bar{X}_{.j.} + \bar{X})^2$$
$$= \sum_i \sum_j \sum_s (X_{ijs} - \bar{X}_{ij.} + \bar{X}_{ij.} - \bar{X}_{i..} - \bar{X}_{.j.} + \bar{X})^2$$
$$= \sum_i \sum_j \sum_s (X_{ijs} - \bar{X}_{ij.})^2 + \sum_i \sum_j \sum_s (\bar{X}_{ij.} - \bar{X}_{i..} - \bar{X}_{.j.} + \bar{X})^2,$$

TWO-WAY ANALYSIS OF VARIANCE WITH INTERACTION

so that we may write

(12) $$F = \frac{\sum_i \sum_j \sum_s (\bar{X}_{ij\cdot} - \bar{X}_{i\cdot\cdot} - \bar{X}_{\cdot j\cdot} + \bar{X})^2}{\sum_i \sum_j \sum_s (X_{ijs} - \bar{X}_{ij\cdot})^2}.$$

Under H_r, the statistic $\{(m-1)ab/[(a-1)(b-1)]\} F$ has the $F((a-1)(b-1), ab(m-1))$ distribution. The likelihood ratio test rejects H_r if

(13) $$\frac{(m-1)ab}{(a-1)(b-1)} \cdot \frac{m \sum_i \sum_j (\bar{X}_{ij\cdot} - \bar{X}_{i\cdot\cdot} - \bar{X}_{\cdot j\cdot} + \bar{X})^2}{\sum_i \sum_j \sum_s (X_{ijs} - \bar{X}_{ij\cdot})^2} > c.$$

Let us write

SS_1 = sum of squares due to factor 1 (row sum of squares)

$\quad = bm \sum_{i=1}^{a} (\bar{X}_{i\cdot\cdot} - \bar{X})^2,$

SS_2 = sum of squares due to factor 2 (column sum of squares)

$\quad = am \sum_{j=1}^{b} (\bar{X}_{\cdot j\cdot} - \bar{X})^2,$

SSI = sum of squares due to interaction

$\quad = m \sum_{i=1}^{a} \sum_{j=1}^{b} (\bar{X}_{ij\cdot} - \bar{X}_{i\cdot\cdot} - \bar{X}_{\cdot j\cdot} + \bar{X})^2,$

and

SSE = sum of squares due to error (residual sum of squares)

$\quad = \sum_{i=1}^{a} \sum_{j=1}^{b} \sum_{s=1}^{m} (X_{ijs} - \bar{X}_{ij\cdot})^2.$

Then we may summarize the above results in the following table.

Two-Way Analysis of Variance Table with Interaction

Source of Variation	Sum of Squares	Degrees of Freedom	Mean Square	F-Ratio
Rows	SS_1	$a-1$	$MS_1 = SS_1/(a-1)$	MS_1/MSE
Columns	SS_2	$b-1$	$MS_2 = SS_2/(b-1)$	MS_2/MSE
Interaction	SSI	$(a-1)(b-1)$	$MSI = SSI/(a-1)(b-1)$	MSI/MSE
Error	SSE	$ab(m-1)$	$MSE = SSE/ab(m-1)$	
Mean	$abm\bar{X}^2$	1	$abm\bar{X}^2$	
Total	$\sum_{i=1}^{a} \sum_{j=1}^{b} \sum_{s=1}^{m} X_{ijs}^2$	abm	$\sum_{i=1}^{a} \sum_{j=1}^{b} \sum_{s=1}^{m} X_{ijs}^2/abm$	

Remark 1. Note that, if $m = 1$, there are no d.f. associated with the SSE. Indeed, the SSE = 0 if $m = 1$. Hence we cannot make tests of hypotheses when $m = 1$, and for this reason we assume $m > 1$. The more general case

in which we take unequal number of observations per cell is quite easily treated along the same lines.

Example 1. To test the effectiveness of three different teaching methods, three instructors were randomly assigned 12 students each. The students were then randomly assigned to the different teaching methods and were taught exactly the same material. At the conclusion of the experiment, identical examinations were given to the students with the following results in regard to grades.

	Instructor		
Teaching Method	I	II	III
1	95	60	86
	85	90	77
	74	80	75
	74	70	70
2	90	89	83
	80	90	70
	92	91	75
	82	86	72
3	70	68	74
	80	73	86
	85	78	91
	85	93	89

From the data the table of means is as follows:

	$\bar{x}_{ij\cdot}$			$\bar{x}_{i\cdot\cdot}$
	82	75	77	78.0
	86	89	75	83.3
	80	78	85	81.0
$\bar{x}_{\cdot j\cdot}$	82.7	80.7	79.0	$\bar{x} = 80.8$

Then

$$SS_1 = \text{sum of squares due to methods}$$
$$= bm \sum_{i=1}^{a} (\bar{x}_{i\cdot\cdot} - \bar{x})^2$$
$$= 3 \times 4 \times 14.13 = 169.56,$$
$$SS_2 = \text{sum of squares due to instructors}$$
$$= am \sum_{j=1}^{b} (\bar{x}_{\cdot j\cdot} - \bar{x})^2$$
$$= 3 \times 4 \times 6.86 = 82.32,$$

SSI = sum of squares due to interaction
$$= m \sum_{i=1}^{3} \sum_{j=1}^{3} (\bar{x}_{ij\cdot} - \bar{x}_{i\cdot\cdot} - \bar{x}_{\cdot j\cdot} + \bar{x})^2$$
$$= 4 \times 140.45 = 561.80$$
SSE = residual sum of squares
$$= \sum_{i=1}^{3} \sum_{j=1}^{3} \sum_{s=1}^{4} (x_{ijs} - \bar{x}_{ij\cdot})^2 = 1830.00.$$

Analysis of Variance

Source	SS	d.f.	MSS	F-Ratio
Methods	169.56	2	84.78	1.25
Instructors	82.32	2	41.16	.61
Interactions	561.80	4	140.45	2.07
Error	1830.00	27	67.78	

With $\alpha = .05$, we see from the tables that $F_{2,27,.05} = 3.35$ and $F_{4,27,.05} = 2.73$, so that we cannot reject any of the three hypotheses that the three methods are equally effective, that the three instructors are equally effective, and that the interactions are all 0.

PROBLEMS 12.6

1. Prove statement (6).
2. Obtain the likelihood ratio test of the null hypothesis $H_\beta : \beta_1 = \beta_2 = \cdots = \beta_b = 0$.
3. Prove statement (10).
4. Suppose that the following data represent the units of production turned out each day by three different machinists, each working on the same machine for three different days:

| | Machinist | | |
Machine	A	B	C
B_1	15,15,17	19,19,16	16,18,21
B_2	17,17,17	15,15,15	19,22,22
B_3	15,17,16	18,17,16	18,18,18
B_4	18,20,22	15,16,17	17,17,17

Using a .05 level of significance, test whether (a) the differences among the machinists are significant, (b) the differences among the machines are significant, and (c) the interactions are significant.

CHAPTER 13

Nonparametric Statistical Inference

13.1 INTRODUCTION

In all the problems of statistical inference considered so far, we assumed that the distribution of the random variable being sampled is known except, perhaps, for some parameters. In practice, however, the functional form of the distribution is seldom, if ever, known. It is therefore desirable to devise some procedures that are free of this assumption concerning distribution. In this chapter we study some procedures that are commonly referred to as *distribution-free* or *nonparametric methods*. The term "distribution-free" refers to the fact that no assumptions are made about the underlying distribution except that the distribution function being sampled is absolutely continuous or purely discrete. The term "nonparametric" refers to the fact that there are no parameters involved in the traditional sense of the term "parameter" used thus far. To be sure, there is a parameter which indexes the family of absolutely continuous df's, but it is not numerical and hence the parameter set cannot be represented as a subset of $\mathcal{R}_n$, for any $n \geq 1$. The restriction to absolutely continuous distribution functions is a simplifying assumption that allows us to use the probability integral transformation (Theorem 5.3.1) and the fact that ties occur with probability 0.

Section 2 of this chapter is devoted to the problem of unbiased (nonparametric) estimation. Sections 3 through 5 deal with some common hypotheses testing problems. In Section 6 we investigate some applications of order statistics in nonparametric methods. Section 7 considers underlying assumptions in some common parametric problems and the effect of relaxation in these assumptions.

13.2 NONPARAMETRIC ESTIMATION

We have already encountered some results in nonparametric estimation. For example, the sample distribution function defined in Section 7.3 as an estimate of the population df is distribution free, and so also are the sample quantiles defined in Section 7.3 as estimates of corresponding population quantiles. To estimate moments of a distribution one can use the method of moments described in Section 8.6. In this section we concentrate mainly on *nonparametric unbiased estimation*. We will see that some obvious generalizations of concepts used in Section 8.4 yield an analogue of the Lehmann-Scheffé theorem.

Let $X_1, X_2, \cdots, X_n$ be iid rv's with common law $\mathscr{L}(X)$, and let $\mathscr{P}$ be the class of all possible distributions of X that consists of the absolutely continuous or discrete distributions, or subclasses of these.

Definition 1. A statistic $T(\mathbf{X})$ is sufficient for the family of distributions $\mathscr{P}$ if the conditional distribution of $\mathbf{X}$, given $T = t$, is the same whatever the true $F \in \mathscr{P}$.

Example 1. Let $X_1, X_2, \cdots, X_n$ be a random sample from an absolutely continuous df, and let $\mathbf{T} = (X_{(1)}, \cdots, X_{(n)})$ be the order statistic. Then

$$f(\mathbf{x} \mid \mathbf{T} = \mathbf{t}) = (n!)^{-1},$$

and we see that $\mathbf{T}$ is sufficient for the family of absolutely continuous distributions on $\mathscr{R}$.

Definition 2. A family of distributions $\mathscr{P}$ is complete if the only unbiased estimate of 0 is the zero function itself, that is,

$$E_F h(\mathbf{X}) = 0 \quad \text{for all } F \in \mathscr{P} \Rightarrow h(\mathbf{x}) = 0$$

for all $\mathbf{x}$ (except for a null set with respect to each $F \in \mathscr{P}$).

Definition 3. A statistic $T(\mathbf{X})$ is said to be complete in relation to a class of distributions $\mathscr{P}$ if the class of induced distributions of T is complete.

We have already encountered many examples of complete statistics or complete families of distributions in Chapter 8.

The following result is stated without proof. For the proof we refer to Fraser [33], pages 27–30, 139–142.

Theorem 1. The order statistic $(X_{(1)}, X_{(2)}, \cdots, X_{(n)})$ is a complete sufficient statistic provided that the iid rv's $X_1, X_2, \cdots, X_n$ are of either the discrete or the continuous type.

Definition 4. A real-valued parameter $g(F)$ is said to be estimable if it has an unbiased estimate, that is, if there exists a statistic $T(\mathbf{X})$ such that

(1) $$E_F T(\mathbf{X}) = g(F) \qquad \text{for all } F \in \mathscr{P}.$$

Example 2. If $\mathscr{P}$ is the class of all distributions for which the second moment exists, $\bar{X}$ is an unbiased estimate of $\mu(F)$, the population mean. Similarly, $\mu_2(F) = \text{var}_F(X)$ is also estimable, and an unbiased estimate is $S^2 = \sum_1^n (X_i - \bar{X})^2 / (n-1)$. We would like to know whether $\bar{X}$ and S^2 are UMVUE's. Similarly, $\bar{X} - \bar{Y}$ is an unbiased estimate of $EX - EY$, and $(1/n)$ · (number of X's $> c$) is an unbiased estimate of $P_F\{X > c\}$, and so on.

Definition 5. The degree $m(m \geq 1)$ of an estimable parameter $g(F)$ is the smallest sample size for which the parameter is estimable, that is, it is the smallest n such that there exists an unbiased estimate $T(X_1, X_2, \cdots, X_n)$ with

$$E_F T = g(F) \qquad \text{for all } F \in \mathscr{P}.$$

Example 3. The parameter $g(F) = P_F\{X > c\}$, where c is a known constant, has degree 1. Also, $\mu(F)$ is estimable with degree 1 (we assume that there is at least one $F \in \mathscr{P}$ such that $\mu(F) \neq 0$), and $\mu_2(F)$ is estimable with degree $m = 2$, since $\mu_2(F)$ cannot be estimated (unbiasedly) by one observation only. At least two observations are needed. Similarly, $\mu^2(F)$ has degree 2.

Definition 6. An unbiased estimate of a parameter based on the minimum sample size (equal to degree m) is called a kernel.

Example 4. X_i is a kernel of $\mu(F)$; $X_i X_j$, $i \neq j$, is a kernel of $\mu^2(F)$; and each

$$T(X_i, X_j) = X_i^2 - X_i X_j, \qquad i = 1, 2, \cdots, n (i \neq j),$$

is a kernel of $\mu_2(F)$.

Lemma 1. There exists a symmetric kernel for every estimable parameter.

Proof. If $T(X_1, X_2, \cdots, X_m)$ is a kernel of $g(F)$, so also is

(2) $$T_s(X_1, X_2, \cdots, X_m) = \frac{1}{m!} \sum_P T(X_{i_1}, X_{i_2}, \cdots X_{i_m}),$$

where the summation P is over all $m!$ permutations of $\{1, 2, \cdots, m\}$.

Example 5. A symmetric kernel for $\mu_2(F)$ is

$$T_s(X_i, X_j) = \tfrac{1}{2}\{T(X_i, X_j) + T(X_j, X_i)\}$$
$$= \tfrac{1}{2}(X_i - X_j)^2, \quad i, j = 1, 2, \cdots, n (i \neq j).$$

Definition 7. Let $g(F)$ be an estimable parameter of degree m, and let $X_1, X_2, \cdots, X_n$ be a sample of size n, $n \geq m$. Corresponding to any kernel $T(X_{i_1}, \cdots, X_{i_m})$ of $g(F)$, we define a U-statistic for the sample by

$$(3) \qquad U(X_1, X_2, \cdots, X_n) = \binom{n}{m}^{-1} \sum_C T_s(X_{i_1}, X_{i_2}, \cdots, X_{i_m}),$$

where the summation C is over all $\binom{n}{m}$ combinations of m integers $(i_1, i_2, \cdots, i_m)$ chosen from $\{1, 2, \cdots, n\}$, and T_s is the symmetric kernel defined in (2).

Clearly the U statistic defined in (3) is symmetric in the X_i's, and

$$(4) \qquad E_F U(\mathbf{X}) = g(F) \qquad \text{for all } F.$$

Example 6. For estimating $\mu(F)$, the U-statistic is $n^{-1}\sum_1^n X_i$. For estimating $\mu_2(F)$, a symmetric kernel is

$$T_s(X_{i_1}, X_{i_2}) = \tfrac{1}{2}(X_{i_1} - X_{i_2})^2, \quad i_1 = 1, 2, \cdots, n (i_1 \neq i_2),$$

so that the corresponding U-statistic is

$$U(\mathbf{X}) = \binom{n}{2}^{-1} \sum_{i_1 < i_2} \tfrac{1}{2}(X_{i_1} - X_{i_2})^2$$
$$= \frac{1}{n-1} \sum_1^n (X_i - \bar{X})^2$$
$$= S^2.$$

Similarly, for estimating $\mu^2(F)$, a symmetric kernel is $T_s(X_{i_1}, X_{i_2}) = X_{i_1} X_{i_2}$, and the corresponding U-statistic is

$$U(\mathbf{X}) = \frac{1}{\binom{n}{2}} \sum_{i<j} X_i X_j = \frac{1}{n(n-1)} \sum_{i \neq j} X_i X_j.$$

For estimating $\mu^3(F)$, a symmetric kernel is $T_s(X_{i_1}, X_{i_2}, X_{i_3}) = X_{i_1} X_{i_2} X_{i_3}$ so that the corresponding U-statistic is

$$U(\mathbf{X}) = \binom{n}{3}^{-1} \sum\sum\sum_{i<j<k} X_i X_j X_k$$
$$= \frac{1}{n(n-1)(n-2)} \sum_{i \neq j \neq k} X_i X_j X_k.$$

The following result shows the importance of the U-statistic.

Theorem 2. Let $\mathscr{P}$ be the class of all absolutely continuous or all purely discrete distribution functions on $\mathscr{R}$. Any estimable function $g(F)$, $F \in \mathscr{P}$, has a unique estimate that is unbiased and symmetric in the observations and has uniformly minimum variance among all unbiased estimates.

Proof. Let $X_1, X_2, \cdots, X_n$ be a random sample from F, $F \in \mathscr{P}$, and let $T(X_1, X_2, \cdots, X_n)$ be an unbiased estimate of $g(F)$. Consider the set of all $n!$ permutations of $\{1, 2, \cdots, n\}$ and suitably index them. Let $\{i_1, i_2, \cdots, i_n\}$ be the ith member in this set, and let

$$T_i = T_i(X_1, X_2, \cdots, X_n) = T(X_{i_1}, X_{i_2}, \cdots, X_{i_n}), \qquad i = 1, 2, \cdots, n!.$$

Let $\bar{T} = \sum_{i=1}^{n!} T_i/n!$. Clearly $E_F \bar{T} = g(F)$, and

$$\begin{aligned}
\operatorname{var}(\bar{T}) &= E\left\{(n!)^{-1} \sum_{i=1}^{n!} T_i\right\}^2 - (g(F))^2 \\
&\leq E\left\{\sum_{i=1}^{n!} (n!)^{-2} \sum_{i=1}^{n!} T_i^2\right\} - (g(F))^2 \\
&= ET^2 - (g(F))^2 \\
&= \operatorname{var}(T).
\end{aligned}$$

The equality holds if and only if

$$T_i(X_1, X_2, \cdots, X_n) = \frac{\alpha}{n!}, \qquad i = 1, 2, \cdots, n!,$$

for all points in the sample space (except perhaps for a null set) where α is a constant. It follows that $T(\mathbf{X})$ is symmetric in the arguments $X_1, X_2 \cdots, X_n$ with probability 1 and $\bar{T}$ is identical with T. Uniqueness follows from Theorem 1 and Problem 4.

Corollary. If $T(X_1, X_2, \cdots, X_n)$ is unbiased for $g(F)$, $F \in \mathscr{P}$, the corresponding U-statistic is an essentially unique UMVUE.

Remark 1. According to Theorem 2, we need only consider estimates that are symmetric in the observations, and all we need to do is to make them unbiased. This procedure leads to an unbiased estimate with the least variance in the class of all unbiased estimates of the parameter. For example, as a consequence of Theorem 2 $\bar{X}$ and S^2 are unique UMVUE's of $\mu(F)$ and $\mu_2(F)$, respectively.

Remark 2. The reader should go back and compare this result (Theorem 2) with the corresponding result in the parametric case (Theorem 8.4.5). See also Remark 8.3.12.

Example 7. Let $\mathscr{P}$ be the class of all absolutely continuous df's, and $X_1, X_2, \cdots, X_n$ be a sample of size n. To estimate $g(F) = P_F\{X_1 > c\}$, where c is a fixed constant, define

$$Y_i = \begin{cases} 1, & X_i > c \\ 0, & X_i \leq c \end{cases}, \quad i = 1, 2, \cdots, n.$$

Consider

$$T(Y_1, Y_2 \cdots, Y_n) = \sum_1^n \alpha_i Y_i,$$

as an estimate of $g(F)$. To find the UMVUE of g we symmetrize T in the Y's. This happens if $\alpha_i = \alpha$, $i = 1, 2, \cdots, n$, and $T(\mathbf{Y}) = \alpha \sum_1^n Y_i$. For T to be unbiased, $E_F T = \alpha \sum_1^n E_F Y_i = \alpha n g(F)$, so that $\alpha = n^{-1}$. Thus $n^{-1} \sum_1^n Y_i$ is the UMVUE;

$$\text{var}_F(T) = \frac{g(F)[1 - g(F)]}{n} \leq \frac{1}{4n}.$$

Also, Y_i is binomial, so that

$$\frac{n^{1/2}[T - g(F)]}{\{g(F)[1 - g(F)]\}^{1/2}} \xrightarrow{L} Z \quad \text{as } n \to \infty,$$

where $Z \sim \mathcal{N}(0, 1)$. This result can be used to find confidence bounds on $g(F)$.

Example 8 (Lehmann [68], 3-24 to 3-26, and Fraser [33], 164-167). Let $\mathscr{P}$ be the class of all absolutely continuous distributions on the real line. For $F, G \in \mathscr{P}$, we define a distance function $\Delta(F, G)$ as follows:

(5) $$\Delta(F, G) = \int_{-\infty}^{\infty} [F(x) - G(x)]^2 \frac{F'(x) + G'(x)}{2} dx.$$

We show that $\Delta(F, G) = 0$ if and only if $F = G$. Clearly, $\Delta(F, F) = 0$. Conversely, let $F(x) \neq G(x)$ for some x_1, and suppose that $F(x_1) - G(x_1) = d > 0$. Since F and G are (absolutely) continuous df's, there exists an $x_0 < x_1$ such that $F(x_0) - G(x_0) = d/2$ and $F(x) - G(x) \geq d/2$ for $x_0 \leq x \leq x_1$. Since F and G are both nondecreasing, at least one of F and G must increase by at least $d/2$ as x varies from x_0 to x_1. Thus

$$\Delta(F, G) \geq \int_{x_0}^{x_1} [F(x) - G(x)]^2 \frac{F'(x) + G'(x)}{2} dx$$

$$\geq \left(\frac{d}{2}\right)^2 \frac{d/2}{2} > 0.$$

Let $X_1, X_2, \cdots, X_m$ be a sample from F, and $Y_1, Y_2, \cdots, Y_n$ be an independent sample from G, where $F, G \in \mathscr{P}$. We wish to find the UMVUE of $\Delta(F, G)$. We first show that

$$g(F, G) = P\{[\max (X_1, X_2) < \min (Y_1, Y_2)] \cup [\max (Y_1, Y_2) < \min (X_1, X_2)]\}$$

(6)
$$= \tfrac{1}{3} + 2\Delta(F, G).$$

We have

$$g(F, G) = P\{\max (X_1, X_2) < \min (Y_1, Y_2)\} + P\{\max (Y_1, Y_2) < \min (X_1, X_2)\}$$

and

$$P\{\max (X_1, X_2) \le x\} = F^2(x),$$
$$P\{\min (Y_1, Y_2) \ge y\} = [1 - G(y)]^2.$$

Let us write $F_1(x) = F^2(x)$ and $G_1(y) = G^2(y)$. Then

$$g(F, G) = \int_{-\infty}^{\infty} [1 - G(y)]^2 F_1'(y) \, dy + \int_{-\infty}^{\infty} [1 - F(x)]^2 G_1'(x) \, dx$$

$$= \int_{-\infty}^{\infty} [1 + G^2(y) - 2G(y)] [2F(y) F'(y)] \, dy$$

$$\qquad + \int_{-\infty}^{\infty} [1 + F^2(x) - 2F(x)] [2G(x) G'(x)] \, dx$$

$$= 2 + \int_{-\infty}^{\infty} 2[G^2(x) F(x) F'(x) + F^2(x) G(x) G'(x)$$
$$\qquad - 2F(x)G(x) (F'(x) + G'(x))] \, dx$$

$$= 2 + \int_{-\infty}^{\infty} d[F^2(x) G^2(x)]$$
$$\qquad - 4 \int_{-\infty}^{\infty} F(x) G(x) [F'(x) + G'(x)] \, dx$$

$$= 3 - 2 \int_{-\infty}^{\infty} \{[F(x) + G(x)]^2 - [F(x) - G(x)]^2\}$$
$$\qquad \cdot \left[\frac{F'(x) + G'(x)}{2} \right] dx$$

$$= 3 - 8 \int_{-\infty}^{\infty} \left[\frac{F(x) + G(x)}{2} \right]^2 \left[\frac{F'(x) + G'(x)}{2} \right] dx + 2\Delta(F, G)$$

$$= 3 - \frac{8}{3} + 2\Delta(F, G),$$

which is (6).

To use Theorem 2, let us define

(7) $\varphi(X_1, X_2, Y_1, Y_2) = \begin{cases} 1 & \text{if } \max (X_1, X_2) < \min (Y_1, Y_2) \text{ or} \\ & \quad \max (Y_1, Y_2) < \min (X_1, X_2), \\ 0 & \text{otherwise.} \end{cases}$

Then $\varphi(X_1, X_2, Y_1, Y_2)$ is an unbiased estimate of $g(F, G)$, and indeed it is a kernel of $g(F, G)$. The corresponding U-statistic therefore must be the UMVUE. We have

(8) $$U(\mathbf{X}, \mathbf{Y}) = \left[\binom{m}{2}\binom{n}{2}\right]^{-1} \sum_{i_1 < i_2} \sum_{j_1 < j_2} \varphi(X_{i_1}, X_{i_2}, Y_{j_1}, Y_{j_2}),$$

so that U is the UMVUE of $g(F, G)$, and the UMVUE of $\Delta(F, G)$ is

(9) $$\hat{\Delta}(\mathbf{X}, \mathbf{Y}) = \tfrac{1}{2} U(\mathbf{X}, \mathbf{Y}) - \tfrac{1}{6}.$$

Example 9 (Lehmann [68], 3-24, and Fraser [33], 162). Let $\mathscr{P}$ be the class of all absolutely continuous df's on the real line, and let X_1, $X_2, \cdots, X_m$ and $Y_1, Y_2, \cdots, Y_n$ be independent random samples from F and G, respectively, $F, G \in \mathscr{P}$. We wish to estimate

$$\rho(F, G) = P\{X < Y\}.$$

Let us define

$$Z_{ij} = \begin{cases} 1, & X_i < Y_j, \\ 0, & X_i \geq Y_j \end{cases}$$

for each pair X_i, Y_j, $i = 1, 2, \cdots, m$, $j = 1, 2, \cdots, n$.

Then $\sum_{i=1}^{m} Z_{ij}$ is the number of X's $< Y_j$, and $\sum_{j=1}^{n} Z_{ij}$ is the number of Y's $> X_i$. Mann and Whitney [78] suggest the use of the estimate U/mn, where

$$U = \sum_{i=1}^{m} \sum_{j=1}^{n} Z_{ij}$$

and

$$EU = mn\, EZ_{ij} = mn\, P\{X < Y\}.$$

Thus

$$\hat{\rho}(F, G) = \frac{U}{mn}$$

is unbiased for ρ. Moreover, $\hat{\rho}$ is symmetric in the X's and Y's, so that it has minimum variance. To compute the minimum variance, we have

$$EU^2 = \sum \sum \sum \sum E(Z_{ij} Z_{hk}),$$

where

$$Z_{ij} Z_{hk} = \begin{cases} 1 & \text{if } X_i < Y_j \text{ and } X_h < Y_k, \\ 0 & \text{otherwise,} \end{cases}$$

so that

$$E(Z_{ij} Z_{hk}) = P\{X_i < Y_j, X_h < Y_k\}$$

$$= \begin{cases} \int F(x)\, G'(x)\, dx & \text{if } i = h, j = k, \\ \int (1 - G(x))^2\, F'(x)\, dx & \text{if } i = h, j \neq k, \\ \int F^2(x)\, G'(x)\, dx & \text{if } i \neq h, j = k, \\ [\int F(x)\, G'(x)\, dx]^2 & \text{if } i \neq h, j \neq k. \end{cases}$$

There are mn terms with $i = h$, $j = k$; $m(m-1)n$ terms with $i \neq h$, $j = k$; $mn(n-1)$ terms with $i = h$, $j \neq k$; and $m(m-1)\,n(n-1)$ terms with $i \neq h$, $j \neq k$. It follows that

$$EU^2 = mn \int F(x)\, G'(x)\, dx + mn(n-1) \int [1 - G(x)]^2\, F'(x)\, dx$$
$$+ m(m-1)n \int F^2(x)\, G'(x)\, dx$$
$$+ mn(m-1)(n-1)[\int F(x)\, G'(x)\, dx]^2,$$

which leads to the variance of U. In particular, if $F = G$, then

$$\text{var } U = \frac{mn(m+n+1)}{12}.$$

PROBLEMS 13.2

1. Let $(\mathcal{R}, \mathcal{B}, P_\theta)$ be a probability space, and let $\mathcal{P} = \{P_\theta: \theta \in \Theta\}$. Let A be a Borel subset of $\mathcal{R}$, and consider the parameter $d(\theta) = P_\theta(A)$. Is d estimable? If so, what is the degree? Find the UMVUE for d, based on a sample of size n, assuming that $\mathcal{P}$ is the class of all continuous distributions.

2. Let $X_1, X_2, \cdots, X_m$ and $Y_1, Y_2, \cdots, Y_n$ be independent random samples from two absolutely continuous df's. Find the UMVUE's of (a) $E\{XY\}$ and (b) var $(X + Y)$.

3. Let $(X_1, Y_1), (X_2, Y_2), \cdots, (X_n, Y_n)$ be a random sample from an absolutely continuous distribution. Find the UMVUE's of (a) $E(XY)$ and (b) var $(X + Y)$.

4. Let $T(X_1, X_2, \cdots, X_n)$ be a statistic that is symmetric in the observations. Show that T can be written as a function of the order statistic. Conversely, if $T(X_1, X_2, \cdots, X_n)$ can be written as a function of the order statistic, T is symmetric in the observations.

13.3 SOME SINGLE-SAMPLE PROBLEMS

In this section we consider some tests of nonparametric hypotheses when a sample of n iid observations $X_1, X_2, \cdots, X_n$ is available.

The Problem of Fit

The problem of fit is to test the hypothesis that the sample of observations $X_1, X_2, \cdots, X_n$ is from some specified distribution against the alternative that it is from some other distribution. Thus the null hypothesis H_0: $X_i \sim F_0$ is simple and is to be tested against the composite alternative H_1: $X_i \sim F$, where $F(x) \ne F_0(x)$ for some x. In Section 10.3 we studied the chi-square test of goodness of fit of H_0.

Another test used for testing H_0 is the *Kolmogorov-Smirnov test*. In Section 7.3 we defined the sample (empirical) df F_n^* of $X_1, X_2, \cdots, X_n$.

Definition 1. Let $X_1, X_2, \cdots, X_n$ be a sample from a df F, and let F_n^* be the corresponding empirical df. The statistic

(1) $$D_n = \sup_x |F_n^*(x) - F(x)|$$

is called the (two-sided) Kolmogorov-Smirnov statistic. We write

(2) $$D_n^+ = \sup_x [F_n^*(x) - F(x)]$$

and

(3) $$D_n^- = \sup_x [F(x) - F_n^*(x)],$$

and call D_n^+, D_n^- the one-sided Kolmogorov-Smirnov statistics.

Theorem 1. The statistics D_n, D_n^-, D_n^+ are distribution free for any continuous df F.

Proof. Clearly, $D_n = \max(D_n^+, D_n^-)$. Let $X_{(1)} \le X_{(2)} \le \cdots \le X_{(n)}$ be the order statistics of $X_1, X_2, \cdots, X_n$, and define $X_{(0)} = -\infty$, $X_{(n+1)} = +\infty$. Then

$$F_n^*(x) = \frac{i}{n} \quad \text{for } X_{(i)} \le x < X_{(i+1)}, \quad i = 0, 1, 2, \cdots, n,$$

and we have

$$\begin{aligned} D_n^+ &= \max_{0 \le i \le n} \sup_{X_{(i)} \le x < X_{(i+1)}} \left\{ \frac{i}{n} - F(x) \right\} \\ &= \max_{0 \le i \le n} \left\{ \frac{i}{n} - \inf_{X_{(i)} \le x < X_{(i+1)}} F(x) \right\} \\ &= \max_{0 \le i \le n} \left\{ \frac{i}{n} - F(X_{(i)}) \right\} \\ &= \max \left\{ \max_{1 \le i \le n} \left[\frac{i}{n} - F(X_{(i)}) \right], 0 \right\}. \end{aligned}$$

Since $F(X_{(i)})$ is the ith-order statistic of a sample from $U(0, 1)$ irrespective of what F is, as long as it is continuous, we see that the distribution of D_n^+ is independent of F. Similarly,

$$D_n^- = \max\left\{\max_{1\le i\le n}\left[F(X_{(i)}) - \frac{i-1}{n}\right], 0\right\},$$

and the result follows.

Without loss of generality, therefore, we assume that F is the df of a $U(0, 1)$ rv.

Theorem 2. If F is continuous, then

(4) $$P\left\{D_n < v + \frac{1}{2n}\right\} = \begin{cases} 0 & \text{if } v \le 0, \\ \int_{(1/2n)-v}^{v+(1/2n)}\int_{(3/2n)-v}^{v+(3/2n)}\cdots\int_{[(2n-1)/2n]-v}^{v+[(2n-1)/2n]} f(u_1, u_2, \cdots, u_n) \cdot \prod_1^n du_i & \\ & \text{if } 0 < v < \frac{2n-1}{2n}, \\ 1 & \text{if } v \ge \frac{2n-1}{2n}, \end{cases}$$

where

(5) $$f(u_1, u_2, \cdots, u_n) = \begin{cases} n!, & 0 < u_1 < \cdots < u_n < 1, \\ 0, & \text{otherwise,} \end{cases}$$

is the joint pdf of an order statistic for a sample of size n from $U(0, 1)$.

We will not prove this result here and instead refer to Gibbons [36], page 77. Let $D_{n,\alpha}$ be the upper α-percent point of the distribution of D_n, that is, $P\{D_n > D_{n,\alpha}\} \le \alpha$. The exact distribution of D_n for selected values of n and α has been tabulated by Miller [81], Owen [88], and Birnbaum [8]. The large-sample distribution of D_n was derived by Kolmogorov [60], and we state it without proof.

Theorem 3. Let F be any continuous df. Then for every $z \ge 0$

(6) $$\lim_{n\to\infty} P\{D_n \le zn^{-1/2}\} = L(z),$$

where

(7) $$L(z) = 1 - 2\sum_{i=1}^{\infty}(-1)^{i-1} e^{-2i^2z^2}.$$

SINGLE-SAMPLE PROBLEMS   541

Theorem 3 can be used to find d_α such that $\lim_{n\to\infty} P\{\sqrt{n}\, D_n \leq d_\alpha\} = 1 - \alpha$. Tables of d_α for various values of α are also available in Owen [88].

The statistics D_n^+ and D_n^- have the same distribution because of symmetry, and their common distribution is given by the following theorem.

Theorem 4. Let F be a continuous df. Then

$$(8) \quad P\{D_n^+ \leq z\} = \begin{cases} 0 & \text{if } z \leq 0, \\ \int_{1-z}^{1} \int_{[(n-1)/n]-z}^{u_n} \cdots \int_{(2/n)-z}^{u_3} \int_{(1/n)-z}^{u_2} f(u_1, u_2, \cdots, u_n) \prod_{i=1}^{n} du_i & \text{if } 0 < z < 1, \\ 1 & \text{if } z \geq 1, \end{cases}$$

where f is given by (5).

Proof. We leave the reader to prove Theorem 4.

Tables for the critical values $D_{n,\alpha}^+$, where $P\{D_n^+ > D_{n,\alpha}^+\} \leq \alpha$, are also available for selected values of n and α; see Birnbaum and Tingey [7]. Table 7 page 661 gives $D_{n,\alpha}^+$ and $D_{n,\alpha}$ for some selected values of n and α. For large samples Smirnov [121] showed that

$$(9) \quad \lim_{n\to\infty} P\{\sqrt{n}\, D_n^+ \leq z\} = 1 - e^{-2z^2}, \quad z \geq 0.$$

In fact, in view of (9), the statistic $V_n = 4n\, D_n^{+2}$ has a limiting $\chi^2(2)$ distribution, for $4n\, D_n^{+2} \leq 4z^2$ if and only if $\sqrt{n}\, D_n^+ \leq z$, $z \geq 0$, and the result follows since

$$\lim_{n\to\infty} P\{V_n \leq 4z^2\} = 1 - e^{-2z^2}, \quad z \geq 0,$$

so that

$$\lim_{n\to\infty} P\{V_n \leq x\} = 1 - e^{-x/2}, \quad x \geq 0,$$

which is the df of a $\chi^2(2)$ rv.

Example 1. Let $\alpha = .01$, and let us approximate $D_{n,\alpha}^+$. We have $\chi^2_{2,.01} = 9.21$. Thus $V_n = 9.21$, yielding

$$D_{n,.01}^+ = \sqrt{\frac{9.21}{4n}} = \frac{3.03}{2\sqrt{n}}.$$

If, for example, $n = 9$, then $D_{n,.01}^+ = 3.03/6 = .50$. Of course, the approximation is better for large n.

The statistic D_n and its one-sided analogues can be used in testing H_0: $X \sim F_0$ against H_1: $X \sim F$, where $F_0(x) \neq F(x)$ for some x.

Definition 2. To test H_0: $F(x) = F_0(x)$ for all x at level α, the Kolmogorov-Smirnov test rejects H_0 if $D_n > D_{n,\alpha}$. Similarly, it rejects $F(x) \geq F_0(x)$ for all x if $D_n^- > D_{n,\alpha}^+$ and rejects $F(x) \leq F_0(x)$ for all x at level α if $D_n^+ > D_{n,\alpha}^+$.

For large samples we can approximate by using Theorem 3 or (9) to obtain an approximate α level test.

Example 2. A die is rolled 15 times with the following results:

Face value:	1	2	3	4	5	6
Frequency:	0	1	4	0	4	6

Let us test H_0 that the die is fair, that is, test H_0: $P\{X = x\} = \frac{1}{6}$, $x = 1, 2, \cdots, 6$, against all alternatives. We have the following:

x:	<1	1	2	3	4	5	6	>6
$F_{15}^*(x)$:	0	0	$\frac{1}{15}$	$\frac{5}{15}$	$\frac{5}{15}$	$\frac{9}{15}$	1	1
$F(x)$:	0	$\frac{1}{6}$	$\frac{2}{6}$	$\frac{3}{6}$	$\frac{4}{6}$	$\frac{5}{6}$	1	1
$\|F_{15}^*(x) - F(x)\|$:	0	$\frac{5}{30}$	$\frac{8}{30}$	$\frac{5}{30}$	$\frac{10}{30}$	$\frac{7}{30}$	0	0

$$D_{15} = \sup_x |F_{15}^*(x) - F(x)| = \frac{10}{30} = .333.$$

Since $D_{15,.10} = .304$, we reject H_0 at level $\alpha = .10$.

Let us now apply the chi-square test to the same data, noting that the number of observations is not large enough. We have the following:

Face value:	1	2	3	4	5	6
$\mathbf{x}$:	0	1	4	0	4	6
$n\mathbf{p}$:	$\frac{15}{6}$	$\frac{15}{6}$	$\frac{15}{6}$	$\frac{15}{6}$	$\frac{15}{6}$	$\frac{15}{6}$

where $\mathbf{x} = (x_1, \cdots, x_6)$, x_i is the number of times i shows up, $i = 1, 2, \cdots, 6$, and $\mathbf{p} = (\frac{1}{6}, \frac{1}{6}, \cdots, \frac{1}{6})$. Let us combine the first two classes, the third and fourth, and the last two to make the expected frequency in each class at least 5. Then $U = 8.4$. Since $\chi^2_{2,.10} = 4.6$, we reject H_0 at level $\alpha = .10$. We emphasize that the application of the chi-square test to this example is somewhat dubious.

Example 3. Let us consider the data in Example 10.3.3, and apply the Kolmogorov-Smirnov test to determine the goodness of the fit. Rearranging the data in increasing order of magnitude, we have the following result:

SINGLE-SAMPLE PROBLEMS

x	$F_0(x)$	$F_{20}^*(x)$	$\left|F_{20}^*(x) - F_0(x)\right|$
−1.787	.0367	$\frac{1}{20}$	.0133
−1.229	.1093	$\frac{2}{20}$	.0093
−0.525	.2998	$\frac{3}{20}$	.1498
−0.513	.3050	$\frac{4}{20}$	.1050
−0.508	.3050	$\frac{5}{20}$	.0550
−0.486	.3121	$\frac{6}{20}$	.0121
−0.482	.3156	$\frac{7}{20}$	.0344
−0.323	.3745	$\frac{8}{20}$	.0255
−0.261	.3974	$\frac{9}{20}$	.0526
−0.068	.4721	$\frac{10}{20}$	.0279
−0.057	.4761	$\frac{11}{20}$	.0739
.137	.5557	$\frac{12}{20}$	.0443
.464	.6772	$\frac{13}{20}$	.0272
.595	.7257	$\frac{14}{20}$	.0257
.881	.8106	$\frac{15}{20}$	.0606
.906	.8186	$\frac{16}{20}$	.0136
1.046	.8531	$\frac{17}{20}$	.0031
1.237	.8925	$\frac{18}{20}$	.0075
1.678	.9535	$\frac{19}{20}$	.0035
2.455	.9931	1	.0069

$$D_{20} = \sup_x \left|F_{20}^*(x) - F_0(x)\right| = .1498.$$

Let us take $\alpha = .05$. Then $D_{20,.05} = .294$. Since $.1498 < .294$, we accept H_0 at the .05 level of significance.

It is worthwhile to compare the chi-square test of goodness of fit and the Kolmogorov-Smirnov test. The latter treats individual observations directly, whereas the former discretizes the data and sometimes loses information through grouping. Moreover, the Kolmogorov-Smirnov test is applicable even in the case of very small samples, but the chi-square test is essentially for large samples.

The chi-square test, on the other hand, can be easily modified to allow estimation of parameters from the data, but the Kolmogorov-Smirnov test does not have this flexibility. The chi-square test can be applied when the data are discrete or continuous, but the Kolmogorov-Smirnov test assumes continuity of the df. This means that the latter test provides a more refined analysis of the data. If the distribution is actually discontinuous, the Kolmogorov-Smirnov test is conservative in that it tends to accept H_0.

We next turn our attention to some other uses of the Kolmogorov-Smirnov statistic. Let $X_1, X_2, \cdots, X_n$ be a sample from a df F, and let F_n^* be the sample df. The estimate F_n^* of F for large n should be close to F. Indeed,

(10) $$P\left\{|F_n^*(x) - F(x)| \le \frac{\lambda\sqrt{F(x)[1-F(x)]}}{\sqrt{n}}\right\} \ge 1 - \frac{1}{\lambda^2},$$

and, since $F(x)[1-F(x)] \le \frac{1}{4}$, we have

(11) $$P\left\{|F_n^*(x) - F(x)| \le \frac{\lambda}{2\sqrt{n}}\right\} \ge 1 - \frac{1}{\lambda^2}.$$

Thus F_n^* can be made close to F with high probability by choosing λ and large enough n. The Kolmogorov-Smirnov statistic enables us to determine the smallest n such that the error in estimate never exceeds a fixed value ε with a large probability $1 - \alpha$. Since

(12) $$P\{D_n \le \varepsilon\} \ge 1 - \alpha,$$

$\varepsilon = D_{n,\alpha}$; and, given ε and α, we can read n from the tables. For large n, we can use the asymptotic distribution of D_n and solve $d_\alpha = \varepsilon\sqrt{n}$ for n.

We can also form confidence bounds for F. Given α and n, we first find $D_{n,\alpha}$ such that

(13) $$P\{D_n > D_{n,\alpha}\} \le \alpha,$$

which is the same as

$$P\{\sup_x |F_n^*(x) - F(x)| \le D_{n,\alpha}\} \ge 1 - \alpha.$$

Thus

(14) $$P\{|F_n^*(x) - F(x)| \le D_{n,\alpha} \text{ for all } x\} \ge 1 - \alpha.$$

Define

(15) $$L_n(x) = \max\{F_n^*(x) - D_{n,\alpha}, 0\}$$

and

(16) $$U_n(x) = \min\{F_n^*(x) + D_{n,\alpha}, 1\}.$$

Then the region between $L_n(x)$ and $U_n(x)$ can be used as a confidence band for $F(x)$ with associated confidence coefficient $1 - \alpha$.

Example 4. For the data on the standard normal distribution of Example 3, let us form a .90 confidence band for the df. We have $D_{20,.10} = .265$. The confidence band is, therefore, $F_{20}^*(x) \pm .265$ as long as the band is between 0 and 1.

The Problem of Location

Let $X_1, X_2, \cdots, X_n$ be a sample of size n from some unknown df F. Let p be a positive real number, $0 < p < 1$, and let $\kappa_p(F)$ denote the quantile of order p for the df F. In the following analysis we assume that F is absolutely continuous. The problem of location is to test H_0: $\kappa_p(F) = \kappa_0$, κ_0 a given number, against one of the alternatives $\kappa_p(F) > \kappa_0$, $\kappa_p < \kappa_0$, and $\kappa_p \neq \kappa_0$. The problem of location and symmetry is to test H_0': $\kappa_{0.5}(F) = \kappa_0$, and F is symmetric against H_1': $\kappa_{0.5}(F) \neq \kappa_0$ or F is not symmetric.

The Sign Test. Let $X_1, X_2, \cdots, X_n$ be iid rv's with common pdf f. The problem is to test

(17) $\qquad H_0: \kappa_p(f) = \kappa_0 \qquad$ against $H_1: \kappa_p(f) > \kappa_0$,

where $\kappa_p(f)$ is the quantile of order p for f, $0 < p < 1$. Let $i(X_1, X_2, \cdots, X_n)$ be the number of positive elements in $X_1 - \kappa_0, X_2 - \kappa_0, \cdots, X_n - \kappa_0$. Note that $P\{X_i = \kappa_0\} = 0$, since X_i are continuous-type rv's. It can be shown (see Fraser [33], 167–170) that a UMP test of H_0 against H_1 is given by

(18) $\qquad \varphi(x_1, x_2, \cdots, x_n) = \begin{cases} 1, & i(x_1, x_2, \cdots, x_n) > c, \\ \gamma, & i(x_1, x_2, \cdots, x_n) = c, \\ 0, & i(x_1, x_2, \cdots, x_n) < c, \end{cases}$

where c and γ are chosen from the size restriction

(19) $\qquad \sum_{i=c+1}^{n} \binom{n}{i(\mathbf{x})} q^{i(\mathbf{x})} p^{n-i(\mathbf{x})} + \gamma \binom{n}{c} q^c p^{n-c} = \alpha$

with $q = 1 - p$. This is so because, under H_0, $\kappa_p(f) = \kappa_0$, so that $P\{X \leq \kappa_0\} = p$, $P\{X \geq \kappa_0\} = 1 - p = q$, and $i(\mathbf{X})$ has a $b(n, q)$ distribution.

The same test is UMP for H_0: $\kappa_p(f) \leq \kappa_0$ against $\kappa_p(f) > \kappa_0$. For the hypothesis H_0: $\kappa_p(f) = \kappa_0$ against the two-sided alternative $\kappa_p(f) \neq \kappa_0$, the two-sided sign test can be shown to be UMP unbiased. (See Fraser [33], 171.)

Example 5. Entering college freshmen have taken a particular high school achievement test for many years, and the upper quartile ($p = .75$) is well established at a score of 195. A particular high school sent 12 of its graduates to college, where they took the examination and obtained scores of 203, 168, 187, 235, 197, 163, 214, 233, 179, 185, 197, 216. Let us test the null hypothesis H_0 that $\kappa_{.75} \leq 195$ against H_1: $\kappa_{.75} > 195$ at the $\alpha = .05$ level.

We have to find c and γ such that

$$\sum_{i=c+1}^{12} \binom{12}{i}\left(\frac{1}{4}\right)^i\left(\frac{3}{4}\right)^{12-i} + \gamma\binom{12}{c}\left(\frac{1}{4}\right)^c\left(\frac{3}{4}\right)^{12-c} = .05.$$

From the table of cumulative binomial distribution (Table 1, page 648) for $n = 12$, $p = \frac{1}{4}$, we see that $c = 6$. Then γ is given by

$$.0142 + \gamma\binom{12}{6}\left(\frac{1}{4}\right)^6\left(\frac{3}{4}\right)^6 = .05.$$

Thus

$$\gamma = \frac{.0358}{.0402} = .89.$$

In our case the number of positive signs, $x_i - 195$, $i = 1, 2, \cdots, 12$, is 7, so that we reject H_0 that the upper quartile is ≤ 195.

Example 6. A random sample of size 8 is taken from a normal population with mean 0 and variance 1. The sample values are $-.465$, $.120$, $-.238$, $-.869$, -1.016, $.417$, $.056$, $.561$. Let us test hypothesis $H_0\colon \mu = -1.0$ against $H_1\colon \mu > -1.0$. We should expect to reject H_0 since we know that it is false. The number of observations, $x_i - \mu_0 = x_i + 1.0$, that are ≥ 0 is 7. We have to find c and γ such that

$$\sum_{i=c+1}^{8} \binom{8}{i}\left(\frac{1}{2}\right)^8 + \gamma \binom{8}{c}\left(\frac{1}{2}\right)^8 = .05, \text{ say,}$$

that is,

$$\sum_{i=c+1}^{8} \binom{8}{i} + \gamma \binom{8}{c} = 12.8.$$

We see that $c = 6$ and $\gamma = .13$. Since the number of positive $x_i - \mu_0$ is > 6, we reject H_0.

Let us now apply the parametric test here. We have

$$\bar{x} = -\frac{1.434}{8} = -.179.$$

Since $\sigma = 1$, we reject H_0 if

$$\bar{x} > \mu_0 + \frac{1}{\sqrt{n}} z_\alpha = -1.0 + \frac{1}{\sqrt{8}} 1.64$$
$$= -.42.$$

Since $-.179 > -.42$, we reject H_0.

The single-sample sign test described above can be easily modified to apply to sampling from a bivariate population. Let $(X_1, Y_1), (X_2, Y_2), \cdots, (X_n, Y_n)$ be a random sample from a bivariate population. Let $Z_i = X_i - Y_i$, $i = 1, 2, \cdots, n$, and assume that Z_i has an absolutely continuous df. Then one can test hypotheses concerning the order parameters of Z by using the sign test. A hypothesis of interest here is that Z has a given median $m_0 (= \kappa_{1/2})$. Without loss of generality let $m_0 = 0$. Then $H_0\colon \text{med}(Z) = 0$, that is, $P\{Z > 0\} = P\{Z < 0\} = \frac{1}{2}$. Note that med (Z) is not necessarily equal to med (X) − med (Y), so that H_0 is not that med (X) = med (Y) but that med $(Z) = 0$. The sign test is UMP against one-sided alternatives and UMP unbiased against two-sided alternatives.

Example 7. We consider an example due to Hahn and Nelson [43], in which two measuring devices take readings on each of 10 test units. Let X and Y, respectively, be the readings on a test unit by the first and second measuring devices. Let $X = A + \zeta$, $Y = A + \eta$, where A, ζ, η, respectively, are the contributions to the readings due to the test unit and to the first and the second measuring devices. Let A, ζ, η be independent with $EA = \mu$, var $(A) = \sigma_a^2$, $E\zeta = E\eta = 0$, var $(\zeta) = \sigma_1^2$, var $(\eta) = \sigma_2^2$, so that X and Y have common mean μ and variances $\sigma_1^2 + \sigma_a^2$ and $\sigma_2^2 + \sigma_a^2$, respectively. Also, the covariance between X and Y is σ_a^2. The data are as follows:

Test unit:	1	2	3	4	5	6	7	8	9	10
First device, X:	71	108	72	140	61	97	90	127	101	114
Second device, Y:	77	105	71	152	88	117	93	130	112	105
$Z = X - Y$:	-6	3	1	-8	-17	-20	-3	-3	-11	9

Let us test the hypothesis H_0: med $(Z) = 0$. The number of Z_i's > 0 is 3. We have

$$P\{\text{Number of } Z_i\text{'s} > 0 \text{ is} \leq 3 | H_0\} = \sum_{k=0}^{3} \binom{10}{k}\left(\frac{1}{2}\right)^{10}$$
$$= .172.$$

Using the two-sided sign test, we have to accept H_0 at level $\alpha = .05$, since $.172 > .025$. The rv's Z_i can be considered to be distributed normally, so that under H_0 the common mean of Z_i's is 0. Using a paired comparison t-test on the data, we can show that $t = -.88$ for 9 d.f., so that we accept the hypothesis of equality of means of X and Y at level $\alpha = .05$.

The Wilcoxon Signed-Ranks Test. The sign test loses information since it ignores the magnitude of the difference between the observations and the hypothesized quantile. The Wilcoxon signed-ranks test provides an alternative test of location (and symmetry) that also takes into account the magnitudes of these differences.

Let $X_1, X_2, \cdots, X_n$ be iid rv's with common absolutely continuous df F, which is symmetric about the median m. The problem is to test $H_0: m = m_0$ against the usual one or two-sided alternatives. Without loss of generality, we assume that $m_0 = 0$. Then $F(-x) = 1 - F(x)$ for all $x \in \mathcal{R}$. To test H_0: $F(0) = \frac{1}{2}$ or $m = 0$, we first arrange $|X_1|, |X_2|, \cdots, |X_n|$ in increasing order of magnitude, and assign ranks $1, 2, \cdots, n$, keeping track of the original signs of X_i. For example, if $n = 4$ and $|X_2| < |X_4| < |X_1| < |X_3|$, the rank of $|X_1|$ is 3, of $|X_2|$ is 1, of $|X_3|$ is 4, and of $|X_4|$ is 2.

Let

(20) $\begin{cases} T^+ = \text{the sum of the ranks of positive } X_i\text{'s,} \\ T^- = \text{the sum of the ranks of negative } X_i\text{'s.} \end{cases}$

Then, under H_0, we expect T^+ and T^- to be the same. Note that

$$T^+ + T^- = \sum_{1}^{n} i = \frac{n(n+1)}{2}, \tag{21}$$

so that T^+ and T^- are linearly related and offer equivalent criteria. Let us define

$$Z_i = \begin{cases} 1 & \text{if } X_i > 0 \\ 0 & \text{if } X_i < 0 \end{cases}, \quad i = 1, 2, \cdots, n, \tag{22}$$

and write $r(|X_i|) = r_i$ for the rank of $|X_i|$. Then $T^+ = \sum_{i=1}^{n} r_i Z_i$ and $T^- = \sum_{i=1}^{n} (1 - Z_i) r_i$. Also,

$$\begin{aligned} T^+ - T^- &= -\sum_{i=1}^{n} r_i + 2 \sum_{i=1}^{n} Z_i r_i \\ &= 2 \sum_{i=1}^{n} r_i Z_i - \frac{n(n+1)}{2}. \end{aligned} \tag{23}$$

The statistic T^+ (or T^-) is known as the *Wilcoxon statistic*. A large value of T^+ (or, equivalently, a small value of T^-) means that most of the large deviations from 0 are positive, and therefore we reject H_0 in favor of the alternative, $H_1: m > 0$.

A similar analysis applies to the other two alternatives. We record the results as follows:

H_0	H_1	Test Reject H_0 if
$m = 0$	$m > 0$	$T^+ > c_1$
$m = 0$	$m < 0$	$T^+ < c_2$
$m = 0$	$m \neq 0$	$T^+ < c_3$ or $T^+ > c_4$

Let us next compute the distribution of T^+. Let

$$Z_{(i)} = \begin{cases} 1 & \text{if the } |X_j| \text{ that has rank } i \text{ is } > 0, \\ 0 & \text{otherwise.} \end{cases} \tag{24}$$

Clearly $T^+ = \sum_{1}^{n} i Z_{(i)}$. The rv's $Z_{(i)}$ are independent Bernoulli rv's but are not necessarily identically distributed. We have

$$\begin{aligned} EZ_{(i)} &= P\{Z_{(i)} = 1\} = P\{[r(|X_j|) = i, X_j > 0] \text{ for some } j\} \\ &= P\{i\text{th-order statistic in } |X_1|, \cdots, |X_n| \text{ corresponds to a positive } X_j\} \\ &= \int_0^\infty n \binom{n-1}{i-1} [F_{|X|}(u)]^{i-1} [1 - F_{|X|}(u)]^{n-i} f(u) \, du \\ &= n \binom{n-1}{i-1} \int_0^\infty [F(u) - F(-u)]^{i-1} [1 - F(u) + F(-u)]^{n-i} f(u) \, du, \end{aligned} \tag{25}$$

where f is the pdf of X. Moreover,

$$\text{var}(Z_{(i)}) = E\{Z_{(i)}(1 - EZ_{(i)})\} \tag{26}$$

and

(27) $$\text{cov}(Z_{(i)}, Z_{(j)}) = 0, \quad i \neq j.$$

Under H_0, X is symmetric about 0 so that $F(0) = \frac{1}{2}$, $F(-u) = 1 - F(u)$ for all $u > 0$. Thus

(28) $$EZ_{(i)} = n\binom{n-1}{i-1} \int_0^\infty [2F(u) - 1]^{i-1} [2 - 2F(u)]^{n-i} f(u)\, du.$$

Letting $v = 2F(u) - 1$, we have

$$EZ_{(i)} = \frac{n}{2}\binom{n-1}{i-1} \int_0^1 v^{i-1}(1-v)^{n-i}\, dv$$

$$= \frac{n}{2}\binom{n-1}{i-1} B(i, n-i+1)$$

(29) $$= \tfrac{1}{2}.$$

The general moments of T^+ are given by

(30) $$ET^+ = \sum_{i=1}^n i EZ_{(i)}$$

and

(31) $$\text{var}(T^+) = \sum_{i=1}^n i^2 \{EZ_{(i)}(1 - EZ_{(i)})\},$$

so that

(32) $$E_{H_0}(T^+) = \frac{1}{2}\sum_1^n i = \frac{n(n+1)}{4}$$

and

(33) $$\text{var}_{H_0}(T^+) = \frac{1}{4}\sum_1^n i^2 = \frac{n(n+1)(2n+1)}{24}.$$

For large n, the Lindeberg condition is satisfied (prove it), so that

(34) $$\frac{T^+ - E_{H_0}T^+}{\sqrt{[n(n+1)(2n+1)]/24}} \xrightarrow{L} Z,$$

where Z is $\mathcal{N}(0, 1)$.

We next compute the distribution of T^+ for small samples. The distribution of T^+ is tabulated by Kraft and Van Eeden [62], pages 221–223.

Note that $T^+ = 0$ if all differences have negative signs, and $T^+ = n(n+1)/2$ if all differences have positive signs. Here a difference means a difference between the observations and the postulated value of the median. T^+ is completely determined by the indicators $Z_{(i)}$, so that the sample space can be considered as a set of 2^n n-tuples $(z_1, z_2, \cdots, z_n)$, where each z_i is 0 or 1. Under H_0, $m = m_0$ and each arrangement is equally likely. Thus

$$P_{H_0}\{T^+ = t\} = \frac{\{\text{number of ways to assign } + \text{ or } - \text{ signs to integers } 1, 2, \cdots, n \text{ so that the sum is } t\}}{2^n}$$

(35)
$$= \frac{n(t)}{2^n}, \text{ say.}$$

Note that every assignment has a conjugate assignment with plus and minus signs interchanged, so that for this conjugate T^+ is given by

(36) $$\sum_{1}^{n} i(1 - Z_{(i)}) = \frac{n(n+1)}{2} - \sum_{1}^{n} iZ_{(i)}.$$

Thus under H_0 the distribution of T^+ is symmetric about the mean $n(n+1)/4$.

Example 8. Let us compute the null distribution for $n = 3$. $E_{H_0}T^+ = n(n+1)/4 = 3$, and T^+ takes values from 0 to $n(n+1)/2 = 6$:

Value of T^+	Ranks Associated with Positive Differences	$n(t)$
6	1, 2, 3	1
5	2, 3	1
4	1, 3	1
3	1, 2; 3	2

so that

(37) $$P_{H_0}\{T^+ = t\} = \begin{cases} \frac{1}{8}, & t = 4, 5, 6, 0, 1, 2, \\ \frac{2}{8}, & t = 3, \\ 0, & \text{otherwise.} \end{cases}$$

Similarly, for $n = 4$, one can show that

(38) $$P_{H_0}\{T^+ = t\} = \begin{cases} \frac{1}{16}, & t = 0, 1, 2, 8, 9, 10, \\ \frac{2}{16}, & t = 3, 4, 5, 6, 7, \\ 0, & \text{otherwise.} \end{cases}$$

An alternative procedure would be to use the mgf technique. Under H_0, the rv's $iZ_{(i)}$ are independent and have the pmf

$$P\{iZ_{(i)} = i\} = P\{iZ_{(i)} = 0\} = \tfrac{1}{2}.$$

Thus

(39) $$M(t) = Ee^{tT^+}$$
$$= \prod_{i=1}^{n}\left(\frac{e^{it} + 1}{2}\right).$$

We express $M(t)$ as a sum of terms of the form $\alpha_j e^{jt}/2^n$. The pmf of T^+ can then be determined by inspection. For example, in the case $n = 4$, we have

$$M(t) = \prod_{i=1}^{4}\left(\frac{e^{it}+1}{2}\right) = \left(\frac{e^{t}+1}{2}\right)\left(\frac{e^{2t}+1}{2}\right)\left(\frac{e^{3t}+1}{2}\right)\left(\frac{e^{4t}+1}{2}\right)$$

(40) $$= \tfrac{1}{4}(e^{3t}+e^{2t}+e^{t}+1)\left(\frac{e^{3t}+1}{2}\right)\left(\frac{e^{4t}+1}{2}\right)$$

(41) $$= \tfrac{1}{8}(e^{6t}+e^{5t}+e^{4t}+2e^{3t}+e^{2t}+e^{t}+1)\left(\frac{e^{4t}+1}{2}\right)$$

(42) $$= \tfrac{1}{16}(e^{10t}+e^{9t}+e^{8t}+2e^{7t}+2e^{6t}+2e^{5t}+2e^{4t}+2e^{3t}+e^{2t}+e^{t}+1).$$

This method gives us the pmf of T^+ for $n=2$, $n=3$, and $n=4$ immediately. Quite simply,

(43) $P_{H_0}\{T^+ = j\}$ = coefficient of e^{jt} in the expansion of $M(t)$, $j = 0, 1, \cdots, n(n+1)/2$.

Example 9. Let us return to the data of Example 6 and test H_0: $m = \mu = -1.0$ against H_1: $m > -1.0$. Ranking $|x_i - m|$ in increasing order of magnitude, we have

.016 < .131 < .535 < .762 < 1.056 < 1.120 < 1.417 < 1.561
  5   4   1   3   7   2   6   8

Thus

$r_1 = 3$, $\quad r_2 = 6$, $\quad r_3 = 4$, $\quad r_4 = 2$, $\quad r_5 = 1$, $\quad r_6 = 7$,
$r_7 = 5$, $\quad r_8 = 8$

and

$$T^+ = 3 + 6 + 4 + 2 + 7 + 5 + 8 = 35.$$

From Table 10 on page 665, H_0 is rejected at level $\alpha = .05$ if $T^+ \geq 31$. Since $35 > 31$, we reject H_0.

Remark 1. The Wilcoxon test statistic can also be used to test for symmetry. Let $X_1, X_2, \cdots, X_n$ be iid observations on an rv with absolutely continuous df F. We set the null hypothesis as

H_0: median, $m = m_0$, and df F is symmetric about m_0.

The alternative is

H_1: $m \neq m_0$ and F symmetric, or F asymmetric.

The test is the same since the null distribution of T^+ is the same.

Remark 2. If we have n independent pairs of observations $(X_1, Y_1), (X_2, Y_2), \cdots, (X_n, Y_n)$ from a bivariate df, we form the differences $Z_i = X_i - Y_i$, $i = 1, 2, \cdots, n$. Assuming that $Z_1, Z_2, \cdots, Z_n$ are (independent) observations from a population of differences with absolutely continuous df F that is symmetric with median m, we can use the Wilcoxon statistic to test H_0: $m = m_0$.

We present some examples.

Example 10. For the data of Example 10.3.3 let us apply the Wilcoxon statistic to test H_0: $m = 0$ and F is symmetric against H_1: $m \neq 0$ and F symmetric or F not symmetric.

The absolute values, when arranged in increasing order of magnitude, are as follows:

$$.057 < .068 < .137 < .261 < .323 < .464 < .482 < .486 < .508 < .513$$
$$\quad 13 \quad\; 5 \quad\;\; 2 \quad\;\; 17 \quad\;\; 4 \quad\;\; 1 \quad\;\; 11 \quad\;\; 15 \quad\;\; 20 \quad\;\; 7$$
$$< .525 < .595 < .881 < .906 < 1.046 < 1.229 < 1.237 < 1.678 < 1.787 < 2.455$$
$$\quad 8 \quad\quad 9 \quad\quad 10 \quad\; 6 \quad\;\; 19 \quad\;\; 14 \quad\;\; 18 \quad\;\; 12 \quad\;\; 16 \quad\quad 3$$

Thus

$r_1 = 6, \quad r_2 = 3, \quad r_3 = 20, \quad r_4 = 5, \quad r_5 = 2, \quad r_6 = 14,$
$r_7 = 10, \quad r_8 = 11, \quad r_9 = 12, \quad r_{10} = 13, \quad r_{11} = 7, \quad r_{12} = 18,$
$r_{13} = 1, \quad r_{14} = 16, \quad r_{15} = 8, \quad r_{16} = 19, \quad r_{17} = 4, \quad r_{18} = 17,$
$r_{19} = 15, \quad r_{20} = 9,$

and

$$T^+ = 6 + 3 + 20 + 14 + 12 + 13 + 18 + 17 + 15 = 118.$$

Since $n = 20$, $E_{H_0} T^+ = 20 \times 21/4 = 105$,

$$\mathrm{var}_{H_0}(T^+) = \frac{20 \cdot 21 \cdot 41}{24} = 717.5.$$

Thus

$$\frac{T^+ - E_{H_0} T^+}{\sqrt{\mathrm{var}_{H_0}(T^+)}} = \frac{118 - 105}{26.8} = \frac{13}{26.8},$$

and $P\{Z > 13/26.6\} = P\{Z > .5\} = .3085$. We therefore accept the null hypothesis.

Example 11. Returning to the data of Example 7, we apply the Wilcoxon test to the differences $Z_i = X_i - Y_i$. The differences are $-6, 3, 1, -8, -17, -20, -3, -3, -11, 9$. To test H_0: $m = 0$ against H_1: $m \neq 0$, we rank the absolute values of z_i in increasing order to get

$$1 < 3 = 3 = 3 < 6 < 8 < 9 < 11 < 17 < 20$$

and

$$T^+ = 1 + 2 + 7 = 10.$$

Here we have assigned ranks 2, 3, 4 to observations $+3, -3, -3$. (If we assign rank 4 to observation 3, then $T^+ = 12$ without appreciably changing the result.)

From the tables, we reject H_0 at $\alpha = .05$ if either $T^+ > 46$ or $T^+ < 9$. Since $T^+ > 5$ and < 50, we accept H_0. Note that hypothesis H_0 was also accepted by the sign test.

PROBLEMS 13.3

1. Prove Theorem 4.
2. Test the goodness of fit for the data of Problem 10.3.5, using the Kolmogorov-Smirnov test.
3. Test the goodness of fit for the data of Problem 10.3.6, using the Kolmogorov-Smirnov test.
4. For the data of Problem 3 find a .95 level confidence band for the distribution function.
5. The following data represent a sample of size 20 from $U[0,1]$: .277, .435, .130, .143, .853, .889, .294, .697, .940, .648, .324, .482, .540, .152, .477, .667, .741, .882, .885, .740. Construct a .90 level confidence band for $F(x)$.
6. In Problem 5 test the hypothesis that the distribution is $U[0,1]$. Take $\alpha = .05$.
7. For the data of Example 3 test, by means of the sign test, the null hypothesis $H_0: \mu = 1.5$ against $H_1: \mu \neq 1.5$.
8. For the data of Problem 5 test the hypothesis that the quantile of order $p = .20$ is .20.
9. For the data of Problem 10.4.8 use the sign test to test the hypothesis of no difference between the two averages.
10. Use the sign test for the data of Problem 10.4.9 to test the hypothesis of no difference in grade-point averages.
11. For the data of Problem 5 apply the signed-rank test to test $H_0: m = .5$ against $H_1: m \neq .5$.
12. For the data of Problems 10.4.8 and 10.4.9 apply the signed-rank test to the differences to test $H_0: m = 0$ against $H_1: m \neq 0$.

13.4 SOME TWO-SAMPLE PROBLEMS

In this section we consider some two-sample tests. Let $X_1, X_2, \cdots, X_m$ and $Y_1, Y_2, \cdots, Y_n$ be independent samples from two absolutely continuous distribution functions F_X and F_Y, respectively. The problem is to test the null hypothesis $H_0: F_X(x) = F_Y(x)$ for all $x \in \mathcal{R}$ against the usual one-sided and two-sided alternatives. First note that $F_X > F_Y$ means

$$P\{X \leq x\} > P\{Y \leq x\}$$

so that

(1) $$P\{X > x\} < P\{Y > x\}.$$

Thus an alternative $F_X > F_Y$ means that the rv Y tends to be larger than the rv X.

Definition 1. We say that a continuous rv Y is stochastically larger than a continuous rv X if inequality (1) is satisfied.

A similar interpretation may be given to the one-sided alternative $F_X < F_Y$. In the special case where both X and Y are normal rv's with means μ_1, μ_2 and common variance σ^2, $F_X = F_Y$ corresponds to $\mu_1 = \mu_2$ and $F_X > F_Y$ corresponds to $\mu_1 < \mu_2$.

Some of the tests developed for single samples may be modified to test the null hypothesis $H_0: F_X = F_Y$.

The Chi-Square Test for Homogeneity

Since the analysis is quite general, let us consider the case of sampling from p populations, $p \geq 2$. We wish to test H_0 that they all have the same distribution. Let $A_1, A_2, \cdots, A_k$ be a partition of the real line, where the A_j's are Borel sets, and let

(2) $$P\{X_j \in A_i\} = p_{ij}, \quad i = 1, 2, \cdots, k; \, j = 1, 2, \cdots, p.$$

If $X_1, X_2, \cdots, X_p$ all have the same df, then

$$p_{i1} = p_{i2} = \cdots = p_{ip}, \quad i = 1, 2, \cdots, k.$$

This is exactly the problem of testing the equality of p independent multinomial distributions. Let $n_1, n_2, \cdots, n_p$ be the number of observations on $X_1, X_2, \cdots, X_p$, respectively. If $n_1, n_2, \cdots, n_p$ are large, the rv

(3) $$V = \sum_{j=1}^{p} \sum_{i=1}^{k} \left[\frac{(X_{ij} - n_j p_{ij})^2}{n_j p_{ij}} \right]$$

is the sum of p independent rv's each of which is asymptotically a chi-square rv with $k - 1$ d.f. Here X_{ij} is the number of observations on X_j that lie in the set A_i. Thus the sum V is a chi-square rv with $p(k - 1)$ d.f., provided that $n_1, n_2, \cdots, n_p$ are sufficiently large. In general, p_{ij}'s will not be known, and we will have to estimate them from the data. Under $H_0: p_{i1} = p_{i2} = \cdots = p_{ip}$, $i = 1, 2, \cdots, k$, and the MLE's for the common probability are

$$\hat{p}_i = \frac{\sum_{j=1}^{p} X_{ij}}{\sum_{j=1}^{p} n_j}, \quad i = 1, 2, \cdots, k.$$

We need to estimate only $k - 1$ of these probabilities, since the last can be estimated from the sum 1. Thus the rv

$$\sum_{j=1}^{p} \sum_{i=1}^{k} \left[\frac{(X_{ij} - n_j \hat{p}_i)^2}{n_j \hat{p}_i} \right]$$

is approximately χ^2 with $p(k - 1) - (k - 1) = (p - 1)(k - 1)$ d.f. We reject H_0 at level α if the computed value of χ^2 is $> \chi^2_{(p-1)(k-1), \alpha}$.

Example 1. Four samples are taken from a table of random normal numbers (Table 6, page 659) with $\mu = 0$ and $\sigma = 1$.

Sample 1	1.221,	−0.439,	1.291,	0.541,	−1.661,	0.665,
Sample 2	1.119,	−0.792,	0.063,	0.484,	1.045,	0.084,
Sample 3	0.004,	−1.275,	−1.793,	−0.986,	−1.363,	−0.880,
Sample 4	−2.015,	−0.623,	−0.699,	0.481,	−0.586,	−0.579,
Sample 1	0.340,	0.008,	0.110,	1.297,	−0.556,	−1.181,
Sample 2	−0.086,	0.427,	−0.528,	−1.433,	2.923,	−1.190,
Sample 3	−0.158,	0.831,	−0.813,	−1.345,	0.500,	−0.318,
Sample 4	−0.120,	0.191,	−0.071,	−3.001,	0.359,	−0.094,
Sample 1	−0.518,	0.843,	0.584,	−0.431,	−0.135,	−0.732,
Sample 2	0.192,	0.942,	1.216,	1.703,	−0.145,	−0.066,
Sample 3	−0.432,	1.045,	0.733,	1.164,	−0.498,	1.006,
Sample 4	1.501,	0.031,	0.402,	0.884,	0.457,	−0.798,
Sample 1	1.049,	2.040,	0.116,	−1.616,	1.381,	−0.394.
Sample 2	1.810,	−0.124.				
Sample 3	2.885,	0.196,	−1.272,	1.262,	−0.281,	1.707,
Sample 4	−0.768.	0.023.				
Sample 3	0.580.					

Here $n_1 = 24$, $n_2 = 20$, $n_3 = 25$, $n_4 = 20$. Let
$A_1 = (-\infty, -.75]$, $A_2 = (-.75, 0]$, $A_3 = (0, .75]$, and $A_4 = (.75, +\infty)$.

The observed frequencies X_{ij} are as follows:

	X_{ij}			
i	X_{i1}	X_{i2}	X_{i3}	X_{i4}
1	3	3	9	4
2	7	5	5	6
3	7	5	5	8
4	7	7	6	2

The following table gives the probabilities p_{ij}, $j = 1, 2, 3, 4$:

i:	1	2	3	4
p_{ij}:	.2266	.2734	.2734	.2266

and the expected frequencies $n_j p_{ij}$ are as follows:

$$n_j p_{ij}$$

i/j	1	2	3	4
1	5.438	4.532	5.665	4.532
2	6.562	5.468	6.835	5.468
3	6.562	5.468	6.835	5.468
4	5.438	4.532	5.665	4.532

Thus

$$v = \sum_{j=1}^{4} \sum_{i=1}^{4} \left[\frac{(x_{ij} - n_j p_{ij})^2}{n_j p_{ij}}\right] = 7.84.$$

Since $\chi^2_{9,.05} = 16.9$, we accept H_0.

If we pool A_1 with A_2 and A_3 with A_4 (because some class frequencies are < 5), we get $v = 1.82$. Since $\chi^2_{3,.05} = 7.81$, we again accept H_0 at the .05 level.

Example 2. Three samples are taken from $U[0, 1]$:

Sample 1 .59, .96, .78, .16, .87, .13, .73, .41, .74, .09, .46, .56,
Sample 2 .25, .91, .26, .77, .55, .72, .47, .43, .31, .56, .50, .96,
Sample 3 .22, .90, .78, .66, .18, .73, .43, .58, .11, .16, .47, .07,

Sample 1 .75, .10, .36, .68, .17, .52, .70, .70, .64, .61, .18, .64, .64.
Sample 2 .87, .46, .26, .60, .59, .11, .02, .33, .66, .70, .32, .02, .01.
Sample 3 .30, .40, .11, .72, .83, .09, .25, .77, .72, .33, .21, .53, .72.

To test the null hypothesis that all the samples came from $U[0, 1]$, let us choose

$A_1 = [0, .25)$, $A_2 = [.25, .50)$, $A_3 = [.50, .75)$, $A_4 = [.75, 1.00]$.

Then the observed frequencies are as follows:

$$X_{ij}$$

i	X_{i1}	X_{i2}	X_{i3}
1	6	4	8
2	3	9	6
3	12	8	7
4	4	4	4

The expected frequencies are these:

$n_j p_{ij}$

i/j	1	2	3
1	6.25	6.25	6.25
2	6.25	6.25	6.25
3	6.25	6.25	6.25
4	6.25	6.25	6.25

Thus

$$v = \sum_{j=1}^{3} \sum_{i=1}^{4} \left[\frac{(X_{ij} - n_j p_{ij})^2}{n_j p_{ij}} \right]$$
$$= 6.3877.$$

Comparing this result with the tabulated value of chi-square for 6 d.f., we see that $6.3877 < \chi^2_{6,.05} = 12.6$, so that we cannot reject H_0.

The Kolmogorov-Smirnov Test

Let $X_1, X_2, \cdots, X_m$ and $Y_1, Y_2, \cdots, Y_n$ be independent random samples from continuous df's F and G, respectively. Let the order statistics be

$$X_{(1)}, X_{(2)}, \cdots, X_{(m)} \quad \text{and} \quad Y_{(1)}, Y_{(2)}, \cdots, Y_{(n)}.$$

Let us write for the empirical df's

(4) $$F_m^*(x) = \begin{cases} 0, & x < X_{(1)}, \\ \dfrac{k}{m}, & X_{(k)} \le x < X_{(k+1)}, \quad k = 1, 2, \cdots, m-1, \\ 1, & k \ge X_{(m)}, \end{cases}$$

and

(5) $$G_n^*(x) = \begin{cases} 0, & x < Y_{(1)}, \\ \dfrac{k}{n}, & Y_{(k)} \le x < Y_{(k+1)}, \quad k = 1, 2, \cdots, n-1, \\ 1, & x \ge Y_{(n)}. \end{cases}$$

In a combined ordered arrangement of m X's and n Y's, F_m^* and G_n^* represent the respective proportions of X and Y values that do not exceed x. Under $H_0: F(x) = G(x)$ for all x, we expect a reasonable agreement between the two sample df's. We define

(6) $$D_{m,n} = \sup_x |F_m^*(x) - G_n^*(x)|.$$

Then $D_{m,n}$ may be used to test H_0 against the two-sided alternative H_1: $F(x) \ne G(x)$ for some x. The test rejects H_0 at level α if

(7) $$D_{m,n} \geq D_{m,n,\alpha},$$

where $P_{H_0}\{D_{m,n} \geq D_{m,n,\alpha}\} \leq \alpha$.

Similarly, one can define the one-sided statistics

(8) $$D^+_{m,n} = \sup_x [F^*_m(x) - G^*_n(x)]$$

and

(9) $$D^-_{m,n} = \sup_x [G^*_n(x) - F^*_m(x)],$$

to be used against the one-sided alternatives

(10) $G(x) \leq F(x)$ for all x and $G(x) < F(x)$ for some x with rejection region $D^+_{m,n} \geq D^+_{m,n,\alpha}$

and

(11) $F(x) \leq G(x)$ for all x and $F(x) < G(x)$ for some x with rejection region $D^-_{m,n} \geq D^-_{m,n,\alpha}$, respectively.

For small samples tables due to Massey [79] are available. In Table 9, page 663, we give the values of $D_{m,n,\alpha}$ and $D^+_{m,n,\alpha}$ for some selected values of m, n, and α. Table 8 gives the corresponding values for the case $m = n$.

For large samples we use the limiting result due to Smirnov [120]. Let $N = mn/(m + n)$. Then

(12) $$\lim_{m,n \to \infty} P\{\sqrt{N} D^+_{m,n} \leq \lambda\} = \begin{cases} 1 - e^{-2\lambda^2}, & \lambda > 0, \\ 0, & \lambda \leq 0, \end{cases}$$

(13) $$\lim_{m,n \to \infty} P\{\sqrt{N} D_{m,n} \leq \lambda\} = \begin{cases} \sum_{j=-\infty}^{\infty} (-1)^j e^{-2j^2\lambda^2}, & \lambda > 0, \\ 0, & \lambda \leq 0. \end{cases}$$

Relations (12) and (13) give the distribution of $D^+_{m,n}$ and $D_{m,n}$, respectively, under $H_0: F(x) = G(x)$ for all $x \in \mathcal{R}$.

Example 3. The following data represent the lifetimes (hours) of batteries for two different brands:

Brand A: 40, 30, 40, 45, 55, 30
Brand B: 50, 50, 45, 55, 60, 40

Are these brands different with respect to average life?

Let us first apply the Kolmogorov-Smirnov test to test H_0 that the population distribution of length of life for the two brands is the same.

x	$F_6^*(x)$	$G_6^*(x)$	$\lvert F_6^*(x) - G_6^*(x) \rvert$
30	$\frac{2}{6}$	0	$\frac{2}{6}$
40	$\frac{4}{6}$	$\frac{1}{6}$	$\frac{3}{6}$
45	$\frac{5}{6}$	$\frac{2}{6}$	$\frac{3}{6}$
50	$\frac{5}{6}$	$\frac{4}{6}$	$\frac{1}{6}$
55	1	$\frac{5}{6}$	$\frac{1}{6}$
60	1	1	0

$$D_{6,6} = \sup_x \lvert F_6^*(x) - G_6^*(x) \rvert = \tfrac{3}{6}.$$

From Table 8, page 662, the critical value for $m = n = 6$ at level $\alpha = .05$ is $D_{6,6,.05} = \tfrac{4}{6}$. Since $D_{6,6} \not> D_{6,6,.05}$, we accept H_0 that the population distribution for the length of life for the two brands is the same.

Let us next apply the two-sample t-test. We have $\bar{x} = 40$, $\bar{y} = 50$, $s_1^2 = 90$, $s_2^2 = 50$, $s_p^2 = 70$. Thus

$$t = \frac{40 - 50}{\sqrt{70}\,\sqrt{\tfrac{1}{6} + \tfrac{1}{6}}} = -2.08.$$

Since $t_{10,.025} = 2.2281$, we accept the hypothesis that the two samples come from the same (normal) population.

The Median Test

A frequently used test, which is essentially a test of the equality of medians of two independent distributions, is the *median test*. This test will tend to accept H_0: $F = G$ even if the shapes of F and G are different as long as the medians are the same.

The test is simple. The combined sample $X_1, X_2, \cdots, X_m, Y_1, Y_2, \cdots, Y_n$ is ordered and a median is found. If $m + n$ is odd, the median is the $[(m + n + 1)/2]$th value in the ordered arrangement. If $m + n$ is even, the median is any number between the two middle values. Let V be the number of observed values of X that are $\leq$ the sample median for the combined sample. If V is large, it is reasonable to conclude that the actual median of X is smaller than the median of Y. One therefore rejects H_0: $F = G$ in favor of H_1: $F(x) \geq G(x)$ for all x and $F(x) > G(x)$ for some x if V is too large, that is, if $V \geq c$. If, however, the alternative is $F(x) \leq G(x)$ for all x and $F(x) < G(x)$ for some x, the median test rejects H_0 if $V \leq c$.

For the two-sided alternative that $F(x) \neq G(x)$ for some x, we use the two-sided test.

We next compute the null distribution of the rv V. If $m + n = 2p$, p a positive integer, then

$$P_{H_0}\{V = v\} = P_{H_0}\{\text{Exactly } v \text{ of the } X_i\text{'s are} \leq \text{combined median}\}$$

(14)
$$= \begin{cases} \dfrac{\binom{m}{v}\binom{n}{p-v}}{\binom{m+n}{p}}, & v = 0, 1, 2, \cdots, m, \\ 0, & \text{otherwise.} \end{cases}$$

Here $0 \leq V \leq \min(m, p)$. If $m + n = 2p + 1$, $p > 0$, is an integer, the $[(m + n + 1)/2]$th value is the median in the combined sample, and

$$P_{H_0}\{V = v\} = P\{\text{Exactly } v \text{ of the } X_i\text{'s are below the } (p + 1)\text{th value in the ordered arrangement}\}$$

(15)
$$= \begin{cases} \dfrac{\binom{m}{v}\binom{n}{p-v}}{\binom{m+n}{p}}, & v = 0, 1, \cdots, \min(m, p), \\ 0, & \text{otherwise.} \end{cases}$$

Remark 1. Under H_0 we expect $(m + n)/2$ observations above the median and $(m + n)/2$ below the median. One can therefore apply the chi-square test with 1 d.f. to test H_0 against the two-sided alternative.

Example 4. For the data of Example 3 the combined sample in ordered form is as follows: $30 \leq 30 \leq 40 \leq 40 \leq 40 \leq 45 \leq 45 \leq 50 \leq 50 \leq 55 \leq 55 \leq 60$. Since $m = n = 6$, $m + n = 12$ is even and the median is 45. Thus

$v =$ number of observed values of X that are less than or equal to 45
$= 5$.

Now

$$P_{H_0}\{V \geq 5\} = \frac{\binom{6}{5}\binom{6}{1}}{\binom{12}{6}} + \frac{\binom{6}{6}\binom{6}{0}}{\binom{12}{6}} \approx .04.$$

Since $P_{H_0}\{V > 5\} > .025$, we accept H_0 that the two samples come from the same population.

The Mann-Whitney-Wilcoxon Test

Let $X_1, X_2, \cdots, X_m$ and $Y_1, Y_2, \cdots, Y_n$ be independent samples from two continuous df's, F and G, respectively. We define

(16)
$$Z_{ij} = \begin{cases} 1, & X_i < Y_j, \\ 0, & X_i > Y_j, \end{cases} \quad i = 1, 2, \cdots, m, \quad j = 1, 2, \cdots, n,$$

and write

(17) $$U = \sum_{i=1}^{m} \sum_{j=1}^{n} Z_{ij}.$$

Note that $\sum_{j=1}^{n} Z_{ij}$ is the number of Y_j's that are larger than X_i, and hence U is the number of values of $X_1, X_2, \cdots, X_m$ that are smaller than each of $Y_1, Y_2, \cdots, Y_n$. The statistic U is called the *Mann-Whitney statistic* and was first encountered in Section 13.2.

Example 5. Let $m = 4$, $n = 3$, and suppose that the combined sample when ordered is as follows:

$$x_2 < x_1 < y_3 < y_2 < x_4 < y_1 < x_3.$$

Then $U = 7$, since there are three values of $x < y_1$, two values of $x < y_2$, and two values of $x < y_3$.

Note that $U = 0$ if all the X_i's are larger than all the Y_j's and $U = mn$ if all the X_i's are smaller than all the Y_j's, because then there are m X's $< Y_1$, m X's $< Y_2$, and so on. Thus $0 \leq U \leq mn$. If U is large, the values of Y tend to be larger than the values of X (Y is stochastically larger than X), and this supports the alternative $F(x) \geq G(x)$ for all x and $F(x) > G(x)$ for some x. Similarly, if U is small, the Y values tend to be smaller than the X values, and this supports the alternative $F(x) \leq G(x)$ for all x and $F(x) < G(x)$ for some x. We summarize these results as follows.

H_0	H_1	Reject H_0 if
$F = G$	$F \geq G$	$U \geq c_1$
$F = G$	$F \leq G$	$U \leq c_2$
$F = G$	$F \neq G$	$U \geq c_3$ or $U \leq c_4$

To compute the critical values we need the null distribution of U. Let

(18) $$p_{m,n}(u) = P_{H_0}\{U = u\}.$$

We will set up a difference equation relating $p_{m,n}$ to $p_{m-1,n}$ and $p_{m,n-1}$. If the observations are arranged in increasing order of magnitude, the largest value can be either an x value or a y value. Under H_0, all $m + n$ values are equally likely, so that the probability that the largest value will be an x value is $m/(m + n)$ and that it will be a y value is $n/(m + n)$.

Now, if the largest value is an x, it does not contribute to U, and the remaining $m - 1$ values of x and n values of y can be arranged to give the observed value $U = u$ with probability $p_{m-1,n}(u)$. If the largest value is a Y, this value is larger than all the m x's. Thus, to get $U = u$, the remaining $n - 1$ values of Y and m values of x contribute $U = u - m$. It follows that

(19) $$p_{m,n}(u) = \frac{m}{m+n} p_{m-1,n}(u) + \frac{n}{m+n} p_{m,n-1}(u-m).$$

If $m = 0$, then for $n \geq 1$

(20) $$p_{0,n}(u) = \begin{cases} 1 & \text{if } u = 0, \\ 0 & \text{if } u > 0. \end{cases}$$

If $n = 0$, $m \geq 1$, then

(21) $$p_{m,0}(u) = \begin{cases} 1 & \text{if } u = 0, \\ 0 & \text{if } u > 0, \end{cases}$$

and

(22) $$p_{m,n}(u) = 0 \quad \text{if } u < 0, \quad m \geq 0, \quad n \geq 0.$$

For small values of m and n one can easily compute the pmf of U. Thus, if $m = n = 1$, then

$$p_{1,1}(0) = \tfrac{1}{2}, \qquad p_{1,1}(1) = \tfrac{1}{2}.$$

If $m = 1$, $n = 2$, then

$$p_{1,2}(0) = p_{1,2}(1) = p_{1,2}(2) = \tfrac{1}{3}.$$

Tables for critical values are available for small values of m and n, $m \leq n$. See, for example, Auble [4] or Mann and Whitney [78]. Table 11 on page 666 gives the values of u_α for which $P_{H_0}\{U > u_\alpha\} \leq \alpha$ for some selected values of m, n, and α.

For large values of m and n, one can use the central limit theorem. We have

(23) $$E_{H_0} U = \sum_{i=1}^m \sum_{j=1}^n E_{H_0} Z_{ij} = \sum_i \sum_j P_{H_0}\{X_i < Y_j\}$$
$$= \frac{mn}{2}.$$

Also

$$E_{H_0} U^2 = \sum_{k=1}^n \sum_{h=1}^m \sum_{j=1}^n \sum_{i=1}^m E_{H_0}(Z_{ij} Z_{hk})$$
$$= \sum_{j=1}^n \sum_{i=1}^m E_{H_0} Z_{ij}^2 + \sum_{\substack{k=1 \\ k \neq j}}^n \sum_{j=1}^n \sum_{i=1}^m E_{H_0}(Z_{ij} Z_{ik})$$

(24) $$+ \sum_{j=1}^n \sum_{\substack{h=1 \\ h \neq i}}^m \sum_{i=1}^m E_{H_0}(Z_{ij} Z_{hj}) + \sum_{k \neq j} \sum_{h \neq i} E_{H_0}(Z_{ij} Z_{hk}).$$

Now

(25) $$E_{H_0} Z_{ij}^2 = P_{H_0}\{X_i < Y_j\} = \tfrac{1}{2},$$

and for $j \neq k$

$$E_{H_0}(Z_{ij}Z_{ik}) = P_{H_0}\{Z_{ij} = 1, Z_{ik} = 1\}$$
$$= P_{H_0}\{X_i < Y_j, X_i < Y_k\},$$
(26) $$= \tfrac{1}{3},$$

since a designated one of three terms is smaller than the other two. Similarly

(27) $$E_{H_0}(Z_{ij}Z_{hj}) = P_{H_0}\{X_i < Y_j, X_h < Y_j\} = \tfrac{1}{3}, \quad i \neq h,$$

and for $k \neq j$, $h \neq i$

$$E_{H_0}(Z_{ij}Z_{hk}) = P_{H_0}\{X_i < Y_j, X_h < Y_k\}$$
$$= P_{H_0}\{X_i < Y_j\}P_{H_0}\{X_h < Y_k\}$$
(28) $$= \tfrac{1}{4}.$$

It follows that

(29) $$E_{H_0}U^2 = \frac{mn}{2} + \frac{mn(n-1)}{3} + \frac{mn(m-1)}{3} + \frac{mn(m-1)(n-1)}{4}$$

and

(30) $$\text{var}_{H_0}(U) = \frac{mn(m+n+1)}{12}.$$

The statistic U is the sum of mn identically distributed (but not independent) rv's. A generalized version of the CLT shows (see, for example, Mann and Whitney [78]) that under H_0 the statistic

$$Z_{m,n} = \frac{U - mn/2}{\sqrt{mn(m+n+1)/12}}$$

tends to the standard normal distribution as $m,n \to \infty$, provided that m/n remains constant. The approximation is fairly good for $m, n \geq 8$.

Example 6. Two samples are as follows:

Values of X_i: 1, 2, 3, 5, 7, 9, 11, 18
Values of Y_i: 4, 6, 8, 10, 12, 13, 14, 15, 19

Thus $m = 8$, $n = 9$, and $U = 3 + 4 + 5 + 6 + 7 + 7 + 7 + 7 + 8 = 54$. Also,

$$E_{H_0}U = \frac{8 \cdot 9}{2} = 36, \quad \text{var}_{H_0}(U) = \frac{8 \cdot 9}{12}(8 + 9 + 1) = 108,$$

and

$$Z = \frac{54 - 36}{\sqrt{108}} = \frac{18}{6\sqrt{3}} = \sqrt{3} = 1.732.$$

Since $1.732 > 1.65$, we reject H_0 that the two samples came from the same population at level $\alpha = .10$.

PROBLEMS 13.4

1. For the data of Example 6 apply the median test.

2. Twelve 4-year-old boys and twelve 4-year-old girls were observed during two 15-minute play sessions, and each child's play during these two periods was scored as follows for incidence and degree of agression:

   Boys: 86, 69, 72, 65, 113, 65, 118, 45, 141, 104, 41, 50
   Girls: 55, 40, 22, 58, 16, 7, 9, 16, 26, 36, 20, 15

   Test the hypothesis that there were sex differences in the amount of aggression shown, using (a) the median test and (b) the Mann-Whitney-Wilcoxon test.

   (Siegel [115])

3. To compare the variability of two brands of tires, the following mileages (1000 miles) were obtained for eight tires of each kind:

   Brand A: 32.1, 20.6, 17.8, 28.4, 19.6, 21.4, 19.9, 30.1
   Brand B: 19.8, 27.6, 30.8, 27.6, 34.1, 18.7, 16.9, 17.9

   Test the null hypothesis that the two samples come from the same population, using the Mann-Whitney-Wilcoxon test.

4. Use the data of Problem 2 to apply the Kolmogorov-Smirnov test.

5. Apply the Kolmogorov-Smirnov test to the data of Problem 3.

13.5 TESTS OF INDEPENDENCE

Let X and Y be two rv's with joint df $F(x, y)$, and let F_1 and F_2, respectively, be the marginal df's of X and Y. In this section we study some tests of the hypothesis of independence, namely,

$$H_0: F(x, y) = F_1(x) F_2(y) \quad \text{for all } (x, y) \in \mathcal{R}_2$$

against the alternative

$$H_1: F(x, y) \neq F_1(x) F_2(y) \text{ for some } (x, y).$$

If the joint distribution function F is bivariate normal, we know that X and Y are independent if and only if the correlation coefficient $\rho = 0$. In this case, the test of independence is to test $H_0: \rho = 0$. (See Remark 12.3.5.)

In the nonparametric situation the most commonly used test of independence is the chi-square test, which we now study.

Chi-square Test of Independence—Contingency Tables

Let X and Y be two rv's, and suppose that we have n observations on (X, Y). Let us divide the space of values assumed by X (the real line) into r mutually exclusive intervals $A_1, A_2, \cdots, A_r$. Similarly, the space of values of Y is divided into c disjoint intervals $B_1, B_2 \cdots, B_c$. As a rule of thumb, we choose the length of each interval in such a way that the probability that $X(Y)$ lies in an interval is approximately $(1/r)(1/c)$. Moreover, it is desirable to have n/r and n/c at least equal to 5. Let X_{ij} denote the number of pairs (X_k, Y_k), $k = 1, 2, \cdots, n$, that lie in $A_i \times B_j$, and let

(1) $\quad p_{ij} = P\{(X, Y) \in A_i \times B_j\} = P\{X \in A_i \text{ and } Y \in B_j\},$

where $i = 1, 2, \cdots, r$, $j = 1, 2, \cdots, c$. If each p_{ij} is known, the quantity

(2) $$\sum_{i=1}^{r} \sum_{j=1}^{c} \left[\frac{(X_{ij} - np_{ij})^2}{np_{ij}} \right]$$

has approximately a chi-square distribution with $rc - 1$ d.f., provided that n is large. (See Theorem 10.3.2.) If X and Y are independent, $P\{(X, Y) \in A_i \times B_j\} = P\{X \in A_i\} P\{Y \in B_j\}$. Let us write $p_{i\cdot} = P\{X \in A_i\}$ and $p_{\cdot j} = P\{Y \in B_j\}$. Then under $H_0: p_{ij} = p_{i\cdot} \cdot p_{\cdot j}$, $i = 1, 2, \cdots, r$, $j = 1, 2, \cdots, c$. In practice, p_{ij} will not be known. We replace p_{ij} by their estimates. Under H_0, we estimate $p_{i\cdot}$ by

(3) $$\hat{p}_{i\cdot} = \frac{\sum_{j=1}^{c} X_{ij}}{n}, \quad i = 1, 2, \cdots, r,$$

and $p_{\cdot j}$ by

(4) $$\hat{p}_{\cdot j} = \sum_{i=1}^{r} \frac{X_{ij}}{n}, \quad j = 1, 2, \cdots, c.$$

Since $\sum_{j=1}^{c} \hat{p}_{\cdot j} = 1 = \sum_{1}^{r} \hat{p}_{i\cdot}$, we have estimated only $r - 1 + c - 1 = r + c - 2$ parameters. It follows (see Theorem 10.3.4) that the rv

(5) $$U = \sum_{i=1}^{r} \sum_{j=1}^{c} \left[\frac{(X_{ij} - n\hat{p}_{i\cdot} \hat{p}_{\cdot j})^2}{n\hat{p}_{i\cdot} \hat{p}_{\cdot j}} \right]$$

is asymptotically distributed as χ^2 with $rc - 1 - (r + c - 2) = (r - 1)(c - 1)$ d.f., under H_0. The null hypothesis is rejected if the computed value of U exceeds $\chi^2_{(r-1)(c-1),\alpha}$.

It is frequently convenient to list the observed and expected frequencies of the rc events $A_i \times B_j$ in an $r \times c$ table, called a *contingency table,* as follows:

	Observed Frequencies, O_{ij}				Expected Frequencies, E_{ij}			
	B_1	$B_2 \cdots B_c$			B_1	B_2	$\cdots \quad B_c$	
A_1	X_{11}	$X_{12} \cdots X_{1c}$	$\sum X_{1j}$	A_1	$np_1 \cdot p_{\cdot 1}$	$np_1 \cdot p_{\cdot 2}$	$\cdots \quad np_1 \cdot p_{\cdot c}$	$np_1.$
A_2	X_{21}	$X_{22} \cdots X_{2c}$	$\sum X_{2j}$	A_2	$np_2 \cdot p_{\cdot 1}$	$np_2 \cdot p_{\cdot 2}$	$\cdots \quad np_2 \cdot p_{\cdot c}$	$np_2.$
.	.	. $\cdots$.	.	.	.	.	$\cdots$	
.	.	. $\cdots$.	.	.	.	.	$\cdots$	
.	.	. $\cdots$.	.	.	.	.	$\cdots$	
A_r	X_{r1}	$X_{r2} \cdots X_{rc}$	$\sum X_{rj}$	A_r	$np_r \cdot p_{\cdot 1}$	$np_r \cdot p_{\cdot 2}$	$\cdots \quad np_r \cdot p_{\cdot c}$	$np_r.$
	$\sum X_{i1}$	$\sum X_{i2} \quad \sum X_{ic}$	n		$np_{\cdot 1}$	$np_{\cdot 2}$	$np_{\cdot c}$	n

Note that the X_{ij}'s in the table are frequencies. Once the category $A_i \times B_j$ is determined for an observation (X, Y), numerical values of X and Y are irrelevant. Next, we need to compute the expected frequency table. This is done quite simply by multiplying the row and column totals for each pair (i, j) and dividing the product by n. Then we compute the quantity

$$\sum_i \sum_j \frac{(E_{ij} - O_{ij})^2}{E_{ij}}$$

and compare it with the tabulated χ^2 value. In this form the test can be applied even to qualitative data. $A_1, A_2, \cdots, A_r$ and $B_1, B_2, \cdots, B_c$ represent the two attributes, and the null hypothesis to be tested is that the attributes A and B are independent.

Example 1. The following are the results for a random sample of 400 employed individuals:

Length of time (years) with the same company	Annual Income (dollars)			
	Less than 8000	8000—15,000	More than 15,000	Total
< 5	50	75	25	150
5—10	25	50	25	100
10 or more	25	75	50	150
	100	200	100	400

If X denotes the length of service with the same company, and Y, the annual income we wish to test the hypothesis that X and Y are independent. The expected frequencies are as follows:

Time (years) with the same company	Expected Frequencies			
	< 8,000	8 − 15,000	≥ 15,000	Total
<5	37.5	75	37.5	150
5—10	25	50	25	100
≥10	37.5	75	37.5	150
	100	200	100	400

Thus
$$U = \frac{(12.5)^2}{37.5} + \frac{0}{25} + \frac{(12.5)^2}{37.5} + 0 + 0 + 0 + \frac{(12.5)^2}{37.5} + 0 + \frac{(12.5)^2}{37.5}$$
$$= 16.66.$$

The number of degrees of freedom is $(3-1)(3-1) = 4$, and $\chi^2_{4,.05} = 9.488$. Since $16.66 > 9.488$, we reject H_0 at level .05 and conclude that length of service with a company is not independent of annual income.

Kendall's tau.

Let $(X_1, Y_1), (X_2, Y_2), \cdots, (X_n, Y_n)$ be a sample from a bivariate population.

Definition 1. For any two pairs (X_i, Y_i) and (X_j, Y_j) we say that the relation is perfect concordance (or agreement) if

(6) $X_i < X_j$ whenever $Y_i < Y_j$ or $X_i > X_j$ whenever $Y_i > Y_j$

and that the relation is perfect discordance (disagreement) if

(7) $X_i > X_j$ whenever $Y_i < Y_j$ or $X_i < X_j$ whenever $Y_i > Y_j$.

Writing π_c and π_d for the probability of perfect concordance and of perfect discordance, respectively, we have

(8) $$\pi_c = P\{(X_j - X_i)(Y_j - Y_i) > 0\}$$

and

(9) $$\pi_d = P\{(X_j - X_i)(Y_j - Y_i) < 0\},$$

and, if the marginal distributions of X and Y are continuous,

$$\pi_c = [P\{Y_i < Y_j\} - P\{X_i > X_j \text{ and } Y_i < Y_j\}]$$
$$+ [P\{Y_i > Y_j\} - P\{X_i < X_j \text{ and } Y_i > Y_j\}]$$
(10) $$= 1 - \pi_d.$$

Definition 2. The measure of association between the rv's X and Y defined by

(11) $$\tau = \pi_c - \pi_d$$

is known as Kendall's tau.

If the marginal distributions of X and Y are continuous, we may rewrite (11), in view of (10), as follows:

(12) $$\tau = 1 - 2\pi_d = 2\pi_c - 1.$$

In particular, if X and Y are independent and continuous rv's, then
$$P\{X_i < X_j\} = P\{X_i > X_j\} = \tfrac{1}{2},$$
since then $X_i - X_j$ is a symmetric rv. Then
$$\begin{aligned}\pi_c &= P\{X_i < X_j\}\,P\{Y_i < Y_j\} + P\{X_i > X_j\}\,P\{Y_i > Y_j\}\\ &= P\{X_i > X_j\}\,P\{Y_i < Y_j\} + P\{X_i < X_j\}\,P\{Y_i > Y_j\}\\ &= \pi_d,\end{aligned}$$
and it follows that $\tau = 0$ for independent continuous rv's.

Note that, in general, $\tau = 0$ does not imply independence. However, for the bivariate normal distribution $\tau = 0$ if and only if the correlation coefficient ρ, between X and Y, is 0, so that $\tau = 0$ if and only if X and Y are independent (Problem 6).

In order to use τ as a test of independence, we first need to find an estimate of τ from the sample. Let us write

(13) $$A_{ij} = \operatorname{sgn}(X_j - X_i)\operatorname{sgn}(Y_j - Y_i),$$

where $\operatorname{sgn} u = -1$ if $u < 0$, $= 0$ if $u = 0$, and $= 1$ if $u > 0$. Then A_{ij} takes values a_{ij} where

(14) $$a_{ij} = \begin{cases} 1 & \text{if pairs } (X_i, Y_i) \text{ and } (X_j, Y_j) \text{ are concordant,}\\ 0 & \text{if pairs } (X_i, Y_i) \text{ and } (X_j, Y_j) \text{ are neither concordant}\\ & \text{nor discordant,}\\ -1 & \text{if pairs } (X_i, Y_i) \text{ and } (X_j, Y_j) \text{ are discordant.}\end{cases}$$

Also

(15) $$P\{A_{ij} = a_{ij}\} = \begin{cases} \pi_c & \text{if } a_{ij} = 1,\\ \pi_d & \text{if } a_{ij} = -1,\\ 1 - \pi_c - \pi_d & \text{if } a_{ij} = 0,\end{cases}$$

and

(16) $$EA_{ij} = \pi_c - \pi_d = \tau.$$

Thus A_{ij} is an unbiased estimate for τ. Note that $a_{ij} = a_{ji}$ and $a_{ii} = 0$, so that

(17) $$T = \sum_{\substack{i=1 \\ i<j}}^{n} \sum_{j=1}^{n} \binom{n}{2}^{-1} A_{ij}$$

is an unbiased estimate of τ.

Definition 3. The statistic T defined in (17) is known as Kendall's sample tau coefficient.

The statistic T may be written in several equivalent forms, but we will put it into a form that is convenient for computation. Let us write

P = number of positive A_{ij}'s,
N = number of negative A_{ij}'s, $\quad 1 \leq i < j \leq n$.

Then

(18) $$T = \frac{P - N}{\binom{n}{2}}.$$

If there are no ties within the X or the Y observations, $A_{ij} \neq 0$ for $i \neq j$ and $P + N = \binom{n}{2}$. In that case

(19) $$T = \frac{2P}{\binom{n}{2}} - 1 = 1 - \frac{2N}{\binom{n}{2}}.$$

Note that $-1 \leq T \leq 1$.

To test H_0: X and Y are independent against H_1: X and Y are dependent, we reject H_0 if $|T|$ is large. Under H_0, $\tau = 0$, so that the null distribution of T is symmetric about 0. Thus we reject H_0 at level α if the observed value of T, t, satisfies $|t| > t_{\alpha/2}$, where $P\{|T| \geq t_{\alpha/2}|H_0\} = \alpha$.

For small values of n the null distribution can be directly evaluated. Values for $4 \leq n \leq 10$ are tabulated by Kendall [57]. Table 12 on page 667 gives the values of S_α for which $P\{S > S_\alpha\} \leq \alpha$, where $S = \binom{n}{2}T$ for selected values of n and α.

For a direct evaluation of the null distribution we note that the numerical value of T is clearly invariant under all order-preserving transformations. It is therefore convenient to order X and Y values and assign them ranks. If we write the pairs from the smallest to the largest according to, say, X values, P is the number of pairs of values of $1 \leq i < j \leq n$ for which $Y_j - Y_i > 0$.

Example 2. Let $n = 4$, and let us find the null distribution of T. There are 4! different permutations of ranks of Y:

Ranks of X values:	1	2	3	4
Ranks of Y values:	a_1	a_2	a_3	a_4

where (a_1, a_2, a_3, a_4) is one of the 24 permutations of 1, 2, 3, 4. Since the distribution is symmetric about 0, we need only compute one half of the distribution.

P	T	Number of Permutations	$P_{H_0}\{T = t\}$
0	-1.00	1	$\frac{1}{24}$
1	$-.67$	3	$\frac{3}{24}$
2	$-.33$	5	$\frac{5}{24}$
3	$.00$	6	$\frac{6}{24}$

Similarly, for $n = 3$ the distribution of T under H_0 is as follows:

P	T	Number of Permutations	$P_{H_0}\{T = t\}$
0	−1.00	1: (3, 2, 1)	$\frac{1}{6}$
1	− .33	2: (2, 3, 1), (3, 1, 2)	$\frac{2}{6}$

Example 3. Two judges rank four essays as follows:

| | \multicolumn{4}{c}{Essay} |
Judge	1	2	3	4
1, X	3	4	2	1
2, Y	3	1	4	2

To test H_0: rankings of the two judges are independent, let us arrange the rankings of the first judge from 1 to 4. Then we have:

| Judge 1, X: | 1 | 2 | 3 | 4 |
| Judge 2, Y: | 2 | 4 | 3 | 1 |

P = number of pairs of rankings for Judge 2 such that for $j > i$, $Y_j - Y_i > 0 = 2$ [the pairs (2, 4) and (2, 3)], and

$$t = \frac{2 \cdot 2}{\binom{4}{2}} - 1 = -.33.$$

Since

$$P_{H_0}\{|T| \geq .33\} = \frac{18}{24} = .75,$$

we accept H_0.

For large $n (\geq 8)$ it can be shown (see Kendall [57]) that the statistic T has an asymptotic normal distribution with mean 0 and variance $2(2n + 5)/[9n(n - 1)]$. Thus, for large n, the statistic

$$(20) \qquad Z = \frac{3\sqrt{n(n - 1)}\, T}{\sqrt{2(2n + 5)}}$$

may be treated approximately as $\mathcal{N}(0, 1)$.

Spearman's Rank Correlation Coefficient

Let $(X_1, Y_1), (X_2, Y_2), \cdots, (X_n, Y_n)$ be a sample from a bivariate population. In Section 7.3 we defined the sample correlation coefficient by

$$(21) \qquad R = \frac{\sum_{i=1}^{n} (X_i - \bar{X})(Y_i - \bar{Y})}{\left\{ \sum_{i=1}^{n} (X_i - \bar{X})^2 \sum_{i=1}^{n} (Y_i - \bar{Y})^2 \right\}^{1/2}},$$

where

$$\bar{X} = n^{-1} \sum_{i=1}^{n} X_i \quad \text{and} \quad \bar{Y} = n^{-1} \sum_{i=1}^{n} Y_i.$$

If the sample values $X_1, X_2, \cdots, X_n$ and $Y_1, Y_2, \cdots, Y_n$ are each ranked from 1 to n in increasing order of magnitude separately, and if the X's and Y's have continuous df's, we get a unique set of rankings. The data will then reduce to n pairs of rankings. Let us write

$$R_i = \text{rank}(X_i), \quad S_i = \text{rank}(Y_i);$$

then R_i and $S_i \in \{1, 2, \cdots, n\}$. Also,

(22) $$\sum_{1}^{n} R_i = \sum_{1}^{n} S_i = \frac{n(n+1)}{2},$$

(23) $$\bar{R} = n^{-1} \sum_{1}^{n} R_i = \frac{n+1}{2}, \quad \bar{S} = n^{-1} \sum_{1}^{n} S_i = \frac{n+1}{2},$$

and

(24) $$\sum_{1}^{n} (R_i - \bar{R})^2 = \sum_{1}^{n} (S_i - \bar{S})^2 = \frac{n(n^2-1)}{12}.$$

Substituting in (21), we obtain

$$R = \frac{12 \sum_{i=1}^{n} (R_i - \bar{R})(S_i - \bar{S})}{n^3 - n}$$

(25) $$= \frac{12 \sum_{1}^{n} R_i S_i}{n(n^2-1)} - \frac{3(n+1)}{n-1}.$$

Writing $D_i = R_i - S_i = (R_i - \bar{R}) - (S_i - \bar{S})$, we have

$$\sum_{i=1}^{n} D_i^2 = \sum_{i=1}^{n} (R_i - \bar{R})^2 + \sum_{i=1}^{n} (S_i - \bar{S})^2 - 2 \sum_{i=1}^{n} (R_i - \bar{R})(S_i - \bar{S})$$

$$= \tfrac{1}{6} n(n^2 - 1) - 2 \sum_{i=1}^{n} (R_i - \bar{R})(S_i - \bar{S}),$$

and it follows that

(26) $$R = 1 - \frac{6 \sum_{i=1}^{n} D_i^2}{n(n^2-1)}.$$

The statistic R defined in (25) or (26) is called *Spearman's rank correlation coefficient*.

From (25) we see that

$$ER = \frac{12}{n(n^2-1)} E\left(\sum_{i=1}^{n} R_i S_i\right) - \frac{3(n+1)}{n-1}$$

$$\text{(27)} \qquad = \frac{12}{n^2 - 1} E(R_i S_i) - \frac{3(n + 1)}{n - 1}.$$

Under H_0, the rv's X and Y are independent, so that the ranks R_i and S_i are also independent. It follows that

$$E_{H_0}(R_i S_i) = ER_i \, ES_i = \left(\frac{n + 1}{2}\right)^2,$$

and

$$\text{(28)} \qquad E_{H_0} R = \frac{12}{n^2 - 1}\left(\frac{n + 1}{2}\right)^2 - \frac{3(n + 1)}{n - 1} = 0.$$

Thus we should reject H_0 if the absolute value of R is large, that is, reject H_0 if

$$\text{(29)} \qquad\qquad\qquad |R| > R_\alpha,$$

where $P_{H_0}\{|R| > R_\alpha\} \le \alpha$. To compute R_α we need the null distribution of R. For this purpose it is convenient to assume, without loss of generality, that $R_i = i$, $i = 1, 2, \cdots, n$. Then $D_i = i - S_i$, $i = 1, 2, \cdots, n$. Under H_0, X and Y being independent, the $n!$ pairs (i, S_i) of ranks are equally likely. It follows that

$$\text{(30)} \qquad P_{H_0}\{R = r\} = (n!)^{-1} \times \text{(number of pairs for which } R = r)$$

$$= \frac{n_r}{n!}, \text{ say}.$$

Note that $-1 \le R \le 1$, and the extreme values can occur only when either the rankings match, that is, $R_i = S_i$, in which case $R = 1$, or $R_i = n + 1 - S_i$, in which case $R = -1$. Moreover, one need compute only one half of the distribution, since it is symmetric about 0 (Problem 8).

In the following example we will compute the distribution of R for $n = 3$ and 4. The exact complete distribution of $\sum_{i=1}^n D_i^2$, and hence R, for $n \le 10$ has been tabulated by Kendall [57]. Table 13 on page 667 gives the values of R_α for some selected values of n and α.

Example 4. Let us first enumerate the null distribution of R for $n = 3$. This is done in the following table:

(s_1, s_2, s_3)	$\sum_{i=1}^n i s_i$	$r = \dfrac{12 \sum_1^n i s_i}{n(n^2 - 1)} - \dfrac{3(n + 1)}{n - 1}$
(1, 2, 3)	14	1.0
(1, 3, 2)	13	.5
(2, 1, 3)	13	.5

Thus

$$P_{H_0}\{R = r\} = \begin{cases} \frac{1}{6}, & r = 1.0, \\ \frac{2}{6}, & r = .5 \\ \frac{2}{6}, & r = -.5, \\ \frac{1}{6}, & r = -1.0. \end{cases}$$

Similarly, for $n = 4$ we have the following:

(s_1, s_2, s_3, s_4)	$\sum_1^n is_i$	r	n_r	$P_{H_0}\{R = r\}$
(1, 2, 3, 4)	30	1	1	$\frac{1}{24}$
{(1, 3, 2, 4), (2, 1, 3, 4), (1, 2, 4, 3)	29	.8	3	$\frac{3}{24}$
(2, 1, 4, 3)	28	.6	1	$\frac{1}{24}$
{(1, 3, 4, 2), (1, 4, 2, 3), (2, 3, 1, 4), (3, 1, 2, 4)	27	.4	4	$\frac{4}{24}$
(1, 4, 3, 2), (3, 2, 1, 4)	26	.2	2	$\frac{2}{24}$
	25	.0	2	$\frac{2}{24}$

The last value is obtained from symmetry.

Example 5. In Example 3, we see that

$$r = \frac{12 \times 23}{4 \times 15} - \frac{3 \times 5}{3} = -.4.$$

Since $P_{H_0}\{|R| \geq .4\} = 18/24 = .75$, we cannot reject H_0 at $\alpha = .05$ or $\alpha = .10$.

For large samples it is possible to use a normal approximation. It can be shown (see, for example, Fraser [33], 247–248) that under H_0 the rv

$$Z = \left(12 \sum_{i=1}^n R_i S_i - 3n^3\right) n^{-5/2}$$

or, equivalently,

$$Z = R\sqrt{n-1}$$

has approximately a standard normal distribution. The approximation is good for $n \geq 10$.

PROBLEMS 13.5

1. A sample of 240 men was classified according to characteristics A and B. Characteristic A was subdivided into four classes A_1, A_2, A_3, and A_4, while B was subdivided into three classes B_1, B_2, and B_3, with the following result:

	A_1	A_2	A_3	A_4	
B_1	12	25	32	11	80
B_2	17	18	22	23	80
B_3	21	17	16	26	80
	50	60	70	60	240

Is there evidence to support the theory that A and B are independent?

2. The following data represent the blood types and ethnic groups of a sample of Iraqi citizens:

Ethnic Group	Blood Type			
	O	A	B	AB
Kurd	531	450	293	226
Arab	174	150	133	36
Jew	42	26	26	8
Turkoman	47	49	22	10
Ossetian	50	59	26	15

Is there evidence to conclude that blood type is independent of ethnic group?

3. In a public opinion poll, a random sample of 500 American adults across the country was asked the following question: "Do you believe that there was a concerted effort to cover up the Watergate scandal? Answer yes, no, or no opinion." The responses according to political beliefs were as follows:

Political Affiliation	Response			
	Yes	No	No Opinion	
Republican	45	75	30	150
Independent	85	45	20	150
Democrat	140	30	30	200
	270	150	80	500

Test the hypothesis that attitude toward the Watergate cover-up is independent of political party affiliation.

4. A random sample of 100 families in Bowling Green, Ohio, showed the following distribution of home ownership by family income:

Residential Status	Annual Income (dollars)		
	Less than 7,500	7,500–12,000	12,000 or Above
Home owner	10	15	30
Renter	8	17	20

Is home ownership in Bowling Green independent of family income?

5. In a flower show the judges agreed that five exhibits were outstanding, and these were numbered arbitrarily from 1 to 5. Three judges each arranged these five exhibits in order of merit, giving the following rankings:

Judge A:	5	3	1	2	4
Judge B:	3	1	5	4	2
Judge C:	5	2	3	1	4

Compute the avarage values of Spearman's rank correlation coefficient R and Kendall's sample tau coefficient T from the three possible pairs of rankings.

6. For the bivariate normally distributed rv (X,Y) show that $\tau = 0$ if and only if X and Y are independent.
(*Hint:* Show that $\tau = (2/\pi) \sin^{-1} \rho$, where ρ is the correlation coefficient between X and Y.)

7. Show that τ is estimable with degree $m = 2$. The estimate A_{12} is unbiased for τ with variance ≤ 1, and the U-statistic is given by (17).

8. Show that the distribution of Spearman's rank correlation coefficient R is symmetric about 0.

9. In Problem 5 test the null hypothesis that rankings of judge A and judge C are independent. Use both Kendall's tau and Spearman's rank correlation tests.

10. A random sample of 12 couples showed the following distribution of heights:

	Height (inches)			Height (inches)	
Couple	Husband	Wife	Couple	Husband	Wife
1	80	72	7	74	68
2	70	60	8	71	71
3	73	76	9	63	61
4	72	62	10	64	65
5	62	63	11	68	66
6	65	46	12	67	67

(a) Compute T.
(b) Compute R.
(c) Test the hypothesis that the heights of husband and wife are independent, using T as well as R. In each case use the normal approximation.

13.6 SOME USES OF ORDER STATISTICS

In this section we consider some applications of order statistics. Here we are mainly interested in three applications, namely, tolerance intervals for distributions, coverages, and confidence interval estimates for quantiles.

Definition 1. Let F be a continuous df. A tolerance interval for F with tolerance coefficient γ is a random interval such that the probability is γ that this random interval covers at least a specified percentage ($100p$) of the distribution.

Let $X_1, X_2, \cdots, X_n$ be a sample of size n from F, and let $X_{(1)}, X_{(2)}, \cdots, X_{(n)}$ be the corresponding set of order statistics. If the end points of the tolerance interval are two order statistics $X_{(r)}, X_{(s)}, r < s$, we have

(1) $$P\{P\{X_{(r)} < X < X_{(s)}\} \geq p\} = \gamma.$$

Since F is continuous, $F(X)$ is $U(0, 1)$, and we have

$$P\{X_{(r)} < X < X_{(s)}\} = P\{X < X_{(s)}\} - P\{X \leq X_{(r)}\}$$
$$= F(X_{(s)}) - F(X_{(r)})$$
(2) $$= U_{(s)} - U_{(r)},$$

where $U_{(r)}, U_{(s)}$ are the order statistics from $U(0, 1)$. Thus (1) reduces to

(3) $$P\{U_{(s)} - U_{(r)} \geq p\} = \gamma.$$

Using the joint distribution of $U_{(r)}$ and $U_{(s)}$ (see Theorem 4.5.4), we have from (3)

(4) $$\gamma = \int_p^1 \int_0^{y-p} \frac{n!}{(r-1)!\,(s-r-1)!\,(n-s)!} x^{r-1} (y-x)^{s-r-1} \cdot (1-y)^{n-s} \, dx \, dy$$

for all $0 < p < 1$ and $r < s$. Unfortunately, (4) is not easy to solve. It is convenient to write $U = U_{(s)} - U_{(r)}$, and let $V = U_{(s)}$. Then the joint pdf of U and V is given by

$$f_{U,V}(u, v) = \begin{cases} \dfrac{n!}{(r-1)!\,(s-r-1)!\,(n-s)!} (v-u)^{r-1} u^{s-r-1} (1-v)^{n-s}, \\ \qquad\qquad\qquad\qquad\qquad\qquad 0 < u < v < 1, \\ 0, \qquad\qquad\qquad\qquad\qquad\qquad \text{otherwise.} \end{cases}$$

Thus

$$f_U(u) = \frac{n!}{(r-1)!\,(s-r-1)!\,(n-s)!} u^{s-r-1} \int_u^1 (v-u)^{r-1} (1-v)^{n-s} \, dv$$

$$= \frac{n!}{(r-1)!\,(s-r-1)!\,(n-s)!} u^{s-r-1} \int_0^1 t^{r-1} [1 - u - t(1-u)]^{n-s} \cdot (1-u)^r \, dt$$

$$= \frac{n!}{(r-1)!\,(s-r-1)!\,(n-s)!} u^{s-r-1} (1-u)^{n-s+r} B(r, n-s+1)$$

(5) $$= \frac{n!}{(s-r-1)!\,(n-s+r)!} u^{s-r-1} (1-u)^{n-s+r}, \quad 0 < u < 1.$$

Using (5) in (3), we may write (3) as

(6) $$\gamma = \int_p^1 n\binom{n-1}{s-r-1} u^{s-r-1} (1-u)^{n-s+r} \, du.$$

Now using (5.3.56), we get

(7) $$\gamma = \sum_{i=0}^{s-r-1} \binom{n}{i} p^i (1-p)^{n-i},$$

which is much easier to use. Given n, p, γ, it may not always be possible to find $s-r$ to yield the exact tolerance coefficient γ.

Example 1. Let $s = n$ and $r = 1$. Then
$$\gamma = \sum_{i=0}^{n-2} \binom{n}{i} p^i (1-p)^{n-i} = 1 - p^n - np^{n-1}(1-p).$$
If $p = .8$, $n = 5$, $r = 1$, then
$$\gamma = 1 - (.8)^5 - 6(.8)^4(.2) = .181.$$
Thus the interval $(X_{(1)}, X_{(5)})$ in this case defines an 18 percent tolerance interval for .80 probability under the distribution (of X).

Example 2. Let X_1, X_2, X_3, X_4, X_5 be a sample from a continuous df F. Let us find r and s, $r < s$, such that $(X_{(r)}, X_{(s)})$ is a 90 percent tolerance interval for .50 probability under F. We have
$$.90 = P\{U \geq \tfrac{1}{2}\} = \sum_{i=0}^{s-r-1} \binom{5}{i} \left(\frac{1}{2}\right)^5.$$
It follows that, if we choose $s - r = 4$, then $\gamma = .81$; and if we choose $s - r = 5$, then $\gamma = .969$. In this case, we must settle for an interval with tolerance coefficient .969, exceeding the desired value .90.

In general, given p, $0 < p < 1$, it is possible to choose a sufficiently large sample size n and a corresponding value of $s - r$ such that with probability $\geq \gamma$ an interval of the form $(X_{(r)}, X_{(s)})$ covers at least $100p$ percent of the distribution. If $s - r$ is specified as a function of n, one chooses the smallest sample size n.

Example 3. Let $p = \tfrac{3}{4}$ and $\gamma = .75$. Suppose that we want to choose the smallest sample size required such that $(X_{(2)}, X_{(n)})$ covers at least 75 percent of the distribution. Thus we want the smallest n to satisfy
$$.75 \leq \sum_{i=0}^{n-3} \binom{n}{i} \left(\frac{3}{4}\right)^i \left(\frac{1}{4}\right)^{n-i}.$$
From Table 1, page 648, of binomial distributions, or by using the normal approximation, we see that $n = 14$.

The statistic $U = U_{(s)} - U_{(r)}$, for $1 \leq r < s \leq n$, defined above is called the *coverage* of the random interval $(X_{(r)}, X_{(s)})$. More precisely, the differences

(8) $$\begin{cases} U_1 = F(X_{(1)}) = U_{(1)}, \\ U_2 = F(X_{(2)}) - F(X_{(1)}) = U_{(2)} - U_{(1)}, \\ \vdots \\ U_n = F(X_{(n)}) - F(X_{(n-1)}) = U_{(n)} - U_{(n-1)}, \\ U_{n+1} = 1 - F(X_{(n)}) = 1 - U_{(n)}, \end{cases}$$

are called *coverages*; $U_{(s)} - U_{(r)} = F(X_{(s)}) - F(X_{(r)})$ is called the *coverage* of the interval $(X_{(r)}, X_{(s)})$. Thus the coverage of a random interval (based on order statistics) is the amount of probability in the continuous df F contained in the random interval. In (5) we determined the pdf of $U_{(s)} - U_{(r)} = U_{r+1} + U_{r+2} + \cdots + U_s$, $1 \le r < s \le n$. Taking $r = i - 1$ and $s = i$, we obtain the marginal pdf of U_i as

(9) $$f_{U_i}(u) = \begin{cases} n(1-u)^{n-1}, & 0 < u < 1, \\ 0, & \text{otherwise.} \end{cases}$$

Thus $EU_i = 1/(n+1)$. This may be interpreted as follows: The order statistics $X_{(1)}, X_{(2)}, \cdots, X_{(n)}$ partition the area under the pdf in $n+1$ parts such that each part has the same expected area.

It is not very difficult to compute the joint pdf of $U_1, U_2, \cdots, U_n$. In fact, the inverse transformations are as follows:

$$U_{(1)} = U_1,$$
$$U_{(2)} = U_1 + U_2,$$
$$\vdots$$
$$U_{(n)} = U_1 + U_2 + \cdots + U_n,$$

so that the Jacobian of the transformation is 1. Also, $u_i \ge 0$ and $\sum_1^n u_i < 1$. Since the joint pdf of $U_{(1)}, U_{(2)}, \cdots, U_{(n)}$ is known to be

$$f(x_1, x_2, \cdots, x_n) = \begin{cases} n! & \text{if } 0 < x_1 < x_2 < \cdots < x_n < 1, \\ 0 & \text{otherwise,} \end{cases}$$

it follows that the joint pdf of $U_1, U_2, \cdots, U_n$ is given by

(10) $$h(u_1, u_2, \cdots, u_n) = \begin{cases} n! & \text{if } u_i \ge 0, \ i = 1, 2, \cdots, n, \ \sum_1^n u_i < 1, \\ 0 & \text{otherwise.} \end{cases}$$

Moreover, h is symmetric in $u_1, u_2, \cdots, u_n$ so that the distribution of every sum of r, $r < n$, of these coverages is the same, and in particular it is the distribution of $U_{(r)} = U_1 + U_2 + \cdots + U_r$. Thus the pdf of $\sum_{i=1}^r U_i (= U_{(r)})$ is given by

(11) $$f_{U_{(r)}}(x) = \begin{cases} n\binom{n-1}{r-1} x^{r-1}(1-x)^{n-r}, & 0 < x < 1, \\ 0, & \text{otherwise.} \end{cases}$$

We have

(12) $$EU_{(r)} = \sum_{i=1}^{r} EU_i = \frac{r}{n+1}.$$

The sum of any r successive coverages $U_{i+1}, U_{i+2}, \cdots, U_{i+r}$ is sometimes referred to as an *r-coverage*. We have

(13) $\quad U_{i+1} + U_{i+2} + \cdots + U_{i+r} = U_{(i+r)} - U_{(i)}, \quad i + r \leq n.$

We next consider the use of order statistics in constructing confidence intervals for population quantiles. Let X be an rv with a continuous df F, $0 < p < 1$. Then the quantile of order p satisfies

(14) $$F(\kappa_p) = p.$$

Let $X_1, X_2, \cdots, X_n$ be n independent observations on X. Then the number of X_i's $< \kappa_p$ is an rv that has a binomial distribution with parameters n and p. Similarly, the number of X_i's that are at least κ_p has a binomial distribution with parameters n and $1 - p$.

Let $X_{(1)}, X_{(2)}, \cdots, X_{(n)}$ be the set of order statistics for the sample. Then

$$P\{X_{(r)} \leq \kappa_p\} = P\{\text{At least } r \text{ of the } X_i\text{'s} \leq \kappa_p\}$$

(15) $$= \sum_{i=r}^{n} \binom{n}{i} p^i (1-p)^{n-i}.$$

Similarly

$$P\{X_{(s)} \geq \kappa_p\} = P\{\text{At least } n - s + 1 \text{ of the } X_i\text{'s} \geq \kappa_p\}$$
$$= P\{\text{At most } s - 1 \text{ of the } X_i\text{'s} < \kappa_p\}$$

(16) $$= \sum_{i=0}^{s-1} \binom{n}{i} p^i (1-p)^{n-i}.$$

It follows from (15) and (16) that

$$P\{X_{(r)} \leq \kappa_p \leq X_{(s)}\} = P\{X_{(s)} \geq \kappa_p\} - P\{X_{(r)} > \kappa_p\}$$
$$= P\{X_{(r)} \leq \kappa_p\} - 1 + P\{X_{(s)} \geq \kappa_p\}$$
$$= \sum_{i=r}^{n} \binom{n}{i} p^i (1-p)^{n-i} + \sum_{i=0}^{s-1} \binom{n}{i} p^i (1-p)^{n-i} - 1$$

(17) $$= \sum_{i=r}^{s-1} \binom{n}{i} p^i (1-p)^{n-i}.$$

It is easy to determine a confidence interval for κ_p from (17), once the confidence level is given. In practice, one determines r and s such that $s - r$ is as small as possible, subject to the condition that the level is $1 - \alpha$.

Example 4. Suppose that we want a confidence interval for the median ($p = \frac{1}{2}$), based on a sample of size 7 with confidence level .90. It suffices to find r and s, $r < s$, such that

$$\sum_{i=r}^{s-1}\binom{7}{i}\left(\frac{1}{2}\right)^{7} \geq .90.$$

By trial and error, using the probability distribution $b(7, \frac{1}{2})$ we see that we can choose $s = 7$, $r = 2$ or $r = 1$, $s = 6$; in either case $s - r$ is minimum ($= 5$), and the confidence level is at least .92.

Example 5. Let us compute the number of observations required for $(X_{(1)}, X_{(n)})$ to be a .95 level confidence interval for the median, that is, we want to find n such that

$$P\{X_{(1)} \leq \kappa_{1/2} \leq X_{(n)}\} \geq .95.$$

It suffices to find n such that

$$\sum_{i=1}^{n-1}\binom{n}{i}\left(\frac{1}{2}\right)^{n} \geq .95.$$

It follows that $n = 6$.

PROBLEMS 13.6

1. Find the smallest values of n such that the intervals (a) $(X_{(1)}, X_{(n)})$, (b) $(X_{(2)}, X_{(n-1)})$ contain the median with probability $\geq .90$.

2. Find the smallest sample size required such that $(X_{(1)}, X_{(n)})$ covers at least 90 percent of the distribution with probability $\geq .98$.

3. Find the relation between n and p such that $(X_{(1)}, X_{(n)})$ covers at least p percent of the distribution with probability $\geq 1 - p$.

4. Given γ, δ, p_0, p_1 with $p_1 > p_0$, find the smallest n such that

$$P\{F(X_{(s)}) - F(X_{(r)}) \geq p_0\} \geq \gamma$$

and

$$P\{F(X_{(s)}) - F(X_{(r)}) \geq p_1\} \leq \delta.$$

Find also $s - r$.
(*Hint:* Use the normal approximation to the binomial distribution.)

5. In Problem 4 find the smallest n and the associated value of $s - r$ if $\gamma = .95$, $\delta = .10$, $p_1 = .75$, $p_0 = .50$.

13.7 ROBUSTNESS

Most of the statistical inference problems treated in this book are parametric in nature. We have assumed that the functional form of the distribution being sampled is known except for a finite number of parameters. It is to be expected that any estimate or test of hypothesis concerning the unknown parameters constructed on this assumption will perform better than the corresponding nonparametric procedure, provided that the underlying assumptions are satisfied. It is therefore of interest to know how well the parametric optimal tests or estimates constructed for one population perform when the basic assumptions are modified. If we can construct tests or estimates that perform well for a variety of distributions, for example, there would be little point in using the corresponding nonparametric method unless the assumptions are seriously violated.

In practice, one makes many assumptions in parametric inference, and any one or all of these may be violated. Thus one seldom has accurate knowledge about the true underlying distribution. Similarly, the assumption of mutual independence or even identical distribution may not hold. Any test or estimate that performs well under modifications of underlying assumptions is usually referred to as *robust*. This subject is receiving considerable attention of late. We refer to Huber [50] for a review of the literature and a historical survey. See also Govindarajulu and Leslie [39]. We content ourselves here by considering some commonly used estimates and test procedures.

The most commonly used estimate for the population mean μ is the sample mean. It has the property of unbiasedness for all populations with finite mean. For many parent populations (normal, Poisson, Bernoulli, gamma, etc.) it is a complete sufficient statistic and hence a UMVUE. Moreover, it is consistent and has asymptotic normal distribution whenever the conditions of the central limit theorem are satisfied. Nevertheless, the sample mean is affected by extreme observations, and a single observation that is either too large or too small may make $\bar{X}$ worthless as an estimate of μ. Suppose, for example, that $X_1, X_2, \cdots, X_n$ is a sample from some normal population. Occasionally something happens to the system, and a wild observation is obtained; that is, suppose one is sampling from $\mathcal{N}(\mu, \sigma^2)$, say, 100α percent of the time and from $\mathcal{N}(\mu, k\sigma^2)$, where $k > 1$, $(1 - \alpha)100$ percent of the time. Here both μ and σ^2 are unknown, and one wishes to estimate μ. In this case one is really sampling from the density function

$$f(x) = \alpha f_0(x) + (1 - \alpha) f_1(x), \tag{1}$$

where f_0 is the pdf of $\mathcal{N}(\mu, \sigma^2)$, and f_1, the pdf of $\mathcal{N}(\mu, k\sigma^2)$. Clearly

$$\bar{X} = \frac{\sum_{1}^{n} X_i}{n} \tag{2}$$

is again unbiased for μ. If α is nearly 1, there is no problem since the underlying distribution is nearly $\mathcal{N}(\mu, \sigma^2)$, and $\bar{X}$ is nearly the UMVUE of μ with variance σ^2/n. If $1 - \alpha$ is large (that is, not nearly 0), then, since one is sampling from f, the variance of X_1 is σ^2 with probability α and is $k\sigma^2$ with probability $1 - \alpha$, and we have

$$\text{(3)} \qquad \text{var}_\sigma(\bar{X}) = \frac{1}{n} \text{var}(X_1) = \frac{\sigma^2}{n} [\alpha + (1 - \alpha)k].$$

If $k(1 - \alpha)$ is large, $\text{var}_\sigma(\bar{X})$ is large and we see that even an occasional wild observation makes $\bar{X}$ subject to a sizable error. The presence of an occasional observation from $\mathcal{N}(\mu, k\sigma^2)$ is frequently referred to as *contamination*. The problem is that we do not know, in practice, the distribution of the wild observations and hence we do not know the pdf f. It is known that the sample median is a much better estimate than the mean in the presence of extreme values. In the contamination model discussed above, if we use $Z_{1/2}$, the sample median of the X_i's, as an estimate of μ (which is the population median), then for large n

$$\text{(4)} \qquad E(Z_{1/2} - \mu)^2 = \text{var}(Z_{1/2}) \approx \frac{1}{4n} \frac{1}{[f(\mu)]^2}.$$

(See Theorem 7.3.7 and Remark 7.3.1.) Since

$$f(\mu) = \alpha f_0(\mu) + (1-\alpha) f_1(\mu)$$
$$= \frac{\alpha}{\sigma\sqrt{2\pi}} + (1 - \alpha) \frac{1}{\sigma\sqrt{2\pi k}} = \left(\alpha + \frac{1-\alpha}{\sqrt{k}} \right) \frac{1}{\sigma\sqrt{2\pi}},$$

we have

$$\text{(5)} \qquad \text{var}(Z_{1/2}) \approx \frac{\pi \sigma^2}{2n} \frac{1}{\{\alpha + [(1-\alpha)/\sqrt{k}]\}^2}.$$

As $k \to \infty$, $\text{var}(Z_{1/2}) \approx \pi\sigma^2/2n\alpha^2$. If there is no contamination, $\alpha = 1$ and $\text{var}(Z_{1/2}) \approx \pi\sigma^2/2n$. Also,

$$\frac{\pi\sigma^2/2n\alpha^2}{\pi\sigma^2/2n} = \frac{1}{\alpha^2},$$

which will be close to 1 if α is close to 1. Thus the estimate $Z_{1/2}$ will not be greatly affected by how large k is, that is, how wild the observations are. We have

$$\frac{\text{var}(\bar{X})}{\text{var}(Z_{1/2})} = \frac{2}{\pi} [\alpha + (1-\alpha)k] [\alpha + \frac{(1-\alpha)}{\sqrt{k}}]^2 \to \infty \qquad \text{as} \quad k \to \infty.$$

Indeed, $\text{var}(\bar{X}) \to \infty$ as $k \to \infty$, whereas $\text{var}(Z_{1/2}) \to \pi\sigma^2/2n\alpha^2$ as $k \to \infty$. One can check that, when $k = 9$ and $\alpha \approx .915$, the two variances are (approximately) equal. As k becomes larger than 9 or α smaller than .915, $Z_{1/2}$ becomes a better estimate of μ than $\bar{X}$.

There are other flaws as well. Suppose, for example, that $X_1, X_2, \cdots, X_n$ is a sample from $U(0, \theta)$, $\theta > 0$. Then both $\bar{X}$ and $T(\mathbf{X}) = (X_{(1)} + X_{(n)})/2$, where $X_{(1)} = \min(X_1, \cdots, X_n)$, $X_{(n)} = \max(X_1, \cdots, X_n)$, are unbiased for $EX = \theta/2$. Also, $\text{var}_\theta(\bar{X}) = \text{var}(X)/n = \theta^2/[12n]$, and one can show that $\text{var}(T) = \theta^2/[2(n+1)(n+2)]$. It follows that the efficiency of $\bar{X}$ relative to that of T is

$$\text{eff}_\theta(\bar{X}|T) = \frac{\text{var}_\theta(\bar{X})}{\text{var}_\theta(T)} = \frac{1}{6n}(n+1)(n+2) > 1 \quad \text{if } n > 2.$$

In fact, $\text{eff}_\theta(\bar{X}|T) \to \infty$ as $n \to \infty$, so that in sampling from a uniform parent $\bar{X}$ is much worse than T, even for moderately large values of n.

Let us next turn our attention to the estimation of standard deviation. Let $X_1, X_2, \cdots, X_n$ be a sample from $\mathcal{N}(\mu, \sigma^2)$. Then the MLE of σ is

(6) $$\hat{\sigma} = \left\{\sum_{i=1}^{n} \frac{(X_i - \bar{X})^2}{n}\right\}^{1/2} = \left(\frac{n-1}{n}\right)^{1/2} S.$$

Note that the lower bound for the variance of any unbiased estimate for σ is $\sigma^2/2n$. Although $\hat{\sigma}$ is not unbiased, the estimate

(7) $$S_1 = \sqrt{\frac{n}{2}} \frac{\Gamma[(n-1)/2]}{\Gamma(n/2)} \hat{\sigma} = \sqrt{\frac{n-1}{2}} \frac{\Gamma[(n-1)/2]}{\Gamma(n/2)} S$$

is unbiased for σ. Also,

$$\text{var}(S_1) = \sigma^2 \left\{\frac{n-1}{2}\left(\frac{\Gamma[(n-1)/2]}{\Gamma(n/2)}\right)^2 - 1\right\}$$

(8) $$= \frac{\sigma^2}{2n} + O\left(\frac{1}{n^2}\right).$$

Thus the efficiency of S_1 (relative to the estimate with least variance $= \sigma^2/2n$) is

$$\frac{\text{var}(S_1)}{\sigma^2/2n} = 1 + \frac{1}{\sigma^2} O\left(\frac{2}{n}\right) \to 1 \quad \text{as } n \to \infty.$$

For small n, the efficiency of S_1 is considerably larger than 1. Thus, for $n = 2$, $\text{eff}(S_1) = 2(\pi - 2) \approx 2.223$, and, for $n = 3$, $\text{eff}(S_1) = (6/\pi)(4 - \pi) \approx 1.636$.

Yet another estimate of σ is the sample mean deviation

(9) $$S_2 = \frac{1}{n}\sum_{i=1}^{n}|X_i - \bar{X}|.$$

Note that

$$E\left\{\sqrt{\frac{\pi}{2}}\frac{1}{n}\sum_{i=1}^{n}|X_i - \mu|\right\} = \sqrt{\frac{\pi}{2}}E|X_i - \mu| = \sigma,$$

and

(10) $$\text{var}\left\{\sqrt{\frac{\pi}{2}}\frac{1}{n}\sum_{i=1}^{n}|X_i - \mu|\right\} = \frac{\pi - 2}{2n}\sigma^2.$$

If n is large enough so that $\bar{X} \approx \mu$, we see that $S_3 = \sqrt{(\pi/2)}S_2$ is nearly unbiased for σ with variance $[(\pi - 2)/2n]\sigma^2$. The efficiency of S_3 is

$$\frac{[(\pi - 2)/2n]\sigma^2}{\sigma^2/2n} = \pi - 2 > 1.$$

For large n, the efficiency of S_1 relative to S_3 is

$$\frac{\text{var}(S_1)}{\text{var}(S_3)} = \frac{(\sigma^2/2n) + O(1/n^2)}{[(\pi - 2)/2n]\sigma^2} = \frac{1}{\pi - 2} + \frac{2}{\pi - 2} O\left(\frac{1}{n}\right)$$
$$\approx .876.$$

Now suppose that there is some contamination. As before, let us suppose that for a proportion α of the time we sample from $\mathcal{N}(\mu, \sigma^2)$ and for a proportion $1 - \alpha$ of the time we get a wild observation from $\mathcal{N}(\mu, k\sigma^2)$, $k > 1$. Assuming that both μ and σ^2 are unknown, suppose that we wish to estimate σ. In the notation used above, let

$$f(x) = \alpha f_0(x) + (1 - \alpha)f_1(x),$$

where f_0 is the pdf of $\mathcal{N}(\mu, \sigma^2)$, and f_1, the pdf of $\mathcal{N}(\mu, k\sigma^2)$. Let us see how even small contamination can make the maximum likelihood estimate $\hat{\sigma}$ of σ quite useless.

If $\hat{\theta}$ is the MLE of θ, and φ is a function of θ, then $\varphi(\hat{\theta})$ is the MLE of $\varphi(\theta)$. By Taylor's expansion, a first approximation is

$$\varphi(\hat{\theta}) = \varphi(\hat{\theta} - \theta + \theta) = \varphi(\theta) + (\hat{\theta} - \theta)\frac{d\varphi(\theta)}{d\theta},$$

so that

(11) $$E\{\varphi(\hat{\theta}) - \varphi(\theta)\}^2 \approx E(\hat{\theta} - \theta)^2 \left[\frac{d\varphi(\theta)}{d\theta}\right]^2.$$

Taking $\theta = \sigma^2$ and $\varphi(\theta) = \sqrt{\theta}$, we get

(12) $$E(\hat{\sigma} - \sigma)^2 \approx \frac{1}{4\sigma^2} E(\hat{\sigma}^2 - \sigma^2)^2.$$

Using Theorem 7.3.5, we see that

(13) $$E(\hat{\sigma}^2 - \sigma^2)^2 \approx \frac{\mu_4 - \mu_2^2}{n}$$

(dropping the other two terms with n^2 and n^3 in the denominator), so that

(14) $$E(\hat{\sigma} - \sigma)^2 \approx \frac{1}{4\sigma^2 n}(\mu_4 - \mu_2^2).$$

For the density f, we see that

(15) $$\mu_4 = 3\sigma^4[\alpha + k^2(1 - \alpha)]$$

and

(16) $$\mu_2 = \sigma^2[\alpha + k(1 - \alpha)].$$

It follows that

(17) $$E\{\hat{\sigma} - \sigma\}^2 \approx \frac{\sigma^2}{4n} \left\{ 3[\alpha + k^2(1 - \alpha)] - [\alpha + k(1 - \alpha)]^2 \right\}.$$

If we are interested in the effect of very small contamination, $\alpha \approx 1$ and $1 - \alpha \approx 0$. Assuming that $k(1 - \alpha) \approx 0$, we see that

$$E\{\hat{\sigma} - \sigma\}^2 \approx \frac{\sigma^2}{4n} \{3[1 + k^2(1 - \alpha)] - 1\}$$

(18) $$= \frac{\sigma^2}{2n} [1 + \tfrac{3}{2}k^2(1 - \alpha)].$$

In the normal case, $\mu_4 = 3\sigma^4$ and $\mu_2^2 = \sigma^4$, so that

$$E\{\hat{\sigma} - \sigma\}^2 \approx \frac{\sigma^2}{2n}.$$

Thus we see that the mean square error due to a small contamination is now multiplied by a factor $[1 + \tfrac{3}{2}k^2(1 - \alpha)]$. If, for example, $k = 10$, $\alpha = .99$, then $1 + \tfrac{3}{2}k^2(1 - \alpha) = \tfrac{5}{2}$. If $k = 10$, $\alpha = .98$, then $1 + \tfrac{3}{2}k^2(1 - \alpha) = 4$, and so on.

A quick comparison with S_3 shows that, although S_1 (or even $\hat{\sigma}$) is a better estimate of σ than S_3 if there is no contamination, the estimate S_3 becomes a much better estimate in the presence of contamination as k becomes large.

One of the most commonly used tests in statistics is Student's t-test for testing the mean of a normal population when the variance is unknown. Let $X_1, X_2, \cdots, X_n$ be a sample from some population with mean μ and finite variance σ^2. As usual, let $\bar{X}$ denote the sample mean, and S^2, the sample variance. If the population being sampled is normal, the t-test rejects H_0: $\mu = \mu_0$ against H_1: $\mu \neq \mu_0$ at level α if $|\bar{x} - \mu_0| > t_{n-1, \alpha/2} (s/\sqrt{n})$. If n is large, we replace $t_{n-1, \alpha/2}$ by the corresponding critical value, $z_{\alpha/2}$, under the standard normal law. If the sample does not come from a normal population, the statistic $T = [(\bar{X} - \mu_0)/S] \sqrt{n}$ is no longer distributed as a $t(n - 1)$ statistic. If, however, n is sufficiently large, we know that T has an asymptotic normal distribution irrespective of the population being sampled, as long as it has a finite variance. Thus, for large n, the distribution of T is independent of the form of the population, and the t-test is robust. The same considerations apply to testing the difference between two means when the two variances are equal. Although we assumed that n is sufficiently large for Cramér's result (Theorem 6.2.15) to hold, empirical investigations have shown that the test based on Student's statistic is robust. Thus a significant value of t may not be interpreted to mean a departure from normality of the observations. Let us next consider the effect of departure from independence on the t-distribution. Suppose that the observations $X_1, X_2, \cdots, X_n$ have a multivariate normal distribution with $EX_i = \mu$, $\mathrm{var}(X_i) = \sigma^2$, and ρ as the common correlation coefficient between any X_i and X_j, $i \neq j$. Then

(19) $$E\bar{X} = \mu, \qquad \mathrm{var}(\bar{X}) = \frac{\sigma^2}{n}[1 + (n-1)\rho],$$

and

$$ES^2 = \frac{1}{n-1} E\left\{\sum_{i=1}^{n} X_i^2 - \frac{1}{n}\sum_{i=1}^{n}\sum_{j=1}^{n} X_i X_j\right\}$$

$$= \frac{1}{n-1}\{n E X_1^2 - n^{-1}[n E X_1^2 + n(n-1)(\rho\sigma^2 + \mu^2)]\}$$

(20) $$= \sigma^2(1 - \rho).$$

For large n, the statistic $\sqrt{n}(\bar{X} - \mu_0)/S$ will be asymptotically distributed as $\mathcal{N}(0, 1 + n\rho/(1 - \rho))$, instead of $\mathcal{N}(0, 1)$. Under H_0, $T^2 = n(\bar{X} - \mu_0)^2/S^2$ is distributed as $F(1, n-1)$ when the observations are independent and identically distributed as $\mathcal{N}(\mu, \sigma^2)$. Consider the ratio

(21) $$\frac{n E(\bar{X} - \mu_0)^2}{ES^2} = \frac{\sigma^2[1 + (n-1)\rho]}{\sigma^2(1 - \rho)} = 1 + \frac{n\rho}{1 - \rho}.$$

The ratio equals 1 if $\rho = 0$, but is > 0 for $\rho > 0$ and $\to \infty$ as $\rho \to 1$. It follows that a large value of T is likely to occur when $\rho > 0$ and is large, even though μ_0 is the true value of the mean. Thus a significant value of t may be due to departure from independence, and the effect can be serious.

Next, consider a test of the null hypothesis $H_0: \sigma = \sigma_0$ against $H_1: \sigma \neq \sigma_0$. Under the usual normality assumptions on the observations $X_1, X_2, \cdots, X_n$, the test statistic used is

(22) $$V = \frac{(n-1)S^2}{\sigma^2} = \frac{\sum_{i=1}^{n}(X_i - \bar{X})^2}{\sigma^2},$$

which has a $\chi^2(n-1)$ distribution under H_0. The usual test is to reject H_0 if

(23) $$V_0 = \frac{(n-1)S^2}{\sigma_0^2} > \chi^2_{n-1,\alpha/2} \qquad \text{or} \qquad V_0 < \chi^2_{n-1,1-\alpha/2}.$$

Let us suppose that $X_1, X_2, \cdots, X_n$ are not normal. It follows from Corollary 2 to Theorem 7.3.5 that

(24) $$\mathrm{var}(S^2) = \frac{\mu_4}{n} + \frac{3-n}{n(n-1)}\mu_2^2,$$

so that

(25) $$\operatorname{var}\left(\frac{S^2}{\sigma^2}\right) = \frac{1}{n}\frac{\mu_4}{\sigma^4} + \frac{3-n}{n(n-1)}.$$

Writing $\gamma_2 = (\mu_4/\sigma^4) - 3$, we have

(26) $$\operatorname{var}\left(\frac{S^2}{\sigma^2}\right) = \frac{\gamma_2}{n} + \frac{2}{n-1}$$

when the X_i's are not normal, and

(27) $$\operatorname{var}\left(\frac{S^2}{\sigma^2}\right) = \frac{2}{n-1}$$

when the X_i's are normal ($\gamma_2 = 0$). Now $(n-1)S^2 = \sum_{i=1}^{n}(X_i - \bar{X})^2$ is the sum of n identically distributed but dependent rv's $(X_j - \bar{X})^2$, $j = 1, 2, \cdots, n$. Using a version of the central limit theorem for dependent rv's (see, for example, Cramér [18], 365), it follows that

$$\left(\frac{n-1}{2}\right)^{-1/2}\left(\frac{S^2}{\sigma^2} - 1\right),$$

under H_0, is asymptotically $\mathcal{N}(0, 1 + (\gamma_2/2))$, and not $\mathcal{N}(0, 1)$ as under the normal theory. As a result the size of the test based on the statistic V_0 will be different from the stated level of significance if γ_2 differs greatly from 0. It is clear that the effect of violation of the normality assumption can be quite serious on inferences about variances, and the chi-square test is not robust.

For a similar examination of some other well-known test criteria, we refer to Scheffé [111], Chapter 10.

In the above discussion we have used somewhat crude calculations to investigate the behavior of the most commonly used estimates and test statistics when one or more of the underlying assumptions are violated. Our purpose here was to indicate that some tests or estimates are robust whereas others are not. The moral is clear: One should check carefully to see that the underlying assumptions are satisfied before using parametric procedures.

PROBLEMS 13.7

1. Let $(X_1, X_2, \cdots, X_n)$ be jointly normal with $EX_i = \mu$, $\operatorname{var}(X_i) = \sigma^2$, and $\operatorname{cov}(X_i, X_j) = \rho\sigma^2$ if $|i - j| = 1$, $i \neq j$, and $= 0$ otherwise.
(a) Show that

$$\operatorname{var}(\bar{X}) = \frac{\sigma^2}{n}\left[1 + 2\rho\left(1 - \frac{1}{n}\right)\right]$$

and
$$E(S^2) = \sigma^2\left(1 - \frac{2\rho}{n}\right).$$

(b) Show that the t-statistic $\sqrt{n}(\bar{X} - \mu)/S$ is asymptotically normally distributed with mean 0 and variance $1 + 2\rho$. Conclude that the significance of t is overestimated for positive values of ρ and underestimated for $\rho < 0$ in large samples.

(c) For finite n, consider the statistic
$$T^2 = \frac{n(\bar{X} - \mu)^2}{S^2}.$$

Compare the expected values of the numerator and the denominator of T^2 and study the effect of $\rho \neq 0$ to interpret significant t values. (Scheffé [111], 338)

2. Let $X_1, X_2, \cdots, X_n$ be a random sample from $G(\alpha, \beta)$, $\alpha > 0$, $\beta > 0$.

(a) Show that
$$\mu_2 = \alpha\beta^2, \qquad \mu_4 = 3\alpha(\alpha + 2)\beta^4.$$

(b) Show that
$$\operatorname{var}\left\{(n - 1)\frac{S^2}{\sigma^2}\right\} \approx (n - 1)\left(2 + \frac{6}{\alpha}\right).$$

(c) Show that the large sample distribution of $(n - 1)S^2/\sigma^2$ is normal.

(d) Compare the large-sample test of $H_0: \sigma = \sigma_0$ based on the asymptotic normality of $(n - 1)S^2/\sigma^2$ with the large-sample test based on the same statistic when the observations are taken from a normal population. In particular, take $\alpha = 2$.

3. Let $X_1, X_2, \cdots, X_m$ and $Y_1, Y_2, \cdots, Y_n$ be two independent random samples from populations with means μ_1 and μ_2, and variances σ_1^2 and σ_2^2, respectively. Let $\bar{X}, \bar{Y}$ be the two sample means, and S_1^2, S_2^2 be the two sample variances. Write $N = m + n$, $R = m/n$, and $\theta = \sigma_1^2/\sigma_2^2$. The usual normal theory test of $H_0: \mu_1 - \mu_2 = \delta_0$ is the t-test based on the statistic
$$T = \frac{\bar{X} - \bar{Y} - \delta_0}{S_p(1/m + 1/n)^{1/2}}$$

where
$$S_p^2 = \frac{(m - 1)S_1^2 + (n - 1)S_2^2}{m + n - 2}.$$

Under H_0, the statistic T has a t-distribution with $N - 2$ d.f., provided that $\sigma_1^2 = \sigma_2^2$.

Show that the asymptotic distribution of T in the nonnormal case is $\mathcal{N}(0, (\theta + R)(1 + R\theta)^{-1})$ for large m and n. Thus, if $R = 1$, T is asymptotically $\mathcal{N}(0, 1)$ as in the normal theory case assuming equal variances, even though the two samples come from nonnormal populations with unequal variances. Conclude that the test is robust in the case of large, equal sample sizes.

(Scheffé [111], 339)

CHAPTER 14

Sequential Statistical Inference

14.1 INTRODUCTION

In all the problems of statistical inference that we have considered so far, the sample size was fixed in advance. We assumed that a sample of fixed size n was available which did not depend on the observations. It is easy to construct examples in which such a procedure is obviously wasteful since it does not take advantage of the information supplied by the observations. Thus, in a coin-tossing experiment where we wish to test $H_0: p = \frac{1}{2}$, if $n - 1$ tosses produce heads the result of the nth trial is not likely to change our decision to reject H_0, provided that n is not too small. This example suggests that it may be disadvantageous in some problems to fix the sample size in advance. An alternative procedure suggests itself: Sample sequentially, that is take one observation after another and make a decision to stop after the nth observation by using some procedure that is based on the observations $x_1, x_2, \cdots, x_n$.

In this chapter we consider some problems of sequential statistical inference. Section 2 introduces some fundamental ideas. Sections 3 and 4 deal with sequential (parametric) point and interval estimation, while Sections 5, 6, and 7 examine the sequential probability ratio test of Wald.

14.2 SOME FUNDAMENTAL IDEAS OF SEQUENTIAL SAMPLING

In all the problems of statistical inference considered so far, we have assumed the availability of a sample of a certain size n. This ignores the fact that sampling is expensive and the taking of each observation involves some cost. The question then arises as to what should be the minimum size required to make a decision. Let us consider some examples.

Example 1. In estimating the mean μ in sampling from a normal population, the sample mean $\bar{X}$ is an unbiased estimate. Let us suppose that each observation costs $\$A$ $(A > 0)$, and that we wish to minimize the size n to make the average loss,

(1) $$E|\bar{X} - \mu|^2 + An,$$

minimum. We have

(2) $$L(n, \mu, \sigma^2) = E|\bar{X} - \mu|^2 + An = \frac{\sigma^2}{n} + nA,$$

which is minimized if we take n to be a solution of

$$\frac{\partial L}{\partial n} = -\frac{\sigma^2}{n^2} + A = 0,$$

that is, $n = n_0$, where

(3) $$n_0 = \frac{\sigma}{\sqrt{A}}.$$

Here we considered n as a continuous variable. It is easy to show that $n = n_0$ minimizes L and that the minimum value of L is

(4) $$L_0 = 2\sigma\sqrt{A} = 2An_0.$$

If σ is known, one takes n_0 observations $\{n_0 = [\sigma/\sqrt{A}] + 1\}$ and estimates μ by $\bar{X}$; in doing so, one minimizes the average loss, as defined in (1). In practice, however, σ will not be known and one cannot compute n_0. How should one proceed in view of the ignorance about σ?

Example 2. Let us suppose that we wish to test H_0: $\mu = \mu_0$ against H_1: $\mu = \mu_1$ ($\mu_1 > \mu_0$) in sampling from a normal population with mean μ (unknown) and known variance σ^2. We would like to make a decision on the basis of the smallest sample size n. Suppose that the probabilities of the two types of error have to be fixed at α and β. We know that the MP test of H_0 rejects H_0 if $\bar{X} > c$. Thus

(5) $$\alpha = P_0\{\bar{X} > c\} = P\left\{\frac{\bar{X} - \mu_0}{\sigma/\sqrt{n}} > \frac{c - \mu_0}{\sigma/\sqrt{n}}\right\},$$

so that

(6) $$\sqrt{n}\,\frac{c - \mu_0}{\sigma} = z_\alpha.$$

Similarly

(7) $$\beta = P_1\{\bar{X} \leq c\} = P\left\{\frac{\bar{X} - \mu_1}{\sigma/\sqrt{n}} \leq \frac{c - \mu_1}{\sigma/\sqrt{n}}\right\},$$

so that

(8) $$\sqrt{n}\,\frac{c-\mu_1}{\sigma} = z_{1-\beta} = -z_\beta.$$

From (6) and (8) we get

(9) $$n = \frac{\sigma^2(z_\alpha + z_\beta)^2}{(\mu_1 - \mu_0)^2}.$$

Thus, if we take $n = n_0$ observations, where n_0 is given by (9), and reject H_0 if

$$\bar{X} > \mu_0 + z_\alpha \frac{\sigma}{\sqrt{n_0}},$$

the test will have the desired probabilities of the two types of error. In practice, however, σ will not be known and one cannot compute n_0. How does one proceed in view of the ignorance of σ?

Example 3. Let $X_1, X_2, \cdots, X_n$ be a sample from $\mathcal{N}(\mu, \sigma^2)$, where both μ and σ^2 are unknown. The shortest confidence interval of level $1 - \alpha$ based on $\bar{X}$ for the parameter μ is

$$\left(\bar{X} - \frac{S}{\sqrt{n}} t_{n-1,\alpha/2},\ \bar{X} + \frac{S}{\sqrt{n}} t_{n-1,\alpha/2}\right),$$

where S^2 is the sample variance. This interval has length $2(S/\sqrt{n})t_{n-1,\alpha/2}$, which is a random variable. In practice, we would like to find a fixed width confidence interval with a given confidence level based on the minimum number of observations. If σ were known, the $1 - \alpha$ level confidence interval $(\bar{X} - (\sigma/\sqrt{n}) z_{\alpha/2},\ \bar{X} + (\sigma/\sqrt{n}) z_{\alpha/2})$ has length $(2\sigma/\sqrt{n}) z_{\alpha/2}$. This length is at most $2d$, say, where $d\,(>0)$ is fixed in advance, provided that we choose n such that

$$d \geq \frac{\sigma}{\sqrt{n}} z_{\alpha/2}$$

or

(10) $$n \geq \frac{\sigma^2}{d^2} z_{\alpha/2}^2.$$

If σ is not known, one does not know this minimum sample size. When σ is unknown, how does one sample to minimize the sample size?

In the examples above we see that it is not always possible to minimize the number of observations required to arrive at a decision that is optimum in some sense. Note that all this discussion is based on the fact that we take a sample of a certain size n which is fixed in advance and may or may not be the smallest number required. An alternative procedure suggests itself: Why not take observations sequentially, that is, one at a time, and use the information provided by the observations to date to determine whether a further observation is required? In such a case we do not need to fix the number of observations in advance of the experiment.

In this section we consider some basic notions of *sequential statistical inference*. Given an infinite sequence $\mathbf{X} = (X_1, X_2, \cdots)$ of rv's, the statistician faces the problem of providing a set of rules that tells the experimenter when to stop sampling. Once the sampling is stopped after taking, say, n observations, the decision problem is treated as a fixed sample size problem.

Let Θ be the parameter space, and $\mathscr{A}$, the space of actions open to the statistician. We assume that the rv's $X_1, X_2, \cdots$ that are observed sequentially are iid. Let $f_\theta(x)$ be the common pdf (pmf) of the X_i's.

Definition 1. A sequential decision procedure has two components.
(a) A stopping rule that specifies whether an element of $\mathscr{A}$ should be chosen without taking any observation. If at least one observation is taken, the rule specifies, for every set of observed values $(x_1, x_2, \cdots, x_n), n \geq 1$, whether to stop sampling and choose a decision in $\mathscr{A}$ or to take another observation X_{n+1}.
(b) A decision rule that specifies the decision $d_0 \in \mathscr{A}$ if no observations are taken, or, if at least one observation is taken, specifies the action $d_n(x_1, x_2, \cdots, x_n) \in \mathscr{A}$ to be taken for each set of observed values $(x_1, x_2, \cdots, x_n)$, after which the sampling might be stopped.

We will assume that the statistician takes at least one observation before making a decision. It is convenient to define *stopping regions* $R_n, n = 1, 2, \cdots$.

Definition 2. Let $R_n \subseteq \mathscr{R}_n, n = 1, 2, \cdots$, be a sequence of Borel-measurable sets such that the sampling is terminated after observing $X_1 = x_1, X_2 = x_2, \cdots, X_n = x_n$ if $(x_1, x_2, \cdots, x_n) \in R_n$. If $(x_1, x_2, \cdots, x_n) \notin R_n$, another observation x_{n+1} is taken. The sets $R_n, n = 1, 2, \cdots$, are called stopping regions.

Definition 3. With every sequential stopping rule we associate a stopping rv N, which takes on the values $1, 2, 3, \cdots$. N is the (random) total number of observations taken before sampling is stopped.

Let $\{N = n\}$ denote the event that sampling is stopped after observing n values $x_1, x_2, \cdots, x_n$, and not before. Thus $\{N = 1\} = R_1$, and

$\{N = n\} = \{(x_1, x_2, \cdots, x_n) \in \mathcal{R}_n:$ sampling is stopped after observing x_n and not before$\}$

$$= (R_1 \cup R_2 \cup \cdots \cup R_{n-1})^c \cap R_n$$

(11)
$$= R_1^c \cap R_2^c \cap \cdots \cap R_{n-1}^c \cap R_n.$$

Note that $\{N = n\}$, and the event $\{N \le n\} = \bigcup_{k=1}^n \{N = k\}$ depend only on the observations $X_1, X_2, \cdots, X_n$, and not on $X_{n+1}, X_{n+2}, \cdots$.

In what follows, we will consider only *closed* sequential sampling procedures, that is, procedures for which sampling eventually terminates with probability 1. Thus we will assume that

(12) $$P\{N < \infty\} = 1$$

or

$$P\{N = \infty\} = 1 - P\{N < \infty\} = 0.$$

Example 4. In Example 1, suppose that σ is not known. Consider the sequential stopping rule T_n:

(13)
$$\begin{cases} \text{Stop after taking } n \ge 2 \text{ observations if } n \text{ is the smallest integer} \\ \text{such that } n \ge s_n/\sqrt{A}, \text{ where} \\ \qquad s_n^2 = \dfrac{\sum_{i=1}^n (x_i - \bar{x})^2}{n-1}, \qquad \bar{x} = \dfrac{\sum_{i=1}^n x_i}{n}. \end{cases}$$

Clearly T_n is a sequential stopping rule that says: After taking each observation, compute the corresponding s_i^2 and check to see whether $i \ge s_i/\sqrt{A}$. If so, stop; otherwise, take another observation x_{i+1}. If $N = n$, that is, sampling stops after observing x_n, then estimate μ unbiasedly by $\bar{x} = \sum_1^n x_i/n$. Here

$$\{N = n\} = \left\{ 2 < \frac{S_2}{\sqrt{A}}, 3 < \frac{S_3}{\sqrt{A}}, \cdots, n-1 < \frac{S_{n-1}}{\sqrt{A}}, n \ge \frac{S_n}{\sqrt{A}} \right\}.$$

Example 5. Suppose that we wish to estimate the probability θ, $0 < \theta \le 1$, of obtaining heads when a given coin is tossed. In fixed sample size procedures of estimation that we have studied, one tosses a coin n times (n fixed in advance) and counts the number of heads. Then the proportion of heads in n trials is a reasonable (unbiased and consistent) estimate of θ. The larger the sample size n, the smaller is the variance of this estimate.

If, instead, we toss the coin until it shows the first head, the procedure is clearly a sequential decision procedure. Let N be the number of trials needed for the first head. Then

$$P\{N = n + 1\} = \theta(1 - \theta)^n, \qquad n = 0, 1, 2, \cdots,$$

and since

$$P\{N < \infty\} = \sum_{n=0}^{\infty} \theta(1-\theta)^n = 1,$$

we note that the sampling terminates with probability 1 (eventually) and the sampling procedure is a closed one. Also,

$$EN = \sum_{n=0}^{\infty}(n+1)\theta(1-\theta)^n$$
$$= 1 + \sum_{n=0}^{\infty} n\theta(1-\theta)^n = \frac{1}{\theta} < \infty \quad \text{for } 0 < \theta \le 1.$$

The MLE of θ is

$$T(N) = \frac{1}{N}.$$

We may write the sequential procedure in terms of each observation as follows. Let $X_i = 1$ if a head shows up on the ith tossing, and $= 0$ otherwise. Then we stop sampling after observing x_{n+1}, if $\sum_{i=1}^{n} x_i = 0$ and $x_{n+1} = 1$. The decision rule is to estimate θ by the reciprocal of the number of observations needed to obtain the first head.

Note that

$$E_\theta T(N) = \theta \sum_{n=0}^{\infty} \frac{1}{n+1}(1-\theta)^n$$
$$= \theta \left[\frac{1}{1-\theta} \sum_{n=0}^{\infty} \frac{(1-\theta)^{n+1}}{n+1} \right]$$
$$= \frac{\theta}{1-\theta} \log \frac{1}{\theta}, \quad 0 < \theta < 1,$$

so that T is not an unbiased estimate of θ.

Example 6. Two players, A and B, play a series of games. A's chance of winning a game is p, and B's chance is $(1-p)$, $0 < p < 1$. Suppose that the loser of each game pays the winner \$1 and that A starts with a capital of \$$a$ and B with \$$b$. The game is finished when either A or B is ruined, that is, when B (or A) has all the money, \$$(a+b)$.

Let

$$p_x = \text{probability of } A\text{'s ruin when he has } \$x \ (1 < x < a+b).$$

At the next game he either has \$ $(x+1)$, if he wins, or \$ $(x-1)$, if he loses. Assuming independence, we have

(14) $$p_x = pp_{x+1} + (1-p)p_{x-1}.$$

The boundary conditions are

(15) $$p_0 = 1, \quad p_{a+b} = 0.$$

It is easy to show that the difference equation (14) subject to (15) has a unique solution:

$$(16) \quad p_x = \begin{cases} \dfrac{[(1-p)/p]^{a+b} - [(1-p)/p]^x}{[(1-p)/p]^{a+b} - 1} & \text{if } p \neq 1-p, \\ \dfrac{a+b-x}{a+b} & \text{if } p = 1-p = \tfrac{1}{2}. \end{cases}$$

(See Problem 4.) Using an analogous argument for B's ruin, we can verify that the series of games terminates with probability 1, that is, the probability of A's ruin plus that of B's ruin is 1, so the procedure is a closed one. In particular,

$$(17) \quad p_a = \begin{cases} \dfrac{[(1-p)/p]^{a+b} - [(1-p)/p]^a}{[(1-p)/p]^{a+b} - 1} & \text{if } p \neq \tfrac{1}{2}, \\ \dfrac{b}{a+b} & \text{if } p = \tfrac{1}{2}. \end{cases}$$

The expected duration of the game is given by

$$(18) \quad E_p N = \begin{cases} \dfrac{a}{1-2p} - \left(\dfrac{a+b}{1-2p}\right) \dfrac{1 - [(1-p)/p]^a}{1 - [(1-p)/p]^{a+b}}, & p \neq \tfrac{1}{2}, \\ ab, & p = \tfrac{1}{2}. \end{cases}$$

(See Problem 5.) We will derive (17) and (18) by a different method in Example 14.6.6.

We conclude this section with some remarks concerning the precision of a sequential procedure. Let $L(\theta, d(\mathbf{X}))$ be the loss incurred when we make decision $d(\mathbf{X})$ and θ is the true parameter value. In sequential sampling another loss enters, namely, the cost $c(N)$ of taking N observations. The statistician's problem therefore is to choose a stopping rule and a decision rule such that the average loss is minimum, or the average loss and cost are minimum, or, simply, the cost is minimum, or the average loss is bounded, and so on. Thus, in the problem of estimating a parameter, the statistician may be interested, for example, in an unbiased estimate of θ such that the squared error loss $L(\theta, d(\mathbf{X})) = [\theta - d(\mathbf{X})]^2$ is minimum, on the average.

PROBLEMS 14.2

1. In Example 1 take the loss to be as follows:

(a) $L(n, \mu, \sigma^2) = B\,E|\bar{X} - \mu|^s + An$, $s > 0, B$ constant.
(b) $L(n, \mu, \sigma^2) = B\,E|\bar{X} - \mu|^s + A \log n$, $s > 0$, B constant.

In each case compute the smallest number of observations required to minimize EL when σ is known. Compute the minimum risk in each case.

2. In Example 5 suppose that we take observations until we observe the rth head. Is this sampling procedure a closed sequential procedure? Let N be the number of trials required to stop. Compute the pmf of N; compute EN. Show that r/N is the MLE estimate of θ. Find an unbiased estimate of θ.

3. Let $X_1, X_2, \cdots$ be observations on a Poisson rv with parameter λ. It is required to estimate λ sequentially. The stopping rule is to stop at the first $N=n$ for which $\sum_{i=1}^n X_i \geq 1$. Let N be the number of observations required. Is this a closed sequential procedure? Find $E_\lambda N$. If the stopping rule is changed as follows: Stop at the first $N=n$ for which $\sum_{i=1}^n X_i \geq 2$, find the distribution of N. Also, find $E_\lambda N$.

4. Prove (16) by writing $(p+q)p_x = pp_{x+1} + qp_{x-1}$, where $q = 1-p$. Thus
$$p_{x+1} - p_x = \frac{q}{p}(p_x - p_{x-1}).$$

5. Prove (18) by an argument similar to the one used in Problem 4. (*Hint*: If E_x is the expected duration of the game, starting with a capital of x, then
$$E_x = 1 + pE_{x+1} + qE_{x-1}, \quad x = 1, 2, \cdots, N-1.$$
Clearly $E_0 = E_{a+b} = 0$.)

14.3 SEQUENTIAL UNBIASED ESTIMATION

In this section we consider some simple problems of sequential estimation. The subject of sequential estimation is not yet very well developed, and the area is under active investigation at this time. Basically no new methods of estimation are needed. One takes observations sequentially and stops according to some well-defined sequential procedure that is a function of the observations to date. Then one estimates the parameter on the basis of the sample thus far obtained by using some standard estimation procedure like the method of unbiased estimation or the method of maximum likelihood estimation. What makes the problem difficult is the fact that the sample space changes with each observation, and one has to solve the distribution problem resulting from the introduction of the stopping variable. We have already considered such a problem in Example 14.2.5, as well as in Example 14.2.1, which we will discuss in some detail in the next section. We first derive some results of general interest concerning the mean and variance of the stopping rv and find an analogue of the Fréchet-Cramér-Rao lower bound for the variance of a sequential estimate.

Let $X_1, X_2, \cdots$ be a sequence of independent rv's with common pdf (pmf) $f_\theta(x)$, $\theta \in \Theta$. Suppose that we wish to estimate θ so that the risk $E_\theta L(\theta, d(\mathbf{X})) = E_\theta(\theta - d(\mathbf{X}))^2$, say, is minimum. If we restrict attention to unbiased estimates of θ, it would be of interest to find a lower bound for the variance of such estimates. In other words an extension of the Fréchet-Cramér-Rao lower bound to the sequential unbiased estimation case is sought. To obtain just such an extension we need some preliminary results.

Theorem 1 (**Wald's Equation**). *Let $X_1, X_2, \cdots$ be iid rv's with $E|X_1| < \infty$. Let N be a stopping variable, and write $S_N = \sum_{k=1}^{N} X_k$. If $EN < \infty$, then*

(1) $$E\{S_N\} = EX_1 EN.$$

Proof. Define rv's

(2) $$Y_i = \begin{cases} 1 & \text{if no decision is reached up to the } (i-1)\text{th stage,} \\ & \text{that is, if } N > (i-1), \\ 0 & \text{otherwise.} \end{cases}$$

Then Y_i is a function of $X_1, X_2, \cdots, X_{i-1}$ alone and is independent of X_i. Consider the rv $\sum_{n=1}^{\infty} X_n Y_n$. Clearly,

$$S_N = \sum_{n=1}^{\infty} X_n Y_n.$$

Thus

(3) $$ES_N = E(\sum_{n=1}^{\infty} X_n Y_n).$$

Now

$$\sum_{n=1}^{\infty} E|X_n Y_n| = \sum_{n=1}^{\infty} E|X_n| E|Y_n|$$
$$= E|X_1| \sum_{n=1}^{\infty} P\{N \geq n\}$$
$$= E|X_1| \sum_{n=1}^{\infty} \sum_{k=n}^{\infty} P\{N = k\}$$
$$= E|X_1| \sum_{n=1}^{\infty} n\, P\{N = n\}$$
$$= E|X_1|\, EN < \infty.$$

We may therefore interchange the expectation and summation signs in (3), obtaining

$$ES_N = \sum_{n=1}^{\infty} EX_n EY_n$$
$$= EX_1 \sum_{n=1}^{\infty} P\{N \geq n\}$$
$$= EX_1\, EN,$$

as asserted.

Remark 1. This result is due to Wald [133]. The proof given above is due to Johnson [53]. An analysis of the proof shows that the rv's X_n do not need to have the same distribution, provided we assume that $EX_n = \mu$ for all n, $E|X_n| \leq A < \infty$ for all n, and $EN < \infty$.

Remark 2. If the rv's $X_1, X_2, \cdots$ are iid and are independent of the iid rv's $Y_1, Y_2, \cdots$, and if M denotes the random number of X's, and N, the random number of Y's, then

$$E|X| < \infty, \qquad E|Y| < \infty, \qquad EM < \infty, \qquad EN < \infty$$

imply

(4) $$E\left(\sum_{i=1}^{M} X_i + \sum_{j=1}^{N} Y_j\right) = EM\ EX + EN\ EY.$$

We leave the reader to complete the proof of (4).

Theorem 2. Let $X_1, X_2, \cdots$ be a sequence of iid rv's with common mean 0 and variance σ^2, $0 < \sigma^2 < \infty$. For any sequential stopping rule with $EN < \infty$, if $E\{(\sum_{k=1}^{N}|X_k|)^2\} < \infty$ then

(5) $$E\left\{\left(\sum_{i=1}^{N} X_i\right)^2\right\} = \sigma^2\ EN.$$

Proof. In the notation of Theorem 1, we have

(6) $$\begin{aligned} ES_N^2 &= E\left(\sum_{i=1}^{\infty} X_i Y_i \sum_{j=1}^{\infty} X_j Y_j\right). \\ &= E\left(\sum_{i=1}^{\infty} X_i^2 Y_i^2 + \sum_{i \neq j} X_i Y_i X_j Y_j\right). \end{aligned}$$

Since

$$E\left(\sum_{i=1}^{\infty} X_i^2 Y_i^2 + \sum_{i \neq j} |X_i X_j| Y_i Y_j\right) = E\left(\sum_{j=1}^{N} |X_j|\right)^2 < \infty,$$

we may interchange the summation and expectation signs in (6). We have from Theorem 1

$$\begin{aligned} E\left(\sum_{i=1}^{\infty} X_i^2 Y_i^2\right) &= E\left(\sum_{i=1}^{N} X_i^2\right) \\ &= EX_i^2\ EN \\ &= \sigma^2\ EN. \end{aligned}$$

Now

$$E\left(\sum_{i \neq j} X_i Y_i X_j Y_j\right) = 2 \sum_{i=2}^{\infty} \sum_{j=1}^{i-1} E(X_i X_j Y_i)$$

$$= 2 \sum_{i=2}^{\infty} \sum_{j=1}^{i-1} E\{Y_i \, E\{X_i X_j | Y_i\}\} \quad \text{(Problem 4.7.5)}$$
$$= 2 \sum_{i=2}^{\infty} \sum_{j=1}^{i-1} E\{Y_i \, EX_i \, E\{X_j | Y_i\}\} \quad (X_j \text{ and } Y_i \text{ are independent of } X_i)$$
$$= 0,$$

and the proof is complete.

Theorem 3 (Wolfowitz [143]). Let $\mathbf{X} = (X_1, X_2, \cdots)$ be an infinite sequence of iid rv's with common pmf (pdf) $f_\theta(x)$, $\theta \in \Theta$. Suppose that the following regularity conditions hold:

(i) Θ is an open interval of the real line which may consist of the entire line or of an entire half-line.

(ii) $f_\theta(x)$ is differentiable with respect to all $\theta \in \Theta$ for all x. Define $\partial \log f_\theta(x) / \partial \theta = 0$ whenever $f_\theta(x) = 0$, so that $\partial \log f / \partial \theta$ is defined for all θ in Θ and all x.

(iii) The integral (the sum) $\int f_\theta(x) \, dx$ ($\sum_x f_\theta(x)$) can be differentiated under the integral (summation) sign.

(iv) $$0 < E_\theta \left[\frac{\partial \log f_\theta(X)}{\partial \theta} \right]^2 < \infty \quad \text{for all } \theta.$$

(v) For each $n = 1, 2, \cdots$ and all θ,
$$E_\theta \left[\sum_{i=1}^{n} \left| \frac{\partial \log f_\theta(X_i)}{\partial \theta} \right| \right]^2 < \infty.$$

Let $\varphi(\theta)$ be an estimable function that is differentiable on Θ. Let $h(X_1, X_2, \cdots)$ be an unbiased estimate of $\varphi(\theta)$ such that

(vi) the equation
$$E_\theta h(X_1, X_2, \cdots, X_n) = \varphi(\theta)$$

can be differentiated under the expectation sign with respect to all θ in Θ and for each $n = 1, 2, \cdots$, and

(vii) the series $\sum_{n=1}^{\infty} d\varphi_n(\theta)/d\theta$ converges uniformly on Θ, where

$$\varphi_n(\theta) = \begin{cases} \int_{\{N=n\}} h(x_1, x_2, \cdots, x_n) \prod_{i=1}^{n} (f_\theta(x_i) \, dx_i), \\ \sum_{\{N=n\}} h(x_1, x_2, \cdots, x_n) f_\theta(x_1) \cdots f_\theta(x_n). \end{cases}$$

Then

(7) $$\operatorname{var}_\theta \{h(X_1, X_2, \cdots, X_N)\} \geq \frac{\{\varphi'(\theta)\}^2}{E\{N\} \, E[\partial \log f_\theta(X)/\partial \theta]^2}$$

for all θ, provided that $EN < \infty$.

Remark 3. Conditions (i) to (iv) are the same as those assumed in Theorem 8.5.1. Conditions (v), (vi), and (vii) are needed because of the sequential nature of the procedure. For (vi) to hold, what we need is the existence, for each integer n, of a nonnegative statistic $T_n(X_1, \cdots, X_n)$ such that

$$\left| h(x_1, x_2, \cdots, x_n) \frac{\partial}{\partial \theta} \prod_{i=1}^{n} f_\theta(x_i) \right| < T_n(x_1, x_2, \cdots, x_n)$$

holds for all $\theta \in \Theta$ and all $(x_1, x_2, \cdots, x_n)$ in $\{N = n\}$ and that $\int_{\{N=n\}} T_n(x_1, \cdots, x_n) \prod_{i=1}^{n} dx_i$ [or the sum $\sum_{\{N=n\}} T_n(x_1, \cdots, x_n)$] is finite.

Proof of the Theorem. Let N be the stopping variable associated with the sequential procedure. The rv's $(\partial/\partial \theta) \log f_\theta(X_i)$, $i = 1, 2, \cdots$, are iid with

$$E_\theta \left[\frac{\partial}{\partial \theta} \log f_\theta(X_i) \right] = 0 \quad \text{for all } i,$$

and

$$0 < E_\theta \left[\frac{\partial}{\partial \theta} \log f_\theta(X_i) \right]^2 < \infty.$$

Since $EN < \infty$, Wald's equation implies that

$$E_\theta \left[\sum_{i=1}^{N} \frac{\partial}{\partial \theta} \log f_\theta(X_i) \right] = E_\theta N \, E_\theta \left[\frac{\partial}{\partial \theta} \log f_\theta(X_i) \right] = 0 \quad \text{for all } \theta.$$

From condition (v) and Theorem 2 we obtain

$$E \left\{ \left[\sum_{i=1}^{N} \frac{\partial \log f_\theta(X_i)}{\partial \theta} \right]^2 \right\} = EN \, E \left[\frac{\partial \log f_\theta(X_1)}{\partial \theta} \right]^2.$$

Now

$$\operatorname{cov}\left[h(X_1, X_2, \cdots, X_N), \sum_{i=1}^{N} \frac{\partial \log f_\theta(X_i)}{\partial \theta} \right]$$

$$= E\left\{ [h(X_1, X_2, \cdots, X_n) - \varphi(\theta)] \sum_{i=1}^{N} \frac{\partial \log f_\theta(X_i)}{\partial \theta} \right\}$$

$$\leq \{E[h(X_1, \cdots, X_N) - \varphi(\theta)]^2\}^{1/2} \left\{ EN \, E\left[\frac{\partial \log f_\theta(X_1)}{\partial \theta} \right]^2 \right\}^{1/2}$$

for all θ. But by conditions (vi) and (vii)

$$\varphi'(\theta) = \frac{\partial}{\partial \theta} E_\theta h(X_1, X_2, \cdots, X_N)$$
$$= E\left[h(X_1, X_2, \cdots, X_N) \sum_{i=1}^{N} \frac{\partial \log f_\theta(X_i)}{\partial \theta}\right],$$

and the result follows.

Remark 4. In the case of a fixed sample size procedure, if there exists an estimate that attains the lower bound of the Fréchet-Cramér-Rao inequality nothing is gained by proceeding sequentially. If we restrict attention to sequential estimation procedures for which $E_\theta N \leq n_0$, and the regularity conditions of Theorem 3 hold, then, for every unbiased estimate $T(\mathbf{X})$,

$$\text{var}_\theta(T) \geq \frac{1}{E_\theta N \, E_\theta[(\partial/\partial\theta) \log f_\theta(X)]^2}$$
$$\geq \frac{1}{n_0 \, E_\theta[(\partial/\partial\theta) \log f_\theta(X)]^2}.$$

Example 1. In Example 14.2.5, conditions (i) through (v) of Theorem 3 are satisfied. Thus for any unbiased estimate $h(X_1, X_2, \cdots, X_N)$ of θ, we have

$$\text{var}_\theta\{h(X_1, X_2, \cdots, X_N)\} \geq \frac{\theta}{E[\partial \log f_\theta(X)/\partial\theta]^2}, \quad \theta \in (0, 1),$$

provided that h satisfies conditions (vi) and (vii) of Theorem 3. Since $f_\theta(X) = \theta^X(1-\theta)^{1-X}$, $\theta \in (0, 1)$,

$$\frac{\partial \log f_\theta(X)}{\partial \theta} = \frac{X - \theta}{\theta(1-\theta)}$$

and

$$E_\theta\left[\frac{\partial \log f_\theta(X)}{\partial \theta}\right]^2 = \frac{\theta(1-\theta)}{\theta^2(1-\theta)^2} = \frac{1}{\theta(1-\theta)}$$

so that

$$\text{var}_\theta\{h(X_1, X_2, \cdots, X_N)\} \geq \theta^2(1-\theta)$$

for any unbiased estimate h of θ satisfying conditions (vi) and (vii) of Theorem 3.

PROBLEMS 14.3

1. Prove result (4), stated in Remark 2.
2. In Problem 14.2.2 find the lower bound for the variance of an unbiased estimate of θ.
3. If X_j and Y_i are independent of X_i, verify that $E\{X_i X_j | Y_i\} = EX_i \, E\{X_j | Y_i\}$.

14.4 SEQUENTIAL ESTIMATION OF THE MEAN OF A NORMAL POPULATION

Let $X_1, X_2, \ldots, X_n, \ldots$ be iid normal rv's with mean μ and variance σ^2, both unknown. In this section we consider the sequential point and interval estimation of μ. We begin with the sequential unbiased estimation procedure suggested by Robbins [100]. As an estimate of μ we choose $\bar{X}_n$, the sample mean. The problem now is to choose n. Let us assume that the loss incurred is $A|\bar{X}_n - \mu|$, where $A > 0$ is a known constant, and let each observation cost one unit. Then we wish to choose n to minimize

(1) $$EL(n) = E\{A|\bar{X}_n - \mu| + n\}.$$

We have

$$E\left|\sqrt{n}\,\frac{\bar{X}_n - \mu}{\sigma}\right| = \sqrt{\frac{2}{\pi}},$$

so that

(2) $$EL(n) = AE\left|\sqrt{n}\,\frac{\bar{X}_n - \mu}{\sigma}\right|\frac{\sigma}{\sqrt{n}} + n$$
$$= A\sqrt{\frac{2}{\pi}}\,\frac{\sigma}{\sqrt{n}} + n.$$

Regarding n as a continuous variable, we differentiate $EL(n)$ and set the derivative equal to 0 to get

(3) $$n_0 = \left(\frac{A\sigma}{\sqrt{2\pi}}\right)^{2/3}$$

as the value of n that minimizes (2). For this value of n (which may not be integral), we have

(4) $$v(\sigma) = EL(n_0) = A\sqrt{\frac{2}{\pi}}\,\sigma\left(\frac{A\sigma}{\sqrt{2\pi}}\right)^{-1/3} + \left(\frac{A\sigma}{\sqrt{2\pi}}\right)^{2/3}$$
$$= 3n_0,$$

so that the loss due to the error of estimate is twice the size of the sample, that is, twice the cost of sampling. Of course, this presupposes knowledge of σ. If we do not know σ, we cannot compute n_0.

For the case where σ is unknown, Robbins [100] suggested the following sequential sampling procedure R:

(5) $\begin{cases} \text{Sample sequentially, observing } x_1, x_2, \ldots \text{. Compute } \bar{x}_n = n^{-1}\sum_{i=1}^{n} x_i \text{ and} \\ s_n^2 = (n-1)^{-1}\sum_{i=1}^{n}(x_i - \bar{x}_n)^2 \text{ successively, and stop sampling the first} \\ \text{time for } n \geq 2 \text{ when} \\ \qquad\qquad n \geq \left(\dfrac{As_n}{\sqrt{2\pi}}\right)^{2/3}. \end{cases}$

We may rewrite this inequality as

(6) $$\sum_{i=1}^{n}(x_i - \bar{x}_n)^2 \leq \frac{2\pi}{A^2}(n-1)n^3.$$

To study the properties of the stopping rule R, we will need some preliminary lemmas. First we note that the sampling procedure is closed.

Lemma 1. Rule R terminates with probability 1.

This result follows from the inequality in (5), since the right-hand side converges to a constant n_0 with probability 1 (see Theorem 6.4.7).

Lemma 2. For any fixed n, $\bar{X}_n$ is independent of the set of rv's $S_2^2, S_3^2, \cdots, S_n^2$, and hence

(7) $$P\{\sqrt{n}\left(\frac{\bar{X}_n - \mu}{\sigma}\right) \leq t \mid s_2^2, s_3^2, \cdots, s_n^2\} = \int_{-\infty}^{t} \frac{1}{\sqrt{2\pi}} e^{-x^2/2} dx.$$

Proof. Define

$$u_i = \frac{x_i - \mu}{\sigma}, \quad i = 1, 2, \cdots, n.$$

Then U_i are iid $\mathcal{N}(0,1)$ rv's. Let us write

$$y_i = \frac{u_1 + u_2 + \cdots + u_i - iu_{i+1}}{\sqrt{i(i+1)}}, \quad i = 1, 2, \cdots, n-1$$

and

$$y_n = \sqrt{n}\,\bar{u},$$

where $\bar{u} = n^{-1}\sum_1^n u_i$. Then the rv's Y_i's are also $\mathcal{N}(0, 1)$ and independent. Now

$$y_n = \frac{\bar{X}_n - \mu}{\sigma}\sqrt{n}$$

and

$$s_i^2 = \sigma^2\left(\frac{y_1^2 + y_2^2 + \cdots + y_{i-1}^2}{i-1}\right), \quad i = 2, 3, \cdots, n.$$

It follows that Y_n is independent of S_i^2 for $i = 2, \cdots, n$. This is the same as saying that $\bar{X}_n$ is independent of $S_2^2, S_3^2, \cdots, S_n^2$.

In fact, we have also shown that the following result holds.

Lemma 3. The joint distribution of the sequence of rv's $\{S_n^2\}$, $n = 2, 3, \cdots$, is the same as that of the sequence

$$\left\{\frac{\sigma^2}{n-1}(Y_1^2 + Y_2^2 + \cdots + Y_{n-1}^2)\right\},$$

where the Y_i's are iid $\mathcal{N}(0, 1)$.

Lemma 4. Let R^* be any sequential stopping rule that terminates with probability 1, and suppose that the sample size N is determined uniquely in such a way that $N = n$ if and only if the point $(s_2^2, s_3^2, \cdots, s_n^2)$ belongs to some set $S_{n-1}^* \subseteq \mathcal{R}_{n-1}$. Then, for any $n = 2, 3, \cdots$,

(8) $$P\left\{\sqrt{N}\,\frac{\bar{X}_N - \mu}{\sigma} \le t \,\Big|\, N = n\right\} = \Phi(t),$$

where

$$\Phi(t) = \frac{1}{\sqrt{2\pi}} \int_{-\infty}^{t} e^{-x^2/2}\, dx,$$

hence the unconditional distribution of $\sqrt{N}(\bar{X}_N - \mu)/\sigma$ is also $\mathcal{N}(0, 1)$.

Proof. The proof is left as an exercise.

Lemma 5. For the sampling procedure R the probability distribution of N may be obtained as follows. Let $\{Y_n\}$, $n = 1, 2, \cdots$, be iid $\mathcal{N}(0, 1)$ rv's, and let

(9) $N = n \Leftrightarrow n \ge 2$ is the first integer such that

$$Y_1^2 + Y_2^2 + \cdots + Y_{n-1}^2 \le \frac{(n-1)n^3}{n_0^3}, \qquad n_0^3 = \frac{A^2\sigma^2}{2\pi}.$$

Lemma 5 follows from Lemma 3 and definitions of n_0, Y_i's, and so forth. Let us now compute the average loss for R. We have

(10) $$L(N) = A\left|\sqrt{N}\,\frac{\bar{X}_N - \mu}{\sigma}\right|\frac{\sigma}{\sqrt{N}} + N,$$

so that

$$\bar{r}(\sigma) = E_R L(N) = \sum_{n=2}^{\infty} P\{N = n\}\, E\{L(N)\,|\,N = n\}$$

$$= \sum_{n=2}^{\infty} P\{N = n\}\left(A\sqrt{\frac{2}{\pi}}\,\frac{\sigma}{\sqrt{n}} + n\right) \qquad \text{by Lemma 4}$$

$$= A\sqrt{\frac{2}{\pi}}\,\sigma\, E(N^{-1/2}) + EN$$

(11) $$= 2n_0^{3/2}\, E(N^{-1/2}) + EN.$$

The risk for the procedure R, given by (11), is to be compared with the risk for the fixed sample size procedure, given by (4). One needs therefore the distribution of N, which is not easy to compute. Ideally, one would like EN to be close to n_0, whatever σ^2 may be. In fact, at least for large n_0, one can show that

$$P\{N \le n_0\} \ge \tfrac{1}{2},$$

so that for at least half the time N is no larger than n_0. Note that, for any fixed integer n,

$$P\{N \le n\} \ge P\left\{Y_1^2 + \cdots + Y_{n-1}^2 \le \frac{(n-1)n^3}{n_0^3}\right\}.$$

It follows that, for $n = n_0$,

$$P\{N \le n_0\} \ge P\{Y_1^2 + \cdots + Y_{n_0-1}^2 \le (n_0 - 1)\}$$
$$= P\{\chi^2(n_0 - 1) \le n_0 - 1\}$$
(12) $$\approx P\{Z \le 0\} = \tfrac{1}{2},$$

where $Z \sim \mathcal{N}(0, 1)$.

Robbins presents some numerical results comparing n_0 and $v(\sigma)$ with values obtained by taking, for each n_0, a thousand sequences $\{y_n\}$ of normal $(0, 1)$ rv's and computing the average values E^*N and $E^*N^{-1/2}$:

n_0:	4	16	64
E^*N:	3.66	15.49	63.84
E_R^*L:	12.49	49.92	193.89
$E_{n_0}L(=3n_0)$:	12.00	48.00	192.00
Regret $= E_R^*L - E_{n_0}L$:	.49	1.92	1.89

These results have been substantiated in subsequent studies by many authors, and the procedure R has been shown to have the following limiting properties.

(a) $P\{\lim_{\sigma \to \infty} n_0^{-1} N = 1\} = 1$.
(b) For $\omega > 0$, $\lim_{\sigma \to \infty} E(N/n_0)^\omega = 1$.
(c) If $\eta(\sigma) = \bar{v}(\sigma)/v(\sigma)$, then $\lim_{\sigma \to \infty} \eta(\sigma) = 1$.
(d) If $w(\sigma) = \bar{v}(\sigma) - v(\sigma)$ is the regret, then $w(\sigma)$ is bounded as $\sigma \to \infty$.

We will not prove these results here, but refer the reader to Starr and Woodroofe [124] for further references. We note that, at least for large-sample values ($\sigma \to \infty$), the sample size required by R will be close to n_0, the average sample number is close to n_0, and the ratio of average loss using R to average loss using a fixed sample size procedure (when σ is known) tends to 1. Moreover, the regret, that is, the average loss due to ignorance of σ, is bounded, at least for large-sample values.

We next consider the sequential interval estimation of the mean μ. Let X_1, $X_2, \cdots$ be independent $\mathcal{N}(\mu, \sigma^2)$ rv's, and suppose that we require a confidence interval for μ of width at most $2d$ ($d > 0$) at confidence level $1 - \alpha$ ($0 < \alpha < 1$). We have seen (Example 14.2.3) that, if σ is known, a confidence interval $I_n = (\bar{X}_n - d, \bar{X}_n + d)$ for μ of width $2d$ and at level $1 - \alpha$ is assured, provided that we choose n to be the (smallest) integer satisfying

$$(13) \qquad n \geq \frac{\sigma^2}{d^2} z_{\alpha/2}^2.$$

If σ is unknown, no procedure based on a fixed number of observations n will satisfy our requirements (see Dantzig [21]). In this situation two choices are open to us. The first option is to take a preliminary sample of size m and estimate σ^2. This procedure (due to Stein [125]), which leads to a sample size that approximately satisfies (13), is simple enough. Take a first sample of size m (fixed) and compute

$$\bar{X}_m = m^{-1} \sum_{i=1}^{m} X_i \quad \text{and} \quad S_m^2 = (m-1)^{-1} \sum_{i=1}^{m} (X_i - \bar{X}_m)^2.$$

Then take a further sample of size $N - m$, where

$$(14) \qquad N \text{ is the smallest integer} \geq \frac{S_m^2}{d^2} t_{m-1,\alpha/2}^2,$$

and use as confidence interval

$$(15) \qquad \left(\bar{X}_N - \frac{S_m}{\sqrt{N}} t_{m-1,\alpha/2}, \bar{X}_N + \frac{S_m}{\sqrt{N}} t_{m-1,\alpha/2} \right),$$

where $\bar{X}_N = N^{-1} \sum_{i=1}^{N} X_i$. The length of this confidence interval is $2t_{m-1,\alpha/2} \cdot S_m / \sqrt{N}$, so that, if we choose N to be the smallest integer satisfying

$$(16) \qquad N \geq \frac{t_{m-1,\alpha/2}^2 S_m^2}{d^2},$$

the confidence interval will have width $\leq 2d$. If $N < m$, the confidence interval based on the first sample itself will have width $\leq 2d$.

We need to show that confidence interval (15) has the required level $(1 - \alpha)$. Note that N is now a random variable. It will be sufficient to show that $\sqrt{N}(\bar{X}_N - \mu)/S_m$ has a t-distribution with $m - 1$ d.f., for then

$$P_{\mu,\sigma} \left\{ \bar{X}_N - \frac{S_m}{\sqrt{N}} t_{m-1,\alpha/2} < \mu < \bar{X}_N + \frac{S_m}{\sqrt{N}} t_{m-1,\alpha/2} \right\}$$

$$= P_{\mu,\sigma} \left\{ \frac{|\bar{X}_N - \mu|}{S_m/\sqrt{N}} < t_{m-1,\alpha/2} \right\}$$

$$= 1 - \alpha.$$

To see that $\sqrt{N}(\bar{X}_N - \mu)/S_m$ has a $t(m - 1)$ distribution, we write

$$\bar{X}_N = \frac{\sum_1^m X_i + \sum_{m+1}^N X_i}{N}$$

(17)
$$= \frac{m}{N} \bar{X}_m + \frac{1}{N} \sum_{i=m+1}^N X_i.$$

Now S_m is independent of $\bar{X}_m$, and since N depends only on S_m, the conditional distribution of $\bar{X}_N$, given $N = k$, is $\mathcal{N}(\mu, \sigma^2/k)$. It follows that $(\bar{X}_N - \mu)\sqrt{N}$, conditional on $N = k$, is $\mathcal{N}(0, \sigma^2)$, and hence the unconditional distribution of $(\bar{X}_N - \mu)\sqrt{N}$ is also $\mathcal{N}(0, \sigma^2)$. Consequently,

$$\frac{(\bar{X}_N - \mu)\sqrt{N}/\sigma}{S_m/\sigma} = \frac{\bar{X}_N - \mu}{S_m} \sqrt{N}$$

has a $t(m - 1)$ distribution, as asserted.

As for the pmf of N, we have

$$P\{N = m\} = P\left\{m \geq t_{m-1,\alpha/2}^2 \frac{S_m^2}{d^2}\right\}$$

$$= P\left\{\frac{S_m^2}{\sigma^2}(m - 1) \leq \frac{m(m-1)}{\sigma^2} \frac{d^2}{t_{m-1,\alpha/2}^2}\right\}$$

(18)
$$= P\left\{\chi^2(m - 1) \leq \frac{m(m-1)}{\sigma^2} \frac{d^2}{t_{m-1,\alpha/2}^2}\right\}.$$

Also, for $k > m$,

$$P\{N = k\} = P\left\{k - 1 < t_{m-1,\alpha/2}^2 \frac{S_m^2}{d^2} \leq k\right\}$$

(19)
$$= P\left\{\frac{(k-1)(m-1)}{t_{m-1,\alpha/2}^2} \frac{d^2}{\sigma^2} < \chi^2(m - 1) \leq \frac{k(m-1)}{t_{m-1,\alpha/2}^2} \frac{d^2}{\sigma^2}\right\},$$

so that the pmf of N depends only on σ. Also,

$$P\{N < \infty\} = P\{N = m\} + \sum_{k=m+1}^{\infty} P\{N = k\}$$

$$= P\left\{\chi^2(m-1) \leq \frac{m(m-1)}{t_{m-1,\alpha/2}^2} \frac{d^2}{\sigma^2}\right\} + \sum_{k=m+1}^{\infty} P\left\{\frac{(k-1)(m-1)}{t_{m-1,\alpha/2}^2} \frac{d^2}{\sigma^2} < \chi^2(m-1) < \frac{k(m-1)}{t_{m-1,\alpha/2}^2} \frac{d^2}{\sigma^2}\right\}$$

$$= P\left\{\chi^2(m-1) \leq \frac{m(m-1)}{t_{m-1,\alpha/2}^2} \frac{d^2}{\sigma^2}\right\}$$

$$+ P\left\{\chi^2(m-1) > \frac{m(m-1)}{t_{m-1,\alpha/2}^2} \frac{d^2}{\sigma^2}\right\}$$

(20)
$$= 1.$$

Moreover,
$$EN = mP\left\{\chi^2(m-1) \le \frac{m(m-1)}{t^2_{m-1,\alpha/2}} \frac{d^2}{\sigma^2}\right\} + \sum_{k=m+1}^{\infty} kP\{N=k\}.$$

Let
$$y = \frac{m(m-1)}{t^2_{m-1,\alpha/2}} \frac{d^2}{\sigma^2}.$$

Then

(21) $$EN = \frac{1}{2^{(m-1)/2}\Gamma[(m-1)/2]}\left\{\int_0^y m e^{-(1/2)x} x^{[(m-1)/2]-1}\,dx \right.$$
$$\left. + \sum_{k=m+1}^{\infty} \int_{[(k-1)/m]y}^{(k/m)y} k e^{-(1/2)x} x^{[(m-1)/2]-1}\,dx\right\}.$$

It follows that
$$\frac{1}{2^{(m-1)/2}\Gamma[(m-1)/2]}\left\{\int_0^y m e^{-(1/2)x} x^{(m-3)/2}\,dx \right.$$
$$\left. + \sum_{k=m+1}^{\infty} \int_{[(k-1)/m]y}^{(k/m)y} \frac{mx}{y} e^{-(1/2)x} x^{(m-3)/2}\,dx\right\}$$
$$< EN$$
$$< \frac{1}{2^{(m-1)/2}\Gamma[(m-1)/2]}\left\{\int_0^y m e^{-(1/2)x} x^{(m-3)/2}\,dx \right.$$
$$\left. + \sum_{k=m+1}^{\infty} \int_{[(k-1)/m]y}^{(k/m)y} \left(\frac{mx}{y}+1\right) e^{-(1/2)x} x^{(m-3)/2}\,dx\right\}.$$

Thus
$$\frac{1}{2^{(m-1)/2}\Gamma[(m-1)/2]}\left\{\int_0^y m e^{-(1/2)x} x^{(m-3)/2}\,dx \right.$$
$$\left. + \int_y^{\infty} e^{-(1/2)x} x^{(m-1)/2} \frac{t^2_{m-1,\alpha/2}\sigma^2}{(m-1)d^2}\,dx\right\}$$
$$< EN$$
$$< \frac{1}{2^{(m-1)/2}\Gamma[(m-1)/2]}\left\{\int_0^y m e^{-(1/2)x} x^{(m-3)/2}\,dx \right.$$
$$\left. + \int_y^{\infty} e^{-(1/2)x} x^{(m-3)/2}\left(\frac{\sigma^2 t^2_{m-1,\alpha/2}}{(m-1)d^2}x + 1\right)dx\right\},$$

which we rewrite as
$$mP\{\chi^2(m-1) < y\} + \frac{\sigma^2}{d^2} t^2_{m-1,\alpha/2} P\{\chi^2(m+1) > y\}$$
(22) $$< EN < mP\{\chi^2(m-1) < y\} + \frac{\sigma^2}{d^2} t^2_{m-1,\alpha/2} P\{\chi^2(m+1) > y\}$$
$$+ P\{\chi^2(m-1) > y\}.$$

If n_0 is the number of observations required for a fixed sample size procedure (when σ is known),

(23) $$EN < m + 1 + n_0,$$

at least when m is large enough so that $t^2_{m-1,\alpha/2} \approx z^2_{\alpha/2}$.

Remark 1. The basic difficulty of Stein's approach is that there is no control on N. If σ is large, N may be very large. Moreover, we lose a lot of information in the estimation of σ^2, since we use only the first sample for this purpose. Another problem is the determination of m: how large or small should it be? Seelbinder [112] and Moshman [83] have considered this problem, and we leave the interested reader to pursue the subject.

Stein's procedure can also be used to test the hypothesis $\mu \leq \mu_0$ against $\mu > \mu_0$ whenever σ is not known. We will not go into the procedure here but refer the reader to Stein [125]. As a by-product of the procedure we obtain the following result.

Theorem 1. $\bar{X}_N$ is an unbiased estimate of μ with variance $\leq [m/(m-2)] \cdot (d^2/t^2_{m-1,\alpha/2})$ for $m > 2$.

Proof. We have seen that

$$\frac{\sqrt{N}(\bar{X}_N - \mu)}{\sigma} \quad \text{and} \quad \frac{(m-1)S^2_m}{\sigma^2}$$

are independent rv's with $\mathcal{N}(0, 1)$ and $\chi^2(m-1)$ distributions, respectively. It follows that $\sqrt{N}(\bar{X}_N - \mu)/S_m \sim t(m-1)$. Now

$$E_{\mu,\sigma}\{\bar{X}_N - \mu\} = E_{\mu,\sigma}\left\{\frac{\sqrt{N}(\bar{X}_N - \mu)}{\sigma} \frac{\sigma}{\sqrt{N}}\right\}$$

$$= E_{\mu,\sigma}\left\{\frac{\sqrt{N}(\bar{X}_N - \mu)}{\sigma}\right\} E_{\mu,\sigma}\left\{\frac{\sigma}{\sqrt{N}}\right\},$$

since N is a function of S_m alone, and $\sqrt{N}[(\bar{X}_N - \mu)/\sigma]$ is independent of S_m. It follows that $E_{\mu,\sigma}(\bar{X}_N - \mu) = 0$, provided that $EN^{-1/2} < \infty$. Moreover,

$$E_{\mu,\sigma}(\bar{X}_N - \mu)^2 = E_{\mu,\sigma}\left\{\frac{N(\bar{X}_N - \mu)^2}{S^2_m} \frac{S^2_m}{N}\right\}$$

$$\leq E_{\mu,\sigma}\left\{\frac{N(\bar{X}_N - \mu)^2}{S^2_m} \frac{d^2}{t^2_{m-1,\alpha/2}}\right\}, \quad \text{using (14),}$$

$$= \frac{m}{m-2} \frac{d^2}{t^2_{m-1,\alpha/2}}, \quad m > 2.$$

We leave the reader to show that $EN^{-1/2} < \infty$ by repeating the argument used in obtaining (22).

Finally, we investigate briefly the second option, namely, to sample sequentially and use all the information in the sample to date to estimate σ^2. The following sequential procedure, Λ, has received wide attention. Proceed sequentially, and at each stage compute

$$\bar{X}_n = n^{-1} \sum_{i=1}^n X_i, \quad \text{and} \quad S_n^2 = (n-1)^{-1} \sum_{i=1}^n (X_i - \bar{X}_n)^2.$$

Stop with observation X_N, where

(24) $\qquad N$ = smallest integer $n \geq n_0$ such that

$$S_n^2 \leq n \frac{d^2}{t_{n-1, \alpha/2}^2},$$

where $n_0 \geq 2$ is a fixed integer. Then form the interval $I_N = (\bar{X}_N - d, \bar{X}_N + d)$. The interval I_N is of fixed length $2d$, and, it is hoped, has level $1 - \alpha$.

We first note that by the strong law $S_n^2 \overset{\text{a.s.}}{\to} \sigma^2$, so that

(25) $\qquad P_{\sigma^2}\{N < \infty\} = 1 - P\{N = \infty\}$

$$= 1 - P\left\{\frac{S_n^2}{n} > \frac{d^2}{t_{n-1, \alpha/2}} \text{ for all } n \geq n_0\right\}$$

$$= 1.$$

Thus the sampling terminates with probability 1. If we write

$$J_n = (\bar{X}_n - t_{n-1, \alpha/2} S_n/n^{1/2}, \bar{X}_n + t_{n-1, \alpha/2} S_n/n^{1/2}),$$

then, for fixed μ, σ, and n,

(26) $\qquad P_{\mu, \sigma}\{J_n \ni \mu\} = P_{\mu, \sigma}\left\{n^{1/2} \frac{|\bar{X}_n - \mu|}{S_n} \leq t_{n-1, \alpha/2}\right\} = 1 - \alpha.$

If $N = n$, then

$$d \geq \frac{S_n}{n^{1/2}} t_{n-1, \alpha/2},$$

and it follows that $I_N \supseteq J_N$. If we can use (26) to show that $P_{\mu, \sigma}\{J_N \ni \mu\} \approx 1 - \alpha$, I_N will have the required confidence level. Since $S_n^2 \overset{\text{a.s.}}{\to} \sigma^2$, and since $t_{n-1, \alpha/2} \to z_{\alpha/2}$ as $n \to \infty$, at least for large N,

$$\sigma^2 \leq N \frac{d^2}{z_{\alpha/2}^2},$$

so that $EN \geq z_{\alpha/2}^2 (\sigma^2/d^2)$, where $z_{\alpha/2}^2 (\sigma^2/d^2)$ is the smallest sample size required if σ is known.

Starr [123] has shown that

(a) $E_\sigma\{N\} < \infty$ for all $\sigma > 0$ and $d > 0$,
(b) $\lim_{d \to 0} P\{I_N \ni \mu\} = 1 - \alpha$,
(c) $\lim_{d \to 0} (E_\sigma N/c) = 1$, $0 < \sigma < \infty$,
(d) $\lim_{d \to 0} (N/c) = 1$ a.s., $0 < \sigma < \infty$,

where c = smallest integer n such that $n \geq z_{\alpha/2}^2 (\sigma^2/d^2)$.

Starr also presents some numerical results which indicate that

$$P\{|\bar{X}_N - \mu| < d\} \geq 1 - \alpha \quad \text{for all } \mu, \sigma^2$$

is nearly achieved. Result (b) has been proved by Chow and Robbins [13] for any rv X with continuous distribution. Simons [117] has also studied the sequential procedure and shows that

$$E_\sigma N \leq c + n_0 + 1 \quad \text{for all } d > 0, \ \sigma > 0,$$

provided that we replace $t_{n-1,\alpha/2}$ by $z_{\alpha/2}$ in the definition of N. Moreover, he shows that the cost of not knowing σ is a finite number of observations in that, with $t_{n-1,\alpha/2}$ replaced by $z_{\alpha/2}$ in the definition of N and taking $n_0 \geq 3$, we have, for some finite integer $k \geq 0$,

$$P\{|\bar{X}_{N+k} - \mu| < d\} \geq 1 - \alpha \quad \text{for all } \mu, \sigma^2, d$$

and

$$E(N + k) \leq c + n_0 + k \quad \text{for all } \mu, \sigma^2, d.$$

We will not pursue this subject any further. The mathematics involved is beyond the scope of this book, and we intended only to give an idea of the type of results obtained. The procedure, in any case, is quite simple and clear. For further details and proofs we refer to the papers cited above and to the book of Zacks [146], Chapter 10.

Remark 2. As we remarked earlier, the subject of sequential estimation is under active investigation. Ideas similar to the one used by Robbins [100] have been employed by several authors in the case of sampling from specific nonnormal populations. We refer, in particular, to the papers by Richter [99], Ghurye and Robbins [35], Simons and Zacks [116], and Zacks [145].

PROBLEMS 14.4

1. In the problem of sequential estimation of the mean μ of a normal distribution with unknown variance σ^2, let us consider the following alternative loss functions:

(a) $L(n) = A|\bar{X} - \mu|^s + n$, where $s > 0$;
(b) $L(n) = A|\bar{X} - \mu|^s + \log n$, where $s > 0$.

In each case define a sequential sampling procedure analogous to (5). Show that Lemmas 1 through 5 hold. Also, compute the average loss for the sequential procedure.

2. Let (X_1, Y_1), (X_2, Y_2), $\cdots$, be independent observations on a bivariate normal rv (X, Y) with mean vector (μ_1, μ_2), and suppose that observations are taken sequentially. How will you estimate $\mu_1 - \mu_2$?

3. Prove Lemma 4.

14.5 THE SEQUENTIAL PROBABILITY RATIO TEST

Let $X, X_1, X_2, \cdots$ be a sequence of iid rv's with common pdf (pmf) $f_\theta(x)$. In this section we study Wald's sequential probability ratio test for testing a simple hypothesis, $H_0\colon X \sim f_{\theta_0}$, against a simple alternative, $H_1\colon X \sim f_{\theta_1}$, when the observations are taken sequentially.

Let f_{0n} and f_{1n} denote the joint pdf's (pmf's) of $X_1, X_2, \cdots, X_n$ under H_0 and H_1, respectively. Then

$$f_{jn}(x_1, x_2, \cdots, x_n) = \prod_{i=1}^{n} f_{\theta_j}(x_i), \quad j = 0,1.$$

Write

(1) $$\lambda_n(x_1, x_2, \cdots, x_n) = \frac{f_{1n}(\mathbf{x})}{f_{0n}(\mathbf{x})},$$

where $\mathbf{x} = (x_1, x_2, \cdots, x_n)$.

Definition 1. The sequential probability ratio test (SPRT) for testing H_0 against H_1 is a rule that states:

(i) if at any stage of the sampling

(2) $$\lambda_n(\mathbf{x}) \geq A,$$

stop and reject H_0;

(ii) if

(3) $$\lambda_n(\mathbf{x}) \leq B,$$

stop and accept H_0 (reject H_1); and

(iii) if

(4) $$B < \lambda_n(\mathbf{x}) < A,$$

continue sampling by taking another observation x_{n+1}.

Here A and $B (A > B)$ are constants which are determined so that the test will have *strength* (α, β), where

(5) $$\alpha = P\{\text{Type I error}\} = P\{\text{Reject } H_0 | H_0\},$$

(6) $\quad\quad\quad \beta = P\{\text{Type II error}\} = P\{\text{Accept } H_0|H_1\}.$

If N is the stopping rv, then

$$\alpha = P_{\theta_0}\{\lambda_N(\mathbf{X}) \geq A\}, \quad \beta = P_{\theta_1}\{\lambda_N(\mathbf{X}) \leq B\}.$$

Remark 1. The SPRT defined by (2), (3), and (4) is due to Wald [134] and is the best procedure in a certain sense to be explained later. The test is suggested by the Neyman-Pearson fundamental lemma. Here both α and β are fixed at a preassigned level, but sample size is not fixed in advance. Note that the assumption of independence and identical distribution on the rv's X_i is not necessary to define the test; it is merely a simplifying assumption.

Let us write

(7) $$Z_i = \log \frac{f_{\theta_1}(X_i)}{f_{\theta_0}(X_i)}.$$

Then the rv's $Z_1, Z_2, \cdots$ are also iid rv's, and

(8) $$\log \lambda_n(\mathbf{X}) = \sum_{i=1}^{n} Z_i = S_n, \text{ say.}$$

In terms of S_n the SPRT (Fig. 1) may be written as follows:

At each stage compute $s_n = \sum_{i=1}^{n} z_i$. If

(9) $$b = \log B < s_n < \log A = a,$$

continue by taking observation Z_{n+1}; if

(10) $$a \leq s_n, \quad \text{reject } H_0,$$

and if

(11) $$b \geq s_n, \quad \text{reject } H_1.$$

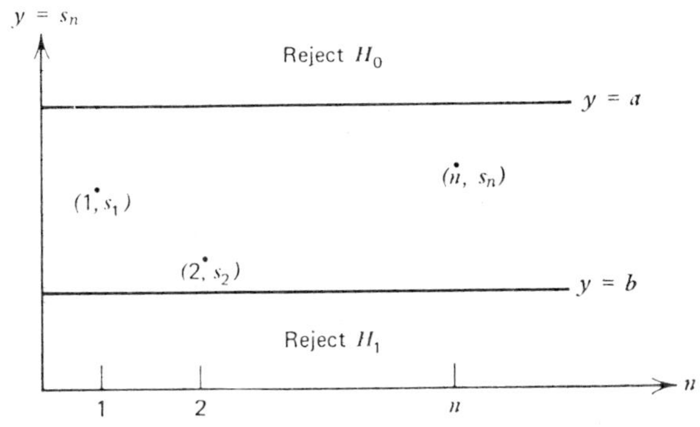

Fig. 1

Remark 2. In a sample where $f_{1n} = 0 = f_{0n}$, we define the ratio $\lambda_n = 1$. If, for some values $x_1, x_2, \cdots, x_n$, $f_{1n} > 0$ but $f_{0n} = 0$, the inequality $\lambda_n \geq A$ is considered fulfilled and H_0 is rejected.

Remark 3. We will ignore the trivial case $P\{Z_i = 0\} = 1$ in the sequel. In this case $f_{\theta_0} = f_{\theta_1}$, and we accept H_0 at the first observation or even without taking any observation.

Remark 4. The same discussion applies to this rule: Reject H_0 if $\lambda_n > A$, reject H_1 if $\lambda_n < B$, and continue if $B < \lambda_n < A$; decide on the boundaries according to some fixed probabilities. If λ_n has an absolutely continuous df, the two rules are clearly equivalent. If λ_n has a discrete distribution, randomization on the boundaries is necessary to achieve a preassigned strength (α, β).

Example 1. Let $X_1, X_2, \cdots$ be iid rv's with a common Bernoulli law with parameter p, $p \in \Theta = \{p_0, p_1\}$, $p_0 \neq p_1$. To test $H_0: X_i \sim b(1, p_0)$ against $H_1: X_i \sim b(1, p_1)$, we compute

$$Z_i = \log \frac{P_{p_1}\{X_i = x\}}{P_{p_0}\{X_i = x\}} = \begin{cases} \log \dfrac{p_1}{p_0} & \text{if } x = 1, \\ \log \dfrac{1 - p_1}{1 - p_0} & \text{if } x = 0. \end{cases}$$

If R is the number of 1's in the sequence of first n observations, then

$$S_n = \sum_{i=1}^{n} Z_i = R \log \frac{p_1}{p_0} + (n - R) \log \frac{1 - p_1}{1 - p_0}.$$

We obtain partial sums s_n cumulatively and accept H_0 if

$$r \log \frac{p_1}{p_0} + (n - r) \log \frac{1 - p_1}{1 - p_0} \leq \log B = b,$$

reject H_0 if

$$r \log \frac{p_1}{p_0} + (n - r) \log \frac{1 - p_1}{1 - p_0} \geq \log A = a,$$

and take another observation otherwise.

Example 2. Let $X_1, X_2, \cdots \sim \mathcal{N}(\theta, 1)$, where $\theta \in \Theta = \{\theta_0, \theta_1\}$. To test $H_0: \theta = \theta_0$ against $H_1: \theta = \theta_1$, we have

$$Z_i = \log \frac{f_{\theta_1}(X_i)}{f_{\theta_0}(X_i)} = (\theta_1 - \theta_0)X_i + \tfrac{1}{2}(\theta_0^2 - \theta_1^2)$$

and

$$\log \lambda_n(\mathbf{X}) = \sum_1^n Z_i = (\theta_1 - \theta_0) \sum_1^n X_i + \frac{n}{2}(\theta_0^2 - \theta_1^2).$$

The quantity $\log \lambda_n(\mathbf{x})$ can be computed cumulatively if, after each observation x_n, we add $(\theta_1 - \theta_0)x_n + \frac{1}{2}(\theta_0^2 - \theta_1^2)$ to the preceding value of $\log \lambda_{n-1}(\mathbf{x})$.

Remark 5. It is instructive to explore somewhat the connection between the SPRT and *random walks*. Let us suppose that a particle at point S_0 undergoes a jump Z_1 at time $n = 1$, a jump Z_2 at time $n = 2$, and so on. Let $Z_1, Z_2, \cdots$ be iid rv's with a given distribution. The particle then moves along the axis, and at time n its position is described by

$$S_n = S_0 + \sum_{i=1}^n Z_i = S_{n-1} + Z_n.$$

In particular, if $Z_1, Z_2, \cdots$ have the common pmf

$$P\{Z_i = 1\} = p, \quad P\{Z_i = -1\} = q, \quad \text{and}$$
$$P\{Z_i = 0\} = 1 - p - q,$$

$0 < p < 1, 0 < q < 1, p + q \leq 1$, we say that S_n performs a *simple random walk*. In practice, it is usual to take $p = 1 - q$. A random walk is therefore just a sequence of rv's $\{S_n\}$. The *state space*, that is, the set of values assumed by S_n, is continuous if the Z_i's are continuous rv's and discrete if the Z_i's are of the discrete type. We say that the *process is in state x at time n* if $S_n = x$. If the particle continues to move indefinitely according to $S_n = S_{n-1} + Z_n$, we say that the random walk is *unrestricted*. The motion of the particle is usually restricted in some manner by the presence of one or more *barriers*. For example, a random walk starting at the origin may be restricted to within a distance a in the positive direction and b in the negative direction from the origin in such a way that, whenever the particle touches or crosses the barrier a or $-b$, it is *absorbed* there and the motion ceases. The states $x \geq a$ and $x \leq -b$ are *absorbing states*, and a and $-b$ are known as *absorbing barriers*.

An example is the problem of Gambler's Ruin, studied in Example 14.2.6. There S_n is player B's cumulative gain at the end of n games. Provided that $-b < S_n < a$, we have $S_n = \sum_1^n Z_i$, where the Z_i's are iid with pmf $P\{Z_i = 1\} = 1 - p$, $P\{Z_i = -1\} = p$, $0 < p < 1$. If at any stage $S_n = a$ (or $-b$), B has gained (lost) all of A's (his) capital and $A(B)$ is ruined. The problem of interest is the probability of A's ruin, which is given by (14.2.16).

The connection between a random walk with two absorbing barriers and the SPRT should now be obvious. Let $Z_1, Z_2, \cdots$ be iid rv's, and $S_n = \sum_{i=1}^n Z_i$. Also, let N be the smallest integer n such that either

$$S_n \geq a \, (= \log A) \quad \text{or} \quad S_n \leq -b \, (= \log B), \quad a, b > 0.$$

Then N is an rv and can be regarded as the time of absorption for the random walk described by S_n with absorbing barriers at a and $-b$. If the SPRT terminates with probability 1, then $1 - \alpha$ and α are the probabilities that absorption happens for the first time when the particle has a position $S_n \leq -b$ or $S_n \geq a$, respectively, given that H_0 is true. A similar interpretation is given to β and $1 - \beta$.

Having defined the test, let us now consider the determination of stopping bounds. The following result holds.

Theorem 1. For the SPRT with stopping bounds A and B, $A > B$, and strength (α, β), we have

(12) $$A \leq \frac{1-\beta}{\alpha}, \quad B \geq \frac{\beta}{1-\alpha}, \quad 0 < \alpha < 1, \quad 0 < \beta < 1.$$

Proof. Let

$$R_n = \{(x_1, x_2, \cdots, x_n): N = n \text{ and } \lambda_N(\mathbf{x}) \geq A\}$$

and

$$S_n = \{(x_1, x_2, \cdots, x_n): N = n \text{ and } \lambda_N(\mathbf{x}) \leq B\}.$$

Then the $\{R_n\}$ are mutually disjoint, and so also are the sets $\{S_n\}$. Also, $\mathbf{x} \in R_n$ means that after taking n observations the SPRT terminates in the rejection of H_0. Thus

$$\begin{aligned}
\alpha &= P\left\{\bigcup_{n=1}^\infty R_n \,\middle|\, H_0\right\} \\
&= \sum_{n=1}^\infty P_{H_0}(R_n) \\
&= \sum_{n=1}^\infty \int_{R_n} f_{0n}(\mathbf{x}) \, d\mathbf{x} \quad \text{if } f_{0n} \text{ is a pdf (a similar argument holds in the discrete case)} \\
&\leq A^{-1} \sum_{n=1}^\infty \int_{R_n} f_{1n}(\mathbf{x}) \, d\mathbf{x} \\
&= A^{-1} P_{H_1}\{\text{Reject } H_0\} \\
&= A^{-1}(1 - \beta).
\end{aligned}$$

Note that we have assumed above (and in what follows) that

$$\sum_{n=1}^\infty P\{N = n\} = \sum_n \int_{R_n \cup S_n} f_{in}(\mathbf{x}) \, d\mathbf{x} = 1, \quad i = 0, 1,$$

that is, the SPRT terminates eventually with probability 1. We will prove this assertion later. Similarly

$$\begin{aligned}\beta &= P\left\{\bigcup_{n=1}^{\infty} S_n \big| H_1\right\} \\ &= \sum_{n=1}^{\infty} \int_{S_n} f_{1n}(\mathbf{x})\, d\mathbf{x} \leq B \sum_{n=1}^{\infty} \int_{S_n} f_{0n}(\mathbf{x})\, d\mathbf{x} \\ &= B P_{H_0}\{\text{Accept } H_0\} = B(1 - \alpha).\end{aligned}$$

Remark 6. Note that $\alpha \leq A^{-1}(1 - \beta) \leq A^{-1}$ and $\beta \leq B(1 - \alpha) \leq B$. Thus the shaded region in Fig. 2 is the region of values of α and β, given A and B.

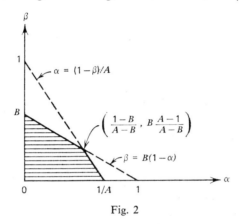

Fig. 2

Inequalities (12) suggest the possibility of approximating the boundaries A and B by

(13) $$A' = \frac{1 - \beta}{\alpha} \quad \text{and} \quad B' = \frac{\beta}{1 - \alpha}.$$

Let (α', β') be the true strength of the SPRT with bounds (A', B'). Note that since α and β are usually small $A' > 1 > B' > 0$. In what follows, we will assume that $0 < B < 1 < A$.

Theorem 2. If for the choice

$$A' = \frac{1 - \beta}{\alpha}, \quad B' = \frac{\beta}{1 - \alpha},$$

the SPRT terminates with probability 1 and the strength is (α', β'), then

(14) $$\alpha' \leq \frac{\alpha}{1 - \beta}, \quad \beta' \leq \frac{\beta}{1 - \alpha}, \quad \text{and} \quad \alpha' + \beta' \leq \alpha + \beta.$$

Proof. The proof is left as an exercise.

Remark 7. Both α' and β' cannot exceed α and β, respectively; at least one of $\alpha' \leq \alpha$, $\beta' \leq \beta$ is true. Generally both are true, and the choice of A', B' leads to a more stringent test then the one with stoppping bounds A and B. In fact, since α, β are small,

$$\alpha' \leq \frac{\alpha}{1-\beta} \approx \alpha(1+\beta)$$

and

$$\beta' \leq \frac{\beta}{1-\alpha} \approx \beta(1+\alpha),$$

so that the increase in the error size in α' (or β') is not beyond a factor $(1+\beta)$ [or $(1+\alpha)$]. For example, if $\alpha = \beta = .05$, then $\alpha' = \beta' \leq .05(1.05) = .0525$.

Remark 8. A' and B' are functions of α, β and can be computed, once and for all, free of the pdf (pmf) f_{θ_i}. Note that in the Neyman-Pearson theory the critical values depend on f_{θ_0} and α. This means that no distribution problem is involved in the SPRT, except when one is interested in the distribution of N, the number of trials required to stop.

Remark 9. The only serious risk in using (A', B') is that (α', β') may be much smaller than required. As a consequence one may need to take a large number of observations. Fortunately, however, there is reason to believe that this effect is also moderate because, if we consider the last step from S_{n-1} to S_n and suppose, for definiteness, that H_0 is rejected at the nth observation, then $S_{n-1} < \log A \approx \log [(1-\beta)/\alpha] \leq S_n$, and the approximation consists in replacing S_n by $\log A'$. Thus the error is due to the last observation Z_n, so that the excess over $\log A'$ is due to Z_n and will usually be moderate.

Remark 10. Theorems 1 and 2 hold even for dependent observations, provided we assume that $P\{N < \infty | H_i\} = 1$, $i = 0, 1$.

Example 3. In Example 1 above we may write the test procedure as follows: Continue sampling by taking another observation as long as

$$\log B - n \log \frac{1-p_1}{1-p_0} < r \left\{ \log \frac{p_1}{p_0} - \log \frac{1-p_1}{1-p_0} \right\}$$
$$< \log A - n \log \frac{1-p_1}{1-p_0}.$$

If $p_1 > p_0$, this is the same as stating: Continue if

$$\frac{\log B + n \log [(1-p_0)/(1-p_1)]}{\log [p_1(1-p_0)/p_0(1-p_1)]} < r$$
$$< \frac{\log A + n \log [(1-p_0)/(1-p_1)]}{\log [p_1(1-p_0)/p_0(1-p_1)]},$$

stop and reject H_0 if

$$r \ge \left\{\log A + n \log \frac{(1-p_0)}{(1-p_1)}\right\} \left\{\log \frac{p_1(1-p_0)}{p_0(1-p_1)}\right\}^{-1},$$

and reject H_1 if

$$r \le \left\{\log B + n \log \frac{(1-p_0)}{(1-p_1)}\right\} \left\{\log \frac{p_1(1-p_0)}{p_0(1-p_1)}\right\}^{-1}.$$

Here r is the number of 1's in the sequence to date.

Let $\alpha = .1$, $\beta = .1$, and $p_0 = \frac{1}{4}$, $p_1 = \frac{3}{4}$. Using the approximation in Theorem 2, we take

$$A' = \frac{1-\beta}{\alpha} = \frac{.9}{.1} = 9 \quad \text{and} \quad B' = \frac{\beta}{1-\alpha} = \frac{1}{9}.$$

The SPRT is to continue if

$$\frac{n}{2} - 1 = \frac{-\log 9 + n \log 3}{\log 9} < r < \frac{n \log 3 + \log 9}{\log 9} = \frac{n}{2} + 1,$$

to reject H_0 if

$$r \ge \frac{n}{2} + 1,$$

and to reject H_1 if

$$r \le \frac{n}{2} - 1.$$

If we observe the sequence 0, 1, 0, 0, then $r = 1$, $n = 4$, so that we reject H_1 since $r = 1 \le \frac{4}{2} - 1 = 1$.

Let us compare this result with the fixed sample size procedure. Note that

$$E_{p_0} R = np_0 = \frac{n}{4}, \quad \text{var}_{p_0}(R) = np_0(1-p_0) = \frac{3n}{16}$$

and

$$E_{p_1} R = np_1 = \frac{3n}{4}, \quad \text{var}_{p_1}(R) = np_1(1-p_1) = \frac{3n}{16}.$$

Using normal approximations, we have

$$.1 = P_{p_0}\{R > k\} \approx P\left\{Z > \frac{k - (n/4)}{\sqrt{3n/16}}\right\}$$

and

$$.1 = P_{p_1}\{R \le k\} \approx P\left\{Z \le \frac{k - (3n/4)}{\sqrt{3n/16}}\right\},$$

where Z is $\mathcal{N}(0, 1)$. From the tables

$$\frac{k - (n/4)}{\sqrt{3n/16}} = 1.28 \quad \text{and} \quad \frac{k - (3n/4)}{\sqrt{3n/16}} = -1.28,$$

giving $n/2 = 2.56(\sqrt{3n}/4)$ or $n \approx 5$. At least in this case, we see that there is a saving of one observation.

PROBLEMS 14.5

1. Find the sequential probability ratio test with stopping bounds $B < 1 < A$ of the simple hypothesis $H_0: \theta = \theta_0$ against $H_1: \theta = \theta_1$ in each of the following cases:
   (a) $X \sim P(\theta)$, $\theta > 0$.
   (b) $X \sim \mathcal{N}(\mu, \theta^2)$, μ known.
   (c) $X \sim G(1, 1/\theta)$, $\theta > 0$.
   (d) X has pdf $f_\theta(x)$, given by
   $$f_\theta(x) = \begin{cases} e^{-(x-\theta)} & \text{if } x \geq \theta, \\ 0 & \text{otherwise.} \end{cases}$$
   (e) $X \sim NB(1; \theta)$, $0 < \theta < 1$.

2. In Problem 1 use the approximation for A and B from Theorem 2 with $\alpha = \beta = .1$. Find the SPRT of $H_0: \theta = \theta_0$ against $H_1: \theta = \theta_1$ in each of the following cases:
   (a) $X \sim P(\theta)$, $\theta > 0$; $\theta_0 = 1$, $\theta_1 = 2$.
   (b) $X \sim \mathcal{N}(\mu, \theta^2)$, $\theta_0 = 9$, $\theta_1 = 4$.
   (c) $X \sim G(1, 1/\theta)$, $\theta_0 = 1$, $\theta_1 = 2$.
   (d) X has pdf $f_\theta(x) = e^{-(x-\theta)}$ if $x \geq \theta$, and $= 0$ otherwise, and $\theta_0 = 0$, $\theta_1 = 1$.
   (e) $X \sim NB(1; \theta)$, $\theta_0 = \frac{1}{4}$, $\theta_1 = \frac{3}{4}$.

3. Prove Theorem 2.

14.6 SOME PROPERTIES OF THE SEQUENTIAL PROBABILITY RATIO TEST

In this section we study some properties of the SPRT. We begin by showing that sampling terminates eventually with probability 1. In the following we write $f_{\theta_0} = f_0$, $f_{\theta_1} = f_1$.

Theorem 1 (Stein [126]). Let $X_1, X_2, \cdots$ be iid rv's with common pdf (pmf) f_i under $H_i (i = 0, 1)$. Let N be the number of observations required for reaching a decision by using the SPRT. With respect to any hypothesis H (not necessarily H_0 or H_1) for which $P\{|Z| > 0|H\} > 0$, where $Z = \log \{f_1(X_1)/f_0(X_1)\}$, the following results hold:

(1) $$P_H\{N < \infty\} = 1,$$
(2) $$E_H\{e^{tN}\} < \infty \quad \text{for } -\infty < t < t_0, \ t_0 > 0.$$

Proof. The rv's $Z_i = \log \{f_1(X_i)/f_0(X_i)\}$ are iid. Let $r = r(m, k)$ be the largest integer $\leq m/k$, where m, k are fixed positive integers, and consider the partial sums S_k, $S_{2k} - S_k$, $\cdots$, $S_{rk} - S_{(r-1)k}$, where $S_n = \sum_{i=1}^{n} Z_i$, $n = 1, 2, \cdots$. If $N > m$, then $S_i \in (b, a)$ for $i = 1, 2, \cdots, m$, $b = \log B$, $a = \log A$, and, in particular, for $i = k, 2k, \cdots, rk$. Hence

$$|T_i| = |S_{ik} - S_{(i-1)k}| < |b| + |a| = c, \text{ say}, \quad (i = 1, 2, \cdots, r).$$

It follows that

(3) $$\begin{aligned} P\{N > m\} &\leq P\{|T_i| < c, \ i = 1, 2, \cdots, r\} \\ &\leq [P\{|T_1| < c\}]^r, \end{aligned}$$

since the T_i are iid. Thus

$$\lim_{m \to \infty} P\{N > m\} \leq \lim_{r \to \infty} [P\{|T_1| < c\}]^r \quad \text{for fixed } k.$$

If $P\{|T_1| < c\} < 1$, this last limit equals 0. This is what we show next. Since $P\{|Z| > 0\} > 0$, we can find an $\varepsilon > 0$ such that either $P\{Z > \varepsilon\} > 0$ or $P\{Z < -\varepsilon\} > 0$ (or both). Then

$$\begin{aligned} P\{|T_1| > c\} &= P\{|Z_1 + \cdots + Z_k| > c\} \\ &= P\{Z_1 + \cdots + Z_k > c\} + P\{-\sum_{1}^{k} Z_i > c\} \\ &\geq P\{Z_i > \frac{c}{k}, \ i = 1, 2, \cdots, k\} + P\{Z_i < -\frac{c}{k}, \ i = 1, 2, \cdots, k\}. \end{aligned}$$

Choose and fix $k = k(\varepsilon)$, an integer such that $c/k < \varepsilon$. Then

$$P\{|T_1| > c\} \geq [P\{Z_1 > \varepsilon\}]^k + [P\{Z_1 < -\varepsilon\}]^k > 0$$

by hypothesis. Thus

$$P\{|T_1| < c\} = \delta < 1,$$

where δ is independent of m. It follows that

$$\lim_{m \to \infty} P\{N > m\} \leq \lim_{m \to \infty} \delta^{r(m)} = 0,$$

where $r = r(m, k) = r(m)$, since k is fixed. Thus

$$P\{N < \infty\} = 1 - P\{N = \infty\} = 1 - \lim_{m \to \infty} P\{N > m\} = 1.$$

To prove (2), we note that we have the estimate

(4) $$P\{N > m\} \leq \delta^{r(m)} \quad \text{for every } m.$$

Thus

$$Ee^{tN} = \sum_{m=1}^{\infty} e^{tm} P\{N = m\}$$
$$\leq \sum_{m=1}^{\infty} e^{tm} P\{N > m - 1\}$$
$$= e^t \sum_{m=0}^{\infty} e^{tm} P\{N > m\}$$
$$\leq e^t \sum_{m=0}^{\infty} e^{tm} \delta^{r(m)}$$
$$= e^t \sum_{m=0}^{\infty} e^{tm} \delta^{m/k} \delta^{r-m/k}$$
$$\leq e^t \delta^{-1/k} \sum_{m=0}^{\infty} (\delta^{1/k} e^t)^m < \infty,$$

provided that $e^t \delta^{1/k} < 1$ or $t < -(\log \delta)/k = t_0$. Since $0 < \delta < 1$, $t_0 > 0$ and the proof is complete.

Remark 1. In the proof of (1) we obtain the rate of convergence of $P\{N > m\}$ to 0 as $m \to \infty$, and show that it is exponential. Clearly, the probability generating function Ez^N of the rv N is finite for $0 < z < 1/\delta^{1/k}$. Moments of all order exist for the stopping variable N.

Corollary to Theorem 1. The SPRT terminates with probability 1 under both H_0 and H_1.

For the proof we need only to show that

$$P\{|Z| > 0 | H_i\} > 0, \quad i = 0, 1.$$

We need the following lemma.

Lemma 1. Let f and g be pdf's (pmf's), and let $C = \{f > g\}$. Then

(5) $$\int_C f(x) \log \frac{f(x)}{g(x)} dx \geq 0$$

if f and g are pdf's, and

(6) $$\sum_{\{x \in C\}} f(x) \log \frac{f(x)}{g(x)} \geq 0$$

if f and g are pmf's, with equality if and only if $f(x) = g(x)$ except for a null set.

We leave the reader to supply the proof of Lemma 1. From Lemma 1, it follows that, unless $P\{f_0(X) = f_1(X)\} = 1$, we have

$$E\{Z | H_0\} < 0.$$

This imples that $P\{Z < 0|H_0\} > 0$ or $P\{|Z| > 0|H_0\} > 0$. Similarly, by taking $f = f_1$, $g = f_0$, we obtain, from Lemma 1, $P\{|Z| > 0|H_1\} > 0$.

We next study the power function (operating characteristic function) of the SPRT. Let us write

(7) $$\beta(\theta) = P_\theta\{\text{Reject } H_0\}.$$

An exact evaluation of $\beta(\theta)$ is, in general, very difficult. A simple approximation can be obtained, however, provided that the equation

(8) $$E_\theta\left\{\left[\frac{f_1(X)}{f_0(X)}\right]^t\right\} = E_\theta e^{tZ} = 1$$

has a nonzero solution, $t = t(\theta)$.

Lemma 2 (Wald [134], 158). If Z is an rv such that

(9) $\quad\quad\quad P\{Z > 0\} > 0 \quad\quad \text{and} \quad\quad P\{Z < 0\} > 0,$

(10) $\quad\quad\quad M(t) = Ee^{tZ}$ exists for any real value t,

and

(11) $$EZ \neq 0,$$

then there exists a $t^* \neq 0$ such that $M(t^*) = 1$. Moreover, if $EZ < 0$, then $t^* > 0$; and if $EZ > 0$, then $t^* < 0$.

Proof. Since $P\{Z > 0\} > 0$, there exists an $\varepsilon > 0$ such that $P\{Z > \varepsilon\} = \delta > 0$. If $t > 0$, then

(12) $$M(t) \geq \int_{z>\varepsilon} e^{tz} f(z)\, dz > e^{t\varepsilon}\, P\{Z > \varepsilon\},$$

where f is the pdf of Z if Z is a continuous rv. Inequality (12) holds also for the case where Z is a discrete rv. It follows that in either case

$$M(t) \to \infty \quad \text{as} \quad t \to \infty.$$

Similarly, we use $P\{Z < 0\} > 0$ to conclude that

$$M(t) \to \infty \quad \text{as} \quad t \to -\infty.$$

Now condition (10) implies that moments of all order of Z exist and may be obtained by differentiating $M(t)$ under the expectation sign as many times as needed. Thus

$$M''(t) = E\{Z^2 e^{tZ}\} > 0 \quad \text{for all real } t,$$

using the same argument as above. It follows that $M(t)$ is a convex function of t; and since $M(t) \to \infty$ as $t \to \pm\infty$, M has a minimum at $t = t_0$, where $M'(t_0) = 0$. Since $M'(t) = E\{Ze^{tZ}\}$, $M'(t)|_{t=0} = EZ \neq 0$ by (11). Thus $t_0 \neq 0$, and since $M(t_0)$ is minimum,

$$M(t_0) < M(0) = 1.$$

Thus M is strictly decreasing over $(-\infty, t_0)$ and strictly increasing over (t_0, ∞). Since $M(0) = 1$ and $M(t_0) < 1$, it follows that there exists exactly one real $t^* \neq 0$ such that $M(t^*) = 1$.

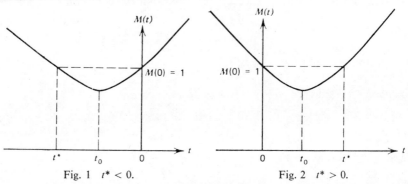

Fig. 1 $t^* < 0$. Fig. 2 $t^* > 0$.

Thus, if $t_0 > 0$ (Fig. 2), so also $t^* > 0$; and if $t_0 < 0$ (Fig. 1), $t^* < 0$. Furthermore, if $t_0 > 0$, then $EZ = M'(t)|_{t=0} < 0$; and if $t_0 < 0$, then $EZ = M'(t)|_{t=0} > 0$. Thus t^* and EZ have opposite signs, and the proof is complete.

Let us use Lemma 2 to approximate $\beta(\theta)$. Let t^* be a nonzero solution of (8). Then

$$(13) \qquad f_\theta^*(x) = \left[\frac{f_1(x)}{f_0(x)}\right]^{t^*} f_\theta(x)$$

is a pdf (pmf). Suppose that $t^* > 0$ (the other case can be treated similarly), and consider the SPRT with stopping bounds A^{t^*}, B^{t^*} for testing f_θ against f_θ^*. The test is to continue as long as

$$B^{t^*} < \frac{\prod_{i=1}^{n} f_\theta^*(x_i)}{\prod_{i=1}^{n} f_\theta(x_i)} < A^{t^*},$$

to reject f_θ if

$$\frac{\prod_{i=1}^{n} f_\theta^*(x_i)}{\prod_{i=1}^{n} f_\theta(x_i)} \geq A^{t^*},$$

and to reject f_θ^* if

$$\frac{\prod_{i=1}^{n} f_\theta^*(x_i)}{\prod_{i=1}^{n} f_\theta(x_i)} \leq B^{t^*}.$$

Let α^* and $1 - \beta^*$ denote the probabilities of rejection (of f_θ), where f_θ and f_θ^*, respectively, are the true densities. Then

(14) $$B^{t^*} \approx \frac{\beta^*}{1 - \alpha^*} \quad \text{and} \quad A^{t^*} \approx \frac{1 - \beta^*}{\alpha^*}.$$

But

(15) $$\frac{\prod_{i=1}^{n} f_\theta^*(x_i)}{\prod_{i=1}^{n} f_\theta(x_i)} = \left[\frac{\prod_{i=1}^{n} f_1(x_i)}{\prod_{i=1}^{n} f_0(x_i)}\right]^{t^*},$$

so that the test is the same as the SPRT of H_0 against H_1 with stopping bounds A and B. It follows that the probability of rejection for the two tests when f_θ is true must be the same. Thus

(16) $$\alpha^* = \beta(\theta) \approx \frac{1 - B^{t^*}}{A^{t^*} - B^{t^*}}.$$

Equation (16) is obtained by solving (14) for α^*.

Note that $t(\theta_0) = 1$ is a solution of

$$E_0\, e^{t(\theta_0) Z} = 1,$$

and $t(\theta_1) = -1$ is a solution of

$$E_1\, e^{t(\theta_1) Z} = 1.$$

Also, $E_\theta\, e^{tZ} = 1$ when $t = 0$. However, $\beta(\theta)$ is not defined for $t = 0$. In this case

(17) $$\beta(\theta)|_{t=0} = -\frac{\log B}{\log A - \log B}$$

by L'Hospital's rule. Also,

$$\beta(\theta_0) = \alpha \quad \text{and} \quad \beta(\theta_1) = P_1\{\text{Reject } H_0\} = 1 - \beta.$$

Finally,

$$\beta(\theta) \approx \left[\frac{1}{A^{t^*}} - \left(\frac{B}{A}\right)^{t^*}\right]\left[1 - \left(\frac{B}{A}\right)^{t^*}\right]^{-1} \to 0$$

as $t^* \to \infty$. Similarly, $\beta(\theta) \to 1$ as $t^* \to -\infty$.

We summarize these values of the power function in the following table:

t:	$-\infty$	-1	0	1	∞
θ:		θ_1		θ_0	
$\beta(\theta)$:	1	$1-\beta$	$-\log B/\log(A/B)$	α	0

Example 1. Let $X_1, X_2, \cdots$ be iid $b(1, \theta)$ rv's. To test $H_0: \theta = \theta_0$ against $H_1: \theta = \theta_1 > \theta_0$, we have

$$E_\theta\left\{\left[\frac{f_{\theta_1}(X)}{f_{\theta_0}(X)}\right]^t\right\} = E_\theta\left\{\left[\frac{\theta_1^X(1-\theta_1)^{1-X}}{\theta_0^X(1-\theta_0)^{1-X}}\right]^t\right\}$$

$$= \theta\left(\frac{\theta_1}{\theta_0}\right)^t + (1-\theta)\left(\frac{1-\theta_1}{1-\theta_0}\right)^t.$$

Setting this equal to 1 and solving for θ, we get

$$(18) \qquad \theta = \frac{1 - [(1-\theta_1)/(1-\theta_0)]^t}{(\theta_1/\theta_0)^t - [(1-\theta_1)/(1-\theta_0)]^t}.$$

Letting $t \to 0$, we get

$$(19) \qquad \theta = -\frac{\log[(1-\theta_1)/(1-\theta_0)]}{\log(\theta_1/\theta_0) - \log[(1-\theta_1)/(1-\theta_0)]}.$$

In particular, if $\theta_0 = .5$, $\theta_1 = .9$, $\alpha = \beta = .05$, then

$$\theta = \frac{5^t - 1}{9^t - 1}.$$

Also, $B = \frac{1}{19}$, $A = 19$, so that

$$\beta(\theta) \approx \frac{1 - 19^{-t}}{19^t - 19^{-t}}.$$

We have

t:	$-\infty$	-1	0	1	∞
θ:	1.0	$.90$	$.733$	$.50$	0
$\beta(\theta)$:	1.0	$.95$	$.50$	$.05$	0

Example 2. Let $X_1, X_2, \cdots$ be iid $\mathcal{N}(\mu, 1)$ rv's. It is required to test $H_0: \mu = \mu_0$ against $H_1: \mu = \mu_1 > \mu_0$. Then

$$Z = \frac{\mu_0^2 - \mu_1^2}{2} + (\mu_1 - \mu_0)X$$

and

$$E_\mu e^{tZ} = \exp\left(\frac{\mu_0^2 - \mu_1^2}{2}t\right)\exp\left\{(\mu_1 - \mu_0)t\mu + (\mu_1 - \mu_0)^2\frac{t^2}{2}\right\}$$

$$(20) \qquad = \exp\left\{(\mu_1 - \mu_0)t\left(\mu - \frac{\mu_0 + \mu_1}{2}\right) + \frac{(\mu_1 - \mu_0)^2}{2}t^2\right\}.$$

Solving
$$E_\mu e^{tZ} = 1$$
with respect to $t(\mu)$, we get

(21) $$t(\mu) = \frac{\mu_0 + \mu_1 - 2\mu}{\mu_1 - \mu_0}.$$

The approximate power function is therefore
$$\beta(\mu) \approx \frac{1 - [\beta/(1-\alpha)]^{t(\mu)}}{[(1-\beta)/\alpha]^{t(\mu)} - [\beta/(1-\alpha)]^{t(\mu)}}$$

whenever $t(\mu) \neq 0$, where $t(\mu)$ is given by (21). It is easy to see that $t(\mu) = 0$ for
$$\mu^* = \frac{\mu_0 + \mu_1}{2},$$

so that from (17)
$$\beta(\mu^*) \approx -\frac{\log B}{\log(A/B)} = \frac{\log[(1-\alpha)/\beta]}{\log[(1-\alpha)(1-\beta)/\alpha\beta]}.$$

We have

t:	$-\infty$	-1	0	1	∞
μ:		μ_1	μ^*	μ_0	
$\beta(\mu)$:	1	$1-\beta$	$\beta(\mu^*)$	α	0

Next we find an approximation for EN. If we approximate S_N by $\log A$ if H_0 is rejected and by $\log B$ if H_1 is rejected, then

(22) $$E_\theta S_N \approx \beta(\theta) \log A + [1 - \beta(\theta)] \log B;$$

and if $E_\theta Z \neq 0$, then

(23) $$E_\theta N = \frac{E_\theta S_N}{E_\theta Z} \approx \frac{\beta(\theta) \log A + [1 - \beta(\theta)] \log B}{E_\theta Z}.$$

Thus

(24) $$E_{\theta_0} N \approx \frac{\alpha \log A + (1-\alpha) \log B}{E_{\theta_0} Z}, \quad \beta(\theta_0) = \alpha,$$

and

(25) $$E_{\theta_1} N \approx \frac{(1-\beta) \log A + \beta \log B}{E_{\theta_1} Z}, \quad \beta(\theta_1) = 1 - \beta.$$

If $E_\theta Z = 0$, we will show in the next section that

(26) $$E_\theta N \approx -\frac{\log A \log B}{E_\theta Z^2}.$$

Example 3. In Example 1,

$$E_\theta Z = \theta \log \frac{\theta_1}{\theta_0} + (1 - \theta) \log \frac{1 - \theta_1}{1 - \theta_0}$$
$$= \theta \log \left[\frac{\theta_1(1 - \theta_0)}{\theta_0(1 - \theta_1)}\right] + \log \left(\frac{1 - \theta_1}{1 - \theta_0}\right).$$

If $\theta_0 = .5$ and $\theta_1 = .9$, then

$$E_\theta Z = \theta \log 9 - \log 5$$

and

$$E_\theta Z^2 = \theta \left(\log \frac{\theta_1}{\theta_0}\right)^2 + (1 - \theta)\left(\log \frac{1 - \theta_1}{1 - \theta_0}\right)^2.$$

We have

t:	$-\infty$	-1	0	1	∞
θ:	1.0	.90	.733	.50	0
$\beta(\theta)$:	1.0	.95	.50	.05	0
$E_\theta Z$:	$\log 1.8$	$.9 \log 9 - \log 5$	0	$\log .6$	$-\log 5$
$E_\theta N$:	5.01	7.2	9.17	5.19	1.83

$E_\theta N$ when $E_\theta Z = 0$ is obtained by using (26).

Example 4. Let $X \sim \mathcal{N}(\theta, 1)$, and $\theta_0 = 0$, $\theta_1 = 1$, $\alpha = \beta = .05$. Then $A = .95/.05 = 19$, $B = .05/.95 = \frac{1}{19}$. also,

$$Z = \log \frac{f_{\theta_1}(X)}{f_{\theta_0}(X)} = X - \tfrac{1}{2}$$

so that $E_\theta Z = \theta - \tfrac{1}{2}$, $E_{\theta_0} Z = -\tfrac{1}{2}$, $E_{\theta_1} Z = \tfrac{1}{2}$. Also,

$$E_\theta Z^2 = E_\theta(X - \tfrac{1}{2})^2 = \theta^2 - \theta + \tfrac{5}{4},$$
$$E_{\theta_0} S_N \approx \alpha \log A + (1 - \alpha) \log B$$
$$= -2.265$$
$$E_{\theta_0} N \approx 5.3$$
$$E_{\theta_1} S_N \approx .95 \log 19 - .05 \log 19$$
$$= .90 \log 19,$$

so that

$$E_{\theta_1} N \approx 5.3$$

Let us consider the fixed sample size procedure which rejects H_0 if $\bar{X}_n > k$. Then

$$\alpha = P_{\theta_0}\{\bar{X}_n > k\} = P\left\{\frac{\bar{X}_n - 0}{1/\sqrt{n}} > \frac{k}{1/\sqrt{n}}\right\} = .05,$$

and
$$\beta = P_{\theta_1}\{\bar{X}_n \le k\} = P\left\{\frac{\bar{X}_n - 1}{1/\sqrt{n}} \le \frac{k - 1}{1/\sqrt{n}}\right\} = .05.$$

It follows that
$$1.65 = \sqrt{n}\, k \quad \text{and} \quad -1.65 = \sqrt{n}\,(k - 1),$$
giving $n = 10.9$.

We have

t:	$-\infty$	-1	0	1	∞
θ:		$\theta_1 = 1$	$.5$	$\theta_0 = 0$	
$\beta(\theta)$:	1	$.95$	$.5$	$.05$	0
$E_\theta Z$:		0.50	0	$-.50$	
$E_\theta N$:		5.3	8.67	5.3	

In Example 4 we see that there is some saving (on the average) if we use the sequential test over the fixed sample size test. Let us examine this possibility a little more closely.

Example 5. Let $X_1, X_2, \cdots$ be iid $\mathcal{N}(\theta, 1)$ rv's, and suppose that we wish to test $H_0: \theta = \theta_0$ against $H_1: \theta = \theta_1 \,(\theta_1 > \theta_0)$. Let us first consider a fixed sample size procedure. The most powerful test rejects H_0 at level α if $\bar{X} > k$. If the probability of type II error is fixed at a preassigned level β, say, we need to choose n, the number of observations, so that
$$P_{\theta_0}\{\bar{X} > k\} = \alpha \quad \text{and} \quad P_{\theta_1}\{\bar{X} \le k\} = \beta.$$

It follows, on standardization, that
$$\frac{k - \theta_0}{1/\sqrt{n}} = z_\alpha \quad \text{and} \quad \frac{k - \theta_1}{1/\sqrt{n}} = -z_\beta.$$

Solving for n, we get
$$(27) \qquad n = \frac{(z_\alpha + z_\beta)^2}{(\theta_1 - \theta_0)^2}.$$

Let us now consider the corresponding SPRT. We have already seen that
$$(28) \qquad t(\theta) = \frac{\theta_1 + \theta_0 - 2\theta}{\theta_1 - \theta_0}$$
is a nonzero root of $E\, e^{tZ} = 1$, where $Z = \log[f_{\theta_1}(X)/f_{\theta_0}(X)]$. Moreover, $t(\theta) = 0$ if $\theta = \theta^*$, where

(29) $$\theta^* = \frac{\theta_1 + \theta_0}{2}.$$

Also,

$$E_\theta Z = E_\theta\{\tfrac{1}{2}[2(\theta_1 - \theta_0)X + \theta_0^2 - \theta_1^2]\}$$

(30) $$= -\frac{(\theta_1 - \theta_0)^2 t(\theta)}{2},$$

and

(31) $$\begin{aligned}E_\theta Z^2 &= \tfrac{1}{4} E_\theta[(\theta_0^2 - \theta_1^2)^2 + 4(\theta_1 - \theta_0)(\theta_0^2 - \theta_1^2)X + 4(\theta_1 - \theta_0)^2 X^2]\\ &= \tfrac{1}{4}[(\theta_1^2 - \theta_0^2)^2 + 4(\theta_1 - \theta_0)(\theta_0^2 - \theta_1^2)\theta + 4(\theta_1 - \theta_0)^2(1 + \theta^2)].\end{aligned}$$

Thus

(32) $$\begin{aligned}E_{\theta^*} Z^2 &= \tfrac{1}{4}[(\theta_1^2 - \theta_0^2)^2 - 2(\theta_1^2 - \theta_0^2)^2 + (\theta_1 - \theta_0)^2(4 + (\theta_0 + \theta_1)^2)]\\ &= (\theta_1 - \theta_0)^2.\end{aligned}$$

Using approximations (24) and (25), we have

(33) $$E_{\theta_0} N \approx -2\,\frac{\alpha \log[(1-\beta)/\alpha] + (1-\alpha)\log[\beta/(1-\alpha)]}{(\theta_1 - \theta_0)^2}$$

and

(34) $$E_{\theta_1} N \approx 2\,\frac{(1-\beta)\log[(1-\beta)/\alpha] + \beta\log[\beta/(1-\alpha)]}{(\theta_1 - \theta_0)^2}.$$

Let us write $r_0 = E_{\theta_0} N/n$, $r_1 = E_{\theta_1} N/n$, where n is given by (27). Then

(35) $$r_0 \approx -2\,\frac{\alpha \log[(1-\beta)/\alpha] + (1-\alpha)\log[\beta/(1-\alpha)]}{(z_\alpha + z_\beta)^2},$$

and

(36) $$r_1 \approx 2\,\frac{(1-\beta)\log[(1-\beta)/\alpha] + \beta\log[\beta/(1-\alpha)]}{(z_\alpha + z_\beta)^2}.$$

Thus r_0 and r_1 depend only on α and β, and not on θ_0 and θ_1, as long as $\theta_1 > \theta_0$. In Tables 1 and 2 we have computed $100 r_0$ and $100 r_1$ for selected values of α and β.

Table 1. $100 r_0$

	α			
β	.01	.03	.05	.10
.01	41.5	48.8	52.5	56.4
.03	38.4	46.2	50.5	55.7
.05	36.6	44.7	49.2	55.2
.10	33.1	41.4	46.4	53.7

Table 2. $100 r_1$

β	α			
	.01	.03	.05	.10
.01	41.5	38.4	36.6	33.1
.03	48.8	46.2	44.7	41.4
.05	52.5	50.5	49.2	46.4
.10	56.4	55.7	55.2	53.7

It is clear that the average sample size required under H_0 and H_1 when the SPRT is used is quite small in comparison to the one needed for a fixed sample size procedure. In other words, the savings in number of observations using SPRT is considerable. Indeed, the SPRT has the following optimal property, which we state without proof.

Theorem 2 (Wald and Wolfowitz [135]). Among all tests (sequential or not) of a simple hypothesis H_0 against a simple alternative H_1 for which

$$P\{\text{Reject } H_0 | H_0\} \le \alpha, \qquad P\{\text{Accept } H_0 | H_1\} \le \beta,$$

and $E_{H_0} N$ and $E_{H_1} N$ are finite, the SPRT with strength (α, β) minimizes both $E_{H_0} N$ and $E_{H_1} N$.

We conclude this section with some remarks and an example.

Remark 2. Several questions have not been answered in Section 14.5 and in this section. For example, given (α, β), do there exist constants A and B? An affirmative answer was provided by Wijsman [138, 139]. Again, granted the existence of stopping bounds A and B, is the SPRT unique in any sense? Several authors investigated this question. The answer is yes, and we refer to Anderson and Friedman [2] for a restricted but readable proof. Wijsman proved uniqueness without any restrictions.

Remark 3. In many special situations there is no "overshooting" of the stopping bounds, and the approximations of this section become exact. The following example illustrates the point.

Example 6. If the rv's X_i take three values, -1, 0, and 1, A and B are taken to be integers, there is no overshooting of bounds A and B. In this case the approximations obtained above are all exact.

Consider, for example, $X_1, X_2, \cdots$ iid rv's with common pmf

$$f_p(x) = \begin{cases} p, & x = 1 \\ 1-p, & x = -1 \end{cases}, \quad 0 < p < 1.$$

It is required to test $H_0: p = p_0$ against $H_1: p = 1 - p_0$, $0 < p_0 < \frac{1}{2}$. Then
$$f_p(x) = p^{(1+x)/2}(1-p)^{(1-x)/2},$$
and
$$\begin{aligned} Z_i &= \log \frac{f_1(X_i)}{f_0(X_i)} \\ &= X_i \log \frac{1-p_0}{p_0} \\ &= zX_i, \end{aligned}$$
where $z = \log(1-p_0)/p_0 > 0$. Thus Z_i takes values $+z$ and $-z$, respectively, with probabilities p and $1-p$. Let us write
$$\log A = a = lz, \quad \log B = -b = -mz,$$
where l and m are positive integers. In this case we get exact results since each step is of size z or $-z$. We have
$$M_p(t) = E_p e^{tZ} = pe^{tz} + (1-p)e^{-tz},$$
so that the nonzero root of $M_p(t) = 1$ is given by
$$t^* = z^{-1} \log \frac{1-p}{p},$$
provided that $p \neq \frac{1}{2}$. Now $\{S_N \geq a\}$ is the event "reject H_0", and $\{S_N \leq -b\}$, the event "reject H_1". Thus from (16) and (17), we have
$$P_p\{\text{Reject } H_0\} = \begin{cases} \dfrac{1 - [(1-p)/p]^{-m}}{[(1-p)/p]^l - [(1-p)/p]^{-m}} & \text{if } p \neq \tfrac{1}{2}, \\ \dfrac{m}{l+m} & \text{if } p = \tfrac{1}{2}. \end{cases}$$

Similarly from (23) and (26) we get
$$E_p N = \begin{cases} \dfrac{1}{2p-1} \dfrac{l\{1 - [(1-p)/p]^{-m}\} - m\{[(1-p)/p]^l - 1\}}{[(1-p)/p]^l - [(1-p)/p]^{-m}} & \text{if } p \neq \tfrac{1}{2}, \\ lm & \text{if } p = \tfrac{1}{2}. \end{cases}$$

The reader will notice that this is the same problem as was discussed in Example 14.2.6. The probability $P_p\{\text{Reject } H_0\}$ is essentially the probability of B's ruin, and $1 - P_p\{\text{Reject } H_0\}$ is the probability of A's ruin.

PROBLEMS 14.6

1. Prove Lemma 1.

[*Hint:* If f is convex and $E|X| < \infty$, then $Ef(X) \geq f(E(X))$].

2. Let $Z_1, Z_2, \cdots$ be independent rv's and a_n, b_n be given constants, $a_n > b_n$, for all n such that $\lim_{n \to \infty} a_n < \infty$ and $\lim_{n \to \infty} b_n > -\infty$. Also, let N be the smallest n such that $S_n = \sum_{k=1}^n Z_k \geq a_n$ or $S_n \leq b_n$. If $\delta > 0$ and $\varepsilon > 0$ can be found such that at least one of $P\{Z_k > \delta\} > \varepsilon$ or $P\{Z_k < -\delta\} > \varepsilon$ holds for all k, then for any $r \geq 0$

$$\lim_{n \to \infty} n^r P\{N > n\} = 0,$$

that is, moments of all orders of N exist.

3. For Problems 14.5.1(b) and 14.5.1(c) find the (approximate) power function and the average sample number (expected sample size.)

14.7 THE FUNDAMENTAL IDENTITY OF SEQUENTIAL ANALYSIS AND ITS APPLICATIONS

In this section we prove the fundamental identity of sequential analysis and study some of its applications.

Theorem 1 (Bahadur [5]). Let $X_1, X_2, \cdots$ be iid rv's with common pdf $f(\cdot | H)$ under hypothesis H. Consider any sequential procedure with a given stopping rule, and let N be the number of observations required to reach a decision. Let $Z_i = Z(X_i)$ be some (Borel-measurable) function of X_i, and write $S_N = \sum_{i=1}^N Z_i$. Assume that $M(t) = E_H e^{tZ_i}$ exists for t in some neighborhood of the origin. Then $P\{N < \infty | H\} = 1$ ensures that for all t

(1) $$E_H\{e^{tS_N}[M(t)]^{-N}\} = P\{N < \infty | H_t\},$$

where, under hypothesis H_t, the pdf (pmf) of X is

(2) $$f(\cdot | H_t) = \frac{e^{tZ(\cdot)}}{M(t)} f(\cdot | H).$$

Proof. Let $R_n \subseteq \mathcal{R}_n$ be the region leading to the termination of the sequential procedure at the nth stage. Then

$$E_H\{e^{tS_N}[M(t)]^{-N}\}$$
$$= \sum_{n=1}^\infty \int_{R_n} [M(t)]^{-n} e^{tS_n} f(\mathbf{x} | H) \, d\mathbf{x}$$
$$= \sum_{n=1}^\infty \int_{R_n} f(\mathbf{x} | H_t) \, d\mathbf{x}$$
$$= P\{N < \infty | H_t\}.$$

Here $\mathbf{x} = (x_1, x_2, \cdots, x_n)$, and we have assumed that the rv's are continuous. The proof for the discrete case is similar.

Corollary (The Fundamental Identity of Sequential Analysis: Wald [134], 159). Consider the SPRT and let

$$Z = Z(X) = \log \frac{f(X|H_1)}{f(X|H_0)}.$$

If H is any hypothesis such that $P\{|Z| > 0 | H\} > 0$, then

(3) $$E_H\{e^{tS_N}[M(t)]^{-N}\} = 1.$$

Proof. We leave the reader to supply the proof.

The fundamental identity has many applications. Assume that both a and b ($a = \log A$, $-b = \log B$) are finite and are the stopping bounds. Then, by Theorem 14.6.1,

(4) $$P\{N < \infty | H_i\} = 1, \quad i = 0, 1.$$

By Lemma 14.6.2, whenever $E_H Z \neq 0$, there exists a $t^* \neq 0$ such that $M(t^*) = 1$ and (3) becomes

(5) $$E_H\{e^{t^*S_N}\} = 1.$$

Let $E_0 = E\{e^{t^*S_N} | S_N \leq -b\}$, and $E_1 = E\{e^{t^*S_N} | S_N \geq a\}$. It follows easily [using (4)] that

(6) $$P\{S_N \leq -b\} = p_{-b} = \frac{E_1 - 1}{E_1 - E_0} \quad \text{and} \quad P\{S_N \geq a\} = p_a = \frac{E_0 - 1}{E_0 - E_1}.$$

If we neglect the excess over the boundaries and write $S_N \approx a$ when $S_N \geq a$, and $S_N \approx -b$ when $S_N \leq -b$, then for $EZ \neq 0$ we have

(7) $$p_a \approx \frac{1 - e^{-t^*b}}{e^{t^*a} - e^{-t^*b}}, \quad p_{-b} \approx \frac{1 - e^{at^*}}{e^{-bt^*} - e^{at^*}}.$$

If $EZ = 0$, then $t^* = 0$; and, letting $t^* \to 0$, we obtain

(8) $$p_a \approx \frac{b}{a + b}, \quad p_{-b} \approx \frac{a}{a + b},$$

which are independent of the particular distribution involved.

Let us next use (3) to obtain the approximate distribution of the stopping rv N. Consider the substitution $z^{-1} = M(t)$. Since $M(t)$ is convex, this equation has two real roots in t, provided that $z^{-1} > M(t_0)$, t_0 being the value at which M achieves a minimum. Let the roots be $\lambda_1(z)$ and $\lambda_2(z)$, where $\lambda_1(1) = 0$ and $\lambda_2(1) = t^*$. Setting $t = \lambda_i(z)$ in (3), we get

(9) $$E\{e^{\lambda_i(z)S_N} z^N\} = 1, \quad i = 1, 2.$$

Now, using the approximation resulting from neglecting the excess over the bounds, we have

(10) $$p_a e^{\lambda_i(z)a} E_a(z^N) + p_{-b} e^{-\lambda_i(z)b} E_{-b}(z^N) \approx 1, \quad i = 1, 2.$$

Equations (10) can be solved for E_a and E_{-b}; and then, using (7) and (8), we can obtain the approximate pgf of N from the following equation:

$$(11) \qquad E(z^N) = p_a E_a(z^N) + p_{-b} E_{-b}(z^N).$$

Here E_a and E_{-b} are conditional expectations, conditional on absorption at a and $-b$, respectively.

To obtain some further interesting results we observe that, if $M(t)$ exists for all real t, the fundamental identity can be differentiated any number of times under the expectation sign. Differentiating once and evaluating at $t = 0$, we obtain Wald's equation (Theorem 14.3.1):

$$(12) \qquad ES_N = EN \cdot EZ, \qquad EZ \neq 0,$$

and if $EZ = 0$, a second derivative yields (Theorem 14.3.2)

$$(13) \qquad EN = ES_N^2 / EZ^2.$$

Ignoring once again the excess over the boundaries, we have

$$(14) \qquad ES_N \approx ap_a - bp_{-b}$$

and

$$(15) \qquad ES_N^2 \approx a^2 p_a + b^2 p_{-b}.$$

It follows from (12), (13), (7), and (8) that

$$(16) \qquad EN \approx \begin{cases} \dfrac{1}{EZ} \dfrac{a + b - ae^{-bt^*} - be^{at^*}}{e^{at^*} - e^{-bt^*}} & \text{if } EZ \neq 0, \\ \dfrac{ab}{EZ^2} & \text{if } EZ = 0. \end{cases}$$

This proves (14.6.26).

Example 1 (Kemperman [56], 70). For $0 < \alpha < 1$, consider the pdf

$$(17) \qquad f(x) = \begin{cases} \alpha \nu e^{-\nu x} & \text{if } x > 0 \quad (\lambda > 0, \nu > 0), \\ (1 - \alpha)\lambda e^{\lambda x} & \text{if } x < 0. \end{cases}$$

The mgf of f is given by

$$(18) \qquad M(t) = \frac{\alpha \nu}{\nu - t} + \frac{\lambda(1 - \alpha)}{\lambda + t},$$

which exists for $-\lambda < t < \nu$. In particular, if $\alpha = \lambda/(\lambda + \nu)$, then

$$(19) \qquad M(t) = \frac{\lambda \nu}{(\nu - t)(\lambda + t)}.$$

In random walk terminology, each step of the particle is the sum of two independent components, one having an exponential distribution an $(0, \infty)$, and the other, an exponential distribution on $(-\infty, 0)$. Alternatively, we may regard each step as the difference between two independent rv's each having an exponential distribution.

If the absorption occurs at the upper barrier a, the step that carries the particle over the boundary must be from the positive component of the mixture. The excess, $S_N - a$, is the excess of the exponentially distributed rv Z_N over the quantity $a - S_{N-1}$, provided that $S_N \geq a$. The excess therefore is due to the last step. Now the exponential distribution has no memory (Theorem 5.3.8). It follows that the distribution of the excess conditional on $S_N \geq a$ is also the same. Thus

$$(20) \qquad E\{e^{t(S_N - a)} | S_N \geq a\} = \frac{\nu}{\nu - t}, \qquad t < \nu,$$

so that

$$E\{e^{tS_N} | S_N \geq a\} = \frac{\nu e^{ta}}{\nu - t},$$

independently of N. Similar remarks apply to the lower boundary.

In this case therefore the excess over the boundaries is exactly accounted for, and we get exact results. The fundamental identity takes the form

$$(21) \qquad p_a \left(\frac{\nu e^{ta}}{\nu - t} \right) E\{[M(t)]^{-N} | S_N \geq a\} \\ + p_{-b} \left(\frac{\lambda e^{-tb}}{\lambda + t} \right) E\{[M(t)]^{-N} | S_N \leq -b\} = 1.$$

The equation

$$M(t) = \frac{\nu \lambda}{(\nu - t)(\lambda + t)} = 1$$

yields $t^* = \nu - \lambda$. It follows from (21) that

$$(22) \qquad \frac{\nu}{\lambda} p_a e^{(\nu - \lambda)a} + \frac{\lambda}{\nu} p_{-b} e^{-b(\nu - \lambda)} = 1.$$

Since $p_a + p_{-b} = 1$, we get

$$(23) \qquad p_a = \begin{cases} \dfrac{1 - (\lambda/\nu) e^{-b(\nu - \lambda)}}{(\nu/\lambda) e^{(\nu - \lambda)a} - (\lambda/\nu) e^{-b(\nu - \lambda)}} & \text{if } \lambda \neq \nu, \\ \dfrac{b + \lambda^{-1}}{a + b + 2\lambda^{-1}} & \text{if } \lambda = \nu. \end{cases}$$

The second result in (23) is obtained by letting $\lambda \to \nu$ in the first expression. The approximate result in this case is

(24) $$p_a \approx \begin{cases} \dfrac{1 - e^{b(\lambda - \nu)}}{e^{a(\nu - \lambda)} - e^{-b(\nu - \lambda)}} & \text{if } \nu \neq \lambda, \\ \dfrac{b}{a + b} & \text{if } \nu = \lambda. \end{cases}$$

Let us specialize (17) a little further by letting $\nu = 1 - \theta$ and $\lambda = 1 + \theta$, $|\theta| < 1$. Then $\alpha = (1+\theta)/2$ and $1 - \alpha = (1-\theta)/2$, so that

(25) $$f_\theta(x) = \begin{cases} \dfrac{1 - \theta^2}{2} e^{-(1-\theta)x} & \text{if } x > 0, \\ \dfrac{1 - \theta^2}{2} e^{(1+\theta)x} & \text{if } x < 0. \end{cases}$$

To test $H_0: \theta = -\tfrac{1}{2}$ against $H_1: \theta = +\tfrac{1}{2}$, we use the SPRT. We have

$$Z_i = \log \frac{f_{1/2}(X_i)}{f_{-1/2}(X_i)}.$$

It is convenient to write

(26) $$f_\theta(x) = \frac{1 - \theta^2}{2} e^{\theta x - |x|}.$$

Then

$$Z_i = \log \{\exp(\tfrac{1}{2} X_i - |X_i| + \tfrac{1}{2} X_i + |X_i|)\}$$
$$= X_i.$$

(27) $$M_Z(t) = Ee^{tX} = \frac{(1 - \theta^2)}{(1 - \theta - t)(1 + \theta + t)} = \frac{1 - \theta^2}{1 - (\theta + t)^2},$$
$$-1 - \theta < t < 1 + \theta,$$

and

(28) $$t^* = \nu - \lambda = (1 - \theta) - (1 + \theta) = -2\theta.$$

From (23)

(29) $$p_a = P_\theta\{S_N \geq a\}$$
$$= \begin{cases} \dfrac{1 - [(1 + \theta)/(1 - \theta)] e^{2b\theta}}{[(1 - \theta)/(1 + \theta)] e^{-2a\theta} - [(1 + \theta)/(1 - \theta)] e^{2b\theta}} & \text{if } \theta \neq 0, \\ \dfrac{1 + b}{a + b + 2} & \text{if } \theta = 0, \end{cases}$$
$$= \begin{cases} \dfrac{1 - \theta^2 - (1 + \theta)^2 e^{2b\theta}}{(1 - \theta)^2 e^{-2a\theta} - (1 + \theta)^2 e^{2b\theta}} & \text{if } \theta \neq 0, \\ \dfrac{1 + b}{a + b + 2} & \text{if } \theta = 0. \end{cases}$$

To find $E_\theta N$, we have, using a similar argument,
$$E_\theta\{S_N - a \,|\, S_N \geq a\} = \frac{1}{\nu}$$
and
$$E_\theta\{S_N + b \,|\, S_N \leq -b\} = -\frac{1}{\lambda}.$$
Thus
$$E_\theta S_N = p_a\left(a + \frac{1}{\nu}\right) - p_{-b}\left(\frac{1}{\lambda} + b\right).$$
Now
$$E_\theta Z = E_\theta X = \frac{2\theta}{1 - \theta^2},$$
which is $\neq 0$ for $\theta \neq 0$. For $\theta \neq 0$,
$$E_\theta N = \frac{1 - \theta^2}{2\theta}\left\{p_a\left(a + \frac{1}{\nu}\right) - p_{-b}\left(\frac{1}{\lambda} + b\right)\right\}$$
$$(30) \qquad = \frac{1 - \theta^2}{2\theta}\left\{p_a\left(a + \frac{1}{1 - \theta}\right) - p_{-b}\left(\frac{1}{1 + \theta} + b\right)\right\}.$$
If $\theta = 0$, we compute $E_\theta Z^2$. We have
$$E_\theta Z^2 = E_\theta X^2$$
$$= \frac{2}{(1 - \theta^2)^2}(3\theta^2 + 1),$$
so that $E_{\theta=0} Z^2 = 2$. Thus
$$E_{\theta=0} S_N^2 = p_a E_0\{S_N^2 \,|\, S_N \geq a\} + p_{-b} E_0\{S_N^2 \,|\, S_N \leq -b\}.$$
Now
$$E_\theta\{(S_N - a)^2 \,|\, S_N \geq a\} = \frac{2}{(1 - \theta)^2},$$
$$E_\theta\{(S_N + b)^2 \,|\, S_N \leq -b\} = \frac{2}{(1 + \theta)^2},$$
so that
$$E_\theta\{S_N^2 \,|\, S_N \geq a\} = \frac{2}{(1 - \theta)^2} + 2a\left(a + \frac{1}{1 - \theta}\right) - a^2$$
and
$$E_\theta\{S_N^2 \,|\, S_N \leq -b\} = \frac{2}{(1 + \theta)^2} + 2b\left(\frac{1}{1 + \theta} + b\right) - b^2.$$

Finally,
$$E_0\{S_N^2 | S_N \geq a\} = 2 + 2a + a^2,$$
and
$$E_0\{S_N^2 | S_N \leq -b\} = 2 + 2b + b^2.$$

It follows that
$$E_0 N = \frac{1}{E_0 Z^2} [p_a(2 + 2a + a^2) + p_{-b}(2 + 2b + b^2)]$$
$$= \frac{1}{2(a + b + 2)} [(b + 1)(2 + 2a + a^2)$$
$$+ (a + 1)(2 + 2b + b^2)]$$
(31)
$$= \tfrac{1}{2}[1 + (a + 1)(b + 1)].$$

Example 2. Let us return to Example 14.6.6 and compute the pgf of the stopping variable N. The equation $x^{-1} = M(t)$ reduces to
$$pxe^{2tz} - e^{tz} + x(1 - p) = 0.$$
The two solutions are
$$\lambda_1(x) = z^{-1} \log \left\{ \frac{1 + \sqrt{1 - 4px^2(1 - p)}}{2xp} \right\}$$
and
$$\lambda_2(x) = z^{-1} \log \left\{ \frac{1 - \sqrt{1 - 4px^2(1 - p)}}{2xp} \right\}.$$
Substituting $t = \lambda_i(x)$ in Wald's fundamental identity, we get
$$E\{e^{\lambda_i(x) S_N} x^N\} = 1, \quad i = 1, 2.$$
Thus
$$E\{x^N | S_N = a\} e^{a\lambda_i(x)} P\{S_N = a\} + E\{x^N | S_N = -b\} e^{-b\lambda_i(x)} = 1,$$
that is,
$$P\{\text{Reject } H_0 | p\} e^{lz\lambda_1(x)} E\{x^N | S_N = lz\}$$
$$+ (1 - P\{\text{Reject } H_0 | p\}) e^{-mz\lambda_1(x)} E\{x^N | S_N = -mz\} = 1$$
and
$$P\{\text{Reject } H_0 | p\} e^{lz\lambda_2(x)} E\{x^N | S_N = lz\}$$
$$+ (1 - P\{\text{Reject } H_0 | p\}) e^{-mz\lambda_2(x)} E\{x^N | S_N = -mz\} = 1.$$
Solving for $E\{x^N | S_N = lz\}$ and $E\{x^N | S_N = -mz\}$, we get

$$E\{x^N | S_N = lz\} = \frac{1}{p^*} \frac{e^{-mz\lambda_2} - e^{-mz\lambda_1}}{e^{lz\lambda_1 - mz\lambda_2} - e^{lz\lambda_2 - m\lambda z\lambda_1}},$$

$$E\{x^N | S_N = -mz\} = \frac{1}{1 - p^*} \frac{e^{lz\lambda_2} - e^{lz\lambda_1}}{e^{-mz\lambda_1 + lz\lambda_2} - e^{-mz\lambda_2 + lz\lambda_1}},$$

where $p^* = P\{\text{Reject } H_0 | p\}$, and we have written $\lambda_1 = \lambda_1(x)$, $\lambda_2 = \lambda_2(x)$. Now

$$\begin{aligned} E_p\{x^N\} &= p^* E\{x^N | S_N = lz\} + (1 - p^*) E\{x^N | S_N = -mz\} \\ &= \frac{(e^{lz\lambda_1} - e^{lz\lambda_2}) - (e^{-mz\lambda_1} - e^{-mz\lambda_2})}{e^{lz\lambda_1 - mz\lambda_2} - e^{lz\lambda_2 - mz\lambda_1}}. \end{aligned}$$

The coefficient of x^n gives the probability $P_p\{N = n\}$, $n = 1, 2, \cdots$.

PROBLEMS 14.7

1. Suppose that $P\{\varepsilon_1 \leq Z_i \leq \varepsilon_2\} = 1$ for all i, $\varepsilon_1 < 0 < \varepsilon_2$. If $S_N = \sum_{j=1}^{N} Z_j$, show that $P\{S_N \geq a\}$ satisfies

$$\frac{1 - e^{-t^*b}}{e^{t^*(a+\varepsilon_2)} - e^{-t^*b}} \leq P\{S_N \geq a\} \leq \frac{1 - e^{t^*(-b+\varepsilon_1)}}{e^{at^*} - e^{t^*(-b+\varepsilon_1)}},$$

where t^* is the unique nonzero root of the equation $M(t) = Ee^{tZ} = 1$.

(Kemperman [56], 64)

2. Suppose that $P_i\{\varepsilon_1 \leq Z_j \leq \varepsilon_2\} = 1$ ($i = 0, 1$). Using $t^* = \pm 1$ in Problem 1, show that (approximately)

$$\frac{\beta}{1 - \alpha} \leq B \leq \frac{\beta e^{-\varepsilon_1}}{1 - \alpha}, \quad e^{-\varepsilon_1} \frac{1 - \beta}{\alpha} \leq A \leq \frac{1 - \beta}{\alpha}.$$

3. For Example 14.6.6 compute $E_p S_N^k$, where k is a positive integer.

4. Prove the corollary to Theorem 1.

References

1. T. W. Anderson, Maximum likelihood estimates for a multivariate normal distribution when some observations are missing, *J. Am. Stat. Assoc.* 52 (1957), 200-203.
2. T. W. Anderson and M. Friedman, A limitation of the optimum property of the sequential probability ratio test, *Contributions to Probability and Statistics,* Stanford University Press, Stanford, Calif., 1960, pages 57-69.
3. T. M. Apostol, *Mathematical Analysis,* Addison-Wesley, Reading, Mass., 1960.
4. J. D. Auble, Extended tables for the Mann-Whitney statistic, *Bull. Inst. Educ. Res.* 1 (1953), i-iii, 1-39.
5. R. R. Bahadur, A note on the fundamental identity of sequential analysis, *Ann. Math. Stat.* 29 (1958), 534-543.
6. D. Bernstein, Sur une propriété charactéristique de la loi de Gauss, *Trans. Leningrad Polytech. Inst.* 3 (1941), 21-22.
7. Z. W. Birnbaum and F. H. Tingey, One-sided confidence contours for probability distribution functions, *Ann. Math. Stat.* 22 (1951), 592-596.
8. Z. W. Birnbaum, Numerical tabulation of the distribution of Kolmogorov's statistic for finite sample size, *J. Am. Stat. Assoc.* 17 (1952), 425-441.
9. D. Blackwell, Conditional expectation and unbiased sequential estimation, *Ann. Math. Stat.* 18 (1947), 105-110.
10. J. Boas, A note on the estimation of the covariance between two random variables using extra information on the separate variables, *Stat. Neerl.* 21 (1967), 291-292.
11. D. G. Chapman and H. Robbins Minimum variance estimation without regularity assumptions, *Ann. Math. Stat.* 22 (1951), 581-586.
12. S. D. Chatterji, Some elementary characterizations of the Poisson distribution, *Am. Math. Month.* 70 (1963), 958-964.
13. Y. S. Chow and H. Robbins, On the asymptotic theory of fixed-width sequential confidence intervals for the mean, *Ann. Math. Stat.* 36 (1965), 457-462.
14. K. L. Chung and P. Erdös, On the application of the Borel-Cantelli lemma, *Trans. Am. Math. Soc.* 72 (1952), 179-186.
15. K. L. Chung, *A Course in Probability Theory,* Harcourt, Brace & World, New York, 1968.
16. R. Courant and F. John, *Introduction to Calculus and Analysis,* John Wiley-Interscience, New York, 1965.
17. H. Cramér, Über eine Eigenschaft der normalen Verteilungsfunktion, *Math. Z.* 41 (1936), 405-414.

18. H. Cramér, *Mathematical Methods of Statistics*, Princeton Univeristy Press, Princeton, N. J., 1946.
19. H. Cramér, A contribution to the theory of statistical estimation, *Skand. Aktuarietidskr.* 29 (1946), 85–94.
20. J. H. Curtiss, A note on the theory of moment generating functions, *Ann. Math. Stat.* 13 (1942), 430–433.
21. G. B. Dantzig, On the nonexistence of tests of Student's hypothesis having power functions independent of σ^2, *Ann. Math. Stat.* 11 (1940), 186–192.
22. D. A. Darmois, Sur diverses propriétés charactéristique de la loi de probabilité de Laplace-Gauss, *Bull. Int. Stat. Inst.* 23 (1951), part II, 79–82.
23. M. M. Desu, A characterization of the exponential distribution by order statistics, *Ann. Math. Stat.* 42 (1971), 837–838.
24. M. M. Desu, Optimal confidence intervals of fixed width, *Am. Stat.* 25 (1971), No. 2, 27–29.
25. S. W. Dharmadhikari, An example in the problem of moments, *Am. Math. Month.* 72 (1965), 302–303.
26. P. Enis, On the relation $EX = E\{E\{X|Y\}\}$, *Biometrika* 60 (1973), 432–433.
27. W. Feller, Über den Zentralen Grenzwertsatz der Wahrscheinlichkeitsrechnung, *Math. Z.* 40 (1935), 521–559; 42(1937), 301–312.
28. W. Feller, *An Introduction to Probability Theory and Its Applications*, Vol 1., 3rd Ed., John Wiley, New York, 1968.
29. W. Feller, *An Introduction to Probability Theory and Its Applications*, Vol 2, 2nd Ed., John Wiley, New York, 1971.
30. T. S. Ferguson, *Mathematical Statistics,* Academic Press, New York, 1967.
31. R. A. Fisher, On the mathematical foundations of theoretical statistics, *Phil. Trans. Royal Soc.* A222 (1922), 309–368.
32. M. Fisz, *Probability Theory and Mathematical Statistics,* 3rd Ed., John Wiley, 1963.
33. D. A. S. Fraser, *Nonparametric Methods in Statistics,* John Wiley, New York, 1965.
34. M. Fréchet, Sur l'extension de certaines evaluations statistiques au cas de petits echantillons, *Rev. Inst. Int. Stat.* 11 (1943), 182–205.
35. S. G. Ghurye and H. Robbins, Two-stage procedures for estimating the difference between means, *Biometrika* 41 (1954), 146–152.
36. J. D. Gibbons, *Nonparametric Inference,* McGraw-Hill, New York, 1971.
37. V. I. Glivenko, Sulla determinazione empirica di una legge di probabilita, *Giorn. Inst. Ital. Attuari* 4 (1933), 1–10.
38. B. V. Gnedenko, Sur la distribution limite du terme maximum d'une série aléatoire, *Ann. Math.* 44 (1943), 423–453.
39. Z. Govindarajulu and R. T. Leslie, *Annotated Bibliography on Robustness Studies of Statistical Procedures,* U. S. Dept. Health, Education, and Welfare, Ser. 2, No. 51, 1971.
40. W. C. Guenther, Shortest confidence intervals, *Am. Stat.* 23 (1969), No. 1, 22–25.
41. W. C. Guenther, Unbiased confidence intervals, *Am. Stat.* 25 (1971), No. 1, 51–53.
42. E. J. Gumbel, Distributions à plusieurs variables dont les marges sont données, *C. R. Acad. Sci. Paris* 246 (1958), 2717–2720.
43. J. H. Hahn and W. Nelson, A problem in the statistical comparison of measuring devices, *Techometrics* 12 (1970), 95–102.

REFERENCES

44. P. R. Halmos and L. J. Savage, Application of the Radon-Nikodym theorem to the theory of sufficient statistics, *Ann. Math. Stat.* 20 (1949), 225–241.
45. P. R. Halmos, *Measure Theory*, Van Nostrand, New York, 1950.
46. H. Hamburger, Uber eine Erweiterung des Stieltjesschen Momentenproblems, *Math. Ann.* 81 (1920), 235-319; 82 (1921), 120-167, 167–187.
47. B. Harris and A. P. Soms, On testing for sufficiency, *Am. Stat.* 27 (1973), 192.
48. J. L. Hodges and E. L. Lehmann, Some problems in minimax point estimation, *Ann. Math. Stat.* 21 (1950), 182–197.
49. D. Hogben, The distribution of the sample variance from a two-point binomial population, *Am. Stat.* 22 (1968), No. 5, 30.
50. P. J. Huber, Robust statistics: A review, *Ann. Math. Stat.* 43 (1972), 1041–1067.
51. V. S. Huzurbazar, The likelihood equation consistency, and maxima of the likelihood function, *Ann. Eugen.* (London) 14 (1948), 185–200.
52. B. Jamison, S. Orey, and W. Pruitt, Convergence of weighted averages of independent random variables, *Z. Wahrschein.* 4 (1965), 40–44.
53. N. L. Johnson, A proof of Wald's theorem on cumulative sums, *Ann. Math. Stat.* 30 (1959), 1245–1247; correction, *Ann. Math. Stat.* 32 (1961), 1344.
54. M. Kac, *Lectures in Probability*, The Mathematical Association of America, 1964–1965.
55. J. F. Kemp, A maximal distribution with prescribed marginals, *Am. Math. Month.* 80 (1973), 83.
56. J. H. B. Kemperman, *The First Passage Problem for a Stationary Markov Chain*, University of Chicago Press, Chicago, 1961.
57. M. G. Kendall, *Rank Correlation Methods*, 3rd Ed., Griffin, London, 1962.
58. J. Kiefer, On minimum variance estimators, *Ann. Math. Stat.* 23 (1952), 627–629.
59. M. S. Klamkin, An expected value, *Siam Rev.* 14 (1972), 378.
60. A. N. Kolmogorov, Sulla determinazione empirica di una legge di distribuzione, *Giorn. Inst. Ital. Attuari* 4 (1933), 83–91.
61. A. N. Kolmogorov and S. V. Fomin, *Elements of the Theory of Functions and Functional Analysis*, Vol 2, Graylock Press, Albany, N. Y., 1961.
62. C. H. Kraft and C. Van Eeden, *A Nonparametric Introduction to Statistics*, Macmillan, New York, 1968.
63. W. Kruskal, Note on a note by C. S. Pillai, *Am. Stat.* 22 (1968), No 5, 24–25.
64. S. Kullback and H. Y. Yasaimaibodi, A property of symmetric (interchangeable) random variables, *Am. Stat.* 21 (1967), No. 1, 32–33.
65. G. Kulldorf, On the conditions for consistency and asymptotic efficiency of maximum likelihood estimates, *Skand. Aktuarietidskr.* 40 (1957), 129–144.
66. J. Lamperti, Density of random variable, *Am. Math. Month.* 66 (1959), 317.
67. J. Lamperti and W. Kruskal, Solution by W. Kruskal to the problem "Poisson Distribution" posed by J. Lamperti, *Am. Math. Month.* 67 (1960), 297–298.
68. E. L. Lehmann, *Theory of Estimation* (mimeographed notes), University of California, 1950.
69. E. L. Lehmann and H. Scheffé, Completeness, similar regions, and unbiased estimation, *Sankhyā*, Ser. A, 10 (1950), 305–340.
70. E. L. Lehmann, *Testing Statistical Hypotheses*, John Wiley, New York, 1959.

71. M. Loève, *Probability Theory*, 3rd Ed., Van Nostrand, Princeton, 1963.
72. E. Lukacs, A characterization of the normal distribution, *Ann. Math. Stat.* 13 (1942), 91–93.
73. E. Lukacs, Characterization of populations by properties of suitable statistics, *Proc. Third Berkeley Symp.* 2 (1956), 195–214.
74. E. Lukacs and R. G. Laha, *Applications of Characteristic Functions*, Hafner, New York, 1964.
75. E. Lukacs, *Characteristic Functions*, 2nd Ed., Hafner, New York, 1970.
76. E. Lukacs, Inequalities and functional equations in probability theory, C. I. M. E., III Ciclo-La Mendola, 1970, pages 235–237.
77. E. Lukacs, *Probability and Mathematical Statistics*, Academic Press, New York, 1972.
78. H. B. Mann and D. R. Whitney, On a test whether one of two random variables is stochastically larger than the other, *Ann. Math. Stat.* 18 (1947), 50–60.
79. F. J. Massey, Distribution table for the deviation between two sample cumulatives, *Ann. Math. Stat.* 23 (1952), 435–441.
80. M. V. Menon, A characterization of the Cauchy distribution, *Ann. Math. Stat.* 33 (1962), 1267–1271.
81. L. H. Miller, Table of percentage points of Kolmogorov statistics, *J. Am. Stat. Assoc.* 51 (1956), 111–121.
82. K. S. Miller, *Advanced Real Calculus*, Harper, New York, 1957.
83. J. Moshman, A method for selecting the size of the initial sample in Stein's two-sample procedure, *Ann. Math. Stat.* 29 (1958), 1271–1275.
84. M. G. Natrella, *Experimental Statistics*, Natl. Bur. Stand. Handbook 91, Washington, D. C., 1963.
85. J. Neyman and E. S. Pearson, On the problem of the most efficient tests of statistical hypotheses, *Phil. Trans. Roy. Soc.* A231 (1933), 289–337.
86. J. Neyman and E. L. Scott, Consistent estimates based on partially consistent observations, *Econometrica* 16 (1948), 1–32.
87. E. H. Oliver, A maximum likelihood oddity, *Am. Stat.* 26 (1972), No. 3, 43–44.
88. D. B. Owen, *Handbook of Statistical Tables*, Addison-Wesley, Reading, Mass., 1962.
89. E. J. G. Pitman and E. J. Williams, Cauchy-distributed functions of Cauchy variates, *Ann. Math. Stat.* 38 (1967), 916–918.
90. J. W. Pratt, Length of confidence intervals, *J. Am. Stat. Assoc.* 56 (1961), 260–272.
91. B. J. Prochaska, A note on the relationship between the geometric and exponential distributions, *Am. Stat.* 27 (1973), 27.
92. P. S. Puri, On a property of exponential and geometric distributions and its relevance to multivariate failure rate, *Sankhyā*, Ser. A, 35 (1973), 61–68.
93. D. A. Raikov, On the decomposition of Gauss and Poisson laws (in Russian), *Izv. Akad. Nauk. SSSR, Ser. Mat.* 2 (1938), 91–124.
94. B. Ramachandaran, On the product of two independent beta random variables, *Ann. Math. Stat.* 33 (1962), 1212.
95. C. R. Rao, Information and the accuracy attainable in the estimation of statistical parameters, *Bull. Calcutta Math. Soc.* 37 (1945), 81–91.

REFERENCES

96. C. R. Rao, Sufficient statistics and minimum variance unbiased estimates, *Proc. Camb. Phil. Soc.* 45 (1949) 213–218.
97. C. R. Rao, *Linear Statistical Inference and Its Applications*, John Wiley, New York, 1965.
98. A. Rényi, *Probability Theory*, North-Holland, Amsterdam, 1970.
99. D. L. Richter, Two-stage experiments for estimating a common mean, *Ann. Math. Stat.* 31(1960), 1164–1173.
100. H. Robbins, Sequential estimation of the mean of a normal population, *Probability and Statistics—The Harald Cramér Volume*, Almquist and Wilsell, Uppsala, 1959, pages 233–245.
101. V. K. Rohatgi, On large deviation probabilities for sums of random variables which are attracted to the normal law, *Commun. Stat.* 2 (1973), 525–533.
102. V. I. Romanovsky, On the moments of the standard deviations and of the correlation coefficient in samples from a normal population, *Metron* 5 (1925), No. 4, 3–46.
103. L. Rosenberg, Nonnormality of linear combinations of normally distributed random variables, *Am. Math. Month.* 72 (1965), 888–890.
104. J. H. Rowland, A counting circuit problem, *Siam Rev.* 10 (1968), 112–113.
105. J. Roy and S. Mitra, Unbiased minimum variance estimation in a class of discrete distributions, *Sankhyā* 18 (1957), 371–378.
106. R. Roy, Y. LePage, and M. Moore, On the power series expansion of the moment generating function, *Am. Stat.* 28 (1974), 58–59.
107. H. L. Royden, *Real Analysis*, Macmillan, New York, 1963.
108. Y. D. Sabharwal, A sequence of symmetric Bernoulli trials, *Siam Rev.* 11 (1969), 406–409.
109. P. A. Samuelson, How deviant can you be? *J. Am. Stat. Assoc.* 63 (1968), 1522–1525.
110. H. Scheffé, A useful convergence theorem for probability distributions, *Ann. Math. Stat.* 18 (1947), 434–438.
111. H. Scheffé, *The Analysis of Variance*, John Wiley, New York, 1961.
112. B. M. Seelbinder, On Stein's two-stage sampling scheme, *Ann. Math. Stat.* 24 (1953), 640–649.
113. D. N. Shanbhag, and I. V. Basawa, On a characterization property of the multinomial distribution, *Ann. Math. Stat.* 42 (1971), 2200.
114. L. Shepp, Normal functions of normal random variables, *Siam Rev.* 4 (1962), 255–256.
115. A. E. Siegel, Film-mediated fantasy aggression and strength of aggression drive, *Child Dev.* 27 (1956), 365–378.
116. G. Simons and S. Zacks, A sequential estimation of tail probabilities in exponential distribution with a proportional closness, Technical Report, Dept. Statistics, Stanford University, 1967.
117. G. Simons, On the cost of not knowing the variance when making a fixed-width confidence interval for the mean, *Ann. Math. Stat.* 39 (1968), 1946–1952.
118. M. Skibinsky, A characterization of hypergeometric distributions, *J. Am. Stat. Assoc.* 65 (1970) 926–929.
119. V. P. Skitovitch, Linear forms of independent random variables and the normal distribution law, *Izv. Akad. Nauk. SSSR, Ser. Mat.* 18 (1954), 185–200.

120. N. V. Smirnov, On the estimation of the discrepancy between empirical curves of distributions for two independent samples (in Russian), *Bull. Moscow Univ.* 2 (1939), 3–16.
121. N. V. Smirnov, Approximate laws of distribution of random variables from empirical data (in Russian), *Usp. Mat. Nauk.* 10 (1944), 179–206.
122. R. C. Srivastava, Two characterizations of the geometric distribution, *J. Am. Stat. Assoc.* 69 (1974), 267–269.
123. N. Starr, The performance of a sequential procedure for the fixed width interval estimation of the mean, *Ann. Math. Stat.* 37 (1966), 36–50.
124. N. Starr and M. B. Woodroofe, Remarks on sequential point estimation, *Proc. Natl. Acad. Sci.* 63 (1969), 285–288.
125. C. Stein, A two-sample test for a linear hypothesis whose power is independent of the variance, *Ann. Math. Stat.* 16 (1945), 243–258.
126. C. Stein, A note on cumulative sums, *Ann. Math. Stat.* 17 (1946), 489–499.
127. S. M. Stigler, Completeness and unbiased estimation, *Am. Stat.* 26 (1972), 28–29.
128. P. T. Strait, A note on the independence and conditional probabilities, *Am. Stat.* 25 (1971), No. 2, 17–18.
129. R. F. Tate and G. W. Klett, Optimum confidence intervals for the variance of a normal distribution, *J. Am. Stat. Assoc.* 54 (1959), 674–682.
130. W. A. Thompson, Jr., *Applied Probability*, Holt, Rinehart and Winston, New York, 1969.
131. H. F. Trotter, An elementary proof of the central limit theorem, *Arch. Math.* 10 (1959), 226–234.
132. H. G. Tucker, *A Graduate Course in Probability*, Academic Press, New York, 1967.
133. A. Wald, On cumulative sums of random variables, *Ann. Math. Stat.* 15 (1944), 283–296.
134. A. Wald, *Sequential Analysis*, John Wiley, New York, 1947.
135. A. Wald and J. Wolfowitz, Optimal character of the sequential probability ratio test, *Ann. Math. Stat.* 19 (1948), 326–339.
136. A. Wald, Note on the consistency of the maximum likelihood estimate, *Ann. Math. Stat.* 20 (1949), 595–601.
137. D. V. Widder, *Advanced Calculus*, 2nd Ed., Prentice-Hall, Englewood Cliffs, N. J., 1961.
138. R. A. Wijsman, On the existence of Wald's sequential test, *Ann. Math. Stat.* 29 (1958), 938–939.
139. R. A. Wijsman, Existence, uniqueness and monotonicity of sequential probability ratio tests, *Ann. Math. Stat.* 34 (1963), 1541–1548.
140. R. A. Wijaman, On the attainment of the Cramér-Rao lower bound. *Ann. Stat.* 1 (1973), 538–542.
141. S. S. Wilks, *Mathematical Statistics*, John Wiley, New York, 1962.
142. J. Wishart, The generalized product-moment distribution in samples from a normal multivariate population, *Biometrika* 20A (1928), 32–52.
143. J. Wolfowitz, The efficiency of sequential estimates and Wald's equation for sequential processes, *Ann. Math. Stat.* 18 (1947), 215–230.

144. C. K. Wong, A note on mutually independent events, *Am. Stat.* 26 (1972), 27.
145. S. Zacks, Sequential estimation of the mean of a log-normal distribution having a prescribed proportional closeness, *Ann. Math. Stat.* 37 (1966), 1688–1696.
146. S. Zacks, *The Theory of Statistical Inference,* John Wiley, New York, 1971.
147. P. W. Zehna, Invariance of maximum likelihood estimation, *Ann. Math. Stat.* 37 (1966), 755.

Table 1. Cumulative Binomial Probabilities, $\sum_{x=0}^{r}\binom{n}{x}p^x(1-p)^{n-x}$, $r = 0, 1, 2, \cdots, n-1$

						p				
n	r	.01	.05	.10	.20	.25	.30	.333	.40	.50
2	0	.9801	.9025	.8100	.6400	.5625	.4900	.4444	.3600	.2500
	1	.9999	.9975	.9900	.9600	.9375	.9100	.8888	.8400	.5000
3	0	.9703	.8574	.7290	.5120	.4219	.3430	.2963	.2160	.1250
	1	.9997	.9928	.9720	.8960	.8438	.7840	.7407	.6480	.5000
	2	1.0000	.9999	.9990	.9920	.9844	.9730	.9629	.9360	.8750
4	0	.9606	.8145	.6561	.4096	.3164	.2401	.1975	.1296	.0625
	1	.9994	.9860	.9477	.8192	.7383	.6517	.5926	.4742	.3125
	2	1.0000	.9995	.9963	.9728	.9492	.9163	.8889	.8198	.6875
	3		1.0000	.9999	.9984	.9961	.9919	.9877	.9734	.9375
5	0	.9510	.7738	.5905	.3277	.2373	.1681	.1317	.0778	.0312
	1	.9990	.9774	.9185	.7373	.6328	.5283	.4609	.3370	.1874
	2	1.0000	.9988	.9914	.9421	.8965	.8370	.7901	.6826	.4999
	3		.9999	.9995	.9933	.9844	.9693	.9547	.9130	.8124
	4		1.0000	1.0000	.9997	.9990	.9977	.9959	.9898	.9686
6	0	.9415	.7351	.5314	.2621	.1780	.1176	.0878	.0467	.0156
	1	.9986	.9672	.8857	.6553	.5340	.4201	.3512	.2333	.1094
	2	1.0000	.9977	.9841	.9011	.8306	.7442	.6804	.5443	.3438
	3		.9998	.9987	.9830	.9624	.9294	.8999	.8208	.6563
	4		.9999	.9999	.9984	.9954	.9889	.9822	.9590	.8907
	5		1.0000	1.0000	.9999	.9998	.9991	.9987	.9959	.9845
7	0	.9321	.6983	.4783	.2097	.1335	.0824	.0585	.0280	.0078
	1	.9980	.9556	.6554	.5767	.4450	.3294	.2633	.1586	.0625
	2	1.0000	.9962	.8503	.8520	.7565	.6471	.5706	.4199	.2266
	3		.9998	.9743	.9667	.9295	.8740	.8267	.7102	.5000
	4		1.0000	.9973	.9953	.9872	.9712	.9547	.9037	.7734
	5			.9998	.9996	.9987	.9962	.9931	.9812	.9375
	6			1.0000	1.0000	.9999	.9998	.9995	.9984	.9922
8	0	.9227	.6634	.4305	.1678	.1001	.0576	.0390	.0168	.0039
	1	.9973	.9427	.8131	.5033	.3671	.2553	.1951	.1064	.0352
	2	.9999	.9942	.9619	.7969	.6786	.5518	.4682	.3154	.1445
	3	1.0000	.9996	.9950	.9437	.8862	.8059	.7413	.5941	.3633
	4		1.0000	.9996	.9896	.9727	.9420	.9120	.8263	.6367
	5			1.0000	.9988	.9958	.9887	.9803	.9502	.8555
	6				1.0000	.9996	.9987	.9974	.9915	.9648
	7					1.0000	.9999	.9998	.9993	.9961
9	0	.9135	.6302	.3874	.1342	.0751	.0404	.0260	.0101	.0020
	1	.9965	.9287	.7748	.4362	.3004	.1960	.1431	.0706	.0196
	2	.9999	.9916	.9470	.7382	.6007	.4628	.3772	.2318	.0899
	3	1.0000	.9993	.9916	.9144	.8343	.7296	.6503	.4826	.2540
	4		.9999	.9990	.9805	.9511	.9011	.8551	.7334	.5001

Table 1 (*Continued*)

n	k									
	5		1.0000	.9998	.9970	.9900	.9746	.9575	.9006	.7462
	6			.9999	.9998	.9987	.9956	.9916	.9749	.9103
	7			1.0000	1.0000	.9999	.9995	.9989	.9961	.9806
	8					1.0000	.9999	.9998	.9996	.9982
10	0	.9044	.5987	.3487	.1074	.0563	.0282	.0173	.0060	.0010
	1	.9958	.9138	.7361	.3758	.2440	.1493	.1040	.0463	.0108
	2	1.0000	.9884	.9298	.6778	.5256	.3828	.2991	.1672	.0547
	3		.9989	.9872	.8791	.7759	.6496	.5592	.3812	.1719
	4		.9999	.9984	.9672	.9219	.8497	.7868	.6320	.3770
	5		1.0000	.9999	.9936	.9803	.9526	.9234	.8327	.6231
	6			1.0000	.9991	.9965	.9894	.9803	.9442	.8282
	7				.9999	.9996	.9984	.9966	.9867	.9454
	8				1.0000	1.0000	.9998	.9996	.9973	.9893
	9						1.0000	.9999	.9999	.9991
11	0	.8954	.5688	.3138	.0859	.0422	.0198	.0116	.0036	.0005
	1	.9948	.8981	.6974	.3221	.1971	.1130	.0752	.0320	.0059
	2	.9998	.9848	.9104	.6174	.4552	.3128	.2341	.1189	.0327
	3	1.0000	.9984	.9815	.8389	.7133	.5696	.4726	.2963	.1133
	4		.9999	.9972	.9496	.8854	.7897	.7110	.5328	.2744
	5		1.0000	.9997	.9884	.9657	.9218	.8779	.7535	.5000
	6			1.0000	.9981	.9924	.9784	.9614	.9007	.7256
	7				.9998	.9988	.9957	.9912	.9707	.8867
	8				1.0000	.9999	.9994	.9986	.9941	.9673
	9					1.0000	.9999	.9999	.9993	.9941
	10						1.0000	1.0000	1.0000	.9995
12	0	.8864	.5404	.2824	.0687	.0317	.0139	.0077	.0022	.0002
	1	.9938	.8816	.6590	.2749	.1584	.0850	.0540	.0196	.0032
	2	.9998	.9804	.8892	.5584	.3907	.2528	.1811	.0835	.0193
	3	1.0000	.9978	.9744	.7946	.6488	.4925	.3931	.2254	.0730
	4	1.0000	.9998	.9957	.9806	.8424	.7237	.6315	.4382	.1939
	5	1.0000	1.0000	.9995	.9961	.9456	.8822	.8223	.6652	.3872
	6			1.0000	.9994	.9858	.9614	.9336	.8418	.6128
	7				.9999	.9972	.9905	.9812	.9427	.8062
	8				1.0000	.9996	.9983	.9962	.9848	.9270
	9					1.0000	.9998	.9995	.9972	.9807
	10						1.0000	.9999	.9997	.9968
	11							1.0000	1.0000	.9998
13	0	.8775	.5134	.2542	.0550	.0238	.0097	.0052	.0013	.0000
	1	.9928	.8746	.6214	.2337	.1267	.0637	.0386	.0126	.0017
	2	.9997	.9755	.8661	.5017	.3326	.2025	.1388	.0579	.0112
	3	1.0000	.9969	.9659	.7473	.5843	.4206	.3224	.1686	.0462
	4		.9997	.9936	.9009	.7940	.6543	.5521	.3531	.1334
	5		1.0000	.9991	.9700	.9198	.8346	.7587	.5744	.2905
	6			.9999	.9930	.9757	.9376	.8965	.7712	.5000
	7			1.0000	.9988	.9944	.9818	.9654	.9024	.7095
	8				.9998	.9990	.9960	.9912	.9679	.8666
	9				1.0000	.9999	.9994	.9984	.9922	.9539

649

Table 1 (*Continued*)

n	k									
	10					1.0000	.9999	.9998	.9987	.9888
	11						1.0000	1.0000	.9999	.9983
	12								1.0000	.9999
14	0	.8687	.4877	.2288	.0440	.0178	.0068	.0034	.0008	.0000
	1	.9916	.8470	.5847	.1979	.1010	.0475	.0274	.0081	.0009
	2	.9997	.9700	.8416	.4480	.2812	.1608	.1054	.0398	.0065
	3	1.0000	.9958	.9559	.6982	.5214	.3552	.2612	.1243	.0287
	4		.9996	.9908	.8702	.7416	.5842	.4755	.2793	.0898
	5		1.0000	.9986	.9562	.8884	.7805	.6898	.4859	.2120
	6			.9998	.9884	.9618	.9067	.8506	.6925	.3953
	7			1.0000	.9976	.9897	.9686	.9424	.8499	.6048
	8				.9996	.9979	.9917	.9826	.9417	.7880
	9				1.0000	.9997	.9984	.9960	.9825	.9102
	10					1.0000	.9998	.9993	.9961	.9713
	11						1.0000	.9999	.9994	.9936
	12							1.0000	.9999	.9991
	13									.9999
15	0	.8601	.4633	.2059	.0352	.0134	.0048	.0023	.0005	.0000
	1	.9904	.8291	.5491	.1672	.0802	.0353	.0194	.0052	.0005
	2	.9996	.9638	.8160	.3980	.2361	.1268	.0794	.0271	.0037
	3	1.0000	.9946	.9444	.6482	.4613	.2969	.2092	.0905	.0176
	4		.9994	.9873	.8358	.6865	.5255	.4041	.2173	.0592
	5		1.0000	.9978	.9390	.8516	.7216	.6184	.4032	.1509
	6			.9997	.9820	.9434	.8689	.7970	.6098	.3036
	7			1.0000	.9958	.9827	.9500	.9118	.7869	.5000
	8				.9992	.9958	.9848	.9692	.9050	.6964
	9				.9999	.9992	.9964	.9915	.9662	.8491
	10				1.0000	.9999	.9993	.9982	.9907	.9408
	11					1.0000	.9999	.9997	.9981	.9824
	12						1.0000	1.0000	.9997	.9963
	13								1.0000	.9995
	14									1.0000

Source. For $n = 2$ through 10, adapted with permission from E. Parzen, *Modern Probability Theory and Its Applications,* John Wiley, New York, 1962. For $n = 11$ through 15, adapted with permission from *Tables of Cumulative Binomial Probability Distribution,* Harvard University Press, Cambridge, Mass., 1955.

Table 2. Tail Probability Under Standard Normal Distribution

This table gives the probability that the standard normal variable Z will exceed a given positive value z, that is, $P\{Z > z_\alpha\} = \alpha$. The probabilities for negative values of z are obtained by symmetry.

z	.00	.01	.02	.03	.04	.05	.06	.07	.08	.09
.0	.5000	.4960	.4920	.4880	.4840	.4801	.4761	.4721	.4681	.4641
.1	.4602	.4562	.4522	.4483	.4443	.4404	.4364	.4325	.4286	.4247
.2	.4207	.4168	.4129	.4090	.4052	.4013	.3974	.3936	.3897	.3859
.3	.3821	.3783	.3745	.3707	.3669	.3632	.3594	.3557	.3520	.3483
.4	.3446	.3409	.3372	.3336	.3300	.3264	.3228	.3192	.3156	.3121
.5	.3085	.3050	.3015	.2981	.2946	.2912	.2877	.2843	.2810	.2776
.6	.2743	.2709	.2676	.2643	.2611	.2578	.2546	.2514	.2483	.2451
.7	.2420	.2389	.2358	.2327	.2297	.2266	.2231	.2206	.2177	.2148
.8	.2119	.2090	.2061	.2033	.2005	.1977	.1949	.1922	.1984	.1867
.9	.1841	.1814	.1788	.1762	.1736	.1711	.1685	.1660	.1635	.1611
1.0	.1587	.1562	.1539	.1515	.1492	.1469	.1446	.1423	.1401	.1379
1.1	.1357	.1335	.1314	.1292	.1271	.1251	.1230	.1210	.1190	.1170
1.2	.1151	.1131	.1112	.1093	.1075	.1056	.1038	.1020	.1003	.0985
1.3	.0968	.0951	.0934	.0918	.0901	.0885	.0869	.0853	.0838	.0823
1.4	.0808	.0793	.0778	.0764	.0749	.0735	.0721	.0708	.0694	.0681
1.5	.0668	.0655	.0643	.0630	.0618	.0606	.0594	.0582	.0571	.0559
1.6	.0548	.0537	.0526	.0516	.0505	.0495	.0485	.0475	.0465	.0455
1.7	.0446	.0436	.0427	.0418	.0409	.0401	.0392	.0384	.0375	.0367
1.8	.0359	.0351	.0344	.0336	.0329	.0322	.0314	.0307	.0301	.0294
1.9	.0287	.0281	.0274	.0268	.0262	.0256	.0250	.0244	.0239	.0233
2.0	.0228	.0222	.0217	.0212	.0207	.0202	.0197	.0192	.0188	.0183
2.1	.0179	.0174	.0170	.0166	.0162	.0158	.0154	.0150	.0146	.0143
2.2	.0139	.0136	.0132	.0129	.0125	.0122	.0119	.0116	.0113	.0110
2.3	.0107	.0104	.0102	.0099	.0096	.0094	.0091	.0089	.0087	.0084
2.4	.0082	.0080	.0078	.0075	.0073	.0017	.0069	.0068	.0066	.0064
2.5	.0062	.0060	.0059	.0057	.0055	.0054	.0052	.0051	.0049	.0048
2.6	.0047	.0045	.0044	.0043	.0041	.0040	.0039	.0038	.0037	.0036
2.7	.0035	.0034	.0033	.0032	.0031	.0030	.0029	.0028	.0027	.0026
2.8	.0026	.0025	.0024	.0023	.0023	.0022	.0021	.0021	.0020	.0019
2.9	.0019	.0018	.0018	.0017	.0016	.0016	.0015	.0015	.0014	.0014
3.0	.0013	.0013	.0013	.0012	.0012	.0011	.0011	.0011	.0010	.0010

Source. Adapted with permission from P.G. Hoel, *Introduction to Mathematical Statistics*, 4th Ed., John Wiley, New York, 1971, page 391.

Table 3. Critical Values under Chi-Square Distribution

For degrees of freedom greater than 30, the expression $\sqrt{2\chi^2} - \sqrt{(2n-1)}$ may be used as a normal deviate with unit variance, where n is the number of degrees of freedom.

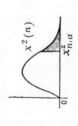

Degrees of freedom	$\alpha=0.99$	0.98	0.95	0.90	0.80	0.70	0.50	0.30	0.20	0.10	0.05	0.02	0.01
1	0.000157	0.000628	0.00393	0.0158	0.0642	0.148	0.455	1.074	1.642	2.706	3.841	5.412	6.635
2	0.0201	0.0404	0.103	0.211	0.446	0.713	1.386	2.408	3.219	4.605	5.991	7.824	9.210
3	0.115	0.185	0.352	0.584	1.005	1.424	2.366	3.665	4.642	6.251	7.815	9.837	11.341
4	0.297	0.429	0.711	1.064	1.649	2.195	3.357	4.878	5.989	7.779	9.488	11.668	13.277
5	0.554	0.752	1.145	1.610	2.343	3.000	4.351	6.064	7.289	9.236	11.070	13.388	15.086
6	0.872	1.134	1.635	2.204	3.070	3.828	5.348	7.231	8.558	10.645	12.592	15.033	16.812
7	1.239	1.564	2.167	2.833	3.822	4.671	6.346	8.383	9.803	12.017	14.067	16.622	18.475
8	1.646	2.032	2.733	3.490	4.594	5.527	7.344	9.524	11.030	13.362	15.507	18.168	20.090
9	2.088	2.532	3.325	4.168	5.380	6.393	8.343	10.656	12.242	14.684	16.919	19.679	21.666
10	2.558	3.059	3.940	4.865	6.179	7.267	9.342	11.781	13.442	15.987	18.307	21.161	23.209
11	3.053	3.609	4.575	5.578	6.989	8.148	10.341	12.899	14.631	17.275	19.675	22.618	24.725
12	3.571	4.178	5.226	6.304	7.807	9.034	11.340	14.011	15.812	18.549	21.026	24.054	26.217
13	4.107	4.765	5.892	7.042	8.634	9.926	12.340	15.119	16.985	19.812	22.362	25.472	27.688
14	4.660	5.368	6.571	7.790	9.467	10.821	13.339	16.222	18.151	21.064	23.685	26.873	29.141
15	5.229	5.985	7.261	8.547	10.307	11.721	14.339	17.322	19.311	22.307	24.996	28.259	30.578
16	5.812	6.614	7.962	9.312	11.152	12.624	15.338	18.418	20.465	23.542	26.296	29.633	32.000
17	6.408	7.255	8.672	10.085	12.002	13.531	16.338	19.511	21.615	24.669	27.587	30.995	33.409

Table 3 (Continued)

18	7.015	7.906	9.390	10.865	12.857	14.440	17.338	20.601	22.760	25.989	28.869	32.346	34.805
19	7.633	8.567	10.117	11.651	13.716	15.352	18.338	21.689	23.900	27.204	30.144	33.687	36.191
20	8.260	9.237	10.851	12.443	14.578	16.266	19.337	22.775	25.038	28.412	31.410	35.020	37.566
21	8.897	9.915	11.591	13.240	15.445	17.182	20.337	23.858	26.171	29.615	32.671	36.343	38.932
22	9.542	10.600	12.338	14.041	16.314	18.101	21.337	24.939	27.301	30.813	33.924	37.659	40.289
23	10.196	11.293	13.091	14.848	17.187	19.021	22.337	26.018	28.429	32.007	35.172	38.968	41.638
24	10.856	11.992	13.848	15.659	18.062	19.943	23.337	27.096	29.553	33.196	36.415	40.270	42.980
25	11.524	12.697	14.611	16.473	18.940	20.867	24.337	28.172	30.675	34.382	37.652	41.566	44.314
26	12.198	13.409	15.379	17.292	19.820	21.792	25.336	29.246	31.795	35.563	38.885	42.856	45.642
27	12.879	14.125	16.151	18.114	20.703	22.719	26.336	30.319	32.912	36.741	40.113	44.140	46.963
28	13.565	14.847	16.928	18.939	21.588	23.647	27.336	31.391	34.027	37.916	41.337	45.419	48.278
29	14.256	15.574	17.708	19.768	22.475	24.577	28.336	32.461	35.139	39.087	42.557	46.693	49.588
30	14.953	16.306	18.493	20.599	23.364	25.508	29.336	33.530	36.250	40.256	43.773	47.962	50.892

Source. Reproduced from *Statistical Methods for Research Workers*, 14th Ed., 1972, with the permission of the Estate of R. A. Fisher, and Hafner Press.

Table 4. Student's t-Distribution

The first column lists the number of degrees of freedom (n). The headings of the other columns give probabilities (α) for t to exceed the entry value. use symmetry for negative t values.

n \ α	.10	.05	0.025	.01	.005
1	3.078	6.314	12.706	31.821	63.657
2	1.886	2.920	4.303	6.965	9.925
3	1.638	2.353	3.182	4.541	5.841
4	1.533	2.132	2.776	3.747	4.604
5	1.476	2.015	2.571	3.365	4.032
6	1.440	1.943	2.447	3.143	3.707
7	1.415	1.895	2.365	2.998	3.499
8	1.397	1.860	2.306	2.896	3.355
9	1.383	1.833	2.262	2.821	3.250
10	1.372	1.812	2.228	2.764	3.169
11	1.363	1.796	2.201	2.718	3.106
12	1.356	1.782	2.179	2.681	3.055
13	1.350	1.771	2.160	2.650	3.012
14	1.345	1.761	2.145	2.624	2.977
15	1.341	1.753	2.131	2.602	2.947
16	1.337	1.746	2.120	2.583	2.921
17	1.333	1.740	2.110	2.567	2.898
18	1.330	1.734	2.101	2.552	2.878
19	1.328	1.729	2.093	2.539	2.861
20	1.325	1.725	2.086	2.528	2.845
21	1.323	1.721	2.080	2.518	2.831
22	1.321	1.717	2.074	2.508	2.819
23	1.319	1.714	2.069	2.500	2.807
24	1.318	1.711	2.064	2.492	2.797
25	1.316	1.708	2.060	2.485	2.787
26	1.315	1.706	2.056	2.479	2.779
27	1.314	1.703	2.052	2.473	2.771
28	1.313	1.701	2.048	2.467	2.763
29	1.311	1.699	2.045	2.462	2.756
30	1.310	1.697	2.042	2.457	2.750
40	1.303	1.684	2.021	2.423	2.704
60	1.296	1.671	2.000	2.390	2.660
120	1.289	1.658	1.980	2.358	2.617
∞	1.282	1.645	1.960	2.326	2.576

Source. P. G. Hoel, *Introduction to Mathematical Statistics,* 4th Ed., John Wiley, New York, 1971, page 393. Reprinted by permission of John Wiley & Sons, Inc.

Table 5. F-Distribution : 5% (Lightface Type) and 1% (Boldface Type) Points for the Distribution of F

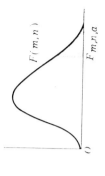

Degrees of freedom for denominator (n)	Degrees of freedom for numerator (m)																					
	1	2	3	4	5	6	7	8	9	10	11	12	14	16	20	24	30	40	50	75	100	200	500	∞
1	161 4052	200 4999	216 5403	225 5625	230 5764	234 5859	237 5928	239 5981	241 6022	242 6056	243 6082	244 6106	245 6142	246 6169	248 6208	249 6234	250 6258	251 6286	252 6302	253 6323	253 6334	254 6352	254 6361	254 6366
2	18.51 98.49	19.00 99.01	19.16 99.17	19.25 99.25	19.30 99.30	19.33 99.33	19.36 99.34	19.37 99.36	19.38 99.38	19.39 99.40	19.40 99.41	19.41 99.42	19.42 99.43	19.43 99.44	19.44 99.45	19.45 99.46	19.46 99.47	19.47 99.48	19.47 99.48	19.48 99.49	19.49 99.49	19.49 99.49	19.50 99.50	19.50 99.50
3	10.13 34.12	9.55 30.81	9.28 29.46	9.12 28.71	9.01 28.24	8.94 27.91	8.88 27.67	8.84 27.49	8.81 27.34	8.78 27.23	8.76 27.13	8.74 27.05	8.71 26.92	8.69 26.83	8.66 26.69	8.64 26.60	8.62 26.50	8.60 26.41	8.58 26.30	8.57 26.27	8.56 26.23	8.54 26.18	8.54 26.14	8.53 26.12
4	7.71 21.20	6.94 18.00	6.59 16.69	6.39 15.98	6.26 15.52	6.16 15.21	6.09 14.98	6.04 14.80	6.00 14.66	5.96 14.54	5.93 14.45	5.91 14.37	5.87 14.24	5.84 14.15	5.80 14.02	5.77 13.93	5.74 13.83	5.71 13.74	5.70 13.69	5.68 13.61	5.66 13.57	5.65 13.52	5.64 13.48	5.63 13.46
5	6.61 16.26	5.79 13.27	5.41 12.06	5.19 11.39	5.05 10.97	4.95 10.67	4.88 10.45	4.82 10.27	4.78 10.15	4.74 10.05	4.70 9.96	4.68 9.89	4.64 9.77	4.60 9.68	4.56 9.55	4.53 9.47	4.50 9.38	4.46 9.29	4.44 9.24	4.42 9.17	4.40 9.13	4.38 9.07	4.37 9.04	4.36 9.02
6	5.99 13.74	5.14 10.92	4.76 9.78	4.53 9.15	4.39 8.75	4.28 8.47	4.21 8.26	4.15 8.10	4.10 7.98	4.06 7.87	4.03 7.79	4.00 7.72	3.96 7.60	3.92 7.52	3.87 7.39	3.84 7.31	3.81 7.23	3.77 7.14	3.75 7.09	3.72 7.02	3.71 6.99	3.69 6.94	3.68 6.90	3.67 6.88
7	5.59 12.25	4.74 9.55	4.35 8.45	4.12 7.85	3.97 7.46	3.87 7.19	3.79 7.00	3.73 6.84	3.68 6.71	3.63 6.62	3.60 6.54	3.57 6.47	3.52 6.35	3.49 6.27	3.44 6.15	3.41 6.07	3.38 5.98	3.34 5.90	3.32 5.85	3.29 5.78	3.28 5.75	3.25 5.70	3.24 5.67	3.23 5.65
8	5.32 11.26	4.46 8.65	4.07 7.59	3.84 7.01	3.69 6.63	3.58 6.37	3.50 6.19	3.44 6.03	3.39 5.91	3.34 5.82	3.31 5.74	3.28 5.67	3.23 5.56	3.20 5.48	3.15 5.36	3.12 5.28	3.08 5.20	3.05 5.11	3.03 5.06	3.00 5.00	2.98 4.96	2.96 4.91	2.94 4.88	2.93 4.86
9	5.12 10.56	4.26 8.02	3.86 6.99	3.63 6.42	3.48 6.06	3.37 5.80	3.29 5.62	3.23 5.47	3.18 5.35	3.13 5.26	3.10 5.18	3.07 5.11	3.02 5.00	2.98 4.92	2.93 4.80	2.90 4.73	2.86 4.64	2.82 4.56	2.80 4.51	2.77 4.45	2.76 4.41	2.73 4.36	2.72 4.33	2.71 4.31

655

Table 5 (Continued)

10	4.96 10.04	4.10 7.56	3.71 6.55	3.48 5.99	3.33 5.64	3.22 5.39	3.14 5.21	3.07 5.06	3.02 4.95	2.97 4.85	2.94 4.78	2.91 4.71	2.86 4.60	2.82 4.52	2.77 4.41	2.74 4.33	2.70 4.25	2.67 4.17	2.64 4.12	2.61 4.05	2.59 4.01	2.56 3.96	2.55 3.93	2.54 3.91
11	4.84 9.65	3.98 7.20	3.59 6.22	3.36 5.67	3.20 5.32	3.09 5.07	3.01 4.88	2.95 4.74	2.90 4.63	2.86 4.54	2.82 4.46	2.79 4.40	2.74 4.29	2.70 4.21	2.65 4.10	2.61 4.02	2.57 3.94	2.53 3.86	2.50 3.80	2.47 3.74	2.45 3.70	2.42 3.66	2.41 3.62	2.40 3.60
12	4.75 9.33	3.88 6.93	3.49 5.95	3.26 5.41	3.11 5.06	3.00 4.82	2.92 4.65	2.85 4.50	2.80 4.39	2.76 4.30	2.72 4.22	2.69 4.16	2.64 4.05	2.60 3.98	2.54 3.86	2.50 3.78	2.46 3.70	2.42 3.61	2.40 3.56	2.36 3.49	2.35 3.46	2.32 3.41	2.31 3.38	2.30 3.36
13	4.67 9.07	3.80 6.70	3.41 5.74	3.18 5.20	3.02 4.86	2.92 4.62	2.84 4.44	2.77 4.30	2.72 4.19	2.67 4.10	2.63 4.02	2.60 3.96	2.55 3.85	2.51 3.78	2.46 3.67	2.42 3.59	2.38 3.51	2.34 3.42	2.32 3.37	2.28 3.30	2.26 3.27	2.24 3.21	2.22 3.18	2.21 3.16
14	4.60 8.86	3.74 6.51	3.34 5.56	3.11 5.03	2.96 4.69	2.85 4.46	2.77 4.28	2.70 4.14	2.65 4.03	2.60 3.94	2.56 3.86	2.53 3.80	2.48 3.70	2.44 3.62	2.39 3.51	2.35 3.43	2.31 3.34	2.27 3.26	2.24 3.21	2.21 3.14	2.19 3.11	2.16 3.06	2.14 3.02	2.13 3.00
15	4.54 8.68	3.68 6.36	3.29 5.42	3.06 4.89	2.90 4.56	2.79 4.32	2.70 4.14	2.64 4.00	2.59 3.89	2.55 3.80	2.51 3.73	2.48 3.67	2.43 3.56	2.39 3.48	2.33 3.36	2.29 3.29	2.25 3.20	2.21 3.12	2.18 3.07	2.15 3.00	2.12 2.97	2.10 2.92	2.08 2.89	2.07 2.87
16	4.49 8.53	3.63 6.23	3.24 5.29	3.01 4.77	2.85 4.44	2.74 4.20	2.66 4.03	2.59 3.89	2.54 3.78	2.49 3.69	2.45 3.61	2.42 3.55	2.37 3.45	2.33 3.37	2.28 3.25	2.24 3.18	2.20 3.10	2.16 3.01	2.13 2.96	2.09 2.89	2.07 2.86	2.04 2.80	2.02 2.77	2.01 2.75
17	4.45 8.40	3.59 6.11	3.20 5.18	2.96 4.67	2.81 4.34	2.70 4.10	2.62 3.93	2.55 3.79	2.50 3.68	2.45 3.59	2.41 3.52	2.38 3.45	2.33 3.35	2.29 3.27	2.23 3.16	2.19 3.08	2.15 3.00	2.11 2.92	2.08 2.86	2.04 2.79	2.02 2.76	1.99 2.70	1.97 2.67	1.96 2.65
18	4.41 8.28	3.55 6.01	3.16 5.09	2.93 4.58	2.77 4.25	2.66 4.01	2.58 3.85	2.51 3.71	2.46 3.60	2.41 3.51	2.37 3.44	2.34 3.37	2.29 3.27	2.25 3.19	2.19 3.07	2.15 3.00	2.11 2.91	2.07 2.83	2.04 2.78	2.00 2.71	1.98 2.68	1.95 2.62	1.93 2.59	1.92 2.57
19	4.38 8.18	3.52 5.93	3.13 5.01	2.90 4.50	2.74 4.17	2.63 3.94	2.55 3.77	2.48 3.63	2.43 3.52	2.38 3.43	2.34 3.36	2.31 3.30	2.26 3.19	2.21 3.12	2.15 3.00	2.11 2.92	2.07 2.84	2.02 2.76	2.00 2.70	1.96 2.63	1.94 2.60	1.91 2.54	1.90 2.51	1.88 2.49
20	4.35 8.10	3.49 5.85	3.10 4.94	2.87 4.43	2.71 4.10	2.60 3.87	2.52 3.71	2.45 3.56	2.40 3.45	2.35 3.37	2.31 3.30	2.28 3.23	2.23 3.13	2.18 3.05	2.12 2.94	2.08 2.86	2.04 2.77	1.99 2.69	1.96 2.63	1.92 2.56	1.90 2.53	1.87 2.47	1.85 2.44	1.84 2.42
21	4.32 8.02	3.47 5.78	3.07 4.87	2.84 4.37	2.68 4.04	2.57 3.81	2.49 3.65	2.42 3.51	2.37 3.40	2.32 3.31	2.28 3.24	2.25 3.17	2.20 3.07	2.15 2.99	2.09 2.88	2.05 2.80	2.00 2.72	1.96 2.63	1.93 2.58	1.89 2.51	1.87 2.47	1.84 2.42	1.82 2.38	1.81 2.36
22	4.30 7.94	3.44 5.72	3.05 4.82	2.82 4.31	2.66 3.99	2.55 3.76	2.47 3.59	2.40 3.45	2.35 3.35	2.30 3.26	2.26 3.18	2.23 3.12	2.18 3.02	2.13 2.94	2.07 2.83	2.03 2.75	1.98 2.67	1.93 2.58	1.91 2.53	1.87 2.46	1.84 2.42	1.81 2.37	1.80 2.33	1.78 2.31
23	4.28 7.88	3.42 5.66	3.03 4.76	2.80 4.26	2.64 3.94	2.53 3.71	2.45 3.54	2.38 3.41	2.32 3.30	2.28 3.21	2.24 3.14	2.20 3.07	2.14 2.97	2.10 2.89	2.04 2.78	2.00 2.70	1.96 2.62	1.91 2.53	1.88 2.48	1.84 2.41	1.82 2.37	1.79 2.32	1.77 2.28	1.76 2.26
24	4.26 7.82	3.40 5.61	3.01 4.72	2.78 4.22	2.62 3.90	2.51 3.67	2.43 3.50	2.36 3.36	2.30 3.25	2.26 3.17	2.22 3.09	2.18 3.03	2.13 2.93	2.09 2.85	2.02 2.74	1.98 2.66	1.94 2.58	1.89 2.49	1.86 2.44	1.82 2.36	1.80 2.33	1.76 2.27	1.74 2.23	1.73 2.21
25	4.24 7.77	3.38 5.57	2.99 4.68	2.76 4.18	2.60 3.86	2.49 3.63	2.41 3.46	2.34 3.32	2.28 3.21	2.24 3.13	2.20 3.05	2.16 2.99	2.11 2.89	2.06 2.81	2.00 2.70	1.96 2.62	1.92 2.54	1.87 2.45	1.84 2.40	1.80 2.32	1.77 2.29	1.74 2.23	1.72 2.19	1.71 2.17

Table 5 (*Continued*)

Degrees of freedom for denominator (n)	Degrees of freedom for numerator (m)																			
	1	2	3	4	5	6	7	8	9	10	11	12	14	16	20	24	30	40	50	75	100	200	500	∞
26	4.22 7.72	3.37 5.53	2.89 4.64	2.74 4.14	2.59 3.82	2.47 3.59	2.39 3.42	2.32 3.29	2.27 3.17	2.22 3.09	2.18 3.02	2.15 2.96	2.10 2.86	2.05 2.77	1.99 2.66	1.95 2.58	1.90 2.50	1.85 2.41	1.82 2.36	1.78 2.28	1.76 2.25	1.72 2.19	1.70 2.15	1.69 2.13
27	4.21 7.68	3.35 5.49	2.96 4.60	2.73 4.11	2.57 3.79	2.46 3.56	2.37 3.39	2.30 3.26	2.25 3.14	2.20 3.06	2.16 2.98	2.13 2.93	2.08 2.83	2.03 2.74	1.97 2.63	1.93 2.55	1.88 2.47	1.84 2.38	1.80 2.33	1.76 2.25	1.74 2.21	1.71 2.16	1.68 2.12	1.67 2.10
28	4.20 7.64	3.34 5.45	2.95 4.57	2.71 4.07	2.56 3.76	2.44 3.53	2.36 3.36	2.29 3.23	3.24 3.11	2.19 3.03	2.15 2.95	2.12 2.90	2.06 2.80	2.02 2.71	1.96 2.60	1.91 2.52	1.87 2.44	1.81 2.35	1.78 2.30	1.75 2.22	1.72 2.18	1.69 2.13	1.67 2.09	1.65 2.06
29	4.18 7.60	3.33 5.52	2.93 4.54	2.70 4.04	2.54 3.73	2.43 3.50	2.35 3.33	2.28 3.20	2.22 3.08	2.18 3.00	2.14 2.92	2.10 2.87	2.05 2.77	2.00 2.68	1.94 2.57	1.90 2.49	1.85 2.41	1.80 2.32	1.77 2.27	1.73 2.19	1.71 2.15	1.68 2.10	1.65 2.06	1.64 2.03
30	4.17 7.56	3.32 5.39	2.92 4.51	2.69 4.02	2.53 3.70	2.42 3.47	2.34 3.30	2.27 3.17	2.21 3.06	2.16 2.98	2.12 2.90	2.09 2.84	2.04 2.74	1.99 2.66	1.93 2.55	1.89 2.47	1.84 2.38	1.79 2.29	1.76 2.24	1.72 2.16	1.69 2.13	1.66 2.07	1.64 2.03	1.62 2.01
32	4.15 7.50	3.30 5.34	2.90 4.46	2.67 3.97	2.51 3.66	2.40 3.42	2.32 3.25	2.25 3.12	2.19 3.01	2.14 2.94	2.10 2.86	2.07 2.80	2.02 2.70	1.97 2.62	1.91 2.51	1.86 2.42	1.82 2.34	1.76 2.25	1.74 2.20	1.69 2.12	1.67 2.08	1.64 2.02	1.61 1.98	1.59 1.96
34	4.13 7.44	3.28 5.29	2.88 4.42	2.65 3.93	2.49 3.61	2.38 3.38	2.30 3.21	2.23 3.08	2.17 2.97	2.12 2.89	2.08 2.82	2.05 2.76	2.00 2.66	1.95 2.58	1.89 2.47	1.84 2.38	1.80 2.30	1.74 2.21	1.71 2.15	1.67 2.08	1.64 2.04	1.61 1.98	1.59 1.94	1.57 1.91
36	4.11 7.39	3.26 5.25	2.86 4.38	2.63 3.89	2.48 3.58	2.36 3.35	2.28 3.18	2.21 3.04	2.15 2.94	2.10 2.86	2.06 2.78	2.03 2.72	1.89 2.62	1.93 2.54	1.87 2.43	1.82 2.35	1.78 2.26	1.72 2.17	1.69 2.12	1.65 2.04	1.62 2.00	1.59 1.94	1.56 1.90	1.55 1.87
38	4.10 7.35	3.25 5.21	2.85 4.34	2.62 3.86	2.46 3.54	2.35 3.32	2.26 3.15	2.19 3.02	2.14 2.91	2.09 2.82	2.05 2.75	2.02 2.69	1.96 2.59	1.92 2.51	1.85 2.30	1.80 2.32	1.76 2.22	1.71 2.14	1.67 2.08	1.63 2.00	1.60 1.97	1.57 1.90	1.54 1.86	1.53 1.84
40	4.08 7.31	3.23 5.18	2.84 4.31	2.61 3.83	2.45 3.51	2.34 3.29	2.25 3.12	2.18 2.99	2.12 2.88	2.07 2.80	2.04 2.73	2.00 2.66	1.95 2.56	1.90 2.49	1.84 2.37	1.79 2.29	1.74 2.20	1.69 2.11	1.66 2.05	1.61 1.97	1.59 1.94	1.55 1.88	1.53 1.84	1.51 1.81
42	4.07 7.27	3.22 5.15	2.83 4.29	2.59 3.80	2.44 3.49	2.32 3.26	2.24 3.10	2.17 2.96	2.11 2.86	2.06 2.77	2.02 2.70	1.99 2.64	1.94 2.54	1.89 2.46	1.82 2.35	1.78 2.26	1.73 2.17	1.68 2.08	1.64 2.02	1.60 1.94	1.57 1.91	1.54 1.85	1.51 1.80	1.49 1.78
44	4.06 7.24	3.21 5.12	2.82 4.26	2.58 3.78	2.43 3.46	2.31 3.24	2.23 3.07	2.16 2.94	2.10 2.84	2.05 2.75	2.01 2.68	1.98 2.62	1.92 2.52	1.88 2.44	1.81 2.32	1.76 2.24	1.72 2.15	1.66 2.06	1.63 2.00	1.58 1.92	1.56 1.88	1.52 1.82	1.50 1.78	1.48 1.75
46	4.05 7.21	3.20 5.10	2.81 4.24	2.57 3.76	2.42 3.44	2.30 3.22	2.22 3.05	2.14 2.92	2.09 2.82	2.04 2.73	2.00 2.66	1.97 2.60	1.91 2.50	1.87 2.42	1.80 2.40	1.75 2.22	1.71 2.13	1.65 2.04	1.62 1.98	1.57 1.90	1.54 1.86	1.51 1.80	1.48 1.76	1.46 1.72
48	4.04 7.19	3.19 5.08	2.80 4.22	2.56 3.74	2.41 3.42	2.30 3.20	2.21 3.04	2.14 2.90	2.08 2.80	2.03 2.71	1.99 2.64	1.96 2.58	1.90 2.48	1.86 2.40	1.79 2.28	1.74 2.20	1.70 2.11	1.64 2.02	1.61 1.96	1.56 1.88	1.53 1.84	1.50 1.78	1.47 1.73	1.45 1.70

Table 5 (Continued)

50	4.03	3.18	2.79	2.56	2.40	2.29	2.20	2.13	2.07	2.02	1.98	1.95	1.90	1.85	1.78	1.74	1.69	1.63	1.60	1.55	1.52	1.48	1.46	1.44
	7.17	5.06	4.20	3.72	3.41	3.18	3.02	2.88	2.78	2.70	2.62	2.56	2.46	2.39	2.26	2.18	2.10	2.00	1.94	1.86	1.82	1.76	1.71	1.68
55	4.02	3.17	2.78	2.54	2.38	2.27	2.18	2.11	2.05	2.00	1.97	1.93	1.88	1.83	1.76	1.72	1.67	1.61	1.58	1.52	1.50	1.46	1.43	1.41
	7.12	5.01	4.16	3.68	3.37	3.15	2.98	2.85	2.75	2.66	2.59	2.53	2.43	2.35	2.23	2.15	2.06	1.96	1.90	1.82	1.78	1.71	1.66	1.64
60	4.00	3.15	2.76	2.52	2.37	2.25	2.17	2.10	2.04	1.99	1.95	1.92	1.86	1.81	1.75	1.70	1.65	1.59	1.56	1.50	1.48	1.44	1.41	1.39
	7.08	4.98	4.13	3.65	3.34	3.12	2.95	2.82	2.72	2.63	2.56	2.50	2.40	2.32	2.20	2.12	2.03	1.93	1.87	1.79	1.74	1.68	1.63	1.60
65	3.99	3.14	2.75	2.51	2.36	2.24	2.15	2.08	2.02	1.98	1.94	1.90	1.85	1.80	1.73	1.68	1.63	1.57	1.54	1.49	1.46	1.42	1.39	1.37
	7.04	4.95	4.10	3.62	3.31	3.09	2.93	2.79	2.70	2.61	2.54	2.47	2.30	2.37	2.18	2.09	2.00	1.90	1.84	1.76	1.71	1.64	1.60	1.56
70	3.98	3.13	2.74	2.50	2.35	2.32	2.14	2.07	2.01	1.97	1.93	1.89	1.84	1.79	1.72	1.67	1.62	1.56	1.53	1.47	1.45	1.40	1.37	1.35
	7.01	4.92	4.08	3.60	3.29	3.07	2.91	2.77	2.67	2.59	2.51	2.45	2.35	2.28	2.15	2.07	1.98	1.88	1.82	1.74	1.69	1.63	1.56	1.53
80	3.96	3.11	2.72	2.48	2.33	2.21	2.12	2.05	1.99	1.95	1.91	1.88	1.82	1.77	1.70	1.65	1.60	1.54	1.51	1.45	1.42	1.38	1.35	1.32
	6.96	4.88	4.04	3.56	3.25	3.04	2.87	2.74	2.64	2.55	2.48	2.41	2.32	2.24	2.11	2.03	1.94	1.84	1.78	1.70	1.65	1.57	1.52	1.49
100	3.94	3.09	2.70	2.46	2.30	2.19	2.10	2.03	1.97	1.92	1.88	1.85	1.79	1.75	1.68	1.63	1.57	1.51	1.48	1.42	1.39	1.34	1.30	1.28
	6.90	4.82	3.98	3.51	3.20	2.99	2.82	2.69	2.59	2.51	2.43	2.36	2.26	2.19	2.06	1.98	1.89	1.79	1.73	1.64	1.59	1.51	1.46	1.43
125	3.92	3.07	2.68	2.44	2.29	2.17	2.08	2.01	1.95	1.90	1.86	1.83	1.77	1.72	1.65	1.60	1.55	1.49	1.45	1.39	1.36	1.31	1.27	1.25
	6.84	4.78	3.94	3.47	3.17	2.95	2.79	2.65	2.56	2.47	2.40	2.33	2.23	2.15	2.03	1.94	1.85	1.75	1.68	1.59	1.54	1.46	1.40	1.37
150	3.91	3.06	2.67	2.43	2.27	2.16	2.07	2.00	1.94	1.89	1.85	1.82	1.76	1.71	1.64	1.59	1.54	1.47	1.44	1.37	1.34	1.29	1.25	1.22
	6.81	4.75	3.91	3.44	3.13	2.92	2.76	2.62	2.53	2.44	2.37	2.30	2.20	2.12	2.00	1.91	1.83	1.72	1.66	1.56	1.51	1.43	1.37	1.33
200	3.89	3.04	2.65	2.41	2.26	2.14	2.05	1.98	1.92	1.87	1.83	1.80	1.74	1.69	1.62	1.57	1.52	1.45	1.42	1.35	1.32	1.26	1.22	1.19
	6.76	4.71	3.88	3.41	3.11	2.90	2.73	2.60	2.50	2.41	2.34	2.28	1.17	2.09	1.97	1.88	1.79	1.69	1.62	1.53	1.48	1.39	1.33	1.28
400	3.86	3.02	2.62	2.39	2.23	2.12	2.03	1.96	1.90	1.85	1.81	1.78	1.72	1.67	1.60	1.54	1.49	1.42	1.38	1.32	1.28	1.22	1.16	1.13
	6.70	4.66	3.83	3.36	3.06	2.85	2.69	2.55	2.46	2.37	2.29	2.23	2.12	2.04	1.92	1.84	1.74	1.64	1.57	1.47	1.42	1.32	1.24	1.19
1000	3.85	3.00	2.61	2.38	2.22	2.10	2.02	1.95	1.89	1.84	1.80	1.76	1.75	1.65	1.58	1.53	1.47	1.41	1.36	1.30	1.26	1.19	1.13	1.08
	6.66	4.62	3.80	3.34	3.04	2.82	2.66	2.53	2.43	2.34	2.26	2.20	2.09	2.01	1.89	1.81	1.71	1.61	1.54	1.44	1.38	1.28	1.19	1.11
	3.84	2.99	2.60	2.37	2.21	2.09	2.01	1.94	1.88	1.83	1.79	1.75	1.69	1.64	1.57	1.52	1.46	1.40	1.35	1.28	1.24	1.17	1.11	1.00
	6.64	4.60	3.78	3.32	3.02	2.80	2.64	2.51	2.41	2.32	2.24	2.18	2.07	1.99	1.87	1.79	1.69	1.59	1.52	1.41	1.36	1.25	1.15	1.00

Source. Reprinted by permission from George W. Snedecor and William G. Cochran, *Statistical Methods*, 6th Ed., (c) 1967 by Iowa State University Press, Ames, Iowa.

Table 6. Random Normal Numbers, $\mu = 0$ and $\sigma = 1$

01	02	03	04	05	06	07	08	09	10
0.464	0.137	2.455	−0.323	−0.068	0.290	−0.288	1.298	0.241	−0.957
0.060	−2.526	−0.531	−0.194	0.543	−1.558	0.187	−1.190	0.022	0.525
1.486	−0.354	−0.634	0.697	0.926	1.375	0.785	−0.963	−0.853	−1.865
1.022	−0.472	1.279	3.521	0.571	−1.851	0.194	1.192	−0.501	−0.273
1.394	−0.555	0.046	0.321	2.945	1.974	−0.258	0.412	0.439	−0.035
0.906	−0.513	−0.525	0.595	0.881	−0.934	1.579	0.161	−1.885	0.371
1.179	−1.055	0.007	0.769	0.971	0.712	1.090	−0.631	−0.255	−0.702
−1.501	−0.488	−0.162	−0.136	1.033	0.203	0.448	0.748	−0.423	−0.432
−0.690	0.756	−1.618	−0.345	−0.511	−2.051	−0.457	−0.218	0.857	−0.465
1.372	0.225	0.378	0.761	0.181	−0.736	0.960	−1.530	−0.260	0.120
−0.482	1.678	−0.057	−1.229	−0.486	0.856	−0.491	−1.983	−2.830	−0.238
−1.376	−0.150	1.356	−0.561	−0.256	−0.212	0.219	0.779	0.953	−0.869
−1.010	0.598	−0.918	1.598	0.065	0.415	−0.169	0.313	−0.973	−1.016
−0.005	−0.899	0.012	−0.725	1.147	−0.121	1.096	0.481	−1.691	0.417
1.393	1.163	−0.911	1.231	−0.199	−0.246	1.239	−2.574	−0.558	0.056
−1.787	−0.261	1.237	1.046	−0.508	−1.630	−0.146	−0.392	−0.627	0.561
−0.105	−0.357	−1.384	0.360	−0.992	−0.116	−1.698	−2.832	−1.108	−2.357
−1.339	1.827	−0.959	0.424	0.969	−1.141	−1.041	0.362	−1.726	1.956
1.041	0.535	0.731	1.377	0.983	−1.330	1.620	−1.040	0.524	−0.281
0.279	−2.056	0.717	−0.873	−1.096	−1.396	1.047	0.089	−0.573	0.932
−1.805	−2.008	−1.633	0.542	0.250	−0.166	0.032	0.079	0.471	−1.029
−1.186	1.180	1.114	0.882	1.265	−0.202	0.151	−0.376	−0.310	0.479
0.658	−1.141	1.151	−1.210	−0.927	0.425	0.290	−0.902	0.610	2.709
−0.439	0.358	−1.939	0.891	−0.227	0.602	0.873	−0.437	−0.220	−0.057
−1.399	−0.230	0.385	−0.649	−0.577	0.237	−0.289	0.513	0.738	−0.300
0.199	0.208	−1.083	−0.219	−0.291	1.221	1.119	0.004	−2.015	−0.594
0.159	0.272	−0.313	0.084	−2.828	−0.430	−0.792	−1.275	−0.623	−1.047
2.273	0.606	0.606	−0.747	0.247	1.291	0.063	−1.793	−0.699	−1.347
0.041	−0.307	0.121	0.790	−0.584	0.541	0.484	−0.986	0.481	0.996
−1.132	−2.098	0.921	0.145	0.446	−1.661	1.045	−1.363	−0.586	−1.023
0.768	0.079	−1.473	0.034	−2.127	0.665	0.084	−0.880	−0.579	0.551
0.375	−1.658	−0.851	0.234	−0.656	0.340	−0.086	−0.158	−0.120	0.418
−0.513	−0.344	0.210	−0.736	1.041	0.008	0.427	−0.831	0.191	0.074
0.292	−0.521	1.266	−1.206	−0.899	0.110	−0.528	−0.813	0.071	0.524
1.026	2.990	−0.574	−0.491	−1.114	1.297	−1.433	−1.345	−3.001	0.479
−1.334	1.278	−0.568	−0.109	−0.515	−0.566	2.923	0.500	0.359	0.326
−0.287	−0.144	−0.254	0.574	−0.451	−1.181	−1.190	−0.318	−0.094	1.114
0.161	−0.886	−0.921	−0.509	1.410	−0.518	0.192	−0.432	1.501	1.068
−1.346	0.193	−1.202	0.394	−1.045	0.843	0.942	1.045	0.031	0.772
1.250	−0.199	−0.288	1.810	1.378	0.584	1.216	0.733	0.402	0.226
0.630	−0.537	0.782	0.060	0.499	−0.431	1.705	1.164	0.884	−0.298
0.375	−1.941	0.247	−0.491	0.665	−0.135	−0.145	−0.498	0.457	1.064

Table 6 (*Continued*)

−1.420	0.489	−1.711	−1.186	0.754	−0.732	−0.066	1.006	−0.798	0.162
−0.151	−0.243	−0.430	−0.762	0.298	1.049	1.810	2.885	−0.768	−0.129
−0.309	0.531	0.416	−1.541	1.456	2.040	−0.124	0.196	0.023	−1.204
0.424	−0.444	0.593	0.993	−0.106	0.116	0.484	−1.272	1.066	1.097
0.593	0.658	−1.127	−1.407	−1.579	−1.616	1.458	1.262	0.736	−0.916
0.862	−0.885	−0.142	−0.504	0.532	1.381	0.022	−0.281	−0.342	1.222
0.235	−0.628	−0.023	−0.463	−0.899	−0.394	−0.538	1.707	−0.188	−1.153
−0.853	0.402	0.777	0.833	0.410	−0.349	−1.094	0.580	1.395	1.298

Source. From tables of the RAND Corporation, by permission.

Table 7. Critical Values of the Kolmogorov-Smirnov One Sample Test Statistics

This table gives the values of $D_{n,\alpha}^+$ and $D_{n,\alpha}$ for which $\alpha \geq P\{D_n^+ > D_{n,\alpha}^+\}$ and $\alpha \geq P\{D_n > D_{n,\alpha}\}$ for some selected values of n and α.

One-Sided Test: $\alpha =$	.10	.05	.025	.01	.005	$\alpha =$	.10	.05	.025	.01	.005
Two-Sided Test: $\alpha =$	.20	.10	.05	.02	.01	$\alpha =$	.20	.10	.05	.02	.01
$n=1$	.900	.950	.975	.990	.995	$n=21$	.226	.259	.287	.321	.344
2	.684	.776	.842	.900	.929	22	.221	.253	.281	.314	.337
3	.565	.636	.708	.785	.829	23	.216	.247	.275	.307	.330
4	.493	.565	.624	.689	.734	24	.212	.242	.269	.301	.323
5	.447	.509	.563	.627	.669	25	.208	.238	.264	.295	.317
6	.410	.468	.519	.577	.617	26	.204	.233	.259	.290	.311
7	.381	.436	.483	.538	.576	27	.200	.229	.254	.284	.305
8	.358	.410	.454	.507	.542	28	.197	.225	.250	.279	.300
9	.339	.387	.430	.480	.513	29	.193	.221	.246	.275	.295
10	.323	.369	.409	.457	.489	30	.190	.218	.242	.270	.290
11	.308	.352	.391	.437	.468	31	.187	.214	.238	.266	.285
12	.296	.338	.375	.419	.449	32	.184	.211	.234	.262	.281
13	.285	.325	.361	.404	.432	33	.182	.208	.231	.258	.277
14	.275	.314	.349	.390	.418	34	.179	.205	.227	.254	.273
15	.266	.304	.338	.377	.404	35	.177	.202	.224	.251	.269
16	.258	.295	.327	.366	.392	36	.174	.199	.221	.247	.265
17	.250	.286	.318	.355	.381	37	.172	.196	.218	.244	.262
18	.244	.279	.309	.346	.371	38	.170	.194	.215	.241	.258
19	.237	.271	.301	.337	.361	39	.168	.191	.213	.238	.255
20	.232	.265	.294	.329	.352	40	.165	.189	.210	.235	.252
Approximation for $n > 40$							$\dfrac{1.07}{\sqrt{n}}$	$\dfrac{1.22}{\sqrt{n}}$	$\dfrac{1.36}{\sqrt{n}}$	$\dfrac{1.52}{\sqrt{n}}$	$\dfrac{1.63}{\sqrt{n}}$

Source. Adapted by permission from Table 1 of Leslie H. Miller, Table of Percentage points of Kolmogorov statistics, *J. Am. Stat. Assoc.* 51 (1956), 111–121.

Table 8. Critical Values of the Kolmogorov-Smirnov Test Statistic for Two Samples of Equal Size

This table gives the values of $D^+_{n,n,\alpha}$ and $D_{n,n,\alpha}$ for which $\alpha \geq P\{D^+_{n,n} > D^+_{n,n,\alpha}\}$ and $\alpha \geq P\{D_{n,n} > D_{n,n,\alpha}\}$ for some selected values of n and α.

One-Sided Test:

$\alpha =$	.10	.05	.025	.01	.005	$\alpha =$	.10	.05	.025	.01	.005

Two-Sided Test:

$\alpha =$	.20	.10	.05	.02	.01	$\alpha =$	.20	.10	.05	.02	.01
$n = 3$	2/3	2/3				$n = 20$	6/20	7/20	8/20	9/20	10/20
4	3/4	3/4	3/4			21	6/21	7/21	8/21	9/21	10/21
5	3/5	3/5	4/5	4/5	4/5	22	7/22	8/22	8/22	10/22	10/22
6	3/6	4/6	4/6	5/6	5/6	23	7/23	8/23	9/23	10/23	10/23
7	4/7	4/7	5/7	5/7	5/7	24	7/24	8/24	9/24	10/24	11/24
8	4/8	4/8	5/8	5/8	6/8	25	7/25	8/25	9/25	10/25	11/25
9	4/9	5/9	5/9	6/9	6/9	26	7/26	8/26	9/26	10/26	11/26
10	4/10	5/10	6/10	6/10	7/10	27	7/27	8/27	9/27	11/27	11/27
11	5/11	5/11	6/11	7/11	7/11	28	8/28	9/28	10/28	11/28	12/28
12	5/12	5/12	6/12	7/12	7/12	29	8/29	9/29	10/29	11/29	12/29
13	5/13	6/13	6/13	7/13	8/13	30	8/30	9/30	10/30	11/30	12/30
14	5/14	6/14	7/14	7/14	8/14	31	8/31	9/31	10/31	11/31	12/31
15	5/15	6/15	7/15	8/15	8/15	32	8/32	9/32	10/32	12/32	12/32
16	6/16	6/16	7/16	8/16	9/16	34	8/34	10/34	11/34	12/34	13/34
17	6/17	7/17	7/17	8/17	9/17	36	9/36	10/36	11/36	12/36	13/36
18	6/18	7/18	8/18	9/18	9/18	38	9/38	10/38	11/38	13/38	14/38
19	6/19	7/19	8/19	9/19	9/19	40	9/40	10/40	12/40	13/40	14/40
Approximation for $n > 40$:							$\dfrac{1.52}{\sqrt{n}}$	$\dfrac{1.73}{\sqrt{n}}$	$\dfrac{1.92}{\sqrt{n}}$	$\dfrac{2.15}{\sqrt{n}}$	$\dfrac{2.30}{\sqrt{n}}$

Source. Adapted by permission from Tables 2 and 3 of Z. W. Birnbaum and R. A. Hall, Small sample distributions for multisample statistics of the Smirnov type, *Ann. Math. Stat.* 31 (1960), 710–720.

Table 9. Critical Values of the Kolmogorov-Smirnov Test Statistic for Two Samples of Unequal Size

This table gives the values of $D^+_{m,n,\alpha}$ and $D_{m,n,\alpha}$ for which $\alpha \geq P\{D^+_{m,n} > D^+_{m,n,\alpha}\}$ and $\alpha \geq P\{D_{m,n} > D_{m,n,\alpha}\}$ for some selected values of $N_1 =$ smaller sample size, $N_2 =$ larger sample size, and α.

One-Sided Test:		$\alpha =$	.10	.05	.025	.01	.005
Two-Sided Test:		$\alpha =$	.20	.10	.05	.02	.01
$N_1 = 1$	$N_2 = 9$		17/18				
	10		9/10				
$N_1 = 2$	$N_2 = 3$		5/6				
	4		3/4				
	5		4/5	4/5			
	6		5/6	5/6			
	7		5/7	6/7			
	8		3/4	7/8	7/8		
	9		7/9	8/9	8/9		
	10		7/10	4/5	9/10		
$N_1 = 3$	$N_2 = 4$		3/4	3/4			
	5		2/3	4/5	4/5		
	6		2/3	2/3	5/6		
	7		2/3	5/7	6/7	6/7	
	8		5/8	3/4	3/4	7/8	
	9		2/3	2/3	7/9	8/9	8/9
	10		3/5	7/10	4/5	9/10	9/10
	12		7/12	2/3	3/4	5/6	11/12
$N_1 = 4$	$N_2 = 5$		3/5	3/4	4/5	4/5	
	6		7/12	2/3	3/4	5/6	5/6
	7		17/28	5/7	3/4	6/7	6/7
	8		5/8	5/8	3/4	7/8	7/8
	9		5/9	2/3	3/4	7/9	8/9
	10		11/20	13/20	7/10	4/5	4/5
	12		7/12	2/3	2/3	3/4	5/6
	16		9/16	5/8	11/16	3/4	13/16
$N_1 = 5$	$N_2 = 6$		3/5	2/3	2/3	5/6	5/6
	7		4/7	23/35	5/7	29/35	6/7
	8		11/20	5/8	27/40	4/5	4/5
	9		5/9	3/5	31/45	7/9	4/5
	10		1/2	3/5	7/10	7/10	4/5
	15		8/15	3/5	2/3	11/15	11/15
	20		1/2	11/20	3/5	7/10	3/4
$N_1 = 6$	$N_2 = 7$		23/42	4/7	29/42	5/7	5/6
	8		1/2	7/12	2/3	3/4	3/4
	9		1/2	5/9	2/3	13/18	7/9

Table 9 (*Continued*)

		10	1/2	17/30	19/30	7/10	11/15
		12	1/2	7/12	7/12	2/3	3/4
		18	4/9	5/9	11/18	2/3	13/18
		24	11/24	1/2	7/12	5/8	2/3
$N_1 = 7$	$N_2 =$	8	27/56	33/56	5/8	41/56	3/4
		9	31/63	5/9	40/63	5/7	47/63
		10	33/70	39/70	43/70	7/10	5/7
		14	3/7	1/2	4/7	9/14	5/7
		28	3/7	13/28	15/28	17/28	9/14
$N_1 = 8$	$N_2 =$	9	4/9	13/24	5/8	2/3	3/4
		10	19/40	21/40	23/40	27/40	7/10
		12	11/24	1/2	7/12	5/8	2/3
		16	7/16	1/2	9/16	5/8	5/8
		32	13/32	7/16	1/2	9/16	19/32
$N_1 = 9$	$N_2 =$	10	7/15	1/2	26/45	2/3	31/45
		12	4/9	1/2	5/9	11/18	2/3
		15	19/45	22/45	8/15	3/5	29/45
		18	7/18	4/9	1/2	5/9	11/18
		36	13/36	5/12	17/36	19/36	5/9
$N_1 = 10$	$N_2 =$	15	2/5	7/15	1/2	17/30	19/30
		20	2/5	9/20	1/2	11/20	3/5
		40	7/20	2/5	9/20	1/2	
$N_1 = 12$	$N_2 =$	15	23/60	9/20	1/2	11/20	7/12
		16	3/8	7/16	23/48	13/24	7/12
		18	13/36	5/12	17/36	19/36	5/9
		20	11/30	5/12	7/15	31/60	17/30
$N_1 = 15$	$N_2 =$	20	7/20	2/5	13/30	29/60	31/60
$N_1 = 16$	$N_2 =$	20	27/80	31/80	17/40	19/40	41/80
Large-sample approximation			$1.07\sqrt{\frac{m+n}{mn}}$	$1.22\sqrt{\frac{m+n}{mn}}$	$1.36\sqrt{\frac{m+n}{mn}}$	$1.52\sqrt{\frac{m+n}{mn}}$	$1.63\sqrt{\frac{m+n}{mn}}$

Source. Adapted by permission from F. J. Massey, Distribution table for the deviation between two sample cumulatives, *Ann. Math. Stat.* 23 (1952), 435–441.

Table 10. Critical Values of the Wilcoxon Signed Ranks Test Statistic

This table gives values of t_α for which $P\{T^+ > t_\alpha\} \leq \alpha$ for selected values of n and α. Critical values in the lower tail may be obtained by symmetry from the equation $t_{1-\alpha} = n(n+1)/2 - t_\alpha$.

		α		
n	.01	.025	.05	.10
3	6	6	6	6
4	10	10	10	9
5	15	15	14	12
6	21	20	18	17
7	27	25	24	22
8	34	32	30	27
9	41	39	36	34
10	49	46	44	40
11	58	55	52	48
12	67	64	60	56
13	78	73	69	64
14	89	84	79	73
15	100	94	89	83
16	112	106	100	93
17	125	118	111	104
18	138	130	123	115
19	152	143	136	127
20	166	157	149	140

Source. Adapted by permission from Table 1 of R. L. McCornack, Extended tables of the Wilcoxon matched pairs signed-rank statistics, *J. Am. Stat. Assoc.* 60 (1965), 864–871.

Table 11. Critical Values of the Mann-Whitney-Wilcoxon Test Statistic

This table gives the values of u_α for which $P\{U > u_\alpha\} \leq \alpha$ for some selected values of m, n, and α. Critical values in the lower tail may be obtained by symmetry from the equation $u_{1-\alpha} = mn - u_\alpha$.

						n				
m	α	2	3	4	5	6	7	8	9	10
2	.01	4	6	8	10	12	14	16	18	20
	.025	4	6	8	10	12	14	15	17	19
	.05	4	6	8	9	11	13	14	16	18
	.10	4	5	7	8	10	12	13	15	16
3	.01		9	12	15	18	20	20	25	28
	.025		9	12	14	16	19	21	24	26
	.05		8	11	13	15	18	20	22	25
	.10		7	10	12	14	16	18	21	23
4	.01			16	19	22	26	29	32	36
	.025			15	18	21	24	27	31	34
	.05			14	17	20	23	26	29	32
	.10			12	15	18	21	24	26	29
5	.01				23	27	31	35	39	43
	.025				22	26	29	33	37	41
	.05				20	24	28	31	35	38
	.10				19	22	26	29	32	36
6	.01					32	37	41	46	51
	.025					30	35	39	43	48
	.05					28	33	37	41	45
	.10					26	30	34	38	42
7	.01						42	48	53	58
	.025						40	45	50	55
	.05						37	42	47	52
	.10						35	39	44	48
8	.01							54	60	66
	.025							50	56	62
	.05							48	53	59
	.10							44	49	55
9	.01								66	73
	.025								63	69
	.05								59	65
	.10								55	61
10	.01									80
	.025									76
	.05									72
	.10									67

Source. Adapted by permission from Table 1 of L. R. Verdooren, Extended tables of critical values for Wilcoxon's test statistic, *Biometrika* 50 (1963), 177–186, with the kind permission of Professor E. S. Pearson, the author, and the *Biometrika* Trustees.

Table 12. Critical Points of Kendall's Tau Test Statistic

This table gives the values of S_α for which $P\{S > S_\alpha\} \leq \alpha$, where $S = \binom{n}{2}T$, for some selected values of α and n. Values in the lower tail may be obtained by symmetry, $S_{1-\alpha} = -S_\alpha$.

		α		
n	.100	.050	.025	.01
3	3	3	3	3
4	4	4	6	6
5	6	6	8	8
6	7	9	11	11
7	9	11	13	15
8	10	14	16	18
9	12	16	18	22
10	15	19	21	25

Source. Adapted by permission from Table 1, page 173, of M. G. Kendall, *Rank Correlation Methods*, 3rd Ed., Griffin, London, 1962. For values of $n \geq 11$, see W. J. Conover, *Practical Nonparametric Statistics*, John Wiley, New York, 1971, page 390.

Table 13. Critical Values of Spearman's Rank Correlation Statistic

This table gives the values of R_α such that $P\{R > R_\alpha\} \leq \alpha$ for some selected values of n and α. Critical values in the lower tail may be obtained by symmetry, $R_{1-\alpha} = -R_\alpha$.

		α		
n	.01	.025	.05	.10
3	1.000	1.000	1.000	1.000
4	1.000	1.000	.800	.800
5	.900	.900	.800	.700
6	.886	.829	.771	.600
7	.857	.750	.679	.536
8	.810	.714	.619	.500
9	.767	.667	.583	.467
10	.721	.636	.552	.442

Source. Adapted by permission from Table 2, Pages 174–175, of M. G. Kendall, *Rank Correlation Methods*, 3rd Ed., Griffin, London, 1962. For values of $n \geq 11$, see W. J. Conover, *Practical Nonparametric Statistics*, John Wiley, New York, 1971, page 391.

Glossary of Some Frequently Used Symbols and Abbreviations

$\Rightarrow$	implies
$\Leftrightarrow$	implies and is implied by
$\rightarrow$	converges to
$\uparrow, \downarrow$	increasing, decreasing
$\Uparrow, \Downarrow$	nonincreasing, nondecreasing
$\Gamma(x)$	gamma function
$\overline{\lim}, \underline{\lim}, \lim$	limit superior, limit inferior, limit
$\mathscr{R}, \mathscr{R}_n$	real line, n-dimensional Euclidean space
$\mathfrak{B}, \mathfrak{B}_n$	Borel σ-field on $\mathscr{R}$, Borel σ-field on $\mathscr{R}_n$
I_A	indicator function of set A
$\varepsilon(x)$	$= 1$ if $x \geq 0$, and $= 0$ if $x < 0$
μ	EX, expected value
m_n	EX^n, $n \geq 0$ integral
β_α	$E|X|^\alpha$, $\alpha > 0$
μ_k	$E(X - EX)^k$, $k \geq 0$ integral
σ^2	$= \mu_2$, variance
f', f'', f'''	first, second, third derivative of f
$\sim$	distributed as
$\approx$	asymptotically (or approximately) equal to
$\xrightarrow{L}$	convergence in law
$\xrightarrow{P}$	convergence in probability
$\xrightarrow{a.s.}$	convergence almost surely
$\xrightarrow{r}$	convergence in rth mean
rv	random variable
df	distribution function
pdf	probability density function
pmf	probability mass function
pgf	probability generating function

GLOSSARY

mgf	moment generating function
d.f.	degress of freedom
BLUE	best linear unbiased estimate
MLE	maximum likelihood estimate
MVUE	minimum variance unbiased estimate
UMA	uniformly most accurate
UMVUE	uniformly minimum variance unbiased estimate
UMAU	uniformly most accurate unbiased
MP	most powerful
UMP	uniformly most powerful
i.o.	infinitely often
iid	independent, identically distributed
SD	standard deviation
MLR	monotone likelihood ratio
MSE	mean square error
WLLN	weak law of large numbers
SLLN	strong law of large numbers
CLT	central limit theorem
SPRT	sequential probability ratio test
$b(1, p)$	Bernoulli with parameter p
$b(n, p)$	binomial with parameters n, p
$NB(r; p)$	negative binomial with parameters r, p
$P(\lambda)$	Poisson with parameter λ
$U[a, b]$	uniform on $[a, b]$
$G(\alpha, \beta)$	gamma with parameters α, β
$B(\alpha, \beta)$	beta with parameters α, β
$\chi^2(n)$	chi-square with d. f. n
$\mathscr{C}(\mu, \theta)$	Cauchy with parameters μ, θ
$\mathscr{N}(\mu, \sigma^2)$	normal with mean μ, variance σ^2
$t(n)$	Student's t with n d. f.
$F(m, n)$	F-distribution with (m, n) d.f.
z_α	$100(1 - \alpha)$th percentile of $\mathscr{N}(0, 1)$
$\chi^2_{n,\alpha}$	$100(1 - \alpha)$th percentile of $\chi^2(n)$
$t_{n,\alpha}$	$100(1 - \alpha)$th percentile of $t(n)$
$F_{m,n,\alpha}$	$100(1 - \alpha)$th percentile of $F(m, n)$

Author Index

Anderson, 387, 631, 641
Apostol, 5, 8-10, 13, 86, 641
Auble, 562, 641

Bahadur, 633, 641
Baswa, 186, 199, 645
Bernstein, 223, 641
Bertrand, 18
Birnbaum, 540, 641, 662
Blackwell, 355, 641
Boas, 360, 641

Chapman, 365, 369, 641
Chatterji, 190, 196, 201, 641
Chow, 611, 641
Chung, 59, 66, 167, 641
Cochran, 658
Conover, 667
Courant, 3-9, 11, 641
Cramér, 222, 253, 309, 327, 331, 361, 373, 384, 385, 446-448, 587, 641, 642
Curtiss, 95, 277, 642

Dantzig, 606, 642
Darmois, 223, 642
Desu, 209, 485, 642
Dharamadhikari, 97, 642

Enis, 170, 642
Erdös, 167, 641

Feller, 87, 116, 180, 222, 283, 288, 291, 642
Ferguson, 414, 428, 429, 642
Fisher, 326, 642, 653
Fisz, 116, 300, 642
Fomin, 21, 643
Fraser, 342, 347, 531, 535, 537, 545, 573, 642
Fréchet, 361, 642
Friedman, 631, 641

Ghurye, 611, 642
Gibbons, 540, 642
Glivenko, 300, 642
Gnedenko, 257, 642
Govindarajulu, 581, 642
Guenther, 479, 492, 642
Gumbel, 119, 642

Hahn, 547, 642
Hall, 662
Halmos, 3, 21, 342, 643
Hamburger, 97, 643
Harris, 402, 643
Hodges, 394, 396, 643

Hoel, 651, 654
Hogben, 307, 643
Huber, 581, 643
Huzurbazar, 386, 643

Jamison, 267, 268, 643
John, 3-9, 11, 641
Johnson, 597, 643

Kac, 48, 643
Kemp, 226, 643
Kemperman, 635, 640, 643
Kendall, 569, 570, 572, 643, 667
Kiefer, 365, 643
Klamkin, 225, 643
Klett, 483, 493, 646
Kolmogorov, 19, 21, 540, 643
Kraft, 549, 643
Kruskal, 166, 196, 643
Kullback, 126, 643
Kulldorf, 385, 643

Laha, 235, 644
Lamperti, 196, 225, 643
Laplace, 18
Lehmann, 238, 342, 347, 356, 394, 396,
  400, 402, 414, 422, 427, 428-430, 434,
  445, 446, 505, 535, 537, 643
LePage, 100, 645
Leslie, 581, 642
Loève, 81, 280, 644
Lukacs, 104, 209, 224, 235, 277, 283, 644

Mann, 537, 562, 563, 644
Massey, 558, 644, 664
McCornack, 665
Menon, 218, 644
Miller, K. S., 12, 644
Miller, L. H., 540, 644, 661
Mitra, 359, 645
Moore, 100, 645
Moshman, 609, 644

Natrella, 450, 451, 456, 457, 644
Nelson, 547, 642
Neyman, 387, 412, 644

Oliver, 380, 644
Orey, 267, 268, 643
Owen, 540, 541, 644

Parzen, 650
Pearson, 412, 644
Pitman, 219, 644
Pratt, 489, 490, 644
Prochaska, 227, 256, 644
Pruitt, 267, 268, 643
Puri, 202, 644

Raikov, 195, 644
Ramachandaran, 215, 644
Rao, 246, 355, 361, 644, 645
Rényi, 283, 645
Richter, 611, 645
Robbins, 365, 369, 602, 611, 641, 642,
  645
Rohatgi, 291, 645
Romanovsky, 326, 645
Rosenberg, 230, 645
Rowland, 126, 645
Roy, J., 359, 645
Roy, R., 100, 645
Royden, 14, 21, 645

Sabharwal, 182, 645
Samuelson, 299, 645
Savage, 342, 643
Scheffé, 15-17, 243, 356, 400, 402, 505,
  587, 588, 643, 645
Scott, 387, 644
Seelbinder, 609, 645
Shanbhag, 186, 199, 645
Shepp, 226, 645
Siegel, 564, 645
Simons, 611, 645
Skibinsky, 194, 645
Skitovitch, 223, 645
Smirnov, 541, 558, 646
Snedecor, 658
Soms, 402, 643
Srivastava, 201, 646
Starr, 605, 610, 646
Stein, 606, 609, 620, 646
Stigler, 357, 646
Strait, 51, 646

Tate, 483, 493, 646
Thompson, 218, 646
Tingey, 541, 641
Trotter, 283, 646
Tucker, 107, 646

Van Eeden, 549, 643
Verdooren, 666

Wald, 386, 597, 613, 623, 631, 633, 646
Whitney, 537, 562, 563, 644
Widder, 13, 14, 95, 96, 162, 217, 646
Wijsman, 371, 631, 646
Wilks, 442, 484, 646
Williams, 219, 644

Wishart, 326, 646
Wolfowitz, 599, 631, 646
Wong, 48, 647
Woodroofe, 605, 646

Yasaimaibodi, 126, 643

Zacks, 611, 645, 647
Zehna, 383, 647

Subject Index

Absolute continuity, 14
Absolutely continuous df, 63, 66
Actions, set of, 388
Analysis of variance, 501
   one-way, 513-518
   two-way, 518-523
   two-way with interaction, 524-529
Asymptotic distribution, of rth order-statistic, 310
   of sample moments, 309
   of sample quantile, 310

Banach's matchbox problem, 188
Bernoulli random variable, 183
Bernoulli trials, 182
Bertrand's paradox, 19, 33
Beta distribution, 213
   bivariate, 118
   mgf, 214
   moments, 213-214
   partial characterization, 215
Bias of an estimate, 350
Binomial, coefficient, 7, 38
   series, 8
Binomial distribution, 132, 184
   bounds for tail probability, 200
   central term, 200
   characterization, 185-186
   coefficient of skewness, 93
   generalized to multinomial, 197
   kurtosis, 93
   as limit of hypergeometric, 201
   mean, 184
   mgf, 185
   pgf, 185
   tail probability as incomplete beta function, 216
   variance, 184
Bonferroni's inequality, 27
Boole's inequality, 28
Borel-Cantelli lemma, 264, 274
Borel-measurable functions, 16, 68
   of an rv, 68, 131
Borel set, 3
Borel σ-field, 2, 3
Buffon's needle problem, 31, 122

Cauchy criterion for a.s. convergence, 270
Cauchy distribution, 79, 170, 217
   bivariate, 118
   characterization, 218
   does not obey WLLN, 260
   mean does not exist, 91
   median, 91
   mgf does not exist, 99
   moments, 217

as ratio of two normal, 224, 236
as stable distribution, 280
truncated, 117
Cauchy-Schwarz inequality, 158, 165
Central limit theorem, 282, 291, 294
   applications of, 292
Change of variables in integrals, 10, 11
Chapman, Robbins and Kiefer inequality, 365
   for discrete uniform, 366
   for normal, 367, 372
   for uniform, 366
Characteristic function, 96
Chebychev's inequality, 100, 221, 269
   improvement of, 101-102
Chi-square distribution, central, 207, 312, 322, 323
   mgf, 208
   moments, 208
   normal approximation, 312
   as square of normal, 224
   noncentral, 314, 319
   mgf, 314
   moments, 315
Chi-square test, 428, 444
   as a goodness of fit test, 448
   for homogeneity, 554
   for independence, 565
   one-tailed, 428
   robustness, 586-587
   for testing equality of proportions, 445-446
   for testing parameters of multinomial, 447
   for testing variance, 444
   two-tailed, 429
Combinatorics, 35-40
Complete, family of distributions, 344, 531
Complete families, binomial, 344
   chi-square, 345
   discrete uniform, 346
   hypergeometric, 350
   uniform, 345
Complete sufficient statistic, 344, 347, 356, 531
   for Bernoulli, 344
   for discrete uniform, 346
   for exponential family, 347
   for normal, 345, 348
   for uniform, 346
Concave function, 4

Conditional, df, 113, 115, 121
   distribution, 167, 339, 390, 531
   pdf, 113, 114
   pmf, 112
   probability, 41, 112
Conditional expectation, 168, 170, 172, 414
   properties of, 168, 169, 171, 172
Confidence, bounds, 467-468
   coefficient, 468
   estimation problem, 467
Confidence interval, 468
   Bayesian, 495
   expected length of, 491
   fixed-width, 469-470
   general method of construction, 471
   level of, 468
   length of, 469
   for the parameter of, Bernoulli, 475
      discrete uniform, 479
      exponential, 478
      normal, 469, 470, 473, 478
      uniform, 476
   shortest-length, 479
   from tests of hypotheses, 486
   UMAU, 491
   UMAU family of, for difference of two normal means, 490
   for normal mean, 489, 491
   for normal variance, 490
   for ratio of two normal variance, 490
   unbiased, 490
   using Chebychev's inequality, 474
   using CLT, 473
   using properties of MLE's, 474
Confidence set, 468
   for mean and variance of normal, 471
   UMA family of, 469
   UMAU family of, 488
   unbiased, 488
Consistent estimate, 335
   for parameter of, Bernoulli, 335
   normal, 336
   uniform, 348
Contaminated normal, 581-582, 584
Contingency table, 565-566
Continuity correction, 293
Continuity theorem, 277, 294
Continuous type distributions, 202-227
Convergence, a.s., 249, 263, 265, 270
   equivalence, 266

Subject Index

in distribution = weak, 240-241, 246, 253, 277, 280
  of mgf's, 277
  modes of, 240, 257
  of moments, 242, 247, 248, 261
  of pdf's, 243
  of pmf's, 242
  in probability, 243, 247, 250, 252, 253, 257, 335, 375
  in rth mean, 247
Convex function, 4, 104
Convolution of df's, 142, 284
Correlation coefficient, 174, 175
Countable additivity, 24
Covariance, 155
  between sample mean and sample variance, 179, 323
  sample, 302
Critical region, 406

Decision function, 388, 592
Degenerate rv, 62, 89, 181
Degrees of freedom when pooling classes, 450
DeMorgan's laws, 1
Density function, probability, 63
Discrete distributions, 181-202
Dispersion matrix = variance - covariance matrix, 233, 236
Distribution, conditional, 112
  of a function of an rv, 68
  a posteriori, 390
  a priori, 390
  of sample mean, 306
  of sample median, 308
  of sample quantile, 151, 308
  of sample range, 142, 308
Distribution function, 56-58, 106, 107-108
  continuity points of a, 64, 280
  of a continuous type rv, 63, 109
  convolution, 142, 284
  decomposition of a, 66
  discontinuity points of a, 56
  of a discrete type rv, 66, 108
  of a function of an rv, 68, 69, 70-71, 73-75, 133, 134-137
  of an rv, 56-58, 106-108
Domain of attraction, 280
  of normal, 291

Efficiency of an estimate, 369
  realtive, 369
Empirical df = sample df, 299, 310
Equivalence lemma, 266
Equivalent rvs, 124, 266
Estimable function, 350, 532
Event, certain, 26
  elementary = simple, 21
  disjoint = mutually exclusive, 24, 46
  independent, 46, 47-48
  null, 21
Expectation, conditional, 168, 170, 172, 414
  properties, 168, 169, 171, 172
Expected value = mean = mathematical expectation, 78-79, 80, 84
  of a function of rv, 80, 154
  of product of rv's, 157-158
  of sum of rv's, 154
Exponential distribution, 137, 207
  characterizations, 209, 212
  memory of, 209
  mgf, 207
  moments, 207
Exponential family, 236, 347
  k-parameter, 238
  one-parameter, 236, 370, 419, 422, 423

Fisher's Z-statistic, 320
Fitting of distribution, binomial, 452
  normal, 449
  Poisson, 449
Fréchet, Cramér, and Rao inequality, 361-363, 382, 386
Fréchet, Cramér, and Rao lower bound for, binomial, 363
  exponential, 372
  geometric, 372
  normal, 372
  Poisson, 363
F-distribution, central, 317, 322, 504
  moments of, 318
  noncentral, 319
  moments of, 320
F-test, 430
  of general linear hypothesis, 502
  as likelihood ratio test, 438
  for testing equality of variances, 457
Fundamental identity of sequential analysis, 633-634

applications, 634-635

Gambler's ruin problem, 594-595, 615
Gamma distribution, 206
   bivariate, 118
   characterizations, 208-209, 225
   mgf, 207
   moments, 207
General linear hypothesis, 498
   canonical form, 503
   estimation in, 499
   likelihood ratio test of, 502
Generating functions, 93
   moment, 95, 161
   probability, 93
Geometric distribution, 94
   characterizations, 189-192, 201
   memory of, 190
   mgf, 188
   moments, 188
Glivenko-Cantelli theorem, 300

Helmert orthogonal matrix, 139, 323
Hölder's inequality, 165
Hypergeometric distribution, 192
   bivariate, 119
   mean and variance, 192

Identically distributed rv's, 123, 126
Implication rule, 28
Improper integral, 12
Independence and correlation, 175, 179, 229, 233
Independence of events, 46
   complete = mutual, 47
   pairwise, 47-48
Independence of rv's, 120, 121, 122, 123, 124, 125, 162-163
   complete = mutual, 122-123
   pairwise, 122
Independent, identically distributed rv's, 123, 142
Indicator function, 54
Infinite, integrals, 12
   product, 8-9
Infinitely often, 264
Infinitesimal rv's, 288
Interchange of limit operations, 11, 12
Invariance, of hypothesis testing problem, 431

principle, 338
Invariant, class of distributions, 337, 430
   estimates, 337-338
   maximal, 431
   tests, 431, 432

Joint, df, 118
   distribution, 120
   pdf, 109, 110
   pmf, 108
Jump, 9
   size of, 9, 57
Jump point, of a df, 57
   of a function, 9

Kendall's sample tau, 568
   distribution of, 569-570
Kendall's tau coefficient, 567
Kendall's tau test, 571
Kolmogorov's, inequality, 268
   strong law of large numbers, 274
Kolmogorov-Smirnov one sample statistic, 539
   for confidence bounds on df, 544
   distribution, 540, 541
   for estimating population df, 544
Kolmogorov-Smirnov test, comparison with chi-square test, 543
   one-sample, 542
   two-sample, 558
Kolmogorov-Smirnov two sample statistic, 557
   distribution, 557
Kronecker lemma, 269
Kurtosis, 93

Laplace distribution, 99
Laplace transform, 13
   bilateral, 13
   convergence differentiation and uniqueness of, 13-14
   unilateral, 13
Least square estimation, 173, 174, 499
   principle, 172, 499
L'Hospital rule, 4
Likelihood, equal, 18
   equation, 377, 382
   function, 376
Likelihood ratio test, 435-436
   asymptotic distribution, 442

for general linear hypothesis, 502
for parameter of, binomial, 436, 444
for simple vs. simple hypotheses, 436
   bivariate normal, 444
   discrete uniform, 443
   exponential, 444
   normal, 437, 438, 443
Limit inferior, 2
   set, 2
   superior, 2, 264
Lindeberg central limit theorem, 282, 291
Lindeberg condition, 282, 288, 290
Linear model, 498
Linear regression model, 500, 506
   confidence intervals, 509, 510, 511
   estimation, 506-507
   testing of hypotheses, 507, 508, 509
Lognormal distribution, 226, 375
Loss function, 388
Lower bound for variance, Chapman, Robbins and Kiefer inequality, 365
   Fréchet, Cramér and Rao inequality, 361
Lyapunov condition, 290
Lyapunov inequality, 103, 167

Maclaurin expansion, 7
   of an mgf, 96, 99
Mann-Whitney statistic, 537, 561
   moments, 537-538, 563-564
   null distribution, 561-562
Mann-Whitney-Wilcoxon test, 561
Marginal, df, 111-112, 118, 120
   pdf, 111, 115
   pmf, 110
Markovian dependence, 115
Markov's inequality, 100, 104
Maximal invariant statistic, 431
   function of, 432
Maximum likelihood estimate, 376
   asymptotic normality, 384-385
   consistency, 384-385
   as a function of sufficient statistic, 381
   invariance property, 383
Maximum likelihood estimation, in case of missing observations, 387
   principle, 376
Maximum likelihood estimation method applied to, Bernoulli, 379-380, 383, 387
   binomial, 386, 387
   bivariate normal, 387
   Cauchy, 387
   discrete uniform, 378, 386
   exponential, 383, 387
   gamma, 381
   geometric, 386
   hypergeometric, 378
   normal, 377, 387
   Poisson, 386
   uniform, 379, 382
Median, 91, 92
Median test, 559
Memory, of exponential, 209
   of geometric, 190
Method of moments, 373
   applied to, beta, 375
   binomial, 374
   gamma, 375
   lognormal, 375
   normal, 375
   Poisson, 374
   uniform, 374
Minimal sufficient statistic, 402
   for Bernoulli, 403
   for beta, 403
   for gamma, 403
   for geometric, 403
   for normal, 402, 403
   for Poisson, 403
   for uniform, 403
Minimax, estimate, 389, 394
   principle, 389
Minimax estimation for parameter of, Bernoulli, 398
   binomial, 394, 398
   hypergeometric, 396
   uniform, 398
Minimum mean square error estimate, 352
   for variance of normal, 358
Minkowski inequality, 165
Missing observations, estimation, 360, 387
Mode, 376
Moment, about origin, 81, 89
   absolute, 81, 86, 103
   central, 89, 155
   condition, 84
   of conditional distribution, 168
   of df, 78
   of functions of random vectors, 155
   inequalities, 100-103

of intercept of sample line of regression, 332
lemma, 86
negative-order, 93
non-existence of order-, 87
of sample covariance, 305
of sample mean, 303
of sample regression coefficient, 331
of sample variance, 304-305
Moment generating function, 95, 161
  continuity theorem for, 277, 294
  differentiation, 96, 99, 162
  expansion, 96, 99
  limiting, 276
  of sample mean, 306
  of sum of independent rv's, 145
  uniqueness, 96, 99
Monotome likelihood ratio, 419
  for hypergeometric, 421
  for one-parameter exponential family, 419
  UMP test for families with, 420
  for uniform, 419
Monotonic function, 9
Most efficient estimate, 369
  asymptotically, 370
  as MLE, 383
  and UMVUE, 370
Most powerful test, 407
  for families with MLR, 420
  as a function of sufficient statistic, 414
  invariant, 432
  Neyman-Pearson, 412
  similar, 426
  unbiased, 425
  uniformly, 408
Multidimensional rv = random vector, 105-106
Multinomial coefficient, 39
Multinomial distribution, 197-198
  characterization, 199
  mgf, 198
  moments, 199

Negative binomial (= Pascal or waiting time) distribution, 186-187
  bivariate, 118, 133
  central term, 200
  mean and variance, 187
  mgf, 187

Neyman-Pearson lemma, 412, 463
Neyman-Pearson lemma applied to,
  Bernoulli, 415
  normal, 415, 417
Nonparametric = distribution-free estimation, 299, 531
  methods, 296-297, 530-588
Nonparametric unbiased estimation, 531
  of distance between two df's, 535
  of population mean, 533
  of population variance, 533
  of tail probability, 535
Norm, 283
Normal approximation, to binomial, 292
  to Poisson, 293
Normal distribution = Gaussian law, 219
  bivariate, 139, 172, 227
  characterizations, 222, 223, 224
  contaminated, 581-582, 584
  as limit of binomial, 281
  as limit of chi-square, 281
  as limit of Poisson, 279
  mgf, 220
  moments, 220-221
  multivariate, 231
  singular, 228
  as stable distribution, 280
  standard, 116, 219
  tail probability, 222
  truncated, 116
Normal equations, 173, 499
Normed measure, 19
Nuisance parameter, 489

Odds, 25
Operator, linear, 284
Ordered samples, 36
Orders of magnitude, o and O notation, 6
Order statistic, 149, 299
  is complete and sufficient, 532
  joint pdf, 150
  joint marginal pdf, 151-152
  marginal pdf, 151
  uses, 576, 579

Parameter, of a distribution, 78
  estimable, 350, 532
  location, 349
  order, 78, 90-91
  scale, 349

space, 334
Parametric statistical hypothesis, 405
  alternative, 405
  composite, 405
  null, 405, 410
  problem of testing, 388, 405-406, 407, 408
  simple, 405
Parametric statistical inference, 296
Pareto distribution, 92
Partition, 399
  coarser, 400
  finer, 400
  minimal sufficient, 400
  reduction of a, 400
  sets, 399
  sub-, 400
  sufficient, 399
Pivot, 480
Point estimate, 333, 334
Point estimation, problem of, 334, 385
Poisson df, as incomplete gamma, 212
Poisson distribution, 69, 130, 194
  central term, 200
  characterizations, 195, 196-197, 201
  coefficient of skewness, 93
  kurtosis, 93
  as limit of binomial, 200, 278
  as limit of negative binomial, 201
  mean and variance, 194-195
  mgf, 195
  moments, 93
  pgf, 94, 195
Polya distribution, 193
Population, 296
Population distribution, 297
Power series, 7
Principle of least squares, 172, 499
Probability, 18, 24
  addition rule, 26
  axioms, 24
  conditional, 41, 112
  continuity of, 30, 35
  countable additivity of, 24
  density function, 63
  distribution, 56
  equally likely, 18
  equally likely assignment, 24, 36
  on finite sample spaces, 35
  generating function, 93, 99, 148
  geometric, 19
  integral transformation, 203
  mass function, 61
  multiplication rule, 42
  posterior and prior, 44
  principle of inclusion-exclusion, 27
  space, 25
  subadditivity, 26
  tail, 83
  total, 42
  uniform assignment of, 24, 36

Quadratic form, 5-6, 103
Quantile of order $p = (100p)$th percentile, 90-91

Random, 18
Random experiment = statistical experiment, 20
Random interval, 467
  coverage of, 577-578
Random sample, 297
  from a finite population, 30, 37
  from a probability distribution, 30, 297
Random sampling, 297
  in geometric probability, 30
Random set, family of, 467
Random variable, 53, 106
  continuous type, 63
  discrete type, 61
  equivalent, 124, 266
  functions of a, 67-75
  interchangeable, 126
  standardized, 90
  symmetric, 142
  symmetrized, 145
  truncated, 116-117, 260
Random vector, 105-106
  continuous type, 109
  discrete type, 108
  functions of, 127-147
Random walk, 615
  and sequential probability ratio test, 616
Range, 142
Regression analysis, 506-512
  coefficient, 174
  linear, 173
  lines, 173, 175, 178
  polynomial, 174
Regularity conditions of FCR inequality, 362

Riesz inequality, 104
Risk function, 388
Robustness, of chi-square test, 586-587
  of sample mean as an estimate, 581-583
  of sample standard deviation as an estimate, 583-585
  of Student's t-test, 585-586, 588
Robust procedure, defined, 581

Sample, 296-297
  correlation coefficient, 302, 310
  covariance, 302
  df, 299, 310
  line of regression, 302
  mean, 298, 301
  median, 302
  mgf, 301
  moments, 301, 302
  point, 21
  quantile of order p, 302
  space, 20-21, 297
  statistic, 298
  variance, 298, 453
Sampling with and without replacement, 36-37
Sampling from bivariate normal, 325, 454
  distribution of sample correlation coefficient, 327
  distribution of sample regression coefficient, 327
  independence of sample mean vector and dispersion matrix, 326
Sampling from unvariate normal, 321, 444, 452-453, 457
  distribution of sample variance, 322
  distribution of $(\bar{X}, S^2)$, 321
  independence of $\bar{X}$ and S^2, 139, 321
Sequential estimation, 593-595, 596, 602
  fixed-with interval, 606, 610
  lower bound for variance, 599-600
  of normal mean, 593, 602
  of parameter of geometric, 593
  unbiased, 596-601, 602
Sequential probability ratio test, 612
  approximating the power function, 624
  approximation of average sample size, 627
  approximation of stopping bounds, 617
  bounds for stopping bounds, 616
  optimal property, 631
  properties, 620
  saving in number of observations, 629-631
  strength, 613
Sequential statistical procedure, 592
  closed, 593
Set function, 3
Shortest-length confidence interval, 479
  for the mean of normal, 480-481
  for the parameter of beta, 485
  for the parameter of exponential, 485
  for the parameter of uniform, 484
  for the variance of normal, 482
σ-field, 2
  choice of, 21
  generated by a class = smallest, 2
Sign test, 545
Similar tests, 425, 426
Single-sample problem, 538-553
  of fit, 539
  of location, 545
  and symmetry, 545
Skewness, coefficient of, 93
Slow variation, 87, 291
Spearman's rank correlation coefficient, 571
  distribution, 572-573
Stable distribution, 218-219, 280
Standard deviation, 89
Standardized rv, 90
Statistic of order k, 149
  marginal pdf, 151
Stein's two-stage procedure, 606
Stirling's approximation, 8, 201
Stochastically larger, 554
Stopping region, rule and rv, 592
Strong law of large numbers, 263, 265, 273, 274
  Borel's, 273
  Kolmogorov's, 274
Student's t-distribution, central, 315, 322, 323, 331, 511
  bivariate, 332
  moments, 316
  noncentral, 317
  moments, 317
Student's t-test, 428-429, 452-455
  as likelihood ratio test, 437
  robustness of, 585-586, 588
Sufficient statistic, 339, 355, 371, 400, 531
  factorization criterion, 341
  joint, 342
Sufficient statistic for, Bernoulli, 340, 343

Subject Index   683

beta, 348
discrete uniform, 343
gamma, 348
lognormal, 349
normal, 339, 343, 348, 349
Poisson, 340
uniform, 344, 349
Support, of a function, 4
line of, 4, 104
Symmetric df or rv, 80, 142-143, 144, 145
Symmetrization, 145
Symmetrized rv, 145
Symmetry, center of, 80

Tail-equivalence, 266
Taylor's expansion, 6
Cauchy and Lagrange form of remainder, 6-7
Test, 406
critical region, 406
of hypothesis, 406, 444, 452
level of significance, 407, 410
most powerful, 407
nonrandomized, 407
power function, 407
randomized, 407
size, 407
statistic, 408
Testing the hypothesis of, equality of several normal means, 514
goodness of fit, 448, 542
homogeneity, 554, 558, 559, 561
independence, 511, 565, 569, 572
Testing hypothesis for parameter of,
Bernoulli, 424, 614, 618
binomial, 436, 443
bivariate normal, 444, 454
discrete uniform, 424
exponential, 424, 444, 465, 620
gamma, 424
geometric, 434, 443, 620
hypergeometric, 421
multinomial, 447
normal, 422, 423, 428, 429, 430, 432, 437, 438, 443, 461, 463, 614, 620
Poisson, 424, 434, 620
several binomial, 445
several normal, 464
two normal, 453, 457
Tests of hypothesis, Bayes, 460

likelihood ratio, 435-436
minimax, 462
Neyman-Pearson, 412
Tests of hypothesis listed, chi-square tests, 428, 429, 444, 445, 447, 448
F-tests, 430, 457
t-tests, 428, 429, 452-455
Tolerance interval, 575
Transformation, 133, 135, 150
of continuous type, 134-136
of discrete type, 131, 133
Helmert, 139, 323
Jacobian of, 135, 136, 150
not one-to-one, 133, 136, 150
one-to-one, 131, 134-136
Transition probability, 115
Triangular distribution, 65
Trinomial distribution, 199
Truncated distribution, 116
Truncation, 116, 260
Two-point distribution, 182
Two-sample problems, 553-573
Types of error in testing hypotheses, 406

Unbiased confidence interval, 488, 490
general method of construction, 492
for mean of normal, 491
for parameter of exponential, 494
for parameter of uniform, 494
for variance of normal, 492
Unbiased estimate, 350, 532
best linear, 352
and complete sufficient statistic, 356
LMV, 352
and sufficient statistic, 355
UMV, 352, 370, 534
Unbiased estimation for parameter of,
Bernoulli, 351, 357, 358
binomial, 363
bivariate normal, 359, 372
discrete uniform, 357, 359
exponential, 372
gamma, 355, 359
hypergeometric, 359
negative binomial, 358
normal, 350, 356-357, 358, 372
Poisson, 351, 354, 358, 359, 363
uniform, 363
Unbiased test, 425
for mean of normal, 426

and similar test, 426
UMP, 425
Uncorrelated rv's, 174
Uniform distribution, 70, 122, 129, 202
 characterization, 205
 discrete, 81-82, 183
 generating samples, 204
 mgf, 146, 203
 moments, 203
 statistic of order k, 216
U-statistic, 533
 for estimating mean and variance, 533
 as UMVUE, 534

Variance, 89
 properties of, 89
 of sum of rv's, 159-160, 161

Wald's equation, 597
Weak law of large numbers, 257, 263, 291
 for weighted sums, 267
Wilcoxon signed-ranks test, 548, 551
Wilcoxon statistic, 548
 distribution, 550-551
 moments, 549

Applied Probability and Statistics (Continued)

FLEISS · Statistical Methods for Rates and Proportions
GALAMBOS · The Asymptotic Theory of Extreme Order Statistics
GIBBONS, OLKIN, and SOBEL · Selecting and Ordering Populations: A New Statistical Methodology
GNANADESIKAN · Methods for Statistical Data Analysis of Multivariate Observations
GOLDBERGER · Econometric Theory
GOLDSTEIN and DILLON · Discrete Discriminant Analysis
GROSS and CLARK · Survival Distributions: Reliability Applications in the Biomedical Sciences
GROSS and HARRIS · Fundamentals of Queueing Theory
GUTTMAN, WILKS, and HUNTER · Introductory Engineering Statistics, *Second Edition*
HAHN and SHAPIRO · Statistical Models in Engineering
HALD · Statistical Tables and Formulas
HALD · Statistical Theory with Engineering Applications
HARTIGAN · Clustering Algorithms
HILDEBRAND, LAING, and ROSENTHAL · Prediction Analysis of Cross Classifications
HOEL · Elementary Statistics, *Fourth Edition*
HOLLANDER and WOLFE · Nonparametric Statistical Methods
HUANG · Regression and Econometric Methods
JAGERS · Branching Processes with Biological Applications
JESSEN · Statistical Survey Techniques
JOHNSON and KOTZ · Distributions in Statistics
    Discrete Distributions
    Continuous Univariate Distributions-1
    Continuous Univariate Distributions-2
    Continuous Multivariate Distributions
JOHNSON and KOTZ · Urn Models and Their Application: An Approach to Modern Discrete Probability Theory
JOHNSON and LEONE · Statistics and Experimental Design in Engineering and the Physical Sciences, Volumes I and II, *Second Edition*
KEENEY and RAIFFA · Decisions with Multiple Objectives
LANCASTER · An Introduction to Medical Statistics
LEAMER · Specification Searches: Ad Hoc Inference with Nonexperimental Data
McNEIL · Interatcive Data Analysis
MANN, SCHAFER, and SINGPURWALLA · Methods for Statistical Analysis of Reliability and Life Data
MEYER · Data Analysis for Scientists and Engineers
OTNES and ENOCHSON · Applied Time Series Analysis: Volume I, Basic Techniques
OTNES and ENOCHSON · Digital Time Series Analysis
PRENTER · Splines and Variational Methods
RAO and MITRA · Generalized Inverse of Matrices and Its Applications
SARD and WEINTRAUB · A Book of Splines
SEARLE · Linear Models
SPRINGER · The Algebra of Random Variables
THOMAS · An Introduction to Applied Probability and Random Processes
UPTON · The Analysis of Cross-Tabulated Data
WHITTLE · Optimization under Constraints